Applied Numerical Analysis

CIS £38.51
AF744255
WITHDRAWN
BOLTON INSTITUTE LIBRARIES
384212

Applied Numerical Analysis

SEVENTH EDITION

Curtis F. Gerald

Patrick O. Wheatley

California Polytechnic State University

BOLTON INSTITUTE OF HIGHER EDUCATION
THE LIBRARY - DEANE CAMPUS
Accession No. 384212
Classification No. 518 GER

PEARSON
Addison Wesley

Boston San Francisco New York
London Toronto Sydney Tokyo Singapore Madrid
Mexico City Munich Paris Cape Town Hong Kong Montreal

.0170069

Publisher: Greg Tobin
Managing Editor: Karen Guardino
Acquisitions Editor: William Hoffman
Associate Editor: RoseAnne Johnson
Production Supervisor: Cindy Cody
Marketing Manager: Pamela Laskey
Marketing Coordinator: Heather Peck
Prepress Supervisor: Caroline Fell
Manufacturing Buyer: Evelyn Beaton
Cover Designer: Dennis Schaefer
Cover Photo Credit: CREATAS/Photography
Compositor: Progressive Information Technologies

If you purchased this book within the United States or Canada you should be aware that it has been wrongfully imported without the approval of the Publisher or the Author.

0-321-19019-X

Copyright © 2004 Pearson Education, Inc.

All rights reserved. No part of this publication may be reproduced, stored in retrieval system, or transmitted, in any form or by any means, electronic, mechanical, photocopying, recording, or otherwise, without the prior written permission of the publisher. Printed in the United States of America.

2 3 4 5 6 7 8 9 10 VH 06 05 04

Contents

5 Numerical Differentiation and Integration 256

6 Numerical Solution of Ordinary Differential Equations 329

7 Optimization 405

8 Partial-Differential Equations 461

9 Finite-Element Analysis 517

Appendixes

Preface

In this seventh edition, we continue on the path established in previous editions. Quoting from the preface of the sixth edition, we "retain the same features that have made the book popular: ease of reading so that the instructor does not have to 'interpret the book' for the student, many illustrative examples that often solve the same problem with different procedures to clarify the comparison of methods, many exercises from which the instructor may choose appropriately for the class, more challenging problems and projects that show practical applications of the material."

We have made substantial improvements on the previous edition. These include:

Theoretical matters that previously were in a separate section near the end of each chapter have been merged with the description of the procedures.

Example computer programs that admittedly were not of professional quality have been deleted, with the idea that this is not normally a programming course anyway. Easy-to-read algorithms have been retained so that students can write programs if they desire.

There is greater emphasis on computer algebra systems; MATLAB is the predominant system, but this is compared with Maple and *Mathematica.* The use of spreadsheets to solve problems is covered as well.

A new chapter on optimization (Chapter 7) has been added that includes multivariable cases as well as single-variable situations. Linear programming has been included, of course, but the treatment is intended to provide a real understanding of the simplex method rather than to merely give a recipe for solving the problem. Nonlinear programming is treated to contrast this with the simpler linear case.

Boundary value problems for ordinary diffferential equations have been separated from those for partial differential equations and are included in the chapter on ordinary differential equations. Partial differential equations that satisfy boundary conditions (elliptic equations) are combined with the other types of partial differential equations in a single chapter.

> Many exercises have been modified or rewritten to provide an even greater variety. New exercises and projects have been added and some of these are more challenging than in the previous edition.
>
> As in previous editions, this book is unique in its inclusion of a thorough survey of numerical methods for solving partial differential equations and an introduction to the finite element method.
>
> Many suggestions from reviewers have allowed us to clarify and extend the treatment of several topics and we have made editorial changes to make the book easier to read and understand.

We again quote from the preface to the sixth edition:

> *Applied Numerical Analysis* is written as a text for sophomores and juniors in engineering, science, mathematics, and computer science. It should be a valuable source book for practicing engineers. Because of its coverage of many numerical methods, the text can serve as a valuable reference.
>
> Although we assume that the student has a good knowledge of calculus, appropriate topics are reviewed in the context of their use. An appendix gives a summary of the most important items that are needed to develop and analyze numerical procedures. We purposely keep the mathematical notation simple for clarity. Furthermore, the answers to exercises marked with a ▶ are found in the back of the text.

Acknowledgements

Many instructors have given valuable suggestions and constructive criticism. We mention those whose thorough reviews have helped make this edition better:

Todd Arbogast, *University of Texas at Austin*
Neil Berger, *University of Illinois at Chicago*
Barbara Bertram, *Michigan Technological Sciences*
Herman Gollwitzer, *Drexel University*
Chenyi Hu, *University of Houston-Downtown*
Tim Sauer, *George Mason University*
Daoqi Yang, *Wayne State University*
Kathie Yerion, *Gonzaga University*

We also want to express our thanks to those at Addison-Wesley who have worked extensively with us to ensure the publication of another quality edition: Greg Tobin, Bill Hoffman, RoseAnne Johnson, Cindy Cody, Pam Laskey, Heather Peck, and Barbara Atkinson.

0

Preliminaries

This book teaches how a computer can be used to solve problems that may not be solvable by the techniques that are taught in most calculus courses. It also shows how those problems that you may have solved before can be solved in a different way. Our emphasis is on problems that exist in the real world, although these examples will be simplified. Many of these simplified examples can be solved analytically, which allows a comparison with the computer-derived solution.

Modern mathematics began when Isaac Newton found mathematical models that matched the empirical laws that Johannes Kepler had reached after about 20 years of observation of the planets. Today, most of applied mathematics is a repetition of what Newton did: to develop mathematical relationships that can be used to simulate some real-world situation and to predict its response to different external factors.

The beauty of mathematics is that it builds on simple cases to arrive at more complex and useful ones. This is true for this book—we start with mathematical applications that are easily understood but that become the basis for other, more important applications of numerical analysis.

Contents of This Chapter

We begin each chapter of this book with a list of the topics that are discussed in that chapter.

0.1 Analysis Versus Numerical Analysis

Describes how numerical analysis differs from analytical analysis and shows where each has special advantages. It briefly lists the topics that will be covered in later chapters.

0.2 Computers and Numerical Analysis
Explains why computers and numerical analysis are intimately related. It describes several ways by which a computer can be employed in carrying out the procedures.

0.3 An Illustrative Example
Tells how a typical problem is solved and uses a special program called a computer algebra system to obtain the solution.

0.4 Kinds of Errors in Numerical Procedures
Examines the important topic of the accuracy of computations and the different sources of errors. Errors that are due to the way that computers store numbers are examined in some detail.

0.5 Interval Arithmetic
Discusses one way to determine the effect of imprecise values in the equations that are used to model a real-world situation.

0.6 Parallel and Distributed Computing
Explains how numerical procedures can sometimes be speeded up by employing a number of computers working together on a problem. Some special difficulties encountered are mentioned.

0.7 Measuring the Efficiency of Numerical Procedures
Tells how one can compare the accuracy of different methods, all of which can accomplish a given task, and how they differ in their use of computing resources.

0.1 Analysis Versus Numerical Analysis

The word *analysis* in mathematics usually means to solve a problem through equations. Of course, the equations must then be reduced to an answer through the procedures of algebra, calculus, differential equations, partial differential equations, or the like. Numerical analysis is similar in that problems are solved, but now the only procedures that are used are arithmetic: add, subtract, multiply, divide, and compare. Since these operations are exactly those that computers can do, numerical analysis and computers are intimately related.

An analytical answer is not always meaningful by itself. Consider this simple cubic equation:

$$x^3 - x^2 - 3x + 3 = 0.$$

It is not hard to find the factors that show that one of the roots is $\sqrt{3}$. That is fine, unless you want to cut a board to that length. But rulers are not graduated in square-root values. So what can you do? Maybe you have a calculator that lets you find the value, or you might use logarithms, or look it up in a table. Numerical analysis has a rich store of methods to find the answer by purely arithmetical operations.

Here's a challenge. You are on a desert island with nothing to work with but a sharp stick that you can use to draw in the sand. You've forgotten everything about mathematics except the four arithmetic operations and you can also compare values (much like a computer). For some reason, maybe because you have nothing more interesting to do, you want to get a good value for the cube root of 2. How would you go about this? One way would be trial and error: You try a set of values to see which one gives a result of 2 when it is multiplied three times, something like this:

$$1.2^3 = 1.728 \quad \text{too small}$$
$$1.4^3 = 2.744 \quad \text{too large}$$
$$1.25^3 = 1.9531 \quad \text{pretty close}$$
$$1.26^3 = 2.0004 \quad \text{really close!}$$

This could go on for some time, but you begin to see that you could interpolate between the last two trials and get an even better answer.

Now you say to yourself, "How good an answer do I really need? Maybe 1.26 is as close as I need. After all, when multiplied, 1.26^3 gives a result that differs from 2.0000 by a very small number, 0.0004."

In this book, we will describe methods that can solve this little problem efficiently and also methods for much more difficult ones. For example, this integral, which gives the length of one arch of the curve $y = \sin(x)$, has no closed form solution:

$$\int_0^{\pi} \sqrt{1 + \cos^2(x)}\, dx.$$

Numerical analysis can compute the length of this curve by standardized methods that apply to essentially any integrand; there is never a need to make a special substitution or to do integration by parts. Further, the only mathematical operations required are addition, subtraction, multiplication, and division, plus doing comparisons.

Another difference between a numerical result and the analytical answer is that the former is always an approximation. Analytical methods usually give the result in terms of mathematical functions that can be evaluated for a specific instance. This also has the advantage that the behavior and properties of the function are often apparent; this is not the case for a numerical answer. However, numerical results can be plotted to show some of the behavior of the solution.

While the numerical result is an approximation, this can usually be as accurate as needed. The necessary accuracy is, of course, determined by the application. The $\sqrt[3]{2}$ example suggests that the accuracy desired depends totally on the context of the problem. (There are limitations to the achievable level of accuracy, because of the way that computers do arithmetic; we will explain these limitations later.) To achieve high accuracy, very many separate operations must be carried out, but computers do them so rapidly without ever making mistakes that this is no significant problem. Actually, evaluating an analytical result to get the numerical answer for a specific application is subject to the same errors.

The analysis of computer errors and the other sources of error in numerical methods is a critically important part of the study of numerical analysis. This subject will occur often throughout this book.

Here are those operations that numerical analysis can do and that are covered in this book:

Find where $f(x) = 0$ for a nonlinear equation or system of equations.

Solve systems of linear equations, even large systems.

Interpolate to find intermediate values from a table of values and fit curves to experimental data.

Approximate functions with polynomials or with a ratio of polynomials.

Approximate values for the derivatives of a function, even if this is known only by a table of function values.

Evaluate the definite integral for any integrand, even if its values are known only from experimental observations.

Solve differential equations when initial values are given; these can be of any order and complexity. Numerical analysis can even solve them if conditions are specified at the boundaries of a region.

Find the minima or maxima of functions, even when subject to constraints.

Solve all types of partial differential equations by several techniques.

0.2 Computers and Numerical Analysis

Numerical methods require such tedious and repetitive arithmetic operations that only when we have a computer to carry out these many separate operations is it practical to solve problems in this way. A human would make so many mistakes that there would be little confidence in the result. Besides, the manpower cost would be more than could normally be afforded. (Once upon a time, military firing tables were computed by hand using desk calculators, but that was a special case of national emergency before computers were available.)

Of course, a computer is essentially dumb and must be given detailed and complete instructions for every single step it is to perform. In other words, a computer program must be written so the computer can do numerical analysis. As you study this book, you will learn enough about the many numerical methods available that you will be able to write programs to implement them. The specific computer language used is not very important; programs can be written in BASIC (many dialects), FORTRAN, Pascal, C, C++, Java, and even assembly language. Most of the methods will be described fully through pseudocode in such a form that translating this code into a program is relatively straightforward.

Actually, writing programs is not always necessary. Numerical analysis is so important that extensive commercial software packages are available. The IMSL (International Mathematical and Statistical Library) MATH/LIBRARY has hundreds of routines, of efficient and of proven performance, written in FORTRAN and C that carry out the methods. Recently, LAPACK (Linear Algebra Package) has been made available at nominal cost. This package of FORTRAN programs incorporates the subroutines that were contained in the earlier packages of LINPACK and EISPACK. Appendix B of this book gives information on these and other programs. The bimonthly newsletter of the Society for Industrial and Applied Mathematics (*SIAM News*) contains discussions and advertisements on some of the latest packages. A set of books, *Numerical Recipes,* lists and discusses numerical analysis programs in a variety of languages: FORTRAN, Pascal, and C.

One important trend in computer operations is the use of several processors working in parallel to carry out procedures with greater speed than can be obtained with a single processor. Some numerical analysis procedures can be carried out this way. Special programming techniques are needed to utilize these fast computer systems. A recent development is to utilize computers that are idle, even personal computers, to carry out computations. If these idle computers are connected in a network, a control computer can send a portion of a large computation to them. After completing its part of the task, the individual computers transmit the results back to the control computer. Such an arrangement is termed *distributed computing.* As you can imagine, coordinating and controlling this distributed system is a difficult task.

An alternative to using a program written in one of the higher-level languages is to use a kind of software sometimes called a computer algebra system (CAS). (This name is not very standardized and not too descriptive.*) This kind of program mimics the way humans solve mathematical problems. Such a program is designed to recognize the type of function (polynomial, transcendental, etc.) presented and then to carry out requested mathematical operations on the function or expression. It does so by looking up in tables the new expressions that result from doing the operation or by using a set of built-in-rules. For example, a program can use the ordinary rules for finding derivatives, employ tables of integrals to do integrations, and factor a polynomial or expand a set of factors. These are only a few of the capabilities. If an analytical answer cannot be given, most of these programs allow the user to get an answer by numerical methods.

In connection with numerical analysis, an important feature of many such programs is the ability to write utility files that are essentially macros: A sequence of the built-in operations is defined to perform a desired larger task or one not inherent in the program. A succession of operations, each of which uses the results of the previous one—a procedure called *iteration*—is also possible. Many numerical analysis procedures are iterative.

Many computer algebra systems are available. We will discuss only three of these: *Mathematica,* MATLAB, and Maple. MATLAB will be used extensively; it will be supplemented and compared to the other two. In this chapter, we will show how MATLAB can plot a function and find where it is a minimum. We anticipate that you will use one of the computer algebra systems as a tool to explore numerical procedures.

One special feature of most of these programs is their ability to carry out many operations with exact arithmetic. An interesting example is to see π displayed to 100 decimal places. Ordinarily, we must be satisfied with a limited number of digits of precision when a normal computer program is employed.

---

* Such programs are also called symbolic algebra systems.

Of particular importance in using such programs is that the plotting of functions, even functions of two independent variables (which require a three-dimensional plot), is built in. In *Mathematica* this graphical capability is especially well developed.

Computer algebra systems, with their ability to perform mathematics symbolically and to carry out numerical procedures with extreme precision, would seem to be almost a preferred tool for the numerical analyst. However, for the large "real-life" problems that a professional analyst often deals with, they do not have the necessary speed. They are good for "small problems" and are an excellent learning environment. However, in many "real world" situations, such as weather prediction or the computation of space vehicle trajectories, the scientist/engineer will employ programs written in FORTRAN or C. And he or she will almost always use the proven routines of IMSL or LAPACK.

Another alternative to writing a computer program to do numerical analysis is to employ a spreadsheet program. Still another way to do numerical procedures is to utilize a programmable calculator. Typcial of these advanced calculators are the TI-89 from Texas Instruments and the HP-48G from Hewlett-Packard. These machines have much of the power of a personal computer to do mathematics. They have limited memory, but built into them are special facilities of interest to the numerical analyst. Programs that are coded in their Read-Only Memory (ROM) can plot functions in two- and three-dimensions, solve for roots of a nonlinear equation, solve systems of linear equations, manipulate matrices, do interpolation, differentiate and integrate (both numerically and analytically), and solve ordinary differential equations as well as perform mathematical and statistical operations. Expressions can include terms like sine, cosine, and other mathematical functions. They not only handle numeric expressions; symbolic manipulations are also possible. We do not discuss programmable calculators in this edition of *Applied Numerical Analysis.*

0.3 An Illustrative Example

We will introduce the subject of numerical analysis by showing a typical problem solved numerically. If you worked for a mining company, Example 0.1 might be a problem you would be asked to solve.

EXAMPLE 0.1 *The Ladder in the Mine.* Two intersecting mine shafts meet at an angle of 123°, as shown in Figure 0.1(a). The straight shaft has a width of 7 ft, and the entrance shaft is 9 ft wide. What is the longest ladder that can negotiate the turn at the intersection of the two shafts? Neglect the thickness of the ladder members, and assume the ladder is not tipped as it is maneuvered around the corner. Provide for the general case in which the angle a is a variable as well as for the widths of the shafts.

Steps in Solving the Problem

Whenever a scientific or engineering problem is to be solved, there are four general steps to follow:

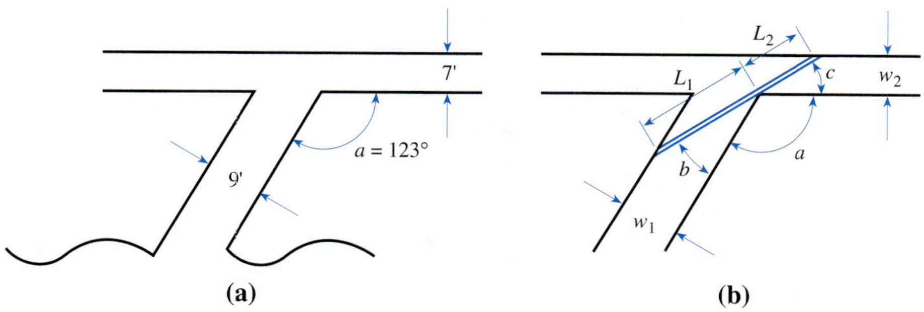

Figure 0.1

1. State the problem clearly, including any simplifying assumptions.
2. Develop a mathematical statement of the problem in a form that can be solved for a numerical answer. This process may involve, as in the present case, the use of calculus. In other situations, other mathematical procedures may be employed. When this statement is a differential equation, appropriate initial conditions and/or boundary conditions must be specified.
3. Solve the equation(s) that result from step 2. Sometimes the method will be algebraic, but frequently more advanced methods will be needed. This text may provide the method that is needed. The result of this step is a numerical answer or set of answers.
4. Interpret the numerical result to arrive at a decision. This will require experience and an understanding of the situation in which the problem is embedded. This interpretation is the hardest part of solving problems and must be learned on the job. This book will emphasize step 3 and will deal to some extent with steps 1 and 2, but step 4 cannot be meaningfully treated in the classroom.

The description of the problem has taken care of step 1. Now for step 2.

Here is one way to analyze our ladder problem. Visualize the ladder in successive locations as we carry it around the corner; there will be a critical position in which the two ends of the ladder touch the walls while a point along the ladder touches the corner where the two shafts intersect (see Fig. 0.1b). Let c be the angle between the ladder and the wall when in this critical position. It is usually preferable to solve problems in general terms, so we work with variables a, b, c, w_1, and w_2.

Consider a series of lines drawn in this critical position—their lengths vary with the angle c, and the following relations hold (angles are expressed in radian measure):

$$L_1 = \frac{w_1}{\sin(b)}, \qquad L_2 = \frac{w_2}{\sin(c)}$$

$$b = \pi - a - c, \tag{0.2}$$

$$L = L_1 + L_2 = \frac{w_1}{\sin(\pi - a - c)} + \frac{w_2}{\sin(c)}$$

The maximum length of ladder that can negotiate the turn is the minimum of L as a function of the angle c. If you were to solve for the minimum of L with respect to c by the methods you learned in calculus, you would first find an expression for dL/dc and then find the value for c that makes this zero. We prefer to use a special function that MATLAB has to get the answer.

MATLAB is command line driven, meaning that we type in commands that invoke operations. It is a large and powerful "computing environment." MathWorks, the developer of the program, calls it "The Language of Technical Computing." In later chapters we will explore many of its capabilities, but for now we will only use it to (1) draw a plot of L versus c (from which we can estimate the minimum point), and (2) find the minimum more accurately with the special MATLAB function.

We start by defining the function L. We know values for w_1 and w_2 from Figure 0.1. Angle c is given as 123° but we want the value in radians. We can ask MATLAB to do the conversion; the value for pi is built into MATLAB, we use "pi" to get it:

```
EDU>>a = 123*2*pi/360
a =
  2.1468
```

We could get more significant figures in the result but this seems good enough. Now we define the function for L. There are other ways to do it, but this is an easy way:

```
EDU>>L = inline('9/sin(pi-2.1468-c)+7/sin(c)')
L =
  Inline function:
  L(c) = 9/sin(pi-2.1468-c)+7/sin(c)
```

We ask MATLAB to plot L versus c:

```
EDU>>fplot(L,[0.4,0.5]);grid on
```

and we see Figure 0.2. (The semicolon before the "grid on" command suppresses the plot until the grid is created.)

From the graph, we can estimate that the minimum point is approximately $L = 34.42$, $c = 0.466$. For this problem, this is perhaps an adequate answer. Still, MATLAB can get the minimum more accurately. We ask for a numerical computation:

```
EDU>>fminbnd(L,0.4,0.5)
ans =
  0.4677
```

Our estimate was really pretty good. But this is the value for c at the minimum—we really want the value for L. So we do:

```
EDU>>L(0.4677)
ans =
  33.4186
```

If we want to see how MATLAB found the c-value at the minimum point, we do:

```
EDU>>fminbnd(L,0.4,0.5,optimset('Display','iter'))
Func-count        x            f(x)        Procedure
     1         0.438197      33.5333       initial
     2         0.461803      33.4231       golden
```

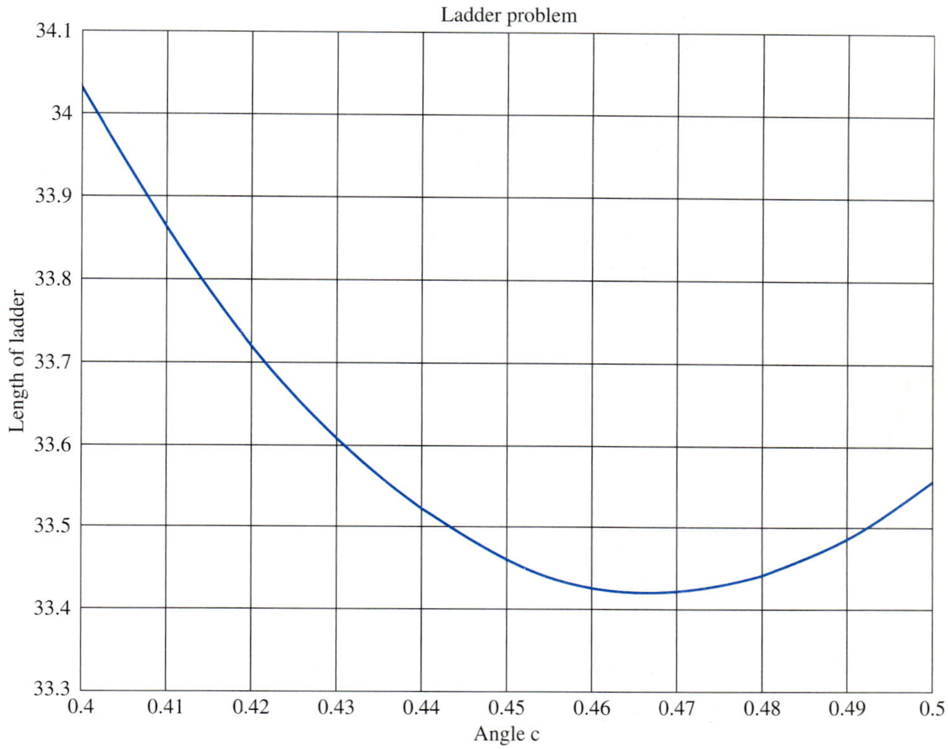

Figure 0.2

```
      3          0.476393      33.4284      golden
      4          0.467721      33.4186      parabolic
      5          0.467688      33.4186      parabolic
      6          0.467654      33.4186      parabolic
Optimization terminated successfully:
the current x satisfies the termination criteria using
OPTIONS.TolX of 1.000000e-004
ans =
   0.4677
```

The table that MATLAB displayed gives the successive steps in finding the minimum (x is used as the name for the independent variable rather than c). In a later chapter, we will explain the different "Procedures" that were used.

You should learn from this example three things about solving problems numerically: (1) There is often more than one way to attack the problem; (2) there are prewritten programs that can help; and (3) the accuracy that is needed in the answer dictates how you should get the solution. When a graph is sufficient, that may be the quickest and best way and it may tell how sensitive the answer is to values of the parameters.

Maple and *Mathematica* are two other computer algebra systems that can solve the ladder problem. If these are available to you, you may want to see how they compare to MATLAB.

0.4 Kinds of Errors in Numerical Procedures

We have mentioned that it is critically important to realize that errors can occur in doing numerical procedures. Some errors are due to the way that a computer does arithmetic but there are other sources of error.

Error in Original Data

Real-world problems, in which an existing or proposed physical situation is modeled by a mathematical equation, will nearly always have coefficients that are imperfectly known. The reason is that the problems often depend on measurements of doubtful accuracy. Further, the model itself may not reflect the behavior of the situation perfectly. We can do nothing to overcome such errors by any choice of method, but we need to be aware of such uncertainties; in particular, we may need to perform tests to see how sensitive the results are to changes in the input information. Because the reason for performing the computation is to reach some decision with validity in the real world, sensitivity analysis is of extreme importance. As Hamming says, "the purpose of computing is insight, not numbers."

Blunders

You will likely always use a computer or at least a programmable calculator in your professional use of numerical analysis. You will probably also use such computing tools extensively while learning the topics covered in this text. Such machines make mistakes very infrequently, but because humans are involved in programming, operation, input preparation, and output interpretation, blunders or gross errors do occur more frequently than we like to admit. The solution here is care, coupled with a careful examination of the results for reasonableness. Sometimes a test run with known results is worthwhile, but it is no guarantee of freedom from foolish error. When hand computation was more common, check sums were usually computed—they were designed to reveal the mistake and permit its correction.

On one occasion, a space flight was lost because someone typed into the program a single value with digits reversed, a common mistake. Human error can be costly!

Truncation Error

The term *truncation error* refers to those errors caused by the method itself (the term originates from the fact that numerical methods can usually be compared to a truncated Taylor series). For instance, we may approximate e^x by the cubic

$$p_3(x) = 1 + \frac{x}{1!} + \frac{x^2}{2!} + \frac{x^3}{3!}.$$

However, we know that to compute e^x really requires an infinitely long series:

$$e^x = p_3(x) + \sum_{n=4}^{\infty} \frac{x^n}{n!}.$$

We see that approximating e^x with the cubic gives an inexact answer. The error is due to truncating the series and has nothing to do with the computer or calculator. For iterative methods, this error can usually be reduced by repeated iterations, but because life is finite and computer time is costly, we must be satisfied with an approximation to the exact analytical answer.

Propagated Error

Propagated error is more subtle than the other errors. By *propagated error* we mean an error in the succeeding steps of a process due to an occurrence of an earlier error—such error is in addition to the local errors. It is somewhat analogous to errors in the initial conditions. Some root-finding methods find additional zeros by changing the function to remove the first root; this technique is called *reducing* or *deflating the equation.* Here the reduced equations reflect the errors in the previous stages. The solution, of course, is to confirm the later results with the original equation.

In examples of numerical methods treated in later chapters, propagated error is of critical importance. If errors are magnified continuously as the method continues, eventually they will overshadow the true value, destroying its validity; we call such a method *unstable.* For a *stable* method—the desirable kind—errors made at early points die out as the method continues. This issue will be covered more thoroughly in later chapters.

Each of these types of error, while interacting to a degree, may occur even in the absence of the other kinds. For example, round-off error can occur even if truncation error is absent, as in an analytical method. Likewise, truncation errors can cause inaccuracies even if we can attain perfect precision in the calculation. The usual error analysis of a numerical method treats the truncation error as though such perfect precision did exist.

Even in the absence of the errors we have discussed, there are errors inherent in the architecture of the computer. We discuss this next.

Round-Off Error

All computing devices represent numbers, except for integers and some fractions, with some imprecision. (MATLAB and similar programs can work with integers and rational fractions to achieve results of higher precision.) Digital computers will nearly always use floating-point numbers of fixed word length; the true values are usually not expressed exactly by such representations. We call the error due to this computer imperfection the *round-off* error. If numbers are rounded when stored as floating-point numbers, the round-off error is less than if the trailing digits were simply chopped off.

Absolute Versus Relative Error, Significant Digits

The accuracy of any computation is always of great importance. There are two common ways to express the size of the error in a computed result: *absolute error* and *relative error.* The first is defined as

$$\text{absolute error} = |\text{true value} - \text{approximate value}|.$$

A given size of error is usually more serious when the magnitude of the true value is small. For example, 1036.52 ± 0.010 is accurate to five significant digits and may be of adequate precision, whereas 0.005 ± 0.010 is a clear disaster.

Using relative error is a way to compensate for this problem. Relative error is defined as

$$\text{relative error} = \frac{\text{absolute error}}{|\text{true value}|}$$

The relative error is more independent of the scale of the value, a desirable attribute. When the true value is zero, the relative error is undefined. It follows that the round-off error due to a finite length of the fractional part of floating-point numbers is more nearly constant when expressed as relative error than when expressed as absolute error. Most people define these errors in terms of magnitudes, in which case the error is always a positive quantity.

Another term that is commonly used to express accuracy is *significant digits,* that is, how many digits in the number have meaning. Extra digits that show up when numbers are shifted to normalize them are meaningless; this is a real problem when there are trailing zeros in a number. We may not know whether they are really zeros or just fillers.

A more formal definition of significant digits follows.

1. Let the true value have digits $d_1 d_2 \ldots d_n d_{n+1} \ldots d_p$.
2. Let the approximate value have $d_1 d_2 \ldots d_n \, e_{n+1} \ldots e_p$.

where $d_1 \neq 0$ and with the first difference in the digits occurring at the $(n + 1)$st digit. We then say that (1) and (2) agree to n significant digits if $|d_{n+1} - e_{n+1}| < 5$. Otherwise, we say they agree to $n - 1$ significant digits.

EXAMPLE 0.7 Let the true value $= 10/3$ and the approximate value $= 3.333$.

The absolute error is $0.000333\ldots = 1/3000$.

The relative error is $(1/3000)/(10/3) = 1/10000$.

The number of significant digits is 4.

Floating-Point Arithmetic

Even though a computer follows exactly the instructions that it is given, when it performs an arithmetic operation it does not get exact answers unless only integers or exact powers of 2 are involved.

A computer stores numbers as floating-point* quantities that resemble scientific notation. For example, 13.524 is the same as the floating-point number $.13524 * 10^2$, which is often displayed as .13524E2. Another example: -0.0442 is the same as $-.442\text{E}-1$. In both of these situations, we have *normalized* the floating-point representation, meaning that we have shifted the decimal point to make the leading digit nonzero.

While not all computers store floating-point numbers in the IEEE standard that we now describe (IBM mainframes are a notable exception), this IEEE standard is by far the most common. A computer number has three parts:

the sign (either + or −),

the fraction part (called the *mantissa*),

the *exponent* part.

The IEEE standard specifies that the number will be stored as a binary quantity. One bit is used for the sign.

There are three levels of precision and these are the number of bits used for mantissa and exponent:

		Number of bits in			
Precision	**Length**	**Sign**	**Mantissa**	**Exponent**	**Range**
Single	32	1	23(+1)	8	$10^{\pm 38}$
Double	64	1	52(+1)	11	$10^{\pm 308}$
Extended	80	1	64	15	$10^{\pm 4931}$

For single and double precision, a clever device is used to get one more bit in the mantissa than the length accommodates: All numbers are normalized so the first bit of the mantissa is always 1. This means that it does not have to be stored and that is why we show the mantissa length as "+1" more than the number of bits actually used.

We need very small numbers as well as large, so the IEEE standard provides for negative exponents. Rather than use one of the bits for the sign of the exponent, exponents are "biased"; a bias value is added to the actual value of the exponent to make all exponents range from zero to a maximum number. For single precision, the bias value is 127 (base 10), so an exponent of -127 is stored as zero; the largest exponent, 128, is stored as 255,

---

* Another name often used for floating-point numbers is *real numbers,* but we reserve the term *real* for the continuous (and infinite) set of numbers on the "number line." When printed as a number with a decimal point, it is called *fixed point.* The essential concept is that these are in contrast to integers.

the largest value that 8 bits can signify. For double and extended precision, the bias values are 1023 and 16383.

It is obvious that only a finite number of different values can be stored in a computer that uses the IEEE standard. Since there is an infinite number of real numbers, it is clear that there must be gaps between the stored values. This is the source of round-off error. Numbers that cannot be stored exactly are approximated. The simplest way to do this is just to chop off the digits beyond those that can be stored. A preferred way is to round to the nearest storable number (with rounding to an even number if there is a tie: 0.1234 becomes 0.123 if we can have only three digits, 0.1235 becomes 0.124).

There is a largest number and smallest (in magnitude) number in the system. Quantities that exceed the maximum cause *overflow;* too small numbers cause *underflow.* How these cases are handled depends on the particular computer. When underflow occurs, many computers replace the value with zero. If there is overflow, they replace the value with a special bit pattern that represents infinity.

In single precision, the smallest and largest storable numbers are:

Smallest: 2.93873 E−39,

Largest: 3.40282 E+38.

The storage of zero is a special case. In the IEEE standard, zero is stored as all zeros: The sign is zero, the mantissa is all zeros, the exponent is all zeros. Obviously, the value for zero cannot be normalized.

Certain mathematical operations are undefined, such as 0/0, $0*\infty$, $\sqrt{-1}$. When a program tries to do any of these, the IEEE standard substitutes another special bit pattern that is displayed as NaN (meaning Not a Number).

eps

The term eps is a shortened form of the Greek letter epsilon; it is used to represent the smallest machine value that can be added to 1.0 that gives a result distinguishable from 1.0. MATLAB can tell what it is for your computer. For the computer used to write this book, here is what MATLAB told us:

```
EDU>> eps
ans =
   2.2204e-016
```

The value of eps depends, however, on the precision of the computer system; MATLAB uses 32 digits and eps is the same as for IEEE double precision; a very small number. In IEEE single precision, eps has the value 1.192E−07 $\approx 2^{-23}$. It is not difficult to see that, if ε is just slightly smaller than eps, $(1 + \varepsilon) + \varepsilon = 1$ but $1 + (\varepsilon + \varepsilon) > 1$.

Round-off Error Versus Truncation Error

We have seen that truncation error is caused by using a procedure that does not give precise results even though the arithmetic is precise. Round off occurs, even when the procedure is exact, due to the imperfect precision of the computer. What we might call *computational error* is the sum of these.

In Chapter 5, we will show how the derivative of a function can be found numerically. Analytically, df/dx is given by

$$f'(x) = \lim \frac{f(x+h) - f(x)}{h} \quad \text{as } h \to 0.$$

We can find an approximate value for $f'(x)$ by computing this ratio with a small value for h. If we make h still smaller, the result is closer to the true value for the derivative (the truncation error is reduced) but at some point, depending on the precision of the computer, round-off errors will dominate and the results become less exact. There is a point where the computational error is least.

Well-Posed and Well-Conditioned Problems

The accuracy of a numerical solution depends not only on the computer's accuracy; the problem itself is a factor. A problem is *well posed* if a solution (a) exists, (b) is unique, and (c) has a solution that varies continuously when the values of its parameters vary continuously. Not all problems have this property. The remedy is to replace the problem with another that has a solution close enough to be useful. A nonlinear problem could be replaced by a linear one; one that extends to infinity might be changed to one that extends to a large but finite extent; a complicated function may be simplified to one that has values that are almost the same.

Some problems are particularly sensitive to changes in the values of the parameters; a small change in the input causes a large change in the output. A *well-conditioned problem* is not so sensitive; the change in the output is not greater than the change in the input (or it could even be little changed). Most applied problems have parameter values that are based on measurements, so these may be not entirely accurate. The values may be numbers based on past experience and today's situation may be different. Modeling and simulation of the system are often used to explore its behavior and the model may be not a really good one. A well-conditioned problem gives useful results in spite of small inaccuracies in the parameters.

The procedure used (the *algorithm*) can sometimes amplify even small errors. In Chapter 6, we will mention a method that seems to have particularly good accuracy but for certain problems it exhibits instability—small initial errors are amplified. In Chapter 2, we discuss how a system of linear equations can be solved and we show that some systems (these are termed *ill-conditioned*) are so affected by round off that the answer is worthless.

Forward and Backward Error Analysis

When we compute a value of a function, $y = f(x)$, we may not get exactly the correct value of y due to computational error. Call this computed value y_{calc}. The *forward error* is defined as

$$E_{\text{fwd}} = y_{\text{calc}} - y_{\text{exact}},$$

where y_{exact} is the function value we would get if computational error were absent.

There is an x-value that will give y_{calc} when there is no computational error; call this x_{calc}:

$$y_{\text{calc}} = f(x_{\text{calc}}).$$

The *backward error* is

$$E_{\text{backw}} = x_{\text{calc}} - x.$$

Here is an example:

Compute $y = x^2$ for $x = 2.37$ and use only two digits. We get $y_{\text{calc}} = 5.6$ while $y_{\text{exact}} = 5.6169$. The forward error is

$$E_{\text{fwd}} = 5.6 - 5.6169 = -0.0169,$$

a relative error of about 0.3%. Because $\sqrt{5.6} = 2.3664\ldots$,

$$E_{\text{backw}} = 2.3664\ldots - 2.37 = -0.0036,$$

a relative error of about 0.15%.

Examples of Computer Numbers

Working with 32- or 64-bit number representation is awkward, so we simplify. Assume that only six bits are available; one bit for the sign, two bits for the exponent, leaving 3(+1) for the mantissa (we normalize). The exponent is biased by one, so we have these translations of the actual exponents:

Actual exponent	Stored in binary as
−1	00
0	01
1	10
2	11

The smallest and largest positive numbers look like this:

Sign	Mantissa	Exponent	Value
0	(1)001	00	$9/16 * 2^{-1} = +9/32$
0	(1)111	11	$15/16 * 2^{2} = +15/4$

Here is how we got the fraction part of these quantities:

$$\text{Smallest} = 1/2 + 0/4 + 0/8 + 1/16 = 9/16,$$
$$\text{Largest} = 1/2 + 1/4 + 1/8 + 1/16 = 15/16.$$

Observe that the smallest cannot have a mantissa of all zeros because that bit pattern is reserved for the number zero. The smallest and largest negative numbers have the same magnitudes; they differ only in the sign bit. These range from $-9/32$ to $-15/4$.

Suppose we draw the number line that shows all possible nonnegative computer numbers in this hypothetical system:

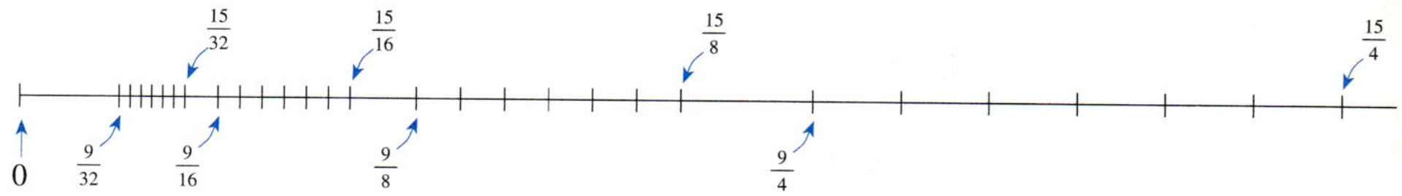

With this very simple computer arithmetic system, the gaps between stored values are very apparent. The gap between zero and the first positive number is extremely large because we have normalized. There is a larger gap between each "decade" as well. In each "decade" there are seven values, so there are $4 * 7 = 28$ positive numbers in all. There are 28 negative numbers, so the total of numbers is $28 + 28 + 1$ (for zero) $= 57$ altogether.

Because of the gaps in this number system, many values cannot be stored exactly. For example, the decimal number 0.601 falls between the first and second numbers in the second "decade." It will be stored as if it were 0.6250 because it is closer to 10/16, an error of about 4%. In the IEEE system, gaps are much smaller but they are still present.

There is a most important consequence to this. When you write a computer program, never use a test such as

$$\text{If } A = B \text{, then} \ldots$$

Instead, you should do something like this:

$$\text{If } |A - B| <= \text{TOL, then} \ldots$$

Here is how numbers appear in the IEEE standard:

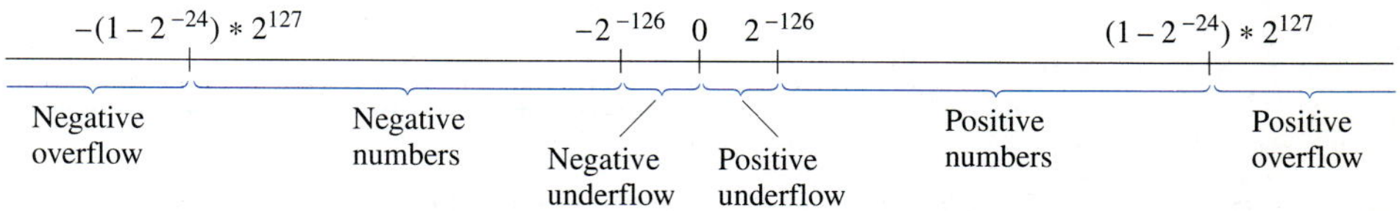

Anomalies with Floating-Point Arithmetic

It may seem surprising that when a set of numbers is added, the order of adding them to the running sum is important. Adding them in the order of smallest in magnitude to the largest gives a more accurate result than if they were added from largest to smallest. (However, there are instances when the opposite is true!)

For some combinations of values, these statements are not true:

$$(X + Y) + Z = X + (Y + Z)$$

$$(X * Y) * Z = X * (Y * Z)$$

$$X * (Y + Z) = (X * Y) + (X * Z)$$

Other peculiar things may happen with floating point. For example, adding 0.0001 one thousand times should equal 1.0 exactly but this is not true with single precision. To see for yourself, try this on your computer.

Here is another unexpected result. If we compute values in single precision of this expression:

$$\frac{(X + Y)^2 - 2XY - Y^2}{X^2} = Z,$$

with different values of X and Y, we get these answers:

X	Y	Z
0.01	1000	1.00000
0.001	1000	0.9999998
0.0001	1000	0.999213
0.00001	1000	1.000444
0.000001	1000	0.68212
0.0000001	1000	−79.58079

The expression for Z can easily be seen to reduce to X^2/X^2, which must equal 1.00000000 (if $X \neq 0$).

Errors When Values Are Converted

The numbers that are input to a computer are normally base-10 numbers. These have to be converted to the base-2 numbers that are stored in the computer. This conversion itself can cause errors. For example, some terminating decimals are nonterminating in base 2: $(0.6)_{10} = (0.100110011001\ldots)_2$.

You may have learned previously how numbers are converted from one base to another, but, to refresh your memory, here are the procedures to convert from base 10 to base 2:

For decimal integers:

Divide repeatedly by 2 and use the remainders in reverse order as the successive base-2 values.

For decimal fractions:

Multiply repeatedly by 2 and use the integer parts as the successive base-2 values.

Examples:

Convert $(327)_{10}$ to binary:
327/2 = 163, r1; 163/2 = 81, r1; 81/2 = 40, r1;
40/2 = 20, r0; 20/2 = 10, r0, 10/2 = 5, r0; 5/2 = 2, r1;
2/2 = 1, r0; 1/2 = 0, rl. Combining remainders gives $(327)_{10} = (101\ 000\ 111)_2$.

Convert $(0.3125)_{10}$ to binary:
0.3125 * 2 = 0.6250 (use 0); 0.6250 * 2 = 1.2500 (use 1);
0.2500 * 2 = 0.5000 (use 0); 0.500 * 2 = 1.000 (use 1).
Combining the integer parts gives
$(0.3125)_{10} = (0.0101)_2$.

0.5 Interval Arithmetic

While there are errors from round off caused by the finite number of bits available for floating-point numbers and errors may occur when the decimal fractions of input values are converted to machine numbers, a major source of error is that the parameters of a mathematical model are from measured quantities that define the parameters. How can we handle such uncertainties?

Interval arithmetic is a relatively new branch of mathematics that allows us to find how parameter errors are propagated through the sequence of computer operations of a program. We discuss here only some elementary concepts.

Interval analysis uses values that fall within a range of numbers. For example, if a measured quantity is reported as 2.4, but this is uncertain by ± 0.05, we should include the uncertainties in the equations. Instead of putting the imperfect number 2.4 into the equation, we should use

$$[2.35, 2.45],$$

an interval that does include the reported 2.4 but also shows the possible range for the true values. The two numbers in brackets show the extreme points of the quantity, the endpoints of the interval, if you will. We always write the lesser value first. (The notation for intervals is not yet well established.)

You can more easily understand the rules for arithmetic operations on intervals if you think of the "worst case" eventuality. We will write interval A as $[a_L, a_R]$. To add interval values A and B, we have:

$$A + B = [a_L, a_R] + [b_L, b_R] = [a_L + b_L, a_R + b_R]$$

(The least sum is $a_L + b_L$, the greatest is $a_R + b_R$.)

$$\text{Example:} \quad [0.5, 0.8] + [-1.2, 0.1] = [-0.7, 0.9]$$

$$\Delta = 0.3 \qquad \Delta = 1.3 \qquad \Delta = 1.6$$

Notice that the width of the sum is the sum of the widths of the terms being added.

For subtraction, we have

$$A - B = [a_L, a_R] - [b_L, b_R] = [a_L - b_R, a_R - b_L]$$

$$\text{Example:} \quad [0.5, 0.8] - [-1.2, 0.1] = [0.4, 2.0]$$

$$\Delta = 0.3 \qquad \Delta = 1.3 \qquad \Delta = 1.6$$

Again, the width of the answer is the sum of the widths of the terms.

Multiplication is more complicated. The definition is easier to explain if we work with a set of four values:

$$S = (a_L * b_L, a_L * b_R, a_R * b_L, a_R * b_R)$$

The members of S are the four possible products of the elements of A and B. Now define

$$S_L = \min(S),$$
$$S_R = \max(S).$$

The product of intervals A and B is then

$$A * B = (S_L, S_R).$$

Example: For [0.5, 0.8]*[−1.2, 0.1], we have

$$\Delta = 0.3 \qquad \Delta = 1.3$$

$$S = (-0.6, 0.05, -0.96, 0.08),$$
$$S_L = -0.96,$$
$$S_R = 0.08,$$

and the product is [−0.96, 0.08].

$$\Delta = 1.04$$

There is no obvious relation between the various widths.

Computing the product of two intervals through this definition may require more operations than the number actually required. Fewer multiplications and comparisons are required if we use alternative definitions for the nine possible cases for intervals A and B. (We have three possible situations for each of A and B: strictly less than zero, containing zero, and strictly greater than zero.) We leave as an exercise the development of the nine definitions.

Division of intervals is reduced to using the rules for multiplication by using the definition of the reciprocal of an interval:

$$1/[a_L, a_R] = [1/a_R, 1/a_R],$$

so that

$$A/B = A * (1/B) = [a_L, a_R] * [1/b_R, 1/b_L].$$

It is most important to remember that if B contains zero (even as an endpoint) division is undefined.

We will not go further into the arithmetic of intervals here, but you may wish to develop the relations for $A > B$, $A < B$, and $A = B$.

Software has been written that performs interval arithmetic. This provides answers in the form of intervals that show the possible range of solutions to the equations that model the applied problems that are solved with numerical procedures. *Mathematica* is one computer algebra system that lets you use interval arithmetic. Here is a sample:

```
In[1]: = a: = Interval[{0.5, 0.8}]
In[2]: = b: = Interval[{−1.2, 0.9}]
In[3]: = a + b
Out[3] = Interval[{−0.7, 1.7}]
In[4]: = a − b
Out[4] = Interval[{−0.4, 2.}]
In[5]: = a * b
Out[5] = Interval[{−0.96, 0.72}]
In[6]: = a/b
Out[6] = Interval[{−Infinity, −0.416667}, {0.555556, Infinity}]
```

Maple can do interval arithmetic too. In this, intervals are called "ranges" and are defined as `a: = INTERVAL(0.5..0.8);`. To perform arithmetic, the command `evalr` is used.

A Visual Example

To see how the imprecision of parameters affects the result of a computation, consider the graph of $y = 2 + x/2$, a straight line of slope 1/2 and y-intercept of 2 if the coefficients are exact. Suppose, though, that the coefficients come from a set of measurement and we know only that the slope is 2 ± 0.2 and the intercept is 0.5 ± 0.1. Figure 0.3 suggests that the plot of y versus x is a band, not a line.

0.6 Parallel and Distributed Computing

Many applications of computers involve tremendously large problems and the solution may be needed almost instantaneously, in so-called real time. Military applications are an example: Victory in war may depend on getting answers quickly. The speed of today's computer, though seemingly very great, is beginning to run into the limits to electron flow in electronic circuits.

Most computer systems run their instructions in sequence—one after another—and this limits their speed. Even though supercomputers are very fast, executing billions of operations per second, in some cases this is not fast enough. The history of computers has seen many techniques to get faster speed.

One of the first techniques to increase the operating speed of a computer was "pipelining"—that is, performing a second instruction within the CPU before the previous instruction is completed. This technique takes advantage of the fact that doing a single "instruction" actually involves several micro-coded steps and that the initial micro-steps can be applied to an additional instruction even though the first sequence of micro-steps has not yet finished. Pipelining permits a speedup by a factor of two or more.

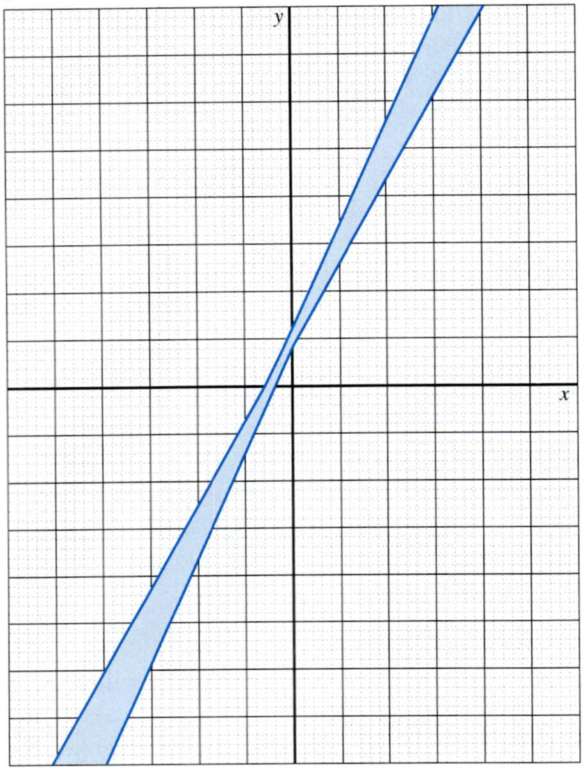

Figure 0.3

Another technique has been to build vector processing operations into the CPU. Because the individual steps required to solve sets of equations involve many multiplications of a vector by another vector, these machines offer significant speed improvements but only by a factor of 5 or 10, not by the factor of 10,000 that is really desired. Further, this feature increases the cost of mainframes considerably. The current trend is to use parallel processing, that is, to put several machines to work on a single problem, dividing the steps of the solution process into many steps that can be performed simultaneously. Not all problems permit such parallel operations, but many important problems of applied mathematics can be so structured. Obtaining many or even several supercomputers is outrageously costly, however. An alternative is to employ a massive number of low-cost microprocessors, of the order of a thousand (1024 is a practical number). Although the individual speed of a microprocessor is not equal to that of a supercomputer, the difference in speed is made up by the larger number of machines that are combined. Intel has been very active in the area of *massively parallel* computers. Their ASCI Red computer consists of over 9000 Pentium Pros and can run at a peak speed of 1.3 teraflops. At the other extreme in this area is the *Beowulf-class* of supercomputers, which are PCs joined together to compete with the dedicated supercomputers. For a description and discussion of one of these, the Loki supercomputer, check the Web sites http://loki-www.lanl.gov/index.html and http://loki-www.lanl.gov/results/. The Loki computer consists of 16 Pentium Pro computers working together to create a modestly priced supercomputer. We can imagine a future of thousands of PCs working together and accessible through the Web.

Massively parallel computers are important in many applications. For example, as stated in the *Atlantic Monthly* for January 1998, "big parallel computers have proved useful for both global climate warming and detailed modeling of ocean circulation" to explain why Europe has winter temperatures about nine to eighteen degrees warmer than comparable latitudes elsewhere.

Recently, much work and interest are found in *distributed computing.* The basic idea here is to connect many different computers, which can work separately on their own tasks as well as in conjunction with each other. In the classification of parallel computers we implicitly assumed a single clock with all the parallel operations in step (synchronous), whereas with distributed computers each machine runs under its own clock; interrupts constantly occur throughout the system to coordinate the actions (asynchronous operations). Moreover, each machine has its separate memory, and the data can flow from one computer to another. Although this seems to complicate the whole business of parallel computing, there are good economic reasons for distributed computing. The hardware is not specialized. One can make use of what is already at hand. The major effort and expense is in software and in connecting the computers and this can be done in a variety of ways.

If many computers are networked together as is common, distributed computing can utilize them when they are idle, such as at night. Not all problems lend themselves to such parceling out of the computations, but one interesting application that uses distributed computing is a program from the University of California, Berkeley, named SETI@home. When a computer's screen saver initiates, a signal is sent to the host computer saying that it is available to join in the Search for Extraterrestrial Intelligence. A chunk of cosmic radio-frequency data is then downloaded for the PC to analyze. Other applications that are being investigated are gene sequencing, weather forecasting, and the decoding of encrypted messages.

Special Problems in Parallel Computing

If parallel computing is to be used to solve a large problem rapidly, several new aspects come into play. Is the data stream provided from a single shared memory, or do the separate units have individual memories? If the memory is distributed, how is communication between the units accomplished? What type of bus provides the data channels to the separate units, and can separate units read and write data at the same time? What sort of intercommunication is there between the individual processors, and can they exchange data without going through memory?

Other questions remain. Do the units operate synchronously, with all controlled by a single clock, or do they run asynchronously? If operation is asynchronous, how does one unit know when to accept data from a prior operation of a different unit, or do all units operate "chaotically"? How can the loads for the separate processors be balanced—will some units sit idle while others are running at capacity? (It would be preferable for all units to run at full loading.) What about the programs for parallel processing? Does the programmer have to be concerned with synchronization and intercommunications? Is the code portable to other machines?

The questions about programming a parallel system are not yet settled. If it were possible to have the compiler recognize parallelism within a conventional program written for sequential operations and have it develop the changed code to be run on the parallel system, the task of programming would be much easier. On the other hand, writing code that specifically takes advantage of the parallel CPUs can be more efficient, but this task is tricky and complicated. It requires a skill that few programmers currently have. This mode would involve knowing exactly how the hardware is organized and what communications problems are involved. Further, it is likely that the best algorithm (solution procedure) for a parallel machine will not always be the optimal one for sequential processing.

Speedup and Efficiency

We do not intend to explain all these many aspects of parallel processing in this book. We must be content to show where parallelism exists for the various kinds of problems that we attack numerically. For example, here is a simple classical problem that exhibits the advantage of parallel processing. Suppose we are to add together n values. We can show the successive steps by a "directed acyclic graph" (dag), as shown in Figure 0.4. Now imagine that we have many separate processors that can be applied to the job. Figure 0.5 shows that the number of time steps can be decreased from seven to three. In both Figures 0.4 and 0.5 the "directed acyclic graphs" (dags) have steps that indicate the sequence of the operations in time. In both cases, step $i + 1$ cannot take place until step i is completed. The flow of operations is from the bottom to the top and one can characterize the dag as having a *height* of 7 in Figure 0.4 and a height of 3 in Figure 0.5. This is consistent with the definition of a "tree." At each level, indicated by a step n, we have the maximum number of processors used at the time. We see from Figure 0.5 that we only need 4 and not 8 processors to speed up the addition of the eight numbers.

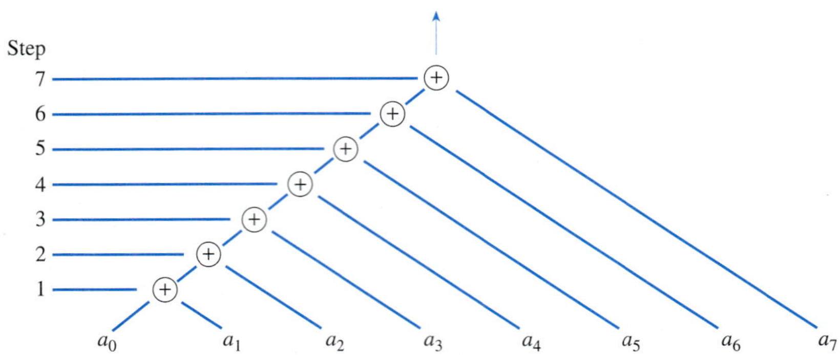

Figure 0.4
Adding eight numbers sequentially

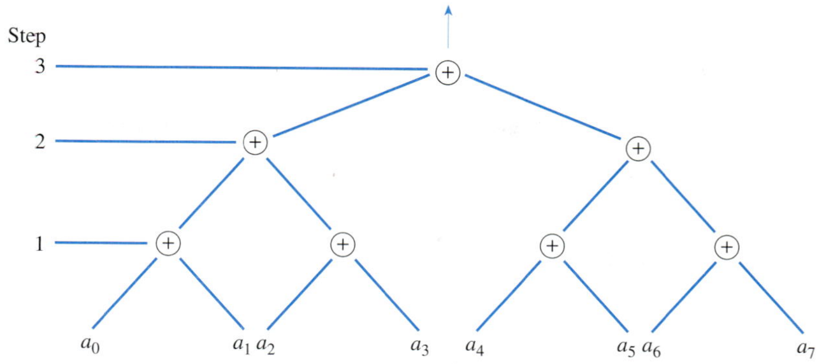

Figure 0.5
Adding eight numbers with parallel processors

The term *speedup* is used to describe the increased performance of a parallel system compared to a single processor. It is the ratio of the execution time for the original sequential process, using a single processor, to the time for the same job using parallel processors. In the preceding simple example, the speedup is 7/3 = 2.333. In computing the speedup for n data items, we use the time for the optimal sequential procedure (or for the best-known procedure if the actual optimal procedure is not known), $T_1(n)$, and for the best (known) parallel algorithm for p processors, $T_p(n)$. With these defined, we can now define speedup:

$$S_p(n) = T_1(n)/T_p(n).$$

In our example, we have $T_1(8) = 7$ and $T_4(8) = 3$, where 7 and 3 are the respective heights of the dags in Figs. 0.4 and 0.5. Another term, the *efficiency,* is based on how the speedup compares to the number of processors used, where

$$E_p(n) = T_1(n)/(pT_p(n)) = S_p(n)/p.$$

Theoretically, if we have p processors, we should be able to do the job p times as fast. In our example, however, $E_4(8) = 2.333/4 = 0.583$.

We have less than an efficiency of 1.00 because some of the processors are idle after the first step. Sometimes the speedup and efficiency are reduced because the size of the problem does not fit to the number of processors. For example, if we were to add only seven numbers in this example, we would still require four processors to get the sum in three steps, but now the speedup would be only 6/3 = 2.000 and the efficiency would drop to 0.5. On the other hand, what if we were limited to four processors but had 15 numbers to add? We would subdivide the problem and, although trivial in this example, it would have a best solution that is not obvious.

0.7 Measuring the Efficiency of Numerical Procedures

It is common that there are several different numerical methods to solve a problem. For example, you will find in the next chapter several techniques to get the roots of $f(x) = 0$, its "zeros." In Chapter 2, we will present several ways to solve a system of linear equations. How can the relative efficiency of different methods be compared?

One comparison is of the number of mathematical operations that are needed to get the answer with a given accuracy. Suppose, on analysis, a method is found to take $f(n)$ multiplies and that f is related to n by

$$f(n) = 1 + 2 + 3 + \ldots + n.$$

An equation that gets this sum is

$$f(n) = \frac{n(n+1)}{2} = \frac{n^2}{2} + \frac{n}{2}.$$

As n gets large, the first term dominates and we say that $f(n)$ is "of order n^2." The common symbol that is used is $f(n) = O(n^2)$. The net effect is with large n, the number of multiplications increases four times if n is doubled. This "order of" measure of operational count occurs often in comparing the efficiency of alternative procedures.

Even though it may seem confusing, there is a second, quite different use of the order relation. Some numerical methods arrive at an answer by varying the size of a parameter. In Chapter 6, we will describe methods for solving a differential equation numerically. The equation is of the form

$$dy/dx = f(x, y),$$

with a value given for y at some value for x. Many of these techniques add together a weighted sum of estimated values for the derivative function at evenly spaced x-values, values that differ by h (a commonly used variable for such spacing). For one method the error in the answer is proportional to the third power of h:

$$\text{Error} = \frac{M}{6}h^3,$$

where M depends on a value for the third derivative of the function $f(x, y)$. Because h is the only parameter of the process that can be chosen by the user, we say that "the error is of order h^3," and this is written as

$$\text{Error} = O(h^3).$$

Even though there are these two uses of the order relation, the context makes the meaning clear.

Taylor Series

The expression for the order of error just given is found by comparison of the procedure with a Taylor series. You will find that the Taylor series is often used in determining the order of error for methods, and the series is itself the basis for some numerical procedures (a particularly good example will be found in Chapter 6).

We remind you that a Taylor series is a power series that can approximate a function, $f(x)$, for values near to $x = a$. Its coefficients use the derivatives of f at $x = a$:

$$f(x) = f(a) + \frac{f'(a)}{1!}(x - a) + \frac{f''(a)}{2!}(x - a)^2 + \frac{f'''(a)}{3!}(x - a)^3 + \ldots$$

In effect, the Taylor series says that if we know the values for all derivatives of $f(x)$ at $x = a$, we can approximate the function as closely as we desire. This implies that $f(x)$ must have derivatives of all orders and that these can be evaluated at $x = a$.

If a Taylor series is truncated while there are still nonzero derivatives of higher order, the truncated power series will not be exact. The error term for a truncated Taylor series can be written in several ways, but the most useful form when the series is truncated after the nth term is

$$\text{Error of TS} = \frac{f^{n+1}(\xi)}{(n + 1)!}$$

where ξ is a value between x and $(x + a)$. The value for ξ is ordinarily not known, so there is some uncertainty in the exact value for the error. Still, this term can give bounds for the error.

Polynomials

We observe that a truncated Taylor series is really just a polynomial in x and that the only arithmetic operations used to compute it are precisely those that a computer can do. We also see that if $f(x)$ has discontinuities, the Taylor series cannot approximate it over the discontinuity.

Polynomials occur frequently in numerical analysis. You will encounter them in several instances in this book: in the development of formulas for interpolation (Chapter 3), for approximation of functions (Chapter 4), and for differentiation and integration (Chapter 5).

One reason for such prevalence is that the only mathematics needed for their evaluation are addition and multiplication and this fits perfectly to a computer. A second reason is the nice behavior of polynomials: They are everywhere continuous and have derivatives of any order. A famous theorem states that any continuous function can be approximated uniformly over a finite interval by a polynomial!

Certain polynomials are especially useful; many of these have a property called *orthogonality*. Typical of these are the Chebyshev polynomials. Using these, we can approximate a function better than with the standard Taylor series. Another set of polynomials that has the orthogonality property is the set of Legendre polynomials, which are involved in a particularly good way to integrate a function numerically.

A ratio of polynomials, a so-called rational function, is also important in numerical analysis. These also can be readily evaluated by a computer, but they may not be continuous everywhere; the denominator polynomials can be zero for some x-values.

When a polynomial is evaluated, it is inefficient to do it in the way that at first seems obvious. Suppose that your polynomial is

$$P(x) = a_0 + a_1x + a_2x^2 + a_3x^3.$$

If you evaluate $P(x)$ as $a_0 + a_1x + a_2x * x + a_3x * x * x$, six multiplications and three additions are required. Putting it into "nested form":

$$P(x) = ((a_3x + a_2)x + a_1x) + a_0,$$

takes only three multiplications and three additions. Nested multiplication is not only faster, but there is less error due to round off.

Exercises

(Answers are given for problems marked with ▶.)

Section 0.2

1. There are many programs and subroutines that have been written to perform numerical analysis. An important Internet resource is "Guide to Available Mathematical Software" (GAMS). It is maintained by a government agency. It can be accessed through the Web address gams.nist.gov/.
Find answers to these questions from that Web site.

a. What is the name of the government agency?

b. Use the link to "Package name" to see a listing of packages from many sources. How many different sources are listed?

c. How many times is Fortran identified as the computer language? How many times is the language C mentioned?

2. Repeat Exercise 1, but now look at "Problem Decision Tree."

a. How many subclasses in the tree?

b. Find a link to a program that gets the characteristics of the floating-point operations of a computer.

3. Make a list of the books in your school's library that deal with parallel computing. How many journal articles can you find that cover this subject?

4. Repeat Exercise 3, but for the topic "Distributed Computing."

Section 0.3

5. If your version of MATLAB allows for symbolic operations, find dL/Dc for the ladder problem of Section 0.3. Plot this derivative function and see where it crosses the x-axis. Does it give the same value for c at the minimum point? (If MATLAB cannot do this for you, use *Mathematica*.)

▶ **6.** Solve this variation to the ladder problem. The height of both the inlet tunnel and the straight shaft are both 6 ft 7 in. If the ladder can be tipped as the corner is negotiated, how much longer can the ladder be and still be taken into the mine?

7. The parameters for the ladder problem must have been determined from measurement; they can hardly be precise for an actual tunnel. Investigate how much the length of the ladder is affected if:

a. The angle a can be in error by as much as 5 degrees.

b. The width of the inlet shaft can be in error by as much as 4 inches.

c. The width of the straight shaft can be in error by as much as 7 inches.

▶ **8.** A circular well is 5.6 ft in diameter and is 14.3 ft deep and has a flat bottom. A ladder that is 17 in. wide (outside measurement) has side rails that are 1 in. by 3 in. What is the longest ladder that can be placed in the well if its top is to be exactly even with the surface of the ground? Get the answer in two different ways.

Section 0.4

9. Develop the Maclaurin series for $\cos(2x)$ up to terms in x^4.

a. What is the greatest truncation error within $x = [-1, 2]$?

b. Plot the truncation error over this range.

10. Repeat Exercise 9 but expand about $x = 1$ (a Taylor series). Also, find the x-values where the truncation error is zero.

11. Express these quantities in the form 0.*xxxx* . . . *x*E*yy*.

a. 1.234567
b. −2.00000111
c. 0.00001325
d. 123456789
e. 0.0000002

12. What is the largest interval between two IEEE numbers in

a. single precision.
b. double precision.

▶13. Can you find examples in single precision where

a. $(X + Y) + Z \neq X + (Y + Z)$?
b. $(X * Y) * Z \neq X * (Y * Z)$?
c. $X * (Y + Z) \neq (X * Y) + (X * Z)$?

14. Evaluate this cubic polynomial for $x = 1.32$, using both rounding and chopping to three digits at each arithmetic operation, getting both the absolute and relative errors:

$$3.12x^3 - 2.11x^2 + 4.01x + 10.33.$$

a. Do it proceeding from left to right.
b. Do it from right to left. Is the answer the same?
c. Repeat part (a) but do it with "nested multiplication." Which takes fewer operations? The nested form is:

$$((3.12x - 2.11)x + 4.01)x + 10.33.$$

▶15. Write a computer program that does the following additions in single precision. What are the absolute and relative errors of each sum?

a. 0.001 added 1000 times
b. 0.0001 added 10,000 times
c. 0.00001 added 100,000 times

16. Are there times when round-off errors tend to cancel in adding series of values?

17. Write a computer program that determines the relative speeds of the four arithmetic operations. Be sure to do enough repetitions so that the intervals between clock ticks do not affect the results; also account for loop overheads.

Section 0.5

▶18. Given these interval numbers, perform the arithmetic. How does the width of the answer compare to the widths of the terms?

$$x = [2.33, 2.54],\ y = [-1.19, 0.11],\ z = [0, 3.45].$$

a. What is $x + y$?
b. What is $x - y + z$?
c. What is $x * z$?
d. What is y/z?

19. If $x = 1.2345$ is stored as a floating-point number, what is the interval that includes it in IEEE single precision?

20. What is the smallest interval between two IEEE numbers?

a. In single precision.
b. In double precision.

21. Plot this function:

$$f(x) = [2.9, 3.1] * x^{[1.95, 2.05]} + [4.1, 4.3]$$

22. Add to the plot of Exercise 21 the plot of $g(x) = 3x^2 + 4.2$. At $x = 3$, how great is the distance between $f(x)$ and $g(x)$?

Section 0.6

23. Under what conditions can parallel processing not be used to speed up a computation?

24. A vector is a quantity that has several values, called its components. An example of a four-component vector is $V = [1.22, 2.33, 3.44, 4.55]$. The inner product of two vectors is the sum of the products of the components taken in order (the vectors must have the same number of components—they must be of the same "size").

a. How can parallel processing speed the computation of the inner product?
b. What is the speedup factor for vectors of size 5?
c. What is the speedup factor for vectors of size n?

▶25. What are the conditions of problems that suggest that distributed processing should be considered?

26. How does distributed computing differ from parallel processing?

Section 0.7

27. When a sequence of n integers is added, the first being 1 and the last being n, the sum is

$$\text{Sum} = n^2/2 + n/2 = O(n^2).$$

For $n = 100$, 1000, and 10,000, how much does the value of Sum differ from $n^2/2$?

28. The numbers of Exercise 27 are called an arithmetic progression and Δ, the difference between successive terms is 1. Find the formula for the following arithmetic progressions, and their order expressions.

a. The sum of odd integers, starting with 1?
b. The sum of even integers, starting with 2?
c. Repeat part (*a*) but for a starting value of *s*.

29. Find a formula for the sum of the squares of integers starting with 1^2 and its order expression. At what number of terms is the formula and the order expression within 1% of each other?

30. Repeat Exercise 29 but for the sum of cubes.

▶31. For a polynomial of degree n, show how many fewer arithmetic operations are needed when it's evaluated in nested form compared to doing it term by term.

32. Repeat Exercise 31 but for a polynomial where some coefficients are zero, say, m zeros in a polynomial of degree n. Are there times when evaluation in nested form has no advantage?

▶33. A rational function is a ratio of two polynomials. Evaluating both the numerator and denominator in nested form should require fewer operations than doing them term by term. How many fewer operations when the degrees are n and d?

Applied Problems and Projects

This group of problems will challenge you more than the exercises do. When you are asked to write a computer program, the language that you use is optional.

APP1. Write a computer program that finds a minimum of $f(x)$ that lies between $x = a$ and $x = b$. It does this by stepping from a toward b in steps of $(b - a)/10$ until the values of $f(x)$ begin to increase. It then reverses the direction with steps one-tenth as large to isolate the minimum more accurately. This is repeated until the minimum is located within x-values that differ by less than 10^{-6}.

APP2. How can APP1 be adapted to find a maximum value? Can the program be modified to permit the user to do either?

a. Critique the procedure of APP1. Consider these questions and others that you think are important:
   What if $f(x)$ is discontinuous in $[a, b]$?
   What if there are multiple minima?
   What if there is no minimum point?

b. Propose a scheme that is more efficient. Define what you think should be the measure of "efficiency."

APP3. Repeat APP1 but now to find where $f(x) = 0$, the point where the function crosses the x-axis. Analyze this as in APP2, parts (a) and (b).

APP4. Do research on the Internet and make a list of at least ten references to parallel computing. Make another list of references to distributed computing.

APP5. Use Maple and/or *Mathematica* to create a plot similar to Figure 0.2.

APP6. Get a formula for the number of mathematical operations needed to evaluate a polynomial of degree n, doing it with:

a. nested multiplication.

b. in standard form.

APP7. Find the Taylor series for $f(x) = 1/x$, expanded about the point $x = 2$. Write a program that displays the computed value at $x = 2.5$, the absolute error, and the relative error, for:

a. A series of three terms.

b. A series of four terms.

c. A series of five terms.

d. Repeat parts (a), (b), and (c) for $x = 3$.

e. For what range of x-values is the infinite series convergent?

APP8. In finding the minimum of L versus c in the ladder problem, MATLAB used two applications of "golden." This refers to the Golden Mean. What is the Golden Mean? Where does this value come from? What other applications is there of this other than in finding a minimum? Why is it called "golden"?

APP9. Write a computer program that converts numbers to/from binary; octal; decimal; hexadecimal.

APP10. Compare the graphs of $(1 - x)^n$ for $n = 2, 4, 6, 8, \ldots$. You will find that, for x-values near 1, the graphs depart less and less from the x-axis as n increases. Determine how accurate the computer must be to get a nonzero value at $x = 0.8$ as a function of n. Then, find the values of x where the departure from zero is just greater than eps when the computation is done in single precision.

APP11. The ABC Manufacturing Company currently ships a product in a cardboard box that measures $6 \times 7.5 \times 2.5$ in. The box is formed from a die-cut pattern using a piece of card stock that is 1/32 in. thick, 12.5 in. wide, and 18 in. high. After the card stock is cut, the unassembled box looks like the figure. Part T forms the top, part B is the bottom, and parts S make the sides. The solid lines represent cuts and the dashed lines represent folds. Flaps F are folded inside the box and are glued to the side pieces. After the box has been filled, the top is folded over and flaps G (which are 1 in. wide) are folded and glued to the outside of the box to seal the box.

You have been asked to lay out the pattern for a new product. Following the same type of design as in the figure, draw the pattern for die-cutting the box material. The box is to have the largest volume that can be made from card stock that measures 15×20 in. Flaps G are still to be 1 in. wide.

Sketch the pattern for the new box. Show how you determined the dimensions to achieve the maximum volume for the box, proving that its volume is the maximum possible.

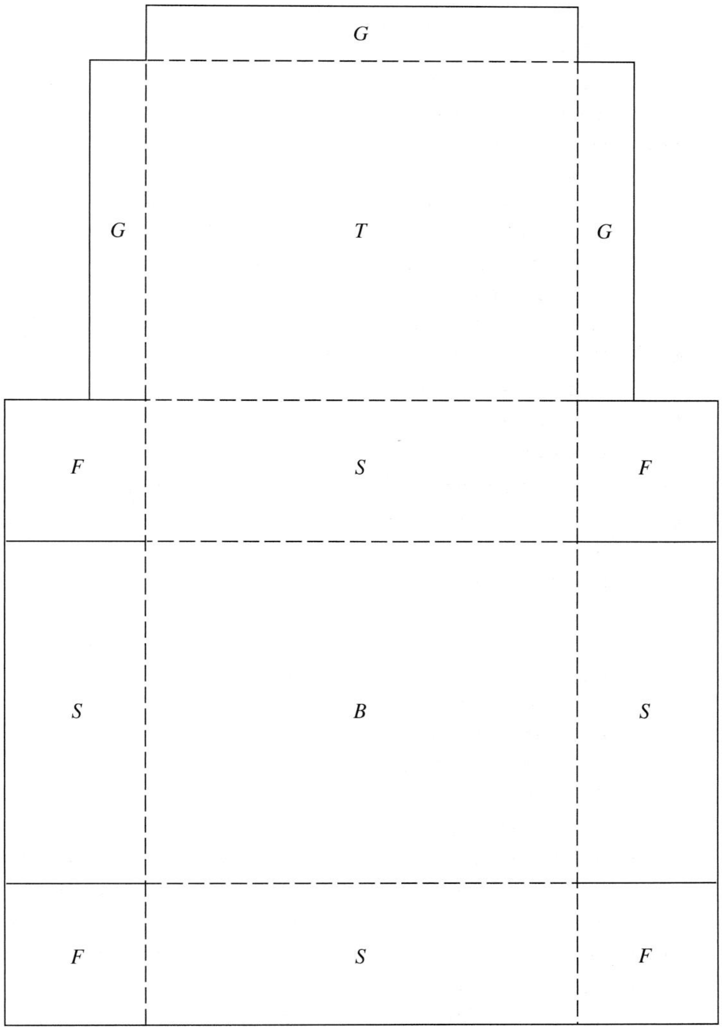

1

Solving Nonlinear Equations

An important problem in applied mathematics is to "solve $f(x) = 0$" where $f(x)$ is a function of x. The values of x that make $f(x) = 0$ are called the *roots* of the equation. They are also called the *zeros* of $f(x)$. This chapter describes some of the many methods for solving $f(x) = 0$ by numerical procedures. We also treat the more complicated case wherein a set of nonlinear equations are to be solved simultaneously:

$$f(x, y, z) = 0,$$
$$g(x, y, z) = 0,$$
$$h(x, y, z) = 0.$$

For example, an engineer might want to find the pressure needed to cause a fluid suspension of particles to flow through a pipe (perhaps in a paper mill). The pressure required depends on the length of the pipe, its diameter, the quantity of fluid that is to flow, and a number called the "friction factor" that has been determined from experiments. This nonlinear equation can compute the friction factor, f:

$$\frac{1}{\sqrt{f}} = \left(\frac{1}{k}\right) \ln(\text{RE}\sqrt{f}) + \left(14 - \frac{5.6}{k}\right).$$

where the parameter k is known and RE, the so-called Reynold's number, can be computed from the pipe diameter, the velocity of flow, and the viscosity of the fluid. The equation for f is not solvable except by the numerical procedures of this chapter.

We show a total of ten root-finding procedures in this chapter. Five of these are described in detail, the others are only mentioned. Of these ten methods, six apply to any type of equation, the others only to polynomials. Why so many? We do this to acquaint you with them, to show that there are often many numerical methods for solving a problem, and to point out why one method may be preferred over another. We even describe other methods for solving nonlinear problems in Chapter 7!

Contents of This Chapter

1.1 Interval Halving (Bisection)

Interval halving (bisection), an ancient but effective method for finding a zero of $f(x)$, is an excellent introduction to numerical methods. It begins with two values for x that bracket a root. It determines that they do in fact bracket a root because the function $f(x)$ changes signs at these two x-values and, if $f(x)$ is continuous, there must be at least one root between the values. A plot of $f(x)$ is useful to know where to start.

The bisection method then successively divides the initial interval in half, finds in which half the root(s) must lie, and repeats with the endpoints of the smaller interval. The test to see that $f(x)$ does change sign between points a and b is to see if $f(a) * f(b) < 0$.

We will compare this method with the others that are described in this chapter by this same function for each:

$$f(x) = 3x + \sin(x) - e^x.$$

It is a good plan to look at a plot of the function to learn where the function crosses the x-axis. MATLAB can do it for us:

```
EDU>> f = inline('3*x + sin(x) - exp(x)')
f =
   Inline function:
   f(x) = 3*x + sin(x) - exp(x)
EDU>> fplot(f,[0 2]);grid on
```

And we see this figure that indicates there are zeros at about $x = 0.35$ and 1.9.

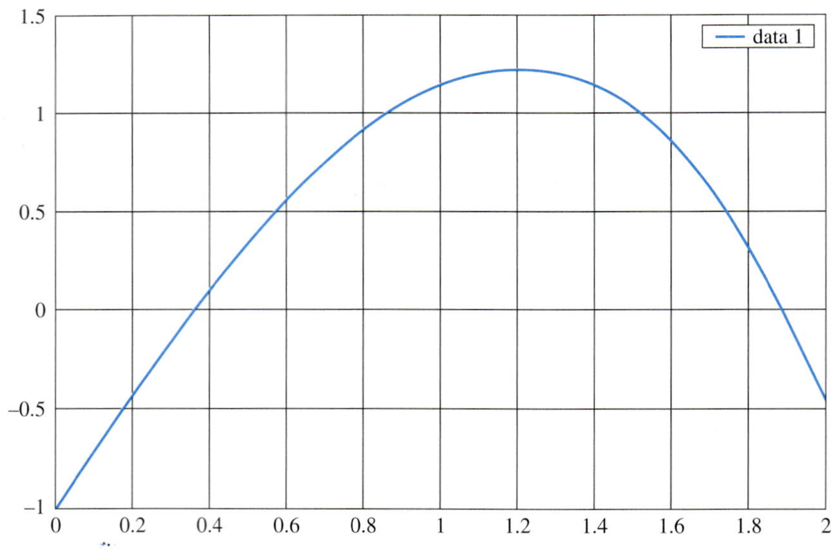

Here is an algorithm for the bisection method:

An Algorithm for Halving the Interval (Bisection)

To determine a root of $f(x) = 0$ that is accurate within a specified tolerance value, given values x_1 and x_2 such that $f(x_1) * f(x_2) < 0$,

Repeat
Set $x_3 = (x_1 + x_2)/2$.

If $f(x_3) * f(x_1) < 0$ Then
Set $x_2 = x_3$
Else Set $x_1 = x_3$ End If.
Until ($|x_1 - x_2|) < 2$ * tolerance value).

The final value of x_3 approximates the root, and it is in error by not more than $|x_1 - x_2|/2$.
Note: The method may produce a false root if $f(x)$ is discontinuous on $[x_1, x_2]$.

A program that implements the method gave the results displayed in Table 1.1. To obtain the true value for the root, which is needed to compute the actual error column, we again used MATLAB:

```
EDU>> solve('3*x + sin(x) - exp(x)')
ans =
.36042170296032440136932951583028
```

which is really more accurate than we need.

MATLAB surely used a more advanced method than bisection to get the answer to the example, but we can write a program in MATLAB that does bisection. We present this to illustrate how you can create a MATLAB program. This is done through a so-called M-file. Clicking on 'File/New/M-file' in MATLAB's toolbar brings up a form into which we enter the commands on the following page:

Table 1.1 The bisection method for $f(x) = 3x + \sin(x) - e^x = 0$, starting from $x_1 = 0$, $x_2 = 1$, using a tolerance value of 1E-4

Iteration	X_1	X_2	X_3	$F(X_3)$	Maximum error	Actual error
1	0.00000	1.00000	0.50000	0.33070	0.50000	0.13958
2	0.00000	0.50000	0.25000	−0.28662	0.25000	−0.11042
3	0.25000	0.50000	0.37500	0.03628	0.12500	0.01458
4	0.25000	0.37500	0.31250	−0.12190	0.06250	−0.04792
5	0.31250	0.37500	0.34375	−0.04196	0.03125	−0.01667
6	0.34375	0.37500	0.35938	−0.00262	0.01563	−0.00105
7	0.35938	0.37500	0.36719	0.01689	0.00781	0.00677
8	0.35938	0.36719	0.36328	0.00715	0.00391	0.00286
9	0.35938	0.36328	0.36133	0.00227	0.00195	0.00091
10	0.35938	0.36133	0.36035	−0.00018	0.00098	−0.00007
11	0.36035	0.36133	0.36084	0.00105	0.00049	0.00042
12	0.36035	0.36084	0.36060	0.00044	0.00024	0.00017
13	0.36035	0.36060	0.36047	0.00013	0.00012	0.00005

```
function rtn = bisec(fx,xa,xb,n)
%bisec does n bisections to approximate
% a root of fx
x = xa; fa = eval(fx);
x = xb; fb = eval(fx);
for i = 1:n
  xc = (xa + xb)/2; x = xc; fc = eval(fx);
  X = [i, xa, xb, xc, fc];
  disp (X)
  if fc*fa < 0
    xb = xc;
  else xa = xc;
  end % of if/else
end % of for loop
```

which we save with the name 'bisec.m.' Now if we enter these commands:

```
EDU>> fx = '3*x + sin(x) - exp(x)'
fx =
  3*x + sin(x) - exp(x)
EDU>> bisec(fx,0,1,13)
```

we see a display similar to Table 1.1, except the iteration numbers are not integers:

1.0000	0	1.0000	0.5000	0.3307
2.0000	0	0.5000	0.2500	−0.2866
3.0000	0.2500	0.5000	0.3750	0.0363
4.0000	0.2500	0.3750	0.3125	−0.1219
5.0000	0.3125	0.3750	0.3438	−0.0420
6.0000	0.3438	0.3750	0.3594	−0.0026
7.0000	0.3594	0.3750	0.3672	0.0169
8.0000	0.3594	0.3672	0.3633	0.0071
9.0000	0.3594	0.3633	0.3613	0.0023
10.0000	0.3594	0.3613	0.3604	−0.0002
11.0000	0.3604	0.3613	0.3608	0.0010
12.0000	0.3604	0.3608	0.3606	0.0004
13.0000	0.3604	0.3606	0.3605	0.0001

It may be interesting for you to see if you can modify the program to produce integers in the first column.

The main advantage of interval halving is that it is guaranteed to work if $f(x)$ is continuous in $[a, b]$ and if the values $x = a$ and $x = b$ actually bracket a root.* Another important advantage that few other root-finding methods share is that the number of interations to achieve a specified accuracy is known in advance. Because the interval $[a, b]$ is halved

* This guarantee can be voided—if the function has a slope very near to zero at the root, the precision of the computations may be inadequate.

each time, the last value of x_3 differs from the true root by less than $\frac{1}{2}$ the last interval. So we can say with surety that

$$\text{error after } n \text{ iterations} < \left|\frac{(b - a)}{2^n}\right|.$$

The major objection of interval halving has been that it is slow to converge. Other methods require fewer iterations to achieve the same accuracy (but then we do not always know a bound on the accuracy).

Observe in Table 1.1 that the estimate of the root may be better at an earlier iteration than at later ones. (The third iterate is closer to the true root than are the next two; we are closer at iteration 6 than at iteration 7.) Of course, in this example we have the advantage of knowing the answer, which is never the case. However, the values of $f(x_3)$ themselves show that these better estimates are closer to the root. (This is not an absolute criterion—some functions may be nearly zero at points not so near the root, but, for smooth functions, a small value of the function is a good indicator that we are near to the root. This is especially true when we are quite close to the root.) The methods we consider in later sections use the values of $f(x)$ to find the root more rapidly.

With speedy computers so prevalent today the slowness of the bisection method is of less concern. When the values of Table 1.1 were computed from a program, the results were seen in less than a second.

When the roots of functions must be computed a great many times (this may be a requirement of some other program that does engineering analysis), the efficiency of interval halving may be inadequate. This will be particularly true if $f(x)$ is not given explicitly but, instead, is developed internally within the other program. In that case, finding values of x that bracket the root may also be a problem.

In spite of arguments that other methods find roots with fewer iterations, interval halving is an important tool in the applied mathematician's arsenal. Bisection is generally recommended for finding an approximate value for the root, and then this value is refined by more efficient methods. The reason is that most other root-finding methods require a starting value near to a root—lacking this, they may fail completely.

Do not overlook other techniques that may seem mundane for getting a first approximation to the root. Graphing the function is always helpful in showing where roots occur, and with programs like MATLAB (or a graphing calculator) that do plots so handily, getting the graph before beginning a root-finding routine is a good practice. Searching methods should also be considered as a preliminary step. Stepping through the interval $[-1, 1]$ and testing whether $f(x)$ changes sign will show whether there are roots in that interval. Roots of larger magnitude can be found by stepping through that same interval with x replaced by $1/y$, because the roots of this modified function are the reciprocals of the roots of the original function. Experience with the particular types of problems that are being solved may also suggest approximate values of roots. Even intuition can be a factor. Acton (1970) gives an especially interesting and illuminating discussion.

When there are multiple roots, interval halving may not be applicable, because the function may not change sign at points on either side of the roots. Here a graph will be most important to reveal the situation. In this case, we may be able to find the roots by working with $f'(x)$, which will be zero at a multiple root.

1.2 Linear Interpolation Methods

Bisection is simple to understand but it is not the most efficient way to find where $f(x)$ is zero.

Most functions can be approximated by a straight line over a small interval. The two methods of this section are based on doing just that.

The Secant Method

The *secant method* begins by finding two points on the curve of $f(x)$, hopefully near to the root we seek. A graph or a few applications of bisection might be used to determine the approximate location of the root. As Figure 1.1 illustrates, we draw the line through these two points and find where it intersects the x-axis. The two points may both be on one side of the root as seen in the figure but they could also be on opposite sides.

If $f(x)$ were truly linear, the straight line would intersect the x-axis at the root. But $f(x)$ will never be exactly linear because we would never use a root-finding method on a linear function! That means that the intersection of the line with the x-axis is not at $x = r$ but that it should be close to it. From the obvious similar triangles we can write

$$\frac{(x_1 - x_2)}{f(x_1)} = \frac{(x_0 - x_1)}{f(x_0) - f(x_1)}$$

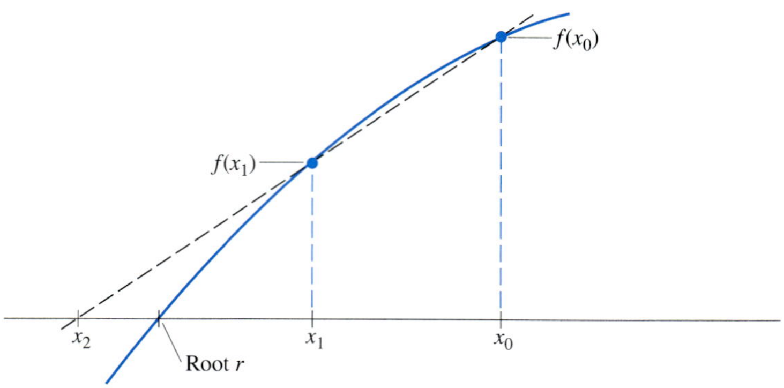

Figure 1.1

and from this solve for x_2:

$$x_2 = x_1 - f(x_1)\frac{(x_0 - x_1)}{f(x_0) - f(x_1)}.$$

Because $f(x)$ is not exactly linear, x_2 is not equal to r, but it should be closer than either of the two points we began with.

If we repeat this, we have:

$$x_{n+1} = x_n - f(x_n)\frac{(x_{n-1} - x_n)}{f(x_{n-1}) - f(x_n)}.$$

Because each newly computed value should be nearer to the root, we can do this easily after the second iterate has been computed, by always using the last two computed points. But after the *first* iteration there aren't "two last computed points." So we make sure to start with x_1 closer to the root than x_0 by testing $f(x_0)$ and $f(x_1)$ and swapping if the first function value is smaller.* The net effect of this rule is to set $x_0 = x_1$ and $x_1 = x_2$ after each iteration. The exceptions to this rule are pathological cases, which we consider next.

The technique we have described is known as the secant method because the line through two points on the curve is called the secant line. Here is pseudocode for the secant method algorithm:

An Algorithm for the Secant Method

To determine a root of $f(x) = 0$, given two values, x_0 and x_1, that are near the root,

If $|f(x_0)| < |f(x_1)|$ Then
  Swap x_0 with x_1.
Repeat
  Set $x_2 = x_1 - f(x_1) * \dfrac{x_0 - x_1}{f(x_0) - f(x_1)}$
  Set $x_0 = x_1$.
  Set $x_1 = x_2$.
Until $|f(x_2)| <$ tolerance value.

Note: If $f(x)$ is not continuous, the method may fail.

An alternative stopping criterion for the secant method is when the pair of points being used are sufficiently close together.

---

* $|f(x_0)| < |f(x_1)|$ does not always mean that x_0 is closer to the root, but that is often the case. When it is, the method is speeded up. In any case, the algorithm still converges to the root when $f(x)$ is continuous and we start near enough to the root.

Table 1.2 Secant method on $f(x) = 3x + \sin(x) - e^x$

Iteration	x_0	x_1	x_2	$f(x_2)$
1	1	0	0.4709896	0.2651588
2	0	0.4709896	0.3722771	2.953367E-02
3	0.4709896	0.3722771	0.3599043	−1.294787E-03
4	0.3722771	0.3599043	0.3604239	5.552969E-06
5	0.3599043	0.3604239	0.3604217	3.554221E-08

At $x = .3604217$, tolerance of .0000001 met!

An Example

Table 1.2 shows the results from the secant method for the same function that was used to illustrate bisection. We know that the root is at 0.3604217. Notice that fewer iterations are required compared to bisection. The efficiency of numerical methods is often measured by how many times a function must be evaluated because that usually is the most time-consuming part of the procedure.

An objection is sometimes raised about the secant method. If the function is far from linear near the root, the successive iterates can fly off to points far from the root, as seen in Figure 1.2.

If the method is being carried out by a program that displays the successive iterates, the user can interrupt the program should such improvident behavior be observed. Also, if the function was plotted before starting the method, it is unlikely that the problem will be encountered, because a better starting value would be used. There are times when this remedy is not possible: when the routine is being used within another program that needs to find a root before it can proceed.

Linear Interpolation (False Position)

A way to avoid such pathology is to ensure that the root is bracketed between the two starting values and remains between the succcessive pairs. When this is done, the method is

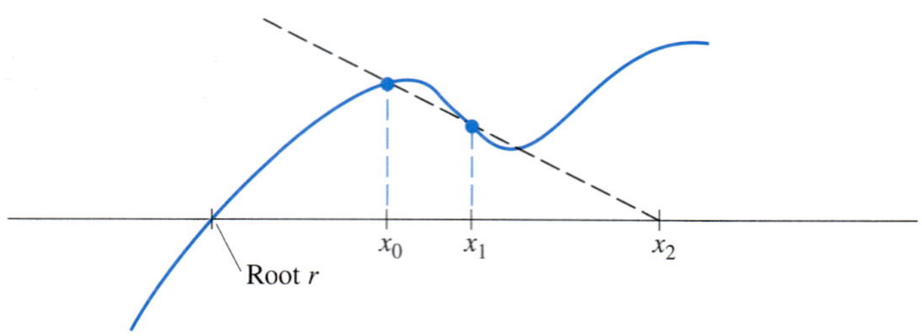

Figure 1.2
A pathological case for the secant method

known as *linear interpolation,* or, more often, as the *method of false position* (in Latin, *regula falsi*). This technique is similar to bisection except the next iterate is taken at the intersection of a line between the pair of x-values and the x-axis rather than at the midpoint. Doing so gives faster convergence than does bisection, but at the expense of a more complicated algorithm.

Here is the pseudocode for *regula falsi* (method of false position):

An Algorithm for the Method of False Position *(regula falsi)*

To determine a root of $f(x) = 0$, given two values of x_0 and x_1 that bracket a root: that is, $f(x_0)$ and $f(x_1)$ are of opposite sign,

Repeat
    Set $x_2 = x_1 - f(x_1) * \dfrac{x_0 - x_1}{f(x_0) - f(x_1)}$
    If $f(x_2)$ is of opposite sign to $f(x_0)$ Then
        Set $x_1 = x_2$
    Else
    Set $x_0 = x_2$
    End If.
Until $|f(x_2)| <$ tolerance value.

Note: If $f(x)$ is not continuous, the method may fail.

Table 1.3 compares the results of three methods—interval halving (bisection), linear interpolation, and the secant method—on $f(x) = 3x + \sin(x) - e^x = 0$. Observe that the speed of convergence is best for the secant method, poorest for interval halving, and

Table 1.3 Comparison of methods, $f(x) = 3x + \sin(x) - e^x = 0, x_0 = 0, x_1 = 1$

	Interval halving		False position		Secant method	
Iteration	x	$f(x)$	x	$f(x)$	x	$f(x)$
1	0.5	0.330704	0.470990	0.265160	0.470990	0.265160
2	0.25	−0.286621	0.372277	0.029533	0.372277	0.029533
3	0.375	0.036281	0.361598	$2.94 * 10^{-3}$	0.359904	$-1.29 * 10^{-3}$
4	0.3125	−0.121899	0.360538	$2.90 * 10^{-4}$	0.360424	$5.55 * 10^{-6}$
5	0.34375	−0.041956	0.360433	$2.93 * 10^{-5}$	0.360422	$3.55 * 10^{-7}$
Error after 5 iterations	0.01667		$-1.17 * 10^{-5}$		$<-1 * 10^{-7}$	

(Exact value of root is 0.360421703.)

intermediate for false position. Notice that false position converges to the root from only one side, slowing it down, especially if that end of the interval is farther from the root. There is a way to avoid this result, called modified linear interpolation. We omit the details of this method.

1.3 Newton's Method

One of the most widely used methods of solving equations is *Newton's method.** Like the previous ones, this method is also based on a linear approximation of the function, but does so using a tangent to the curve. Figure 1.3 gives a graphical description. Starting from a single initial estimate, x_0, that is not too far from a root, we move along the tangent to its intersection with the x-axis, and take that as the next approximation.† This is continued until either the successive x-values are sufficiently close or the value of the function is sufficiently near zero.**

The calculation scheme follows immediately from the right triangle shown in Figure 1.3, which has the angle of inclination of the tangent line to the curve at $x = x_0$ as one of its acute angles:

$$\tan\theta = f'(x_0) = \frac{f(x_0)}{x_0 - x_1}, \qquad x_1 = x_0 - \frac{f(x_0)}{f'(x_0)}.$$

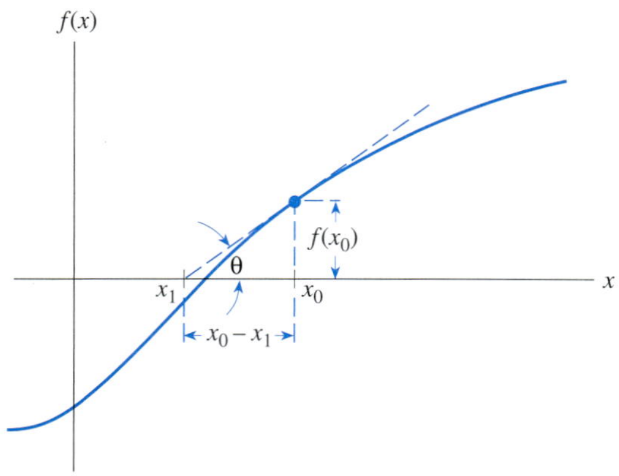

Figure 1.3

* Newton did not publish an extensive discussion of this method, but he solved a cubic polynomial in *Principia* (1687). The version given here is considerably improved over his original example.

† The algorithm for Newton's method can be derived from a Taylor series. We suggest that you do it this way.

** Which criterion should be used often depends on the particular physical problem to which the equation applies. Customarily, agreement of successive x-values to a specified tolerance is required.

We continue the calculation scheme by computing

$$x_2 = x_1 - \frac{f(x_1)}{f'(x_1)},$$

or, in more general terms,

$$x_{n+1} = x_n - \frac{f(x_n)}{f'(x_n)}, \qquad n = 0, 1, 2, \ldots.$$

Newton's algorithm is widely used because, at least in the near neighborhood of a root, it is more rapidly convergent than any of the methods discussed so far. We show in a later section that the method is quadratically convergent, by which we mean that the error of each step approaches a constant K times the square of the error of the previous step. The net result of this is that the number of decimal places of accuracy nearly doubles at each iteration. However, there is the need for two function evaluations at each step, $f(x_n)$ and $f'(x_n)$, and we must obtain the derivative function at the start.*

When Newton's method is applied to $f(x) = 3x + \sin x - e^x = 0$, we have the following calculations:

$$f(x) = 3x + \sin x - e^x,$$
$$f'(x) = 3 + \cos x - e^x$$

There is little need to use MATLAB to get this simple derivative, but, for practice, here is how to do it:

```
EDU>> fx = '3*x + sin(x) - exp(x)'
fx =
3*x + sin(x) - exp(x)

EDU>> dfx = diff(fx)
dfx =
3 + cos(x) - exp(x)
```

If we begin with $x_0 = 0.0$, we have

$$x_1 = x_0 - \frac{f(x_0)}{f'(x_0)} = 0.0 - \frac{-1.0}{3.0} = 0.33333;$$

$$x_2 = x_1 - \frac{f(x_1)}{f'(x_1)} = 0.33333 - \frac{-0.068418}{2.54934} = 0.36017;$$

$$x_3 = x_2 - \frac{f(x_2)}{f'(x_2)} = 0.36017 - \frac{-6.279 \times 10^{-4}}{2.50226} = 0.3604217.$$

* Finding $f'(x)$ may be difficult. Computer algebra systems can be a real help.

After three iterations, the root is correct to seven digits; convergence is much more rapid than any previous method. In fact, the error after an iteration is about one-third of the square of the previous error. In comparing numerical methods, however, we usually count the number of times functions must be evaluated. Because Newton's method requires two function evaluations per step, the comparison is not as one-sided in favor of Newton's method as at first appears; the three iterations with Newton's method required six function evaluations. Five iterations with the previous methods also required six evaluations. If a difficult problem requires many iterations to converge, the number of function evaluations with Newton's method may be many more than with linear iteration methods because Newton always uses two per iteration whereas the others take only one (after the first step that takes two).

A more formal statement of the algorithm for Newton's method, suitable for implementation in a computer program, is shown here:

Newton's Method

To determine a root of $f(x) = 0$, given x_0 reasonably close to the root,

Compute $f(x_0), f'(x_0)$.
If $(f(x_0) \neq 0)$ And $(f'(x_0) \neq 0)$ Then
Repeat
  Set $x_1 = x_0$.
  Set $x_0 = x_0 - f(x_0)/f'(x_0)$.
  Until $(|x_1 - x_0| <$ tolerance value 1) Or
  $|f(x_0)| <$ tolerance value 2).
  End If.

Note: The method may converge to a root different from the expected one or diverge if the starting value is not close enough to the root.

When Newton's method is applied to polynomial functions, special techniques facilitate such application. We consider these in a later section of this chapter.

In some cases Newton's method will not converge. Figure 1.4 illustrates this situation. Starting with x_0, one never reaches the root r because $x_6 = x_1$ and we are in an endless loop. Observe also that if we should ever reach the minimum or maximum of the curve, we will fly off to infinity. We will develop the analytical condition for this in a later section and show that Newton's method is quadratically convergent in most cases.

Relating Newton's Method to Other Methods

It is of interest to notice that the previous interpolation methods are closely related to Newton's method. For linear interpolation, whose algorithm we can write as

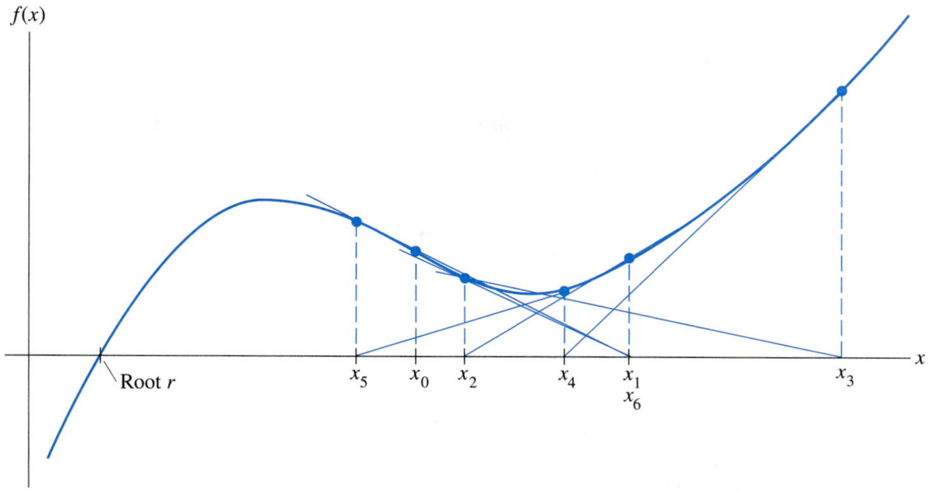

Figure 1.4

$$x_{n+1} = x_n - \frac{f(x_n)}{\dfrac{f(x_n) - f(x_{n-1})}{x_n - x_{n-1}}}$$

we see that the denominator of the fractional term is exactly the definition of the derivative except not taken to the limit as the two x-values approach each other. This *difference quotient* is an approximation to the derivative, as we will explain in detail in a later chapter. Because the denominator of the fractional term is an approximation to the derivative of f, we see the close resemblance to Newton's method.

The secant method has exactly this same resemblance to Newton's method because it is just linear interpolation without the requirement that the two x-values bracket the root. Because these two values usually are closer together than for linear interpolation, the approximation to the derivative is even better.

From this we see that there is an alternative way to get the derivative for Newton's method. If we compute $f(x)$ at two closely spaced values for x and divide the difference in the function values by the difference in x-values, we have the derivative (nearly) without having to differentiate. Although this sounds like spending an extra function evaluation, we avoid having to evaluate the derivative function and so it breaks even. (Convergence will not usually be as fast, however.)

Complex Roots

Newton's method works with complex roots if we give it a complex value for the starting value. Here is an example.

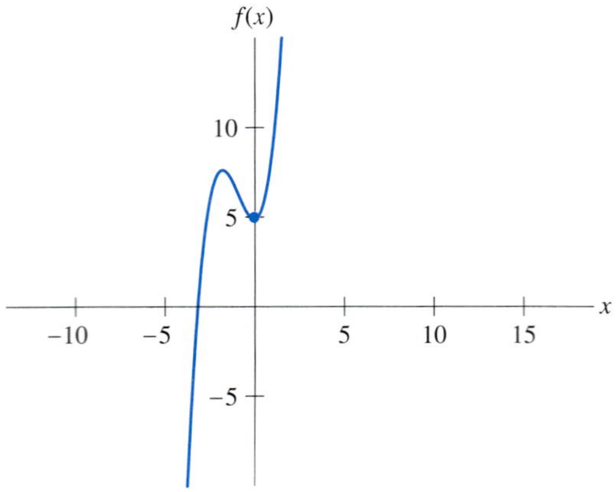

Figure 1.5
Plot of $f(x) = x^3 + 2x^2 - x + 5$

EXAMPLE 1.1 Use Newton's method on $f(x) = x^3 + 2x^2 - x + 5$.

Figure 1.5 shows the graph of $f(x)$. It has a real root at about $x = -3$, whereas the other two roots are complex because the x-axis is not crossed again.

If we begin Newton's method with $x_0 = 1 + i$ (we used this in the lack of knowledge about the complex root), we get these successive iterates:

1. $0.486238 + 1.04587i$
2. $0.448139 + 1.23665i$
3. $0.462720 + 1.22242i$
4. $0.462925 + 1.22253i$
5. $0.462925 + 1.22253i$

Because the fourth and fifth iterates agree to six significant figures, we are sure that we have an estimate good to at least that many figures. The second complex root is the conjugate of this: $0.462925 - 1.22253i$. If we begin with $x_0 = 1 - i$, the method converges to the conjugate.

If we begin with a real starting value—say, $x_0 = -3$—we get convergence to the root at $x = -2.92585$.

Newton's Method for Polynomials

We have already pointed out that polynomials are of great importance in numerical analysis because of their "nice" behavior and because they can be evaluated using only arithmetic operations. Descartes's rule of signs (see Appendix A) lets us predict the number of positive roots. Any root-finding method can get their roots but there are some special techniques with Newton's method.

As we have said, evaluating a polynomial at some x-value is best done by nested multiplication. The name for this is *Horner's Method* and MATLAB has a built-in function to rearrange a polynomial into nested form:

```
EDU>> P = poly2sym([2 1 -3 -3])
P =
2x^3 + x^2 - 3*x - 3
EDU>> horner(P)
ans =
((2x + 1) * x - 3) * x - 3
```

In the above, the first command created a symbolic representation of the polynomial from the vector of coefficients and the second put this into nested form.

Evaluating a polynomial in nested form can be done in a computer program by *synthetic division.* This procedure was done by hand before the advent of computers and perhaps you have seen it before. While you may not do hand computations very often, synthetic division is a good way to start a discussion of the computer algorithm.

Suppose we want to find the value at $x = 2$ of

$$P(x) = 2x^3 + x^2 - 3x - 3.$$

Write the coefficients in a row and follow this pattern:

$$\begin{array}{c|cccc} x = 2 & 2 & 1 & -3 & -3 \\ & & 4 & 10 & 14 \\ \hline & 2 & 5 & 7 & \textcircled{11} \end{array}$$

Here is what was done to get the tableau: Copy the first coefficient below the line, multiply this times the x-value and add to the second coefficient, multiply that result by the x-value and add to the third coefficient, and do the same for the last coefficient. The last row of numbers is the coefficients of the *reduced polynomial* and the remainder from the division.

The final result, 11, which has been circled, is the value of the polynomial at $x = 2$! This is also the remainder from the division:

$$\frac{2x^3 + x^2 - 3x - 3}{x - 2} = 2x^2 + 5x + 7 + \frac{11}{x - 2}.$$

If you study the steps in synthetic division, you will see that these are exactly what is done if the polynomial is evaluated in nested form: Horner's method and synthetic division are precisely the same.

The value of synthetic division in getting a root by Newton's method is that, if the reduced polynomial is divided by $(x - 2)$, the remainder from this is the value of the derivative at $x = 2$:

$$\begin{array}{c|ccc} x = 2 & 2 & 5 & 7 \\ & & 4 & 18 \\ \hline & 2 & 9 & \textcircled{25} \end{array}$$

where the circled 25 is $P'(2)$.

With the values of $P(2)$ and $P'(2)$ available, we can use them in Newton's method to estimate a root starting with $x_1 = 2$:

$$x_2 = 2 - 11/25 = 1.56,$$

which is closer to a root of $P(x)$, which MATLAB tells us is at $x = 1.3782$:

```
EDU>> p = [2 1 -3 -3]
p =
   2 1 -3 -3
EDU>> r = roots(p)
r =
     1.3782
   -0.9391 + 0.4545i
   -0.9391 - 0.4545i
```

MATLAB also told us that there are two complex-valued roots.

To divide two polynomials using MATLAB, we first define them as arrays of the coefficients, then use the command 'deconv' (which really means to get the inverse of the convolution of two vectors, which is the equivalent of multiplying the polynomials). So, to divide $2x^3 + x^2 - 3x - 3$ by $(x - 2)$, we do:

```
EDU>> N = [2 1 -3 -3]; D = [1 -2];
EDU>> [q,r] = deconv (N, D)
q =
     2     5     7
r =
     0     0     0    11
```

which is MATLAB's way of telling us that N/D is $(2x^2 + 5x + 7)$ plus the remainder, $11/(x - 2)$, exactly as the synthetic division gave us. A second division of the reduced polynomial in the same fashion will give us $P'(2)$.

Parallel Processing

Horner's method for evaluating a polynomial is one of the classic examples where we can speed up a computation by using *parallel processors.* The directed acyclic graphs (dags) for the sequential and parallel algorithms are shown in Figure 1.6. Although we have more operations (five multiplies and three adds) with the parallel scheme (compared to three multiplies and three adds), the time required to produce the result is reduced from six steps to four steps. The time savings comes from doing some operations in parallel rather than in succession, of course. Observe that the most efficient method for sequential processing (Horner's method) is not used in parallel processing.

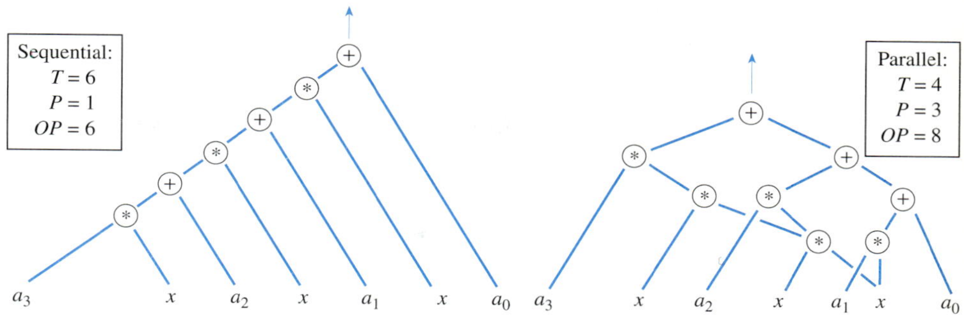

Figure 1.6
Dags for evaluating a polynomial of degree 3

An Algorithm for Synthetic Division and the Remainder Theorem

We can develop an algorithm for synthetic division and show that the remainders are the value of the polynomial and its derivative by writing the nth-degree polynomial as

$$P_n(x) = a_n x^n + a_{n-1} x^{n-1} + \cdots + a_1 x + a_0.$$

We wish to divide this by the factor $(x - x_1)$, giving a reduced polynomial $Q_{n-1}(x)$ of degree $n - 1$, and a remainder, R, which is a constant:

$$\frac{P_n(x)}{x - x_1} = Q_{n-1}(x) + \frac{R}{x - x_1}.$$

Rearranging yields

$$P_n(x) = (x - x_1)Q_{n-1}(x) + R.$$

Note that at $x = x_1$,

$$P_n(x_1) = (0)[Q_{n-1}(x_1)] + R = R,$$

which is the remainder theorem: The remainder on division by $(x - x_1)$ is the value of the polynomial at $x = x_1$, $P_n(x_1)$.

If we differentiate $P_n(x)$, we get

$$P'_n(x) = (x - x_1)Q'_{n-1}(x) + (1)Q_{n-1}(x) + 0.$$

Letting $x = x_1$, we have

$$P'_n(x_1) = Q_{n-1}(x_1).$$

We evaluate the Q-polynomial at x_1 by a second division whose remainder equals $Q_{n-1}(x_1)$. This verifies that the second remainder from synthetic division yields the value for the derivative of the polynomial.

We now develop the synthetic division algorithm, writing $Q_{n-1}(x)$ in form similar to $P_n(x)$:

$$\begin{aligned} P_n(x) &= a_n x^n + a_{n-1}x^{n-1} + \cdots + a_1 x + a_0 \\ &= (x - x_1)Q_{n-1}(x) + R \\ &= (x - x_1)(b_{n-1}x^{n-1} + b_{n-2}x^{n-2} + \cdots + b_1 x + b_0) + R. \end{aligned}$$

Multiplying out and equating coefficients of like terms in x, we get

$$\left.\begin{aligned} \text{coef. of } x^n: \quad & a_n = b_{n-1} \\ x^{n-1}: \quad & a_{n-1} = b_{n-2} - x_1 b_{n-1} \\ x^{n-2}: \quad & a_{n-2} = b_{n-3} - x_1 b_{n-2} \\ & \vdots \\ x: \quad & a_1 = b_0 - x_1 b_1 \\ \text{const:} \quad & a_0 = R - x_1 b_0 \end{aligned}\right\} \quad \text{or} \quad \left\{\begin{aligned} & b_{n-1} = a_n \\ & b_{n-2} = a_{n-1} + x_1 b_{n-1} \\ & b_{n-3} = a_{n-2} + x_1 b_{n-2} \\ & \vdots \\ & b_0 = a_1 + x_1 b_1 \\ & R = a_0 + x_1 b_0 \end{aligned}\right.$$

The general form is $b_i = a_{i+1} + x_1 b_{i+1}$, by which all the b's may be calculated, provided that we first set $b_n = 0$. If this is compared to the preceding synthetic divisions, it is seen to be identical, except that we now have a vertical array. The horizontal layout is easier for hand computation. For evaluation of the derivative, a set of c-values is computed from the b's in the same way in which the b's are computed from the a's.

1.4 Muller's Method

Most of the root-finding methods that we have considered so far have approximated the function in the neighborhood of the root by a straight line. Obviously, this is never true; if the function were linear, finding the root would take practically no effort. *Muller's method* is based on approximating the function in the neighborhood of the root by a quadratic polynomial. This gives a much closer match to the actual curve.

A second-degree polynomial is made to fit three points near a root, at x_0, x_1, x_2, with x_0 between x_1 and x_2. The proper zero of this quadratic, using the quadratic formula, is used as the improved estimate of the root. The process is then repeated using the set of three points nearest the root being evaluated.

The procedure for Muller's method is developed by writing a quadratic equation that fits through three points in the vicinity of a root, in the form $av^2 + bv + c$. (See Fig. 1.7.) The development is simplified if we transform axes to pass through the middle point, by letting $v = x - x_0$.

Let $h_1 = x_1 - x_0$ and $h_2 = x_0 - x_2$. We evaluate the coefficients by evaluating $p_2(v)$ at the three points:

$$\begin{aligned} v = 0: \quad & a(0)^2 + b(0) + c = f_0; \\ v = h_1: \quad & ah_1^2 + bh_1 + c = f_1; \\ v = -h_2: \quad & ah_2^2 - bh_2 + c = f_2. \end{aligned}$$

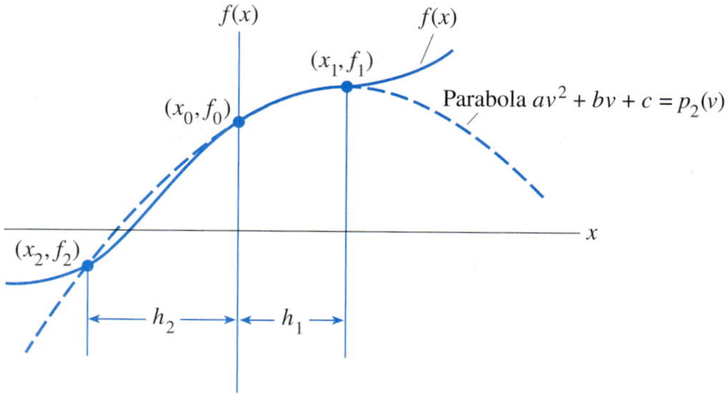

Figure 1.7

From the first equation, $c = f_0$. Letting $h_2/h_1 = \gamma$, we can solve the other two equations for a and b:

$$a = \frac{\gamma f_1 - f_0(1+\gamma) + f_2}{\gamma h_1^2(1+\gamma)}, \qquad b = \frac{f_1 - f_0 - ah_1^2}{h_1}.$$

After computing a, b, and c, we solve for the root of $av^2 + bv + c = 0$ by the quadratic formula, choosing the root nearest to the middle point x_0. This value is

$$\text{root} = x_0 - \frac{2c}{b \pm \sqrt{b^2 - 4ac}},$$

with the sign in the denominator taken to give the largest absolute value of the denominator (that is, if $b > 0$, choose plus; if $b < 0$, choose minus; if $b = 0$, choose either). The reason for using this somewhat unusual form of the quadratic formula is to make the next iterate closer to the root.

We take the root of the polynomial as one of a set of three points for the next approximation, taking the three points that are most closely spaced (that is, if the root is to the right of x_0, take x_0, x_1, and the root; if to the left, take x_0, x_2, and the root). We always reset the subscripts to make x_0 be the middle of the three values.

An algorithm for Muller's method is shown here:

Mueller's Method

Given the points x_2, x_0, x_1 in increasing value,

Evaluate the corresponding function values: f_2, f_0, f_1.

Repeat

(Evaluate the coefficients of the parabola, $ax^2 + bx + c$, determined by the three points. $\{(x_2, f_2), (x_0, f_0), (x_1, f_1)\}$.)

Set $h_1 = x_1 - x_0$; $h_2 = x_0 - x_2$; $\gamma = h_2/h_1$.
Set $c = f_0$.
Set $a = \dfrac{\gamma f_1 - f_0(1 + \gamma) + f_2}{\gamma h_1^2(1 + \gamma)}$
Set $b = \dfrac{f_1 - f_0 - ah_1^2}{h_1}$

(Next, compute the roots of the polynomial.)

Set root $= x_0 - \dfrac{2c}{b \pm \sqrt{b^2 - 4ac}}$

Choose root, x_r, closest to x_0 by making the denominator as large as possible; i.e. if $b > 0$, choose plus; otherwise, choose minus.

If $x_r > x_0$,
  Then rearrange to: x_0, x_1, and the root
  Else rearrange to: x_0, x_2, and the root
End If.

(In either case, reset subscripts so that x_0, is in the middle.)

Until $|f(x_r)| <$ Ftol.

Muller's method, like Newton's, will find a complex root if given complex starting values. Of course, the computations must use complex arithmetic.

Muller's method can fail under some conditions. We leave as a challenge to the student to find when this will happen. (*Hint:* What will make the denominator of the equation for the root of the quadratic zero or nearly so?)

Experience shows that Muller's method converges at a rate that is similar to that for Newton's method.* It does not require the evaluation of derivatives, however, and (after we have obtained the starting values) needs only one function evaluation per iteration. There is an initial penalty in that one must evaluate the function three times, but this is frequently overcome by the time the required precision is attained. (We do have to evaluate the coefficients a, b, and c each time, of course.)

* Atkinson (1978) shows that each error is about proportional to the previous error to the 1.85th power.

Here is an example of the use of Muller's method.

EXAMPLE 1.2 Find a root between 0 and 1 of the same transcendental function as before: $f(x) = 3x + \sin(x) - e^x$. Let

$$\begin{aligned} x_0 &= 0.5, & f(x_0) &= 0.330704 & h_1 &= 0.5, \\ x_1 &= 1.0, & f(x_1) &= 1.123489 & h_2 &= 0.5, \\ x &= 0.0, & f(x_2) &= -1 & \gamma &= 1.0. \end{aligned}$$

Then

$$a = \frac{(1.0)(1.123189) - 0.330704(2.0) + (-1)}{1.0(0.5)^2(2.0)} = -1.07644,$$

$$b = \frac{1.123189 - 0.330704 - (-1.07644)(0.5)^2}{0.5} = 2.12319,$$

$$c = 0.330704,$$

and

$$\text{root} = 0.5 - \frac{2(0.330704)}{2.12319 + \sqrt{(2.12319)^2 - 4(-1.07644)(0.330704)}}.$$
$$= 0.354914.$$

For the next iteration, we have

$$\begin{aligned} x_0 &= 0.354914, & f(x_0) &= -0.0138066 & h_1 &= 0.145086, \\ x_1 &= 0.5, & f(x_1) &= 0.330704 & h_2 &= 0.354914, \\ x_2 &= 0, & f(x_2) &= -1 & \gamma &= 2.44623. \end{aligned}$$

Then

$$a = \frac{(2.44623)(0.330704) - (-0.0138066)(3.44623) + (-1)}{2.44623(0.145086)^2(3.44623)} = -0.808314,$$

$$b = \frac{0.330704 - (-0.0138066) - (-0.808314)(0.145086)^2}{0.145086} = 2.49180,$$

$$c = -0.0138066,$$

$$\text{root} = 0.354914 - \frac{2(-0.0138066)}{2.49180 + \sqrt{(2.49180)^2 - 4(-0.808314)(-0.0138066)}}$$
$$= 0.360465.$$

After a third iteration, we get 0.3604217 as the value for the root, which is identical to that from Newton's method after three iterations.

1.5 Fixed-Point Iteration: $x = g(x)$ Method

The method known as *fixed-point iteration* [we also call it the $x = g(x)$ method] can be a useful way to get a root of $f(x) = 0$. This method is also the basis for some important theory. To use the method, we rearrange $f(x)$ into an equivalent form $x = g(x)$, which usually can be done in several ways. Observe that if $f(r) = 0$, where r is a root of $f(x)$, it follows that $r = g(r)$. Whenever we have $r = g(r)$, r is said to be a fixed point for the function g.

Under suitable conditions that we explain later, the iterative form

$$x_{n+1} = g(x_n) \qquad n = 0, 1, 2, 3, \ldots,$$

converges to the fixed point r, a root of $f(x)$.

Here is a simple example:

$$f(x) = x^2 - 2x - 3 = 0.$$

$f(x)$ is easy to factor to show roots at $x = -1$ and $x = 3$. (We pretend that we don't know this.)

Suppose we rearrange to give this equivalent form:

$$x = g_1(x) = \sqrt{2x + 3}.$$

If we start with $x = 4$ and iterate with the fixed-point algorithm, successive values of x are

$$\begin{aligned}
x_0 &= 4,\\
x_1 &= \sqrt{11} = 3.31662,\\
x_2 &= \sqrt{9.63325} = 3.10375,\\
x_3 &= \sqrt{9.20750} = 3.03439,\\
x_4 &= \sqrt{9.06877} = 3.01144,\\
x_5 &= \sqrt{9.02288} = 3.00381,
\end{aligned}$$

and it appears that the values are converging on the root at $x = 3$.

Other Rearrangements

Another rearrangement of $f(x)$ is

$$x = g_2(x) = \frac{3}{(x - 2)}.$$

Let us start the iterations again with $x_0 = 4$. Successive values then are

$$\begin{aligned}
x_0 &= 4,\\
x_1 &= 1.5,\\
x_2 &= -6,
\end{aligned}$$

$$x_3 = -0.375,$$
$$x_4 = -1.263158,$$
$$x_5 = -0.919355,$$
$$x_6 = -1.02762,$$
$$x_7 = -0.990876,$$
$$x_8 = -1.00305,$$

and it seems that we now converge to the other root, at $x = -1$. We also see that the convergence is oscillatory rather than monotonic as we saw in the first case.

Consider a third rearrangement:

$$x = g_2(x) = \frac{(x^2 - 3)}{2}.$$

Starting again with $x_0 = 4$, we get

$$x_0 = 4,$$
$$x_1 = 6.5,$$
$$x_2 = 19.625,$$
$$x_3 = 191.070,$$

and the iterates are obviously diverging.

This difference in behavior of the three rearrangements is interesting and worth further study. First, though, let us look at the graphs of the three cases. The fixed point of $x = g(x)$ is the intersection of the line $y = x$ and the curve $y = g(x)$ plotted against x. Figure 1.8 shows the three cases.

Observe that we always get the successive iterates by this construction: Start on the x-axis at the initial x_0, go vertically to the curve, then horizontally to the line $y = x$, then vertically to the curve, and again horizontally to the line. Repeat this process until the points on the curve converge to a fixed point or else diverge. It appears that the different behaviors depend on whether the slope of the curve is greater, less, or of opposite sign to the slope of the line (which equals $+1$).

Here is pseudocode for the fixed-point ($x = g(x)$) method:

Iteration Algorithm with the Form $x = g(x)$

To determine a root of $f(x) = 0$, given a value x_1 reasonably close to the root,

Rearrange the equation to an equivalent form $x = g(x)$.

Repeat
Set $x_2 = x_1$.
Set $x_1 = g(x_1)$
Until $|x_1 - x_2| <$ tolerance value

Note: The method may converge to a root different from the expected one, or it may diverge. Different rearrangements will converge at different rates.

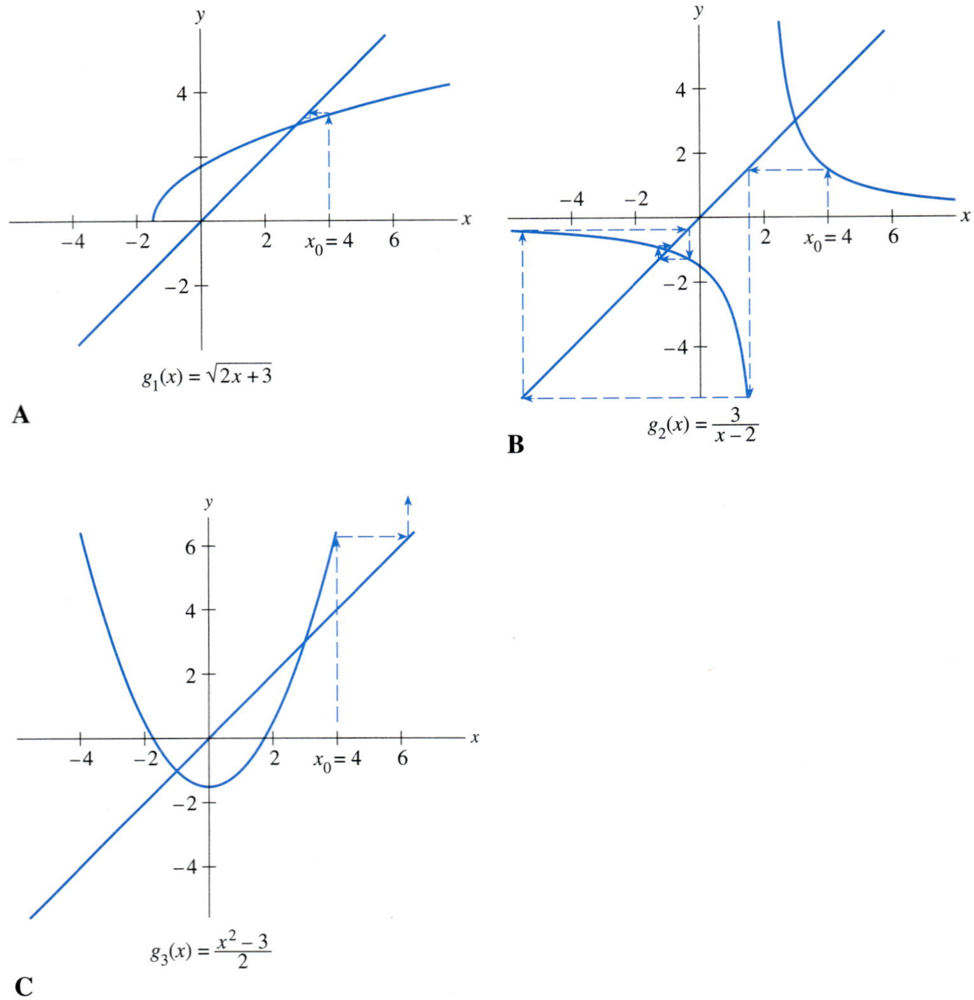

Figure 1.8

Order of Convergence

The fixed-point method converges at a linear rate; it is said to be *linearly convergent,* meaning that the error at each successive iteration is a constant fraction of the previous error. (Actually, this is true only as the errors approach zero.) If we tabulate the errors after each step in getting the roots of the polynomial and its ratio to the previous error, we find:

Iteration	If $g(x) = \sqrt{2x + 3}$ Error	Ratio	If $g(x) = 3/(x - 2)$ Error	Ratio
1	0.31662	0.31662	2.50000	0.50000
2	0.10375	0.32767	−5.00000	−2.00000
3	0.03439	0.33143	0.62500	−0.12500
4	0.01144	0.33270	−0.26316	−0.42105
5	0.00381	0.33312	0.08065	−0.30645
6			−0.02762	−0.34254
7			0.00912	−0.33029
8			−0.00305	−0.33435

Notice that the magnitudes of the ratios seem to be leveling out at 0.3333. In fact, if the iterations are continued, they become exactly one-third.

Accelerating Convergence

For any iterative process where the errors decrease proportionally, we can speed the convergence by a technique known as *Aitken acceleration.* Based on the assumption that each error is a constant times the previous error, we can write

$$\frac{e_{n+2}}{e_{n+1}} = \frac{e_{n+1}}{e_n},$$

or, because $e_n = R - x_n$, where R is the true value for the root,

$$\frac{R - x_{n+2}}{R - x_{n+1}} = \frac{R - x_{n+1}}{R - x_n},$$

giving $(R - x_{n+2})(R - x_n) = (R - x_{n+1})^2$, which we can solve for R:

$$R = \frac{x_n * x_{n+2} - x_{n+1}^2}{x_n - 2x_{n+1} + x_{n+2}}.$$

Let us apply this to the first three computations from $x = g(x)$, where $g(x) = \sqrt{2x + 3}$. Substituting the values for x_0, x_1, and x_2, we get $R = 3.00744$, closer to the true value of 3.0.

There is a better way to do this extrapolation, called the Δ^2 process, that uses fewer arithmetic operations. Define $\Delta x_n = x_{n+1} - x_n$ and $\Delta^2 x_n = \Delta x_{n+1} - \Delta x_n$, and the equation for R can be written as

$$R = x_n - \frac{(\Delta x_n)^2}{\Delta^2 x_n}.$$

Let us apply this to the first three computations from $x = g(x)$, where $g(x) = \sqrt{2x + 3}$.

x	Δx	$\Delta^2 x$
4		
	0.68338	
3.31662		0.47051
	0.21287	
3.10375		

Extrapolating, $R = 4 - (0.68338)^2/0.47051 = 4 - 0.99256 = 3.00744$, which is closer to the true value than the fifth iterate when we didn't extrapolate. We could proceed from this point to do two more iterations and then extrapolate again. For this simple definition of $g(x)$ it may not seem worth the effort; when $g(x)$ is expensive to compute, it certainly is.

Some Theory

The above demonstrated that fixed-point iterations seem to converge linearly. We now show when this is true. We have

$$x_{n+1} = g(x_n),$$

and we can write this relation for the error after iteration $n + 1$, where R is the true value of the root:

$$R - x_{n+1} = R - g(x_n) = g(R) - g(x_n)$$

because, when $x = R$, $R = g(R)$. Multiplying and dividing by $(R - x_n)$:

$$R - x_{n+1} = \frac{(g(R) - g(x_n))}{(R - x_n)}(R - x_n),$$

we can use the mean-value theorem* [if $g(x)$ and $g'(x)$ are continuous] to say that

$$R - x_{n+1} = g'(\xi_n) * (R - x_n),$$

where ξ_n lies between x_n and R.

Writing e_n for the error of the nth iterate, we have

$$|e_{n+1}| = |g'(\xi_n)| * |e_n|$$

because e_n, the error in x_n is $R - x_n$. (We take absolute values because the successive iterates may oscillate around the root.)

Now suppose that $|g'(\xi_n)| < K < 1$ where $K < 1$ on some interval of size h around the root R. If we begin with an x-value in this interval, fixed-point iterations will converge because

$$|e_n| = K|e_{n-1}| = K^2|e_{n-2}| = K^3|e_{n-3}| = \cdots = K^n|e_0|.$$

* This theorem is covered in Appendix A.

This proves that the fixed-point method is linearly convergent in the limit as x_n approaches R, provided that we start within the interval where $|K| < 1$.

Convergence of Newton's Method

Newton's method uses iterations that resemble fixed point:

$$x_{n+1} = x_n - \frac{f(x_n)}{f'(x_n)} = g(x_n).$$

Successive iterates will converge if $|g'(x)| < 1$, and, doing the differentiation, we see that the method converges if

$$|g'(x)| = \left| \frac{f(x) * f''(x)}{[f'(x)]^2} \right| < 1, \quad \textbf{(1.1)}$$

which requires that $f(x)$ and its derivatives exist and be continuous. Newton's method is shown to be quadratically convergent by the following: As before,

$$R - x_{n+1} = g(R) - g(x_n).$$

Now we expand $g(x_n)$ as a Taylor series in terms of $(R - x_n)$, with the second derivative term as the remainder, getting

$$g(x_n) = g(R) + g'(R) * (R - x_n) + (g''(\xi)/2) * (R - x_n)^2, \quad \textbf{(1.2)}$$

where ξ lies within $[x_n, R]$. However, from Eq. (1.1),

$$|g'(R)| = \left| \frac{f(R) * f''(R)}{[f'(R)]^2} \right| = 0$$

because $f(R) = 0$ at the root and Eq. (1.2) reduces to

$$g(x_n) = g(R) + (g''(\xi)/2) * (R - x_n)^2. \quad \textbf{(1.3)}$$

Using $e_n = R - x_n$ for the error of the nth iterate, Eq. (1.3) becomes

$$e_{n+1} = R - x_{n+1} = g(R) - g(x_n) = (g''(\xi)/2) * (e_n)^2,$$

proving that Newton's method is quadratically convergent.

Convergence of the Secant Method and False Position

Both of the secant method and false position use iterations that can be written as

$$x_{n+1} = \frac{f(x_n)}{f(x_n) - f(x_{n-1})} (x_n - x_{n-1}),$$

which is similar to $x = g(x)$, except that $x = g(x_n, x_{n-1})$. When we apply Taylor series, the derivatives are pretty complicated; we omit the details. It turns out that the error relation is

$$e_{n+1} = g(\xi_1, \xi_2)/2 * e_n * e_{n-1},$$

showing that the error is proportional to the product of the two previous errors. We can conclude that the convergence is better than linear but poorer than quadratic.

Pizer (1975) shows that the order of convergence of the secant method is $(1 + \sqrt{5})/2 = 1.62$.

1.6 Multiple Roots

A function can have more than one root of the same value. Look at Figure 1.9. The curve on the left has a triple root at $x = -1$ [the function is $(x + 1)^3$]. The curve on the right has a double root at $x = 2$ [the function is $(x - 2)^2$]. If there were more than two or three roots, the plots would be similar except they would be flatter near the x-axis and rise more steeply away from the root.

The methods we have described do not work well for multiple roots. For example, Newton's method is only linearly convergent at a double root. $f(x) = (x - 1)\,(e^{(x-1)} - 1)$ has a double root at $x = 1$, as seen in Figure 1.10. Table 1.4 gives the errors of successive iterates and the convergence is clearly linear.

When Newton's method is applied to a triple root, convergence is still linear, as seen in Table 1.5. With a triple root, the ratio of errors is larger, about $\frac{2}{3}$, compared to $\frac{1}{2}$ for the double root of Table 1.4.

In addition to a slow convergence, there is another disadvantage to using these methods to find multiple roots: imprecision. Because the curve is "flat" in the neighborhood of the root—$f'(x)$ will always be zero at a multiple root, as is apparent from Figure 1.9—there is a "neighborhood of uncertainty" around the root where values of $f(x)$ are very small. Thus, the imprecise arithmetic of almost all computational devices will find $f(x)$ "equal" to zero throughout this neighborhood; that is, the program cannot distinguish which x-value is really the root. Using double precision will decrease the neighborhood of uncertainty. In

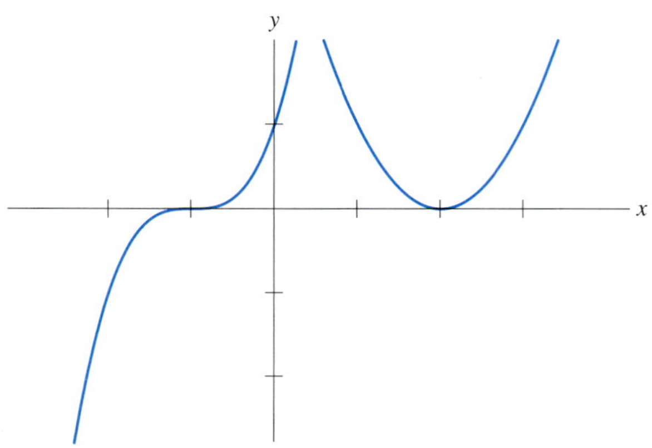

Figure 1.9

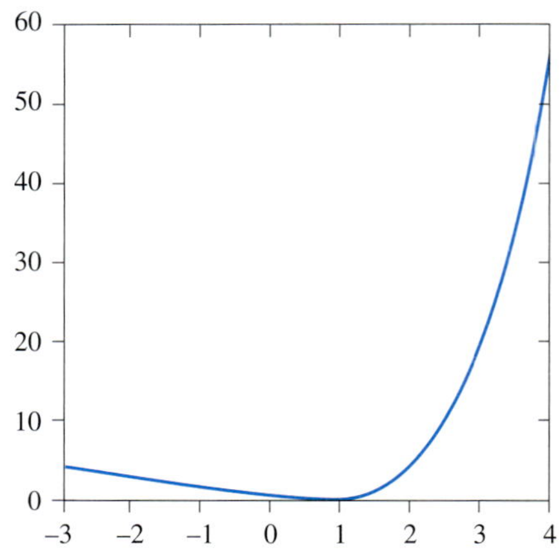

Figure 1.10
Plot of $(x - 1)\,(e^{(x-1)} - 1)$

Table 1.4 Errors when finding a double root

Iteration	Error	Ratio
1	0.3679	
2	0.1666	0.453
3	0.0798	0.479
4	0.0391	0.490
5	0.0193	0.494
6	0.0096	0.497
7	0.0048	0.500
8	0.0024	0.500

fact, MATLAB's 'vpa' command can give as much precision as desired, even to 100 significant figures, so this "neighborhood" can be very small.

Remedies for Multiple Roots with Newton's Method

When $f(x)$ has only one zero at $x = R$, we saw in Section 1.5 that Newton's method is quadratically convergent. We did this by examining the Taylor expansion for $g(x)$ about $(x - R)$ where

$$x_{n+1} = x_n - \frac{f(x_n)}{f'(x_n)} = g(x_n). \tag{1.4}$$

That series was

$$g(x_n) = g(R) + g'(R) * (R - x_n) + (g''(\xi)/2) * (R - x_n)^2, \tag{1.5}$$

and we saw that $g'(R)$ was zero.

However, if $f(x)$ has a root of multiplicity k at $x = R$, we can factor out $(x - R)^k$ from $f(x)$ to get

$$f(x) = (x - R)^k Q(x) \tag{1.6}$$

Table 1.5 Successive errors with Newton's method, for $f(x) = (x + 1)^3 = 0$

Iteration	Error	Iteration	Error
0	0.5	6	0.0439
1	0.3333	7	0.0293
2	0.2222	8	0.0195
3	0.1482	9	0.0130
4	0.0988	10	0.00867
5	0.0658		

where $Q(x)$ has no root at $x = R$. That means that $Q(R)$ is nonzero, even though $f(R)$, $f'(R), f''(R), \ldots f^{(k-1)}(R)$ are all zero, as is readily found by differentiating Eq. (1.6). We then see that the denominator and numerator in Eq. (1.4) are both zero. While this is an indeterminant form, we cannot say that $g'(R)$ is zero, confirming that Newton's method with multiple roots is only linearly convergent.

Look now at a different formulation of Newton's method:

$$x_{n+1} = x_n - k * \frac{f(x_n)}{f'(x_n)} = g_k(x_n). \tag{1.7}$$

As before, at $f(R) = 0$, $g_k(R) = R$. Using the reformulation of $f(x)$ as given by Eq. (1.6), and differentiate, we get

$$g'(x) = \frac{(R - x)\{k(r - x)QQ'' + Q'[2kQ - (k - 1)(R - x)Q']\}}{[(R - x)Q' + kQ]^2},$$

and we see that $g'(R) = 0$. From the preceding argument, then, the modified Newton's method now converges quadratically at a multiple root. (It also does so at a simple root with $k = 1$, of course.) Using this method to get the root of $f(x) = (x - 1) * (e^{(x-1)} - 1)$, we find that the third iterate is $x = 1.00088$ with $f(x) = 0.00000$. We also find that $e_{n+1} = 0.24 * e_n^2$, confirming quadratic convergence.

This algorithm would seem to solve the problem of multiple roots using Newton's method, but we don't know the multiplicity of the root in advance! (This objection is a little academic as the following argument shows.)

We might guess at the value for k and see whether we get quadratic convergence, or we could try several values and see what happens. Better yet, we could compare a graph of $f(x)$ with the plots of $(x - R)^k$, using an approximate value for R and various values for k. The "flatness" of the curves will be the same for $f(x)$ and the plot of equivalent multiplicity. We wonder, though, whether all such effort is justified—why not just live with the linear convergence? We will find the root with sufficient accuracy from that operation long before we complete the alternative explorations.

Another solution to multiple roots is tempting to consider. We can divide $f(x)$ by $(x - R)$ and deflate the function, reducing the multiplicity by one. The problem here is that we don't know R. However, dividing by $(x - s)$, where s is an approximation of R does almost the same thing. We suggest that you might want to explore this idea. Be warned that the division creates an indeterminate form at $x = R$ and a strong discontinuity at $x = s$.

Acton (1970) gives another technique by which we may obtain a multiple root with quadratic convergence. If $f(x)$ has a root of multiplicity k at $x = R$, we have $f(x) = (R - x)^k * Q(x)$. Let $S(x)$ be $f(x)/f'(x)$, so that

$$S(x) = \frac{(x - R)^k Q(x)}{k(x - R)^{k-1}Q(x) + (x - R)^k Q'(x)} = \frac{(x - R)Q(x)}{kQ(x) + (x - R)Q'(x)}$$

which has a simple root at $x = R$. When S(x) is used in the Newton formula, we get

$$x_{n+1} = x_n - \frac{S(x_n)}{S'(x_n)} = x_n - \frac{f(x_n) * f'(x_n)}{[f'(x_n)]^2 - f(x_n) * f''(x_n)}$$

and we see that we need to evaluate three functions at each iteration: $f(x_n)$, $f'(x_n)$, and $f''(x_n)$. Acton also points out that there are nearly equal quantities being subtracted in the denominator, a source of arithmetic error.

Nearly Multiple Roots

A problem related to multiple roots is a function that has two or more roots very close together. If these roots are all within the region of uncertainty (which is a function of the arithmetic precision we are using), they are effectively multiple roots, because for all of them $f(x)$ is computationally equal to zero.

Newton's method is again essentially linearly convergent when we have nearly equal roots, provided that we start outside the interval that holds the roots. Unfortunately, modifying the method by considering them to be multiple roots doesn't work; often an infinite loop occurs. If we are so unlucky as to start between two almost equal roots, Newton's method can fly off to "outer space," as we previously observed.

Whenever we want to find roots that are near $f'(x) = 0$, we are in trouble. We strongly recommend that you graph the function, before jumping into a root-finding routine, to see in advance whether such problems will arise.

1.7. Nonlinear Systems

When we have a system of simultaneous nonlinear equations, the situation is more difficult. In fact, some sets have no real solutions. Consider this example of a pair of equations:

$$x^2 + y^2 = 4,$$
$$e^x + y = 1.$$

Graphically, the solution to this system is represented by the intersections of the circle $x^2 + y^2 = 4$ with the curve $y = 1 - e^x$. Figure 1.11 shows that these are near $(-1.8, 0.8)$ and $(1, -1.7)$.

Newton's method can be applied to systems as well as to a single nonlinear equation. We begin with the forms

$$f(x, y) = 0,$$
$$g(x, y) = 0.$$

Let $x = r$, $y = s$ be a root, and expand both functions as a Taylor series about the point (x_i, y_i) in terms of $(r - x_i)$, $(s - y_i)$, where (x_i, y_i) is a point near the root:

$$f(r, s) = 0 = f(x_i, y_i) + f_x(x_i, y_i)(r - x_i) + f_y(x_i, y_i)(s - y_i) + \cdots,$$
$$g(r, s) = 0 = g(x_i, y_i) + g_x(x_i, y_i)(r - x_i) + g_y(x_i, y_i)(s - y_i) + \cdots.$$

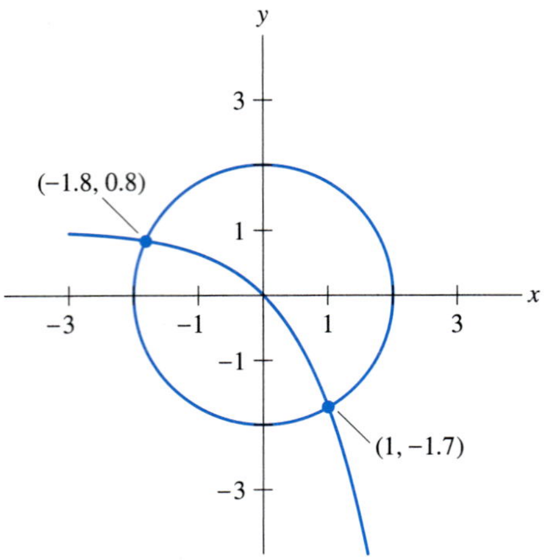

Figure 1.11

Truncating both series gives

$$0 = f(x_i, y_i) + f_x(x_i, y_i)(r - x_i) + f_y(x_i, y_i)(s - y_i),$$
$$0 = g(x_i, y_i) + g_x(x_i, y_i)(r - x_i) + g_y(x_i, y_i)(s - y_i),$$

which we can rewrite as

$$f_x(x_i, y_i)\,\Delta x_i + f_y(x_i, y_i)\Delta y_i = -f(x_i, y_i),$$
$$g_x(x_i, y_i)\,\Delta x_i + g_y(x_i, y_i)\Delta y_i = -g(x_i, y_i), \qquad \textbf{(1.8)}$$

where Δx_i and Δy_i are used as increments to x_i and y_i, so that $x_{i+1} = x_i + \Delta x_i$ and $y_{i+1} = y_i + \Delta y_i$ are improved estimates of the (x, y) values. We repeat this until both $f(x, y)$ and $g(x, y)$ are close to zero.

The extension to more than two simultaneous equations is straightforward, but solving larger sets of equations requires information from the next chapter.

We illustrate by solving the example at the beginning of this section:

$$f(x, y) = 4 - x^2 - y^2 = 0,$$
$$g(x, y) = 1 - e^x - y = 0.$$

The partial derivatives are

$$f_x = -2x, \qquad f_y = -2y,$$
$$g_x = -e^x, \qquad g_y = -1.$$

Beginning with $x_0 = 1$, $y_0 = -1.7$, where

$$f_x = -2, \qquad f_y = 3.4,$$
$$g_x = -2.7183, \qquad g_y = -1.0,$$
$$f(1, -1.7) = 0.1100, \qquad g(1, -1.7) = -0.0183,$$

we solve

$$-2\,\Delta x_0 + 3.4\Delta y_0 = -0.1100,$$
$$-2.7183\,\Delta x_0 - 1.0\,\Delta y_0 = 0.0183.$$

This gives $\Delta x_0 = 0.0043$, $\Delta y_0 = -0.0298$, from which $x_1 = 1.0043$, $y_1 = -1.7298$. These agree with the true value within 2 in the fourth decimal place. Repeating the process once more produces $x_2 = 1.004169$, $y_2 = -1.729637$. The function values at this second iteration are approximately -0.0000001 and -0.00000001.

Newton's method, as you would expect, converges quadratically when we are near the solution but notice that it is rather expensive. For even this 2×2 system there are six function evaluations at each step. For a 3×3 system, there are twelve. For a $n \times n$ system, the number is $n^2 + n$.

For larger systems, the number of function evaluations can be reduced by not recomputing the partials at every step but only after n steps, reusing the same values n times (n being the number of equations). We then only need to evaluate the n function values until we again update the partials. We do sacrifice quadratic convergence thereby but it is usually better than linear. Unfortunately, this modification of Newton's method for a system can diverge unless we are close to the roots.

With MATLAB, getting the solution to a system is easy:

```
EDU>> [x, y] = solve ('x^2 + y^2 = 4', 'exp(x) + y = 1')
x =
        -1.8162640688251505742443123715859
y =
        1 - exp(-1.8162640688251505742443123715859)
```

but this is the leftmost intersection! We can get the one near $(1, -1.7)$ with

```
EDU>> [x, y] = solve('abs(x^2) + y^2 = 4', 'exp(x) + y = 1')
x =
        1.0041687384746591657874315472901
y =
        -1.7296372870258699313633129362508
```

which reproduces the result from Newton's method with many more digits.

Solving a System by Iteration

There is another way to attack a system of nonlinear equations. Consider this pair of equations:

$$e^x - y = 0,$$
$$xy - e^x = 0.$$

We know how to solve a single nonlinear equation by fixed-point iterations—we rearrange it to solve for the variable in a way that successive computations may reach a solution. Sometimes we can do the same for a system. Let us solve the first of the pair for x and the second for y:

$$x = \ln(y),$$
$$y = e^x/x.$$

To start, we guess at a value for y, say, $y = 2$. We enter this into the first rearranged equation and get an x-value that we use in the second. This gives a new value for y from which we get a new value for x, and repeat. Here is what we get:

y-value	x-value
2	0.69315
2.88539	1.05966
2.72294	1.00171
2.71829	1.00000
2.71828	1.00000

which are precisely the correct results.

Here is another example for the pair of equations whose plot is Figure 1.11:

$$x^2 + y^2 = 4,$$
$$e^x + y = 1.$$

If we will try this rearrangement:

$$y = -\sqrt{(4 - x^2)},$$
$$x = \ln(1 - y),$$

and begin with $x = 1.0$, the successive values for y and x are:

$$x = 1.0051, \qquad y = -1.7291,$$
$$x = 1.00398, \qquad y = -1.72975,$$
$$x = 1.00421, \qquad y = -1.72961,$$
$$x = 1.00416, \qquad y = -1.72964,$$
$$x = 1.00417, \qquad y = -1.72963,$$

and we are converging to the solution in an oscillatory manner.

Other rearrangements are possible. You may wish to see that this one diverges from the starting point $(1, -1.7)$:

$$y = (4 - x^2)/y, \qquad x = \ln(1 - y).$$

You may also want to see that both rearrangements diverge when used to find the intersection in the fourth quadrant, and to discover rearrangements that will converge to this point.

Exercises

Section 1.1

1. The function $f(x) = 2 * \sin(x) - e^x/4 - 1$ is zero for two values near $x = -5$. Use bisection to find both, starting with $[-7, -5]$ and $[-5, -3]$. How many iterations are needed to get results that agree to five significant figures?

2. The quadratic $(x - 0.3) * (x - 0.5)$ obviously has zeros at 0.3 and 0.5.
   a. Why is the interval [0.1, 0.6] not a satisfactory starting interval for bisection?
   b. What are good starting intervals for each root?
   c. If you start with [0, 0.49] which root is reached with bisection?

   Which root from [0.31, 1.0]?

▶3. Where do the curves of $y = \cos(x)$ and $y = x^3 - 1$ intersect? Use bisection.

4. The function $f(x) = x * \sin((x - 2)/(x - 1))$ has many zeros, especially near $x = 1$ where the function is discontinuous. Find the four zeros nearest to $x = 0.95$ by bisection, correct to five significant figures. How can you find good starting intervals?

5. Suppose that your computing device has only 3 bits (plus one hidden bit) for the fraction part. There are large gaps between the numbers that can be stored, as indicated by a sketch in Section 0.4. The relation $3 * \sin(x) = x^2 - 2$ is true at a point very near to $x = 2.13$. If you begin bisection with a starting interval of [1, 3], what will be the successive x-values that are used in finding the solution?

6. Exercise 5 is an extreme example that shows that every computer that uses a finite number of bits to represent the fraction part has gaps between the machine numbers. This limits the accuracy when finding where $f(x)$ is zero. What is the limit to the accuracy of getting a zero by bisection if the number of fraction bits (including the hidden bit) is m?

▶7. How many iterations of bisection will be required to attain an accuracy of 10^{-4} if the starting interval is $[a, b]$?

Section 1.2

8. Repeat Exercise 1 but use the secant method. How many fewer iterations are required?

▶9. Repeat Exercise 1 but now use *regula falsi*. Compare the number of iterations with Exercises 1 and 8.

10. Explain why the secant method usually converges to reach a given stopping tolerance faster than either bisection or linear interpolation.

11. In bisection and the method of false position, one tests to see that a function changes sign between $x = a$ and $x = b$. If this is done by seeing if $f(a) * f(b) < 0$, underflow may occur. Is there an alternative way to make the test that avoids this problem?

Section 1.3

12. Solve Exercise 3 with Newton's method.

13. The function $f(x) = 4x^3 - 1 - \exp(x^2/2)$ has values of zero near $x = 1.0$ and $x = 3.0$.
    a. What is the derivative of f?
    b. If you begin Newton's method at $x = 2$, which root is reached? How many iterations to achieve an error less than 10^{-5}?
    c. Begin Newton's method at another starting point to get the other zero.
    d. For both parts (b) and (c), tabulate the number of correct digits at each iteration.

▶14. Apply Newton's method to the equation $x^2 = N$ to derive this algorithm for getting the square root of N:

$$x_{n+1} = \frac{1}{2}\left(x_n + \frac{N}{x_n}\right).$$

15. Find algorithms for getting the third and fourth roots of N that are similar to that in Exercise 14 for the square root. Can this be generalized for the nth root?

16. If the algorithm of Exercise 14 is applied twice, show that

$$(A * B)^{1/2} \approx \frac{A + B}{4} + \frac{A * B}{A + B}.$$

17. Show that the error of the approximation in Exercise 14 is nearly equal to

$$\frac{(A - B)^4}{16\,(A + B)^2}.$$

18. $f(x) = (x - 1)^2\,(x + 1)$ obviously has roots at $+1$ and -1. Using starting values that differ from the roots by 0.2, compare the number of repetition of Newton's

method required to reach the roots within 0.0001. Explain the difference.

19. Newton's method will find complex roots. Find the roots of these relations including the complex ones:

a. $x^3 + 2 = 0$.
b. $2x^3 - 3x^2 = 1$.
c. $x^2 = (e^{-2x} - 2)/x$.

▶**20.** The sum of two numbers is 20, the square root of their product is 9. What are the numbers?

21. A fourth-degree polynomial, $P_4(x)$, is

$$P_4(x) = 1.1x^4 + 4.6x^3 + 6.6x^2 - 12x - 16.$$

a. Use synthetic division (by hand) to get $P(-1)$ and $P'(-1)$.
b. One of the real roots of $P(x)$ is near to $x = -1$. Use the results of part (a) to get a first estimate of the root. Then continue until you have the root to within 0.001. How does each error compare to the previous error?
c. A second real root of $P(x)$ is near 1.5. Perform Newton's method with synthetic division on $P(x)$ to get this second root. Again compare the successive errors.
d. Use the result of part (b) to deflate $P(x)$ to obtain a cubic polynomial. Then repeat part (b) on this cubic. Do you get the same answer as you did in part (c)?
e. What are the last two roots?

22. $P_4(x) = (x - 1.1)(x - 2.2)(x - 3.3)(x - 4.4)$ has four positive roots, of course. Expanded, $P(x)$ is

$$x^4 - 11.0x^3 + 42.35x^2 - 66.55x + 35.1384.$$

a. Use Newton's method to get the roots, each correct to only three significant digits. Do this by deflating the polynomial after getting a root and then getting the next from the deflated polynomial. Start each computation with an x-value 10% greater than the actual root.
b. Which gives more accuracy on the successive roots? (1) Begin with the largest root and work down to the smallest, or (2) work up from the smallest root.

23. Inaccuracies in the coefficients of a polynomial can sometimes have a very great influence on the values of the roots.

a. How do the roots of Exercise 22 change if the coefficient of x^3 were -11.11 (a 1% change)? Are any of the roots relatively unaffected?
b. Investigate the effect of a 1% change in the other coefficients. In which of the coefficients does a change by this amount cause the greatest change in the computed roots?
c. What if all the coefficients are increased by 1%? Does this cause an even greater change in the roots?

▶**24.** Synthetic division finds $P(a)$ by dividing $P(x)$ by $(x - a)$, then gets $P'(a)$ by synthetic division of the reduced polynomial. Does this mean that we can evaluate $P''(a), P'''(a), \ldots$ by repeated divisions?

25. This polynomial obviously has roots at $x = 2$ and at $x = 4$; one is a double root, the other a triple root:

$$\begin{aligned} P(x) &= (x - 2)^3 (x - 4)^2 \\ &= x^5 - 14x^4 + 76x^3 - 200x^2 + 256x - 128. \end{aligned}$$

a. Which root can you get with bisection? Which root can't you get?
b. Repeat part (a) with the secant method.
c. If you begin with the interval [1, 5], which root will you get with (1) bisection, (2) the secant method, (3) false position?

26. Use Newton's method on the polynomial of Exercise 25 with $x_0 = 3$. Does it converge? To which root? Is convergence quadratic?

27. An equation in Section 1.6 shows how to restore quadratic convergence when Newton's method is used for multiple roots. Use this device to restore quadratic convergence in getting both roots of the polynomial in Exercise 25.

▶**28.** Apply Newton's method to the derivative of the polynomial of Exercise 25 to get the double root. Show that the convergence is now quadratic. How can this technique be applied to get the triple roots with quadratic convergence?

29. If $P(x)$ is divided by $P'(x)$, the resulting polynomial is effectively deflated and the multiplicity of roots is reduced.

a. Plot the $P(x)$ of Exercise 25.
b. Plot $P(x)/P'(x)$. What is the multiplicity of the double root of $P(x)$?
c. What is the multiplicity of the triple root?
d. If $P(x)$ is divided by $P'(x)$ and this quotient is divided by $P''(x)$, what is the result?

▶**30.** This quadratic has two nearly equal roots:

$$P_2(x) = x^2 - 4x + 3.9999.$$

a. Which root do you get with Newton's method starting at $x = 2.1$? Is convergence quadratic?

b. Repeat part (a) but starting with $x = 1.9$.

c. What happens with Newton's method starting from $x = 2.0$?

d. Repeat part (c) but change $P(x)$ so it has roots at 2.01 and 2.03. If you start with $x_0 = 2.02$, are the results similar to those of part (c)? If not, explain.

Section 1.4

31. Use Muller's method to find roots of these equations.

a. $4x^3 - 3x^2 + 2x - 1 = 0$, root near $x = 0.6$.

b. $x^2 + e^x = 5$, roots near $x = 1, x = -2$.

c. $\sin(x) = x^2$, root near 0.9. What are other roots?

d. $\tan(x) + 3x^2 - 1$, root near 0.8 and three others near $x = 0$.

▶**32.** Muller's method can be started in a "self-starting" way. One automatically begins with $[-0.5, 0, 0.5]$ rather than with x-values near the root. Use this technique on a function with three roots near to zero. It has been said that the root nearest $x = 0$ will be found. Is this true?

33. After one root of $f(x)$ is found, another root can be found from the *deflated* function. To deflate a function, we form $g(x) = f(x)/(x - r)$, where r is the first root; $g(x)$ has all the roots of $f(x)$ except r. To see that this is true, compare the graphs of

$$f(x) = x(2x - 2)(3x - 3),$$

$$g(x) = (2x - 2)(3x - 3),$$

$$h(x) = x(2x - 2),$$

$$k(x) = x(3x - 3).$$

34. When will Muller's method fail? Are there times when the quadratic that is formed does not have a real solution? If this occurs, what can you do to remedy the situation?

35. Muller's method can find complex roots if complex arithmetic is used. Do this to find the complex roots of

$$f(x) = x^3 - x - 2.$$

Section 1.5

36. Most functions can be rearranged in several ways to give $x = g(x)$ with which to begin the fixed-point method. For $f(x) = e^x - 2x^2$, one $g(x)$ is

$$x = \pm\sqrt{(e^x/2)}.$$

a. Show that this converges to the root near 1.5 if the positive value is used and to the root near -0.5 if the negative is used.

b. There is a third root near 2.6. Show that we do not converge to this root even though values near to the root are used to begin the iterations. Where does it converge if $x_0 = 2.5$? If $x_0 = 2.7$?

c. Find another rearrangement that does converge correctly to the third root.

▶**37.** Here are three different $g(x)$ functions. All are rearrangements of the same $f(x)$. What is $f(x)$?

a. $(4 + 2x^3)/(x^2) - 2x$.

b. $\sqrt{(4/x)}$.

c. $(16 + x^3)/(5x^2)$.

d. Which of these converge? What x-value is obtained? Are there starting values for which one or more diverge? Which diverge?

38. If $f(x) = x^2 + 2x - 1 = 0$, one form of $g(x)$ is $1/(x + 2)$.

a. How many iterations are needed to attain a tolerance value of 1. 0E-5, starting with $x_0 = 1.0$?

b. If Aitken acceleration is used, is this speeded up?

c. The function has a second root. Will this $g(x)$ converge to it? If not, find another rearrangement that does. See if Aitken acceleration will speed convergence.

39. In Exercise 38, for what ranges of starting values does the $g(x)$ function converge? [If a division by zero occurs, as it does with the first $g(x)$ when $x = -2$, do not stop but continue with the next iterate equal to zero.]

▶**40.** The cubic $x^3 - 2x^2 - x + 1$ has three real roots; one is negative.

a. Find a rearrangement that converges to the negative root. Will this converge to either of the positive roots?

b. Find a rearrangement that will converge to the larger positive root. Will this same rearrangement converge to the other roots?

c. Find a different rearrangement that converges to the smaller positive root. Does it work for the other roots?

41. Can $\sqrt{5}$ be approximated through fixed-point iteration? Define a $f(x)$ and a $g(x)$ to do this if it can. Can you find several forms of $g(x)$?

Section 1.6

▶**42.** When a polynomial, $P(x)$, has coefficients that are all real numbers and a root that is complex, this root comes with a companion, its complex conjugate. The product of these two roots is a quadratic with real coefficients. Bairstow's method is a technique that uses synthetic division of $P(x)$ by a trial quadratic, and from the remainder of this division, a monomial, gets values that are used in a two-dimensional Newton's method to close in on the correct quadratic factor. The algorithm for this synthetic division is somewhat complicated (previous editions of this book give the details). However, MATLAB makes it easy to divide $P(x)$ by the trial quadratic. The command '$[p, r] =$ deconv (P, Q)' divides polynomial P by quadratic Q to give the reduced polynomial p and the remainder monomial r.

a. Use MATLAB to divide P by Q to get $p(x)$ and $r(x)$ where

$$P(x) = x^3 + x^2 + x - 3, \qquad Q(x) = x^2 + x + 1.$$

b. Part (a) gives a remainder that has nonzero coefficients, so you know that $Q(x)$ is not a factor. Repeat part (a), but now use $Q(x) = x^2 + 2x + 3$. Do you get a remainder? A zero remainder means that the trial quadratic is an exact factor. The coefficients of q are those of the other factor (a monomial, in this instance).

43. Graeffe's root-squaring method has the advantage that no initial estimate for the roots is necessary. It is based on the fact that when $P(x)$ is multiplied by $P(-x)$, the result has roots that are the squares of the roots of $P(x)$ and these are spread farther apart than those of the original polynomial. (Of course, the signs of the roots are lost in the process.) If this is repeated n times, the magnitudes of the roots of $P(x)$ are given by

$$|(a_j/a_{j+1})| \text{ for } j = 0 \ldots (n-1).$$

a. MATLAB's command 'conv (A, B)' gets the product of polynomials A and B so this can be used to do the multiplying. The trick is to get $P(-x)$ easily. Use Graeffe's method to show that the roots of this polynomial are 1, −2, and 4:

$$P(x) = x^3 - 3x^2 - 6x + 8.$$

b. What will the method do if there are complex roots?

c. Will the method work if some of the coefficients are themselves complex numbers?

44. Laquerre's method starts with an estimate of a root of $P(x)$, say, $x = a$, which we hope is near to the desired root of $P(x)$. (One authority calls this a "sure-fire method." It does make a pretty rash assumption about all other roots than the one we seek, but the assumption is valid.) To use the method, two quantities are first computed:

$$A = P'(a)/P(a), \qquad B = A^2 - P''(a)/P(a).$$

From these two quantities, we compute another quantity, d, by

$$d = \frac{n}{A \pm \sqrt{(n-1)(nB - A^2)}},$$

where the positive sign is used if A is positive and the minus sign if it is negative. If this value, d, is subtracted from the original estimate, a, $a - d$ is an improved estimate. The procedure is continued until the adjustment, d, is negligible.

a. For $P(x) = x^3 - 8.6x^2 + 22.41x - 16.236$, use Laguerre's method, starting with $a = 1.0$, to find a root at $x = 1.2$.

b. Repeat part (a), but start with $a = 5.0$ to find a second root.

c. From the results of parts (a) and (b), get the third root without doing further approximations.

45. Do some research of the literature to find out about the following methods. Neither of them require an initial approximation.

a. Lehmer's method.

b. The QD algorithm.

c. Use one of the methods on a polynomial of degree 4.

d. Compare the efficiency of these methods to Newton's method. How do they work for complex roots?

Section 1.7

46. For this system of two equations

$$y = \cos^2(x),$$
$$x^2 + y^2 - x = 2,$$

a. Plot the two equations and observe two intersections that occur near $x = 2$ and $x = -1$.

b. Substitute y from the first equation into the second, getting an equation in only x. Solve this equation for the x-value(s) by any of the methods in this chapter to find where the function is zero, then substitute this x-value in either of the original equations to get y-values.

c. Repeat part (b) but now solve for x from one equation and use it to eliminate x from the other. You should get the same solutions, of course, but which is easier to use?

d. Compute the partial derivatives of the equations and use them to find the solutions by Newton's method for a system, starting with the points [2, 0.2] and [−1, 0.3].

e. When is the technique of part (b) not a good option compared to using Newton's method?

47. MATLAB finds six solutions to this system and two are complex valued. Two of the real solutions are near (1, 1, 1) and (1.3, 0.9, −1.2).

$$x - 3y - z^2 = -3,$$
$$2x^3 + y - 5z^2 = -2,$$
$$4x^2 + y + z = 7.$$

a. What are the partial derivatives that would be used in Newton's method?

b. The matrix of partial derivatives is called the Jacobian matrix. For the starting vector [1, 1, 1], what are its elements?

c. Complete getting the two solutions with Newton's method. Find starting values that converge. Is convergence quadratic?

48. Repeat Exercise 47, part (c), but now only recompute the elements of the Jacobian after every third iteration rather than each time. How does the rate of convergence compare to that of Exercise 47?

49. Compare the number of function evaluations with those needed in both Exercises 47 and 48 to perform five iterations.

▶**50.** Can the system of Exercise 46 be solved by iteration? Do this if you think it is possible, or explain why it cannot be done this way.

51. Repeat Exercise 50, but for the equations in Exercise 47.

52. Use MATLAB to get the solutions referred to in Exercise 47.

Applied Problems and Projects

APP1. If an initial amount of money is invested and earns interest compounded annually at the rate of i %, the "Rule of 72" says that the money will double in about $72/i$ years. This is only approximately true, the exact final amount is given by

$$S = P\,(1 + i/100)^n,$$

where S is the final amount, P is the initial amount, and n is the number of years. Make a table comparing the exact values for the number of years for P to double with that from the rule. Do for i from 2% to 12%.

a. At what interest rate is the Rule of 72 exact?

b. Interest is often added to the account more frequently than annually. This makes the growth faster. What would be a good "Rule" value if interest is compounded:
quarterly?
monthly?
daily?
continuously?

APP2. Given are

$$x'' + x + 2y' + y = f(t), \quad x'' - x + y = g(t), \quad x\,(0) = x'(0) = y(0) = 0.$$

In solving this pair of simultaneous second-order differential equations by the Laplace transform method, it becomes necessary to factor the expression.

$$(S^2 + 1)(S) - (2S + 1)(S^2 - 1) = -S^3 - S^2 + 3S + 1,$$

so that partial fractions can be used in getting the inverse transform. What are the factors?

APP3. DeSantis (1976) has derived a relationship for the compressibility factor of real gases of the form

$$z = \frac{1 + y + y^2 - y^3}{(1 - y)^3},$$

where $y = b/(4v)$, b being the van der Waals correction and v the molar volume. If $z = 0.892$, what is the value of y?

APP 4. In studies of solar-energy collection by focusing a field of plane mirrors on a central collector, one researcher obtained this equation for the geometrical concentration factor C:

$$C = \frac{\pi(h/\cos A)^2 F}{0.5\pi D^2(1 + \sin A - 0.5 \cos A)}$$

where A is the rim angle of the field, F is the fractional coverage of the field with mirrors, D is the diameter of the collector, and h is the height of the collector. Find A if $h = 300$, $C = 1200$, $F = 0.8$, and $D = 14$.

APP5. Lee and Duffy (1976) relate the friction factor for flow of a suspension of fibrous particles to the Reynolds number by this empirical equation:

$$\frac{1}{\sqrt{f}} = \left(\frac{1}{k}\right) \ln (\text{RE}\sqrt{f}) + \left(14 - \frac{5.6}{k}\right).$$

In their relation, f is the friction factor, RE is the Reynolds number, and k is a constant determined by the concentration of the suspension. For a suspension with 0.08% concentration, $k = 0.28$. What is the value of f if RE $= 3750$?

APP6. Based on the work of Frank–Kamenetski in 1955, temperatures in the interior of a material with embedded heat sources can be determined if we solve this equation:

$$e^{-(1/2)t} \cosh^{-1} (e^{(1/2)t}) = \sqrt{\frac{1}{2} L_{cr}}.$$

Given that $L_{cr} = 0.088$, find t.

APP7. Suppose we have the 555 Timer Circuit

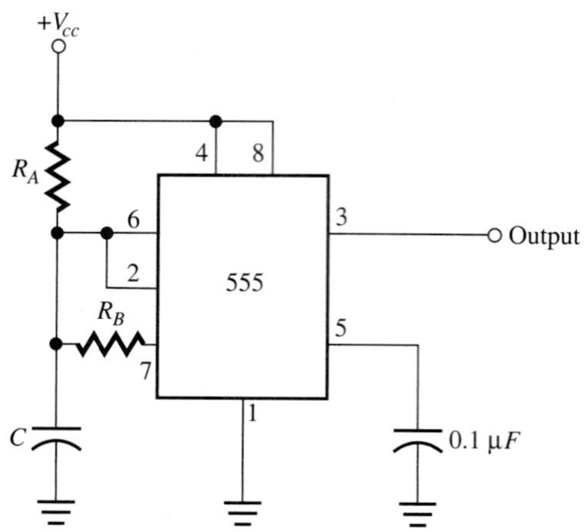

whose output waveform is

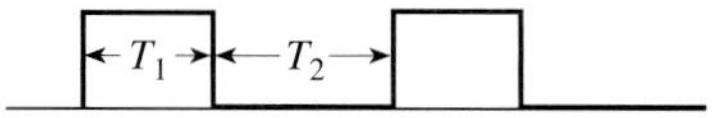

where

$$T_1 + T_2 = \frac{1}{f}$$

$$f = \text{frequency}$$

$$\text{Duty cycle} = \frac{T_1}{T_1 + T_2} \times 100\%.$$

It can be shown that

$$T_1 = R_A C \ln(2)$$

$$T_2 = \frac{R_A R_B C}{R_A + R_B} * \ln\left(\left|\frac{R_A - 2R_B}{2R_A - R_B}\right|\right).$$

Given that $R_A = 8670$, $C = 0.01 \times 10^{-6}$, $T_2 = 1.4 \times 10^{-4}$,

a. Find T_1, f, and the duty cycle.
b. Find R_B using any program you have written.
c. Select an f and duty cycle, then find T_1 and T_2.

APP8. In solving this boundary-value problem by the Fourier series method,

$$y'' + \lambda y = 0, \quad y(0) = 0, \quad y(1) = y'(1),$$

we must find the values of z where $\tan(z) = z$. Find three positive solutions where the equation is satisfied. Make a graph to get approximate values. Compare several methods to see which is faster.

APP9. In Chapter 5, a particularly efficient method for numerical integration of a function, called *Gaussian quadrature,* is discussed. In the development of formulas for this method, it is necessary to evaluate the zeros of Legendre polynomials. Find the zeros of the Legendre polynomial of sixth order:

$$P_6(x) = \frac{1}{48}(693x^6 - 945x^4 + 315x^2 - 15).$$

(*Note:* All the zeros of the Legendre polynomials are less than one in magnitude and, for polynomials of even order, are symmetrical about the origin.)

APP10. The Legendre polynomials of APP9 are one set of a class of polynomials known as *orthogonal* polynomials. Another set are the *Laguerre* polynomials. Find the zeros of the following:

a. $L_3(x) = x^3 - 9x^2 + 18x - 6$
b. $L_4(x) = x^4 - 16x^3 + 72x^2 - 96x + 24$

APP11. Still another set of orthogonal polynomials are the *Chebyshev* polynomials. (We will use these in Chapter 4.) Find the roots of

$$T_6(x) = 32x^6 = 48x^4 + 18x^2 - 1 = 0.$$

(Note the symmetry of this function. All the roots of Chebyshev polynomials are also less than one in magnitude.)

APP12. A sphere of density d and radius r weighs $\frac{4}{3}\pi r^3 d$. The volume of a spherical segment is $\frac{1}{3}\pi(3rh^2 - h^3)$.

Find the depth to which a sphere of density 0.6 sinks in water as a fraction of its radius. (See the accompanying figure.)

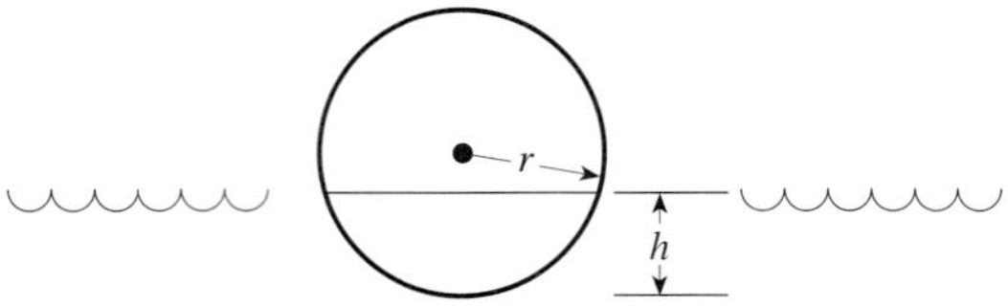

APP13. For several functions that have multiple roots; investigate whether Aitken acceleration improves the rate of convergence. Do this for several methods.

APP14. Make experimental comparisons of the rates of conversion for Newton's method, for Newton's method with the derivative estimated numerically, and for the secant method. Make a table that shows how the errors decrease for each method, then make a log plot of the errors.

APP15. When two alternative machines, A and B, are considered for purchase, the choice is often based on what is termed the "break-even time." If the machine will be used for less than this time, machine A is purchased, if greater, it is B. Suppose these costs and benefits apply:

	Machine A	**Machine B**
Initial cost	\$3250	\$5680
Annual expenses	510	830
Annual return	860	1070

In finding the answers, you should reduce all costs to their "present worth."

a. What is the break-even time?

b. Find the values of costs/benefits for several years before and after the break-even time.

c. Which is preferred if the machine will be used for only four years?

APP16. Muller's method is said to converge with an order of convergence equal to 1.85. Verify this experimentally. Is it true if there is a multiple root?

APP17. Spreadsheet programs can perform iterated computations. Devise and test a spreadsheet program that implements fixed-point iterations.

APP18. Repeat APP17, but for the

a. Bisection method,

b. Newton's method.

APP19. Fixed-point iterations sometimes converge (a) by "walking up a staircase" (Fig. 1.8a) or (b) "spirally" (Fig. 1.8b), or they may diverge (Fig. 1.8c). The conditions for these cases are discussed in Section 1.5, and we have shown that convergence is of order 1. This means that Aitken acceleration applies. Jones (1982) discusses other ways to accelerate the convergence; so does Acton (1970).

Consider this problem: Where do the curves for $e^y + 1 = e^x$ and $x^2 + y^2 = 1$ cross? One intersection is near (0.9, 0.4). Find rearrangements of the form $x = g(x)$ that converge to this intersection and compare how fast they converge. Apply several acceleration techniques to this.

APP20. a. The rate of flow of water through a stream is often measured by installing a weir. This amounts to building a dam across the stream with a vee-shaped notch near the center (the point of the notch is down). If the upstream velocity is neglected, the flow Q (ft^3/sec) is related to distance h (ft) from the surface of the upstream water to the point of the vee and to the angle θ (degrees) between the sides of the notch by this formula:

$$Q = 0.59 * \left(\frac{8}{15}\right) * \tan\left(\frac{\theta}{2}\right) * \sqrt{(2g)} * h^{2.5},$$

where g is the gravitational constant, 32.2 ft/sec^2. If $Q = 200$, make a table that shows how h is related to θ for values of θ between 20 and 130 degrees. Can this be done without using one of the methods of this chapter?

b. More often, weirs have a rectangular notch. Look up formulas for this case, but now the formula should allow for the effects of the velocity of the incoming water, v. Make a table that shows how h varies with v for several values of the width of the notch. Can you do this without using a method of this chapter? (You might want to repeat this for other notch configurations.)

APP16. It once was thought that the planets revolved around the sun in circular orbits. Johannes, Kepler (1571–1630), using the observations of Tycho Brahe, found that the orbits are really ellipses. He further found that their speed was not constant; they move faster when nearer the sun than when farther away. Kepler's equation relates time, t, and the central angle, A, measured from the sun (which is at one of the foci of the ellipse):

$$2\pi t = P(A - \varepsilon * \sin(A)),$$

where P is the period of revolution (the planet's "year,") and ε is the eccentricity of the ellipse.

You will remember that the equation of an ellipse with the center as the origin is

$$x^2/a^2 + y^2/b^2 = 1,$$

where a and b are one-half the major and minor diameters. The eccentricity, ε is c/a, where c is the distance from the origin to the foci.

The earth's orbit is almost circular, $\varepsilon = 0.02$, while the planet Mercury's is much flatter, $\varepsilon = 0.21$. (The orbit of Halley's comet has an eccentricity of 0.97!) Mercury has a year of 88 earth days and is 29 million miles from the sun at its closest point, the perigee.

Plot the ellipse for Mercury. Then solve Kepler's equation for angle A to superimpose the positions of Mercury at ten equispaced intervals during its year. Be sure to remember that Kepler's equations has the sun at the focal point nearest the perigee. (You can save some computations by noticing that the orbit is symmetrical.)

2

Solving Sets of Equations

Solving sets of linear equations is the most frequently used numerical procedure when real-world situations are modeled. Linear equations are the basis for mathematical models of economics, weather prediction, heat and mass transfer, statistical analysis, and a myriad of other applications. The methods for solving ordinary and partial-differential equations depend on them. In this book, almost every chapter uses the algorithms that we discuss here.

It is almost impossible to discuss systems of more than two or three equations without using matrices and vectors, so we cover some of the concepts of these at the start. Other aspects of the characteristics of a matrix are described in later chapters.

Contents of This Chapter

It shows when a matrix cannot be inverted and tells of situations where no unique solution exists to a system of equations.

2.4 Ill-Conditioned Systems

Explores systems for which getting the solution with accuracy is very difficult. A number, the condition number, is a measure of such difficulty; a property of a matrix, called its norm, is used to compute its condition number. A way to improve an inaccurate solution is described.

2.5 Iterative Methods

This section describes how a linear system can be solved in an entirely different way, by beginning with an initial estimate of the solution and performing computations that eventually arrive at the correct solution. It tells how the convergence can be accelerated. An iterative method is particularly important in solving systems that have few nonzero coefficients.

2.6 Parallel Processing

Tells how parallel computing can be applied to the solution of linear systems. An algorithm is developed that allows a significant reduction in processing time.

2.1 Matrices and Vectors

When a system of equations has more than two or three equations, it is difficult to discuss them without using *matrices* and *vectors*. While you may already know something about them, it is important that we review this topic in some detail.

A *matrix* is a rectangular array of numbers in which not only the value of the number is important but also its position in the array. The size of the matrix is described by the number of its rows and columns. A matrix of n rows and m columns is said to be $n \times m$. The elements of the matrix are generally enclosed in brackets, and double-subscripting is the common way of indexing the elements. The first subscript also denotes the row, and the second denotes the column in which the element occurs. Capital letters are used to refer to matrices. For example,

$$A = \begin{bmatrix} a_{11} & a_{12} \dots a_{1m} \\ a_{21} & a_{22} \dots a_{2m} \\ \vdots & \\ a_{n1} & a_{n2} \dots a_{nm} \end{bmatrix} = [a_{ij}], \quad i = 1, 2, \dots, n, \quad j = 1, 2, \dots, m.$$

Enclosing the general element a_{ij} in brackets is another way of representing matrix A, as just shown. Sometimes we will enclose the name of the matrix in brackets, $[A]$, to emphasize that A is a matrix.

Two matrices of the same size may be added or subtracted. The sum of

$$A = [a_{ij}] \qquad \text{and} \qquad B = [b_{ij}]$$

is the matrix whose elements are the sum of the corresponding elements of A and B:

$$C = A + B = [a_{ij} + b_{ij}] = [c_{ij}].$$

Similarly, we get the *difference* of two equal-sized matrices by subtracting corresponding elements. If two matrices are not equal in size, they cannot be added or subtracted. Two matrices are equal if and only if each element of one is the same as the corresponding element of the other. Obviously, equal matrices must be of the same size. Some examples will help make this clear.

If

$$A = \begin{bmatrix} 4 & 7 & -5 \\ -4 & 2 & 12 \end{bmatrix} \qquad \text{and} \qquad B = \begin{bmatrix} 1 & 5 & 4 \\ 2 & -6 & 3 \end{bmatrix},$$

we say that A is 2×3 because it has two rows and three columns. B is also 2×3. Their sum C is also 2×3:

$$C = A + B = \begin{bmatrix} 5 & 12 & -1 \\ -2 & -4 & 15 \end{bmatrix}.$$

The difference D of A and B is

$$D = A - B = \begin{bmatrix} 3 & 2 & -9 \\ -6 & 8 & 9 \end{bmatrix}.$$

Multiplication of two matrices is defined as follows, when A is $n \times m$ and B is $m \times r$:

$$[a_{ij}] * [b_{ij}] = [c_{ij}]$$
$$= \begin{bmatrix} (a_{11}b_{11} + a_{12}b_{21} + \cdots + a_{1m}b_{m1}) & \ldots & (a_{11}b_{1r} + \cdots + a_{1m}b_{mr}) \\ (a_{21}b_{11} + a_{22}b_{21} + \cdots + a_{2m}b_{m1}) & \ldots & (a_{21}b_{1r} + \cdots + a_{2m}b_{mr}) \\ \vdots & & \\ (a_{n1}b_{11} + a_{n2}b_{21} + \cdots + a_{mn}b_{m1}) & \ldots & (a_{n1}b_{1r} + \cdots + a_{nm}b_{mr}) \end{bmatrix},$$

$$c_{ij} = \sum_{k=1}^{m} a_{ik}b_{kj}, \qquad i = 1, 2, \ldots, n, \qquad j = 1, 2, \ldots, r.$$

It is simplest to select the proper elements if we count across the rows of A with the left hand while counting down the columns of B with the right. Unless the number of columns of A equals the number of rows of B (so the counting comes out even), the matrices cannot be multiplied. Hence, if A is $n \times m$, B must have m rows or else they are said to be "nonconformable for multiplication" and their product is undefined. In general $AB \neq BA$, so the order of factors must be preserved in matrix multiplication.

If a matrix is multiplied by a scalar (a pure number), the product is a matrix, each element of which is the scalar times the original element. We can write

$$\text{If } kA = C, \qquad c_{ij} = ka_{ij}.$$

A matrix with only one column, $n \times 1$ in size, is termed a *column vector,* and one of only one row, $1 \times m$ in size, is called a *row vector*. When the unqualified term *vector* is used, it nearly always means a *column* vector. Frequently the elements of vectors are only singly subscripted.

Some examples of matrix multiplication follow.

$$\text{Suppose } A = \begin{bmatrix} 3 & 7 & 1 \\ -2 & 1 & -3 \end{bmatrix}, \qquad B = \begin{bmatrix} 5 & -2 \\ 0 & 3 \\ 1 & -1 \end{bmatrix}, \qquad x = \begin{bmatrix} -3 \\ 1 \\ 4 \end{bmatrix}, \qquad y = \begin{bmatrix} y_1 \\ y_2 \\ y_3 \end{bmatrix}.$$

$$A * B = \begin{bmatrix} 16 & 14 \\ -13 & 10 \end{bmatrix}; \qquad B * A = \begin{bmatrix} 19 & 33 & 11 \\ -6 & 3 & -9 \\ 5 & 6 & 4 \end{bmatrix};$$

$$A * x = \begin{bmatrix} 2 \\ -5 \end{bmatrix}; \qquad A * y = \begin{bmatrix} 3y_1 + 7y_2 + y_3 \\ -2y_1 + y_2 - 3y_3 \end{bmatrix}.$$

Because A is 2×3 and B is 3×2, they are conformable for multiplication and their product is 2×2. When we form the product of $B * A$, it is 3×3. Observe that not only is $AB \neq BA$; AB and BA are not even the same size. The product of A and the vector x (a 3×1 matrix) is another vector, one with two components. Similarly, Ay has two components. We cannot multiply B times x or B times y; they are nonconformable.

The product of the scalar number 2 and A is

$$2A = \begin{bmatrix} 6 & 14 & 2 \\ -4 & 2 & -6 \end{bmatrix}.$$

Because a vector is just a special case of a matrix, a column vector can be multiplied by a matrix, as long as they are conformable in that the number of columns of the matrix equals the number of elements (rows) in the vector. The product in this case will be another column vector. The size of a product of two matrices, the first $m \times n$ and the second $n \times r$, is $m \times r$. An $m \times n$ matrix times an $n \times 1$ vector gives an $m \times 1$ product.

The general relation for $Ax = b$ is

$$b_i = \sum_{k=1}^{\text{No. of cols.}} a_{ik}x_k, \qquad i = 1, 2, \ldots, \text{No. of rows.}$$

This definition of matrix multiplication permits us to write the set of linear equations

$$\begin{aligned} a_{11}x_1 + a_{12}x_2 + \cdots + a_{1n}x_n &= b_1, \\ a_{21}x_1 + a_{22}x_2 + \cdots + a_{2n}x_n &= b_2, \\ &\vdots \\ a_{n1}x_1 + a_{n2}x_2 + \cdots + a_{nn}x_n &= b_n, \end{aligned}$$

much more simply in matrix notation, as $Ax = b,$ where

$$A = \begin{bmatrix} a_{11} & a_{12} \ldots a_{1n} \\ a_{21} & a_{22} \ldots a_{2n} \\ \vdots & \\ a_{n1} & a_{n2} \ldots a_{nn} \end{bmatrix}, \qquad x = \begin{bmatrix} x_1 \\ x_2 \\ \vdots \\ x_n \end{bmatrix}, \qquad b = \begin{bmatrix} b_1 \\ b_2 \\ \vdots \\ b_n \end{bmatrix}.$$

For example,

$$\begin{bmatrix} 3 & 2 & 4 \\ 1 & -2 & 0 \\ -1 & 3 & 2 \end{bmatrix} * x = \begin{bmatrix} 14 \\ -7 \\ 2 \end{bmatrix}$$

is the same as the set of equations

$$\begin{aligned} 3x_1 + 2x_2 + 4x_3 &= 14, \\ x_1 - 2x_2 \qquad &= -7, \\ -x_1 + 3x_2 + 2x_3 &= 2. \end{aligned}$$

Two vectors, each with the same number of components, may be added or subtracted. Two vectors are equal if each component of one equals the corresponding component of the other.

A very important special case is the multiplication of two vectors. The first must be a row vector if the second is a column vector, and each must have the same number of components. For example,

$$[1 \quad 3 \quad -2] * \begin{bmatrix} 4 \\ -1 \\ 3 \end{bmatrix} = [-5]$$

gives a "matrix" of one row and one column. The result is a pure number, a scalar. This product is called the *scalar product* of the vectors, also called the *inner product*.

If we reverse the order of multiplication of these two vectors, we obtain

$$\begin{bmatrix} 4 \\ -1 \\ 3 \end{bmatrix} * [1 \quad 3 \quad -2] = \begin{bmatrix} 4 & 12 & -8 \\ -1 & -3 & 2 \\ 3 & 9 & -6 \end{bmatrix}.$$

This product is called the *outer product*. Although not as well known as the inner product, the outer product is very important in nonlinear optimization problems.*

A vector whose length is one is called a *unit vector*.† A vector that has all of its elements equal to zero is the *zero vector*. If all elements are zero except one, it is a *unit basis vector*. There are three distinct unit basis vectors of order-3:

$$\begin{bmatrix} 1 \\ 0 \\ 0 \end{bmatrix}, \quad \begin{bmatrix} 0 \\ 1 \\ 0 \end{bmatrix}, \quad \text{and} \quad \begin{bmatrix} 0 \\ 0 \\ 1 \end{bmatrix}.$$

Some Special Matrices and Their Properties

Square matrices are particularly important when a system of equations is to be solved. Square matrices have some special properties.

The elements on the *main diagonal* are those from the upper-left corner to the lower-right corner. These are commonly referred to just as the *diagonal elements;* most often, just the word *diagonal* is used. If all elements except those on the diagonal are zero, the matrix is called a *diagonal matrix*.

If the nonzero elements of a diagonal matrix all are equal to one, the matrix is called the *identity matrix of order n* where n equals the number of row and columns. The usual symbol for this identity matrix is I_n and it has properties similar to unity. For example, the order-4 identity matrix is

$$\begin{bmatrix} 1 & 0 & 0 & 0 \\ 0 & 1 & 0 & 0 \\ 0 & 0 & 1 & 0 \\ 0 & 0 & 0 & 1 \end{bmatrix} = I_4.$$

The subscript is omitted when the order is clear from the context.

An important property of an identity matrix, I, is that for any $n \times n$ matrix, A, it is always true that

$$I * A = A * I = A.$$

If two rows of an identity matrix are interchanged, it is called a *transposition matrix*. (We also get a transposition matrix by interchanging two columns.) The name is appropriate because, if transposition matrix P_1 is multiplied with a square matrix of the same size, A, the product $P_1 * A$ will be the A matrix but with the same two rows interchanged. Here is an example:

$$P_1 = \begin{bmatrix} 1 & 0 & 0 & 0 \\ 0 & 0 & 0 & 1 \\ 0 & 0 & 1 & 0 \\ 0 & 1 & 0 & 0 \end{bmatrix}, \quad A = \begin{bmatrix} 9 & 6 & 2 & 13 \\ 4 & 2 & 8 & 11 \\ 0 & 7 & 1 & 9 \\ 3 & 2 & 6 & 8 \end{bmatrix}, \quad P_1 * A = \begin{bmatrix} 9 & 6 & 2 & 13 \\ 3 & 2 & 6 & 8 \\ 0 & 7 & 1 & 9 \\ 4 & 2 & 8 & 11 \end{bmatrix}.$$

*Another important product of three-component vectors is the *vector product,* also known as the *cross product*.

† The length of a vector is the square root of the sum of the squares of its components, an extension of the idea of the length of a two-component vector drawn from the origin.

However, if the two matrices are multiplied in reverse order, $A * P_1$, the result will be matrix A but with the columns of A interchanged. You should check this for yourself with the example matrices.

A *permutation matrix* is obtained by multiplying several transposition matrices.

A square matrix is called a *symmetric matrix* when the pairs of elements in similar positions across the diagonal are equal. Here is an example:

$$\begin{bmatrix} 1 & x & y \\ x & 2 & z \\ y & z & 3 \end{bmatrix}$$

The transpose of a matrix is the matrix obtained by writing the rows as columns or by writing the columns as rows. (A matrix does not have to be square to have a transpose.) The symbol for the transpose of matrix A is A^T. Example 2.1 illustrates.

EXAMPLE 2.1

$$A = \begin{bmatrix} 3 & -1 & 4 \\ 0 & 2 & -3 \\ 1 & 1 & 2 \end{bmatrix}; \quad A^T = \begin{bmatrix} 3 & 0 & 1 \\ -1 & 2 & 1 \\ 4 & -3 & 2 \end{bmatrix}.$$

It should be clear that $A = A^T$ if A is symmetric, and that for any matrix, the transpose of the transpose, $(A^T)^T$, is just A itself. It is also true, though not so obvious, that

$$(A * B)^T = B^T * A^T.$$

When a matrix is square, a quantity called its *trace* is defined. The trace of a square matrix is the sum of the elements on its main diagonal. For example, the traces of the previous matrices are

$$\text{tr}(A) = \text{tr}(A^T) = 3 + 2 + 2 = 7.$$

It should be obvious that the trace remains the same if a square matrix is transposed.

If all the elements above the diagonal are zero, a matrix is called *lower-triangular;* it is called *upper-triangular* when all the elements below the diagonal are zero. For example, these order-3 matrices are lower- and upper-triangular:

$$L = \begin{bmatrix} 1 & 0 & 0 \\ 4 & 6 & 0 \\ -2 & 1 & -4 \end{bmatrix}, \quad U = \begin{bmatrix} 1 & -3 & 3 \\ 0 & -1 & 0 \\ 0 & 0 & 1 \end{bmatrix}.$$

Triangular matrices are of special importance, as will become apparent later in this chapter and in several other chapters.

Tridiagonal matrices are those that have nonzero elements only on the diagonal and in the positions adjacent to the diagonal; they will be of special importance in certain partial-differential equations. An example of a tridiagonal matrix is

$$\begin{bmatrix} -4 & 2 & 0 & 0 & 0 \\ 1 & -4 & 1 & 0 & 0 \\ 0 & 1 & -4 & 1 & 0 \\ 0 & 0 & 1 & -4 & 1 \\ 0 & 0 & 0 & 2 & -4 \end{bmatrix}.$$

For a tridiagonal matrix, only the nonzero values need to be recorded, and that means that the $n \times n$ matrix can be compressed into a matrix of 3 columns and n rows. For this example, we can write the matrix as

$$\begin{bmatrix} x & -4 & 2 \\ 1 & -4 & 1 \\ 1 & -4 & 1 \\ 1 & -4 & 1 \\ 2 & -4 & x \end{bmatrix}.$$

(The x entries are not normally used; they might be entered as zeros.)

In some important applied problems, only a few of the elements are nonzero. Such a matrix is termed a *sparse matrix* and procedures that take advantage of this sparseness are of value.

Examples of Operations with Matrices

Here are some examples of matrix operations:

$$3 * \begin{bmatrix} 1 & 2 \\ 3 & 4 \end{bmatrix} = \begin{bmatrix} 3 & 6 \\ 9 & 12 \end{bmatrix}.$$

$$\begin{bmatrix} 1 & 3 & 2 \\ -1 & 0 & 4 \end{bmatrix} + \begin{bmatrix} -1 & 0 & 2 \\ 4 & 1 & -3 \end{bmatrix} = \begin{bmatrix} 0 & 3 & 4 \\ 3 & 1 & 1 \end{bmatrix}.$$

$$\begin{bmatrix} 2 & 1 \\ 0 & -4 \\ 7 & 2 \end{bmatrix} - \begin{bmatrix} 3 & -2 \\ 4 & 1 \\ 0 & -2 \end{bmatrix} = \begin{bmatrix} -1 & 3 \\ -4 & -5 \\ 7 & 4 \end{bmatrix}.$$

$$\begin{bmatrix} 2 & 0 & -1 \\ 3 & 2 & 6 \end{bmatrix} * \begin{bmatrix} -1 \\ 2 \\ 1 \end{bmatrix} = \begin{bmatrix} -3 \\ 7 \end{bmatrix}, \text{ but } \begin{bmatrix} 6 & -1 \\ 3 & -2 \end{bmatrix} * \begin{bmatrix} -1 \\ 2 \\ 3 \end{bmatrix} \text{ is undefined.}$$

$$\begin{bmatrix} 1 & 3 \\ 2 & -1 \end{bmatrix} * \begin{bmatrix} 0 & 3 \\ -1 & 1 \end{bmatrix} = \begin{bmatrix} -3 & 6 \\ 1 & 5 \end{bmatrix}, \text{ but } \begin{bmatrix} 0 & 3 \\ -1 & 1 \end{bmatrix} * \begin{bmatrix} 1 & 3 \\ 2 & -1 \end{bmatrix} = \begin{bmatrix} 6 & -3 \\ 1 & -4 \end{bmatrix}.$$

Division of matrices is not defined, but we will discuss the *inverse* of a matrix later in this chapter.

The *determinant* of a square matrix is a number. For a 2×2 matrix, the determinant is computed by subtracting the product of the elements on the minor diagonal (from upper right to lower left) from the product of terms on the major diagonal. For example,

$$A = \begin{bmatrix} -5 & 3 \\ 7 & 2 \end{bmatrix}, \qquad \det(A) = (-5)(2) - (7)(3) = -31;$$

$\det(A)$ is the usual notation for the determinant of A. Sometimes the determinant is symbolized by writing the elements of the matrix between vertical lines (similar to representing the absolute value of a number).

For a 3×3 matrix, you may have learned a crisscross way of forming products of terms (we call it the "spaghetti rule") that probably should be forgotten, for it applies only to the special case of a 3×3 matrix; it won't work for larger systems. The general rule that applies in all cases is to expand in terms of the minors of some row or column. The *minor* of any term is the matrix of lower order formed by striking out the row and column in which the term is found. The determinant is found by adding the product of each term in any row or column by the determinant of its minor, with signs alternating + and −. We expand each of the determinants of the minor until we reach 2×2 matrices. For example,

$$\text{Given } A = \begin{bmatrix} 3 & 0 & -1 & 2 \\ 4 & 1 & 3 & -2 \\ 0 & 2 & -1 & 3 \\ 1 & 0 & 1 & 4 \end{bmatrix},$$

$$\det(A) = 3\begin{vmatrix} 1 & 3 & -2 \\ 2 & -1 & 3 \\ 0 & 1 & 4 \end{vmatrix} - 0\begin{vmatrix} 4 & 3 & -2 \\ 0 & -1 & 3 \\ 1 & 1 & 4 \end{vmatrix}$$

$$+ (-1)\begin{vmatrix} 4 & 1 & -2 \\ 0 & 2 & 3 \\ 1 & 0 & 4 \end{vmatrix} - 2\begin{vmatrix} 4 & 1 & 3 \\ 0 & 2 & -1 \\ 1 & 0 & 1 \end{vmatrix}$$

$$= 3\left\{(1)\begin{vmatrix} -1 & 3 \\ 1 & 4 \end{vmatrix} - (3)\begin{vmatrix} 2 & 3 \\ 0 & 4 \end{vmatrix} + (-2)\begin{vmatrix} 2 & -1 \\ 0 & 1 \end{vmatrix}\right\}$$

$$- 0 + (-1)\left\{(4)\begin{vmatrix} 2 & 3 \\ 0 & 4 \end{vmatrix} - (1)\begin{vmatrix} 0 & 3 \\ 1 & 4 \end{vmatrix} + (-2)\begin{vmatrix} 0 & 2 \\ 1 & 0 \end{vmatrix}\right\}$$

$$- 2\left\{(4)\begin{vmatrix} 2 & -1 \\ 0 & 1 \end{vmatrix} - (1)\begin{vmatrix} 0 & -1 \\ 1 & 1 \end{vmatrix} + (3)\begin{vmatrix} 0 & 2 \\ 1 & 0 \end{vmatrix}\right\}$$

$$= 3\{(1)(-7) - (3)(8) + (-2)(2)\} - 0 + (-1)\{(4)(8) - (1)(-3) + (-2)(-2)\}$$
$$- 2\{(4)(2) - (1)(1) + (3)(-2)\}$$
$$= 3(-7 - 24 - 4) - 0 + (-1)(32 + 3 + 4) - 2(8 - 1 - 6)$$
$$= 3(-35) - 0 + (-1)(39) - 2(1) = -146.$$

In computing the determinant, the expansion can be about the elements of any row or column. To get the signs, give the first term a plus sign if the sum of its column number and row number is even; give it a minus if the sum is odd, with alternating signs thereafter. (For example, in expanding about the elements of the third row we begin with a plus; the first element a_{31} has $3 + 1 = 4$, an even number.) Judicious selection of rows and columns with many zeros can hasten the process, but this method of calculating determinants is a lot of work if the matrix is of large size. Methods that triangularize a matrix, as described in Section 2.2, are much better ways to get the determinant.

If a matrix, B, is triangular (either upper or lower), its determinant is just the product of the diagonal elements: $\det(B) = \Pi B_{ii}$, $i = 1, \ldots, n$. It is easy to show this if the determinant of the triangular matrix is expanded by minors. The following example illustrates this:

$$\det\begin{bmatrix} 4 & 0 & 0 \\ 6 & -2 & 0 \\ 1 & -3 & 5 \end{bmatrix} = 4 * \det\begin{bmatrix} -2 & 0 \\ -3 & 5 \end{bmatrix} + 0 + 0$$

$$= 4 * (-2) * |5| = 4 * (-2) * 5 = -40.$$

Determinants can be used to obtain the *characteristic polynomial* and the *eigenvalues* of a matrix, which are the roots of that polynomial. In Chapter 6, you will see that these are important in solving certain differential equations. The Greek symbol λ is commonly used to represent an eigenvalue. (Eigenvalue is a German word, the corresponding English term is *characteristic value,* but it is less frequently used.)

The two terms, eigenvalue and characteristic polynomial are interrelated: For matrix A, $P_A(\lambda) = \det(A - \lambda I)$.

For example, if

$$A = \begin{bmatrix} 1 & 3 \\ 4 & 5 \end{bmatrix},$$

then

$$P_A(\lambda) = |A - \lambda I| = \det\begin{bmatrix} (1-\lambda) & 3 \\ 4 & (5-\lambda) \end{bmatrix}$$

$$= (1-\lambda)(5-\lambda) - 12$$

$$= \lambda^2 - 6\lambda - 7.$$

(The characteristic polynomial is always of degree n if A is $n \times n$.) If we set the characteristic polynomial to zero and solve for the roots, we get the eigenvalues of A. For this example, these are $\lambda_1 = 7$, $\lambda_2 = -1$, or, in more symbolic mathematical notation,

$$\Lambda(A) = \{7, \quad -1\}.$$

We also mention the notion of an *eigenvector* corresponding to an eigenvalue. The eigenvector is a nonzero vector w such that

$$Aw = \lambda w, \qquad \text{that is,} \qquad (A - \lambda I)w = 0. \tag{2.1}$$

In the current example, the eigenvectors are

$$w_1 = \begin{bmatrix} 1 \\ 2 \end{bmatrix}, \qquad w_2 = \begin{bmatrix} -3 \\ 2 \end{bmatrix}.$$

We leave it as an exercise to show that these eigenvectors satisfy Eq. (2.1).

Observe that the trace of A is equal to the sum of the eigenvalues: $\text{tr}(A) = 1 + 5 = \lambda_1 + \lambda_2 = 7 + (-1) = 6$. This is true for any matrix: The sum of its eigenvalues equals its trace.

For now, we limit the finding of eigenvalues and eigenvectors to small matrices because getting these through the characteristic polynomial is not recommended for matrices larger than 4×4. In Chapter 6 we examine other, more efficient ways to get these important quantities.

If a matrix is triangular, its eigenvalues are equal to the diagonal elements. This follows from the fact that its determinant is just the product of the diagonal elements and its characteristic polynomial is the product of the terms $(a_{i,i} - \lambda)$ with i going from 1 to n, the number of rows of the matrix. This simple example illustrates for a 3×3 matrix, A:

$$A = \begin{bmatrix} 1 & 2 & 3 \\ 0 & 4 & 5 \\ 0 & 0 & 6 \end{bmatrix}, \ \det(A - \lambda I) = \det \begin{bmatrix} (1-\lambda) & 2 & 3 \\ 0 & (4-\lambda) & 5 \\ 0 & 0 & (5-\lambda) \end{bmatrix}$$
$$= (1-\lambda)(4-\lambda)(6-\lambda),$$

whose roots are clearly 1, 4, and 6. It does not matter if the matrix is upper- or lower-triangular.

Using Computer Algebra Systems

MATLAB can do matrix operations. We first define two matrices and a vector, A, B, and v:

```
EDU>> A = [4 1 -2; 5 1 3; 4 0  -1]
A =
   4    1    -2
   5    1     3
   4    0    -1
EDU>> B = [3 3 1; -2 1 5; 2 2 0]
B =
     3    3    1
    -2    1    5
     2    2    0
EDU>> v = [-2 3 1]
v =
   -2    3    1
```

Now we so some operations:

```
EDU>> 3*A
ans =
     12     3    -6
     15     3     9
     12     0    -3
```

```
EDU>> A + B
ans =
   7   4   -1
   3   2    8
   6   2   -1
EDU>> B - A
ans =
   -1   2   3
   -7   0   2
   -2   2   1
EDU>> A*B
ans =
    6    9    9
   19   22   10
   10   10    4
EDU>> B*v
??? Error using ==> *
Inner matrix dimensions must agree.
```

We can't multiply by the row vector, but we could with a column vector—so we transpose the vector:

```
EDU>> vt = v'
vt =
  -2
   3
   1
EDU>> B*Vt
ans =
   4
  12
   2
```

and now it works. Here are some other operations:

```
EDU>> det(A)
ans =
  21
EDU>> v*vt
ans =
  14
EDU>> vt*v
ans =
    4   -6   -2
   -6    9    3
   -2    3    1
```

```
EDU>> trace(A)
ans =
   4
```

We can get the characteristic polynomial:

```
EDU>> poly(A)
ans =
   1.0000    -4.0000    2.0000    -21.0000
```

where the coefficients are given. This represents

$$x^4 - 4x^3 + 2x - 21.$$

Using Maple

Maple and MATLAB are interrelated and MATLAB commands can be invoked in Maple and vice versa, but Maple can do matrix manipulations on its own. Here are a few—Maple's commands are somewhat different. Most need to be preceded by `with(linalg)`.

```
matadd (A, B)     does A + B
multiply (A, B) does A * B
   (but evalm ( . . . ) is more versatile, does all arithmetic
     operations with matrices, vectors, and scalars.)
trace (A)         gets the trace
transpose (A)     transposes
det (A)           gets the determinant
```

2.2 Elimination Methods

We now discuss numerical methods that are used to solve a set of linear equations. The term *linear equation* means an equation in several variables where all of the variables occur to the first power and there are no terms involving transcendental functions such as sine and cosine.

It used to be that students were taught to use Cramer's rule, in which a system can be solved through the use of determinants. However, Cramer's rule is inefficient and is almost impossible to use if there are more than two or three equations. As we have said, most applied problems are modeled with large systems and we present methods that work well with them. Even so, we use small systems as illustrations.

Suppose we have a system of equations that is of a special form, an *upper-triangular system,* such as

$$\begin{aligned} 5x_1 + 3x_2 - 2x_3 &= -3, \\ 6x_2 + x_3 &= -1, \\ 2x_3 &= 10. \end{aligned}$$

Whenever a system has this special form, its solution is very easy to obtain. From the third equation we see that $x_3 = 5$. Substituting this value into the second equation quickly gives $x_2 = -1$. Then substituting both values into the first equation reveals that $x_1 = 2$; now we have the solution: $x_1 = 2, x_2 = -1, x_3 = 5$.*

The first objective of the *elimination method* is to change the matrix of coefficients so that it is upper triangular. Consider this example of three equations:

$$\begin{aligned} 4x_1 - 2x_2 + x_3 &= 15, \\ -3x_1 - x_2 + 4x_3 &= 8, \\ x_1 - x_2 + 3x_3 &= 13. \end{aligned} \tag{2.2}$$

Multiplying the first equation by 3 and the second by 4 and adding these will eliminate x_1 from the second equation. Similarly, multiplying the first by -1 and the third by 4 and adding eliminates x_1 from the third equation. (We prefer to multiply by the negatives and add, to avoid making mistakes when subtracting quantities of unlike sign.) The result is

$$\begin{aligned} 4x_1 - 2x_2 + x_3 &= 15, \\ -10x_2 + 19x_3 &= 77, \\ -2x_2 + 11x_3 &= 37. \end{aligned}$$

We now eliminate x_2 from the third equation by multiplying the second by 2 and the third by -10 and adding to get

$$\begin{aligned} 4x_1 - 2x_2 + x_3 &= 15, \\ -10x_2 + 19x_3 &= 77, \\ -72x_3 &= -216. \end{aligned}$$

Now we have a triangular system and the solution is readily obtained; obviously $x_3 = 3$ from the third equation, and *back-substitution* into the second equation gives $x_2 = -2$. We continue with back-substitution by substituting both x_2 and x_3 into the first equation to get $x_1 = 2$.

The essence of any elimination method is to reduce the coefficient matrix to a triangular matrix and then use back-substitution to get the solution.

We now present the same problem, solved in exactly the same way, in matrix notation:

$$\begin{bmatrix} 4 & -2 & 1 \\ -3 & -1 & 4 \\ 1 & -1 & 3 \end{bmatrix} \begin{bmatrix} x_1 \\ x_2 \\ x_3 \end{bmatrix} = \begin{bmatrix} 15 \\ 8 \\ 13 \end{bmatrix}.$$

The arithmetic operations that we have performed affect only the coefficients and the right-hand-side terms, so we work with the matrix of coefficients *augmented* with the

---

* A system that is *lower-triangular* is equally easy to solve. We then do *forward-substitution* rather than *back-substitution*.

right-hand-side vector:

$$A|b = \left[\begin{array}{rrr:r} 4 & -2 & 1 & 15 \\ -3 & -1 & 4 & 8 \\ 1 & -1 & 3 & 13 \end{array}\right]$$

(The dashed line is usually omitted.)

We perform elementary row transformations* to convert A to *upper-triangular* form:

$$\begin{bmatrix} 4 & -2 & 1 & 15 \\ -3 & -1 & 4 & 8 \\ 1 & -1 & 3 & 13 \end{bmatrix}, \quad \begin{array}{r} \\ 3R_1 + 4R_2 \rightarrow \\ (-1)R_1 + 4R_3 \rightarrow \end{array} \begin{bmatrix} 4 & -2 & 1 & 15 \\ 0 & -10 & 19 & 77 \\ 0 & -2 & 11 & 37 \end{bmatrix},$$

$$\begin{array}{r} \\ \\ 2R_2 - 10R_3 \rightarrow \end{array} \begin{bmatrix} 4 & -2 & 1 & 15 \\ 0 & -10 & 19 & 77 \\ 0 & 0 & -72 & -216 \end{bmatrix}. \tag{2.3}$$

The steps here are to add 3 times the first row to 4 times the second row and to add -1 times the first row to 4 times the third row. The next and final phase (in order to get a triangular system) is to add 2 times the second row to -10 times the third row.

The array in Eq. (2.3) represents the equations

$$\begin{aligned} 4x_1 - 2x_2 + x_3 &= 15, \\ -10x_2 + 19x_3 &= 77, \\ -72x_3 &= -216. \end{aligned} \tag{2.4}$$

The back-substitution step can be performed quite mechanically by solving the equations of Eq. (2.4) in reverse order. That is,

$$\begin{aligned} x_3 &= -216/(-72) = 3, \\ x_2 &= (77 - 3(19))/(-10) = -2, \\ x_1 &= (15 - 1(3) - (-2)(-2))/4 = 2. \end{aligned}$$

Thinking of the procedure in terms of matrix operations, we transform the augmented coefficient matrix by elementary row operations until a triangular matrix is created on the left. After back-substitution, the x-vector stands as the rightmost column.†

These operations, which do not change the relationships represented by a set of equations, can be applied to an augmented matrix, because this is only a different notation for the equations. (We need to add one proviso: Because round-off error is related to the magnitude of the values when we express them in fixed-word-length computer

---

* Elementary row operations are arithmetic operations that obviously are valid rearrangements of a set of equations: (1) Any equation can be multiplied by a constant; (2) the order of the equations can be changed; (3) any equation can be replaced by its sum with another of the equations.

† Making the matrix triangular by row operations is not a way to get its eigenvalues; the row operations change them.

representations, some of our operations may have an effect on the accuracy of the computed solution.)

An alternative to converting the system to upper-triangular is to make it lower-triangular. For example, if we have this set of equations (or an augmented matrix):

$$\begin{aligned} 6x_1 \qquad\qquad &= 18 \\ -2x_1 + 5x_2 \qquad &= 2 \\ 3x_1 - 4x_2 + 3x_3 &= 11 \end{aligned} \quad \text{or} \quad \begin{bmatrix} 6 & 0 & 0 \\ -2 & 5 & 0 \\ 3 & -4 & 3 \end{bmatrix} \begin{bmatrix} x_1 \\ x_2 \\ x_3 \end{bmatrix} = \begin{bmatrix} 18 \\ 2 \\ 11 \end{bmatrix}.$$

Here, we would solve for the variables in this order: x_1, then x_2, and finally x_3, with the same number of computations as in the case of a lower-triangular system. Both the lower- and upper-triangular systems play an important part in the development of algorithms in the following sections, because these systems require fewer multiplications/divisions than the general system. We shall also show that we can often write a general matrix A as the product LU, a lower-triangular matrix times an upper-triangular matrix.

Note that there exists the possibility that the set of equations has no solution, or that the prior procedure will fail to find it. During the triangularization step, if a zero is encountered on the diagonal, we cannot use that row to eliminate coefficients below that zero element. However, in that case, we can continue by interchanging rows and eventually achieve an upper-triangular matrix of coefficients. The real stumbling block is finding a zero on the diagonal after we have triangularized. If that occurs, the back-substitution fails, for we cannot divide by zero. It also means that the determinant is zero: There is no solution.

It is worthwhile to explain in more detail what we mean by the *elementary row operations* that we have used here, and to see why they can be used in solving a linear system. There are three of these operations:

1. We may multiply any row of the augmented coefficient matrix by a constant.
2. We can add a multiple of one row to a multiple of any other row.
3. We can interchange the order of any two rows (this was not used earlier).

The validity of these row operations is intuitively obvious if we think of them applied to a set of linear equations. Certainly, multiplying one equation through by a constant does not change the truth of the equality. Adding equal quantities to both sides of an equality results in an equality, and this is the equivalent of the second transformation. Obviously, the order of the set is arbitrary, so rule 3 is valid.

Gaussian Elimination

The procedure just described has a major problem. While it may be satisfactory for hand computations with small systems, it is inadequate for a large system. Observe that the transformed coefficients can become very large as we convert to a triangular system. The method that is called *Gaussian elimination* avoids this by subtracting a_{i1}/a_{11} times the first equation from the ith equation to make the transformed numbers in the first column equal to zero. We do similarly for the rest of the columns.

We must always guard against dividing by zero. Observe that zeros may be created in the diagonal positions even if they are not present in the original matrix of coefficients. A useful strategy to avoid (if possible) such zero divisors is to rearrange the equations so as to put the coefficient of largest magnitude on the diagonal at each step. This is called *pivoting*. Complete pivoting may require both row and column interchanges. This is not frequently done. Partial pivoting, which places a coefficient of larger magnitude on the diagonal by row interchanges only, will guarantee a nonzero divisor if there is a solution to the set of equations, and will have the added advantage of giving improved arithmetic precision. The diagonal elements that result are called *pivot elements*. (When there are large differences in magnitude of coefficients in one equation compared to the other equations, we may need to *scale* the values; we consider this later.)

We repeat the example of the previous section, incorporating these ideas and carrying four significant digits in our work. We begin with the augmented matrix.

$$\left[\begin{array}{ccc:c} 4 & -2 & 1 & 15 \\ -3 & -1 & 4 & 8 \\ 1 & -1 & 3 & 13 \end{array}\right]$$

$$\begin{array}{r} \\ R_2 - (-3/4)R_1 \rightarrow \\ R_3 - (1/4)R_1 \rightarrow \end{array} \left[\begin{array}{ccc:c} 4 & -2 & 1 & 15 \\ 0 & -2.5 & 4.75 & 19.25 \\ 0 & -0.5 & 2.75 & 9.25 \end{array}\right]$$

$$\begin{array}{r} \\ \\ R_3 - (-0.5/-2.5)R_2 \rightarrow \end{array} \left[\begin{array}{ccc:c} 4 & -2 & 1 & 15 \\ 0 & -2.5 & 4.75 & 19.25 \\ 0 & 0.0 & 1.80 & 5.40 \end{array}\right]$$

[The notation used here means to subtract $(-3/4)$ times row 1 from row 2 and to subtract $(1/4)$ times row 1 from row 3 in reducing in the first column; to subtract $(-0.5/-2.5)$ times row 2 from row 3 in the third column.]

The method we have just illustrated is called *Gaussian elimination*. (In this example, no pivoting was required to make the largest coefficients be on the diagonal.) Back-substitution, as presented with Eq. (2.4), gives us, as before, $x_3 = 3$, $x_2 = -2$, $x_1 = 2$. We have come up with the exact answer to this problem. Often it will turn out that we shall obtain answers that are just close approximations to the exact answer because of round-off error. When there are many equations, the effects of round-off (the term is applied to the error due to chopping as well as when rounding is used) may cause large effects. In certain cases, the coefficients are such that the results are particularly sensitive to round off; such systems are called *ill-conditioned.*

In the example just presented, the zeros below the main diagonal show that we have reduced the problem [Eq. (2.3)] to solving an upper-triangular system of equations as in Eqs. (2.4). However, at each stage, if we had stored the ratio of coefficients in place of zero (we show these in parentheses), our final form would have been

$$\left[\begin{array}{ccc|c} 4 & -2 & 1 & 15 \\ (-0.75) & -2.5 & 4.75 & 19.25 \\ (0.25) & (0.20) & 1.80 & 5.4 \end{array}\right].$$

Then, in addition to solving the problem as we have done, we find that the original matrix

$$A = \begin{bmatrix} 4 & -2 & 1 \\ -3 & -1 & 4 \\ 1 & -1 & 3 \end{bmatrix}$$

can be written as the product:

$$\underbrace{\begin{bmatrix} 1 & 0 & 0 \\ -0.75 & 1 & 0 \\ 0.25 & 0.20 & 1 \end{bmatrix}}_{L} * \underbrace{\begin{bmatrix} 4 & -2 & 1 \\ 0 & -2.5 & 4.75 \\ 0 & 0 & 1.80 \end{bmatrix}}_{U}. \qquad (2.5)$$

This procedure is called a *LU decomposition* of A. In this case,

$$A = L * U,$$

where L is *lower-triangular* and U is *upper-triangular*. As we shall see in the next example, usually $L * U = A'$, where A' is just a permutation of the rows of A due to row interchange from pivoting.

Finally, because the determinant of two matrices, $B * C$, is the product of each of the determinants, for this example we have

$$\det(L * U) = \det(L) * \det(U) = \det(U),$$

because L is triangular and has only ones on its diagonal so that $\det(L) = 1$. Thus, for the example given in Eq. (2.5), we have

$$\det(A) = \det(U) = (4) * (-2.5) * (1.8) = -18,$$

because U is upper-triangular and its determinant is just the product of the diagonal elements.

From this example, we see that Gaussian elimination does the following:

1. It solves the system of equations.
2. It computes the determinant of a matrix very efficiently.
3. It can provide us with the LU decomposition of the matrix of coefficients, in the sense that the product of the two matrices, $L * U$, may give us a permutation of the rows of the original matrix.

With regard to item 2, when there are row interchanges,

$$\det(A) = (-1)^m * u_{11} * \cdots * u_{nn},$$

where the exponent m represents the number of row interchanges.

We summarize the operations of Gaussian elimination in a form that will facilitate the writing of a computer program. Note that in the actual implementation of the algorithm, the L and U matrices are actually stored in the space of the original matrix A.

Gaussian Elimination

To solve a system of n linear equations: $Ax = b$.

```
For j = 1 To (n − 1)
    pvt = |a[ j, j ]|
    pivot [ j ] = j
    ipvt_temp = j

    For i = j + 1 To n        (Find pivot row)
        IF |a [i, j]| > pvt   Then
            pvt = |a[i, j]|
            ipvt_temp = i
        End IF
    End For i

    (Switch rows if necessary)
    IF pivot [ j ]  < >  ipvt_temp
        [switch_rows(rows j and ipvt_temp)]

    For i = j + 1 to n        (Store multipliers)
        a[i, j] = a[i, j ]/a[ j, j ]
    End For i

    (Create zeros below the main diagonal)
    For i = j + 1 To n
        For k = j + 1 To n
            a[i, k] = a[i, k] − a[i, j] * a[j, k]
        End For k
        b [i] = b[i] − a[i, j] * b[j]
    End For i

End For j;

(Back Substitution Part)
x[n] = b[n]/a[n, n]
For j = n − 1 Down To 1
    x[ j ] = b[ j ]
    For k = n Down To j + 1
        x[ j] = x[ j ] − x[k] * a[j, k]
    End For k
    x[ j] = x[ j ]/a[ j, j]

End For j.
```

Interchanging rows in a large matrix can be expensive; there is a better way. We keep track of the order of the rows in an *order vector* and, when a row interchange is indicated, we only interchange the corresponding elements in the order vector. This vector then tells which rows are to be worked on. Using an order vector saves computer time because only two numbers of this vector are interchanged; we do not have to switch all the elements of the two rows. However, we do not do this here in order to keep our explanations simple. You will later see an example that uses an order vector.

The algorithm for Gaussian elimination will be clarified by an additional numerical example. Solve the following system of equations using Gaussian elimination. In addition, compute the determinant of the coefficient matrix and the *LU* decomposition of this matrix.

Given the system of equations, solve

$$\begin{aligned} 2x_2 \qquad\quad + x_4 &= 0, \\ 2x_1 + 2x_2 + 3x_3 + 2x_4 &= -2, \\ 4x_1 - 3x_2 \qquad\quad + x_4 &= -7, \\ 6x_1 + x_2 - 6x_3 - 5x_4 &= 6. \end{aligned} \tag{2.6}$$

The augmented coefficient matrix is

$$\begin{bmatrix} 0 & 2 & 0 & 1 & 0 \\ 2 & 2 & 3 & 2 & -2 \\ 4 & -3 & 0 & 1 & -7 \\ 6 & 1 & -6 & -5 & 6 \end{bmatrix}. \tag{2.7}$$

We cannot permit a zero in the a_{11} position because that element is the pivot in reducing the first column. We could interchange the first row with any of the other rows to avoid a zero divisor, but interchanging the first and fourth rows is our best choice. This gives

$$\begin{bmatrix} 6 & 1 & -6 & -5 & 6 \\ 2 & 2 & 3 & -2 & -2 \\ 4 & -3 & 0 & 1 & -7 \\ 0 & 2 & 0 & 1 & 0 \end{bmatrix}. \tag{2.8}$$

We make all the elements in the first column zero by subtracting the appropriate multiple of row one:

$$\begin{bmatrix} 6 & 1 & -6 & -5 & 6 \\ 0 & 1.6667 & 5 & 3.6667 & -4 \\ 0 & -3.6667 & 4 & 4.3333 & -11 \\ 0 & 2 & 0 & 1 & 0 \end{bmatrix}. \tag{2.9}$$

We again interchange before reducing the second column, not because we have a zero divisor, but because we want to preserve accuracy.* Interchanging the second and third rows

---

* A numerical example that demonstrates the improved accuracy when partial pivoting is used will be found in Section 2.4.

puts the element of largest magnitude on the diagonal. (We could also interchange the fourth column with the second, giving an even larger diagonal element, but we do not do this.) After the interchange, we have

$$\begin{bmatrix} 6 & 1 & -6 & -5 & 6 \\ 0 & -3.6667 & 4 & 4.3333 & -11 \\ 0 & 1.6667 & 5 & 3.6667 & -4 \\ 0 & 2 & 0 & 1 & 0 \end{bmatrix}. \tag{2.10}$$

Now we reduce in the second column

$$\begin{bmatrix} 6 & 1 & -6 & -5 & 6 \\ 0 & -3.6667 & 4 & 4.3333 & -11 \\ 0 & 0 & 6.8182 & 5.6364 & -9.0001 \\ 0 & 0 & 2.1818 & 3.3636 & -5.9999 \end{bmatrix}.$$

No interchange is indicated in the third column. Reducing, we get

$$\begin{bmatrix} 6 & 1 & -6 & -5 & 6 \\ 0 & -3.6667 & 4 & 4.3333 & -11 \\ 0 & 0 & 6.8182 & 5.6364 & -9.0001 \\ 0 & 0 & 0 & 1.5600 & -3.1199 \end{bmatrix}. \tag{2.11}$$

Back-substitution gives

$$x_4 = \frac{-3.1199}{1.5600} = -1.9999,$$

$$x_3 = \frac{-9.0001 - 5.6364(-1.9999)}{6.8182} = 0.33325,$$

$$x_2 = \frac{-11 - 4.3333(-1.9999) - 4(0.33325)}{-3.6667} = 1.0000,$$

$$x_1 = \frac{6 - (-5)(-1.9999) - (-6)(0.33325) - (1)(1.0000)}{6} = -0.50000.$$

The correct answers are -2, $\frac{1}{3}$, 1, and $-\frac{1}{2}$ for x_4, x_3, x_2, and x_1. In this calculation we have carried five significant figures and rounded each calculation. Even so, we do not have five-digit accuracy in the answers. The discrepancy is due to round off. The question of the accuracy of the computed solution to a set of equations is a most important one, and at several points in the following discussion we will discuss how to minimize the effects of round off and avoid conditions that can cause round-off errors to be magnified.

In this example, if we had replaced the zeros below the main diagonal with the ratio of coefficients at each step, the resulting augmented matrix would be

$$\begin{bmatrix} 6 & 1 & -6 & -5 & 6 \\ (0.66667) & -3.6667 & 4 & 4.3333 & -11 \\ (0.33333) & (-0.45454) & 6.8182 & 5.6364 & -9.0001 \\ (0.0) & (-0.54545) & (0.32) & 1.5600 & -3.1199 \end{bmatrix}. \quad \textbf{(2.12)}$$

This gives a *LU* decomposition as

$$\begin{bmatrix} 1 & 0 & 0 & 0 \\ 0.66667 & 1 & 0 & 0 \\ 0.33333 & -0.45454 & 1 & 0 \\ 0.0 & -0.54545 & 0.32 & 1 \end{bmatrix} * \begin{bmatrix} 6 & 1 & -6 & -5 \\ 0 & -3.6667 & 4 & 4.3333 \\ 0 & 0 & 6.8182 & 5.6364 \\ 0 & 0 & 0 & 1.5600 \end{bmatrix}. \quad \textbf{(2.13)}$$

It should be noted that the product of the matrices in Eq. (2.13) produces a permutation of the original matrix, call it A', where

$$A' = \begin{bmatrix} 6 & 1 & -6 & -5 \\ 4 & -3 & 0 & 1 \\ 2 & 2 & 3 & 2 \\ 0 & 2 & 0 & 1 \end{bmatrix},$$

because rows 1 and 4 were interchanged in Eq. (2.8) and rows 2 and 3 in Eq. (2.10). The determinant of the original matrix of coefficients—the first four columns of Eq. (2.7)—can be easily computed from Eq. (2.11) or Eq. (2.12) according to the formula

$$\det(A) = (-1)^2 * (6) * (-3.6667) * (6.8182) * (1.5600) = -234.0028,$$

which is close to the exact solution: −234.* The exponent 2 is required, because there were two row interchanges in solving this system. To summarize, you should note that the Gaussian elimination method applied to Eq. (2.6) produces the following:

1. The solution to the four equations.
2. The determinant of the coefficient matrix

$$\begin{bmatrix} 0 & 2 & 0 & 1 \\ 2 & 2 & 3 & 2 \\ 4 & -3 & 0 & 1 \\ 6 & 1 & -6 & -5 \end{bmatrix}.$$

---

* The difference is because the computed values have been rounded to four decimal places.

3. A LU decomposition of the matrix, A', which is just the original matrix, A, after we have interchanged its rows in the process.

MATLAB can get the matrices of Eq. (2.13) with its `lu` command:

```
EDU>> A = [0 2 0 1 0; 2 2 3 2 -2; 4 -3 0 1 -7; 6 1 -6 -5 6]
A =
   0     2     0     1     0
   2     2     3     2    -2
   4    -3     0     1    -7
   6     1    -6    -5     6
EDU>> [L,U,P] = lu(A)
L =
   1.0000        0          0          0
   0.6667      1.0000       0          0
   0.3333     -0.4545     1.0000       0
   0          -0.5455     0.3200     1.0000
U =
   6.0000     1.0000     -6.0000     -5.0000      6.0000
   0         -3.6667      4.0000      4.3333    -11.0000
   0          0           6.8182      5.6364     -9.0000
   0          0           0           1.5600     -3.1200
P =
   0     0     0     1
   0     0     1     0
   0     1     0     0
   1     0     0     0
```

In this, matrix P is the permutation matrix that was used to put the largest magnitude coefficient in the pivot position. Observe that MATLAB got a more accurate solution.

We really desire the solution to $Ax = b$. MATLAB gets this with a simple command. We define A and b (trailing semicolons suppress the outputs); the apostrophe on b gets the transpose:

```
EDU>> A = [0 2 0 1; 2 2 3 2; 4 -3 0 1; 6 1 -6 -5];
EDU>> b = [0 2 -7 6]';
EDU>> A\b
ans =
   -0.5000
    1.0000
    0.3333
   -2.0000
```

Again, we obtained a more accurate solution.

Operational Count

The efficiency of a numerical procedure is ordinarily measured by counting the number of arithmetic operations that are required. In the past, only multiplications and divisions were counted because they used to take much longer to perform than additions and subtractions. In today's computers using math coprocessors, all four of these take about the same time, so we should count them all.

In a system of n equations with just one right-hand side, we compute the number of operations as follows. The augmented matrix is $n \times n + 1$ in size.

To reduce the elements below the diagonal in column 1, we first compute $(n - 1)$ multiplying factors [takes $(n - 1)$ divides]. We multiply each of these by all the elements in row 1 except the first element [takes (n) multiplies)] and subtract these products from the n elements in each of the $n - 1$ rows below row 1, ignoring the first elements because these are known to become zero [takes $n * (n - 1)$ multiplies and the same number of subtracts]. In summary:

$$\text{Divides} = (n - 1),$$
$$\text{Multiplies} = n * (n - 1),$$
$$\text{Subtracts} = n * (n - 1).$$

In the other columns, we do similarly except each succeeding column has one fewer element. So, we have for column i:

$$\text{Divides: } (n - i),$$
$$\text{Multiplies: } (n - i + 1) * (n - i),$$
$$\text{Subtracts: } (n - i + 1)(n - i).$$

We add these quantities together for the reduction in columns 1 through $n - 1$ to get:

$$\text{Divides} = \sum_{i=1}^{n-1} (n - i) = \sum_{i=1}^{n-1} i = n^2/2 - n/2,$$

$$\text{Multiplies} = \sum_{i=1}^{n-1} (n - i + 1)(n - i) = \sum_{i=1}^{n-1} i(i + 1) = n^3/3 - n/3.$$

Subtracts are the same as multiplies $= n^3/3 - n/3$. If we add these together, we get, for the triangularization part, $2n^3/3 + n^2/2 - 7n/6$ total operations. In terms of the order relation discussed in Chapter 0, this is $O(n^3/3)$. We still need to do the back-substitutions. A little reflection shows that this requires

$$\text{Multiplies} = \sum_{i=1}^{n-1} i = n^2/2 - n/2,$$

$$\text{Subtracts} = \text{same as number of multiplies},$$

$$\text{Divides} = n,$$

so the back substitution requires a total of n^2 operations.

If we add the operations needed for the entire solution of a system of n equations, we get:

$$2n^3/3 + 3n^2/2 - 7n/6.$$

Multiple Right-Hand Sides

Gaussian elimination can readily work with more than one right-hand-side vector. We just append the additional column vectors within the augmented matrix and apply the reduction steps to these new columns exactly as we do to the first column. If row interchanges are made, the entire augmented matrix is included.

The Gauss–Jordan Method

There are many variants to the Gaussian elimination scheme. The back-substitution step can be performed by eliminating the elements above the diagonal after the triangularization has been finished, using elementary row operations and proceeding upward from the last row. This technique is equivalent to the procedures described in the following example. The diagonal elements may all be made ones as a first step before creating zeros in their column; this performs the divisions of the back-substitution phase at an earlier time.

One variant that is sometimes used is the *Gauss–Jordan* scheme. In it, the elements above the diagonal are made zero at the *same time* that zeros are created below the diagonal. Usually, the diagonal elements are made ones at the same time that the reduction is performed; this transforms the coefficient matrix into the identity matrix. When this has been accomplished, the column of right-hand sides has been transformed into the solution vector. Pivoting is normally employed to preserve arithmetic accuracy.

The previous example, solved by the Gauss–Jordan method, gives this succession of calculations. The original augmented matrix is

$$\begin{bmatrix} 0 & 2 & 0 & 1 & 0 \\ 2 & 2 & 3 & 2 & -2 \\ 4 & -3 & 0 & 1 & -7 \\ 6 & 1 & -6 & -5 & 6 \end{bmatrix}.$$

Interchanging rows 1 and 4, dividing the new first row by 6, and reducing the first column gives

$$\begin{bmatrix} 1 & 0.1667 & -1 & -0.8333 & 1 \\ 0 & 1.6667 & 5 & 3.3667 & -4 \\ 0 & -3.6667 & 4 & 4.3334 & -11 \\ 0 & 2 & 0 & 1 & 0 \end{bmatrix}.$$

Interchanging rows 2 and 3, dividing the new second row by -3.6667, and reducing the second column (operating above the diagonal as well as below) gives

$$\begin{bmatrix} 1 & 0 & -0.8182 & -0.6364 & 0.5 \\ 0 & 1 & -1.0909 & -1.1818 & 3 \\ 0 & 0 & 6.8182 & 5.6364 & -9 \\ 0 & 0 & 2.1818 & 3.3636 & -6 \end{bmatrix}.$$

No interchanges now are required. We divide the third row by 6.8182 and zero the other elements in the third column:

$$\begin{bmatrix} 1 & 0 & 0 & 0.04 & -0.58 \\ 0 & 1 & 0 & -0.280 & 1.56 \\ 0 & 0 & 1 & 0 & -1.32 \\ 0 & 0 & 0 & 1.5599 & -3.12 \end{bmatrix}.$$

We complete by dividing the fourth row by 1.5599 and create zeros above:

$$\begin{bmatrix} 1 & 0 & 0 & 0 & -0.5 \\ 0 & 1 & 0 & 0 & 1.0001 \\ 0 & 0 & 1 & 0 & 0.3333 \\ 0 & 0 & 0 & 1 & -2 \end{bmatrix}.$$

The fourth column is now the solution. It differs slightly from that obtained with Gaussian elimination; round-off errors have been entered in a different way.

While the Gauss–Jordan method might seem to require the same effort as Gaussian elimination, it really requires almost 50% more operations. As an exercise, you should show that it takes $(n^2 - n)/2$ divides, $(n^3 - n)/2$ multiplies, and $(n^3 - n)/2$ subtracts for a total of $n^3 + n^2 - 2n$ altogether. It is $O(n^3)$, compared to $O(2n^3/3)$.

Scaled Partial Pivoting*

There are times when the partial pivoting procedure is inadequate. When some rows have coefficients that are very large in comparison to those in other rows, partial pivoting may not give a correct solution. The answer to this problem is *scaling,* which means that we adjust the coefficients to make the largest in each row of the same magnitude.

Coefficients may differ in magnitude for several reasons. It might be caused by relations where the quantities are in widely different units: microvolts versus kilovolts, seconds versus years, for example. It could be due to inherently large numbers in just one equation. Here is a simple example to show how partial pivoting may not be enough:

$$\text{Given: } A = \begin{bmatrix} 3 & 2 & 100 \\ -1 & 3 & 100 \\ 1 & 2 & -1 \end{bmatrix}, \quad b = \begin{bmatrix} 105 \\ 102 \\ 2 \end{bmatrix}.$$

whose correct answer obviously is $x = [1.00, 1.00, 1.00]^T$.

---

* Sometimes this is called *virtual scaling.*

If we solve this using partial pivoting but with only three digits of precision to emphasize round-off error, we get this augmented matrix after triangularization:

$$\begin{bmatrix} 3 & 2 & 100 & 105 \\ 0 & 3.67 & 133 & 135 \\ 0 & 0 & -82.4 & -82.6 \end{bmatrix},$$

from which the solution vector is readily found to be the erroneous value of $[0.939, 1.09, 1.00]^T$.

The trouble here is that, while no pivoting appears to be needed during the solution, the coefficients in the third equation are much smaller than those in the other two. We can do scaled partial pivoting by first dividing each equation by its coefficient of largest magnitude. Doing so with the original equations gives (still using just three digits):

$$\begin{bmatrix} 0.0300 & 0.01 & 1.00 & 1.05 \\ -0.0100 & 0.03 & 1.00 & 1.02 \\ 0.500 & 1.00 & -0.500 & 1.00 \end{bmatrix}.$$

We now see that we should interchange row 1 with row 3 before we begin the reduction.

We could now solve this scaled set of equations but there is a better way that uses the original equations, eliminating the rounding off that may occur in obtaining the scaled equations. This method begins by computing a *scaling vector* whose elements are the elements in each row of largest magnitude. Calling the scaling vector S, we have for this example:

$$S = [100, 100, 2].$$

Before reducing the first column, we divide each element by the corresponding element of S to get $R = [0.0300, -0.0300, 0.500]$ in which the largest element is the third. This shows that the third equation should be the pivot and that the third equation should be interchanged with the first. (As you will see below, we do not have to actually interchange equations.) In preparation for further reduction steps, we do interchange the elements of S to get:

$$S' = [2, 100, 100].$$

The reduced matrix after reducing in the first column (still using only three digits of precision) is

$$\begin{bmatrix} 1 & 2 & -1 & 2 \\ 0 & 5 & 99 & 104 \\ 0 & -4 & 103 & 99 \end{bmatrix}.$$

We now are ready to reduce in column 2. We divide the elements of this column by the elements of S to get $R = [1, 0.0500, -0.0400]$. We ignore the first element of R, and see that no interchange is needed. Doing the reduction we get this final matrix:

$$\begin{bmatrix} 1 & 2 & -1 & 2 \\ 0 & 5 & 99 & 104 \\ 0 & 0 & 182 & 182 \end{bmatrix},$$

and back substitution gives the correct answer: $x = [1.00, 1.00, 1.00]$.

Using an Order Vector

We now give an example of using an order vector to avoid the actual interchange of rows. Our system in augmented matrix form is

$$\begin{bmatrix} 4 & -3 & 0 & -7 \\ 2 & 2 & 3 & -2 \\ 6 & 1 & -6 & 6 \end{bmatrix}.$$

Initially, we set the order vector to [1, 2, 3]. Looking at column 1, we see that $A(3, 1)$ should be the pivot; we exchange elements in the order vector to get [3, 2, 1]. In the reduction of column 1, we use row 3 as the pivot row to get

$$\begin{bmatrix} (0.6667) & -3.667 & 4 & -11 \\ (0.3333) & 1.667 & 5 & -4 \\ 6 & 1 & -6 & 6 \end{bmatrix}.$$

From here, we ignore row 3. We see that $A(1, 2)$ should be the next pivot. We exchange elements in the order vector to get [3, 1, 2] so as to use row 1 as the next pivot row. Reducing column 2, we then get

$$\begin{bmatrix} (0.6667) & -3.667 & 4 & -11 \\ (0.3333) & (-0.4545) & 6.8182 & -9 \\ 6 & 1 & -6 & 6 \end{bmatrix},$$

and "back-substituting" from the final set of equations in the order given by the final order vector: first 2, then 1, then 3, gives the solution: $x_2 = 1.5600$, $x_1 = -0.5800$, $x_3 = -1.3200$.

Using the *LU* Matrix for Multiple Right-Hand Sides

Many physical situations are modeled with a large set of linear equations: an example is determining the internal temperatures in a nuclear reactor, and knowing the maximum temperature is critical. The equations will depend on the geometry and certain external factors that will determine the right-hand sides. If we want the solution for many different values of these right-hand sides, it is inefficient to solve the system from the start with each one of the right-hand-side values—using the *LU* equivalent of the coefficient matrix is preferred.

Of course, getting the solutions for a problem with several right-hand sides can be done in ordinary Gaussian elimination by appending the several right-hand vectors to the coefficient matrix. However, when these vectors are not known in advance, we might think we would have to start from the beginning.

We can use the *LU* equivalent of the coefficient matrix to avoid this if we solve the system in two steps. Suppose we have solved the system $Ax = b$ by Gaussian elimination—we now know the *LU* equivalent of A: $A = L * U$. Consider now that we want to solve $Ax = b$ with some new b-vector. We can write

$$Ax = b = L * U * x = b.$$

The product of U and x is a vector, call it y. Now, we can solve for y from $Ly = b$ and this is readily done because L is lower-triangular and we get y by "forward-substitution." Call the solution $y = b'$.

Going back to the original $LUx = b$, we see that, from $Ux = y = b'$, we can get x from $Ux = b'$, which is again readily done by back-substitution (U is upper-triangular). The operational count for either forward- or back-substitution is exactly n^2 operations, so solving $Ax = b$ will take only $2n^2$ operations if the LU equivalent of A is already known, which is significantly less than the $2n^3/3 + 3n^2/2 - 7n/6$ operations required to solve $Ax = b$ directly.

What if we had reordered the rows of matrix A by pivoting? This is no problem if we save the order vector that tells the final order of the rows of the matrix. We then use this to rearrange the elements of the b-vector, or perhaps use it to select the proper elements during the forward- and back-substitutions.

AN EXAMPLE Solve $Ax = b$, where we already have its L and U matrices:

$$L = \begin{bmatrix} 1.0000 & 0 & 0 & 0 \\ 0.6667 & 1.0000 & 0 & 0 \\ 0.3333 & -0.4545 & 1.0000 & 0 \\ 0 & -0.5455 & 0.3200 & 1.0000 \end{bmatrix},$$

$$U = \begin{bmatrix} 6.0000 & 1.0000 & -6.0000 & -5.0000 \\ 0 & -3.6667 & 4.0000 & 4.3333 \\ 0 & 0 & 6.8182 & 5.6364 \\ 0 & 0 & 0 & 1.5600 \end{bmatrix}.$$

Suppose that the b-vector is $[6, -7, -2, 0]^T$. We first get $y = Ux$ from $Ly = b$ by forward-substitution:

$$y = [6, -11, -9, -3.12]^T,$$

and use it to compute x from $Ux = y$:

$$x = [-0.5, 1, 0.3333, -2]^T.$$

[This is the same system as Eq. (2.6) but with the equations reordered so pivoting is not needed.]

Now, if we want the solution with a different b-vector:

$$bb = [1 \;\; 4 \;\; -3 \;\; 1]^T,$$

we just do $Ly = bb$ to get

$$y = [1, 3.3333, -1.8182, 3.4]^T,$$

and then use this y in $Ux = y$ to find the new x:

$$x = [0.0128, -0.5897, -2.0684, 2.1795]^T.$$

Tridiagonal Systems

There are some applied problems where the coefficient matrix is of a special form. Most of the coefficients are zero; only those on the diagonal and adjacent to the diagonal are non-zero. Such a matrix is called *tridiagonal.* Here is an example from Chapter 6:

$$\begin{bmatrix} -2.70192 & 2 & 0 & 0 & 0 & 0 & 0 & 0 & -14.0385 \\ 1 & -2 & 1 & 0 & 0 & 0 & 0 & 0 & 0 \\ 0 & 1 & -2 & 1 & 0 & 0 & 0 & 0 & 0 \\ 0 & 0 & 1 & -2 & 1 & 0 & 0 & 0 & 0 \\ 0 & 0 & 0 & 1 & -2 & 1 & 0 & 0 & 0 \\ 0 & 0 & 0 & 0 & 1 & -2 & 1 & 0 & 0 \\ 0 & 0 & 0 & 0 & 0 & 1 & -2 & 1 & 0 \\ 0 & 0 & 0 & 0 & 0 & 0 & 1 & -2 & -100.0 \end{bmatrix}.$$

When this system is solved by Gaussian elimination, the zeros do not enter into the solution; only the non-zero coefficients are used. That means that there is no need to store the zeros; we can compress the coefficients into an array of three columns and put the right-hand-side terms into a fourth column. The number of arithmetic operations is reduced significantly.

Here is an algorithm that carries out the solution of the problem:

Gaussian Elimination for a Tridiagonal System

Given the $n \times 4$ matrix that has the right-hand side as its fourth column,

```
(LU decomposition phase)

For i = 2 To n
    A[i, 1] = A[i, 1]/A[i − 1, 2]
    A[i, 2] = A[i, 2] − A[i, 1] * A[i − 1, 3]
    A[i, 4] = A[i, 4] − A[i, 1] * A[i − 1, 4]

End For i

(Back-substitution)

A[n, 4] = A[n, 4]/A[n, 2]
For i = (n − 1) Down To 1
    A[i, 4] = (A[i, 4] − A[i, 3] * A[i + 1, 4])/A[i, 2]µ
End For i
```

2.3 The Inverse of a Matrix and Matrix Pathology

Division by a matrix is not defined but the equivalent is obtained from the *inverse* of the matrix. If the product of two square matrices, $A * B$, equals the identity matrix, I, B is said to be the inverse of A (and A is the inverse of B). The usual notation for the inverse of matrix A is A^{-1}. We have said that matrices do not commute on multiplication but inverses are an exception: $A * A^{-1} = A^{-1} * A$.

One way to find the inverse of matrix A is to employ the minors of its determinant but this is not efficient. The better way is to use an elimination method. We augment the A matrix with the identity matrix of the same size and solve. The solution is A^{-1}. This is equivalent to solving the system with n right-hand sides, each column being one of the n unit basis vectors in turn. Here is an example:

EXAMPLE 2.2 Given matrix A, find its inverse. First use the Gauss–Jordan method with exact arithmetic.

$$A = \begin{bmatrix} 1 & -1 & 2 \\ 3 & 0 & 1 \\ 1 & 0 & 2 \end{bmatrix}.$$

Augment A with the identity matrix and then reduce:

$$\begin{bmatrix} 1 & -1 & 2 & 1 & 0 & 0 \\ 3 & 0 & 1 & 0 & 1 & 0 \\ 1 & 0 & 2 & 0 & 0 & 1 \end{bmatrix} \rightarrow \begin{bmatrix} 1 & -1 & 2 & 1 & 0 & 0 \\ 0 & 3 & -5 & -3 & 1 & 0 \\ 0 & 1 & 0 & -1 & 0 & 1 \end{bmatrix}$$

$$\overset{(1)}{\rightarrow} \begin{bmatrix} 1 & -1 & 2 & 1 & 0 & 0 \\ 0 & 1 & 0 & -1 & 0 & 1 \\ 0 & 0 & -5 & 0 & 1 & -3 \end{bmatrix} \overset{(2)}{\rightarrow} \begin{bmatrix} 1 & -1 & 0 & 1 & \frac{2}{5} & -\frac{6}{5} \\ 0 & 1 & 0 & -1 & 0 & 1 \\ 0 & 0 & 1 & 0 & -\frac{1}{5} & \frac{3}{5} \end{bmatrix}$$

$$\rightarrow \begin{bmatrix} 1 & 0 & 0 & 0 & \frac{2}{5} & -\frac{1}{5} \\ 0 & 1 & 0 & -1 & 0 & 1 \\ 0 & 0 & 1 & 0 & -\frac{1}{5} & \frac{3}{5} \end{bmatrix}.$$

We confirm the fact that we have found the inverse by multiplication:

$$\begin{bmatrix} 1 & -1 & 2 \\ 3 & 0 & 1 \\ 1 & 0 & 2 \end{bmatrix} \begin{bmatrix} 0 & \frac{2}{5} & -\frac{1}{5} \\ -1 & 0 & 1 \\ 0 & -\frac{1}{5} & \frac{3}{5} \end{bmatrix} = \begin{bmatrix} 1 & 0 & 0 \\ 0 & 1 & 0 \\ 0 & 0 & 1 \end{bmatrix}.$$

It is more efficient to use Gaussian elimination. We show only the final triangular matrix; we used pivoting:

---

(1) Interchange the third and second rows before eliminating from the third row.

(2) Divide the third row by -5 before eliminating from the first row.

$$\begin{bmatrix} 1 & -1 & 2 & 1 & 0 & 0 \\ 3 & 0 & 1 & 0 & 1 & 0 \\ 1 & 0 & 2 & 0 & 0 & 1 \end{bmatrix} \rightarrow \begin{bmatrix} 3 & 0 & 1 & 0 & 1 & 0 \\ (0.333) & -1 & 1.667 & 1 & -0.333 & 0 \\ (0.333) & (0) & 1.667 & 0 & -0.333 & 1 \end{bmatrix}.$$

After doing the back-substitutions, we get

$$\begin{bmatrix} 3 & 0 & 1 & 0 & 0.4 & -0.2 \\ (0.333) & -1 & 1.667 & -1 & 0 & 1 \\ (0.333) & (0) & 1.667 & 0 & -0.2 & 0.6 \end{bmatrix}.$$

If we have the inverse of a matrix, we can use it to solve a set of equations, $Ax = b$, because multiplying by A^{-1} gives the answer:

$$A^{-1}Ax = A^{-1}b,$$
$$x = A^{-1}b.$$

This would seem like a good way to solve equations, and there are many references to it. But this is not the best way to solve a system—getting the LU equivalent of A first and then using the L and U to solve $Ax = b$ requires only two back-substitutions and that requires exactly the same work as multiplying the vector b by a matrix. Finding A^{-1} means solving a system with three right-hand sides.

The real importance of the inverse is to develop theory and is essential to understanding many things in applied mathematics. For example, does every square matrix have an inverse? The answer is no, and we look now at when this is true.

Pathological Systems

When a real physical situation is modeled by a set of linear equations, we can anticipate that the set of equations will have a solution that matches the values of the quantities in the physical problem, at least as far as the equations truly do represent it.* Because of round-off errors, the solution vector that is calculated may imperfectly predict the physical quantity, but there is assurance that a solution exists, at least in principle. Consequently, it must always be theoretically possible to avoid divisions by zero when the set of equations has a solution.

An arbitrary set of equations may not have such a guaranteed solution, however. There are several such possible situations, which we term "pathological." In each case, there is *no unique solution* to the set of equations.

* There are certain problems for which values of interest are determined from a set of equations that do not have a unique solution; these are called *eigenvalue* problems and are discussed in Chapter 6.

First, here is an example of a matrix that has no inverse: What is the *LU* equivalent of

$$A = \begin{bmatrix} 1 & -2 & 3 \\ 2 & 4 & -1 \\ -1 & -14 & 11 \end{bmatrix}?$$

MATLAB can find this:

```
EDU>> A = [1 -2 3; 2 4 -1;-1 -14 11]
A =
     1    -2     3
     2     4    -1
    -1   -14    11
EDU>> lu(A)
ans =
     2.0000        4.0000       -1.0000
    -0.5000      -12.0000       10.5000
     0.5000        0.3333        0
```

It is obvious that we cannot ordinarily solve a system $Ax = b$ that has this A matrix, for the zero in element $A(3, 3)$ cannot be used as a divisor in the back-substitution. That means that we cannot solve a system with the identity matrix as the right-hand sides. And that would have to be done to find A^{-1}. What does MATLAB say if we ask it to find the inverse?

```
EDU>> inv(A)
Warning: Matrix is singular to working precision.
ans =
   Inf     Inf     Inf
   Inf     Inf     Inf
   Inf     Inf     Inf
```

and we ask ourselves what the term *singular* means. Actually, the definition of a singular matrix is a matrix that does not have an inverse!

Are there other ways to see if a matrix is singular without trying to triangularize it? Yes, here are five other tests.

1. A singular matrix has a determinant of zero. This follows directly from the above result, where we saw that element $A(3, 3)$ was zero, and $\det(A) = (2)(-12)(0) = 0$.
2. The rank of the matrix is less than n, the number of rows. The rank is not as easy to find, but MATLAB can find it:

```
EDU>> rank(A)
ans =
     2
```

The rank of a matrix is really determined by the next two properties of a singular matrix.

3. A singular matrix has rows that are linearly dependent vectors. A set of vectors is said to be *linearly dependent* if a weighted sum equals zero without using all weighting factors equal to zero. For matrix A above,

$$-3[1, -2, 3] + 2[2, 4, -1] + 1[-1, -14, 11] = [0, 0, 0].$$

4. A singular matrix has columns that are linearly dependent vectors. For matrix A,

$$-10[1, 2, -1]^T + 7[-2, 4, -14]^T + 8[3, -1, 11]^T = [0, 0, 0]^T.$$

5. A set of equations with this coefficient matrix has no unique solution.

Redundant and Inconsistent Systems

Even though a matrix is singular, it may still have a solution. Consider again the same singular matrix:

$$A = \begin{bmatrix} 1 & -2 & 3 \\ 2 & 4 & -1 \\ -1 & -14 & 11 \end{bmatrix}.$$

Suppose we solve the system $Ax = b$ where the right-hand side is $b = [5, 7, 1]^T$. MATLAB then gives

```
EDU>> Ab = [1 -2 3 5; 2 4 -1 7; -1 -14 11 1]
Ab =
    1      -2      3     5
    2       4     -1     7
   -1     -14     11     1
EDU>> lu(Ab)
ans =
    2.0000     4.0000    -1.0000    7.0000
   -0.5000   -12.0000    10.5000    4.5000
    0.5000     0.3333    0          0
```

and the back-substitution cannot be done. The display suggests that x_3 can have any value. Suppose we set it equal to 0. We can solve the first two equations with that substitution; that gives $[17/4, -3/8, 0]^T$. We get the same result from solving any other combination of two equations.

Suppose we set x_3 to 1 and repeat. This gives $[3, 1/2, 1]^T$, and this is another solution. In fact, any linear combination of these two is a solution! While that may not be a satisfactory answer, we must agree that we have found a solution, actually, an infinity of them. The reason for this is that the system is redundant: The third equation, given by the third row, is just a linear combination of the first two:

$$-3[1, -2, 3, 5] + 2[2, 4, -1, 7] = -1[-1, -14, 11, 1].$$

Of course, this means that any one equation is a linear combination of the other two. What we have here is not truly three linear equations but only two independent ones. The system is called *redundant*.

What if the right-hand vector is different, say, $[5, 7, 2]^T$? Solving with this vector, we find this final array:

2.0000	4.0000	−1.0000	7.0000
(−0.5000)	−12.0000	10.5000	5.5000
(0.5000)	(0.3333)	0	−0.3333

We now say, for this system, that it is *inconsistent;* there is no solution that satisfies the equations. In either case, there is no unique solution to a system with a singular coefficient matrix.

Here is a comparison of singular and nonsingular matrices:

For Singular Matrix A:	**For Nonsingular Matrix A:**
It has no inverse, A^{-1}	It has an inverse, A^{-1} exists
Its determinant is zero	The determinant is nonzero
There is no unique solution to the system $Ax = b$	There is a unique solution to the system $Ax = b$
Gaussian elimination cannot avoid a zero on the diagonal	Gaussian elimination does not encounter a zero on the diagonal
The rank is less than n	The rank equals n
Rows are linearly dependent	Rows are linearly independent
Columns are linearly dependent	Columns are linearly independent

2.4 Ill-Conditioned Systems

We have seen that a system whose coefficient matrix is singular has no unique solution. What if the matrix is "almost singular"? Here is an example:

$$A = \begin{bmatrix} 3.02 & -1.05 & 2.53 \\ 4.33 & 0.56 & -1.78 \\ -0.83 & -0.54 & 1.47 \end{bmatrix}.$$

The LU equivalent has a very small element in position (3, 3):

$$LU = \begin{bmatrix} 4.33 & 0.56 & -1.78 \\ 0.6975 & -1.4406 & 3.7715 \\ -0.1917 & 0.3003 & -0.0039 \end{bmatrix}.$$

Let's look at the inverse:

$$\text{inv}(A) = \begin{bmatrix} 5.6611 & -7.2732 & -18.5503 \\ 200.5046 & -268.2570 & -669.9143 \\ 76.8511 & -102.6500 & -255.8846 \end{bmatrix},$$

and we see that this has elements very large in comparison to A.

Both of these results suggest that matrix A is nonsingular but is "almost singular." Suppose we solve the system $Ax = b$, with b equal to $[-1.61, 7.23, -3.38]^T$. It is clear, if you do the math, that the solution is $x = 1.0000, 2.0000, -1.0000$.

Now suppose that we make a small change in just the first element of the b-vector: $b = [-1.60, 7.23, -3.38]$. Now solve again; we get $x = [1.0566, 4.0051, -0.2315]$. What a difference! Let us try another small change in the b-vector: $b = [-1.61, 7.22, -3.38]^T$. The solution now is $x = [1.07271, 4.6826, 0.0265]$ which also differs much from our first answer.

A system whose coefficient matrix is nearly singular is called *ill-conditioned.* When a system is ill-conditioned, the solution is very sensitive to changes in the right-hand vector. It is also sensitive to small changes in the coefficients. If $A(1, 1)$ is changed from 3.02 to 3.00 and we solve the system again with the original b-vector, we now find a large change in the solution: $x = [1.1277, 6.5221, 0.7333]^T$. This means that it is also very sensitive to round-off error.

This phenomenon shows up even more pointedly in large systems. But even this system of only two equations shows the effect of near singularity:

$$\begin{bmatrix} 1.01 & 0.99 \\ 0.99 & 1.01 \end{bmatrix} \begin{bmatrix} x \\ y \end{bmatrix} = \begin{bmatrix} 2.00 \\ 2.00 \end{bmatrix}.$$

The solution is clearly seen to be $x = 1.00$, $y = 1.00$.

However, if we make a small change to the b-vector, to $[2.02, 1.98]^T$, the solution now is $x = 2$, $y = 0$. If we had another slightly different b-vector: $[1.98, 2.02]^T$, we would have $x = 0$, $y = 2$!

It is helpful to think of the system, $Ax = b$, as a *linear system solver machine.* In this view, we have *inputs* to the machine, the b-vector, and *outputs,* the x-values. For an ill-conditioned system, small changes in the input make large changes in the output.

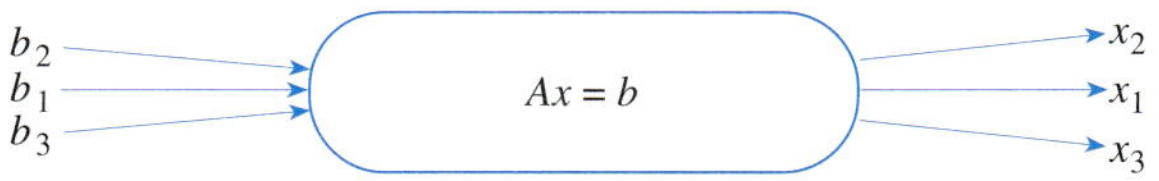

Even though the three inputs are "close together"—$b_1 = (2, 2)^T$, $b_2 = (2.02, 1.98)^T$, and $b_3 = (1.98, 2.02)^T$—we get very "distant" outputs—$x_1 = (1, 1)^T$, $x_2 = (2, 0)^T$, $x_3 = (0, 2)^T$. This modest example shows the basic idea of an ill-conditioned system: *For small changes in the input, we get large changes in the output.*

In some situations, one can combat ill-conditioning by transforming the problem into an equivalent set of equations that are not ill-conditioned. The efficiency of this scheme is related to the relative amount of computation required for the transformation, compared to the cost of doing the calculations in higher precision.

An interesting phenomenon of an ill-conditioned system is that we cannot test for the accuracy of the computed solution merely by substituting it into the equations to see whether the right-hand sides are reproduced. Consider again the ill-conditioned example we have previously examined:

$$A = \begin{bmatrix} 3.02 & -1.05 & 2.53 \\ 4.33 & 0.56 & -1.78 \\ -0.83 & -0.54 & 1.47 \end{bmatrix}, \qquad b = \begin{bmatrix} -1.61 \\ 7.23 \\ -3.38 \end{bmatrix}.$$

If we compute the vector Ax, using the exact solution $x = (1, 2, -1)^T$, we of course get

$$Ax = (-1.61, 7.23, -3.38)^T = b.$$

However, if we substitute a clearly erroneous vector

$$\bar{x} = (0.880, -2.34, -2.66)^T,$$

we get $A\bar{x}$ $(-1.6152, 7.2348, -3.3770)^T$, which is very close to b.

Effect of Precision

We have mentioned that it is difficult to get an accurate solution when a system is ill-conditioned and have demonstrated that small changes in either the coefficients or the right-hand side make large changes in the solution. The solution is also dependent on the accuracy of the arithmetic computations. We use here the computer algebra system, Maple, to show this. We begin by invoking the linalg "package" that enables many operations:

```
with (linalg)
```

Now we define the matrix:

```
>A: = matrix (3,4,[3.02,−1.05,2.53,−1.61,4.33,0.56,−1.78,7.23,
−0.83,−0.54,1.47,−3.38]);
```

$$\text{A:} = \begin{bmatrix} 3.02 & -1.05 & 2.53 & -1.61 \\ 4.33 & .56 & -1.78 & 7.23 \\ -.83 & -.54 & 1.47 & -3.38 \end{bmatrix}.$$

We already have seen that this system is ill-conditioned. What does Maple say the solution is?

```
>rref(A);
```

$$\begin{bmatrix} 1 & 0 & 0 & 1.000000037 \\ 0 & 1 & 0 & 2.000001339 \\ 0 & 0 & 1 & -.9999994882 \end{bmatrix},$$

which is pretty close to the exact solution, $x = [1, 2, -1]$. Maple used its default precision of ten digits in getting this answer. If we change the precision to 20, we get a more accurate solution but it is still not exact. What if we change it to only three digits?

```
Digits: = 3; rref(A);
  Digits: = 3
```

$$\begin{bmatrix} 1 & 0 & -.073 & 0 \\ 0 & 1 & -2.62 & 0 \\ 0 & 0 & 0 & 1 \end{bmatrix}$$

and Maple tells us that the coefficient matrix is singular at that precision. If we change to four digits:

```
Digits: = 4; rref(A);
  Digits: = 4
```

$$\begin{bmatrix} 1 & 0 & 0 & .9824 \\ 0 & 1 & 0 & 1.346 \\ 0 & 0 & 1 & -1.250 \end{bmatrix}$$

we get a poor approximation to $x = [1, 2, -1]$, the exact solution.

The 'rref' command can be used in MATLAB also. It rather obviously uses the Gauss–Jordan method.

Condition Numbers and Norms

The degree of ill-conditioning of a matrix is measured by its *condition number*. This is defined in terms of its *norms,* a measure of the magnitude of the matrix. We discuss norms before discussing condition numbers.

The magnitude of a single number is just its distance from zero: $|-4.2| = 4.2$. But a matrix is not a single number; its norm is different. For any norm, these properties are essential:

1. The norm must always have a value greater than or equal to zero, and must be zero only when the matrix is the zero matrix (one with all elements equal to zero).
2. The norm must be multiplied by k if the matrix is multiplied by the scalar k.
3. The norm of the sum of two matrices must not exceed the sum of the norms.
4. The norm of the product of two matrices must not exceed the product of the norms.

More formally, we can state these conditions, using $\|A\|$ to represent the *norm of matrix A:*

1. $\|A\| \geq 0$ and $\|A\| = 0$ if and only if $A = 0$.
2. $\|kA\| = |k|\,\|A\|$.
3. $\|A + B\| \leq \|A\| + \|B\|$.
4. $\|AB\| \leq \|A\|\,\|B\|$.

(2.14)

The third relationship is called the *triangle inequality*. The fourth is important when we deal with the product of matrices.

For the special kind of matrices that we call vectors, our past experience can help us. For vectors in two- or three-space, the length satisfies all four requirements and is a good value to use for the norm of a vector. This norm is called the *Euclidean norm,* and is computed by $\sqrt{x_1^2 + x_2^2 + x_3^2}$.

We compute the Euclidean norm of vectors with more than three components by generalizing:

$$\|x\|_e = \sqrt{x_1^2 + x_2^2 + \cdots + x_n^2} = \left(\sum_{i=1}^{n} x_i^2\right)^{1/2}.$$

This is not the only way to compute a vector norm, however. The sum of the absolute values of the x_i can be used as a norm; the maximum value of the magnitudes of the x_i will also serve. These three norms can be interrelated by defining the p-norm as.

$$\|x\|_p = \left(\sum_{i=1}^{n} |x_i|^p\right)^{1/p}.$$

From this its is readily seen that

$$\|x\|_1 = \sum_{i=1}^{n} |x_i| = \text{sum of magnitudes;}$$

$$\|x\|_2 = \left(\sum_{i=1}^{n} x_i^2\right)^{1/2} = \text{Euclidean norm;}$$

$$\|x\|_\infty = \max_{1 \le i \le n} |x_i| = \text{maximum-magnitude norm.}$$

Which of these vector norms is best to use may depend on the problem. In most cases, satisfatory results are obtained with any of these measures of the "size" of a vector.

EXAMPLE 2.3 Compute the 1-, 2-, and ∞-norms of the vector x, if $x = (1.25, 0.02, -5.15, 0)$.

$$\|x\|_1 = |1.25| + |0.02| + |-5.15| + |0| = 6.42.$$
$$\|x\|_2 = [(1.25)^2 + (0.02)^2 + (-5.15)^2 + (0)^2]^{1/2} = 5.2996.$$
$$\|x\|_\infty = |-5.15| = 5.15.$$

Matrix Norms

The norms of a matrix are similar to the norms of a vector. Two norms that are closely related are

$$\|A\|_1 = \max_{1 \le j \le n} \sum_{i=1}^{n} |a_{ij}| = \text{maximum column sum};$$

$$\|A\|_\infty = \max_{1 \le j \le n} \sum_{j=1}^{n} |a_{ij}| = \text{maximum row sum}.$$

The matrix norm $\|A\|_2$ that corresponds to the 2-norm of a vector is not readily computed. It is defined in terms of the eigenvalues of the matrix $A^T * A$. Suppose r is the largest eigenvalue of $A^T * A$. Then $\|A\|_2 = r^{1/2}$, the square root of r. This is called the *spectral norm* of A, and $\|A\|_2$ is always less than (or equal to) $\|A\|_1$ and $\|A\|_\infty$.

For an $m \times n$ matrix, the Frobenius norm is defined as

$$\|A\|_f = \left(\sum_{i=1}^{m} \sum_{j=1}^{n} a_{ij}^2 \right)^{1/2}.$$

EXAMPLE 2.4 Compute the Frobenius norms of A, B, and C, and the ∞-norms, given that

$$A = \begin{bmatrix} 5 & 9 \\ -2 & 1 \end{bmatrix}; \quad B = \begin{bmatrix} 0.1 & 0 \\ 0.2 & 0.1 \end{bmatrix}; \quad \text{and} \quad C = \begin{bmatrix} 0.2 & 0.1 \\ 0.1 & 0 \end{bmatrix}.$$

$$\|A\|_f = \sqrt{25 + 81 + 4 + 1} = \sqrt{111} = 10.54; \qquad \|A\|_\infty = 14.0;$$

$$\|B\|_f = \sqrt{0.01 + 0 + 0.04 + 0.01} = \sqrt{0.06} = 0.2449; \qquad \|B\|_\infty = 0.3;$$

$$\|C\|_f = \sqrt{0.04 + 0.01 + 0.01 + 0} = \sqrt{0.06} = 0.2449; \qquad \|C\|_\infty = 0.3.$$

The results of our examples look quite reasonable; certainly A is "larger" than B or C. Although $B \neq C$, both are equally "small." The Frobenius norm is a good measure of the magnitude of a matrix.

We see then that there are a number of ways that the norm of a matrix can be expressed. Which way is preferred? There are certainly differences in their cost; for example, some will require more extensive arithmetic than others. The spectral norm is usually the most "expensive." Which norm is best? The answer to this question depends in part on the use for the norm. In most instances, we want the norm that puts the smallest upper bound on the magnitude of the matrix. In this sense, the spectral norm is usually the "best." We observe, in the next example, that not all the norms give the same value for the magnitude of a matrix.

MATLAB can compute all of the norms of a matrix:

```
A =
     5  -5  -7
    -4   2  -4
    -7  -4   5
```

```
EDU>> norm(A,'fro')
ans =
  15
EDU>> norm(A,inf)
ans =
  17
EDU>> norm(A,1)
ans =
  16
EDU>> norm(A)
ans =
  12.0301
EDU>> norm(A,2)
ans =
  12.0301
```

We observe that the 2-norm, the spectral norm, is the norm we get if we just ask for the "norm." The smallest norm of the matrix is the spectral norm, it is the "tightest" measure.

Errors in the Solution and Norms

When we solve a system of equations, we hope that the result has little error but, as we have seen, that is not always true. If the coefficient matrix is ill-conditioned, we cannot check the accuracy by just substituting our answer into the equations, but we can use norms to see how great the error is.

Let $\bar{x}$ be the computed solution, an approximation to the true solution. Define the *residual, r*, as $r = b - A\bar{x}$, the difference between the b-vector and what we get when the approximate $\bar{x}$ is substituted into the equations. Let e be the error in $\bar{x}$ and x be the true solution to the system (which we don't know), $e = x - \bar{x}$. Because $Ax = b$, we have

$$r = b - A\bar{x} = Ax - A\bar{x} = A(x - \bar{x}) = Ae.$$

Hence,

$$e = A^{-1}r.$$

Taking norms and recalling Eq. (2.14), line 4, for a product, we write

$$\|e\| \leq \|A^{-1}\|\,\|r\|. \tag{2.15}$$

From $r = Ae$, we also have $\|r\| \leq \|A\|\,\|e\|$, which combines with Eq. (2.15) to give

$$\frac{\|r\|}{\|A\|} \leq \|e\| \leq \|A^{-1}\|\;\|r\|. \tag{2.16}$$

Applying the same reasoning to $Ax = b$ and $x = A^{-1}b$, we get

$$\frac{\|b\|}{\|A\|} \leq \|x\| \leq \|A^{-1}\|\;\|b\|. \tag{2.17}$$

Taking Eqs. (2.16) and (2.17) together, we reach a most important relationship:

$$\frac{1}{\|A\|\,\|A^{-1}\|}\frac{\|r\|}{\|b\|} \le \frac{\|e\|}{\|x\|} \le \|A\|\,\|A^{-1}\|\frac{\|r\|}{\|b\|}.$$

Condition Number of a Matrix

The product of the norm of A and the norm of its inverse is called the *condition number* of matrix A and is the best measure of ill-conditioning. A small number means good-conditioning, a large one means ill-conditioning. So, we usually write the previous equation as

$$\frac{1}{(\text{Condition no.})}\frac{\|r\|}{\|b\|} \le \frac{\|e\|}{\|x\|} \le (\text{Condition no.})\frac{\|r\|}{\|b\|}. \quad \textbf{(2.18)}$$

Equation (2.18) shows that the relative error in the computed solution vector $\bar{x}$ can be as great as the relative residual multiplied by the condition number. Of course it can also be as small as the relative residual divided by the condition number. Therefore, when the condition number is large, the residual gives little information about the accuracy of $\bar{x}$. Conversely, when the condition number is near unity, the relative residual is a good measure of the relative error of $\bar{x}$.

When we solve a linear system, we are normally doing so to determine values for a physical system for which the set of equations is a model. We use the measured values of the parameters of the physical system to evaluate the coefficients of the equations, so we expect these coefficients to be known only as precisely as the measurements. When these are in error, the solution of the equations will reflect these errors. We have already seen that an ill-conditioned system is extremely sensitive to small changes in the coefficients. The condition number lets us relate the change in the solution vector to such errors in the coefficients of the set of equations $Ax = b$.

Assume that the errors in measuring the parameters cause errors in the coefficients of A so that the actual set of equations being solved is $(A + E)\bar{x} = b$, where $\bar{x}$ represents the solution of the perturbed system and A represents the true (but unknown) coefficients. We let $\bar{A} = A + E$ represent the perturbed coefficient matrix. We desire to know how large $x - \bar{x}$ is.

Using $Ax = b$ and $\bar{A}\bar{x} = b$, we can write

$$\begin{aligned} x = A^{-1}b = A^{-1}(\bar{A}\bar{x}) &= A^{-1}(A + \bar{A} - A)\bar{x} \\ &= [I + A^{-1}(\bar{A} - A)]\bar{x} \\ &= \bar{x} + A^{-1}(\bar{A} - A)\bar{x}\,. \end{aligned}$$

Because $\bar{A} - A = E$, we have

$$x - \bar{x} = A^{-1}E\bar{x}.$$

Taking norms, we get

$$\|x - \bar{x}\| \leq \|A^{-1}\|\,\|E\|\,\|\bar{x}\| = \|A^{-1}\|\,\|A\|\frac{\|E\|}{\|A\|}\|\bar{x}\|,$$

so that

$$\frac{\|x - \bar{x}\|}{\|\bar{x}\|} \leq (\text{Condition no.})\frac{\|E\|}{\|A\|}.$$

This says that the error of the solution relative to the norm of the computed solution can be as large as the relative error in the coefficients of A multiplied by the condition number. The net effect is that, if the coefficients of A are known to only four-digit precision and the condition number is 1000, the computed vector x may have only one digit of accuracy.

Iterative Improvement

When the solution to the system $Ax = b$ has been computed, and, because of round-off error, we obtain the approximate solution vector $\bar{x}$, it is possible to apply iterative improvement to correct $\bar{x}$ so that it more closely agrees with x. Define $e = x - \bar{x}$. Define $r = b - A\bar{x}$.

$$Ae = r. \tag{2.19}$$

If we could solve this equation for e, we could apply this as a correction to $\bar{x}$. Furthermore, if $\|e\|/\|\bar{x}\|$ is small, it means that $\bar{x}$ should be close to x. In fact, if the value of $\|e\|/\|\bar{x}\|$ is 10^{-p}, we know that $\bar{x}$ is probably correct to p digits.

The process of iterative improvement is based on solving Eq. (2.19). Of course this is also subject to the same round-off error as the original solution of the system for $\bar{x}$, so we actually get $\bar{e}$, an approximation to the true error vector. Even so, unless the system is so ill-conditioned that $\bar{e}$ is not a reasonable approximation to e, we will get an improved estimate of x from $\bar{x} + \bar{e}$. One special caution is important to observe: The computation of the residual vector r must be as precise as possible. One always uses double-precision arithmetic; otherwise, iterative improvement will not be successful. An example will make this clear.

We are given

$$A = \begin{bmatrix} 4.23 & -1.06 & 2.11 \\ -2.53 & 6.77 & 0.98 \\ 1.85 & -2.11 & -2.32 \end{bmatrix}, \qquad b = \begin{bmatrix} 5.28 \\ 5.22 \\ -2.58 \end{bmatrix},$$

whose true solution is

$$x = \begin{bmatrix} 1.000 \\ 1.000 \\ 1.000 \end{bmatrix}.$$

If inadequate precision is used, we might get this approximate solution vector: $\bar{x} = (0.991, 0.997, 1.000)^T$. Using double precision, we compute $A\bar{x}$, storing this product in a register that holds six digits, then we get the residual.

$$A\bar{x} = \begin{bmatrix} 5.24511 \\ 5.22246 \\ -2.59032 \end{bmatrix}, \qquad r = \begin{bmatrix} 0.0349 \\ -0.00246 \\ 0.0103 \end{bmatrix}.$$

We now solve $A\bar{e} = r$, again using three-digit precision, and get

$$\bar{e} = \begin{bmatrix} 0.00822 \\ 0.00300 \\ -0.00000757 \end{bmatrix}.$$

Finally, correcting $\bar{x}$ with $\bar{x} + \bar{e}$ gives almost exactly the correct solution:

$$\bar{x} + \bar{e} = \begin{bmatrix} 0.999 \\ 1.000 \\ 1.000 \end{bmatrix}.$$

In the general case, the iterations are repeated until the corrections are negligible. Because we want to make the solution of Eq. (2.19) as economical as possible, we should use an *LU* method to solve the original system and apply the *LU* to Eq. (2.19).

Pivoting and Precision

We have previously said that pivoting reduces the errors due to round off. We examine this further with a small system, one with only two equations:

$$\varepsilon x + By = C,$$
$$Dx + Ey = F,$$

with ε a very small number. If this is solved without pivoting and the first column is reduced ($-D/\varepsilon$ is the multiplier), we get

$$\varepsilon x + By = C,$$
$$(E - DB/\varepsilon)y = F - CD\backslash\varepsilon.$$

and, solving for y, we see that

$$y = \frac{F - CD/\varepsilon}{E - DB/\varepsilon} \approx \frac{CD}{DB} = \frac{C}{B} \text{ if } \varepsilon \text{ is very small.}$$

Substituting this for y in the first equation, we find

$$x = \frac{C - B(C/B)}{\varepsilon} = \frac{C - C}{\varepsilon} = 0!$$

showing that $x \approx 0$ for any values of C and F if ε is small enough. Now, suppose that $F = D + E$ and $C = \varepsilon + B$ and we do pivot by interchanging the equations:

$$Dx + Ey = F = D + E,$$
$$\varepsilon x + By = C = \varepsilon + B.$$

(Obviously, the correct solution must be $x = 1$, $y = 1$.) Now, reducing the first column ($-\varepsilon/D$ is the multiplier) we have:

$$Dx + Ey = D + E$$
$$(B - (\varepsilon/D)E)y = \varepsilon + B - (\varepsilon/D)\,(D + E) = \frac{\varepsilon D + BD - \varepsilon D - \varepsilon E}{D} = \frac{BD - \varepsilon E}{D},$$

so that

$$y = \frac{(BD - \varepsilon E)/D}{(BD - \varepsilon E)/D} = 1.$$

We get x by substituting $y = 1$ into the first equation, so that

$$x = \frac{D + E - E(1)}{D} = 1,$$

which demonstrates how pivoting may be very necessary.

Here is a numerical example of the same thing. The augmented matrix is

$$\begin{bmatrix} 0.02 & 10 & 10.02 \\ 10 & 10 & 20 \end{bmatrix},$$

which must yield $x = 1$, $y = 1$. If precision is infinite, pivoting is not required. Let's reduce in the first column without pivoting ($-10/0.02 = -500$ is the multiplier):

$$\begin{bmatrix} 0.02 & 10 & 10.02 \\ 0 & -4990 & -4490 \end{bmatrix},$$

which gives $y = (-4990)/(-4990) = 1$.

Now, substituting $y = 1$ into the first equation,

$$x = \frac{10.02 - 10}{0.02}.$$

The result for x is 1 if the numerator equals 0.02. But suppose we have only three digits of precision—that means the numerator does not equal 0.02 but is 0.00 and x is zero!

2.5 Iterative Methods

Gaussian elimination and its variants are called *direct methods*. An entirely different way to solve many systems is through *iteration*. In this, we start with an initial estimate of the solution vector and proceed to refine this estimate. There are times when this is preferred over a direct method. This is especially true when the coefficient matrix is sparse.

The two methods for solving $Ax = b$ that we shall discuss in this section are the *Jacobi method* and the *Gauss–Seidel method*. These methods not only can solve a system of equations but they are also the basis for other accelerated methods that we shall introduce in later chapters of this book. When the system of equations can be ordered so that each diagonal entry of the coefficient matrix is larger in magnitude than the sum of the magnitudes of the other coefficients in that row—such a system is called *diagonally dominant*—the iteration will converge for any starting values. Formally, we say that an $n \times n$ matrix A is diagonally dominant if and only if for each $i = 1, 2, \ldots, n$,

$$|a_{ii}| > \sum_{\substack{j=1 \\ j\neq i}}^{n} |a_{i,j}|, \qquad i = 1, 2, \ldots, n.$$

Although this may seem like a very restrictive condition, it turns out that there are very many applied problems that have this property (steady-state and transient heat transfer are two). Our approach is illustrated with the following simple example of a system.

$$\begin{aligned} 6x_1 - 2x_2 + \ \ x_3 &= 11, \\ x_1 + 2x_2 - 5x_3 &= -1, \\ -2x_1 + 7x_2 + 2x_3 &= \ \ 5. \end{aligned}$$

The solution is $x_1 = 2, x_2 = 1, x_3 = 1$. However, before we begin our iterative scheme we must first reorder the equations so that the coefficient matrix is diagonally dominant.

$$\begin{aligned} 6x_1 - 2x_2 + \ \ x_3 &= 11, \\ -2x_1 + 7x_2 + 2x_3 &= \ \ 5, \\ x_1 + 2x_2 - 5x_3 &= -1. \end{aligned} \tag{2.20}$$

The iterative methods depend on the rearrangement of the equations in this manner:

$$x_i = \frac{b_i}{a_{i,i}} - \sum_{\substack{j=1 \\ j\neq i}}^{n} \frac{a_{ij}}{a_{ii}} x_j, \qquad i = 1, 2, \ldots, n.$$

Each equation is now solved for the variables in succession:

$$\begin{aligned} x_1 &= 1.8333 \qquad\qquad\qquad + 0.3333x_2 - 0.1667x_3, \\ x_2 &= 0.7143 + 0.2857x_1 \qquad\qquad\quad - 0.2857x_3, \\ x_3 &= 0.2000 + 0.2000x_1 + 0.4000x_2. \end{aligned}$$

We begin with some initial approximation to the value of the variables. (Each component might be taken equal to zero if no better initial estimates are at hand.) Substituting these approximations into the right-hand sides of the set of equations generates new approximations that, we hope, are closer to the true value. The new values are substituted in the right-hand sides to generate a second approximation, and the process is repeated until successive values of each of the variables are sufficiently alike. We indicate the iterative process on Eq. (2.20), as follows, by putting superscripts on variables to indicate successive iterates. Thus our set of equations becomes

$$\begin{aligned} x_1^{(n+1)} &= 1.8333 + 0.3333x_2^{(n)} - 0.1667x_3^{(n)}, \\ x_2^{(n+1)} &= 0.7143 + 0.2857x_1^{(n)} - 0.2857x_3^{(n)}, \\ x_3^{(n+1)} &= 0.2000 + 0.2000x_1^{(n)} + 0.4000x_2^{(n)}. \end{aligned} \tag{2.21}$$

Starting with an initial vector of $x^{(0)} = (0, 0, 0)$, we get

Successive estimates of solution (Jacobi method)

	First	Second	Third	Fourth	Fifth	Sixth	. . .	Ninth
x_1	0	1.833	2.038	2.085	2.004	1.994	. . .	2.000
x_2	0	0.714	1.181	1.053	1.001	0.990	. . .	1.000
x_3	0	0.200	0.852	1.080	1.038	1.001	. . .	1.000

Note that this method is exactly the same as the method of fixed-point iteration for a single equation that was discussed in Chapter 1, but it is now applied to a set of equations; we see this if we write Eq. (2.21) in the form of

$$x^{(n+1)} = G(x^{(n)}) = b' - Bx^{(n)},$$

which is identical to $x_{n+1} = g(x_n)$ as used in Chapter 1.

In the present context, of course, $x^{(n)}$ and $x^{(n+1)}$ refer to the nth and $(n + 1)$st iterates of a vector rather than a simple variable, and g is a linear transformation rather than a nonlinear function. For the preceding example, we restate Eq. (2.20) in matrix form:

$$Ax = b, \quad \begin{bmatrix} 6 & -2 & 1 \\ -2 & 7 & 2 \\ 1 & 2 & -5 \end{bmatrix} \begin{bmatrix} x_1 \\ x_2 \\ x_3 \end{bmatrix} = \begin{bmatrix} 11 \\ 5 \\ -1 \end{bmatrix}. \tag{2.22}$$

Now, let $A = L + D + U$, where

$$L = \begin{bmatrix} 0 & 0 & 0 \\ -2 & 0 & 0 \\ 1 & 2 & 0 \end{bmatrix}, \quad D = \begin{bmatrix} 6 & 0 & 0 \\ 0 & 7 & 0 \\ 0 & 0 & -5 \end{bmatrix}, \quad U = \begin{bmatrix} 0 & -2 & 1 \\ 0 & 0 & 2 \\ 0 & 0 & 0 \end{bmatrix}.$$

Then Eq. (2.22) can be rewritten as

$$\begin{aligned} Ax &= (L + D + U)x = b, \quad \text{or} \\ Dx &= -(L + U)x + b, \quad \text{which gives} \\ x &= -D^{-1}(L + U)x + D^{-1}b. \end{aligned}$$

From this we have, identifying x on the left as the new iterate,

$$x^{(n+1)} = -D^{-1}(L + U)x^{(n)} + D^{-1}b. \tag{2.23}$$

In Eqs. (2.21) we see that

$$b' = D^{-1}b = \begin{bmatrix} 1.8333 \\ 0.7143 \\ 0.2000 \end{bmatrix},$$

$$B = D^{-1}(L + U) = \begin{bmatrix} 0 & -0.3333 & 0.1667 \\ -0.2857 & 0 & 0.2857 \\ -0.2000 & -0.4000 & 0 \end{bmatrix}.$$

The procedure we have just described is known as the *Jacobi method,* also called "the method of simultaneous displacements" because each of the equations is simultaneously changed by using the most recent set of x-values.

We can write the algorithm for the Jacobi iterative method as follows:

Algorithm for Jacobi Iteration

We assume that the system $Ax = b$ has been rearranged so that the matrix A is diagonally dominant. That is, for each row of A:

$$|a_{i,i}| > \sum_{\substack{j=i \\ j\neq i}}^{n} |a_{i,j}|, \qquad i = 1, 2, \ldots, n.$$

This is a sufficient condition for convergence both for this method and for the one that we discuss next. We begin with an initial approximation to the solution vector, which we store in the vector: *old_x*.

```
For i = 1 To n
   b[i] = b[i]/a[i, i]
   new_x[i] = old_x[i]
   a[i, j] = a[i, j]/a[i, i];  j = 1 . . . n and i <> j
End For i

Repeat

   For i = 1 To n
      old_x[i] = new_x[i]
      new_x[i] = b[i]
   End For i

   For i = 1 To n
      For j = 1 To n
```

```
            If (j <> i) Then
                new_x[i] = new_x[i] - a[i, j] * old_x[j]
            End For j
        End For i
    Until new_x and old_x converge to each other.
```

Gauss–Seidel Iteration

Observe that we never use the values *new-x* in the algorithm for Jacobi iteration until we have found all of its components. Even though we have *new_x* [1] available, we do not use it to compute *new_x* [2] even though in nearly all cases the new values are better than the old and ought to be used instead. When this is done, the procedure known as *Gauss–Seidel iteration* results.

We begin exactly as with the Jacobi method by rearranging the equations, solving each equation for the variable whose coefficient is dominant in terms of the others. We proceed to improve each x-value in turn, using always the most recent approximations of the other variables. The rate of convergence is more rapid than for the Jacobi method, as shown by reworking the previous example [Eq. (2.20)].

Successive estimates of solution (Gauss–Seidel method)

	First	Second	Third	Fourth	Fifth	Sixth
x_1	0	1.833	2.069	1.998	1.999	2.000
x_2	0	1.238	1.002	0.995	1.000	1.000
x_3	0	1.062	1.015	0.998	1.000	1.000

These values were computed by using this iterative scheme:

$$x_1^{(n+1)} = 1.8333 + 0.3333x_2^{(n)} - 0.1667x_3^{(n)},$$
$$x_2^{(n+1)} = 0.7143 + 0.2857x_1^{(n+1)} - 0.2857x_3^{(n)},$$
$$x_3^{(n+1)} = 0.2000 + 0.2000x_1^{(n+1)} + 0.4000x_2^{(n+1)},$$

beginning with $x^{(1)} = (0, 0, 0)^T$.

The algorithm for the Gauss–Seidel iteration is as follows:

Algorithm for Gauss–Seidel Iteration

We assume as we did in the previous algorithm that the system $Ax = b$ has been rearranged so that the coefficient matrix, A, is diagonally dominant. As before, we begin with an initial approximation to the solution vector, which we store in the vector: x.

```
For i = 1 To n
    b[i] = b[i]/a[i, i]
    a[i, j] = a[i, j]/a[i, i]; j = 1 . . . n and i <> j
End for i;

While Not (yet convergent) Do
    For i = 1 To n
        x[i] = b[i];
        For j = 1 To n
            If (j <> i) Then
                x[i] = x[i] − a[i, j] * x[j]
        End For j
    End For i
End While
```

The matrix formulation for the Gauss–Seidel method is almost the same as the one given in Eq. (2.23). For Gauss–Seidel, $Ax = b$ can be rewritten as

$$(L + D)x = -Ux + b, \tag{2.24}$$

and from this we get

$$x^{(n+1)} = -(L + D)^{-1}Ux^{(n)} + (L + D)^{-1}b. \tag{2.25}$$

The usefulness of this matrix notation will become apparent in Chapter 6 where the eigenvalues of matrices $D^{-1}(L + U)$ of Eq. (2.23) and $(L + D)^{-1}\,U$ of Eq. (2.25) will be studied. The eigenvalues of the two matrices indicate how fast the iterations will converge. We emphasize, however, that without diagonal dominance, neither Jacobi nor Gauss–Seidel is sure to converge. (Some authors use the term *row diagonal dominance* for our term *diagonal dominance*. Their term is perhaps more accurate.)

There are some instances of the system $Ax = b$ where the coefficient matrix does not have (row) diagonal dominance but still both Jacobi and Gauss–Seidel methods do converge. It can be shown that, if the coefficient matrix, A, is symmetric and positive definite,* the Gauss–Seidel method will converge from any starting vector. In another class of problems, where matrix A has diagonal elements that are all positive and off-diagonal elements that are all negative, both Jacobi and Gauss–Seidel methods will either converge or diverge. When both methods converge, the Gauss–Seidel method converges faster. Datta (1995) discusses this and gives examples.

For a general coefficient matrix, there is little that can be said. In fact, there are examples where Jacobi converges and Gauss–Seidel diverges from the same starting vector! Still, returning to the focus of this section, we can say that, given row diagonal dominance in the coefficient matrix, the Gauss–Seidel method is often the better choice. Having said

---

* Matrix A is positive definite if $x * Ax > 0$ for all nonzero vectors x.

that, we may still prefer the Jacobi method if we are running the program on parallel processors because all n equations can be solved simulaneously at each iteration.

Accelerating Convergence

Convergence in the Gauss–Seidel method can be speeded if we do what is called *overrelaxing*. The term comes from an old hand method where a set of "residuals" (the right-hand-side values for a rearrangement of the equations when the unknowns were given certain values) were "relaxed" to zero. Overelaxation will be encountered again in Chapter 8, where we solve partial-differential equations.

The standard relationship for Gauss–Seidel iteration for the set of equations $Ax = b$, for variable x_i, can be written

$$x_i^{(k+1)} = \frac{1}{a_{ii}}\left(b_i - \sum_{j=1}^{i-1} a_{ij}x_j^{(k+1)} - \sum_{j=i+1}^{n} a_{ij}x_j^{(k)}\right), \qquad \textbf{(2.26)}$$

where the superscript $(k + 1)$ indicates that this is the $(k + 1)$st iterate. On the right side we use the most recent estimates of the x_j, which will be either $x_j^{(k)}$ or $x_j^{(k+1)}$.

An algebraically equivalent form for Eq. (2.26) is

$$x_i^{(k+1)} = x_i^{(k)} + \frac{1}{a_{ii}}\left(b_i - \sum_{j=1}^{i-1} a_{ij}x_j^{(k+1)} - \sum_{j=i}^{n} a_{ij}x_j^{(k)}\right),$$

because $x_i^{(k)}$ is both added to and subtracted from the right side. Overrelaxation can be applied to Gauss–Seidel if we will add to $x_i^{(k)}$ some multiple of the second term. It can be shown that this multiple should never be more than 2 in magnitude (to avoid divergence), and the optimal overrelaxation factor lies between 1.0 and 2.0. Our iteration equations take this form, where w is the *overrelaxation factor:*

$$x_i^{(k+1)} = x_i^{(k)} + \frac{w}{a_{ii}}\left(b_i - \sum_{j=1}^{i-1} a_{ij}x_j^{(k+1)} - \sum_{j=1}^{n} a_{ij}x_j^{(k)}\right).$$

Table 2.1 shows how the convergence rate is influenced by the value of w for the system

$$\begin{bmatrix} -4 & 1 & 1 & 1 \\ 1 & -4 & 1 & 1 \\ 1 & 1 & -4 & 1 \\ 1 & 1 & 1 & -4 \end{bmatrix} x = \begin{bmatrix} 1 \\ 1 \\ 1 \\ 1 \end{bmatrix},$$

Table 2.1 Acceleration of convergence of Gauss–Seidel iteration

w, the overrelaxation factor	Number of iterations to reach error $<1 \times 10^{-5}$	
1.0	24	
1.1	18	
1.2	13	
1.3	11	←Minimum
1.4	14	of iterations
1.5	18	
1.6	24	
1.7	35	
1.8	55	
1.9	100+	

starting with an initial estimate of $x = 0$. The exact solution is

$$x_1 = -1, \qquad x_2 = -1, \qquad x_3 = -1, \qquad x_4 = -1.$$

Sparse Matrices and Banded Matrices

It has already been said that many applied problems are solved with systems whose coefficient matrix is sparse—only a fraction of the elements are nonzero. In Chapters 6 and 8 you will see several instances. The tridiagonal system of Section 2.2 is the prime example. In other applications the coefficient matrix may be sparse and have elements situated in selected positions. A *banded matrix* is one where the nonzero elements lie on diagonals parallel to the main diagonal. A tridiagonal matrix is obviously banded.

Some sparse matrices are not as compact as a tridiagonal one. It often happens that the nonzero coefficients lie on diagonals but some or all of these bands are not adjacent to the main diagonal. Algorithms similar to the one for a tridiagonal matrix can be developed. However, one usually finds that the nonzero elements between the bands and the main diagonal do not stay zero and more arithmetic operations are needed to get a solution or to find an LU equivalent to the coefficient matrix.

Fortunately, most sparse systems have a main diagonal that is dominant so it is easy to set up the rearranged equations that can be solved quickly by iteration. And these iterations can be speeded up by overrelaxation. Here is an example of a small system:

$$\begin{bmatrix} 3 & 1 & 0 & -1 & 0 & 0 & 2.05 \\ 1 & 4 & 2 & 0 & 2 & 0 & 3.33 \\ 0 & 2 & 4 & 1 & 0 & 3 & -6.21 \\ 2 & 0 & -1 & 3 & 3 & 0 & 5.25 \\ 0 & 3 & 0 & 1 & 5 & 2 & 8.92 \\ 0 & 0 & 1 & 0 & -1 & 2 & 10.87 \end{bmatrix}$$

whose solution is

$$x = [-1.7040, 17.9737, -19.1286, 10.8118, -14.3019, 7.8483]^T.$$

which we found by Gaussian elimination. It would be good practice to get this solution by an iterative method.

Iteration Is Minimizing

Getting successive improvements to an initial x-vector, x_0, that converge to the solution to the system $Ax = b$ can be considered to be minimizing the errors in the x-vectors, the *residuals:*

$$r_n = b - Ax_n.$$

For a special class of problems, those whose coefficient matrix is symmetric and positive definite, there is a method that is extremely rapidly convergent, the *conjugate gradient method.* In Chapter 7 we discuss this method of finding the minimum of a function of several variables.

When matrix A is multiplied with vector x, a new transformed vector results. Because the product of matrix A with vector x depends on x, we can say that the product is a function of x because it changes when x is varied—Ax is then a "function of x." Our statement that iteration is minimizing makes sense.

We will not give a full explanation of the conjugate gradient method at this point, only give one example where it works and another where it doesn't.

Consider this small system whose coefficient matrix is symmetric and positive definite:*

$$A = \begin{bmatrix} 4 & -3 & -1 \\ -3 & 5 & 2 \\ -1 & 2 & 3 \end{bmatrix}, \qquad b = \begin{bmatrix} 7 \\ 2 \\ -3 \end{bmatrix},$$

whose solution is $x = [3.9167, 3.5833, -2.0833]^T$. If we start with $x_0 = [0, 0, 0]^T$, Gauss–Seidel iterations converge in 20 iterations. With the Jacobi method, there is no convergence; the successive x-vectors after 34 iterations oscillate about the true answer:

$$\text{Iteration 34: } x = [3.5833, 3.9166, -1.7500]^T,$$
$$\text{Iteration 35: } x = [4.2500, 3.2500, -2.4166]^T,$$

whose averages are exactly the solution.

If the conjugate gradient method is applied, again with $x_0 = [0, 0, 0]^T$, we get these results:

$$x_1 = [2.4520, 0.7006, -1.0508]^T,$$
$$x_2 = [4.0670, 3.4771, -1.6197]^T,$$
$$x_3 = [3.9167, 3.5833, -2.0833]^T,$$

and we obtain the exact solution in three tries!

---

* A matrix is positive definite if and only if the determinants of all its leading minors are positive. The leading minors are the submatrices whose upper-left elements are the diagonal elements of the matrix. This matrix is clearly symmetric. It is positive definite because the determinants of it leading minors are 24, 11, and 3.

The conjugate gradient method will always converge in n tries with a system of n-equations; it is the preferred iterative method for systems that have the necessary conditions. Still, each iteration of the conjugate gradient method is more expensive than Jacobi or Gauss–Seidel.

If we attempt to use the method when the coefficient matrix is not symmetric, it fails. With this set of two equations:

$$\begin{bmatrix} 2 & 1 & 3 \\ -1 & 3 & 2 \end{bmatrix}$$

which obviously is solved with $x = [1,\ 1]^T$, the conjugate gradient method actually diverges from $x_0 = [0, 0]^T$, while Gauss–Seidel converges in 7 iterations and Jacobi in 12.

2.6 Parallel Processing

We have mentioned that the operation of many numerical methods can be speeded up by the proper use of parallel processing or distributed systems. In this section, we describe how vector/matrix operations, Gaussian elimination, and Jacobi iteration can be efficiently performed in a parallel or distributed processing environment. We shall show how much the performance can be improved depending on the topology of the network in each case and pay special attention to the implementation of Gaussian elimination.

Vector/Matrix Operations

For inner products, a very elementary case, we assume we have two vectors, v, u, of length n, and an equal number of parallel processors, proc(i), $i = 1 \ldots n$, where each proc(i) contains the components, v_i, u_i. Then the multiplication of all the $v_i * u_i$ can be done in parallel in one time unit. In Section 0.6, we found that if the processors are connected suitably we can actually do the addition part in log(n) time units. Thus, we can estimate the time for an inner product as $1 + \log(n) = O(\log(n))$.

This assumes a high degree of connectivity between the processors. There has been much study of such connectivity. These different designs are referred to as the topologies of the systems. In our present example, we assume the topology of a hypercube. However, before we describe that design, we shall introduce the simpler topology of the linear array. Suppose our processors were only connected as a linear array in which the communication send/receive is just between two adjacent processors:

$$P_1 \leftrightarrow P_2 \leftrightarrow \cdots \leftrightarrow P_{n-1} \leftrightarrow P_n.$$

Then our addition of the n elements would be $n/2$ time steps, because we could do an addition at each end in parallel and proceed to the middle.

The n-dimensional hypercube is a graph with 2^n vertices in which each vertex has n edges (is connected to n other vertices). This graph can be easily defined recursively because there is an easy algorithm to determine the order in which the vertices are connected.

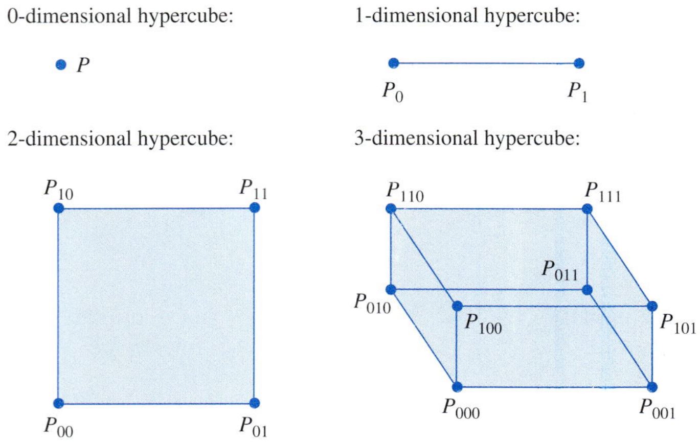

Two vertices are *adjacent* if and only if the indices differ in exactly one bit. We can get to the $(n + 1)$-dimensional hypercube by making two copies of the n-dimensional hypercube and then adding a zero to the leftmost bit of the first n-dimensional cube and then doing the same with a 1 to the second cube. It was this kind of connectivity that allowed us to make the addition of n numbers in $\log(n)$ time steps in Section 0.6. There are many other designs for connecting the processors. Such designs include names like star, ring, torus, mesh, and others. However, for the rest of this section, we shall assume that we have the processors optimally connected.

For the *matrix/vector product, Ax*, we assume that processor proc(i) contains the ith row of A as well as the vector, x. Because one processor is performing this dot product of row i of A and of vector x, this could be done in $2n$ units of time. However, because the other processors are proceeding in parallel, the whole operation will only take $O(n)$ units of time. We would have had a less efficient algorithm had we made use of the inner product algorithm above on the individual rows of A and the vector x.

For a linear array of processors, Bertsekas and Tsitsiklis give the following time for our more simplified case. The time is

$$\alpha n + (n - 1)(\beta + \gamma),$$

where α is the time for an addition or multiplication, β is the time for a transmission of the product along the link, and γ is a positive constant.

For the *matrix/matrix product, AB*, for two $n \times n$ matrices, we suppose that we have n^2 processors. Before we start our computations, each processor, $P_{i,j}$, will have received the values for row i of A and column j of B. Based on our previous discussion, we can expect the time to be $O(n)$. For n^3 processors, this can be reduced to just $O(\log(n))$ time. Here each processor, proc(i,k,j), would store the elements, $A_{i,k}$, $B_{k,j}$. Then all the multiplications can be done in parallel, and the additional $(\log(n))$ time units are for the additions.

Gaussian Elimination

Recall how we achieve a solution to a system of linear equations through Gaussian elimination. We first perform a sequence of row reductions on the augmented matrix $(A : b)$ until the coefficient matrix A is in upper-triangular form. Then we employ back-substitution to find the solution.

To see how this can be done in a parallel-processing environment, we must examine the row-reduction phase and the back-substitution phase in some detail.

We begin with the row-reduction phase: Consider the following example of the first stage of row reduction of a 4×4 system with one right-hand side:

$$\begin{bmatrix} 1 & 2 & 1 & 3 & 4 \\ 2 & 5 & 4 & 3 & 4 \\ 1 & 4 & 2 & 3 & 3 \\ 3 & 2 & 4 & 1 & 8 \end{bmatrix} \begin{matrix} \\ R_2 - (2/1) * R_1 \\ R_3 - (1/1) * R_1 \\ R_4 - (3/1) * R_1 \end{matrix} \rightarrow \begin{bmatrix} 1 & 2 & 1 & 3 & 4 \\ 0 & 1 & 2 & -3 & -4 \\ 0 & 2 & 1 & 0 & -1 \\ 0 & -4 & 1 & -8 & -4 \end{bmatrix}.$$

Although each of these row reductions depends on the elements of row 1, they are completely independent of one another. For example, the elements of rows 2 and 4 play no part in the row operations performed on row 3. Thus, the row reductions on rows 2, 3, and 4 can be computed simultaneously.

If we are computing in a parallel processing environment, we can take advantage of this independence by assigning each row-reduction task to a different processor:

$$\begin{bmatrix} 1 & 2 & 1 & 3 & 4 \\ 2 & 5 & 4 & 3 & 4 \\ 1 & 4 & 2 & 3 & 3 \\ 3 & 2 & 4 & 1 & 8 \end{bmatrix} \begin{matrix} \\ \rightarrow \text{Processor 1: } R_2 - (2/1) * R_1 \\ \rightarrow \text{Processor 2: } R_3 - (1/1) * R_1 \rightarrow \cdots \\ \rightarrow \text{Processor 3: } R_4 - (3/1) * R_1 \end{matrix}$$

Suppose each row assignment statement requires 4 time units, one for each element in a row. Then the sequential algorithm performs this stage of the row reduction in 12 time units, whereas we need only 4 time units for the parallel algorithm. This example of parallel processing on the first stage of row reduction of a 4×4 system matrix generalizes to any row reduction in stage j of an $n \times n$ system matrix.

Recall that there are $n - 1$ row-reduction stages in Gaussian elimination, one for each of the n columns of the coefficient matrix except for the last column. This suggests that we need $n - 1$ processors to do the reduction in parallel.* Also recall that each row-reduction stage j creates zeros in every cell below the diagonal in the jth column. The following two pseudocodes compare the use of a single processor with the use of n processors to perform the entire row-reduction phase of Gaussian elimination.

* Even so, we will need n processors in the final algorithm, as will be seen.

Algorithms for Row Reduction in Gaussian Elimination

Sequential Processing (without pivoting)

```
For j = 1 To (n − 1)
    For i = ( j + 1) To n
        For k = j To (n + 1)
            a[i, k] = a[i, k] − a[i, j]/a[j, j] * a[j, k]
        End For k
    End For i
End For j
```

Parallel Processing

```
For i = 1 To (n − 1)          (Counts stages = columns)
    For k = i To (n + 1)      (On Processor j = (i + 1) To n)
        a[i, k] = a[i, k] − a[i, j]/a[j, j] * a[j, k]
    End For k
End For i
```

If we total the arithmetic operations to carry out the reduction of an $n \times n$ coefficient matrix to upper-triangular form, we find that the sequential algorithm requires $O(n^3)$ successive operations and that the parallel algorithm with n processors accomplishes the same task in $O(n^2)$ successive operations.

What happens if we have more than n processors? As indicated in our earlier discussion, we can speed up the reduction process even more. Suppose we increase the number of processors from n to, say, $n^2 + n$. We can effectively use this extra power for Gaussian elimination just as we did for the matrix/vector operations earlier. The complete algorithm for the row-reduction phase of Gaussian elimination on these processors runs only $O(n)$ successive steps, and each step requires just three time steps for one subtraction, one division, and one multiplication. If, as before, we label each processor as proc(i,j), $i = 1, \ldots, n$, $j = 1, \ldots, n + 1$, proc(i,j) is responsible for each element a_{ij} of the matrix $[A{:}\ b]$. We can now rewrite the algorithm for parallel processing to reflect this improvement:

```
For i = 2 To n
    {On Processor ( j, k)}
    a[ j, k] = a[ j, k] − a[ j, i]/a[i, i] * a[ j, k]
End For i
```

The row-reduction phase of Gaussian elimination leaves us with an upper-triangular coefficient matrix and an appropriately adjusted right-hand side. In the sequential algorithm we now find a solution using back-substitution. Before we consider parallelization of back-substitution, let us examine the activity of the processors during row reduction in greater detail.

As we have observed, the processors responsible for computations on the elements of row 1 sit idle during row reduction, because those elements of the matrix never change. In addition, after the first row-reduction stage in which zeros are placed in the first column, the processors for row 2 also sit idle. In fact, each stage of row reduction frees $n + 1$ processors.

It is natural to wonder if these idle processors could be employed in our algorithm. Indeed, they can. We use them to perform row reductions above the diagonal at the same time that corresponding row reductions occur below the diagonal. Thus at each stage j of the reduction, zeros appear in all but the diagonal element of the jth column.

This diagram illustrates our improved procedure, continuing the simple 4×4 example examined before and doing stages 3 and 4:

$$\rightarrow \begin{bmatrix} 1 & 0 & -3 & 9 & 12 \\ 0 & 1 & 2 & -3 & -4 \\ 0 & 0 & -3 & 6 & 7 \\ 0 & 0 & 9 & -20 & -20 \end{bmatrix} \begin{matrix} R_1 - (3/3) * R_3 \\ R_2 - (-2/3) * R_3 \\ \\ R_4 - (9/-3) * R_3 \end{matrix} \rightarrow \begin{bmatrix} 1 & 0 & 0 & 3 & 5 \\ 0 & 1 & 0 & 1 & \frac{2}{3} \\ 0 & 0 & -3 & 6 & 7 \\ 0 & 0 & 0 & -2 & 1 \end{bmatrix}$$

$$\begin{matrix} R_1 - (3/-2) * R_4 \\ R_2 - (1/-2) * R_4 \\ R_3 - (6/-2) * R_4 \\ \\ \end{matrix} \begin{bmatrix} 1 & 0 & 0 & 0 & \frac{13}{2} \\ 0 & 1 & 0 & 0 & \frac{7}{6} \\ 0 & 0 & -3 & 0 & 10 \\ 0 & 0 & 0 & -2 & 1 \end{bmatrix} \rightarrow x = \begin{bmatrix} \frac{13}{2} \\ \frac{7}{6} \\ -\frac{10}{3} \\ -\frac{1}{2} \end{bmatrix}.$$

The result of n such reductions—one for each column of the coefficient matrix A—is $[D : b']$, where D is a diagonal matrix and b' is an appropriately adjusted right-hand-side vector.

Specifically, the solution x for $Dx = b'$ also satisfies $Ax = b$. This solution is the vector x whose elements are $x_i = b_i'/d_{ii}$ for $i = 1, \ldots, n$. We can use n processors to perform these n divisions simultaneously. Notice that the back-substitution phase of Gaussian elimination is no longer necessary! We find that the Gauss–Jordan procedure is preferred when doing parallel processing!

The parallel algorithm for n^2 processors required n time units for row reduction, and one additional time unit for division. Recall that the sequential algorithm required $O(n^3)$ time units. To understand the magnitude of the improvement in running time, consider that a solution achieved in 10 seconds via the parallel algorithm would require around 15 minutes via the sequential algorithm.*

Our final parallel algorithm for solving a system of linear equations more closely resembles the Gauss–Jacobi solution technique than it does Gaussian elimination. This is not surprising. It is not uncommon for good parallel algorithms to differ dramatically from their speediest sequential counterparts.

---

* This neglects the overhead of interprocessor communications.

Problems in Using Parallel Processors

It is essential to mention some important concerns that have been neglected in the preceding discussion. When we actually implement this parallel algorithm, we must worry about four issues.

1. The algorithm described here does not pivot. Thus, our solution may not be as numerically stable as one obtained via a sequential algorithm with partial pivoting. In fact, if a zero appears on the diagonal at any stage of the reduction, we are in big trouble. Bertsekas and Tsitsiklis observe that Gaussian elimination with pivoting can have an upper bound of $O(n \log(n))$ time when $n^2 + n$ processors are used, and still $O(n^2)$ time in the case of n processors.
2. The coefficient matrix A is assumed to be nonsingular. It is easy to check for singularity at each stage of the row reduction, but such error-handling will more than double the running time of the algorithm.
3. We have ignored the communication and overhead time costs that are involved in parallelization. Because of these costs, it is probably more efficient to solve small systems of equations using a sequential algorithm.
4. Other, perhaps faster, parallel algorithms exist for solving systems of linear equations. One technique, which is easily derived from ours, involves computing A^{-1} via row operations and simply multiplying the right-hand side to get the solution $x = A^{-1}b$. Another technique requires computing the coefficients of the characteristic polynomial and then applying these coefficients in building A^{-1} from powers of A. This method finds a solution in only $[2 \log_2 n + O(\log n)]$ time units, but it requires $n^4/2$ processors to do so. In addition, it often leads to numeric instability.*

Despite these concerns, our algorithm is an effective approach to solving systems of linear equations in a parallel environment

Iterative Solutions—The Jacobi Method

The method of simultaneous displacements (the Jacobi method) that was discussed in Section 2.5 is adapted very simply to a parallel environment. Recall that at each iteration of the algorithm a new solution vector $x^{(n+1)}$ is computed using only the elements of the solution vector from the previous iteration, $x^{(n)}$. In fact, the elements of the vector $x^{(n)}$ can be considered fixed with respect to the iteration $(n + 1)$. Thus, though each element $x_i^{(n+1)}$ in the vector $x^{(n+1)}$ depends on the elements in $x^{(n)}$, these $x_i^{(n+1)}$ are independent of one another and can be computed simultaneously.

Suppose the solution vector x has m elements. Then each iteration of the Jacobi algorithm in a sequential environment requires m assignment statements. If we have m processors in parallel, these m assignment statements can be performed simultaneously, thereby reducing the running time of the algorithm by a factor of m.

* JaJa (1992) describes these alternative algorithms in some detail.

Notice also that each assignment statement is a summation over approximately m terms. As demonstrated in Section 0.6, this summation can be performed in $\log_2 m$ time units with m parallel processors, compared to m time units for sequential addition. If m^2 processors are available, we can employ both of these parallelizations and reduce the time for each iteration of the Jacobi algorithm to $\log_2 m$ time units. This is a significant speedup over the sequential algorithm, which requires m^2 time units per iteration.

As seen in Section 2.5, the actual running time of the algorithm (the number of iterations) depends on the degree of diagonal dominance of the coefficient matrix. Parallelization decreases only the time required for each iteration.

Because Gauss–Seidel iteration requires that the new iterates for each variable be used after they have been obtained, this method cannot be speeded up by parallel processing. Again, the preferred algorithm for sequential processing is not the best for parallel processing.

Exercises

Section 2.1

1. For these four matrices:

$$A = \begin{bmatrix} 3 & 2 & -1 \\ 2 & 1 & 3 \\ -3 & -2 & 1 \end{bmatrix}, \quad B = \begin{bmatrix} 2 & 4 \\ -1 & -2 \\ 3 & 0 \end{bmatrix},$$

$$C = \begin{bmatrix} 3 & -1 & 3 \\ 1 & 1 & -2 \\ 4 & -2 & 0 \end{bmatrix}, \quad D = \begin{bmatrix} 0 & -2 & 3 \\ -1 & 5 & 0 \\ -2 & -1 & 6 \end{bmatrix},$$

a. Which pairs can be added? Find the sums.
b. Which pairs can be subtracted? Get their differences, then repeat in opposite order.
c. Which pairs can be multiplied? Find the products.
d. Which of these has a trace? Compute the traces.

2. Get the transpose for each matrix in Exercise 1. Then repeat each part of Exercise 1 with the transposes.

▶**3.** For these vectors:

$$v1 = \begin{bmatrix} -2 \\ 3 \\ 4 \end{bmatrix}, \qquad v2 = [3 \;\; 4 \;\; -1], \qquad v3 = \begin{bmatrix} 4 \\ 2 \\ -3 \end{bmatrix},$$

a. Which pairs can be multiplied? Find the products.
b. Using the matrices in Exercise 1, which matrices can multiply these vectors? Compute the products.
c. Can any of these vectors multiply with one of the matrices of Exercise 1 in the order $v * M$? Get the products for those that do.
d. Find the product of each vector times its transpose. Repeat with the transpose as the first factor.

4. Given the matrices:

$$A = \begin{bmatrix} -2 & 1 & 2 \\ 2 & 3 & -2 \\ 1 & -2 & -3 \end{bmatrix}, \quad B = \begin{bmatrix} -2 & 3 & 5 \\ 2 & 1 & -4 \\ 4 & -1 & 6 \end{bmatrix},$$

a. Find BA, B^3, AA^T.
b. Get det (A) and det (B).
c. A square matrix can always be expressed as the sum of a lower-triangular matrix L and an upper-triangular matrix U. Find two different combinations of L and U for both A and B.

▶**5.** Let

$$A = \begin{bmatrix} 2 & 9 \\ 3 & -10 \end{bmatrix}, \quad B = \begin{bmatrix} 1 & 6 & 2 \\ 4 & -1 & 3 \\ 1 & -3 & -1 \end{bmatrix}.$$

a. Find the characteristic polynomials of both A and B.
b. Find the eigenvalues of both A and B.
c. Is [0.2104, 0.8401] an eigenvector of A?

6. Write this as a set of equations:

$$\begin{bmatrix} 4 & 2 & -2 & -1 \\ 0 & 4 & 1 & 2 \\ 3 & -2 & 1 & 2 \\ 2 & 0 & 3 & -5 \end{bmatrix} \begin{bmatrix} x_1 \\ x_2 \\ x_3 \\ x_4 \end{bmatrix} = \begin{bmatrix} 7 \\ 10 \\ 2 \\ 3 \end{bmatrix}.$$

7. Write these equations in matrix form:

$$\begin{aligned} 6x - 2y + 3z &= 12, \\ x + y - 4z &= 8, \\ -2x - 3y &= 12. \end{aligned}$$

8. It is true that $(A * B)^T = B^T * A^T$.

a. Test this statement with two 3×3 matrices of your choice.
b. Is this true if A is 3×2 and B is 2×3.
c. Prove the statement.

9. For matrix A, write the transposition matrices that perform the interchanges.

$$A = \begin{bmatrix} 3 & 5 & -2 & -1 & 0 \\ -2 & 3 & 4 & -5 & 3 \\ 5 & 2 & -1 & 3 & -6 \\ 2 & -3 & 4 & 2 & 0 \\ -5 & 5 & 3 & -3 & 4 \end{bmatrix}.$$

a. Row 3 with row 5.
b. Column 2 with column 1.
▶ c. Both row 3 with row 5 and column 4 with column 2.
d. Multiply A with each of your matrices in parts (a), (b), and (c) to confirm that the correct interchanges occur.
e. What happens if the transposition matrix of part (a) is used to postmultiply with A rather than to premultiply?

10. Confirm that $P^T * P = I$ and that $P^T = P$ for each transposition matrix of Exercise 9.

Section 2.2

11. a. Solve by back-substitution:

$$\begin{aligned} 3x_1 + 3x_2 + x_3 &= 12, \\ -4x_2 - 3x_3 &= -10, \\ 2x_3 &= 4. \end{aligned}$$

b. Solve by forward-substitution:

$$\begin{aligned} 3x_1 &= 15, \\ 2x_1 - x_2 &= 10, \\ 5x_1 + x_2 - 2x_3 &= 5. \end{aligned}$$

12. The first procedure described in Section 2.2 is sometimes called "Naive Gaussian Elimination." Use it to solve Exercises 13 and 14.

13. Solve the following (given as the augmented matrix):

$$\left[\begin{array}{rrr|r} 3 & 1 & -4 & 7 \\ -2 & 3 & 1 & -5 \\ 2 & 0 & 5 & 10 \end{array}\right].$$

14. Here is a system of equations that is called "ill-conditioned," meaning that the solution is not easy to get accurately. Section 2.4 discusses this; here, we give a "taste" of the problem. This is the system as an augmented matrix:

$$\begin{bmatrix} 3 & 2 & 4 & 9 \\ 8 & -6 & -8 & -6 \\ -1 & 2 & 3 & 4 \end{bmatrix}.$$

You can see that $x = [1, 1, 1]^T$ is the solution.

a. Confirm the solution by doing naive Gaussian elimination using exact arithmetic (use fractions throughout).
b. Now get the solution using only three significant figures in your computations. Observe that the solution is different.
c. Compute the solution when the system is changed only slightly: Change the coefficient in the first column of the first row to 3.1. Use more precise computations, perhaps single or even double precision. Observe that this makes a large change in the solution.

▶**15.** Use Gaussian elimination with partial pivoting to solve the equations of Exercise 13. Are any row interchanges needed?

16. In which column(s) are row interchanges needed to solve the equations in Exercise 6 by Gaussian elimination with partial pivoting?

17. Solve this system by Gaussian elimination with partial pivoting:

$$\begin{bmatrix} 1 & -2 & 4 & 6 \\ 8 & -3 & 2 & 2 \\ -1 & 10 & 2 & 4 \end{bmatrix}.$$

a. How many row interchanges are needed?
b. Solve again but use only three significant digits of precision.
c. Repeat part (b) without any row interchanges. Do you get the same results?

18. Solve the system

$$\begin{aligned} 2.51x + 1.48y + 4.53z &= 0.05, \\ 1.48x + 0.93y - 1.30z &= 1.03, \\ 2.68x + 3.04y - 1.48z &= -0.53. \end{aligned}$$

a. Use Gaussian elimination, but use only three significant digits and do no interchanges. Observe the small divisor in reducing the third column. The correct solution is $x = 1.45310$, $y = -1.58919$, $z = -0.27489$.
b. Repeat part (a) but now do partial pivoting.
c. Repeat part (b) but now chop the numbers rather than rounding.
d. Substitute the solutions found in (a), (b), and (c) into the equations. How well do these match the original right-hand sides?

19. Use the Gauss–Jordan method to solve the equations of Exercise 17.

20. Use the Gauss–Jordan method to solve the equations of Exercise 18.

▶21. Confirm that the Gauss–Jordan method requires $O(n^3)$ total arithmetic operations.

22. What if we solve a system with m right-hand sides rather than just one? How many total operations are then required for both Gaussian elimination and the Gauss–Jordan method?

23. Suppose that multiplication takes twice as long to do as an addition/subtraction and that division takes three times as long (which used to be true). For a system of ten equations, how much longer does it take to get a solution compared to when each operation takes the same amount of time? Do this for both Gaussian elimination and for Gauss–Jordan.

24. Write an algorithm for the Gauss–Jordan method. Provide for partial pivoting.

a. When there is only one right-hand side.
b. When there are m right-hand sides.

25. Modify the algorithm for Gaussian elimination to incorporate scaled partial pivoting.

26. Repeat Exercise 25 but now employ an order vector to avoid actually interchanging the rows.

27. Use scaled partial pivoting to solve:

$$\begin{bmatrix} 4.13 & -2.20 & 0.95 & 3.02 \\ 6.14 & 4.45 & -1.45 & -4.02 \\ 1.03 & 1.86 & 0.44 & 5.22 \end{bmatrix}.$$

a. Employ six significant digits.
b. Repeat with only three significant digits. Is the solution much different?

▶28. If a comparison takes one-half as long as an addition/subtraction and to interchange two numbers takes twice times as long, how much time is saved by using an order vector rather than doing the actual row interchanges? Express the answer in terms of addition times for a system of n equations.

29. A system of two equations can be solved by graphing the two lines and finding where they intersect. (Graphing three equations could be done, but locating the intersection of the three planes is difficult.) Graph this system; you should find the intersection at (6, 2).

$$0.1x + 51.7y = 104,$$
$$5.1x - 7.3y = 16.$$

a. Now, solve using three significant digits of precision and no row interchanges. Compare the answer to the correct value.
b. Repeat part (a) but do partial pivoting.
c. Repeat part (a) but use scaled partial pivoting. Which of part (a) or (b) does this match, if any?
d. Complete pivoting chooses the largest of all of the coefficients at the current stage as the pivot element. Repeat part (a) with complete pivoting. How does this answer compare to those of parts (a), (b), and (c)?

30. The determinant of a matrix can be found by expanding in terms of its minors. Compare the number of arithmetic operations when done this way with the number if the matrix is reduced to a triangular one by Gaussian elimination. Do this for a 4×4 matrix. Then find a relation for an $n \times n$ matrix.

▶31. When you solved Exercise 17, you could have saved the row multipliers and obtained a LU equivalent of the coefficient matrix. Use this LU to solve Exercise 17 but with right-hand sides of:

a. $[1, -3, 5]^T$.
b. $[-3, 7, -2]^T$.

32. Repeat Exercise 17, but now use the LU.

33. Repeat Exercise 27, but now use the LU.

▶34. Given this tridiagonal system:

$$\begin{bmatrix} 4 & -1 & 0 & 0 & 0 & 0 & 100 \\ -1 & 4 & -1 & 0 & 0 & 0 & 200 \\ 0 & -1 & 4 & -1 & 0 & 0 & 200 \\ 0 & 0 & -1 & 4 & -1 & 0 & 200 \\ 0 & 0 & 0 & -1 & 4 & -1 & 200 \\ 0 & 0 & 0 & 0 & -1 & 4 & 100 \end{bmatrix},$$

a. Solve the system using the algorithm for a compacted system matrix that has n rows but only four columns.
b. How many arithmetic operations are needed to solve a tridiagonal system of n equations in this compacted arrangement? How does this compare to solving such a system with Gaussian elimination without compacting?

35. The system of Exercise 34 is an example of a symmetric matrix. Because the elements at opposite positions across the diagonal are exactly the same, it can be stored as a matrix with n rows but only three columns.

a. Write an algorithm for solving a symmetric tridiagonal system that takes advantage of such compacting.

b. Use the algorithm from part (a) to solve the system in Exercise 34.
c. How many arithmetic operations are needed with this algorithm for a system of n equations?

▶**36.** Write the algorithm for LU reduction that puts ones on the diagonal of U.

37. When are row interchanges absolutely required in forming the LU equivalent of matrix A?

38. Given system A:

$$A = \begin{bmatrix} 2 & -1 & 3 & 2 \\ 2 & 2 & 0 & 4 \\ 1 & 1 & -2 & 2 \\ 1 & 3 & 4 & -1 \end{bmatrix}.$$

Find the LU equivalent of matrix A that has 2's in each diagonal position of L rather than 1's.

39. Repeat Exercise 38, but now make the diagonal elements of L equal to [1, 2, 3, 4].

40. If you were asked to create a LU reduction of matrix A that has at least one zero on a diagonal,

a. When can you do this, putting the zero(s) on the diagonal of L?
b. When can you do this, putting the zero(s) on the diagonal of U?
c. Give examples where A is 3×3.

Section 2.3

▶**41.** Which of these matrices are singular?

a. $$\begin{bmatrix} -2 & 1 & -1 \\ -3 & 4 & -6 \\ 2 & 7 & 15 \end{bmatrix}.$$

b. $$\begin{bmatrix} 2 & 3 & 1 & 1 \\ 1 & 3 & 4 & 1 \\ 7 & 0 & -4 & 2 \\ 4 & -6 & -9 & 0 \end{bmatrix}.$$

c. $$\begin{bmatrix} 2 & 3 & 1 & 1 \\ 1 & 3 & 4 & 1 \\ 7 & 0 & -4 & 2 \\ 4 & -6 & 9 & 0 \end{bmatrix}.$$

42. For this matrix:

$$A = \begin{bmatrix} 3 & 5 & 1 \\ -1 & 3 & 2 \\ a & b & -1 \end{bmatrix},$$

a. Find values for a and b that make A singular.
b. Find values for a and b that make A nonsingular.

43. The matrix in Exercise 41, part (b), is singular.

a. That means its rows form vectors that are linearly dependent. Find the weighting factors for the rows that makes their sum zero.
b. Repeat part (a), but with the columns.

44. Do these equations have a solution? Find the solution if it exists. Explain why when it doesn't.

a. $$\begin{aligned} -2x + 3y + z &= 2, \\ -3x + y + z &= 5, \\ x + y - z &= -5, \\ 3y + z &= 0. \end{aligned}$$

b. $$\begin{bmatrix} 0 & 1 & 1 & 1 \\ 1 & 1 & 0 & -2 \\ 1 & 0 & 1 & 0 \\ 1 & 1 & 1 & 4 \end{bmatrix}.$$

c. $$\begin{bmatrix} 2 & -1 & 6 & 1 \\ 1 & 0 & 2 & 0 \\ 3 & 2 & 2 & 0 \end{bmatrix}.$$

▶**45.** The Hilbert matrix is a classic case of the pathological situation called "ill-conditioning." The 4×4 Hilbert matrix is

$$H_4 = \begin{bmatrix} 1 & \frac{1}{2} & \frac{1}{3} & \frac{1}{4} \\ \frac{1}{2} & \frac{1}{3} & \frac{1}{4} & \frac{1}{5} \\ \frac{1}{3} & \frac{1}{4} & \frac{1}{5} & \frac{1}{6} \\ \frac{1}{4} & \frac{1}{5} & \frac{1}{6} & \frac{1}{7} \end{bmatrix}.$$

For the system $Hx = b^T$, with $b^T = [25/12, 77/60, 57/60, 319/420]$, the exact solution is $x^T = [1, 1, 1, 1]$.

a. Show that the matrix is ill-conditioned by showing that it is nearly singular.
b. Using only three significant digits (chopped) in your arithmetic, find the solution to $Hx = b$. Explain why the answers are so poor.
c. Using only three significant digits, but rounding, again find the solution and compare it to that obtained in part (b).

46. For this system of equations

$$\begin{aligned} ax + 4y + z &= 6, \\ 2ax - y + 2z &= 3, \\ x + 3y + az &= 5, \end{aligned}$$

a. What value of a gives a unique solution to the system?
b. What value of a makes the system have no solution?

c. What value of a makes the system have an infinity of solutions?

47. Solve this pair of equations by Gaussian elimination:

$$0.2205x + 0.1254y = 0.6606,$$
$$0.4457x + 0.2506y = 0.8897.$$

a. Use only four significant digits in the solution.
b. Compare the solution using seven significant digits with that of part (a). Explain why the solutions are different.

48. Use the Gaussian elimination method to triangularize this matrix and from that get its determinant:

$$A = \begin{bmatrix} 3 & -1 & 2 \\ 1 & 1 & 3 \\ -3 & 0 & 5 \end{bmatrix}.$$

49. Repeat Exercise 48 but convert matrix A to a LU that has ones on the diagonal of U rather than on L.

▶**50.** Change the element in row 3, column 3 of Exercise 48 from +5 to −5 and repeat Exercise 48. Explain why this causes the determinant to become smaller.

51. For this matrix:

$$\begin{bmatrix} 4 & -2 & 3 & -5 \\ 3 & 3 & 5 & -8 \\ -6 & -1 & 4 & 3 \\ -4 & 2 & -3 & 5 \end{bmatrix},$$

a. Show that the matrix is singular.
b. Change the element in row 4 column 4 from 5.0 to 5.1 and get its determinant. Even though this matrix is larger than the matrix in Exercise 48 and most of its elements are greater, why is the determinant a smaller number than for Exercise 48?

52. First show that det $(A * B) = \det(A) * \det(B)$ for two 4×4 matrices that you compose, then prove that this will always be true for any two square matrices of the same size.

53. If some of the elements of a matrix are very small in magnitude and others are very large, will the value of its determinant be large or small? What if only one element is very large and the rest very small? What if only one is very small and the rest very large? Are there situations where the magnitudes of the elements are not important?

▶**54.** Get the inverse of the matrix in Exercise 48.

a. Do it through Gaussian elimination.
b. Repeat, but with the Gauss–Jordan method.
c. How many arithmetic operations are used in parts (a) and (b)?

55. Find the inverse of the matrix in part (b) of Exercise 51.

a. Do this using only three significant digits of precision.
b. Repeat part(a), but now use seven digits. Why are the results different?

56. Repeat Exercise 55, but for the matrix in Exercise 48. Why are the results with three digits the same as those with seven when rounded?

57. Find the determinant of matrix A and the determinant of its inverse.

$$A = \begin{bmatrix} -2 & 3 & 2 & 8 \\ 1 & 3 & -2 & 6 \\ 5 & -1 & 3 & 9 \\ 2 & 3 & 8 & -1 \end{bmatrix}.$$

Section 2.4

▶**58.** Evaluate the 1-, 2- and ∞-norms of these vectors:

a. [3.06, −2.11, 8.12, −4.45].
b. [−5, −3, 2, 7].

59. Verify the relations of Eq. (2.14) for each of the definitions of a vector norm.

60. Which vector norm usually gives the smallest value? Is there an instance when all vector norms have the same value?

▶**61.** Evaluate the 1-, 2-, and ∞-norms of these matrices:

$$A = \begin{bmatrix} 5 & -9 & 6 \\ 2 & -7 & 4 \\ 1 & 5 & 8 \end{bmatrix},$$

$$B = \begin{bmatrix} 10.2 & 2.4 & 4.5 \\ -2.3 & 7.7 & 11.1 \\ -5.5 & -3.2 & 0.9 \end{bmatrix}.$$

62. Is the spectral norm of a matrix always the smallest norm? Is there a case where all matrix norms have the same value?

63. For the matrices of Exercise 61, compare these norms:

a. norm $(A + B)$ with norm (A) + norm(B).
b. norm $(A * B)$ with norm (A) * norm(B).
c. norm (A^2) with norm (A) * norm(A).
d. What conclusion do you draw from these results?

▶**64.** Find the ∞-norm of the Hilbert matrix of Exercise 45.

65. If a matrix is nearly singular, how does its norm compare to the norm of its inverse?

66. Given this system of equations:

$$\begin{bmatrix} 6.03 & 1.99 & 3.01 & 1 \\ 4.16 & -1.23 & 1.27 & 1 \\ -4.81 & 9.34 & 0.987 & 1 \end{bmatrix},$$

a. Solve with a precision of ten significant digits.
b. Solve again with a precision of four significant digits.
c. What happens if you solve with a precision of only three significant digits?
d. Let x be the solution from part (a) and let ξ be the solution from part (b). Let $e = x - \xi$. What are the norms of e?
e. Is the system ill-conditioned? What is the condition number of the coefficient matrix? Compute this for each definition of condition number.

67. Repeat Exercise 66 after changing element a_{32} to -9.34. Why are the results so different?

68. What if we discover that one of the coefficients in Exercise 66 is slightly in error due to measuring errors? Specifically, suppose that a_{13} should be 3.02 rather than 3.01. How does this affect the answers to parts (a) and (b) of Exercise 66?

69. What are the residuals for the imperfect solutions of Exercises 66, 67, and 68?

▶**70.** What is the condition number for the coefficient matrix of Exercise 67. Why is it so different from that for Exercise 66?

71. Verify Eq. (2.16) with the residuals from Exercises 66 and 67.

72. Verify Eq. (2.18) with the residuals from Exercises 66 and 67.

73. Apply iterative improvement to the solution from Exercise 66, part (b).

74. Compare the condition numbers for the Hilbert matrix of order-4:

a. Using exact numbers (use fractional numbers throughout).
b. Using floating-point values with only three significant digits.

▶**75.** Prove that cond $(A) \geq 1$ for any square matrix. Are there any exceptions to this?

76. For what values of a does this matrix have a condition number greater than 100?

$$\begin{bmatrix} a & 2 & 2 \\ 2 & 2 & 2 \\ 2 & 2 & a \end{bmatrix}$$

77. Find a 2×2 matrix whose condition number is exactly 289 using infinity norms.

Section 2.5

▶**78.** Solve this system with the Jacobi method. First rearrange to make it diagonally dominant if possible. Use [0, 0, 0] as the starting vector. How many iterations to get the solution accurate to five significant digits?

$$\begin{bmatrix} 7 & -3 & 4 & 6 \\ -3 & 2 & 6 & 2 \\ 2 & 5 & 3 & -5 \end{bmatrix}.$$

▶**79.** Repeat Exercise 78 with the Gauss–Seidel method. Are fewer iterations required?

80. Is convergence faster in Exercises 78 and 79 if the starting vector is $[-0.26602, -0.26602, -0.26602]$ which is the average value of the elements of the solution vector?

81. Solve this system of equations, starting with the initial vector of [0, 0, 0]:

$$\begin{aligned} 4.63x_1 - 1.21x_2 + 3.22x_3 &= 2.22, \\ -3.07x_1 + 5.48x_2 + 2.11x_3 &= -3.17, \\ 1.26x_1 + 3.11x_2 + 4.57x_3 &= 5.11. \end{aligned}$$

a. Solve using the Jacobi method.
b. Solve using the Gauss–Seidel method.

82. The coefficient matrix of Exercise 81 is diagonally dominant. If the value of the element in position (2, 2) is smaller in magnitude than 5.48, it is no longer diagonally dominant. How small can it be and still converge to a solution by iterating with

a. The Jacobi method?
b. The Gauss–Seidel method?

83. This 2×2 matrix is obviously singular and is almost diagonally dominant. If the right-hand-side vector is [0, 0], the equations are satisfied by any pair where $x = y$.

$$\begin{bmatrix} 2 & -2 \\ -2 & 2 \end{bmatrix}.$$

a. What happens if you use the Jacobi method with these starting vectors: [1, 1], [1, −1], [−1, 1], [2, 5], [5, 2]?
b. What happens if the Gauss–Seidel method is used with the same starting vectors as in part (a)?
c. If the elements whose values are −2 in the matrix are changed slightly, to −1.99, the matrix is no longer singular but is almost singular. Repeat parts (a) and (b) with these new matrix.

84. For the system of equations in Exercise 78, find the matrices that correspond to Eqs. (2.23) and (2.25). For which method is the norm of the multiplier of $x^{(n)}$ a smaller number?

85. What is the optimal values of the overrelaxation factors that speed the solutions of Exercise 81?

Section 2.6

86. Section 2.6 says that the time to compute the inner product of two n-component vectors is proportional to $1 + \log(n)$, when n processors are available and each processor holds just one component of each vector. However, we can multiply an $n \times n$ matrix times a vector in $2n$ units of time if each processor holds one entire row of the matrix as well as the vector. Make a table that compares the times for these alternative methods for values of $n = 10 * e$ for $e = 1$ to 5.

▶87. Develop an algorithm for inverting an $n \times n$ matrix by parallel processing with approximately n^2 processors.

88. The final algorithm developed in Section 2.6 used $n^2 + n$ processors. Show that this can be further improved so that only $(n + 1)(n - 1) = n^2 - 1$ processors are required.

89. Develop an algorithm for doing Jacobi iterations to solve a system of n linear equations using n^2 processors.

Applied Problems and Projects

APP1. In considering the movement of space vehicles, it is frequently necessary to transform coordinate systems. The standard inertial coordinate system has the N-axis pointed north, the E-axis pointed east, and the D-axis pointed toward the center of the earth. A second system is the vehicle's local coordinate system (with the i-axis straight ahead of the vehicle, the j-axis to the right, and the k-axis downward). We can transform the vector whose local coordinates are (i, j, k) to the inertial system by multiplying by transformation matrices:

$$\begin{bmatrix} n \\ e \\ d \end{bmatrix} = \begin{bmatrix} \cos a & -\sin a & 0 \\ \sin a & \cos a & 0 \\ 0 & 0 & 1 \end{bmatrix} \begin{bmatrix} \cos b & 0 & \sin b \\ 0 & 1 & 0 \\ -\sin b & 0 & \cos b \end{bmatrix} \begin{bmatrix} 1 & 0 & 0 \\ 0 & \cos c & -\sin c \\ 0 & \sin c & \cos c \end{bmatrix} \begin{bmatrix} i \\ j \\ k \end{bmatrix}.$$

Transform the vector $[2.06, -2.44, -0.47]^T$ to the inertial system if $a = 27°$, $b = 5°$, $c = 72°$.

APP2. Exercise 45 showed the pattern for a Hilbert matrix. The $n \times n$ Hilbert matrix can be defined more formally as:

$$H_n = (1/(i + j + 1)).\ i, j = 0, 1, \ldots, n - 1.$$

a. Use this in a program that displays the Hilbert matrix of order-5.

b. What is the condition number of the 9×9 Hilbert matrix, H_9?

c. Solve $H_9 x = [1, 1, 1, 1, 1, 1, 1, 1, 1]^T$. Then change the first component of the right-hand side to 1.01 and solve again. Which component of x is most changed?

APP3. Electrical engineers often must find the currents flowing and voltages existing in a complex resistor network. Here is a typical problem.

Seven resistors are connected as shown, and voltage is applied to the circuit at points 1 and 6 (see Fig. 2.1) You may recognize the network as a variation on a Wheatstone bridge.

Although we are especially interested in finding the current that flows through the ammeter, the computational method can give the voltages at each numbered point (these are called *nodes*) and the current through each of the branches of the circuit. Two laws are involved:

Kirchhoff's law: The sum of all currents flowing into a node is zero.

Ohm's law: The current through a resistor equals the voltage across it divided by its resistance.

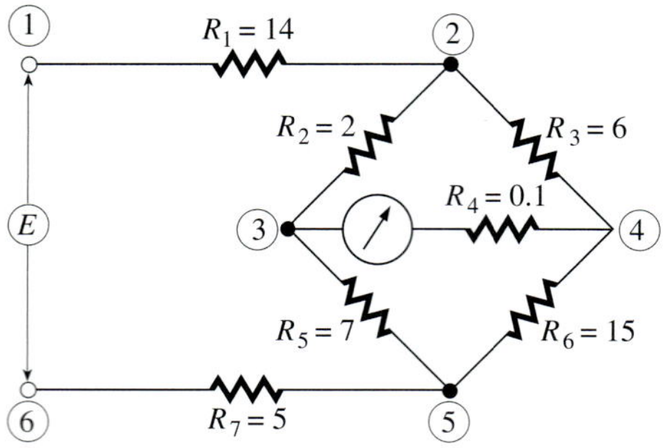

Figure 2.1

We can set up 11 equations using these laws and from these solve for 11 unknown quantities (the four voltages and seven currents). If $V_1 = 5$ volts and $V_6 = 0$ volts, set up the 11 equations and solve to find the voltage at each other node and the currents flowing in each branch of the circuit.

APP4. A square matrix can be partitioned into submatrices. We write

$$M = \begin{bmatrix} A & B \\ C & D \end{bmatrix},$$

where A and D are square. Suppose that

$$M^{-1} = \begin{bmatrix} E & F \\ G & H \end{bmatrix}.$$

Ralston shows that we can get M^{-1} in the following way:

1. Invert A to get A^{-1}.
2. Compute $D - C * A^{-1} * B$.
3. Invert $D - C * A^{-1} * B$, which gives H.
4. Compute $-A^{-1} * B * H$, which gives F.
5. Compute $-H * C * A^{-1}$, which gives G.
6. Compute $A^{-1} - A^{-1} * B * G$, which gives E.

We then get the inverse of M by inverting two smaller matrices and doing some arithmetic operations on matrices that are smaller than M. In this, matrices A and D must not be singular.

Choose some 4×4 matrix. Call it M. Then,

a. Partition M into submatrices. This can be done in three different ways.
b. Get the inverse of M with Ralston's technique. Do all partitionings give the same result?
c. Is Ralston's technique more or less efficient than inverting M directly? Does the difference in operational count depend on the size of M? Does it depend on the way that M is partitioned?

APP5. Mass spectrometry analysis gives a series of peak height readings for various ion masses. For each peak, the height h_j is contributed to by the various constituents. These make different contributions c_{ij} per unit concentration p_i so that the relation

$$h_j = \sum_{i=1}^{n} c_{ij} p_i$$

Table 2.2

Peak number	Component				
	CH_4	C_2H_4	C_2H_6	C_3H_6	C_3H_8
1	0.165	0.202	0.317	0.234	0.182
2	27.7	0.862	0.062	0.073	0.131
3		22.35	13.05	4.420	6.001
4			11.28	0	1.110
5				9.850	1.684
6					15.94

holds, with n being the number of components present. Carnahan (1964) gives the values shown in Table 2.2 for c_{ij}.

If a sample had measured peak heights of $h_1 = 5.20$, $h_2 = 61.7$, $h_3 = 149.2$, $h_4 = 79.4$, $h_5 = 89.3$, and $h_6 = 69.3$, calculate the values of p_i for each component. The total of all the p_i values was 21.53.

APP6. Figure 2.2 shows a structure that might support a bridge (a "truss"). The support at point a is constrained so that it cannot move; the one at point f can move horizontally. There are two external loads, at joints b and d.

In analyzing a truss, the members are assumed not to bend, so the forces within them act only in the direction of the member; these are considered to act from the joint toward the center.

This truss has nine members, so there are nine member forces, F_i, $i = 1, \ldots, 9$. If we set the sum of all forces acting either vertically or horizontally within each member, nine equations can be written. Solving these equations gives the values for the nine forces, the F_i.

a. Set up the equations and solve.
b. The matrix is sparse. Is it banded?
c. Can the band width be reduced by reordering the equations?

APP7. The truss in APP6, Figure 2.2, is called *statically determinant,* because nine linearly independent equations can be set up to solve for the nine forces. If a tenth member is added to give better stability to the structure, as shown in Figure 2.3, there are ten member forces to be determined but only nine force equations can be written. This truss is called *statically indeterminant.*

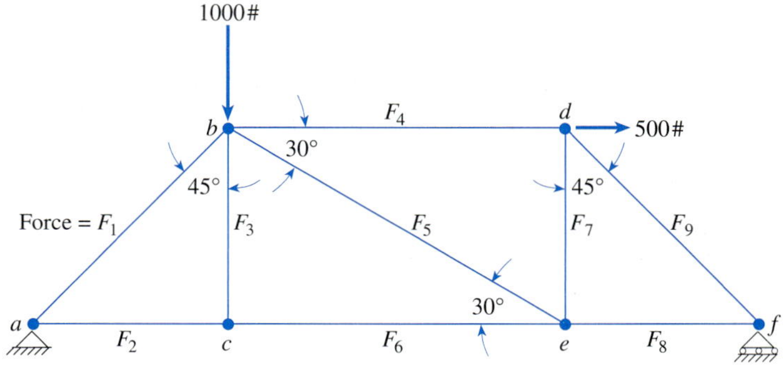

Figure 2.2

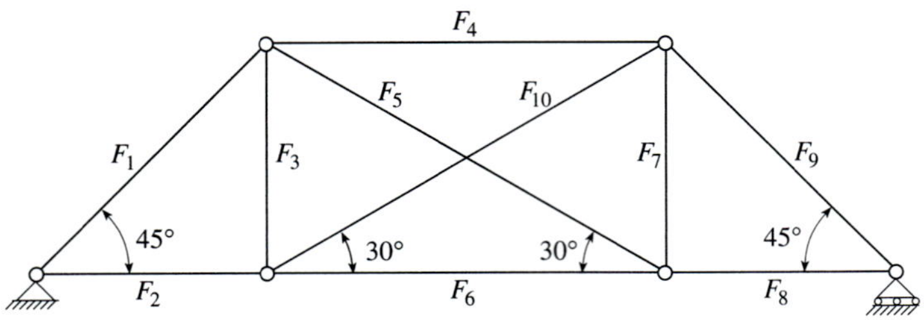

Figure 2.3

A solution can be found if the stretching or compression of the members is considered. We need to solve a set of equations that gives the displacement of each joint; these are of the form $ASA^Tx = P$. We get the tensions in the members, f, by the matrix multiplication $SA^Tx = f$

The required matrices and vectors are

$$A = \begin{bmatrix} 0.7071 & 0 & 0 & -1 & -0.8660 & 0 & 0 & 0 & 0 & 0 \\ 0.7071 & 0 & 1 & 0 & 0.5 & 0 & 0 & 0 & 0 & 0\ \mathrm{p} \\ 0 & 1 & 0 & 0 & 0 & -1 & 0 & 0 & 0 & -0.8660 \\ 0 & 0 & -1 & 0 & 0 & 0 & 0 & 0 & 0 & -0.5 \\ 0 & 0 & 0 & 0 & 0 & 0 & 1 & 0 & 0.7071 & 0.5 \\ 0 & 0 & 0 & 1 & 0 & 0 & 0 & 0 & -0.7071 & 0.8660 \\ 0 & 0 & 0 & 0 & 0.8660 & 1 & 0 & -1 & 0 & 0 \\ 0 & 0 & 0 & 0 & -0.5 & 0 & -1 & 0 & 0 & 0 \\ 0 & 0 & 0 & 0 & 0 & 0 & 0 & 1 & 0.7071 & 0 \end{bmatrix}$$

S is a diagonal matrix with values (from upper left to lower right) of

$$4255,\ 6000,\ 6000,\ 3670,\ 3000,$$
$$3670,\ 6000,\ 6000,\ 4255,\ 3000.$$

(These quantities are the values of aE/L, where a is the cross-sectional area of a member, E is the Young's modulus for the material, and L is the length.)

Solve the system of equations to determine the values of f for each of three loading vectors:

$$P_1 = [0, -1000, 0, 0, 500, 0, 0, -500, 0]^T,$$
$$P_2 = [1000, 0, 0, -500, 0, 1000, 0, -500, 0]^T,$$
$$P_3 = [0, 0, 0, -500, 0, 0, 0, -500, 0]^T.$$

APP8. For turbulent flow of fluids in an interconnected network (see Fig. 2.4) the flow rate V from one node to another is about proportional to the square root of the difference in pressures at the nodes. (Thus, fluid flow differs from flow of electrical current in a network in that nonlinear equations result.) For the conduits in Figure 2.4, find the pressure at each node. The values of b represent conductance factors in the relation $v_{ij} = b_{ij}(p_i - p_j)^{1/2}$.

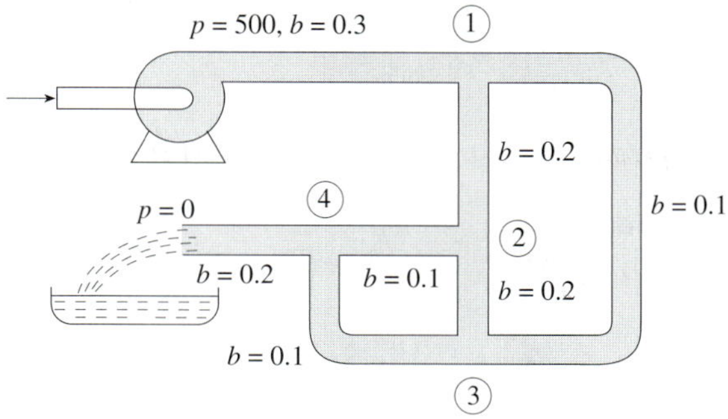

Figure 2.4

These equations can be set up for the pressures at each node:

$$\text{At node 1: } 0.3\sqrt{500 - p_1} = 0.2\sqrt{p_1 - p_2} + 0.1\sqrt{p_1 - p_3};$$

$$\text{node 2: } 0.2\sqrt{p_1 - p_2} = 0.1\sqrt{p_2 - p_4} + 0.2\sqrt{p_2 - p_3};$$

$$\text{node 3: } 0.1\sqrt{p_1 - p_3} + 0.2\sqrt{p_2 - p_3} = 0.1\sqrt{p_3 - p_4};$$

$$\text{node 4: } 0.1\sqrt{p_2 - p_4} + 0.1\sqrt{p_3 - p_4} = 0.2\sqrt{p_4 - 0}.$$

APP9. a. The elements in the matrix equation $Ax = B$ may be complex valued. Write a program to do Gaussian elimination in some computer language that permits complex values and solve a few examples. How will you determine the proper pivot rows in this program? Does the coefficient matrix have an inverse? If so, multiply this inverse by the original coefficient matrix but, before doing this, try to predict the result.

b. It is not necessary to use complex arithmetic to solve a system that has complex-valued elements. How can this be done? Solve the examples that you used in part (a) in this way. You should get the same solutions; do you?

APP10. Electrical circuits always have some capacitance and inductance in addition to resistance. Suppose that a 500 μF capacitor is added to the network of APP3 between nodes 1 and 2 and a 4 mH inductance is added between nodes 5 and 6. Of course, if the voltage source E is a direct current source, no current will flow after the capacitance becomes saturated, but if E is an alternating voltage source, there will be continuous (though fluctuating) current in the network. Set up the equations that can be solved for the voltages at the nodes and the currents in each branch of the network. You may need to consult a reference to handle this mixture of resistors, capacitors, and inductance.

APP11. We have shown how a tridiagonal system is especially advantageous in that it can be solved with fewer arithmetic operations than a full $n \times n$ system. A banded matrix is similarly advantageous, and this is particularly true if the coefficient matrix is symmetric. What are the number of multiplies and divides for a symmetrical system of n equations that has m elements to the right and to the left of the diagonal? Your answer should be expressed in terms of n and m.

APP12. It has been claimed that the National Weather Service uses extremely large sets of equations to forecast the weather. Do research to see if this is true and if these equations are linear or nonlinear. There are several models in use. Five of these are

1. NGM (Nested Grid Model—also called RAFS or Regional Analysis Forecast System).
2. ETA–forecast out to 48 hours.
3. MESO–ETA—forecast to 33 hours.
4. AVN–aviation model to 72 hours.
5. MRF—medium range forecast.

You may find the answer to this question from NWS/CIO (the Office of the Chief Information Officer); the Internet may provide a link to this office. If this is a group project, one member might send an inquiry to:

National Weather Service, NOAA

1325 West-West Highway

Silver Spring, MD 20910

APP13. MATLAB has commands that give you quantitative information on sparse matrices. A tridiagonal matrix is sparse if there are many equations. Generate a large tridiagonal matrix in MATLAB, name it *A*, and then use these commands to investigate it. What is the information given by each command?

```
nnz (A)
nonzeros (A)
nzmax (A)
spy(A)
[i, j, s) = find (A);
  [m, n) = size (A);
    B = sparse (i, j, s, m, n)
```

APP14. Can Gaussian elimination be used to solve a system where there are inequalities in addition to equalities? Try to do it with this small system. (4, 0) is a solution; what other points are a solution? (Chapter 7 discusses this type of problem in detail.) A graphical solution is easy.

$$\begin{aligned} 5x_1 - 3x_2 &\geq 12 \\ 2x_1 + 4x_2 &\leq 15 \\ x_1 + 3x_2 &= 4 \end{aligned}$$

3

EAGLE LIBRARY
Bolton Institute of Higher
Education - EAGLE CAMPUS

Interpolation and Curve Fitting

It once was the case that students found values for sines, logarithms, and other nonalgebraic functions from tables rather than getting the values using a computer or calculator, as one commonly does today. Those earlier tables had values of the function at uniformly spaced values of the argument. One most often interpolated linearly: The value for $x = 0.125$ was computed as at the half-way point between $x = 0.12$ and $x = 0.13$. If the function does not vary too rapidly and the tabulated points are close enough together, this linearly estimated value would be accurate enough.

If you don't have to use tables and interpolate from them, why do we devote a lengthy chapter to a topic that may seem obsolete? There are four reasons: (1) Interpolation methods are the basis for many other procedures that you will study in this course, such as numerical integration and differentiation; (2) they are behind the ways that we use to solve ordinary and partial-differential equations; (3) they demonstrate important theory about polynomials and the accuracy of numerical methods; and (4) they are one of the more important ways that curves are drawn on your computer screen. In addition, history itself may hold a special fascination for some.

There is a rich history behind interpolation. It really began with the early studies of astronomy when the motion of heavenly bodies was determined from periodic observations. The names of many famous mathematicians are associated with interpolation: Gauss, Newton, Bessel, Stirling.

An application of interpolation that you see every day is in weather forecasting. When you watch the weather forecasts on television, you may wonder where these (usually) correct projections come from. The weather service people collect information on temperature, wind speed and direction, humidity, and barometric pressure from hundreds of weather stations around the United States. Added to these are cloud data from satellites that are in elevated orbits above the earth. All of these data items are entered into a massive computer program that models the weather. Up to a million pieces of data are involved.

There is a problem, however. The locations where the data is collected are not uniformly distributed. These places are at various towns and cities and some remote locations where the data are obtained and transmitted automatically. It is a complex problem, one of several dimensions because the various weather stations are also at different elevations. The problem is that the computer models are massive, partial-differential equations that require the data to be at points on a uniform grid.

In this chapter, you will learn how such scattered data can be interpolated to estimate values at uniformly positioned grid points. This chapter will also compare several ways that one can do interpolation and contrast these with other techniques for fitting functions to imprecise data and for drawing smooth curves.

Contents of This Chapter

3.1 Interpolating Polynomials

Describes a straightforward but computationally awkward way to fit a polynomial to a set of data points so that an interpolated value can be computed. The cost of getting the interpolant with a desired accuracy is facilitated by a variant, Neville's method.

3.2 Divided Differences

These provide a more efficient way to construct an interpolating polynomial—one that allows one to readily change the degree of the polynomial. If the data are at evenly spaced x-values, there is some simplification.

3.3 Spline Curves

Using special polynomials, *splines,* one can fit polynomials to data more accurately than with an interpolating polynomial. At the expense of added computational effort, some important problems that one has with interpolating polynomials is overcome.

3.4 Bezier Curves and B-Splines

Are modern techniques for constructing smooth curves. They are used widely in computer graphics. They are not interpolating polynomials but are closely related.

3.5 Interpolating on a Surface

When the function has two independent variables, the points lie on a surface. Interpolating at points on the surface between the given points is more difficult but the previous techniques can be applied.

3.6 Least-Squares Approximations

Are methods by which polynomials and other functions can be fitted to data that are subject to errors likely in experiments. These approximations are widely used to analyze experimental observations.

3.1 Interpolating Polynomials

It has been mentioned that weather prediction requires that scattered data must be interpolated to estimate values at uniformly positioned grid points. That is a multidimensional problem—we start with simpler problems that have only one dimension, where y is a function of x.

In this simpler problem, for example, we have a table of x and y-values. Two entries in this table might be $y = 2.36$ at $x = 0.41$ and $y = 3.11$ at $x = 0.52$. If we desire an estimate for y at $x = 0.43$, we would use the two table values for that estimate. The quickest and easiest way to get this estimate would be to use the value at the point closest to $x = 0.43$, which would be $y = 2.36$. You are thinking, "Yes, that is quick and easy but surely not the best estimate. Why not interpolate as if $y(x)$ was linear between the two x-values?" That is a good suggestion. We will explore this and other even better ways to interpolate in this chapter.

We will be most interested in techniques adapted to situations where the data are far from linear. The basic principle is to fit a polynomial curve to the data. The reason for using polynomials has already been stated—they are "nice" functions and their evaluation requires only those arithmetic operations that computers can do.

In this section through Section 3.5, we assume that the tabulated data are exact. In Section 3.6, we consider the case where the data may have errors of measurement, which is true for most experimental results.

Fitting a Polynomial to Data

Suppose that we have the following data pairs—x-values and $f(x)$-values—where $f(x)$ is some unknown function:

x	$f(x)$
3.2	22.0
2.7	17.8
1.0	14.2
4.8	38.3
5.6	51.7

First, we need to select the points that determine our polynomial. (The maximum degree of the polynomial is always one less than the number of points.) Suppose we choose the first four points. If the cubic is $ax^3 + bx^2 + cx + d$, we can write four equations involving the unknown coefficients a, b, c, and d:

$$\begin{aligned}
\text{when } x = 3.2&: a(3.2)^3 + b(3.2)^2 + c(3.2) + d = 22.0,\\
\text{if } x = 2.7&: a(2.7)^3 + b(2.7)^2 + c(2.7) + d = 17.8,\\
\text{if } x = 1.0&: a(1.0)^3 + b(1.0)^2 + c(1.0) + d = 14.2,\\
\text{if } x = 4.8&: a(4.8)^3 + b(4.8)^2 + c(4.8) + d = 38.3.
\end{aligned}$$

Solving these equations by the methods of the previous chapter gives us the polynomial. We can then estimate the values of the function at some value of x—say, $x = 3.0$—by substituting 3.0 for x in the polynomial.

For this example, the set of equations gives

$$a = -0.5275,$$
$$b = 6.4952,$$
$$c = -16.1177,$$
$$d = 24.3499,$$

and our polynomial is

$$-0.5275x^3 + 6.4952x^2 - 16.1177x + 24.3499.$$

At $x = 3.0$, the estimated value is 20.212.

We seek a better and simpler way of finding such interpolating polynomials. This procedure is awkward, especially if we want a new polynomial that is also made to fit at the point (5.6, 51.7), or if we want to see what difference it would make to use a quadratic instead of a cubic. Furthermore, this technique leads to an ill-conditioned system of equations.*

Lagrangian Polynomials

We will first look at one very straightforward approach—the Lagrangian polynomial. The Lagrangian polynomial is perhaps the simplest way to exhibit the existence of a polynomial for interpolation with unevenly spaced data. Data where the x-values are not equispaced often occur as the result of experimental observations or when historical data are examined.

Suppose we have a table of data with four pairs of x- and $f(x)$-values, with x_i indexed by variable i:

i	x	$f(x)$
0	x_0	f_0
1	x_1	f_1
2	x_2	f_2
3	x_3	f_3

Here we do not assume uniform spacing between the x-values, nor do we need the x-values arranged in a particular order. The x-values must all be distinct, however. Through these four data pairs we can pass a cubic. The Lagrangian form for this is

* For this example, the condition number is about 2700. If a quartic were fitted to all five points, it would be about 62,000!

$$P_3(x) = \frac{(x-x_1)(x-x_2)(x-x_3)}{(x_0-x_1)(x_0-x_2)(x_0-x_3)}f_0 + \frac{(x-x_0)(x-x_2)(x-x_3)}{(x_1-x_0)(x_1-x_2)(x_1-x_3)}f_1 + \frac{(x-x_0)(x-x_1)(x-x_3)}{(x_2-x_0)(x_2-x_1)(x_2-x_3)}f_2 + \frac{(x-x_0)(x-x_1)(x-x_2)}{(x_3-x_0)(x_3-x_1)(x_3-x_2)}f_3 \quad \textbf{(3.1)}$$

Note that this equation is made up of four terms, each of which is a cubic in x; hence the sum is a cubic. The pattern of each term is to form the numerator as a product of linear factors of the form $(x - x_i)$, omitting one x_i in each term, the omitted value being used to form the denominator by replacing x in each of the numerator factors. In each term, we multiply by the f_i corresponding to the x_i omitted in the numerator factors. The Lagrangian polynomial for other degrees of interpolating polynomials employs this same pattern of forming a sum of polynomials all of the desired degree; it will have $n + 1$ terms when the degree is n.

It is easy to see that the Lagrangian polynomial does in fact pass through each of the points used in its construction. For example, in the preceding equation for $P_3(x)$, let $x = x_2$. All terms but the third vanish because of a zero numerator, while the third term becomes just $(1) * f_2$. Hence, $P_3(x_2) = f_2$. Similarly, $P_3(x_i) = f_i$ for $i = 0, 1, 3$.

EXAMPLE 3.1 Fit a cubic through the first four points of the preceding table and use it to find the interpolated value for $x = 3.0$.

$$\begin{aligned} P_3(3.0) = {} & \frac{(3.0-2.7)(3.0-1.0)(3.0-4.8)}{(3.2-2.7)(3.2-1.0)(3.2-4.8)}(22.0) \\ & + \frac{(3.0-3.2)(3.0-1.0)(3.0-4.8)}{(2.7-3.2)(2.7-1.0)(2.7-4.8)}(17.8) \\ & + \frac{(3.0-3.2)(3.0-2.7)(3.0-4.8)}{(1.0-3.2)(1.0-2.7)(1.0-4.8)}(14.2) \\ & + \frac{(3.0-3.2)(3.0-2.7)(3.0-1.0)}{(4.8-3.2)(4.8-2.7)(4.8-1.0)}(38.3). \end{aligned}$$

Carrying out the arithmetic, $P_3(3.0) = 20.21$.

Observe that we get the same result as before. The arithmetic in this method is tedious, although hand calculators are convenient for this type of computation. Writing a computer program that implements the method is not hard to do. Both MATLAB and *Mathematica* can get interpolating polynomials of any degree (but high degrees are usually undesirable).

Using MATLAB

MATLAB gets interpolating polynomials readily. The cubic fitted to the first four points, in Example 3.1 is done by:

```
EDU>> x = [3.2 2.7 1.0 4.8]; y = [22.0 17.8 14.2 38.3];
EDU>> p = polyfit (x, y, 3)
P =
   -0.5275     6.4952     -16.1177     24.3499
```

which are the coefficients of the same cubic as before:

$$-0.5275x^3 + 6.4952x^2 - 16.1177x + 24.3499.$$

If we want the value of the polynomial at $x = 3.0$:

```
EDU>> xval = polyval(p, 3.0)
xval =
   20.2120
```

exactly what we got in Example 3.1.

Error of Interpolation

When we fit a polynomial $P_n(x)$ to some data points, it will pass exactly through those points, but between those points $P_n(x)$ will not be precisely the same as the function $f(x)$ that generated the points (unless the function is that polynomial). How much is $P_n(x)$ different from $f(x)$? How large is the error of $P_n(x)$?

We begin the development of an expression for the error of $P_n(x)$, an nth-degree interpolating polynomial, by writing the error function in a form that has the known property that it is zero at the $n + 1$ points, from x_0 through x_n, where $P_n(x)$ and $f(x)$ are the same. We call this function $E(x)$:

$$E(x) = f(x) - P_n(x) = (x - x_0)(x - x_1) \cdots (x - x_n)g(x).$$

The $n + 1$ linear factors give $E(x)$ the zeros we know it must have, and $g(x)$ accounts for its behavior at values other than at $x_0, x_1, \ldots, x_n$. Obviously, $f(x) - P_n(x) - E(x) = 0$, so

$$f(x) - P_n(x) - (x - x_0)(x - x_1) \cdots (x - x_n)g(x) = 0. \tag{3.2}$$

To determine $g(x)$, we now use the interesting mathematical device of constructing an auxiliary function (the reason for its special form becomes apparent as the development proceeds). We call this auxiliary function $W(t)$, and define it as

$$W(t) = f(t) - P_n(t) - (t - x_0)(t - x_1) \cdots (t - x_n)g(x).$$

Note in particular that x has *not* been replaced by t in the $g(x)$ portion. (W is really a function of both t and x, but we are only interested in variations of t.) We now examine the zeros of $W(t)$.

Certainly at $t = x_0, x_1, \ldots, x_n$, the W function is zero ($n + 1$ times), but it is also zero if $t = x$ by virtue of Eq. (3.2). There are then a total of $n + 2$ values of t that make $W(t) = 0$. We now impose the necessary requirements on $W(t)$ for the *law of mean value* to hold. $W(t)$ must be continuous and differentiable. If this is so, there is a zero to its derivative $W'(t)$ between each of the $n + 2$ zeros of $W(t)$, a total of $n + 1$ zeros. If $W''(t)$ exists, and we suppose it does, there will be n zeros of $W''(t)$, and likewise $n - 1$ zeros of $W'''(t)$, and so on, until we reach $W^{(n+1)}(t)$, which must have at least one zero in the interval that has x_0, x_n, or x as endpoints. Call this value of $t = \xi$. We then have

$$W^{(n+1)}(\xi) = 0 = \frac{d^{n+1}}{dt^{n+1}}[f(t) - P_n(t) - (t - x_0) \cdots (t - x_n)g(x)]_{t=\xi} \tag{3.3}$$
$$= f^{(n+1)}(\xi) - 0 - (n+1)!g(x).$$

The right-hand side of Eq. (3.3) occurs because of the following arguments. The $(n + 1)$st derivative $f(t)$, evaluated at $t = \xi$, is obvious. The $(n + 1)$st derivative of $P_n(t)$ is zero because every time any polynomial is differentiated its degree is reduced by one, so that the nth derivative is of degree zero (a constant) and its $(n + 1)$st derivative is zero. We apply the same argument to the $(n + 1)$st-degree polynomial in t that occurs in the last term—its $(n + 1)$st derivative is a constant that results from the t^{n+1} term and is $(n + 1)!$. Of course, $g(x)$ is independent of t and goes through the differentiations unchanged. The form of $g(x)$ is now apparent:

$$g(x) = \frac{f^{(n+1)}(\xi)}{(n+1)!}, \quad \xi \text{ between } (x_0, x_n, x).$$

The conditions on $W(t)$ that are required for this development (continuous and differentiable $n + 1$ times) will be met if $f(x)$ has these same properties, because $P_n(x)$ is continuous and differentiable. We now have our error term:

$$E(x) = (x - x_0)(x - x_1) \cdots (x - x_n) \frac{f^{(n+1)}(\xi)}{(n+1)!}, \tag{3.4}$$

with ξ on the smallest interval that contains $\{x, x_0, x_1, \ldots, x_n\}$.

The expression for error given in Eq. (3.4) is interesting but is not always extremely useful. This is because the actual function that generates the x_i, f_i values is often unknown; we obviously then do not know its $(n + 1)$st derivative. We can conclude, however, that if the function is "smooth," a low-degree polynomial should work satisfactorily. (The smaller the higher derivatives of a function, the smoother it is. For example, for a straight line, all derivatives above the first are zero.) On the other hand, a "rough" function can be expected to have larger errors when interpolated. We can also conclude that extrapolation (applying the interpolating polynomial outside the range of x-values employed to construct it) will have larger errors than for interpolation. It also follows that the error is smaller if x is centered within the x_i, because this makes the product of the $(x - x_i)$ terms smaller.

Here is an algorithm for interpolation with a Lagrangian polynomial of degree N.

An Algorithm for Interpolation from a Lagrange Polynomial

Given a set of $n + 1$ points $[(x_i, f_i), i = 0, \ldots, n]$ and a value for x at which the polynomial is to be evaluated:

```
Set Sum = 0.
For i = 0 To n Step 1 Do
  Set P = 1.
  For j = 0 to n step 1 Do
    If (j ≠ i) Then
      Set P = P * (x − x_j)/(x_i − x_j)
    End If.
   End Do j.
  Set Sum = Sum + P * f_i
End Do i.
```

Sum is the interpolated value at x.

A Word of Caution

It is most important that you never fit a polynomial of a degree higher than 4 or 5 to a set of points. If you need to fit to a set of more than six points, be sure to break up the set into subsets and fit separate polynomials to these. Figure 3.18, a part of Applied Problem 11, illustrates why this is so necessary. A still better way to fit a large number of data points is to use *spline curves,* as described in Section 3.3.

Recognize also that you cannot fit a function that is discontinuous or one whose derivative is discontinuous with a polynomial. This is because every polynomial is everywhere continuous and has continuous derivatives. A Fourier series that we discuss in Chapter 4 can approximate such functions.

Is Our Function a Cubic Polynomial?

In Example 3.1, we have fitted a cubic polynomial that matches the table exactly at four points. Is $f(x)$, a function whose form is not given, really a polynomial of degree-3? If it is, the error of the Lagrangian polynomial would be zero because the fourth derivative term in Eq. (3.4) would be zero. How can we tell? One way is to see if the fifth point (5.6, 51.7) is on the cubic. MATLAB says it is not:

```
EDU>> x2 = polyval (f, 5.6)
x2 =
  45.1473
```

and we see that $P_3(5.6)$ is not equal to $f(5.6) = 51.7$. Another way would be to use (5.6, 51.7) as one of four points with any three others to see if that interpolating polynomial is the same. Still another technique would be to plot the interpolating polynomial and see if the fifth point lies on the curve.

This discussion also points out that extrapolation with an interpolating polynomial incurs a larger error than does interpolation, a fact that can be observed from Eq. (3.2).

Neville's Method

The trouble with the standard Lagrangian polynomial technique is that we do not know which degree of polynomial to use. If the degree is too low, the interpolating polynomial does not give good estimates of $f(x)$. If the degree is too high, undesirable oscillations in polynomial values can occur. (More on this later, in the section on spline curves.)

Neville's method can overcome this difficulty. It essentially computes the interpolated value with polynomials of successively higher degree, stopping when the successive values are close together.* The successive approximations are actually computed by linear interpolation from the previous values. The Lagrange formula for linear interpolation to get $f(x)$ from two data pairs, (x_1, f_1) and (x_2, f_2), is

$$f(x) = \frac{(x - x_2)}{(x_1 - x_2)} f_1 + \frac{(x - x_1)}{(x_2 - x_1)} f_2,$$

which can be written more compactly as

$$f(x) = \frac{(x - x_2) * f_1 + (x_1 - x) * f_2}{x_1 - x_2}. \tag{3.5}$$

We will use Eq. (3.5) in Neville's method.

If we examine Eq. (3.4) for the error term of Lagrangian interpolation, we see that the smallest error results when we use data pairs where the x_i's are closest to the x-value we are interpolating. Neville's method begins by arranging the given data pairs, (x_i, f_i), so the successive values are in order of the closeness of the x_i to x.

EXAMPLE 3.2 Suppose we are given these data:

x	$f(x)$
10.1	0.17537
22.2	0.37784
32.0	0.52992
41.6	0.66393
50.5	0.63608

and we want to interpolate for $x = 27.5$. We first rearrange the data pairs in order of closeness to $x = 27.5$:

i	$\lvert x - x_i \rvert$	x_i	$f_i = P_{i0}$
0	4.5	32.0	0.52992
1	5.3	22.2	0.37784
2	14.1	41.6	0.66393
3	17.4	10.1	0.17537
4	23.0	50.5	0.63608

* Neville's method is not the most efficient method to compute an interpolated value. It is better to obtain the interpolating polynomial by the procedures of the next section and then evaluate it for the desired x-value.

Neville's method begins by renaming the f_i as P_{i0}. We build a table by first interpolating linearly between pairs of values for $i = 0, 1$, $i = 1, 2$, $i = 2, 3$, and so on. These values are written in a column to the right of the first P of each pair. The next column of the table is created by linearly interpolating from the previous column for $i = 0, 2$, $i = 1, 3$, $i = 2, 4$, and so on. The next column after this uses values for $i = 0, 3$, $i = 0, 4, \ldots$, and continues until we run out of data pairs.

Here is the Neville table for the preceding data:

i	x_i	P_{i0}	P_{i1}	P_{i2}	P_{i3}	P_{i4}
0	32.0	0.52992	0.46009	0.46200	0.46174	0.45754
1	22.2	0.37784	0.45600	0.46071	0.47901	
2	41.6	0.66393	0.44524	0.55843		
3	10.1	0.17537	0.37379			
4	50.5	0.63608				

The general formula for computing entries into the table is

$$P_{i,j} = \frac{(x - x_i) * P_{i+1,j-1} + (x_{i+j} - x) * P_{i,j-1}}{x_{i+j} - x_i}. \quad (3.6)$$

Thus, the values of P_{01} and P_{11} are computed by

$$P_{01} = \frac{(27.5 - 32.0) * 0.37784 + (22.2 - 27.5) * 0.52992}{22.2 - 32.0} = 0.46009,$$

$$P_{11} = \frac{(27.5 - 22.2) * 0.66393 + (41.6 - 27.5) * 0.37784}{41.6 - 22.2} = 0.45600.$$

Once we have the column of P_{i1}'s, we compute the next column. For example,

$$P_{22} = \frac{(27.5 - 41.6) * 0.37379 + (50.5 - 27.5) * 0.44524}{50.5 - 41.6} = 0.55843.$$

The remaining columns are computed similarly by using Eq. (3.6).

The top line of the table represents Lagrangian interpolates at $x = 27.5$ using polynomials of degree equal to the second subscript of the P's. Each of these polynomials uses the required number of data pairs, taking them as a set starting from the top of the table. (An exercise asks you to prove that the top line does represent Lagrangian interpolates with polynomials of increasing degree.)

The preceding data are for sines of angles in degrees and the correct value for $x = 27.5$ is 0.46175. Observe that the top line values get better and better until the last, when it diverges. This divergence becomes apparent when we notice that the successive values get closer to a constant value until the last one. (If the table is not arranged to center the x-value within the x_i, the convergence is not as quick.)

If we instead do this computation by hand, we can save computing time by computing, not the entire table, but only as much as required to get convergence to the desired number of decimal places. We therefore do only the computations needed to compute those top row values that are required. In a computer program it is hardly worth the added programming complications because the entire table is computed so quickly.

Parallel Processing

The several terms of a Lagrange polynomial, as shown in Eq. (3.1), can all be computed simultaneously with parallel processing. Each entry in the successive columns of the table for Neville's method can be computed simultaneously. An exercise asks you to determine the number of time steps that are saved.

3.2 Divided Differences

There are two disadvantages to using the Lagrangian polynomial or Neville's method for interpolation. First, it involves more arithmetic operations than does the divided-difference method we now discuss. Second, and more importantly, if we desire to add or subtract a point from the set used to construct the polynomial, we essentially have to start over in the computations. Both the Lagrangian polynomials and Neville's method also must repeat all of the arithmetic if we must interpolate at a new x-value. The divided-difference method avoids all of this computation.

Actually, we will not get a polynomial different from that obtained by Lagrange's technique. As we will show later on, every nth-degree polynomial that passes through the same $n + 1$ points is identical. Only the way that the polynomial is expressed is different.

Our treatment of divided-difference tables assumes that the function, $f(x)$, is known at several values for x:

x_0	f_0
x_1	f_1
x_2	f_2
x_3	f_3

We do not assume that the x's are evenly spaced or even that the values are arranged in any particular order (but some ordering may be advantageous).

Consider the nth-degree polynomial written in a special way:

$$\begin{aligned} P_n(x) = a_0 + (x - x_0)a_1 + (x - x_0)(x - x_1)a_2 + \cdots \\ + (x - x_0)(x - x_1) \cdots (x - x_{n-1})a_n. \end{aligned} \tag{3.7}$$

If we chose the a_i so that $P_n(x) = f(x)$ at the $n + 1$ known points, (x_i, f_i), $i = 0, \ldots, n$, then $P_n(x)$ is an interpolating polynomial. We will show that the a_i's are readily determined by using what are called the *divided differences of the tabulated values*.

A special standard notation for divided differences is

$$f[x_0, x_1] = \frac{f_1 - f_0}{x_1 - x_0},$$

called the *first divided difference between* x_0 *and* x_1. The function

$$f[x_1, x_2] = \frac{f_2 - f_1}{x_2 - x_1}$$

is the first divided difference between x_1 and x_2. (We use $f[x_0] = f_0 = f(x_0)$.)

In general,

$$f[x_s, x_t] = \frac{f_t - f_s}{x_t - x_s}$$

is the first divided difference between x_s and x_t. Observe that the order of the points is immaterial:

$$f[x_s, x_t] = \frac{f_t - f_s}{x_t - x_s} = \frac{f_s - f_t}{x_s - x_t} = f[x_t, x_s].$$

Second- and higher-order differences are defined in terms of lower-order differences. For example,

$$f[x_0, x_1, x_2] = \frac{f[x_1, x_2] - f[x_0, x_1]}{x_2 - x_0};$$

$$f[x_0, x_1, \ldots, x_n] = \frac{f[x_1, x_2, \ldots, x_n] - f[x_0, x_1, \ldots, x_{n-1}]}{x_n - x_0}$$

The concept is even extended to a zero-order difference:

$$f[x_s] = f_s.$$

Using the standard notation, a divided-difference table is shown in symbolic form in Table 3.1. Table 3.2 shows specific numerical values. (These data are the same as in the first table of Section 3.1.)

Table 3.1

x_i	f_i	$f[x_i, x_{i+1}]$	$f[x_i, x_{i+1}, x_{i+2}]$	$f[x_i, x_{i+1}, x_{i+2}, x_{i+3}]$
x_0	f_0	$f[x_0, x_1]$	$f[x_0, x_1, x_2]$	$f[x_0, x_1, x_2, x_3]$
x_1	f_1	$f[x_1, x_2]$	$f[x_1, x_2, x_3]$	$f[x_1, x_2, x_3, x_4]$
x_2	f_2	$f[x_2, x_3]$	$f[x_2, x_3, x_4]$	
x_3	f_3	$f[x_3, x_4]$		
x_4	f_4			

Table 3.2

x_i	f_i	$f[x_i, x_{i+1}]$	$f[x_i, \ldots, x_{i+2}]$	$f[x_i, \ldots, x_{i+3}]$	$f[x_i, \ldots, x_{i+4}]$
3.2	22.0	8.400	2.856	−0.528	0.256
2.7	17.8	2.118	2.012	0.0865	
1.0	14.2	6.342	2.263		
4.8	38.3	16.750			
5.6	51.7				

We are now ready to establish that the a_i of Eq. (3.7) are given by these divided differences. We write Eq. (3.7) with x set equal to $x_0, x_1, \ldots, x_n$ in succession, giving

$$\begin{aligned}
x = x_0: &\qquad P_n(x_0) = a_0,\\
x = x_1: &\qquad P_n(x_1) = a_0 + (x_1 - x_0)a_1,\\
x = x_2: &\qquad P_n(x_2) = a_0 + (x_2 - x_0)a_1 + (x_2 - x_0)(x_2 - x_1)a_2,\\
&\vdots\\
x = x_n: &\qquad P_n(x_n) = a_0 + (x_n - x_0)a_1 + (x_n - x_0)(x_n - x_1)a_2 + \cdots\\
&\qquad\qquad + (x_n - x_0) \ldots (x_n - x_{n-1})\, a_n.
\end{aligned}$$

If $P_n(x)$ is to be an interpolating polynomial, it must match the table for all $n + 1$ entries:

$$P_n(x_i) = f_i \qquad \text{for } i = 0, 1, 2, \ldots, n.$$

If the $P_n(x_i)$ in each equation is replaced by f_i, we get a triangular system, and each a_i can be computed in turn.

From the first equation,

$$a_0 = f_0 = f[x_0] \qquad \text{makes} \qquad P_n(x_0) = f_0.$$

If $a_1 = f[x_0, x_1]$, then

$$P_n(x_1) = f_0 + (x_1 - x_0)\,\frac{f_1 - f_0}{x_1 - x_0} = f_1.$$

If $a_2 = f[x_0, x_1, x_2]$, then

$$\begin{aligned}
P_n(x_2) &= f_0 + (x_2 - x_0)\,\frac{f_1 - f_0}{x_1 - x_0}\\
&\qquad + (x_2 - x_0)(x_2 - x_1)\,\frac{(f_2 - f_1)/(x_2 - x_1) - (f_1 - f_0)/(x_1 - x_0)}{x_2 - x_0}\\
&= f_2.
\end{aligned}$$

One can show in similar fashion that each $P_n(x_i)$ will equal f_i if $a_i = f[x_0, x_1, \ldots, x_i]$.

We then can write:

$$P_n(x) = f[x_0] + (x - x_0)f[x_0, x_1] + (x - x_0)(x - x_1)f[x_0 \ldots x_2]$$
$$+ (x - x_0)(x - x_1)(x - x_2)f[x_0 \ldots x_3] + \cdots \quad \textbf{(3.8)}$$
$$+ (x - x_0)(x - x_1) \ldots (x - x_{n-1})f[x_0 \ldots x_n].$$

EXAMPLE 3.3 Write the interpolating polynomial of degree-3 that fits the data of Table 3.2 at all points from $x_0 = 3.2$ to $x_3 = 4.8$.

$$P_3(x) = 22.0 + 8.400(x - 3.2) + 2.856(x - 3.2)(x - 2.7)$$
$$-0.528(x - 3.2)(x - 2.7)(x - 1.0).$$

What is the fourth-degree polynomial that fits at all five points? We only have to add one more term to $P_3(x)$:

$$P_4(x) = P_3(x) + 0.256(x - 3.2)(x - 2.7)(x - 1.0)(x - 4.8).$$

When this method is used for interpolation, we observe that nested multiplication can be used to cut down on the number of arithmetic operations, for example, for $x = 3$:

$$P_3(3) = \{[-0.528(3 - 1.0) + 2.856](3 - 2.7) + 8.400\}(3 - 3.2) + 22.0$$
$$= 20.2120.$$

If we compute the interpolated value at $x = 3.0$ for each of the third-degree polynomials in Section 3.1, we get the same result: $P_3(3.0) = 20.2120$. This is not surprising, because all third-degree polynomials that pass through the same four points are identical. They may look different but they can all be reduced to the same form.

An algorithm for constructing a divided-difference table is

An Algorithm for Interpolation from a Divided Difference Table

Given a set of $n + 1$ points $[(x_i, f_i), i = 0, \ldots, n]$ and a value $x = u$ at which the interpolating polynomial is to be evaluated:

We first find the coefficients of the interpolating polynomial. These are stored in vector dd.

For $i = 0$ To n Step 1 Do
    Set dd[i] = f[i]
End For i.
For $j = 1$ To n Step 1 Do
    Set temp1 = dd[$j - 1$].

```
    For k = j To n Step 1 Do
        Set temp2 = dd[k].
        Set dd[k] = (dd[k] − temp1)/(x[k] − x[k − j]).
        temp1 = temp2
    End For k.
End For j.
```

Now we compute the value of the polynomial at u. We do this by nested multiplication from the highest term.

```
Set sum = 0
For i = n DownTo 1 Step 1 Do
    Set sum = (sum + dd[i]) * (u − x[i − 1])
    Set sum = sum + dd[0]
End For i.
ddvalue = sum.
```

ddvalue is the value of the polynomial at u, $p_n(u)$.

Observe that parallel processing can compute all entries in the successive columns simultaneously. If there are $N + 1$ data pairs and a full table is constructed, the number of time steps equals the number of new columns, N. Sequential processing would require $N(N + 1)/2$ steps.

Divided Differences for a Polynomial

It is of interest to look at the divided differences for $f(x) = P_n(x)$. Suppose that $f(x)$ is the cubic

$$f(x) = 2x^3 - x^2 + x - 1.$$

Here is its divided-difference table:

x_i	$f[x_i]$	$f[x_i, x_{i+1}]$	$f[x_i \ldots x_{i+2}]$	$f[x_i \ldots x_{i+3}]$	$f[x_i \ldots x_{i+4}]$	$f[x_i \ldots x_{i+5}]$
0.30	−0.7360	2.4800	3.0000	2.0000	0.0000	0.0000
1.00	1.0000	3.6800	3.6000	2.0000	0.0000	
0.70	−0.1040	2.2400	5.4000	2.0000		
0.60	−0.3280	8.7200	8.2000			
1.90	11.0080	21.0200				
2.10	15.2120					

Observe that the third divided differences are all the same. (It then follows that all higher divided differences will be zero.) We can take advantage of this fact by not using differences beyond the column where the values are essentially constant, because this indicates that the function behaves nearly like a polynomial of that degree.

It is most important to also observe that the third derivative of a cubic polynomial is also a constant. (In this instance, $P^{(3)}(x) = 2 * 3! = 12$.) The relationship between divided differences and derivatives will be explored in detail in Chapter 5. For now, we just state that for an nth-degree polynomial, $P_n(x)$, whose highest-power term has the coefficient a_n, the nth divided differences will always be equal to a_n. Because the nth derivative of this polynomial is equal to $a_n * n!$, the relationship between derivatives and divided differences seems to involve $n!$. We exploit this later.

Identical Polynomials

The interpolating polynomials obtained by the Lagrangian method and through divided differences look different but they are really identical. We will explore other methods for constructing polynomials. It is important to recognize that every polynomial of degree n that has the same value at $n + 1$ distinct points is exactly the same.

When a polynomial of the nth degree is developed from $n + 1$ data points, we have exactly enough data to determine the $n + 1$ coefficients, so the conclusion that any resulting polynomial is the same is intuitively true. Further, every expression of the polynomial can be reduced to standard form and this must always be identical.

A more formal and compelling proof is by contradiction:

> Suppose there are two different polynomials of degree n that agree at $n + 1$ distinct points. Call these $P_n(x)$ and $Q_n(x)$, and write their difference:
>
> $$D(x) = P_n(x) - Q_n(x),$$
>
> where $D(x)$ is a polynomial of at most degree n. But because P and Q match at the $n + 1$ points, their difference $D(x)$ is equal to zero at all $n + 1$ of these x-values; that is, $D(x)$ is a polynomial of degree n at most but has $n + 1$ distinct zeros. However, this is impossible unless $D(x)$ is identically zero. Hence $P_n(x)$ and $Q_n(x)$ are not different—they must be the same polynomial.

A most important consequence of this uniqueness property of interpolating polynomials is that their error terms are also identical (though we may want to express the error terms in different forms). We only have to derive the error term for one form of interpolating polynomial to have the error term for all forms of interpolating polynomials.

Error of Interpolation from Differences

The error term for an interpolating polynomial derived from a divided-difference table is identical to that for the equivalent Lagrangian polynomial because, as we have just observed, all polynomials of degree n that match at $n + 1$ points are identical. That means that the error term associated with the nth-degree polynomial $P_n(x)$ of Eq. (3.8) is simply Eq. (3.4), which we repeat here:

$$E(x) = (x - x_0)(x - x_1) \cdots (x - x_n)\frac{f^{(n+1)}(\xi)}{(n + 1)!}.$$

It is still not convenient to use this error expression, because the derivative of f that appears is unknown. However, if $f(x)$ is almost the same as some polynomial of degree n (and we will know that this is true because the nth divided differences will be almost constant), interpolating with an nth-degree polynomial should be nearly exact. The reason is that the $(n + 1)$st derivative of $f(x)$ will be nearly zero and the error of the nth-degree interpolating polynomial will be very small.

What if we use a lower-degree polynomial? The error should be larger. If $f(x)$ is a known function, we can use Eq. (3.4) to bound the error. Here is an example.

EXAMPLE 3.4 Here is a divided difference table for $f(x) = x^2e^{-x/2}$:

x_i	$f[x_i]$	$f[x_i, x_{i+1}]$	$f[x_i \ldots x_{i+2}]$	$f[x_i \cdots x_{i+3}]$	$f[x_i \cdots x_{i+4}]$
1.10	0.6981	0.8593	−0.1755	0.0032	0.0027
2.00	1.4715	0.4381	−0.1631	0.0191	
3.50	2.1287	−0.0511	−0.0657		
5.00	2.0521	−0.2877			
7.10	1.4480				

Find the error of the interpolates for $f(1.75)$ using polynomials of degrees-1, -2, and -3.

The results are shown in Table 3.3, for which Eq. (3.8) was used to do the interpolations. (MATLAB helped in finding the derivatives and evaluating the maximum and minimum values within the intervals.) The error formula does bracket the actual errors, as expected. In this case, observe that the use of a cubic polynomial does not improve the accuracy. In part, this is because we do not have the x-value well centered within the tabulated values; also, the value of the derivative is not decreasing. ■

Table 3.3 Errors of interpolation for $f(1.75)$

Degree	Interpolated value	Actual error	$f^{(n+1)}$ maximum	$f^{(n+1)}$ minimum	Upper bound	Lower bound
1	1.25668	0.01996	−0.3679	0.0594	0.0299	−0.00483
2	1.28520	−0.00856	−0.8661	0.1249	0.0059	−0.0408
3	1.28611	−0.00947	1.1398	−0.0359	0.0014	−0.0439

Error Estimation When $f(x)$ Is Unknown—The Next-Term Rule

Occasionally, almost always when dealing with experimental data, the function is unknown. Still, there is a way to estimate the error of the interpolation. This is because the nth-order divided difference is itself an approximation for $f^{(n)}(x)/n!$, as will be demonstrated in Chapter 5. What this means is that the error of the interpolation is given approximately by the value of the next term that would be added!

This most valuable rule for estimating the error of interpolation we call the *next-term rule.* It is easy to state and to use:

> $E_n(x)$ = (approximately) the value of the next term that would be added to $P_n(x)$.

Here is how it works for the preceding example:

Degree	Exact error	Estimate from next-term rule
1	0.01996	0.02852
2	0.00856	0.00091
3	−0.00947	−0.00249

As you can see, the agreement is at least fair.

Interpolation Near the End of a Table

Thus far, we have assumed that the entries are indexed from the top to the bottom of the table. This would appear to indicate that our formulas do not work well for constructing polynomials from divided differences at the end of the table. Remember, however, that the ordering of the points is immaterial. We can just as well begin at the bottom and number the entries going upward, with no adjustment of Eq. (3.8) required. The table is really not changed at all, just the symbols that we use. We now use Eq. (3.8) with the newly indexed values.

Table 3.4a Conventional divided-difference table

x_0	$f[x_0]$	$f[x_0, x_1]$	$f[x_0 \ldots x_2]$	$f[x_0 \ldots x_3]$	$f[x_0 \ldots x_4]$
x_1	$f[x_1]$	$f[x_1, x_2]$	$f[x_1 \ldots x_3]$	$f[x_1 \ldots x_4]$	
x_2	$f[x_2]$	$f[x_2, x_3]$	$f[x_2 \ldots x_4]$		
x_3	$f[x_3]$	$f[x_3, x_4]$			
x_4	$f[x_4]$				

Table 3.4b Divided-difference table indexed upwardly

x_4	$f[x_4]$				
x_3	$f[x_3]$	$f[x_3, x_4]$			
x_2	$f[x_2]$	$f[x_2, x_3]$	$f[x_2 \ldots x_4]$		
x_1	$f[x_1]$	$f[x_1, x_2]$	$f[x_1 \ldots x_3]$	$f[x_1 \ldots x_4]$	
x_0	$f[x_0]$	$f[x_0, x_1]$	$f[x_0 \ldots x_2]$	$f[x_0 \ldots x_3]$	$f[x_0 \ldots x_4]$

Tables 3.4a and b compare the two different numbering schemes. The entries in the rows of Table 3.4b are exactly the same numbers as in the upward diagonals of Table 3.4a.

Evenly Spaced Data

If the x-values are evenly spaced, getting an interpolating polynomial is considerably simplified. Instead of using divided differences, "ordinary differences" are used; the differences in f-values are not divided by the differences in x-values. A delta symbol is used to write them and, for a table of $N + 1$ $(x, f(x))$ pairs, differences up the Nth order can be computed.

We suppose that the table has entries indexed from 0 to N. First-order differences are then written as Δf_i and are computed as $\Delta f_i = f_{i+1} - f_i$, $i = 0, \ldots, (N - 1)$. Second-order differences, $\Delta^2 f_i$, are the differences of the first-order differences: $\Delta^2 f_i = \Delta(\Delta f_{i+1} - \Delta f_i)$, which is easily shown to be $\Delta^2 f_i = f_{i+2} - 2f_{i+1} + f_i$, $i = 0, \ldots, (N - 2)$. Higher-order differences are again the differences of the next lower-order differences. They can be computed from the original f-values:

$$\Delta^n f_i = f_{i+n} - n f_{i+n-1} + \frac{n(n-1)}{2!} f_{i+n-2} - \ldots \pm f_i, \qquad i = 0, \ldots, (N - n).$$

Observe that the coefficients are the familiar binomial coefficients.

An interpolating polynomial of degree n can be written in terms of these ordinary differences, with x evaluated at x_s:

$$P_n(x_s) = f_0 + s\Delta f_0 + \frac{s(s-1)}{2!}\Delta^2 f_0 + \frac{s(s-1)(s-2)}{3!}\Delta^3 f_0 + \ldots$$
$$+ \frac{s(s-1)\ldots(s-n+1)}{n!}\Delta^n f_0,$$

where $s = (x - x_0)/h$, with $h = \Delta x$, the uniform spacing in x-values. Observe again that the coefficients are the familiar binomial coefficients.

This form of the interpolating polynomial is called the Newton–Gregory forward polynomial. We will use this type of interpolating polynomial several times in later chapters. Several other forms of interpolating polynomials can be written in terms of the differences of the table. We do not pursue this topic further because the divided difference formulas apply to evenly spaced data, although earlier editions of this text go into considerable detail.

The next-term rule applies to this Newton–Gregory polynomial: The error of interpolation is approximated by the next term that would be added. Here is an example.

Given this table of x, $f(x)$ values, and the columns of differences, find $f(0.73)$ from a cubic interpolating polynomial.

x	$f(x)$	Δf	$\Delta^2 f$	$\Delta^3 f$	$\Delta^4 f$
0.0	0.000	0.203	0.017	0.024	0.020
0.2	0.203	0.220	0.041	0.044	0.052
0.4	0.423	0.261	0.085	0.096	0.211
0.6	0.684	0.346	0.181	0.307	
0.8	1.030	0.527	0.488		
1.0	1.557	1.015			
1.2	2.572				

In order to center the x-values around $x = 0.73$, we must use the four entries beginning with $x = 0.4$. That makes $x_0 = 0.4$ and $s = (0.73 - 0.4)/0.2 = 1.65$. Inserting the proper values into the expression for the Newton–Gregory polynomial, we get

$$f(0.73) = 0.423 + (1.65)(0.261) + \frac{(1.65)(0.65)}{2!}(0.085) + \frac{(1.65)(0.65)(-0.35)}{3!}(0.096)$$

$$= 0.423 + 0.4306 + 0.0456 - 0.0060 = 0.893.$$

The function is actually for $f(x) = \tan(x)$, so we know that the true value of $f(0.73)$ is 0.895; the error is 0.002. The next-term rule estimates the error as 0.004. This estimate is very good.

One nice feature of a table of ordinary differences is that an error in an entry for $f(x)$ can be readily detected. Such an error causes a disruption to the regular progression of values in the columns of differences. For example, if the entry for $x = 0.6$ has two digits reversed (0.648 rather than 0.684) and the table is recomputed, the columns for $\Delta^2 f$ and $\Delta^3 f$ lose their regularity.

MATLAB's 'diff' command gets the differences between the elements of a vector, so the columns of the above table are generated readily. We exhibit these as rows to save space:

```
EDU>> f = [0.0 0.203 0.423 0.684 1.030 1.557 2.572];
EDU>> diff(f)
  ans =
    0.2030  0.2200  0.2610  0.3460  0.5270  1.0150
EDU>> diff(ans)
  ans =
    0.0170  0.0410  0.0850  0.1810  0.4880
EDU>> diff(ans)
  ans =
    0.0240  0.0440  0.0960  0.3070
EDU>> diff(ans)
  ans =
    0.0200  0.0520  0.2110
```

Table 3.5a Table of function differences for $f(x) = 2x^3$, $h = 0.5$

x_i	f_i	Δf_i	$\Delta^2 f_i$	$\Delta^3 f_i$	$\Delta^4 f_i$	$\Delta^5 f_i$
0.00	0.00	0.25	1.50	1.50	0.00	0.00
0.50	0.25	1.75	3.00	1.50	0.00	0.00
1.00	2.00	4.75	4.50	1.50	0.00	
1.50	6.75	9.25	6.00	1.50		
2.00	16.00	15.25	7.50			
2.50	31.25	22.75				
3.00	54.00					

Function Differences Versus Divided Differences

Obviously, the table of function differences that we have been discussing is closely related to the table of divided differences. Except for dividing function differences by a difference of x-values in the latter, these two tables are the same when the x-values are evenly spaced. To make this crystal clear, compare the tables for the simple case of $f(x) = 2x^3$ with $h = 0.5$, as shown in Tables 3.5a and b.

As expected, the columns of third differences are constant in both tables. For divided differences, this constant is equal to just 2, the coefficient of x^3. For the difference table, it is equal to that coefficient times $(3!)(h^3)$, or $2 * 6 * 0.5^3 = 1.5$.

For first differences, the divided differences are equal to the function differences divided by h (0.5 here). Second divided differences are equal to second function differences divided by $(h)(2h)$ (0.5 in this example). Third divided differences are equal to third function differences divided by $(h)(2h)(3h)$ (0.75 in this instance). The pattern should now be clear:

$$f[x_i \ldots x_{i+n}] = \frac{\Delta^n f_i}{n!h^n}.$$

Table 3.5b Table of divided differences for $f(x) = 2x^3$, $h = 0.5$

x_i	$f[x_i]$	$f[x_i, x_{i+1}]$	$f[x_i \ldots x_{i+2}]$	$f[x_i \ldots x_{i+3}]$	$f[x_i \ldots x_{i+4}]$	$f[x_i \ldots x_{i+5}]$
0.00	0.00	0.50	3.00	2.00	0.00	0.00
0.50	0.25	3.50	6.00	2.00	0.00	
1.00	2.00	9.50	9.00	2.00		
1.50	6.75	18.50	12.00	2.00		
2.00	16.00	30.50	15.00			
2.50	31.25	45.50				
3.00	54.00					

If the values are not evenly spaced, a comparison is impossible because the table of function differences is not defined.

This difference between the two kinds of tables has a great effect on the relation between differences and derivatives, a topic that we explore in Chapter 5.

3.3 Spline Curves

There are times when fitting an interpolating polynomial to data points is very difficult. Here is an example where we try to fit to data pairs from a known function. Figure 3.1a is a plot of $f(x) = \cos^{10}(x)$ on the interval $[-2, 2]$. It is a nice, smooth curve but has a pronounced maximum at $x = 0$ and is near to the x-axis for $|x| > 1$. The curves of Figure 3.1b, c, d, and e are for polynomials of degrees-2, -4, -6, and -8 that match the function at evenly spaced points. None of the polynomials is a good representation of the function. In particular, observe how the eighth-degree polynomial deviates widely near $|x| = 2$. Polynomials

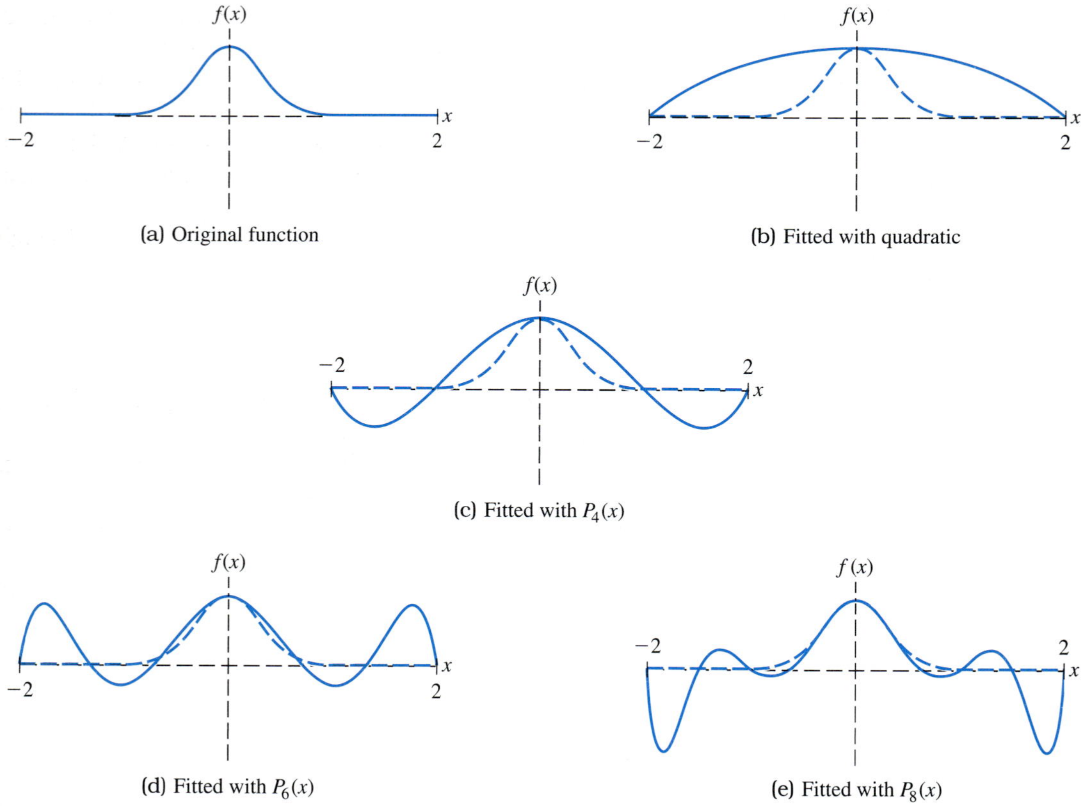

(a) Original function

(b) Fitted with quadratic

(c) Fitted with $P_4(x)$

(d) Fitted with $P_6(x)$

(e) Fitted with $P_8(x)$

Figure 3.1

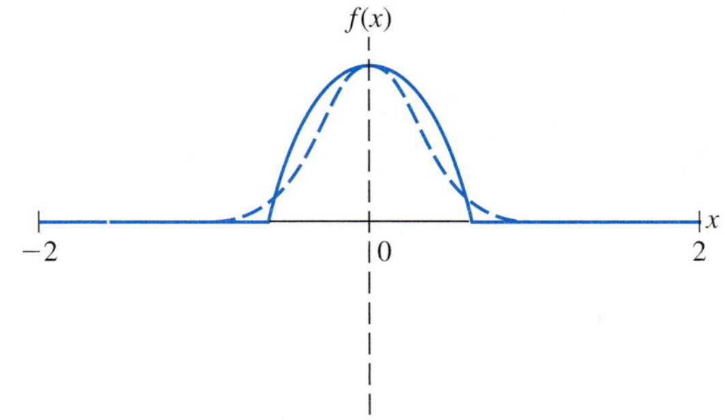

Figure 3.2

of degree higher than 8 will exhibit even greater deviations because, when we try to match $f(x)$ where it is flat, the polynomials must have many zeros.

One might think that a solution to the problem would be to break up the interval $[-2, 2]$ into subintervals and fit separate polynomials to the function in these smaller intervals. Figure 3.2 shows a much better fit if we use a quadratic between $x = -0.65$ and $x = 0.65$, with $P(x) = 0$ outside that interval. That is better (and one could further improve the fit), but there are discontinuities in the slope where the separate polynomials join.

An answer to the dilemma is to use *spline curves.* It borrows from the idea of a device used in drafting. A draftsman fits curves such as our example by bending a flexible rod to conform to the curve; the rod is held in place by placing weights on it. This device is better than using a French curve, for how the French curve is moved to conform is very subjective and is not effective where the curvature is great.

Spline curves may be of varying degrees. Suppose that we have a set of $n + 1$ points (which do not have to be evenly spaced):

$$(x_i, y_i), \qquad \text{with } i = 0, 1, 2, \ldots, n.$$

A spline fits a set of nth-degree polynomials, $g_i(x)$, between each pair of points, from x_i to x_{i+1}. The points at which the splines join are called knots.

If the polynomials are all of degree-1, we have a *linear spline* and the "curve" would appear as in the accompanying figure. As you can see, the slopes are discontinuous where the segments join. Splines of degree greater than 1 do not have this problem. Most often cubic splines are used.

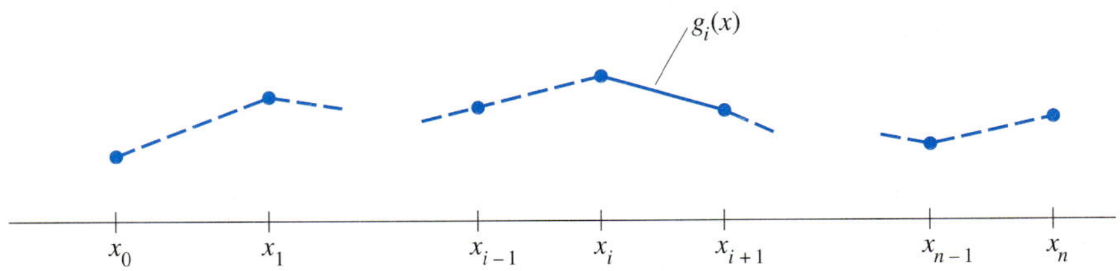

The Equation for a Cubic Spline

The drafting spline, from which the concept of spline curves is taken, bends according to the laws of beam flexure, so both the slope and curvature are everywhere continuous. Our mathematical spline curve must have this same behavior, requiring that they be of at least degree-3.

We will create a succession of cubic splines over successive intervals of the data. (There is no requirement that the points be evenly spaced.) Each spline must join with its neighboring cubic polynomials at the knots where they join with the same slope and curvature. (The end splines have only one neighbor, so their slope and curvature is not so constrained. This factor will be covered later.)

We write the equation for a cubic polynomial, $g_i(x)$, in the ith interval, between points (x_i, y_i), (x_{i+1}, y_{i+1}). It looks like the solid curve shown here. The dashed curves are other cubic spline polynomials. It has this equation:

$$g_i(x) = a_i(x - x_i)^3 + b_i(x - x_i)^2 + c_i(x - x_i) + d_i. \tag{3.9}$$

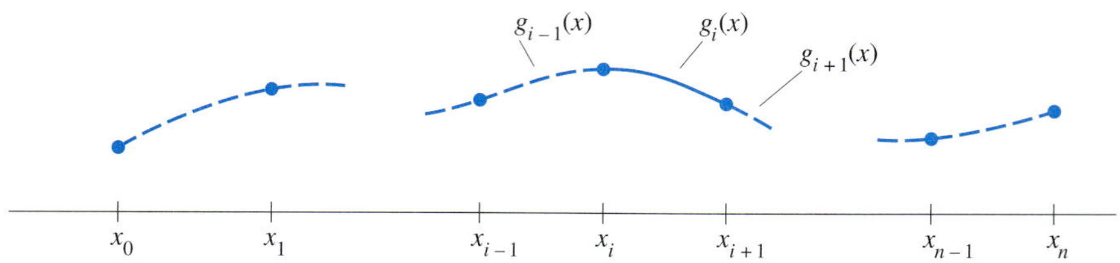

Thus, the cubic spline function we want is of the form

$$g(x) = g_i(x) \text{ on the interval } [x_i, x_{i+1}], \quad \text{for } i = 0, 1, \ldots, n-1$$

and meets these conditions:

$$g_i(x_i) = y_i, \quad i = 0, 1, \ldots, n-1 \quad \text{and} \quad g_{n-1}(x_n) = y_n; \tag{3.10a}$$

$$g_i(x_{i+1}) = g_{i+1}(x_{i+1}), \quad i = 0, 1, \ldots, n-2; \tag{3.10b}$$

$$g_i'(x_{i+1}) = g_{i+1}'(x_{i+1}), \quad i = 0, 1, \ldots, n-2; \tag{3.10c}$$

$$g_i''(x_{i+1}) = g_{i+1}''(x_{i+1}), \quad i = 0, 1, \ldots, n-2. \tag{3.10d}$$

[Equation (3.10) says that the cubic spline fits to each of the points (3.10a), is continuous (3.10b), and is continuous in slope and curvature (3.10c) and (3.10d), throughout the region spanned by the points.]

If there are $n + 1$ points, the number of intervals and the number of $g_i(x)$'s are n. Thus, there are four times n unknowns, which are the $\{a_i, b_i, c_i, d_i\}$ for $i = 0, 1, \ldots, n-1$. Equation (3.10a) immediately gives

$$d_i = y_i, \quad i = 0, 1, \ldots, n-1. \tag{3.11}$$

Equation (3.10 b) then gives

$$y_{i+1} = g_{i+1}(x_{i+1}) = g_i(x_{i+1}) = a_i(x - x_{i+1})^3 + b_i(x - x_{i+1})^2 + c_i(x - x_{i+1}) + y_i$$
$$= a_ih_i^3 + b_ih_i^2 + c_ih_i + y_i, \qquad i = 0, 1, \ldots, n - 1. \tag{3.12}$$

[In the last part of Eq (3.12), we used $h_i = (x_{i+1} - x_i)$, the width of the ith interval.]

To relate the slopes and curvatures of the joining splines, we differentiate Eq. (3.9):

$$g_i'(x) = 3a_i(x - x_i)^2 + 2b_i(x - x_i) + c_i, \tag{3.13}$$
$$g_i''(x) = 6a_i(x - x_i) + 2b_i, \qquad \text{for } i = 0, 1, \ldots, n - 1. \tag{3.14}$$

Observe that the second derivative of a cubic is linear, so $g''(x)$ is linear within $[x_i, x_{i+1}]$.

The development is simplified if we write the equations in terms of the second derivative—that is, if we let $S_i = g''(x_i)$ for $i = 0, 1, \ldots, n - 1$ and $S_n = g''_{n-1}(x_n)$.

From Eqs. (3.10d) and (3.14), we have

$$S_i = 6a_i(x_i - x_i) + 2b_i$$
$$= 2b_i;$$
$$S_{i+1} = 6a_i(x_{i+1} - x_i) + 2b_i$$
$$= 6a_ih_i + 2b_i.$$

Hence we can write

$$b_i = \frac{S_i}{2}, \tag{3.15}$$

$$a_i = \frac{S_{i+1} - S_i}{6h_i}. \tag{3.16}$$

We substitute the relations for a_i, b_i, d_i given by Eqs. (3.11), (3.15) and (3.16) into Eq. (3.9) and then solve for c_i:

$$y_{i+1} = \left(\frac{S_{i+1} - S_i}{6h_i}\right)h_i^3 + \frac{S_i}{2}h_i^2 + c_ih_i + y_i;$$

$$c_i = \frac{y_{i+1} - y_i}{h_i} - \frac{2h_iS_i + h_iS_{i+1}}{6}.$$

We now invoke the condition that the slopes of the two cubics that join at (x_i, y_i) are the same. For the equation in the ith interval, Eq. (3.10c) becomes, with $x = x_i$,

$$y_i' = 3a_i(x_i - x_i)^2 + 2b_i(x_i - x_i) + c_i = c_i.$$

In the previous interval, from x_{i-1} to x_i, the slope at its right end will be

$$y_i' = 3a_{i-1}(x_i - x_{i-1})^2 + 2b_{i-1}(x_i - x_{i-1}) + c_{i-1}$$
$$= 3a_{i-1}h_{i-1}^2 + 2b_{i-1}h_{i-1} + c_{i-1}.$$

Equating these, and substituting for a, b, c, d their relationships in terms of S and y, we get

$$y_i' = \frac{y_{i+1} - y_i}{h_i} - \frac{2h_iS_i + h_iS_{i+1}}{6}$$
$$= 3\left(\frac{S_i - S_{i-1}}{6h_{i-1}}\right)h_{i-1}^2 + 2\left(\frac{S_{i-1}}{2}\right)h_{i-1} + \frac{y_i - y_{i-1}}{h_{i-1}} - \frac{2h_{i-1}S_{i-1} + h_{i-1}S_i}{6}.$$

Simplifying this equation, we get

$$h_{i-1}S_{i-1} + (2h_{i-1} + 2h_i)S_i + h_iS_{i+1} = 6\left(\frac{y_{i+1} - y_i}{h_i} - \frac{y_i - y_{i-1}}{h_{i-1}}\right)$$
$$= 6(f[x_i, x_{i+1}] - f[x_{i-1}, x_i]). \qquad \textbf{(3.17)}$$

The last part of Eq. (3.17) involves divided differences.

Equation (3.17) applies at each internal point, from $i = 1$ to $i = n - 1$, there being $n + 1$ total points. This gives $n - 1$ equations relating the $n + 1$ values of S_i. We get two additional equations involving S_0 and S_n when we specify conditions pertaining to the end intervals of the whole curve. To some extent, these end conditions are arbitrary. Four* alternative choices are often used: Observe that the fourth end condition is "not a knot condition."

1. Take $S_0 = 0$ and $S_n = 0$. This makes the end cubics approach linearity at their extremities. This condition, called a *natural spline,* matches precisely to the drafting device. This technique is used very frequently.
2. Another often used condition is to force the slopes at each end to assume specified values. When that information is not known, the slope might be estimated from the points. If $f'(x_0) = A$ and $f'(x_n) = B$, we use these relations (note that divided differences are employed):

   At left end: $2h_0S_0 + h_0S_1 = 6(f[x_0, x_1] - A)$.

   At right end: $h_{n-1}S_{n-1} + 2h_{n-1}S_n = 6(B - f[x_{n-1}, x_n])$.
3. Take $S_0 = S_1$, $S_n = S_{n-1}$. This is equivalent to assuming that the end cubics approach parabolas at their extremities.
4. Take S_0 as a linear extrapolation from S_1 and S_2, and S_n as a linear extrapolation from S_{n-1} and S_{n-2}. Only this condition gives cubic spline curves that match exactly to $f(x)$ when $f(x)$ is itself a cubic. For condition 4, we use these relations:

*A fifth condition is sometimes encountered—a function is periodic and the data cover a full period. In this case, $S_0 = S_n$ and the slopes are also the same at the first and last points.

At left end: $\dfrac{S_1 - S_0}{h_0} = \dfrac{S_2 - S_1}{h_1}, \qquad S_0 = \dfrac{(h_0 + h_1)S_1 - h_0 S_2}{h_1}.$

At right end: $\dfrac{S_n - S_{n-1}}{h_{n-1}} = \dfrac{S_{n-1} - S_{n-2}}{h_{n-2}},$ **(3.18)**

$$S_n = \frac{(h_{n-2} + h_{n-1})S_{n-1} - h_{n-1}S_{n-2}}{h_{n-2}}.$$

This is called "not a knot condition."

Relation 1, where $S_0 = 0$ and $S_n = 0$, is called a *natural spline*. It is often felt that this flattens the curve too much at the ends; in spite of this, it is frequently used. Relation 4 frequently suffers from the other extreme, giving too much curvature in the end intervals. Probably the best end condition to use is condition 2, provided reasonable estimates of the derivative are available.

If we write the equation of $S_1, S_2, \ldots, S_{n-1}$ [Eq. (3.17)] in matrix form, we get

$$\begin{bmatrix} h_0 & 2(h_0 + h_1) & h_1 & & & & \\ & h_1 & 2(h_1 + h_2) & h_2 & & & \\ & & h_2 & 2(h_2 + h_3) & h_3 & & \\ & & & \ddots & & & \\ & & & & h_{n-2} & 2(h_{n-2} + h_{n-1}) & h_{n-1} \end{bmatrix} \begin{bmatrix} S_0 \\ S_1 \\ S_2 \\ S_3 \\ \vdots \\ S_{n-1} \\ S_n \end{bmatrix}$$

$$= 6 \begin{bmatrix} f[x_1, x_2] - f[x_0, x_1] \\ f[x_2, x_3] - f[x_1, x_2] \\ f[x_3, x_4] - f[x_2, x_3] \\ \vdots \\ f[x_{n-1}, x_n] - f[x_{n-2}, x_{n-1}] \end{bmatrix}.$$

In this matrix array there are only $n - 1$ equations, but $n + 1$ unknowns. We can eliminate two unknowns (S_0 and S_n) using the relations that correspond to the end-condition assumptions. In the first three cases, this reduces the S vector to $n - 1$ elements, and the coefficient matrix becomes square, of size $(n - 1 \times n - 1)$. Furthermore, the matrix is always tridiagonal (even in case 4), and hence is solved speedily and can be stored economically.

For each end condition, the coefficient matrices become

Condition 1 $\quad S_0 = 0, S_n = 0$:

$$\begin{bmatrix} 2(h_0 + h_1) & h_1 & & & \\ h_1 & 2(h_1 + h_2) & h_2 & & \\ & h_2 & 2(h_2 + h_3) & h_3 & \\ & & & \ddots & \\ & & & h_{n-2} & 2(h_{n-2} + h_{n-1}) \end{bmatrix}.$$

Condition 2 $\quad f'(x_0) = A$ and $f'(x_n) = B$:

$$\begin{bmatrix} 2h_0 & h_0 & & & \\ h_0 & 2(h_0+h_1) & h_1 & & \\ & h_1 & 2(h_1+h_2) & h_2 & \\ & & & \ddots & \\ & & & h_{n-1} & 2h_{n-1} \end{bmatrix}.$$

Condition 3 $\quad S_0 = S_1,\ S_n = S_{n-1}$:

$$\begin{bmatrix} (3h_0+2h_1) & h_1 & & & \\ h_1 & 2(h_1+h_2) & h_2 & & \\ & h_2 & 2(h_2+h_3) & h_3 & \\ & & & \ddots & \\ & & & h_{n-2} & (2h_{n-2}+3h_{n-1}) \end{bmatrix}.$$

Condition 4 $\quad S_0$ and S_n are linear extrapolations:

$$\begin{bmatrix} \dfrac{(h_0+h_1)(h_0+2h_1)}{h_1} & \dfrac{h_1^2-h_0^2}{h_1} & & & \\ h_1 & 2(h_1+h_2) & h_2 & & \\ & h_2 & 2(h_2+h_3) & h_3 & \\ & & & \ddots & \\ & & & \dfrac{h_{n-2}^2-h_{n-1}^2}{h_{n-2}} & \dfrac{(h_{n-1}+h_{n-1})(h_{n-1}+2h_{n-2})}{h_{n-2}} \end{bmatrix}.$$

With condition after 4, solving the set of equations, we must compute S_0 and S_n using Eq. (3.18). For conditions 1, 2, and 3, no computations are needed. For each of the first three cases, the right-hand-side vector is the same; it is given in Eq. (3.17). If the data are evenly spaced, the matrices reduce to a simple form.

After the S_i values are obtained, we get the coefficients a_i, b_i, c_i, and d_i for the cubics in each interval. From these we can compute points on the interpolating curve.

$$a_i = \frac{S_{i+1} - S_i}{6h_i};$$

$$b_i = \frac{S_i}{2};$$

$$c_i = \frac{y_{i+1} - y_i}{h_i} - \frac{2h_iS_i + h_iS_{i+1}}{6};$$

$$d_i = y_i.$$

EXAMPLE 3.5 Fit the data of Table 3.6 with a natural cubic spline curve, and evaluate the spline values $g(0.66)$ and $g(1.75)$. [The true relation is $f(x) = 2e^x - x^2$.] We see that $h_0 = 1.0$, $h_1 = 0.5$, and $h_2 = 0.75$. The divided differences that we can use to get the right-hand sides of our equations are $f[0, 1] = 2.4366$, $f[1, 1.5] = 4.5536$, and $f[1.5, 2.25] = 9.5995$.

Table 3.6

x	$f(x)$
0.0	2.0000
1.0	4.4366
1.5	6.7134
2.25	13.9130

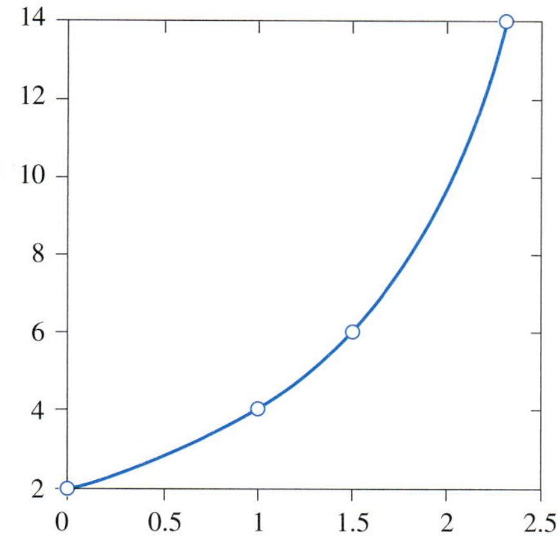

Figure 3.3

For a natural cubic spline, we use end condition 1 and solve

$$\begin{bmatrix} 3.0 & 0.5 \\ 0.5 & 2.5 \end{bmatrix} \begin{bmatrix} S_1 \\ S_2 \end{bmatrix} = \begin{bmatrix} 12.7020 \\ 30.2754 \end{bmatrix},$$

giving $S_1 = 2.2920$ and $S_2 = 11.6518$. ($S_0 = S_3 = 0$, of course.) Using these S's, we compute the coefficients of the individual cubic splines to arrive at

i	Interval	$g_i(x)$
0	[0.0, 1.0]	$0.3820(x-0)^3 + 0(x-0)^2 + 2.0546(x-0) + 2.0000$
1	[1.0, 1.5]	$3.1199(x-1)^3 + 1.146(x-1)^2 + 3.2005(x-1) + 4.4366$
2	[1.5, 2.25]	$-2.5893(x-1.5)^3 + 5.8259(x-1.5)^2 + 6.6866(x-1.5) + 6.7134$

Figure 3.3 shows the cubic spline curve. (You should verify that these equations satisfy all the conditions that were given for cubic spline curves.)

We use g_0 to find $g(0.66)$: It is 3.4659. (True = 3.4340)

We use g_2 to find $g(1.75)$: It is 8.7087. (True = 8.4467)

Some observations on this example: (a) We were given four points that define three intervals, (b) on each of the three intervals a $g(x)$ is defined, and (c) because each g has four coefficients, we must evaluate 12 unknown coefficients. However, by introducing the S's, we only had to solve two equations!

Using MATLAB

If you have access to MATLAB's spline toolbox, you can construct a spline curve that is almost exactly the same as Figure 3.3 by

```
EDU>> x = [0.0 1.0 1.5 2.25];
EDU>> y = [2.0 4.4366 6.7134 13.9139];
EDU>> cs = csapi(x, y);
EDU>> fnplt(cs); hold on; plot(x, y, 'o')
```

We can interpolate to get y-values from the spline at $x = 0.66$ and $x = 1.75$ by

```
EDU>> csapi(x, y, .66)
ans =
  3.5115
EDU>> csapi(x, y, 1.75)
ans =
  8.4994
```

which are not identical to those in Example 3.5. The reason is that MATLAB uses a different end condition, called the "not a knot" end condition. (It is the same as condition 4.) The value at $x = 0.66$ is less accurate but that at $x = 1.75$ is better. MATLAB has another command that works the same and is avaliable from the student edition without the spline toolbox:

```
EDU>> yi = interp1(x, y, 0.66, 'spline')
yi =
  3.5114
EDU>> yi = interp1(x, y, 1.75, 'spline')
yi =
  8.4993
```

Mathematica and Maple can get spline fits to data as well. Maple uses the natural cubic spline.

Here is another example in which we compare using all four end conditions. The data appear to be periodic. As a project, see if you can develop the equations and solve the example for a periodic spline. (Is there much difference in interpolated values?)

EXAMPLE 3.6 The data in the following table are from astronomical observations of a type of variable star called a *Cepheid variable* and represent variations in its apparent magnitude with time:

Time	0.0	0.2	0.3	0.4	0.5	0.6	0.7	0.8	1.0
Apparent magnitude	0.302	0.185	0.106	0.093	0.240	0.579	0.561	0.468	0.302

Use each of the four end conditions to compute cubic splines, and compare the values interpolated from each spline function at intervals of time of 0.05.

The augmented matrices whose solutions give values for $S_1, S_2, \ldots, S_7$ are shown in Table 3.7. A computer program was used to obtain the results shown in Table 3.8.

The results from the four end conditions between $x = 0.2$ and $x = 0.8$ are nearly identical; they differ by less than 0.001. Only in the end portions is there some difference.

Table 3.7

Condition 1

Matrix coefficients are

—	0.60	0.10	−1.23
0.10	0.40	0.10	3.96
0.10	0.40	0.10	9.60
0.10	0.40	0.10	11.52
0.10	0.40	0.10	−21.42
0.10	0.40	0.10	−4.50
0.10	0.60	—	0.60

Condition 2

Matrix coefficients are

—	0.40	0.10	0.00
0.20	0.60	0.10	−1.23
0.10	0.40	0.10	3.96
0.10	0.40	0.10	9.60
0.10	0.40	0.10	11.52
0.10	0.40	0.10	−21.42
0.10	0.40	0.10	−4.50
0.10	0.60	0.20	0.60
0.10	0.40	—	0.00

Condition 3

Matrix coefficients are

—	0.80	0.10	−1.23
0.10	0.40	0.10	3.96
0.10	0.40	0.10	9.60
0.10	0.40	0.10	11.52
0.10	0.40	0.10	−21.42
0.10	0.40	0.10	−4.50
0.10	0.80	—	0.60

Condition 4

Matrix coefficients are

—	1.20	−0.30	−1.23
0.10	0.40	0.10	3.96
0.10	0.40	0.10	9.60
0.10	0.40	0.10	11.52
0.10	0.40	0.10	−21.42
0.10	0.40	0.10	−4.50
−0.30	1.20	—	0.60

Figure 3.4 shows the points for the four conditions from $x = 0.0$ to $x = 0.2$. Condition 4 gives values that are significantly different from the others.

Fitting Splines to a Hump

At the beginning of the section, it was pointed out that a function with a sharp rise from a base line does not lend itself to being fitted with interpolating polynomials and that cubic splines are preferred. The example function was $f(x) = \cos^{10}(x)$ between $x = -2$ and $x = 2$ (Fig. 3.1a). Example 3.7 shows that a cubic spline gets a very good fit.

EXAMPLE 3.7 Fit cubic splines to $f(x) = \cos^{10}(x)$ with knots at -2, -1, -0.5, 0, 0.5, 1, and 2. Figure 3.5 shows the points superimposed on the spline function, and Table 3.9 compares the values from the splines with the true values for the function at several points. The agreement is excellent. (The figure and table are on p. 179.)

Table 3.8

t	Values, condition 1	Values, condition 2*	Values, condition 3	Values, condition 4
0.00	**0.302**	**0.302**	**0.302**	**0.302**
0.05	0.278	0.276	0.282	0.297
0.10	0.252	0.250	0.256	0.271
0.15	0.222	0.221	0.224	0.231
0.20	**0.185**	**0.185**	**0.185**	**0.185**
0.25	0.143	0.143	0.142	0.141
0.30	**0.106**	**0.106**	**0.106**	**0.106**
0.35	0.087	0.087	0.088	0.088
0.40	**0.093**	**0.093**	**0.093**	**0.093**
0.45	0.133	0.133	0.133	0.133
0.50	**0.240**	**0.240**	**0.240**	**0.240**
0.55	0.424	0.424	0.424	0.424
0.60	**0.579**	**0.579**	**0.579**	**0.579**
0.65	0.608	0.608	0.608	0.608
0.70	**0.561**	**0.561**	**0.561**	**0.561**
0.75	0.511	0.511	0.511	0.511
0.80	**0.468**	**0.468**	**0.468**	**0.468**
0.85	0.426	0.426	0.426	0.430
0.90	0.385	0.385	0.384	0.392
0.95	0.343	0.343	0.343	0.350
1.00	**0.302**	**0.302**	**0.302**	**0.302**

* Note that in the values for condition 2 we used forward and backward differences to approximate the slope at either end of the curve; that is, $V'(0.0) = -0.585$ and $V'(1.0) = -0.830$.

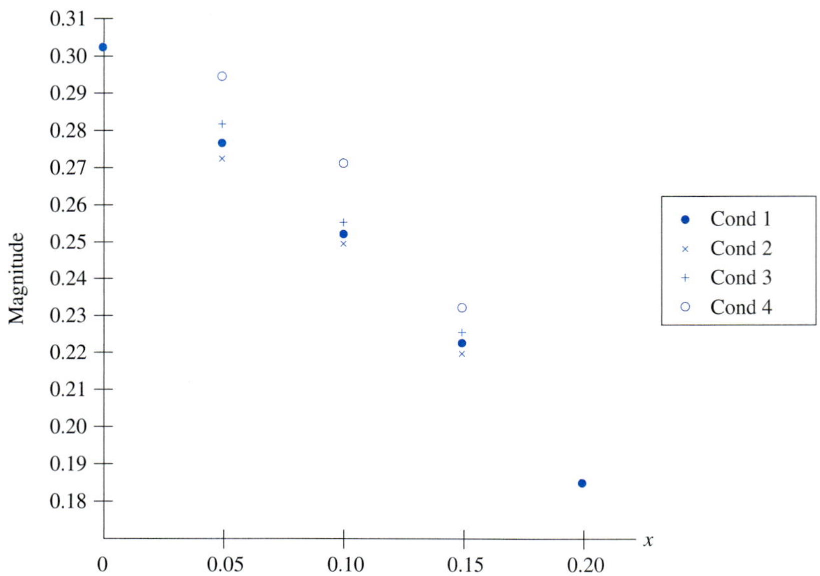

Figure 3.4

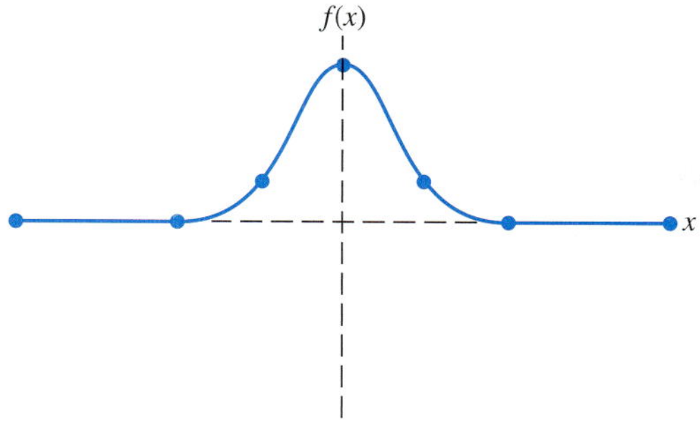

Figure 3.5

3.4 Bezier Curves and B-Spline Curves

In addition to the splines we have studied in the previous section, there are others that are important. In particular, Bezier curves and B-splines are widely used in computer graphics and computer-aided design. B-splines are often used to numerically integrate and differentiate functions that are defined only through a set of data points. These two types of curves are not really interpolating splines, because the curves do not normally pass through all of the points. In this respect, they show some similarity to least-squares curves, which are discussed in a later section. However, both Bezier curves and B-splines have the important property of staying within the polygon determined by the given points. We will be more explicit about this property later. In addition, these two new spline curves have a nice

Table 3.9 A cubic spline fitted to the function $f(x) = \cos^{10}(x)$, end condition 1

x-value	Spline value	$f(x)$	Error
−2.00	0.0002	0.0002	0.0000
−1.75	−0.0046	0.0000	0.0046
−1.50	−0.0073	0.0000	0.0073
−1.25	−0.0058	0.0000	0.0058
−1.00	0.0021	0.0021	−0.0000
−0.75	0.0467	0.0440	−0.0027
−0.50	0.2709	0.2709	−0.0000
−0.25	0.7283	0.7292	0.0009
0.00	1.0000	1.0000	0.0000
0.25	0.7283	0.7292	0.0009
0.50	0.2709	0.2709	−0.0000
0.75	0.0467	0.0440	−0.0027
1.00	0.0021	0.0021	−0.0000
1.25	−0.0058	0.0000	0.0058
1.50	−0.0073	0.0000	0.0073
1.75	−0.0046	0.0000	0.0046
2.00	0.0002	0.0002	−0.0000

geometric property in that in changing one of the points we change only one portion of the curve, a "local" effect. For the cubic spline curve of the previous section, changing just one point has a "global" effect in that the entire curve from the first to the last point is affected. Finally, for the cubic splines just studied, the points were given data points. For the two curves we study in this section, the points in question are more likely "control" points that we select to determine the shape of the curve we are working on.

For simplicity, we consider mainly the cubic version of these two curves. In what follows, we will express $y = f(x)$ in parametric form. The parametric form represents a relation between x and y by two other equations, $x = F_1(u)$, $y = F_2(u)$. The independent variable u is called the *parameter*. For example, the equation for a circle can be written, with θ as the parameter, as

$$x = r\cos(\theta),$$
$$y = r\sin(\theta).$$

If we express x and y in terms of a parameter, u, the point (x, y) becomes $(x(u), y(u))$. We will use this with values of the parameter u between 0 and 1.

We discuss Bezier curves first. Bezier curves are named after the French engineer P. Bezier of the Renault Automobile Company. He developed them in the early 1960s to fill a need for curves whose shape can be readily controlled by changing a few parameters. Bezier's application was to construct pleasing surfaces for car bodies.

Suppose we are given a set of control points, $p_i = (x_i, y_i)$, $i = 0, 1, \ldots, n$. (These points are also referred to as *Bezier points*.) Figure 3.6 is an example.

These points could be chosen on a computer screen, using a pointing device. The points do not necessarily progress from left to right. We treat the coordinates of each point as a two-component vector,

$$p_i = \begin{bmatrix} x_i \\ y_i \end{bmatrix}.$$

The set of points, in parametric form, is

$$P(u) = \begin{bmatrix} x(u) \\ y(u) \end{bmatrix}, \qquad 0 \le u \le 1.$$

The nth-degree Bezier polynomial determined by $n + 1$ points is given by

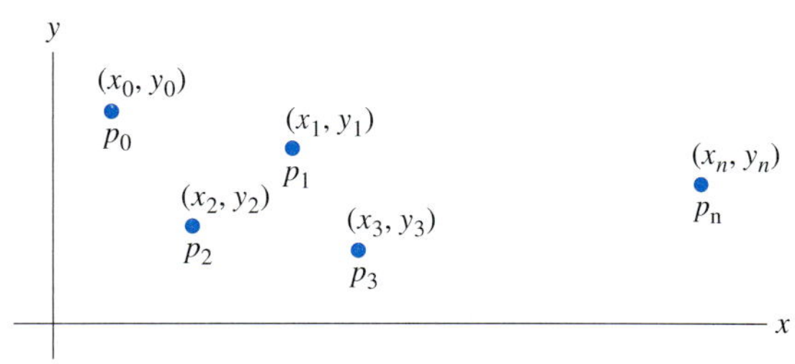

Figure 3.6

$$P(u) = \sum_{i=0}^{n} \binom{n}{i} (1 - u)^{n-i} u^i p_i,$$

where

$$\binom{n}{i} = \frac{n!}{i!(n - i)!}.$$

$P(u)$ is actually a *Bernstein polynomial.* Bernstein showed in 1912 that a weighted sum of these polynomials will converge uniformly to any continuous function on the interval [0, 1] as n approaches infinity. (Maple knows the Bernstein polynomials. The command is in a library: 'bernstein'.)

When $n = 2$, $P(u)$ is a quadratic equation defined by three points, p_0, p_1, and p_2:

$$P(u) = (1)(1 - u)^2 p_0 + 2(1 - u)(u)p_1 + (1)u^2 p_2,$$

because, for $n = 2$ and $i = 0, 1, 2$, we have $\binom{2}{0} = 1$, $\binom{2}{1} = 2$, $\binom{2}{2} = 1$. The preceding equation represents the pair of equations

$$x(u) = (1 - u)^2 x_0 + 2(1 - u)u\, x_1 + u^2 x_2,$$
$$y(u) = (1 - u)^2 y_0 + 2(1 - u)u\, y_1 + u^2 y_2.$$

Observe that, if $u = 0$, $x(0)$ is identical to x_0 and similarly for $y(0)$. If $u = 1$, the point referred to is (x_2, y_2). As u takes on values between 0 and 1, a curve is traced that goes from the first point to the third of the set. Ordinarily the curve will not pass through the central point of the three. (If the points are collinear, the curve is the straight line through them all.) In effect, the points of the second-degree Bezier curve have coordinates that are weighted sums of the coordinates of the three points that are used to define it. From another point of view, one can think of the Bezier equations as weighted sums of three polynomials in u, where the weighting factors are the coordinates of the three points.

In one of the exercises, you are to find the Bezier curve for seven points, with $(x(0), y(0)) = P_0$, $(x(1), y(1)) = P_3$, and $(x(2), y(2)) = P_6$.

Applying the general defining equation for $n = 3$, we get the cubic Bezier polynomial that we now consider in some detail. The properties of other Bezier polynomials are the same as for the cubic. Here is the Bezier cubic:

$$x(u) = (1 - u)^3 x_0 + 3(1 - u)^2 u x_1 + 3(1 - u)u^2 x_2 + u^3 x_3,$$
$$y(u) = (1 - u)^3 y_0 + 3(1 - u)^2 u y_1 + 3(1 - u)u^2 y_2 + u^3 y_3.$$

Observe again that $(x(0), y(0)) = p_0$ and $(x(1), y(1)) = p_3$, and that the curve will not ordinarily go through the intermediate points. As illustrated in the example curves in Figure 3.7, changing the intermediate "control" points changes the shape of the curve. The

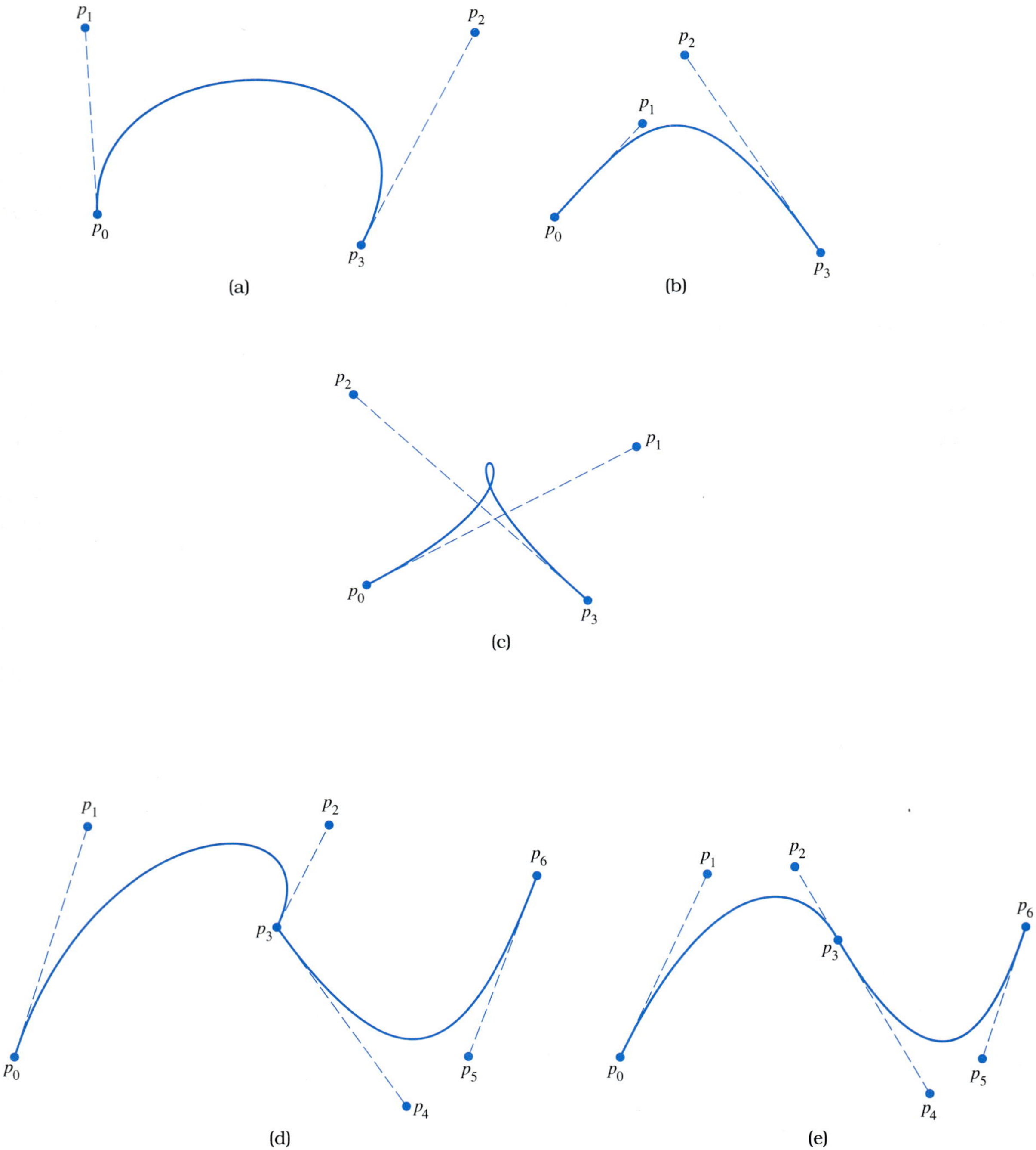

Figure 3.7
Bezier curves defined by four and seven points

examples are in Figure 3.7a through e. The first three of these show Bezier curves defined by one group of four points.

Figure 3.7d and e demonstrate how cubic Bezier curves can be continued beyond the first set of four points; one just subdivides seven points (p_0 to p_6) into two groups of four, with the central one (p_3) belonging to both sets. Figure 3.7e shows that p_2, p_3, and p_4 must be collinear to avoid a discontinuity in the slope at p_3.

It is of interest to list the properties of Bezier cubics:

1. $P(0) = p_0, \qquad P(1) = p_3.$
2. Because $dx/du = 3(x_1 - x_0)$ and $dy/du = 3(y_1 - y_0)$ at $u = 0$, the slope of the curve at $u = 0$ is $dy/dx = (y_1 - y_0)/(x_1 - x_0)$, which is the slope of the secant line between p_0 and p_1. Similarly, the slope at $u = 1$ is the same as the secant line between the last two points. This is indicated in the figures by dashed lines.
3. The Bezier curve is contained in the convex hull determined by the four points.

The *convex hull* of a set of points is the smallest convex set that contains the points. A set, C, is *convex* if and only if the line segment between any two points in the set lies entirely in set C. The following sketches show examples of the convex hull of four points.

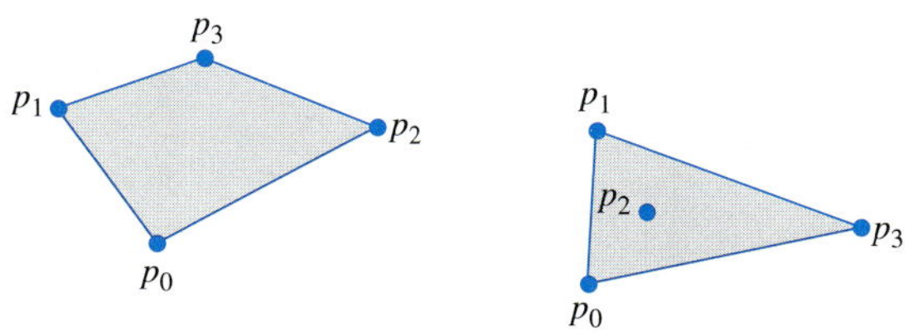

It is often convenient to represent the Bezier curve in matrix form. For Bezier cubics, this is

$$P(u) = [u^3, u^2, u, 1] \begin{bmatrix} -1 & 3 & -3 & 1 \\ 3 & -6 & 3 & 0 \\ -3 & 3 & 0 & 0 \\ 1 & 0 & 0 & 0 \end{bmatrix} \begin{bmatrix} p_0 \\ p_1 \\ p_2 \\ p_3 \end{bmatrix}$$
$$= u^T M_2 p.$$

Mathematica can draw Bezier curves as well as the splines of the previous section. In what follows, the first command defines some x, y pairs, the second invokes a graphics package. One must be sure to use *back quotes* in this. Then, in [3], a spline curve is set up which is displayed by [4].

```
In[1]: =
  spdata = {{1,1},{2,4},{3,3},{4,4}}
Out[1]: =
```

```
  {{1,1},{2,4},{3,3},{4,4}}
In[2]: =
  <<Graphics 'Spline'
In[3]: =
  splin = Spline[spdata, cubic]
Out[3]: =
  Spline[{{1,1},{2,4),{3,3},{4,4}},Cubic,<>]
In[4]: =
  Show{Graphics[{Line[spdata],splin}]]
```

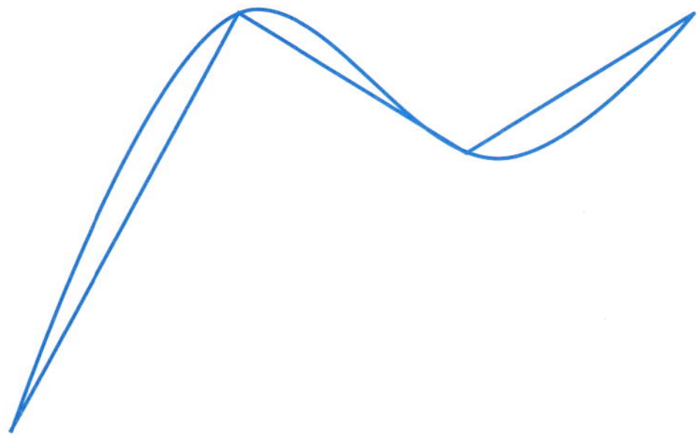

We use the same data to draw a Bezier curve:

```
In[5]: =
  Show[Graphics [{Line[spdata],Spline[spdata, Bezier]}]]
```

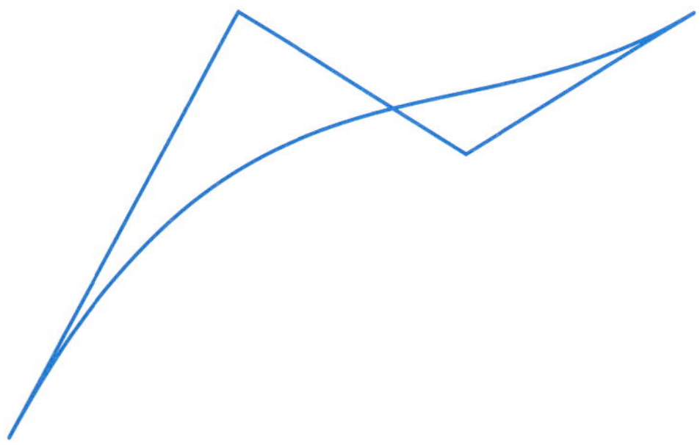

B-Spline Curves

We now discuss B-splines. These curves are like Bezier curves in that they do not ordinarily pass through the given data points. (The least-squares curves that are described in Section 3.6 are similar in this respect.) They can be of any degree, but we will concentrate on the cubic form. Cubic B-splines resemble the ordinary cubic splines of the previous section in that a separate cubic is derived for each pair of points in the set. However, the B-spline need not pass through any points of the set that are used in its definition.

We begin the description by stating the formula for a cubic B-spline in terms of parametric equations whose parameter is u.

Given the points $p_i = (x_i, y_i)$, $i = 0, 1, \ldots, n$, the cubic B-spline for the interval (p_i, p_{i+1}), $i = 1, 2, \ldots, n-1$, is

$$B_i(u) = \sum_{k=-1}^{2} b_k p_{i+k}, \quad \text{where}$$

$$\begin{aligned}
b_{-1} &= \frac{(1-u)^3}{6}, \\
b_0 &= \frac{u^3}{2} - u^2 + \frac{2}{3}, \\
b_1 &= -\frac{u^3}{2} + \frac{u^2}{2} + \frac{u}{2} + \frac{1}{6}, \\
b_2 &= \frac{u^3}{6}, \qquad 0 \le u \le 1.
\end{aligned} \tag{3.19}$$

As before, p_i refers to the point (x_i, y_i); it is a two-component vector. The coefficients, the b_k's, serve as a basis and do not change as we move from one set of points to the next. Observe that they can be considered weighting factors applied to the coordinates of a set of four points. The weighted sum, as u varies from 0 to 1, generates the B-spline curve.

If we write out the equations for x and y from Eq. (3.19), we get

$$\begin{aligned}
x_i(u) &= \frac{1}{6}(1-u)^3 x_{i-1} + \frac{1}{6}(3u^3 - 6u^2 + 4)x_i \\
&\quad + \frac{1}{6}(-3u^3 + 3u^2 + 3u + 1)x_{i+1} + \frac{1}{6}u^3 x_{i+2}; \\
y_i(u) &= \frac{1}{6}(1-u)^3 y_{i-1} + \frac{1}{6}(3u^3 - 6u^2 + 4)y_i \\
&\quad + \frac{1}{6}(-3u^3 + 3u^2 + 3u + 1)y_{i+1} + \frac{1}{6}u^3 y_{i+2}.
\end{aligned}$$

Note the notation here: $x_i(u)$ and $y_i(u)$ are functions (of u) and x_i, y_i are components of the point p. (The end portions are a special situation that we discuss later.)

As we have said, the u-cubics act as weighting factors on the coordinates of the four successive points to generate the curve. For example, at $u = 0$, the weights applied are 1/6, 2/3, 1/6, and 0. At $u = 1$, they are 0, 1/6, 2/3, and 1/6. These values vary throughout the interval from $u = 0$ to $u = 1$. As an exercise, you are asked to graph these factors. This will give you a visual impression of how the weights change with u.

Let us now examine two B-splines determined from a set of exactly four points. Figure 3.8a and b shows the effect of varying just one of the points. As you would expect, when p_2 is moved upward and to the left, the curve tends to follow; in fact, it is pulled to the opposite side of p_1. You may be surprised to see that the curve is never very close to the two

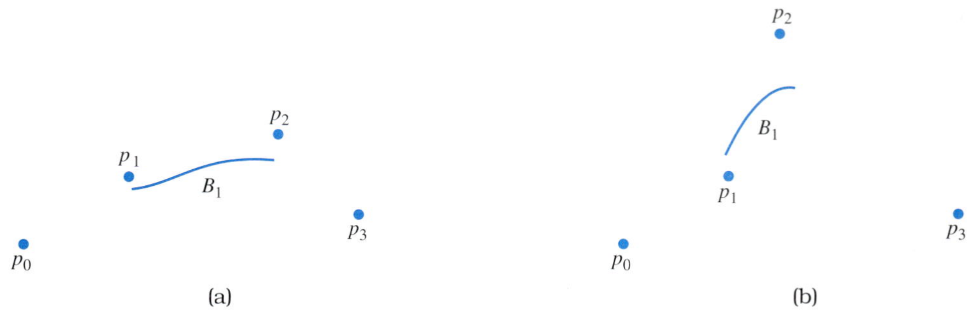

Figure 3.8

intermediate points, though it begins and ends at positions somewhat adjacent. It will be helpful to think of the curve generated from the defining equation for B_1 as associated with a curve that goes from near p_1 to p_2. It is also helpful to remember that points p_0, p_1, p_2, and p_3 are used to get B_1.

Because a set of four points is required to generate only a portion of the B-spline, that associated with the two inner points, we must consider how to get the B-spline for more than four points as well as how to extend the curve into the region outside of the middle pair. We use a method analogous to the cubic splines of Section 3.3 marching along one point at a time, forming new sets of four. We abandon the first of the old set when we add the new one.

The conditions that we want to impose on the B-spline are exactly the same as for ordinary splines: continuity of the curve and its first and second derivatives. It turns out that the equations for the weighting factors (the u-polynomials, the b_k) are such that these requirements are met. Figure 3.9 shows how three successive parts of a B-spline might look.

We can summarize the properties of B-splines as follows:

1. Like the cubic splines of Section 3.4, B-splines are pieced together so they agree at their joints in three ways:

   a. $B_i(1) = B_{i+1}(0) = \dfrac{p_i + 4p_{i+1} + p_{i+2}}{6}$,

   b. $B_i'(1) = B_{i+1}'(0) = \dfrac{-p_i + p_{i+2}}{2}$,

   c. $B_i''(1) = B_{i+1}''(0) = p_i - 2p_{i+1} + p_{i+2}$.

   The subscripts here refer to the portions of the curve and the points in Figure 3.9.

2. The portion of the curve determined by each group of four points is within the convex hull of these points.

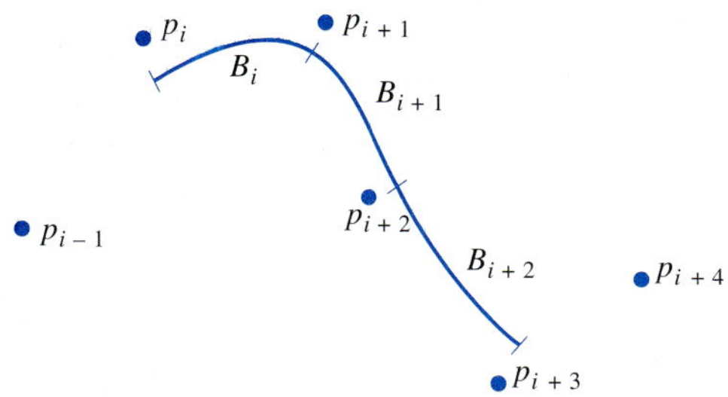

Figure 3.9
Successive B-splines joined together

Now we consider how to generate the ends of the joined B-spline. If we have points from p_0 to p_n, we already can construct B-splines B_1 through B_{n-2}. We need B_0 and B_{n-1}. Our problem is that, using the procedure already defined, we would need additional points outside the domain of the given points. We probably also want to tie down the curve in some way—having it start and end at the extreme points of the given set seems like a good idea. How can we do this?

First, we can add more points without creating artificiality by making the added points coincide with the given extreme points. If we add not just a single fictitious point at each end of the set, but two at each end, we will find that the new curves not only join properly with the portions already made, but start and end at the extreme points as we wanted. (It looks like we have added two extra portions, but reflection shows these are degenerate, giving only a single point.)

In summary: We add fictitious points p_{-2}, p_{-1}, p_{n+1}, and p_{n+2}, with the first two identical with p_0 and the last two identical with p_n. (There are other methods to handle the starting and ending segments of B-splines that we do not cover.)

The matrix formulation for cubic B-splines is helpful. Here it is:

$$B_i(u) = \frac{1}{6}[u^3, u^2, u, 1]\begin{bmatrix} -1 & 3 & -3 & 1 \\ 3 & -6 & 3 & 0 \\ -3 & 0 & 3 & 0 \\ 1 & 4 & 1 & 0 \end{bmatrix}\begin{bmatrix} p_{i-1} \\ p_i \\ p_{i+1} \\ p_{i+2} \end{bmatrix} \quad (3.20)$$

$$= \frac{u^T M_b p}{6}.$$

This applies on the interval [0, 1] and for the points (p_i, p_{i+1}).

B-splines differ from Bezier curves in three ways:

1. For a B-spline, the curve does not begin and end at the extreme points.
2. The slopes of the B-splines do not have any simple relationship to lines drawn between the points.
3. The endpoints of the B-splines are in the vicinity of the two intermediate given points, but neither the x- nor the y-coordinates of these endpoints normally equal the coordinates of the intermediate points.

An algorithm for drawing a B-spline curve is as follows:

An Algorithm for Drawing a B-spline Curve

Given a set of $n + 1$ points, $P_i = (x_i, y_i)$, $i = 0, \ldots, n$:

Set $p_{-2} = p_{-1} = p_0$.
Set $p_{n+1} = p_{n+2} = p_n$.

For $i = 0$ To n Step 1 Do
    For $u = [0, \ldots, 1]$
    Compute

$$x(u) = (1 - u)^3 x_{i-1}/6 + (3u^3 - 6u^2 + 4)x_i/6 + (-3u^3 + 3u^2 + 3u + 1)\, x_{i+1}/6 + u^3 x_{i+2}/6.$$

$$y(u) = (1 - u)^3 y_{i-1}/6 + (3u^3 - 6u^2 + 4)y_i/6 + (-3u^3 + 3u^2 + 3u + 1)y_{i+1}/6 + u^3 y_{i+2}/6.$$

    Plot $(x(u), y(u))$
    End For u.
End For i.

We conclude this section by looking at several examples of B-splines. The five parts of Figure 3.10 show B-splines that are defined by the same sets of points as the Bezier curves in Figure 3.7. (Fictitious points have been added to complete the end portions of these B-splines.) There are significant differences.

3.5 Interpolating on a Surface

In the opening of this chapter, we mentioned an interpolation problem that is faced by the National Weather Service, that of interpolating from scattered data to get values at points on a uniform grid. (The chapter on partial-differential equations will tell why this is

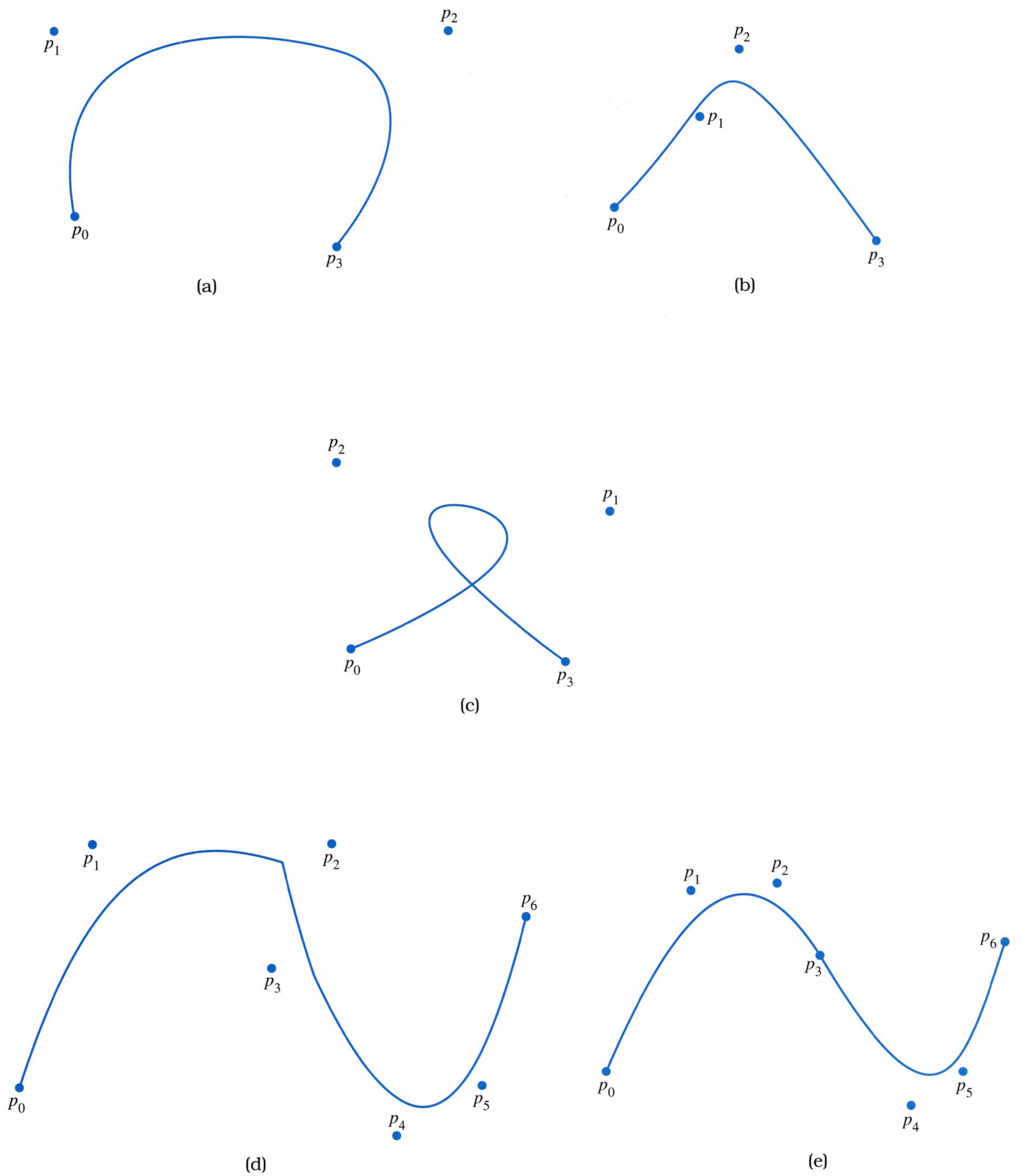
p_1
p_2
p_0
p_3
(a)
p_2
p_1
p_0
p_3
(b)
p_2
p_1
p_0
p_3
(c)
p_1
p_2
p_6
p_3
p_0
p_4
p_5
(d)
p_1
p_2
p_6
p_3
p_0
p_5
p_4
(e)

Figure 3.10

important.) This very multidimensional problem is not an easy one. We begin therefore with the simpler case where data values, $z = f(x, y)$, are taken from a table that has the independent variables, x and y, spaced uniformly.

When a function z is a polynomial function* of two variables x and y—say, of degree-3 in x and degree-2 in y—we would have

$$z = f(x, y) = a_0 + a_1x + a_2y + a_3x^2 + a_4xy + a_5y^2 + a_6x^3 + a_7x^2y + a_8xy^2 + a_9x^3y + a_{10}x^2y^2 + a_{11}x^3y^2. \tag{3.21}$$

Such a function describes a surface; (x, y, z) is a point on it. The functional relation is seen to involve many terms. If we are concerned with four independent variables (three space dimensions plus time, say), even low-degree polynomials would be quite intractable. Except for special purposes, such as when we need an explicit representation, perhaps to permit ready differentiation at an arbitrary point, we can avoid such complications by handling each variable separately. We will treat only this case.

Note the immediate simplification of Eq. (3.21) if we let y take on a constant value, say, $y = c$. Combining the y factors with the coefficients, we get

$$z|_{y=c} = b_0 + b_1x + b_2x^2 + b_3x^3.$$

This suggests that we can interpolate for z at $(x, y) = (a, b)$ by holding one of the independent variables constant, say, $y = y_1$, making a table in which there is only one independent variable, x. Any procedure that we have explained previously can then be used. We can repeat this at other values for y, $y = y_2, y_3, \ldots, y_n$ so that we develop a table for z at $x = a$ and various y-values. From this we interpolate for $y = b$.

Example 3.8 illustrates this attack.

EXAMPLE 3.8 Estimate $f(1.6, 0.33)$ from the values in Table 3.10. Use quadratic interpolation in the x-direction and cubic interpolation for y. We select one of the variables to hold constant, say, x. (This choice is arbitrary because we would get the same result, except for differences due to round off, if we had chosen to hold y constant.) We decide to interpolate

Table 3.10 Tabulation of a function of two variables $z = f(x, y)$

x \ y	**0.1**	**0.2**	**0.3**	**0.4**	**0.5**	**0.6**
0.5	0.165	0.428	0.687	0.942	1.190	1.431
1.0	0.271	0.640	1.003	1.359	1.703	2.035
1.5	0.447	0.990	1.524	2.045	2.549	3.031
2.0	0.738	1.568	2.384	3.177	3.943	4.672
2.5	1.216	2.520	3.800	5.044	6.241	7.379
3.0	2.005	4.090	6.136	8.122	10.030	11.841
3.5	3.306	6.679	9.986	13.196	16.277	19.198

* We approximate a nonpolynomial function by a polynomial that agrees with the function, just as we have done with a function of one variable.

Table 3.11 Tabulations at three x-values

	y	z	Δz	$\Delta^2 z$	$\Delta^3 z$
$x = 1.0$	0.2	0.640	0.363	−0.007	−0.005
	0.3	1.003	0.356	−0.012	
	0.4	1.359	0.344		
	0.5	1.703			
$x = 1.5$	0.2	0.990	0.534	−0.013	−0.004
	0.3	1.524	0.521	−0.017	
	0.4	2.045	0.504		
	0.5	2.549			
$x = 2.0$	0.2	1.568	0.816	−0.023	−0.004
	0.3	2.384	0.793	−0.027	
	0.4	3.177	0.766		
	0.5	3.943			

for y within the three rows of the table at $x = 1.0, 1.5$, and 2.0, because the desired value at $x = 1.6$ is most nearly centered within this set. We choose y-values of 0.2, 0.3, 0.4, and 0.5 so that $y = 0.33$ is centralized.

Because the x-values are evenly spaced, we elect to use Newton–Gregory forward polynomials. Table 3.11 shows the ordinary differences that we need.

We need the subtables from $y = 0.2$ to $y = 0.5$, because, for a cubic interpolation, four points are required. Using any convenient formula (remember that all cubics that agree at four points are identical), we get Table 3.12. In the last tabulation we carry one extra decimal to guard against round-off errors. Interpolating again, we get $z = 1.8406$, which we report as $z = 1.841$.

The function tabulated in Table 3.10 is $f(x, y) = e^x \sin y + y - 0.1$, so the true value is $f(1.6, 0.33) = 1.8350$. Our error of 0.0056 occurs because quadratic interpolation for x is inadequate in view of the large second difference. In retrospect, it would have been better to use quadratic interpolation for y, because the third differences of the y-subtables are small, and let x take on a third-degree relationship. (You may want to verify that this reduces the error to 0.0022.)

Table 3.12 Tabulation at $y = 0.33$

	x	z	Δz	$\Delta^2 z$
$y = 0.33$	1.0	1.1108	0.5710	0.3717
	1.5	1.6818	0.9427	
	2.0	2.6245		

It is instructive to observe which of the values in Table 3.10 entered into our computation. The shaded rectangle covers these values. This is the "region of fit" for the interpolating polynomial that we have used. The principle of choosing values so that the point at which the interpolating polynomial is used is centered in the region of fit obviously applies here in exact analogy to the one-way table situation. It also applies to tables of three and four variables in the same way. Of course, the labor of interpolating in such multidimensional cases soon becomes burdensome.

A rectangular region of fit is not the only possibility. We may change the degree of interpolation as we subtabulate the different rows or columns. Intuitively, it would seem best to use higher-degree polynomials for the rows near the interpolating point, decreasing the degree as we get farther away. The coefficient of the error term, when this is done, will be found to be minimized thereby, although for multidimensional interpolating polynomials the error term is quite complex. The region of fit will be diamond-shaped when such tapered degree functions are used.

We may adapt the Lagrangian form of interpolating polynomial to the multidimensional case also. It is perhaps easiest to employ a process similar to the preceding example. Holding one variable constant, we write a series of Lagrangian polynomials for interpolation at the given value of the other variable, and then combine these values in a final Lagrange form. The net result is a Lagrangian polynomial in which the function factors are replaced by Lagrangian polynomials. The resulting expression for the previous example would be

$$
\begin{aligned}
&\frac{(y-0.3)(y-0.4)(y-0.5)}{(0.2-0.3)(0.2-0.4)(0.2-0.5)} \\
&\times\left[\frac{(x-1.5)(x-2.0)}{(1.0-1.5)(1.0-2.0)}(0.640)+\frac{(x-1.0)(x-2.0)}{(1.5-1.0)(1.5-2.0)}(0.990)+\frac{(x-1.0)(x-1.5)}{(2.0-1.0)(2.0-1.5)}(1.568)\right] \\
&+\frac{(y-0.2)(y-0.4)(y-0.5)}{(0.3-0.2)(0.3-0.4)(0.3-0.5)} \\
&\times\left[\frac{(x-1.5)(x-2.0)}{(1.0-1.5)(1.0-2.0)}(1.003)+\frac{(x-1.0)(x-2.0)}{(1.5-1.0)(1.5-2.0)}(1.524)+\frac{(x-1.0)(x-1.5)}{(2.0-1.0)(2.0-1.5)}(2.384)\right] \\
&+\frac{(y-0.2)(y-0.3)(y-0.5)}{(0.4-0.2)(0.4-0.3)(0.4-0.5)} \qquad (3.22)\\
&\times\left[\frac{(x-1.5)(x-2.0)}{(1.0-1.5)(1.0-2.0)}(1.359)+\frac{(x-1.0)(x-2.0)}{(1.5-1.0)(1.5-2.0)}(2.045)+\frac{(x-1.0)(x-1.5)}{(2.0-1.0)(2.0-1.5)}(3.177)\right] \\
&+\frac{(y-0.2)(y-0.3)(y-0.4)}{(0.5-0.2)(0.5-0.3)(0.5-0.4)} \\
&\times\left[\frac{(x-1.5)(x-2.0)}{(1.0-1.5)(1.0-2.0)}(1.703)+\frac{(x-1.0)(x-2.0)}{(1.5-1.0)(1.5-2.0)}(2.549)+\frac{(x-1.0)(x-1.5)}{(2.0-1.0)(2.0-1.5)}(3.943)\right].
\end{aligned}
$$

The equation is easy to write, but its evaluation by hand is laborious. If one is writing a computer program for interpolation in such multivariate situations, the Lagrangian form is recommended. There is a special advantage in that equal spacing in the table is not required. The Lagrangian form is also perhaps the most straightforward way to write out the polynomial as an explicit function.

When the given points are not evenly spaced, Lagrangian polynomials or the method of divided differences should be used for interpolation. With the latter, exactly the same principle is involved: Hold one variable constant while subtables of divided differences are constructed, then combine the interpolated values from these subtables into a new table.

Parallel processing can save many time steps in the preceding computations. Each value in the column of differences of Tables 3.11 and 3.12 can be computed at the same time. (We must wait for the interpolations from Table 3.11 to be completed before we do Table 3.12, of course.) Every factor of Eq. (3.22) can be evaluated in parallel.

Interpolation for the Weather Service

Here is a simplified form of interpolation that might be used to estimate a predicted value for the temperature at a grid point from data from weather stations located in its neighborhood. Suppose that the stations where the temperature is known are as in Table 3.13. The coordinates of the known temperatures are relative to our desired grid point (so the origin is there). Because weather comes generally from the west, the data from stations in that direction are given double weight. One way to give them this weight is to consider them to be duplicated.

If these data were to be entered into a table like Table 3.9, we would find entries only along the diagonal, so we cannot solve this in the same manner as Example 3.8. Still, all methods for interpolation are really finding a weighted average of data. [Examination of Eq. (3.1) shows this clearly; the ratios of x-values are weights applied to the f-values.] So this simplified weather problem gets the predicted temperature at the grid point [coordinates of (0, 0)] by

$$[56(2) + 62(2) + 59(2) + 64 + 61]/8 = 59.9°\text{F}.$$

The Weather Service must have more sophisticated ways of doing this.

Table 3.13 Temperature data at weather stations

Station 1:	Coordinates:	−14.2, 25.6, Temperature 56°F
Station 2:	Coordinates:	−22.7, −12.1, Temperature 62°F
Station 3:	Coordinates:	−33.6, −2.5, Temperature 59°F
Station 4:	Coordinates:	4.7, −8.3, Temperature 64°F
Station 5:	Coordinates:	13.4, 15.7, Temperature 61°F

Using Cubic Splines, Bezier Surfaces, and B-Spline Surfaces

Another alternative is to use cubic splines for interpolation in multivariate cases. Here again, it is perhaps best to hold one variable constant while constructing one-way splines, then combine the results from these in the second phase. The computational effort would be significant, however.

Interpolating for values of functions of two independent variables can also be thought of as constructing a surface that is defined by the given points. Rather than finding values on a surface that contains the given points, we can construct surfaces that are analogous to Bezier curves and B-spline curves where the surface does not normally contain the given points.

So far, we have been able to interpolate on simple surfaces where we are given z as a function of x and y. Suppose now we are given a set of points, $p_i = \{(x_i, y_i, z_i), i = 0, \ldots, n\}$, and we wish to fit a surface to those points. This would be the case if we were trying to draw a mountain, an airplane, or a teapot. But first we consider the representation of more general surfaces. Let $p = (x, y, z)$ be any point on the surface. Then the coordinates of each point are represented as the equations

$$x = x(u, v),$$
$$y = y(u, v),$$
$$z = z(u, v),$$

where u, v are the independent variables that range over a given set of values and x, y, z are the dependent variables. This is a slight change of notation from the first part of this section.

An example of this would be the equations of a sphere of radius r about the origin: (0, 0, 0). Here any point on the surface of the sphere is given by

$$x = r\cos(u)\sin(v),$$
$$y = r\sin(u)\sin(v),$$
$$z = r\cos(v),$$

where u ranges in value from 0 to 2π and v ranges from 0 to π. Figure 3.11 illustrates this.

MATLAB can interpolate on a surface, $z = f(x, y)$. One of four methods can be specified: 'nearest,' 'linear,' 'cubic,' and 'spline.' The 'linear' method is the default. The methods do interpolations in the following ways:

'nearest'—nearest neighbor interpolation
'linear' —bilinear interpolation
'cubic' —bicubic interpolation
'spline' —spline interpolation

Here is an example that uses a known function: $z = 2xy + e^{(x-y)}$, so we can see how good the interpolated results are. We will estimate $z(1.7, 2.0)$.

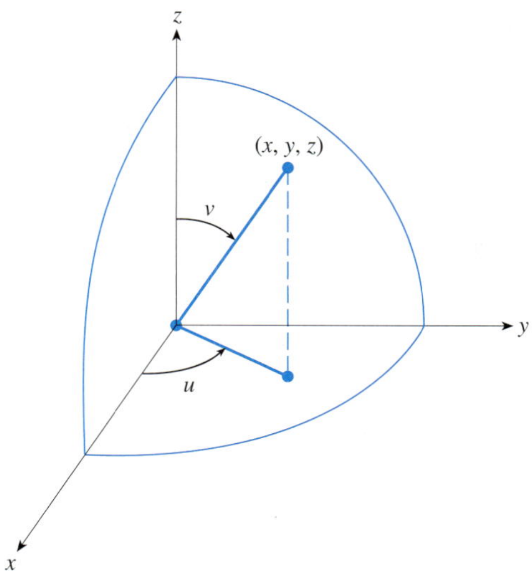

Figure 3.11

We begin by creating a table of z-values for a set of x-values: $x = [0, 1.2, 2.5, 2.9]$ with $y = [0, 0.9, 1.8, 3.2]$. We will get a matrix of z-values in a very simple way—we first define the y-vector and compute rows of the table in turn for each x-value.

```
EDU>> y = [0 0.9 1.8 3.2];

EDU>> x = 0;

EDU>> z1 = 2*x*y + exp(x - y)
      z1 =
           1.0000    0.4066    0.1653    0.0408

EDU>> x = 1.2;

EDU>> z2 = 2*x*y + exp(x - y)
      z2 =
           3.3201    3.5099    4.8688    7.8153

EDU>> x = 2.5;

EDU>> z3 = 2*x*y + exp(x - y)
      z3 =
           12.1825   9.4530    11.0138   16.4966

EDU>> x = 2.9;

EDU>> z4 = 2*x*y + exp(x - y)
      z4 =
           18.1741   12.6091   13.4442   19.3008
```

Now we join these to get a matrix of z-values:

```
EDU>> z = [z1; z2; z3; z4]
      z =
           1.0000    0.4066    0.1653    0.0408
           3.3201    3.5099    4.8688    7.8153
          12.1825    9.4530   11.0138   16.4966
          18.1741   12.6091   13.4442   19.3008
```

We need the x-values in a vector:

```
EDU>> x = [0 1.2 2.5 2.9]
      x =
             0      1.2000      2.5000      2.9000
```

and we are ready to do the interpolation. First, using the 'linear' method (the default method):

```
EDU>> zil = interp2(x,y,z,1.7,2.0)
      zil =
            10.4643
```

which does not match well to the correct value of $z = 7.5408$. Will the cubic interpolant be better?

```
EDU>> zic = interp2(x,y,z,1.7,2.0, 'cubic')
      zic =
            9.0978
```

which is better but still not very good. We would get results closer to the true value if the table were more closely spaced. A plot of the function shows that the z-values change rapidly at $(x, y) = (1.7, 2.0)$.

Creating a B-Spline Surface

We now describe constructing a B-spline surface. [A most interesting and informative description of Bezier surfaces can be found in Crow (1987). See also Pokorny and Gerald (1989).]

From the previous section, we know that a cubic B-spline curve segment starting near the point p_i to near the point p_{i+1} is determined by the four points

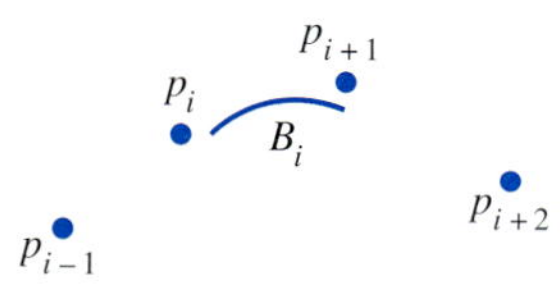

where $p_i(u) = (x_i(u), y_i(u))$ in two dimensions, or $p_i(u) = (x_i(u), y_i(u), z_i(u))$ if we had been working in three dimensions. The segment was then extended by introducing p_{i+3}, deleting p_{i-1}, and generating the curve for $0 \le u \le 1$.

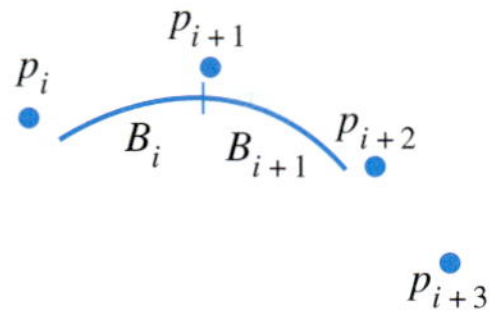

The process is continued until we have B_{n-2}. Finally, the first and last segments are generated by starting with p_0, p_0, p_0, p_1 and ending with p_{n-1}, p_n, p_n, p_n.

In an analogous manner the interpolating B-spline surface patch depends on 16 points, as Figure 3.12 shows. Here $p_{i,j} = (x_{i,j}, y_{i,j}, z_{i,j})$, a point in E^3. This patch is generated by computing the points $p_{i,j}(u, v)$, for $0 \le u \le 1$ and $0 \le v \le 1$. Here we have changed the subscripts on the points $p_{i,j}$ so as to fit into matrix notation.

For simplicity, we will consider only the x-coordinate in detail. Comparable formulations hold for the y- and z-coordinates. The simplest formulation for $x_{ij}(u, v)$ is based on the matrix formulation of Eq. (3.20) and is given by

$$x_{ij}(u, v) = \frac{1}{36}[u^3, u^2, u, 1]M_b X_{i,j} M_b^T \begin{bmatrix} v^3 \\ v^2 \\ v \\ 1 \end{bmatrix}, \tag{3.23}$$

where $X_{i,j}$ is the 4×4 matrix

$$\begin{bmatrix} x_{i-1,j-1} & x_{i-1,j} & x_{i-1,j+1} & x_{i-1,j+2} \\ x_{i,j-1} & x_{i,j} & x_{i,j+1} & x_{i,j+2} \\ x_{i+1,j-1} & x_{i+1,j} & x_{i+1,j+1} & x_{i+1,j+2} \\ x_{i+2,j-1} & x_{i+2,j} & x_{i+2,j+1} & x_{i+2,j+2} \end{bmatrix},$$

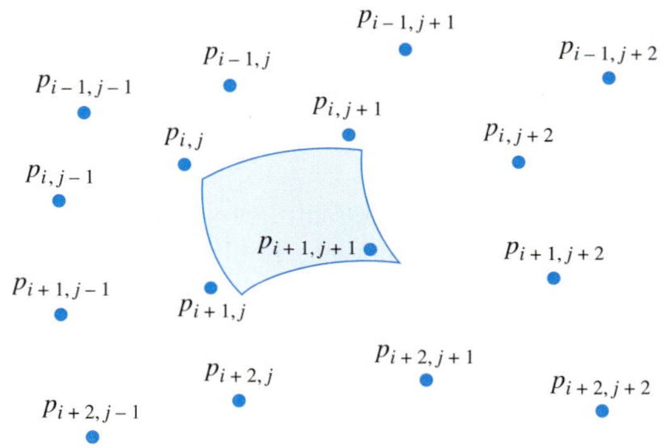

Figure 3.12

which are just the x-coordinates of the 16 points of Figure 3.12. The matrix M_b is the matrix we saw before in Eq. (3.20)

$$M_b = \begin{bmatrix} -1 & 3 & -3 & 1 \\ 3 & -6 & 3 & 0 \\ -3 & 0 & 3 & 0 \\ 1 & 4 & 1 & 0 \end{bmatrix}.$$

The y and z equations are then obtained merely by substituting the corresponding matrices $Y_{i,j}$ and $Z_{i,j}$, which are formed from the y and z components of the 16 points. Because each of these equations is cubic in u and v, they are referred to as *bicubic equations.* The coordinates of the points on a patch are given by

$$x(u, v) = \frac{1}{36}[u^3, u^2, u, 1]M_b X_{i,j} M_b^T [v^3, v^2, v, 1]^T,$$

$$y(u, v) = \frac{1}{36}[u^3, u^2, u, 1]M_b Y_{i,j} M_b^T [v^3, v^2, v, 1]^T,$$

$$z(u, v) = \frac{1}{36}[u^3, u^2, u, 1]M_b Z_{i,j} M_b^T [v^3, v^2, v, 1]^T,$$

as u and v range between 0 and 1. It is easily verified that the weights applied to each of the 16 points are

$$\begin{bmatrix} 1 & 4 & 1 & 0 \\ 4 & 16 & 4 & 0 \\ 1 & 4 & 1 & 0 \\ 0 & 0 & 0 & 0 \end{bmatrix} \quad \text{At } p_{i,j}(u, v) \text{ (for } u = 0, v = 0\text{), and}$$

$$\begin{bmatrix} 0 & 0 & 0 & 0 \\ 0 & 1 & 4 & 1 \\ 0 & 4 & 16 & 4 \\ 0 & 1 & 4 & 1 \end{bmatrix} \quad \text{At } p_{i,j}(u, v) \text{ (for } u = 1, v = 1\text{)}$$

where each (i, j)th element is the coefficient for the corresponding point in Figure 3.12. In effect, these matrices are templates that overlay the points shown in Figure 3.12.

The surface patch is extended by adding another row or column of points and deleting a corresponding row or column of points. One should verify that the current and previous patches are connected smoothly along the edge where they join. An initial or final patch can be obtained by repeating a corner, as was suggested for the B-spline curve. This will ensure that the patch actually starts or ends at a point. For the surface, we would repeat a point nine times, instead of three times as was done for the curve.

For a more detailed and informative discussion of interpolating curves and surfaces, the reader should consult Pokorny and Gerald (1989).

3.6 Least-Squares Approximations

Until now, in this chapter we have assumed that the data are accurate, but when these values are derived from an experiment, there is some error in the measurements. This section explains the usual method of treating such inaccurate data. We begin with a simple example. Some students are assigned to find the effect of temperature on the resistance of a metal wire. They have recorded the temperature and resistance values in a table and have plotted their findings, as seen in Figure 3.13. The graph suggest a linear relationship. If so, then

$$R = aT + b, \tag{3.24}$$

and values for the parameters, a and b, can be obtained from the plot.

If someone else were given the data and asked to draw the line, it is not likely that they would draw exactly the same line and they would get different values for a and b.

We would like a way of fitting a line to experimental data that is unambiguous and that, in some sense, minimizes the deviations of the points from the line. The usual method for doing this is called the *least-squares method.* The deviations are determined by the distances between the points and the line. How these distances are measured depends on whether there are experimental errors in both variables or in just one of them.

In analyzing the data from the students' experiments, we will assume that the temperature values are accurate and that the errors are only in the resistance numbers; we then will use the vertical distances. (If both measurements were in error, we might use the perpendicular distances and would modify the following. If this is done, the problem becomes more complicated. We treat only the simpler case.)

We might first suppose we could minimize the deviations by making their sum a minimum, but this is not an adequate criterion. Consider the case of only two points (Fig. 3.14).

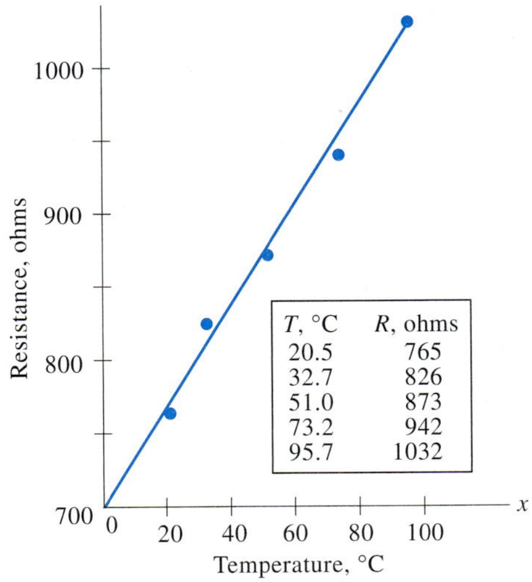

Figure 3.13

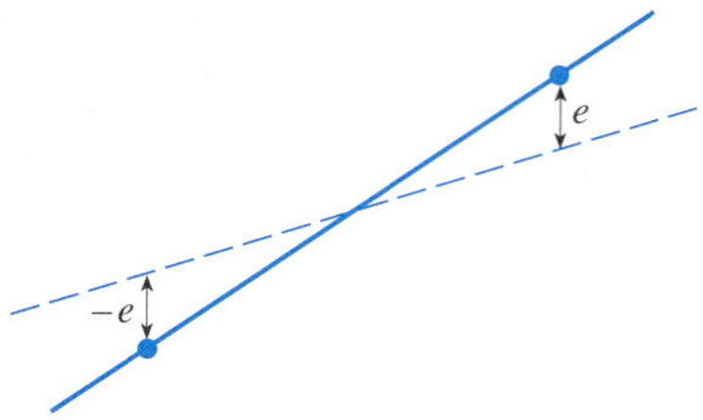

Figure 3.14

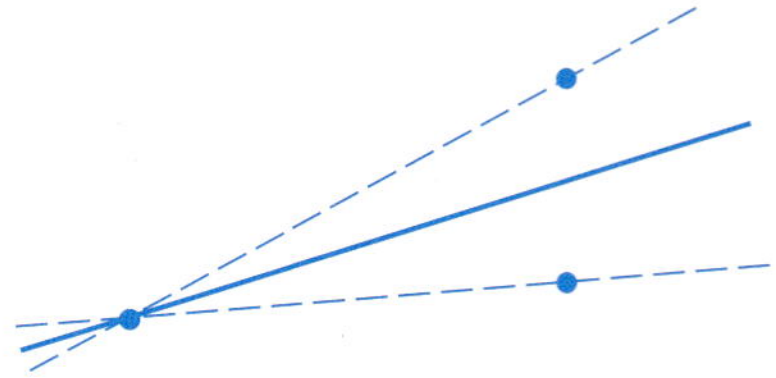

Figure 3.15

Obviously, the best line passes through each point, but any line that passes through the midpoint of the segment connecting them has a sum of errors equal to zero.

Then what about making the sum of the magnitudes of the errors a minimum? This also is inadequate, as the case of three points shows (Fig. 3.15). Assume that two of the points are at the same x-value (which is not an abnormal situation, as frequently experiments are duplicated). The best line will obviously pass through the average of the duplicated tests. However, any line that falls between the dotted lines shown will have the same sum of the magnitudes of the vertical distances. We wish an unambiguous result, so we cannot use this as a basis for our work.

We might accept the criterion that we make the magnitude of the maximum error a minimum (the so-called *minimax* criterion), but for the problem at hand this is rarely done. This criterion is awkward because the absolute-value function has no derivative at the origin, and it also is felt to give undue importance to a single large error. The usual criterion is to minimize the sum of the *squares* of the errors, the "least-squares" principle.*

In addition to giving a unique result for a given set of data, the least-squares method is also in accord with the *maximum-likelihood* principle of statistics. If the measurement errors have a so-called normal distribution and if the standard deviation is constant for all the data, the line determined by minimizing the sum of squares can be shown to have values of slope and intercept that have maximum likelihood of occurrence.

Let Y_i represent an experimental value, and let y_i be a value from the equation

$$y_i = ax_i + b,$$

where x_i is a particular value of the variable assumed to be free of error. We wish to determine the best values for a and b so that the y's predict the function values that correspond to x-values. Let $e_i = Y_i - y_i$. The least-squares criterion requires that

$$\begin{aligned} S &= e_1^2 + e_2^2 + \cdots + e_N^2 \\ &= \sum_{i=1}^{N} e_i^2 \\ &= \sum_{i=1}^{N} (Y_i - ax_i - b)^2 \end{aligned}$$

* The various criteria for a "best fit" can be described by minimizing a norm of the error vector. Relate each criterion to its corresponding vector norm to review the definition of such norms.

be a minimum. N is the number of (x, Y)-pairs. We reach the minimum by proper choice of the parameters a and b, so they are the "variables" of the problem. At a minimum for S, the two partial derivatives $\partial S/\partial a$ and $\partial S/\partial b$ will both be zero. Hence, remembering that the x_i and Y_i are data points unaffected by our choice of values for a and b, we have

$$\frac{\partial S}{\partial a} = 0 = \sum_{i=1}^{N} 2(Y_i - ax_i - b)(-x_i),$$

$$\frac{\partial S}{\partial b} = 0 = \sum_{i=1}^{N} 2(Y_i - ax_i - b)(-1).$$

Dividing each of these equations by -2 and expanding the summation, we get the so-called *normal equations*

$$a \sum x_i^2 + b \sum x_i = \sum x_i Y_i,$$
$$a \sum x_i + bN = \sum Y_i. \qquad \textbf{(3.25)}$$

All the summations in Eq. (3.25) are from $i = 1$ to $i = N$. Solving these equations simultaneously gives the values for slope and intercept a and b.

For the data in Figure 3.13 we find that

$$N = 5, \quad \Sigma T_i = 273.1, \quad \Sigma T_i^2 = 18{,}607.27, \quad \Sigma R_i = 4438, \quad \Sigma T_i R_i = 254{,}932.5.$$

Our normal equations are then

$$18{,}607.27a + 273.1b = 254{,}932.5,$$
$$273.1a + 5b = 4438.$$

From these we find $a = 3.395$, $b = 702.2$, and hence write Eq. (3.24) as

$$R = 702 + 3.39T.$$

MATLAB gets a least-squares polynomial with its 'polyfit' command, the same one that fits an interpolating polynomial to data defined in vectors x and y:

```
eq = polyfit (x, y, N).
```

When the numbers of points (the size of x) is greater than the degree plus one, the polynomial is the least squares fit. So, to solve for the equation for the least squares line with the data of Figure 3.13, we do

```
EDU>> x = [20.5 32.7 51.0 73.2 95.7];
EDU>> y = [765 826 873 942 1032];
EDU>> eq = polyfit (x, y, 1)
eq =
   3.3949   702.1721
```

which give us the coefficients of the equation with somewhat more precision.

Nonlinear Data

In many cases, of course, data from experimental tests are not linear, so we need to fit to them some function other than a first-degree polynomial. Popular forms that are tried are the exponential form

$$y = ax^b$$

or

$$y = ae^{bx}.$$

We can develop normal equations for these analogously to the preceding development for a least-squares line by setting the partial derivatives equal to zero. Such nonlinear simultaneous equations are much more difficult to solve than linear equations. Thus, the exponential forms are usually linearized by taking logarithms before determining the parameters:

$$\ln y = \ln a + b \ln x$$

or

$$\ln y = \ln a + bx.$$

We now fit the new variable $z = \ln y$ as a linear function of $\ln x$ or x as described earlier. Here we do not minimize the sum of squares of the deviations of Y from the curve, but rather the deviations of $\ln Y$. In effect, this amounts to minimizing the squares of the percentage errors, which itself may be a desirable feature. An added advantage of the linearized forms is that plots of the data on either log-log or semilog graph paper show at a glance whether these forms are suitable by whether a straight line represents the data when so plotted.

In cases when such linearization of the function is not desirable, or when no method of linearization can be discovered, graphical methods are frequently used; one merely plots the experimental values and sketches in a curve that seems to fit well. Special forms of graph paper, in addition to log-log and semilog, may be useful (probability, log-probability, and so on). Transformation of the variables to give near linearity, such as by plotting against $1/x$, $1/(ax + b)$, $1/x^2$, and other polynomial forms of the argument may give curves with gentle enough changes in slope to allow a smooth curve to be drawn. S-shaped curves are not easy to linearize; the Gompertz relation

$$y = ab^{c^x}$$

is sometimes employed. The constants a, b, and c are determined by special procedures. Another relation that fits data to an S-shaped curve is

$$\frac{1}{y} = a + be^{-x}.$$

In awkward cases, subdividing the region of interest into subregions with a piecewise fit in the subregions can be used.

The objection to the graphical technique is its *lack of uniqueness.* Two individuals will usually not draw the same curve through the points. One's judgment is frequently distorted by one or two points that deviate widely from the remaining data. Often one tends to pay too much attention to the extremities in comparison to the points in the central parts of the region of interest.

Further problems are caused if we wish to integrate or differentiate the function. Our discussion of least-squares polynomials is one solution to these difficulties.

Least-Squares Polynomials

Because polynomials can be readily manipulated, fitting such functions to data that do not plot linearly is common. We now consider this case. It will turn out that the normal equations are linear for this situation, which is an added advantage. In the development, we use n as the degree of the polynomial and N as the number of data pairs. Obviously, if $N = n + 1$, the polynomial passes exactly through each point and the methods discussed earlier in this chapter apply, so we will always have $N > n + 1$ in the following.

We assume the functional relationship

$$y = a_0 + a_1 x + a_2 x^2 + \cdots + a_n x^n, \tag{3.26}$$

with errors defined by

$$e_i = Y_i - y_i = Y_i - a_0 - a_1 x_i - a_2 x_i^2 - \cdots - a_n x_i^n.$$

We again use Y_i to represent the observed or experimental value corresponding to x_i, with x_i free of error. We minimize the sum of squares,

$$S = \sum_{i=1}^{N} e_i^2 = \sum_{i=1}^{N} (Y_i - a_0 - a_1 x_1 - a_2 x_i^2 - \cdots - a_n x_i^n)^2.$$

At the minimum, all the partial derivatives $\partial S/\partial a_0$, $\partial S/\partial a_1, \ldots, \partial S/\partial a_n$ vanish. Writing the equations for these gives $n + 1$ equations:

$$\frac{\partial S}{\partial a_0} = 0 = \sum_{i=1}^{N} 2(Y_i - a_0 - a_1 x_i - \cdots - a_i x_i^n)(-1),$$

$$\frac{\partial S}{\partial a_1} = 0 = \sum_{i=1}^{N} 2(Y_i - a_0 - a_1 x_i - \cdots - a_i x_i^n)(-x_i),$$

$$\vdots \qquad \vdots$$

$$\frac{\partial S}{\partial a_n} = 0 = \sum_{i=1}^{N} 2(Y_i - a_0 - a_1 x_i - \cdots - a_n x_i^n)(-x_i^n).$$

Dividing each by -2 and rearranging gives the $n + 1$ normal equations to be solved simultaneously:

$$
\begin{aligned}
a_0 N + a_1 \sum x_i + a_2 \sum x_i^2 + \cdots + a_n \sum x_i^n &= \sum Y_i, \\
a_0 \sum x_i + a_1 \sum x_i^2 + a_2 \sum x_i^3 + \cdots + a_n \sum x_i^{n+1} &= \sum x_i Y_i, \\
a_0 \sum x_i^2 + a_1 \sum x_i^3 + a_2 \sum x_i^4 + \cdots + a_n \sum x_i^{n+2} &= \sum x_i^2 Y_i, \\
\vdots \qquad\qquad & \qquad \vdots \\
a_0 \sum x_i^n + a_1 \sum x_i^{n+1} + a_2 \sum x_i^{n+2} + \cdots + a_n \sum x_i^{2n} &= \sum x_i^n Y_i.
\end{aligned}
\tag{3.27}
$$

Putting these equations in matrix form shows an interesting pattern in the coefficient matrix.

$$
\begin{bmatrix}
N & \sum x_i & \sum x_i^2 & \sum x_i^3 & \dots & \sum x_i^n \\
\sum x_i & \sum x_i^2 & \sum x_i^3 & \sum x_i^4 & \dots & \sum x_i^{n+1} \\
\sum x_i^2 & \sum x_i^3 & \sum x_i^4 & \sum x_i^5 & \dots & \sum x_i^{n+2} \\
 & & \vdots & & & \vdots \\
\sum x_i^n & \sum x_i^{n+1} & \sum x_i^{n+2} & \sum x_i^{n+3} & \dots & \sum x_i^{2n}
\end{bmatrix}
[a] =
\begin{bmatrix}
\sum Y_i \\ \sum x_i Y_i \\ \sum x_i^2 Y_i \\ \vdots \\ \sum x_i^n Y_i
\end{bmatrix}.
\tag{3.28}
$$

All the summations in Eqs. (3.27) and (3.28) run from 1 to N. (We will let B stand for the coefficient matrix.)

Equation (3.28) represents a linear system; how this can be solved was covered in Chapter 2. However, you need to know that this system is ill-conditioned and round-off errors can distort the solution: the a's of Eq. (3.26). Up to degree-3 or -4, the problem is not too great. Special methods that use *orthogonal* polynomials are a remedy. We do not pursue this because degrees higher than 4 are used very infrequently. It is often better to fit a series of lower-degree polynomials to subsets of the data.

Matrix B of Eq. (3.28) is called the *normal matrix* for the least-squares problem. There is another matrix that corresponds to this, called the *design matrix*. It is of the form

$$
A = \begin{bmatrix}
1 & 1 & 1 & \dots & 1 \\
x_1 & x_2 & x_3 & \dots & x_N \\
x_1^2 & x_2^2 & x_3^2 & \dots & x_N^2 \\
\vdots & & & & \vdots \\
x_1^n & x_2^n & x_3^n & \dots & x_N^n
\end{bmatrix}.
$$

It is easy to show that AA^T is just the coefficient matrix of Eq. (3.28). It is also easy to see that Ay, where y is the column vector of Y-values, gives the right-hand side of Eq. (3.28) (You ought to try this for, say, a 3×3 case to reassure yourself.) This means that we can rewrite Eq. (3.28) in matrix form, as

$$AA^Ta = Ba = Ay.$$

We can use Gaussian elimination to solve the system (but only for low-degree polynomials). However, because B has special properties, another method can be used that avoids the problem of ill-conditioning.

1. The matrix $B = AA^T$ is symmetric and positive definite. An $n \times n$ matrix, M, is said to be positive semidefinite if, for every n-component vector, $x^TMx \geq 0$. If we add the condition that $x^TMx = 0$ only if x is the zero vector, M is said to be positive definite. (You should show that B is positive definite and symmetric.)
2. In linear algebra, it is shown that B can be diagonalized by an orthogonal matrix P:

$$PBP^T = PAA^TP^T = D,$$

   where the diagonal elements of D are the eigenvalues of B. Note that orthogonality implies that $PP^T = I$, the identity matrix.
3. B is positive definite, so all of its eigenvalues are nonnegative. This means that we can define a matrix S as

$$S = \sqrt{D}, \qquad \text{or} \qquad S^2 = D.$$

   The diagonal elements of S are called the singular values of A.
4. We can rewrite Eq. (3.28) and its solution as follows:

$$AA^Ta = P^TDPa = (SP)^T(SP)a = Ay,$$
$$a = P^TD^{-1}PAy.$$

This last eliminates having to multiply out AA^T and, by extending this approach, leads to an important method for solving Eq. (3.28) called *singular-value decomposition*. [See Press, *Numerical Recipes* (1992) on this topic.]

MATLAB has a command `'(U, S, V] = svd(A)'` that computes the singular value decomposition of matrix A. The combination `U * S * V` is equal to A and the singular values of A are on the diagonal of S. *Mathematica* can do the same. (When A is symmetric and semidefinite, the singular values are the eigenvalues.) We do not pursue this idea further.

We illustrate the use of Eqs. (3.27) to fit a quadratic to the data of Table 3.14. Figure 3.16 shows a plot of the data. (The data are actually a perturbation of the relation $y = 1 - x + 0.2x^2$.

Table 3.14 Data to illustrate curve fitting

x_i	0.05	0.11	0.15	0.31	0.46	0.52	0.70	0.74	0.82	0.98	1.171
Y_i	0.956	0.890	0.832	0.717	0.571	0.539	0.378	0.370	0.306	0.242	0.104

$\Sigma x_i = 6.01$	$N = 11$
$\Sigma x_i^2 = 4.6545$	$\Sigma Y_i = 5.905$
$\Sigma x_i^3 = 4.1150$	$\Sigma x_i Y_i = 2.1839$
$\Sigma x_i^4 = 3.9161$	$\Sigma x_i^2 Y_i = 1.3357$

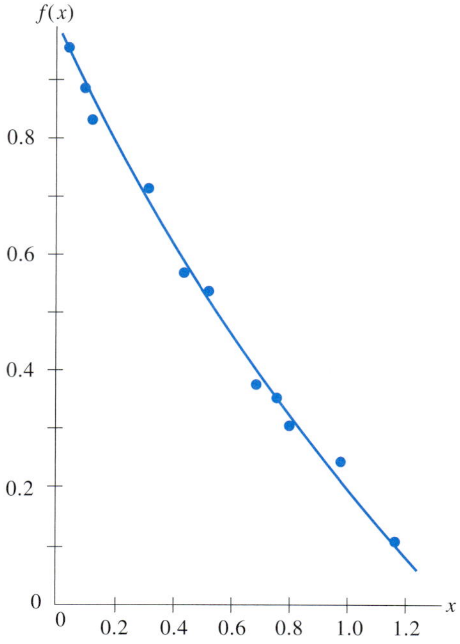

Figure 3.16

It will be of interest to see how well we approximate this function.) To set up the normal equations, we need the sums tabulated in Table 3.14. The equations to be solved are:

$$\begin{aligned} 11a_0 + \quad 6.01a_1 + 4.6545a_2 &= 5.905, \\ 6.01a_0 + 4.6545a_1 + 4.1150a_2 &= 2.1839, \\ 4.6545a_0 + 4.1150a_1 + 3.9161a_2 &= 1.3357. \end{aligned}$$

The result is $a_0 = 0.998$, $a_1 = -1.018$, $a_2 = 0.225$, so the least squares method gives

$$y = 0.998 - 1.018x + 0.225x^2,$$

which we compare to $y = 1 - x + 0.2x^2$. Errors in the data cause the equations to differ.

Use of Orthogonal Polynomials

We have mentioned that the system of normal equations for a polynomial fit is ill-conditioned when the degree is high. Even for a cubic least-squares polynomial, the condition number of the coefficient matrix can be large. In one experiment, [Atkinson (1985), p. 263] a cubic polynomial was fitted to 21 data points. When the data were put into the coefficient matrix of Eq. (3.28), its condition number (using 2-norms) was found to be 22,000! This means that small differences in the y-values will make a large difference in the solution. In fact, if the four right-hand-side values are each changed by only 0.01 (about 0.1%), the solution for the parameters of the cubic were changed significantly, by as much as 44%!

However, if we fit the data with *orthogonal polynomials** such as the *Chebyshev polynomials* that are described in the next chapter, the condition number of the coefficient matrix is reduced to about 5 and the solution is not much affected by the perturbations. We will postpone further discussion of orthogonal polynomials.

What Degree of Polynomial Should Be Used?

In the general case, we may wonder what degree of polynomial should be used. As we use higher-degree polynomials, we of course will reduce the deviations of the points from the curve until, when the degree of the polynomial, n, equals $N - 1$, there is an exact match (assuming no duplicate data at the same x-value) and we have an interpolating polynomial. The answer to this problem is found in statistics. One increases the degree of approximating polynomial as long as there is a statistically significant decrease in the variance, σ^2, which is computed by

$$\sigma^2 = \frac{\sum e_i^2}{N - n - 1}. \tag{3.29}$$

For the preceding example, when the degree of the polynomial made to fit the points is varied from 1 to 7, we obtain the results shown in Table 3.15.

The criterion of Eq. (3.29) chooses the optimal degree as 2. This is no surprise, in view of how the data were constructed. It is important to realize that the numerator of Eq. (3.29), the *sum of the deviations squared* of the points from the curve, should continually decrease as the degree of the polynomial is raised. It is the denominator of Eq. (3.29) that makes σ^2 increase as we go above the optimal degree. In this example, the smallest value for σ^2 is at degree-2 as we expect. The small value when the degree is 5 may be due to ill-conditioning, even though double precision was used to get the values in Table 3.15.

Before leaving this section, we illustrate how to apply these methods to a more complicated function.

Table 3.15

Degree	**Equation**	**σ^2 (Eq. 3.27)**	**$\sum e^2$**
1	$y = 0.95228 - 0.76041x$	0.00106	0.00915
2	$y = 0.99800 - 1.0180x + 0.22468x^2$	0.00023	0.00187
3	$y = 1.0037 - 1.0794x + 0.35137x^2 - 0.06894x^3$	0.00026	0.00181
4	$y = 0.98810 - 0.83690x - 0.52680x^2 + 1.0461x^3 - 0.45635x^4$	0.00027	0.00165
5	$y = 1.0369 - 1.8241x + 4.8953x^2 - 10.753x^3 + 10.537x^4 - 3.6594x^5$	0.00013	0.00067

* A sequence of polynomials is said to be orthogonal with respect to the interval $[a, b]$ if

$$\int_a^b P_n(x) * P_m(x)\, dx = 0 \text{ when } n \neq m.$$

EXAMPLE 3.9 The results of a wind tunnel experiment on the flow of air on the wing tip of an airplane provide the following data:

R/C: 0.73, 0.78, 0.81, 0.86, 0.875, 0.89, 0.95, 1.02, 1.03, 1.055, 1.135, 1.14, 1.245, 1.32, 1.385, 1.43, 1.445, 1.535, 1.57, 1.63, 1.755;

V_θ/V_∞: 0.0788, 0.0788, 0.064, 0.0788, 0.0681, 0.0703, 0.0703, 0.0681, 0.0681, 0.079, 0.0575, 0.0681, 0.0575, 0.0511, 0.0575, 0.049, 0.0532, 0.0511, 0.049, 0.0532, 0.0426;

where R is the distance from the vortex core, C is the aircraft wing chord, V_θ is the vortex tangential velocity, and V_∞ is the aircraft free-stream velocity. Let $x = R/C$ and $y = V_\theta/V_\infty$. We would like our curve to be of the form

$$g(x) = \frac{A}{x}(1 - e^{-\lambda x^2}),$$

and our least-squares equation becomes

$$\begin{aligned} S &= \sum_{i=1}^{21}(Y_i - g(x_i))^2 \\ &= \sum_{i=1}^{21}\left(Y_i - \frac{A}{x_i}(1 - e^{-\lambda x_i^2})\right)^2. \end{aligned}$$

Setting $S_A = S_\lambda = 0$ gives the following equations:

$$\sum_{i=1}^{21}\left(\frac{1}{x_i}\right)(1 - e^{-\lambda x_i^2})\left(Y_i - \frac{A}{x_i}(1 - e^{-\lambda x_i^2})\right) = 0,$$

$$\sum_{i=1}^{21} x_i(e^{-\lambda x_i^2})\left(Y_i - \frac{A}{x_i}(1 - e^{-\lambda x_i^2})\right) = 0.$$

When this system of nonlinear equations is solved, we get

$$g(x) = \frac{0.07618}{x}(1 - e^{-2.30574x^2}).$$

For these values of A and λ, $S = 0.000016$. The graph of this function is presented in Figure 3.17.

Here is an algorithm for obtaining a least-squares polynomial:

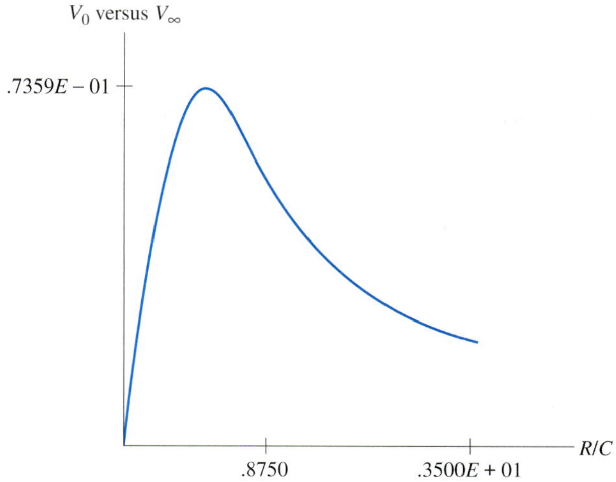

Figure 3.17

Given N data pairs, (x_i, Y_i) $i = 1, \ldots, N$, obtain an nth-degree least-squares polynomial by the following:
Form the coefficient matrix, M, with $n + 1$ rows (r) and $n + 1$ columns (c), by

$$\text{Set } M_{rc} = \sum_{i=1}^{N} x_i^{r+c-2}.$$

Form the right-hand-side vector b, with $n + 1$ rows (r), by:

$$\text{Set } b_r = \sum_{i=1}^{N} x_i^{r-1} Y_i,$$

Solve the linear system $Ma = b$ to get the coefficients in

$$y = a_0 + a_1 x + a_2 x^2 + \cdots + a_n x^n,$$

which is the desired polynomial that fits the data.

Exercises

Section 3.1

1. Write out the Lagrangian polynomial from this table:

x	y
2.1	−12.4
4.1	7.3
7.1	10.1

a. Confirm that it reproduces the y's for each x-value.
b. Interpolate with it to estimate y at $x = 3$.
c. Extrapolate for $x = 8$.
d. Plot the polynomial between $x = 0$ and $x = 10$ together with the original points and the interpolates in parts (b) and (c).

2. Suppose in Exercise 1 that the y-value for $x = 4.1$ is mistakenly entered as 7.2 rather than 7.3. Repeat Exercise 1 with this incorrect value. How much difference does this make to the answers for parts (b) and (c)?

▶ **3.** Multiply out the Lagrangian polynomials in Exercises 1 and 2 to get the quadratics in the form $ax^2 + bx + c$. How different are the values for a, b, and c?

4. Given the four points (2, 1), (4, 3), (3, 5), (8, 9),

a. Find the cubic that passes through them, in Lagrangian form.
b. Multiply out to express as $ax^3 + bx^2 + cx + d$.
c. How many arithmetic operations (addition, subtraction, multiplication, division) are required to use the Lagrangian polynomial to interpolate at $x = 6.5$?
d. How many operations are required to convert the Lagrangian form to the cubic equation in part (b)?
e. How many operations are used to interpolate at $x = 6.5$ with the cubic of part (b)?
f. Interpolating as in part (c) requires more operations than in part (e), but to use the polynomial as in part (d) means that part (b) must be done first. But if we must interpolate many times, using the cubic of part (b) would be more efficient. How many interpolations does it take to overcome the overhead work?

5. Plot the coefficients of each term of the Lagrangian polynomial of Exercise 4 to see their form. Then superimpose each of these together with the plot of the cubic of Exercise 4, part (b).

▶ **6.** If $e^{1.3}$ is approximated by Lagrangian interpolation from the values for $e^0 = 1$, $e^1 = 2.7183$, and $e^2 = 7.3891$, what are the minimum and maximum estimates for the error? Compare to the actual error.

7. Repeat Exercise 6, but now extrapolate to get $e^{2.7}$.

8. Use the following data to construct the Neville table that interpolates at $x = 0.6$.

x	$f(x)$
0.3	0.404958
0.5	0.824361
0.7	1.40963
0.9	2.21364
1.1	3.30458

a. Compare the result to those when only the first three data points are used.
b. Repeat part (a), but with the three points from $x =$ 0.5 to 0.9.
c. Repeat with the first four points.
d. Repeat again with the last four points.
e. The table is for $f(x) = x * e^x$. Which of the answers is closest to the correct result, $f(0.6) = 1.09327$? Why this one?

9. Repeat Exercise 8, but after rounding the function values to four significant digits. Does this make a noticeable difference in the answers?

▶ **10.** If parallel processing with n processors is used to interpolate at some x-value from a Lagrangian polynomial constructed from n $(x, f(x))$ values, how much computing time can be saved compared to sequential processing? Express in terms of T, the time it takes to compute one term of the polynomial by itself. If we have $2n$ processors, can the solution be obtained more rapidly?

11. Repeat Exercise 10, but now for Neville's method where T is the time to compute one term in a column. What should be the order in which terms are computed in getting the table? Assume that there are enough processors to give maximum speedup. How many processors can be usefully employed? Can this take advantage of the fact that not all n points may be needed?

Section 3.2

12. Construct the divided-difference table from these data:

x	−0.2	0.3	0.7	−0.3	0.1
$f(x)$	1.23	2.34	−1.05	6.51	−0.06

13. Repeat Exercise 4, but now do it with divided differences (use the divided-difference polynomial wherever the Lagrangian polynomial is mentioned). Omit part (e).

14. Use the divided-difference table from Exercise 12 to interpolate for $f(0.4)$

a. Using the first three points.
b. Using the last three points.
c. Using the best set of three points. Which points should be used?
d. Using the best set of four points.
e. Using all of the points.
f. Explain why the results are not all the same.

▶ **15.** Is this table of values from a polynomial? If so, how do you know that $f(x)$ is truly a polynomial? What is its degree? What is the polynomial? Use divided differences to answer the questions.

x	$f(x)$
−1	8
3	0
2	−1
−2	15
4	3

16. After you have solved Exercise 15, you are told that $f(0) = 5$. With this new information, repeat Exercise 15.

▶**17.** Complete the table of ordinary differences from these data:

x	1.20	1.25	1.30	1.35	1.40	1.45	1.50
$f(x)$	0.1823	0.2231	0.2624	0.3001	0.3365	0.3716	0.4055

a. What degree of polynomial is required to exactly fit all seven points to within three significant digits?
b. What polynomial of lesser degree will almost fit to the same precision?
c. Justify your answer to part (b).

18. Using the data in Exercise 17,

a. Compute $f[x_0, x_1, x_2, x_3]$ directly from the data without making a divided-difference table if x_0 is 1.30.
b. Repeat part (a), but now get $\Delta^3 f_0$. Verify from the divided-difference table you created in Exercise 17.
c. Construct the divided-difference table for the data in Exercise 17 and verify your answer to part (a).

19. Estimate the value of $f(1.33)$ from the data in Exercise 17 using the third-degree Newton–Gregory interpolating polynomial. Use the best starting point. Estimate the error from the next-term rule.

20. Repeat Exercise 19, but now estimate $f(0.67)$. Is the estimated error greater than in Exercise 19? If so, explain.

▶**21.** The function tabulated in Exercise 17 is not known but you can still use the data to interpolate. Estimate $f(1.32)$ using the best set of three points. Estimate the error from the next-term rule.

22. What are the bounds to the error in the result from Exercise 7? Compare to the actual error.

23. Repeat Exercise 8, but now from a divided-difference table. Compare the error of the estimate to the bounds on the error.

24. Show how nested multiplication can be used to evaluate the polynomial $P_n(x)$ of Eq. (3.8). How many fewer operations are needed when $n = 2$? When $n = 3$? When $n = 4$?

25. Prove that the divided difference of order n is always a constant if $f(x)$ is a polynomial of degree n.

▶**26.** Suppose you have a table of x, $f(x)$ values that has seven entries.

a. How many computer operations are used in computing the divided-difference table up to the third order?
b. How many operations are used to compute the ordinary differences?

27. Given three points from which polynomial $P_2(x)$ has been found. You know the function that gives the three points.

a. Show that $f''(x) \approx 2 * f[x_1, x_2, x_3]$ for any three distinct points.
b. What is the relation if the points are evenly spaced and you use the table of ordinary differences?

28. Given the $f(x)$ values at three distinct x-values, x_1, x_2, x_3. There are six different ways in which the points can be ordered. Show that the second-order divided differences are identical for all permutations of the ordering.

29. In a table of (x, y) values, one point is a duplicate of another point. What happens when the divided-difference table is constructed? What happens with an ordinary difference table?

30. Use this ordinary difference table:

x	$f(x)$	Δf	$\Delta^2 f$	$\Delta^3 f$	$\Delta^4 f$
0.12	0.79168	−0.01834	−0.01129	0.00134	0.00038
0.24	0.77334	−0.02963	−0.00995	0.00172	0.00028
0.36	0.74371	−0.03958	−0.00823	0.00200	
0.48	0.70413	−0.04781	−0.00623		
0.60	0.65632	−0.05404			
0.72	0.60228				

a. Estimate $f(0.231)$ from the Newton–Gregory polynomial of degree-2 with $x_0 = 0.12$.
b. Add one term to part (a) to get $f(0.231)$ from the third-degree polynomial.
c. Estimate the errors of both parts (a) and (b).
d. Is it better to start with $x_0 = 0.24$ or with $x_0 = 0.36$ when getting $f(0.42)$ from a quadratic? Justify your answer.

e. Demonstrate that the precision of the data can have a large effect on the table by recomputing it with the function values chopped after three decimal places.

▶**31.** You have these values for x and $y(x)$:

x	0.1	0.3	0.5	0.7	0.9	1.1	1.3
y	0.003	0.067	0.148	0.248	0.370	0.518	0.697

a. Find $y(0.54)$ from a cubic that starts from $x = 0.1$.
b. Repeat part (a) but start from $x = 0.3$. Should this be a better value?
c. What is the minimum degree of polynomial that fits all the data?
d. Construct the difference table, then construct the divided difference table. How do these differ?

Section 3.3

32. Fit this function with interpolating polynomials that match $f(x)$ at equal intervals in $[-1, 1]$:

$$f(x) = \begin{cases} 0, & -1 < x < -0.5 \\ 1 - |2x|, & -0.5 < x < \ \ 0.5 \\ 0, & 0.5 < x < \ \ 1.0 \end{cases}$$

a. A polynomial of degree 2.
b. A polynomial of degree 3.
c. A polynomial of degree 4.
d. Plot the function and each of the polynomials.

33. Fit the function in Exercise 32 with a natural cubic spline that matches to $f(x)$ at five evenly spaced points in $[-1, 1]$. Plot the spline curve together with $f(x)$.

34. Repeat Exercise 33 but now use end conditions 3 and 4 as defined in Eq. (3.18). Which end condition gives the best fit to the function?

35. Repeat Exercise 33, but now force the slopes at the ends to be zero. Which spline fits better, this one or one from Exercises 33 or 34?

▶**36.** Find the coefficient matrix and the right-hand-side vector for the cubic spline that fits to these data:

x	0.15	0.27	0.76	0.89	1.07	2.11
$f(x)$	0.1680	0.2974	0.7175	0.7918	0.8698	0.9972

37. Solve the equations of Exercise 36 to get the equations for the cubics. Use this to interpolate at $x = 0.31$, $x = 0.85$, and $x = 2.05$. Compare these interpolates with the values of $ERF(x)$, the so-called error function, which you can find in tables.

38. Develop the equations to get the coefficients of quadratic splines. What end conditions are appropriate for these quadratic splines?

39. Repeat Exercise 33 with the equations of Exercise 38. Use the end conditions that you provided for in Exercise 38.

▶**40.** A cubic spline with end condition 1 is called a "free spline." If the slopes at the ends are specified (condition 2), it is called a "clamped spline." Suppose that $f(x)$ is a third-degree polynomial:

a. Show that $f(x)$ is its own clamped spline.
b. Show that $f(x)$ is not its own free spline.

41. When data for a periodic function are tabulated, the first and last points are identical. Develop the equations for fitting a cubic spline to such periodic data, taking into account the matching of endpoints.

42. What if end condition 1 is used at one end ($S_0 = 0$) but end condition 4 is used at the other end? What are the first and last rows of the coefficient matrix for such a spline?

43. Is $f(x)$ a linear spline?

$$f(x) = \begin{cases} 1 - x, & -1 \le x \le 1 \\ 2(x - 1), & 1 \le x \le 2 \\ (x + 2)/2, & 2 \le x \le 4 \end{cases}$$

▶**44.** For the function of Exercise 43, fit the four points $f(-1), f(1), f(2)$, and $f(4)$ with a cubic spline. What is the maximum deviation of this spline from $f(x)$ in the interval $[-1, 4]$? At what x-value does this occur?

Section 3.4

45. Show that the matrix form of the equations for the Bezier curve is equivalent to the algebraic form in Section 3.4.

46. Repeat Exercise 45 for B-splines.

▶**47.** What is the matrix form for a Bezier curve of order-4?

48. Compute the connected Bezier curve from this set of points:

	Point #									
	0	1	2	3	4	5	6	7	8	9
x	10	50	75	90	105	150	180	190	160	130
y	10	15	60	100	140	200	140	120	100	80

a. Draw the graph determined by the ten points.
b. Why is the graph smoothly connected at points 3 and 6?
c. Rewrite the Bezier equations so that the parameter u

is defined on [0, 1] for points 0 to 3, on [1, 2] for points 3 to 6, and on [2, 3] for points 6 to 9.

49. Repeat Exercise 48 for a B-spline curve.

50. Plot the weighting factors that produce a cubic Bezier curve by letting u vary between zero and one.

51. Repeat Exercise 50, but for a cubic B-spline.

▶**52.** Prove that the slopes at the ends of a cubic B-spline curve are the same as the slope between the two endpoints.

53. If these four points are connected in order by straight lines, a zigzag line is created:

$$(0, 0), (1, 0.3), (2, 1.7), (3, 1.5).$$

a. Using the two interior points as controls, find the cubic Bezier curve. Plot this together with the zigzag line.
b. Use this cubic equation to find interpolates at $x = 0.5$, $x = 0.75$, and $x = 2.5$. How close are these to the zigzag line?
c. If the second and third points (the control points) are moved, the Bezier curve will change. If these are moved vertically, where should they be located so that the Bezier curve passes through all of the original four points?

54. Repeat Exercise 53, but for B-spline curves. Add fictitious points at the end so the end portions are completed.

▶**55.** If one of the points used in constructing a connected B-spline curve is changed, what parts of the curve are affected? Is there a change that does not affect the curve? Do the terms *local control* and *global control* apply to what you observe?

56. Repeat Exercise 55 for a connected Bezier curve.

57. A fourth-degree B-spline is a natural extension of the cubic B-spline. Can the degree be reduced to two? What assumptions are reasonable for such a quadratic?

58. The function $y = x/(x^3 + 1/5)$ is discontinuous near $x = -0.6$, is zero at $x = 0$, and has a sharp maximum near $x = 0.5$. Find the seven evenly spaced points near $x = 0.5$ that define a B-spline that goes through the endpoints and matches to the maximum within ± 0.002.

Section 3.5

▶**59.** In Section 3.5, it is asserted that interpolation can be done by making subtables with x held constant or, alternatively, from subtables with y held constant. Example 3.8 did it the first way. Recompute $f(1.6, 0.33)$ in the second way. Do you get the same result?

60. After Example 3.8 was completed, it was observed that a cubic in x and a quadratic in y might be preferred. Use this preferred attack to estimate $f(1.6, 0.33)$. How does this compare to the answer in Exercise 59? How much is it in error from the true value, 1.8350? Was the "preferred attack" really better?

61. In Example 3.8 and in Exercises 59 and 60, a rectangular set of points from the table was used. Is it advantageous to use a more nearly circular set of points? Estimate $f(1.6, 0.33)$ from a set of the 12 points closest to $(x, y) = (1.6, 0.33)$ and compare with the results from Example 3.8, and from Exercises 59 and 60.

▶**62.** From this table, estimate $z(x, y)$ for $x = 2.8$ and $y = 0.54$ using an array of nine points nearest to the point of interpolation to construct interpolating polynomials. (There may be several ways to choose these points; try them all.) The function whose values are tabulated is $z = x + e^y$.

$x\backslash y$	0.2	0.4	0.5	0.7	0.9
1.3	2.521	2.792	2.949	3.314	3.760
2.5	3.721	3.992	4.149	4.514	4.960
3.1	4.321	4.592	4.749	5.114	5.560
4.7	5.921	6.192	6.349	6.714	7.160
5.5	6.721	6.992	7.149	7.514	7.960

63. Using the data from Exercise 62, construct the B-spline surface from the rectangular array of 16 points nearest to (2.8, 0.54) and find $z(2.8, 0.54)$. Compare to the result of Exercise 62.

64. Repeat Exercise 63, but now for a Bezier surface.

65. Repeat Exercise 63, but now use cubic splines.

Section 3.6

66. Figure 3.13 plots data that appear to be linear and the least-squares line that fits is $R = 702.2 + 3.395T$. A line drawn by eye that also seems to fit the data is $R = 700 + 3.5T$.

a. Draw both lines on a copy of Figure 3.13 to confirm that both equations are reasonable representations.
b. Compute the deviations of the R-values of the data from each of these lines. Find the sum of squares of the deviations and compare. Which sum is smaller? By how much?

c. Compare the maximum of the deviations for each equation. Are they much different? How do the averages of the errors compare?

▶**67.** Show that (X, Y) where X is the average of the x-values and Y is the average of the y-values for any set of data points is a point on the least-squares line that fits to the data.

68. Fit these six (x, y)-values with a straight line:

x	1.2	3.1	5.6	6.2	8.8	9.1
y	4.957	12.909	23.404	25.981	36.907	38.212

a. Do this assuming that the x-values are free of errors, given $y = f(x)$.
b. Repeat but now assume that the y-values are error free.
c. Part (b) gives $x = g(y)$. Translate this to $y = h(x)$. Is $h(x)$ the same as the result of part (a)?
d. For which line is the sum of the squares of the deviations smaller?

69. The equation of a plane is $z = ax + by + c$. We can fit experimental data to such a plane using the least-squares technique. Here are some data for $z = f(x, y)$:

x	0.40	1.2	3.4	4.1	5.7	7.2	9.3
y	0.70	2.1	4.0	4.9	6.3	8.1	8.9
z	0.031	0.933	3.058	3.349	4.870	5.757	8.921

a. Develop the normal equations to fit the (x, y) data to a plane.
b. Use these equations to fit $z = ax + by + c$.
c. What is the sum of the squares of the deviations of the points from the plane?

▶**70.** Plot the line between (2, 5) and (6, −1) and get its equation. Now add a third point at $x = 4$ and find the least-squares line for the three points. For which y-value does the line shift the most

a. If the y-value at the third point is 5?
b. If the y-value is 0?
c. If the y-value is 4?
d. Find the equations of the least-squares lines for each.

71. These data are measured solubilities of n-butane in liquid anhydrous hydrofluoric acid. Fit to the equation $S = a * \exp(b * T)$ using least squares. (This is important in a process that converts n-butane to i-butane, which gives a higher octane number to gasoline.)

T, °F	77	100	185	239	285
S, wt. %	2.4	3.4	7.0	1.1	19.6

72. Plot the data of Exercise 71:

a. On ordinary graph paper. Observe the points do not fall on a line.
b. On semi-log paper. Observe that the points fall near to a line.
c. Part (b) suggest that fitting a line to

$$\ln(S) = \ln(a) + b * T$$

will give the same results as in Exercise 71. Do this to confirm.

73. $y = ax^2 + bx + c$ is a quadratic equation, of course. Compute $z = y$ + a random number within the range [0, .2] for six x-values chosen randomly within the range [2, 7].

a. Fit the least-squares line to these points.
b. Fit the least-squares quadratic to them.
c. Fit the least-squares cubic to them.
d. Compare the sum of squares of the deviations for each part.

74. If A is the design matrix defined in Section 3.6,

a. Show that $A * A^T$ gives the coefficient matrix of Eq. (3.28).
b. Show that $A * y$, where y is the column vector of y-values, gives the right-hand side of Eq. (3.28).
c. Is it more economical to compute the values for Eq. (3.28) by using the design matrix rather than computing them with Eq. (3.28)?

▶**75.** From theoretical considerations, it is suspected that the rate of flow from a fire hose is proportional to some power of the pressure at the nozzle. Do these data confirm that? Get the least-squares values for the exponent and the proportionality factor.

Flow rate	94	118	147	180	230
Pressure	10	16	25	40	60

76. If the data of Exercise 75 are plotted on log-log paper, the points appear to be nearly linear with a slope of 2. That means that a quadratic, $F = aP^2 + bP + c$, should fit the data.

a. Get the coefficients of the quadratic by least squares.
b. Is the sum of squares of the deviations less than for the relation of Exercise 75?

▶**77.** Fit a polynomial of optimal degree to these points:

x	1.1	1.6	11.4	4.1	5.3	17.5	9.4	11.5	12.1
$f(x)$	7.9	24.8	−28.8	42.6	29.6	−34.6	−3.1	−28.7	−39.6

78. Repeat Exercise 77,

a. Using only every other point.
b. Using the other half of the points.
c. Compare the results of parts (a) and (b) with that of Exercise 77.

79. Suppose you want to use least squares to fit the data of Exercise 77 with this equation:

$$f(x) = a + b * \sin(c * x).$$

a. What difficulties will be experienced if the normal equations for a polynomial are used?
b. If it were known that $c = \pi/10$, would it then be easier to get values for a and b?
c. Does part (b) suggest that it would be preferred to obtain least-squares values for a and b using a succession of c-values and thus finding good values for a, b, and c by seeing when the sum of squares of deviations is smallest?

80. In Section 3.6 it was pointed out that the coefficient matrix in Eq. (3.28) is ill-conditioned if the degree of the polynomial is more than 3 or 4. If experimental data are available at ten evenly spaced x-values from $x = 3$ to $x = 7$, it is possible to find least-squares polynomials $P_n(x)$ for n from 1 to 8.

a. What is the condition number for $P_4(x)$?
b. For $P_6(x)$?
c. For $P_8(x)$?

Applied Problems and Projects

APP1. In Section 3.2, we described how the Newton–Gregory interpolating polynomial can be constructed from a table of ordinary differences. There are other ways to get interpolating polynomials from such a table and these bear the names of famous mathematicians—Gauss–Forward, Gauss–Backward, Stirling, Bessel. There is even a Newton–Gregory backward polynomial. Do research to find how these differ from one another.

APP2. The cost of government welfare programs adds significantly to our taxes. The table below gives data for several years:

Year	Expenditures in billions of dollars
1985	731
1986	782
1987	833
1988	886
1989	956
1990	1049
1991	1159
1992	1267
1993	1367
1994	1436
1995	1505

Use the data between 1991 to 1994 to estimate what the value would be in 1995 and compare to the value in the table. Do this

a. From a cubic interpolating polynomial.
b. From the least-squares line.
c. From the least-squares quadratic.
d. From a cubic spline.

From each of these, project to find what one would anticipate the expenditures for the year 2000 might be; then find what the actual expenditures were for comparison.

APP3. Use the data of APP2 with several approaches to extrapolate backward to estimate the expected expenditure for 1980. How do these values compare to 492 billion, the amount actually spent?

APP4. S. H. P. Chen and S. C. Saxena report experimental data for the emittance of tungsten as a function of temperature [*Ind. Eng. Chem. Fund.* 12, 220 (1973)]. Their data follow. They found that the equation

$$e(T) = 0.02424\left(\frac{T}{303.16}\right)^{1.27591}$$

correlated the data for all temperatures accurately to three digits. What degree of interpolating polynomial is required to match to their correlation at points midway between the tabulated temperatures? Discuss the pros and cons of polynomial interpolation in comparison to using their correlation.

T, °K	300	400	500	600	700	800	900	1000	1100
e	0.024	0.035	0.046	0.058	0.067	0.083	0.097	0.111	0.125

T, °K	1200	1300	1400	1500	1600	1700	1800	1900	2000
e	0.140	0.155	0.170	0.186	0.202	0.219	0.235	0.252	0.269

APP5. In studies of radiation-induced polymerization, a source of gamma rays was employed to give measured doses of radiation. However, the dosage varied with position in the apparatus, with these figures being recorded:

Position, in. from base point	0	0.5	1.0	1.5	2.0	3.0	3.5	4.0
Dosage, 10^5 rads/hr	1.90	2.39	2.71	2.98	3.20	3.20	2.98	2.74

For some reason, the reading at 2.5 in. was not reported, but the value of radiation there is needed. Fit interpolating polynomials of various degrees to the data to supply the missing information. What do you think is the best estimate for the dosage level at 2.5 in.?

APP6. Studies of the kinetics of elution of copper compounds from ion-exchange resins gave the following data. The normality of the leaching liquid was the most important factor in determining the diffusivity. The data were obtained at convenient values of normality; we desire a table of D for integer values of normality (N = 0.0, 1.0, 2.0, 3.0, 4.0, 5.0). Use the data to construct such a table.

N	$D \times 10^6$, cm²/sec	N	$D \times 10^6$, cm²/sec
0.0521	1.65	0.9863	3.12
0.1028	2.10	1.9739	3.06
0.2036	2.27	2.443	2.92
0.4946	2.76	5.06	2.07

APP7. When the steady-state heat-flow equation is solved numerically, temperatures $u(x, y)$ are calculated at the nodes of a gridwork constructed in the domain of interest. (This is the content of Chapter 8.) When a certain problem was solved, the values given in the following table were obtained. This procedure does not give the temperatures at points other than the nodes of the grid; if they are desired, one can interpolate to find them. Use the data to estimate the values of the temperature at the points (0.7, 1.2), (1.6, 2.4), and (0.65, 0.82).

$x \backslash y$	**0.0**	**0.5**	**1.0**	**1.5**	**2.0**	**2.5**
0.0	0.0	5.00	10.00	15.00	20.00	25.00
0.5	5.00	7.51	10.05	12.70	15.67	20.00
1.0	10.00	10.00	10.00	10.00	10.00	10.00
1.5	15.00	12.51	9.95	7.32	4.33	0.0
2.0	20.00	15.00	10.00	5.00	0.00	−5.00

APP8. The interpolating polynomials that have been described in this chapter have all fit the polynomial to match certain function values. One can also fit a polynomial that fits not just to values of the function but also to values of its derivative. Such an interpolating polynomial is called a *Hermite polynomial*.

Develop the relations to construct a cubic Hermite polynomial from $[x_1, f(x_1)]$, $[x_1, f'(x_1)]$, $[x_2, f(x_2)]$, and $[x_2, f'(x_2)]$. Then use your formula to find the cubic polynomial that interpolates from these data

x	$f(x)$	$f'(x)$
1	2.71828	0
3	6.69518	4.46345

to estimate $f(1.5)$, $f(2.0)$, and $f(2.5)$.

The data are for the function $f(x) = e^x/x$. How great are the errors of the interpolants? Are these errors less than those from the cubic interpolating polynomial that fits the function at $x = 1, 1.5, 2.4,$ and 3.0?

Superimpose the graphs of (a) the function, (b) the Hermite polynomial, and (c) the interpolating polynomial.

APP9. Exercise 44 asked you to fit a spline to the four points where the function has changes in its slope. Experiment with fitting the spline to four other points on the function to find a set that matches better to the function throughout its range. Can you conclude from this how a broken-line function should be fitted?

If you have the graph of a function whose derivative is continuous, where should points be chosen to get the best fit with a cubic spline?

APP10. Star S in the Big Dipper (Ursa Major) has a regular variation in its apparent magnitude. Leon Campbell and Laizi Jacchia give data for the mean light curve of this star in their book *The Story of Variable Stars* (Blakeston, 1941). A portion of these data is given here.

Phase	−110	−80	−40	−10	30	80	110
Magnitude	7.98	8.95	10.71	11.70	10.01	8.23	7.86

The data are periodic in that the magnitude for phase = −120 is the same as for phase = +120. The spline functions discussed in Section 3.4 do not allow for periodic behavior. For a periodic function, the slope and second derivatives are the same at the two endpoints. Taking this into account, develop a spline that interpolates the preceding data.

Other data given by Campbell and Jacchia for the same star are

Phase	−100	−60	−20	20	60	100
Magnitude	8.37	9.40	11.39	10.84	8.53	7.89

How well do interpolants based on your spline function agree with this second set of observations?

APP11. A fictitious chemical experiment produces seven data points:

t	−1	−0.96	−0.86	−0.79	0.22	0.5	0.930
y	−1	−0.151	0.894	0.986	0.895	0.5	−0.306

a. Plot the points and interpolate a smooth curve by intuition.
b. Plot the unique sixth-degree polynomial that interpolates these points.

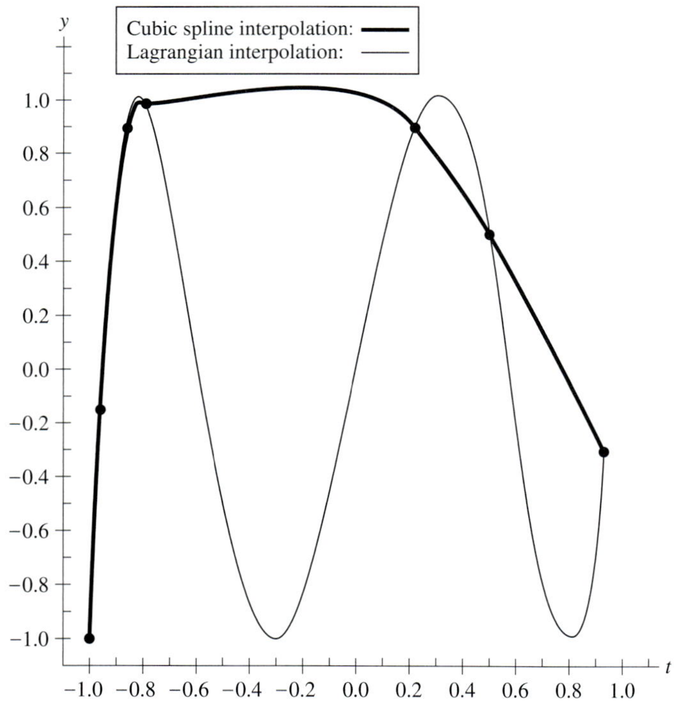

Figure 3.18

c. Use a spline program to evaluate enough points to plot this curve.

d. Compare your results with the graph in Figure 3.18.

APP12. In Exercise 30, what if the entry at $x = 0.36$ was mistakenly entered as 0.73471, rather than the correct value of 0.74371? How does this affect the computations of that exercise? How do values in the difference table change? Is there a pattern to the changes in the table?

APP13. The figure below is the profile of a pretty girl. (It is a tracing from a photograph of a daughter-in-law of one of the authors, taken by Elsie F. Gerald.) What is the best way to construct a sequence of polynomials that essentially duplicate the outline? Where will you choose the points where the polynomials join or what "knots" or "control points" will you specify? Pay particular attention to the portions of the figure at the lips and teeth.

You will want to trace the drawing on graph paper to find the coordinates of the outline. Do this and use your result to reproduce the figure.

4

Approximation of Functions

When your software program asks the computer to get the value of sin(2.113) or $e^{-3.5}$, have you wondered how it can get the values if the most powerful functions it can compute are polynomials? It doesn't look these up in tables and interpolate! Rather, the computer approximates every function other than polynomials from some polynomial that is tailored to give the values very accurately.

This chapter describes how such approximating polynomials are developed. We want the approximation to be efficient in that it obtains the values with the smallest error in the least number of arithmetic operations. Our approach will be gradual, building toward the more efficient methods from a less powerful starting point.

This chapter includes a second topic of great importance in applied mathematics—representing a function with a series of sine and cosine terms. In view of the above, this may seem a very roundabout way of doing the job. Still, such a series, a *Fourier series,* is usually the best way to represent a periodic function, something that cannot be done with a polynomial or a Taylor series. A Fourier series can even approximate functions with discontinuities and discontinuous derivatives.

The origin of the subject began with studies of vibrating strings. It reached fuller development with Jean Baptiste Joseph Fourier (1768–1830), who used them in solving heat conduction problems. The theory behind Fourier series has been extended to methods for solving other partial-differential equations.

One of the most important applications is in the analysis of the vibrational modes of a structure to determine which frequencies are of most importance. If external forces act with one of these frequencies, severe damage can occur.

Contents of This Chapter

4.1 Chebyshev Polynomials and Chebyshev Series

Chebyshev polynomials are *orthogonal polynomials* that are the basis for fitting nonalgebraic functions with maximum efficiency. They can be used to modify a Taylor series so that there is greater efficiency. A series of such polynomials converges more rapidly than a Taylor series.

4.2 Rational Function Approximations
Are the ratio of two polynomials that can be developed from a Taylor series; the result is a *Padé approximation,* a better match to the function being approximated. If the rational function is developed from a Chebyshev series, an even better approximation results. Even more improved approximations are mentioned—a *minimax approximation.*

4.3 Fourier Series
These are series of sine and cosine terms that can be used to approximate a function within a given interval very closely, even functions with discontinuities. Fourier series are important in many areas, particularly in getting an analytical solution to partial-differential equations.

4.1 Chebyshev Polynomials and Chebyshev Series

If we want to represent a known function as a polynomial, one way to do it is with a Taylor series. This you learned in your calculus course: Given a function, $f(x)$, we write

$$P_n(x) = a_0 + a_1(x - a) + a_1(x - a)^2 + a_2(x - a)^3 + \cdots + a_n(x - a)^n + \cdots$$

where $a_i = f^{(i)}(a)/i!$ (we remember that $f^{(0)}$ is just $f(a)$). Unless $f(x)$ is itself a polynomial, the series may have an infinite number of terms. Terminating the series incurs an error, the truncation error. The error after the $(x - a)^n$ term can be written in different ways but a most useful form is

$$\text{Error} = \frac{(x - a)^{n+1}}{(n + 1)!} f^{(n+1)}(\xi), \ \xi \text{ in } [a, x].$$

A problem with using the Taylor series to get polynomial approximations to a transcendental function is that the error grows rapidly as x-values depart from $x = a$.

For $f(x) = e^x$, the Taylor series is easy to write because the derivatives are so simple: $f^{(n)}(a) = e^a$ for all orders and we have, for $a = 0$ (which is then called a Maclaurin series),

$$e^x \approx 1 + 1(x - 0) + 1/2(x - 0)^2 + 1/6(x - 0)^3$$

if we use only terms through x^3; the error term shows that the error of this will grow about proportional to x^4 as x-values depart from zero. There is a way to combat this rapid growth of the errors, and that is to write the polynomial approximation to $f(x)$ in terms of *Chebyshev polynomials*. Chebyshev was a Russian mathematician; an older spelling of his name is Tschebycheff.

Chebyshev Polynomials

A Maclaurin series can be thought of as representing $f(x)$ as a weighted sum of polynomials. The kind of "polynomials" that are used are just the successive powers of x: $1, x, x^2, x^3, \ldots$. Chebyshev polynomials are not as simple; the first 11 of these are

$$\begin{aligned}
T_0(x) &= 1,\\
T_1(x) &= x,\\
T_2(x) &= 2x^2 - 1,\\
T_3(x) &= 4x^3 - 3x,\\
T_4(x) &= 8x^4 - 8x^2 + 1,\\
T_5(x) &= 16x^5 - 20x^3 + 5x,\\
T_6(x) &= 32x^6 - 48x^4 + 18x^2 - 1,\\
T_7(x) &= 64x^7 - 112x^5 + 56x^3 - 7x,\\
T_8(x) &= 128x^8 - 256x^6 + 160x^4 - 32x^2 + 1,\\
T_9(x) &= 256x^9 - 576x^7 + 432x^5 - 120x^3 + 9x,\\
T_{10}(x) &= 512x^{10} - 1280x^8 + 1120x^6 - 400x^4 + 50x^2 - 1.
\end{aligned} \tag{4.1}$$

The members of this series of polynomials can be generated from the two-term recursion formula

$$T_{n+1}(x) = 2xT_n(x) - T_{n-1}(x), \qquad T_0(x) = 1, \qquad T_1(x) = x. \tag{4.2}$$

(Using the symbol T for these derives from the older spelling of Chebyshev.)

Note that the coefficient of x^n in $T_n(x)$ is always 2^{n-1}. In Figure 4.1 we plot the first four polynomials of Eq. (4.1).

These polynomials have some unusual properties. They form an orthogonal set, in that

$$\int_{-1}^{1} \frac{T_n(x)T_m(x)}{\sqrt{1 - x^2}}\,dx = \begin{cases} 0, & n \neq m,\\ \pi, & n = m = 0,\\ \dfrac{\pi}{2}, & n = m \neq 0. \end{cases} \tag{4.3}$$

The orthogonality of these functions will not be of immediate concern to us.

The Chebyshev polynomials are also terms of a Fourier series,* because

$$T_n(x) = \cos n\theta, \tag{4.4}$$

where $\theta = \arccos x$. Observe that $\cos 0 = 1$, $\cos \theta = \cos(\arccos x) = x$.

---

* We discuss Fourier series later in this chapter.

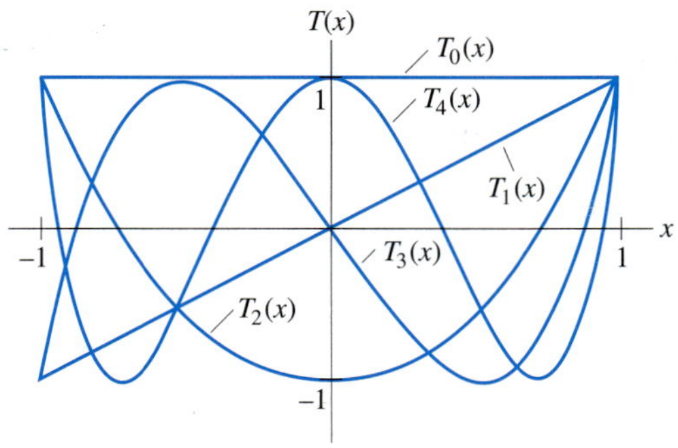

Figure 4.1

To demonstrate the equivalence of Eq. (4.4) to Eqs. (4.1) and (4.2), we recall some trigonometric identities, such as

$$\begin{aligned}\cos 2\theta &= 2\cos^2\theta - 1,\\ T_2(x) &= 2x^2 - 1;\\ \cos 3\theta &= 4\cos^3\theta - 3\cos\theta,\\ T_3(x) &= 4x^3 - 3x;\\ \cos(n+1)\theta + \cos(n-1)\theta &= 2\cos\theta\cos n\theta,\\ T_{n+1}(x) + T_{n-1}(x) &= 2xT_n(x).\end{aligned}$$

Because of the relation $T_n(x) = \cos(n\theta)$, the Chebyshev polynomials will have a succession of maxima and minima of alternating signs, as Figure 4.1 shows. It follows from Eq. (4.4), because $|\cos n\theta| = 1$ for $n\theta = 0, \pi, 2\pi, \ldots$, and because θ varies from 0 to π as x varies from 1 to -1, that $T_n(x)$ assumes a maximum magnitude of one $n + 1$ times on the interval $[-1, 1]$. For example, as seen in Figure 4.1, $T_4(x)$ has three maxima and two minima, a total of five, and all are of magnitude 1.

It is most important that, of all polynomials of degree n that have a coefficient of one on x^n, the polynomial

$$\frac{1}{2^{n-1}} T_n(x)$$

has a smaller upper bound to its magnitude in the interval $[-1, 1]$ than any other. Because the maximum magnitude of $T_n(x)$ is one, the upper bound is $1/2^{n-1}$ (we must make the coefficient of x^n equal to one). This is important because we will be able to write power function approximations to functions whose maximum errors are given in terms of this upper bound.

Error Bounds for Chebyshev Polynomials

We have asserted that, of all polynomials of degree n whose highest power of x has a coefficient of one, $T_n(x)/2^{n-1}$ has the smallest error bounds on $[-1, 1]$. The proof is by contradiction.

Let $P_n(x)$ be a polynomial whose leading term* is x^n and suppose that its maximum magnitude on $[-1, 1]$ is less than that of $T_n(x)/2^{n-1}$. Write

$$\frac{T_n(x)}{2^{n-1}} - P_n(x) = P_{n-1}(x),$$

where $P_{n-1}(x)$ is a polynomial of degree $n - 1$ or less, as the x^n terms cancel. The polynomial $T_n(x)$ has $n + 1$ extremes (counting endpoints), each of magnitude 1, so $T_n(x)/2^{n-1}$ has $n + 1$ extremes each of magnitude $1/2^{n-1}$, and these successive extremes alternate in sign. By our supposition about $P_n(x)$, at each of these maxima or minima, the magnitude of $P_n(x)$ is less than $1/2^{n-1}$; hence, $P_{n-1}(x)$ must change its sign at least for every extreme of $T_n(x)$, which is then at least $n + 1$ times. Hence, $P_{n-1}(x)$ crosses the axis at least n times and would have n zeros. However, this is impossible if $P_{n-1}(x)$ is only of degree $n - 1$, unless it is identically zero. The premise must then be false and $P_n(x)$ has a larger magnitude than the polynomial we are testing or, alternatively, $P_n(x)$ is exactly the same polynomial.

Using Computer Algebra Systems

The computer algebra systems that we have described in earlier chapters can get Chebyshev polynomials. Suppose we want $T_5(x)$.

In Maple, the command to get a Chebyshev polynomial is included in Maple's orthopoly package. That package provides several related commands. To use them, we must first invoke the package; its commands are then available:

```
with (orthopoly);
T(5,x)
  16x^5 − 20x^3 + 5x
```

and we see the fifth-degree polynomial exactly as in Eq. (4.1). *Mathematica* can also do this with a built-in function:

```
ChebyshevT[5, x]
  5x − 20x^3 + 16x^5
```

which is the same except the terms are in reverse order.

MATLAB has no commands for these polynomials but this M-file will compute them:

```
function T = Tch(n)
if n == 0
  disp('1')
elseif n == 1
  disp('x')
else
```

* We restrict the polynomials to those whose leading term is x^n so that all are scaled alike.

```
    t0 = '1';
    t1 = 'x';
    for i = 2:n
      T = symop('2*x','*',t1,'-',t0);
      t0 = t1;
      t1 = T;
    end % for
end % for
```

We invoke this by

```
EDU>> Tch(5)
ans =
   2*x*(2*x*(2*x*(2*x^2 - 1)-x) - 2*x^2 + 1) - 2*x*(2*x^2 - 1) + x
```

which is hard to read, but we can input

```
EDU>> collect(ans)
ans =
   16*x^5 - 20*x^3 + 5*x
```

to see the polynomial.

Economizing a Power Series

We begin a search for better power series representations of functions by using Chebyshev polynomials to "economize" a Maclaurin series. This example will give a modification of the Maclaurin series that produces a fifth-degree polynomial whose errors are only slightly greater than those of a sixth-degree Maclaurin series. We start with a Maclaurin series for e^x:

$$e^x = 1 + x + \frac{x^2}{2} + \frac{x^3}{6} + \frac{x^4}{24} + \frac{x^5}{120} + \frac{x^6}{720} + \cdots.$$

If we would like to use a truncated series to approximate e^x on the interval [0, 1] with a precision of 0.001, we will have to retain terms through that in x^6, because the error after the term in x^5 will be more than 1/720. Suppose we subtract

$$\left(\frac{1}{720}\right)\left(\frac{T_6}{32}\right)$$

from the truncated series. We note from Eq. (4.1) that this will exactly cancel the x^6 term and at the same time make adjustments in other coefficients of the Maclaurin series. Because the maximum value of T_6 on the interval [0, 1] is unity, this will change the sum of the truncated series by only

$$\frac{1}{720} * \frac{1}{32} < 0.00005,$$

which is small with respect to our required precision of 0.001. Performing the calculations, we have

$$e^x \approx 1 + x + \frac{x^2}{2} + \frac{x^3}{6} + \frac{x^4}{24} + \frac{x^5}{120} + \frac{x^6}{720}$$
$$- \frac{1}{720}\left(\frac{1}{32}\right)(32x^6 - 48x^4 + 18x^2 - 1), \tag{4.5}$$
$$e^x \approx 1.000043 + x + 0.499219x^2 + \frac{x^3}{6} + 0.043750x^4 + \frac{x^5}{120}.$$

The resulting fifth-degree polynomial approximates e^x on [0, 1] nearly as well as the sixth-degree Maclaurin series: Its maximum error (at $x = 1$) is 0.000270, compared to 0.000226 for the Maclaurin polynomial. We "economize" in that we get about the same precision with a lower-degree polynomial.

By subtracting $\frac{1}{120}(T_5/16)$ we can economize further, getting a fourth-degree polynomial that is almost as good as the economized fifth-degree one. It is left as an exercise to do this and to show that the maximum error is now 0.000781, so that we have found a fourth-degree power series that meets an error criterion that requires us to use two additional terms of the original Maclaurin series. Because of the relative ease with which they can be developed, such economized power series are frequently used for approximations to functions and are much more efficient than power series of the same degree obtained by merely truncating a Taylor or Maclaurin series. Table 4.1 compares the errors of these power series.

Observe in Table 4.1 that even the economized polynomial of degree-4 is more accurate than a fifth-degree Maclaurin series. Also notice that near $x = 0$, the economized polyno-

Table 4.1 Comparison of economized series with Maclaurin series

		Maclaurin of degree			**Economized of degree**	
x	e^x	**6**	**5**	**4**	**5**	**4**
0.0	1.00000	1.00000	1.00000	1.00000	1.00004	1.00004
0.2	1.22140	1.22140	1.22140	1.22140	1.22142	1.22098
0.4	1.49182	1.49182	1.49182	1.49173	1.49178	1.49133
0.6	1.82212	1.82212	1.82205	1.82140	1.82208	1.82212
0.8	2.22554	2.22549	2.32513	2.22240	2.22553	2.22605
1.0	2.71828	2.71806	2.71667	2.70833	2.71801	2.71749
Maximum error		0.00023	0.00162	0.00995	0.00027	0.00078

mials are less accurate; in effect, we permit small errors at points within the range but get a smaller maximum error. We return to this later. Also notice that the desired accuracy of a maximum error less than 0.001 is met with the economized fourth-degree polynomial.

Computer Algebra Systems Can Economize a Series

All three of the computer algebra systems can get Maclaurin series and Chebyshev polynomials. It follows that they should be able to economize the Maclaurin one. In this demonstration, we omit the intermediate results.

Maple gets the Maclaurin series for e^x with

```
taylor(exp(x), x = 0);
```

and if `Order` is set at 7, we get the sixth-degree polynomial that we have been working with. However, this includes the error term. We can remove it with

```
p: = convert(%,polynom);
```

where `'%'` refers to the previous answer. Now, doing

```
p - orthopoly[T](6,x)/6!/2^5;
```

produces Eq. (4.5) but with the coefficients expressed as ratios of integers:

$$\frac{23041}{23040} + x + \frac{639}{1280}x^2 + \frac{1}{6}x^3 + \frac{7}{160}x^4 + \frac{1}{120}x^5$$

Now, if we do `evalf(%)` we get the coefficients in floating point:

$$1.000043403 + x + .4992187500x^2 + .1666666667x^3 + .04375000000x^4 + .008333333333x^5$$

Mathematica is similar but we do not have to remove the error term:

```
Series[Exp[x],{x,0,6}] - ChebyshevT[6,x]/6!/2^5
```

$$\frac{23041}{23040} + x + \frac{639x^2}{1280} + \frac{x^3}{6} + \frac{7x^4}{160} + \frac{x^5}{120} + O[x]^7$$

We can get the economized series with MATLAB by employing our M-file for the Chebyshev series. We must start with x as a symbolic variable, then get the Maclaurin series and subtract the proper multiple of the Chebyshev series:

```
EDU>> syms x
EDU>> ts = taylor(exp(x),7)
   1 + x + 1/2*x^2 + 1/6*x^3 + 1/24*x^4 + 1/120*x^5 + 1/720*x^6
EDU>> cs = Tch(6);
EDU>> es = ts-cs/factorial(6)/2^5
es =
   23041/23040 + x + 639/1280*x^2 + 1/6*x^3 + 7/160*x^4 + 1/120*x^5
```

which duplicates the others. If we prefer to see this in floating point with seven digits:

```
EDU>> vpa(cs,7)
ans =
1.000043 + x + .4992188*x^2 + .1666667*x^3 + .4375000e - 1*x^4
   + .8333333e - 2*x^5
```

Chebyshev Series

By rearranging the Chebyshev polynomials, we can express powers of x in terms of them:

$$\begin{aligned}
1 &= T_0,\\
x &= T_1,\\
x^2 &= \frac{1}{2}(T_0 + T_2),\\
x^3 &= \frac{1}{4}(3T_1 + T_3),\\
x^4 &= \frac{1}{8}(3T_0 + 4T_2 + T_4),\\
x^5 &= \frac{1}{16}(10T_1 + 5T_3 + T_5), \qquad\qquad (4.6)\\
x^6 &= \frac{1}{32}(10T_0 + 15T_2 + 6T_4 + T_6),\\
x^7 &= \frac{1}{64}(35T_1 + 21T_3 + 7T_5 + T_7),\\
x^8 &= \frac{1}{128}(35T_0 + 56T_2 + 28T_4 + 8T_6 + T_8),\\
x^9 &= \frac{1}{256}(126T_1 + 84T_3 + 36T_5 + 9T_7 + T_9).
\end{aligned}$$

By substituting these identities into an infinite Taylor series and collecting terms in $T_i(x)$, we create a Chebyshev series. For example, we can get the first four terms of a Chebyshev series by starting with the Maclaurin expansion for e^x. Such a series converges more rapidly than does a Taylor series on $[-1, 1]$:

$$e^x = 1 + x + \frac{x^2}{2} + \frac{x^3}{6} + \frac{x^4}{24} + \cdots.$$

Replacing terms by Eq. (4.6), but omitting polynomials beyond $T_3(x)$ because we want only four terms,* we have

$$\begin{aligned} e^x &= T_0 + T_1 + \frac{1}{4}(T_0 + T_2) + \frac{1}{24}(3T_1 + T_3) + \frac{1}{192}(3T_0 + 4T_2 + \cdots) \\ &\quad + \frac{1}{1920}(10T_1 + 5T_3 + \cdots) + \frac{1}{23{,}040}(10T_0 + 15T_2 + \cdots) + \cdots \\ &= 1.2661T_0 + 1.1302T_1 + 0.2715T_2 + 0.0443T_3 + \cdots. \end{aligned}$$

To compare the Chebyshev expansion with the Maclaurin series, we convert back to powers of x, using Eq. (4.1):

$$\begin{aligned} e^x &= 1.2661 + 1.1302(x) + 0.2715(2x^2 - 1) + 0.0443(4x^3 - 3x) + \cdots. \\ e^x &= 0.9946 + 0.9973x + 0.5430x^2 + 0.1772x^3 + \cdots. \end{aligned} \tag{4.7}$$

Table 4.2 and Figure 4.2 compare the error of the Chebyshev expansion, Eq. (4.7), with the Maclaurin series, using terms through x^3 in each case. The figure shows how the Chebyshev expansion attains a smaller maximum error by permitting the error at the origin to increase. The errors can be considered to be distributed more or less uniformly throughout the interval. In contrast to this, the Maclaurin expansion, which gives very small errors near the origin, allows the error to bunch up at the ends of the interval.

Table 4.2 Comparison of Chebyshev series for e^x with Maclaurin series:
$e^x = 0.9946 + 0.9973x + 0.5430x^2 + 0.1772x^3$;
$e^x = 1 + x + 0.5x^2 + 0.1667x^3$

x	e^x	Chebyshev	Error	Maclaurin	Error
−1.0	0.3679	0.3631	0.0048	0.3333	0.0346
−0.8	0.4493	0.4536	−0.0042	0.4346	0.0147
−0.6	0.5488	0.5534	−0.0046	0.5440	0.0048
−0.4	0.6703	0.6712	−0.0009	0.6693	0.0010
−0.2	0.8187	0.8154	0.0033	0.8187	0.0001
0	1.0000	0.9946	0.0054	1.0000	0.0000
0.2	1.2214	1.2172	0.0042	1.2213	0.0001
0.4	1.4918	1.4917	0.0001	1.4907	0.0012
0.6	1.8221	1.8267	−0.0046	1.8160	0.0061
0.8	2.2255	2.2307	−0.0051	2.2054	0.0202
1.0	2.7183	2.7121	0.0062	2.6667	0.0516

* The number of terms that are employed determines the accuracy of the computed values, of course.

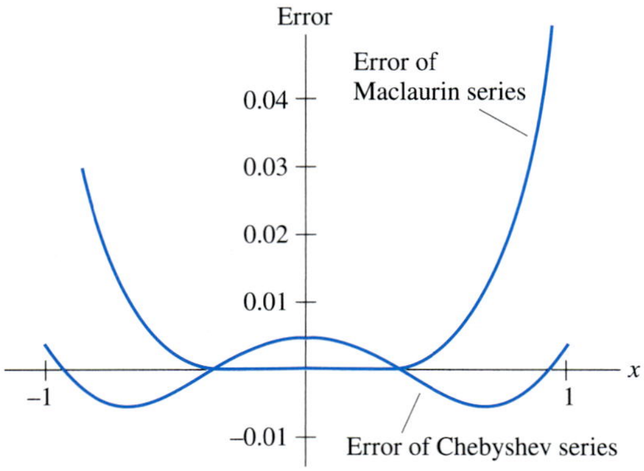

Figure 4.2

If the function is to be expressed directly as an expansion in Chebyshev polynomials, the coefficients can be obtained by integration. Based on the orthogonality property, the coefficients are computed from*

$$a_i = \frac{2}{\pi}\int_{-1}^{1} \frac{f(x)T_i(x)}{\sqrt{1-x^2}}\,dx,$$

and the series is expressed as

$$f(x) = \frac{a_0}{2} + \sum_{i=1}^{\infty} a_i T_i(x).$$

A change of variable will be required if the desired interval is other than $[-1, 1]$. In some cases, the definite integral that defines the coefficients can be profitably evaluated by the numerical procedures that we discuss in the next chapter.

Because the coefficients of the terms of a Chebyshev expansion usually decrease even more rapidly than the terms of a Maclaurin expansion, we can get an estimate of the magnitude of the error from the next nonzero term after those that were retained. For the truncated Chebyshev series given by Eq. (4.7), the $T_4(x)$ term would be

$$\frac{1}{192}(T_4) + \frac{1}{23{,}040}(6T_4) + \cdots = 0.00525T_4.$$

Because the maximum value of $T_4(x)$ on $[-1, 1]$ is 1.0, we estimate the maximum errors of Eq. (4.7) to be 0.00525. The maximum error in Table 4.2 is 0.0062. This good agreement is caused by the very rapid decrease in coefficients in this example.

Only Maple has a built-in command to get a Chebyshev series:

```
with (numapprox);
chebyshev(exp(x), x = -1 .. 1);
```

---

* The integration is not easy because the integrand is infinite at the endpoints.

which produces a series up to terms in $T_9(x)$. The first four terms of this match Eq. (4.7):

```
1.266065878 T(0, x) + 1.130318208 T(1, x) + .2714953396 T(2, x)
 + .04433684985 T(3, x) + .005474240443 T(4, x)
 + .0005429263119 T(5, x) + .00004497732296 T(6, x)
 + .3198436462 10^-5 T(7, x) + .1992124806 10^-6 T(8, x)
```

The computational economy to be gained by economizing a Maclaurin series, or by using a Chebyshev series, is even more dramatic when the Maclaurin series is slowly convergent. The previous example for $f(x) = e^x$ is a case in which the Maclaurin series converges rapidly. The power of the methods of this section is better demonstrated in the following example.

EXAMPLE 4.1 A Maclaurin series for $(1 + x)^{-1}$ is

$$(1 + x)^{-1} = 1 - x + x^2 - x^3 + x^4 - \cdots \qquad (-1 < x < 1).$$

Table 4.3 compares the accuracy of truncated Maclaurin series with the economized series derived from them.

In Table 4.3, we see that the error of the Maclaurin series is small for $x = 0.2$, and this also would be true for other values near $x = 0$, whereas the economized polynomial has less accuracy. At $x = 0.8$, the situation is reversed, however. Economized polynomials of degrees 8 and 6, derived from truncated Maclaurin series of degrees 10 and 8, actually have smaller errors than their precursors. Further economization, giving polynomials of degrees 6 and 4, have lesser or only slightly greater errors than their precursors, at

Table 4.3 Comparison of errors: Maclaurin, Chebyshev, and economized series for $1/(1 + x)$

		Maclaurin		Chebyshev		Economized*	
	Degree	Value	Error	Value	Error	Value	Error
At $x = 0.2$	2	0.840000	0.006667	0.841035	0.007702	0.758600	−0.074733
	4	0.833600	0.000267	0.833316	−0.000017	0.764594	−0.068739
	6	0.833344	$11 * 10^{-6}$	0.833359	0.000026	0.803646	−0.029687
	8	0.833334	$1 * 10^{-6}$	0.833365	0.000032	0.822786	−0.010547
	10	0.833333	0	0.833364	0.000031		
At $x = 0.8$	2	0.840000	0.284445	0.549866	−0.005899	0.812600	0.257045
	4	0.737600	0.182045	0.555518	−0.000038	0.738314	0.122759
	6	0.672064	0.116509	0.555561	0.000006	0.628558**	0.073003
	8	0.630121	0.074566	0.555568	0.000012	0.602106***	0.046551
	10	0.603277	0.047722	0.555568	0.000012		

* Economized series were derived from Maclaurin series whose degree is greater by 2.

** A series of degree equal to 4 has a value of 0.658246 with an error of 0.102691.

*** A series of degree equal to 6 has a value of 0.598199 with an error of 0.042644.

significant savings of computational effort and with smaller storage requirements in a computer's memory for the coefficients of the polynomials.

4.2 Rational Function Approximations

Approximating a known function with a Chebyshev series is much better than with a Taylor series in that it has a smaller maximum error in the interval $[-1, 1]$. (We would have to make transformations to the function to allow us to translate the x-value into this interval). Still, there is a way to improve even further.

As has been mentioned several times, the most complicated function that a computer can directly evaluate is a polynomial. This means that it can also evaluate a ratio of polynomials. Using a *rational function* permits this further improvement.

We approach this topic in stages: the first is *Padé approximations.*

Padé Approximations

A Padé approximation is a rational function, the quotient of two polynomials, the numerator of degree n and the denominator of degree m, which we can write as

$$f(x) \approx R_N(x) = \frac{a_0 + a_1 x + a_2 x^2 + \cdots + a_n x^n}{1 + b_1 x + b_2 x^2 + \cdots + b_m x^m}, \qquad N = n + m.$$

The constant term in the denominator can be taken as unity without loss of generality, because we can always convert to this form by dividing numerator and denominator by b_0. The constant b_0 will generally not be zero, for, in that case, the fraction would be undefined at $x = 0$. The most useful of the Padé approximations are those with the degree of the numerator equal to, or one more than, the degree of the denominator. Note that the number of constants in $R_N(x)$ is $N + 1 = n + m + 1$.

The Padé approximations are related to Maclaurin expansions in that the coefficients are determined in a similar fashion to make $f(x)$ and $R_N(x)$ agree at $x = 0$ and also to make the first N derivatives agree at $x = 0$.*

We begin with the Maclaurin series for $f(x)$ (we use only terms through x^N) and write

$$f(x) - R_N(x) \approx (c_0 + c_1 x + c_2 x^2 + \cdots + c_N x^N) - \frac{a_0 + a_1 x + \cdots + a_n x^n}{1 + b_1 x + \cdots + b_m x^m} \tag{4.8}$$

$$= \frac{(c_0 + c_1 x + \cdots + c_N x^N)(1 + b_1 x + \cdots + b_m x^m) - (a_0 + a_1 x + \cdots + a_n x^n)}{1 + b_1 x + \cdots + b_m x^m}$$

* A similar development can be derived for the expansion about a nonzero value of x, but the manipulations are not as easy. By a change of variable we can always make the region of interest contain the origin.

The coefficients c_i are $f^{(i)}(0)/(i!)$ of the Maclaurin expansion. Now if $f(x) = R_N(x)$ at $x = 0$, the numerator of Eq. (4.8) must have no constant term. Hence

$$c_0 - a_0 = 0.$$

For the first N derivatives of $f(x)$ and $R_N(x)$ to be equal at $x = 0$, the coefficients of the powers of x up to and including x^N in the numerator must all be zero also. This gives N additional equations for the a's and b's. The first n of these involve a's, the rest only b's and c's:

$$\begin{aligned}
&b_1c_0 + c_1 - a_1 = 0,\\
&b_2c_0 + b_1c_1 + c_2 - a_2 = 0,\\
&b_3c_0 + b_2c_1 + b_1c_2 + c_3 - a_3 = 0,\\
&\quad\vdots\\
&b_mc_{n-m} + b_{m-1}c_{n-m+1} + \cdots + c_n - a_n = 0,\\
&\\
&b_mc_{n-m+1} + b_{m-1}c_{n-m+2} + \cdots + c_{n+1} = 0,\\
&b_mc_{n-m+2} + b_{m-1}c_{n-m+3} + \cdots + c_{n+2} = 0,\\
&\quad\vdots\\
&b_mc_{N-m} + b_{m-1}c_{N-m+1} + \cdots + c_N = 0.
\end{aligned} \tag{4.9}$$

Note that, in each equation, the sum of the subscripts on the factors of each product is the same, and is equal to the exponent of the x-term in the numerator. The $N + 1$ equations of Eqs. (4.8) and (4.9) give the required coefficients of the Padé approximation. We illustrate by an example.

EXAMPLE 4.2 Find $\arctan(x) \approx R_{10}(x)$. Use degree-5 in both numerator and denominator.

The Maclaurin series through x^{10} is

$$\arctan(x) \approx x - 1/3x^3 + 1/5x^5 - 1/7x^7 + 1/9x^9. \tag{4.10}$$

We form $f(x) - R_{10}(x)$:

$$\frac{(x - 1/3x^3 + 1/5x^5 - 1/7x^7 + 1/9x^9)(1 + b_1x + b_2x^2 + b_3x^3 + b_4x^4 + b_5x^5) - (a_0 + a_1x + a_2x^2 + a_3x^3 + a_4x^4 + a_5x^5)}{(1 + b_1x + b_2x^2 + b_3x^3 + b_4x^4 + b_5x^5)} \tag{4.11}$$

If we multiply out in the numerator and set the coefficients of the x-terms through x^{10} to zero, we get for the a's:

$$\begin{aligned}
a_0 &= 0,\\
a_1 &= 1,\\
a_2 &= b_1,\\
a_3 &= -1/3 + b_2,\\
a_4 &= -1/3b_1 + b_3,\\
a_5 &= 1/5 - 1/3b_2 + b_4.
\end{aligned}$$

and for the b's:

$$\begin{aligned} 1/5b_1 - 1/3b_3 &= 0, \\ -1/7 + 1/5b_2 - 1/3b_4 &= 0, \\ -1/7b_1 + 1/5b_3 - 1/3b_5 &= 0, \\ 1/9 - 1/7b_2 + 1/5b_4 &= 0, \\ 1/9b_1 - 1/7b_3 + 1/5b_5 &= 0. \end{aligned}$$

We solve the last five equations for the b's. The matrix is

$$\begin{bmatrix} 1/5 & 0 & -1/3 & 0 & 0 \\ 0 & 1/5 & 0 & -1/3 & 0 \\ -1/7 & 0 & 1/5 & 0 & -1/3 \\ 0 & -1/7 & 0 & 1/5 & 0 \\ 1/9 & 0 & -1/7 & 0 & 1/5 \end{bmatrix} [b] = \begin{bmatrix} 0 \\ 1/7 \\ 0 \\ -1/9 \\ 0 \end{bmatrix},$$

whose solution is

$$b_1 = 0, b_2 = 10/9, b_3 = 0, b_4 = 5/21, b_5 = 0.$$

We get the a's from the first six equations:

$$a_0 = 0, a_1 = 1, a_2 = 0, a_3 = 7/9, a_4 = 0, a_5 = 64/945.$$

A rational function that approximates arctan x is then

$$\arctan x = \frac{x + \frac{7}{9}x^3 + \frac{64}{945}x^5}{1 + \frac{10}{9}x^2 + \frac{5}{21}x^4}. \qquad \textbf{(4.12)}$$

In Table 4.4 we compare the errors for this Padé approximation (Eq. 4.12) to the Maclaurin series expansion (Eq. 4.10). Enough terms are available in the Maclaurin series to give five-decimal precision at $x = 0.2$ and 0.4, but at $x = 1$ (the limit for convergence of the series) the error is sizable. Even though we used no more information in establishing it, the Padé formula is surprisingly accurate, having an error only 1/275 as large at $x = 1$. It is then particularly astonishing to realize that the Padé approximation is still not the best one of its form, for it violates the minimax principle. If the extreme precision near $x = 0$ is relaxed, we can make the maximum error smaller in the interval.

Figure 4.3 shows how closely the Padé approximation matches arctan(x), especially on $[-1, 1]$.

The error of a Padé approximation can often be roughly estimated by computing the next nonzero term in the numerator of Eq. (4.12). For Example 4.2, the coefficient of x^{10} is

Table 4.4 Comparison of Padé approximation to Maclaurin series for arctan x

x	True value	Padé (Eq. 4.12)	Error	Maclaurin (Eq. 4.10)	Error
0.2	0.19740	0.19740	0.00000	0.19740	0.00000
0.4	0.38051	0.38051	0.00000	0.38051	0.00000
0.6	0.54042	0.54042	0.00000	0.54067	−0.00025
0.8	0.67474	0.67477	−0.00003	0.67982	−0.00508
1.0	0.78540	0.78558	−0.00018	0.83492	−0.04952

zero, and the next term is

$$\left(-\frac{1}{7}b_4 + \frac{1}{9}b_2 - \frac{1}{11}\right)x^{11} = \left[-\frac{1}{7}\left(\frac{5}{21}\right) + \frac{1}{9}\left(\frac{10}{9}\right) - \frac{1}{11}\right]x^{11}$$
$$= -0.0014x^{11}.$$

Dividing by the denominator, we have

$$\text{Error} = \frac{-0.0014x^{11}}{1 + 1.1111x^2 + 0.2381x^4}.$$

At $x = 1$ this estimate gives -0.00060, which is about three times too large, but still of the correct order of magnitude. It is not unusual that such estimates are rough; analogous estimates of error by using the next term in a Maclaurin series behave similarly. The validity of the rule of thumb that "next term approximates the error" is poor when the coefficients do not decrease rapidly.

The preference for Padé approximations with the degree of the numerator the same as or one more than the degree of the denominator rests on the empirical fact that the errors are usually less for these. There are, however, even more efficient rational functions.

Here is how we can get a Padé approximation to arctan(x) with fifth-degree polynomials in both numerator and denominator from Maple:

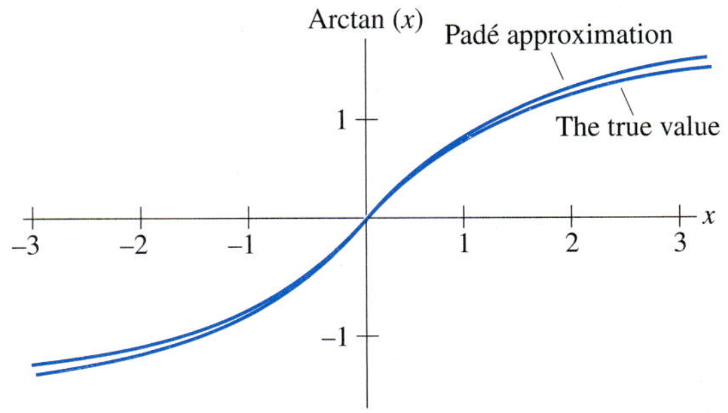

Figure 4.3

```
numapprox[padé] (arctan(x), x = 0, [5,5]);
```

to give a result very much like Eq. (4.12):

$$\frac{\frac{64}{945}x^5 + 7/9x^3 + x}{1 + 10/9x^2 + 5/21x^4}$$

Mathematica can do it too, if we first load a package called

```
<<Calculus'Padé';
```

and then do

```
Padé[ArcTan[x],{x,0,5,5}]
```

which gives a result exactly like Eq. (4.12):

$$\frac{x + \frac{7x^3}{9} + \frac{64x^5}{945}}{1 + \frac{10x^2}{9} + \frac{5x^4}{21}}$$

MATLAB does not have a command to get a Padé approximation.

A special feature of rational approximations like a Padé ratio is that these can approximate a function that is discontinuous. For example, a Padé approximation to tan(x) of degree-3 in both numerator and denominator is

$$\tan(x) \approx \frac{x - x^3/15}{1 - 2x^2/5}.$$

If you compare the plot of this with that for tan(x), you will find an excellent match between $x = -1.8$ and $x = 1.8$, even though the tangent function is infinite at $x = \pm\pi/2 = \pm 1.5708$; the portions of the branches beyond $x = \pm\pi$ are well fitted.

Continued Fractions

Because small differences in computational effort accumulate for a frequently used function, it is of interest to see if we can reduce the effort to evaluate Eq. (4.12). If we evaluate it as it stands, using nested form for the polynomials, we have

$$\text{Numerator} = (0.0677x^2 + 0.7778)x^3 + x$$
$$\text{Denominator} = (0.2381x^2 + 1.1111)x^2 + 1$$

We need to count adds, subtracts, multiplies, and divides because today's computers take about the same time to process each of these operations. For the numerator, there are three multiplies plus one for x^2 and two adds. The denominator takes two multiplies and two adds (x^2 is reused). There is one division. The total is

$$\text{Multiplies} = 3 + 1 + 2 = 6,$$
$$\text{Adds} = 2 + 2 = 4,$$
$$\text{Divides} = 1;$$
$$\text{Total} = 11.$$

Using the nested form of the Maclaurin series takes six multiplies and four add/subtracts for a total of ten. The divide required for the Padé approximation means it takes one more operation but it gives greater precision.

However, by doing a number of successive divisions, we can express Eq. (4.12) in *continued fraction* form:*

$$\frac{0.0677x^5 + 0.7778x^3 + x}{0.2381x^4 + 1.1111x^2 + 1} = \frac{0.2844x^5 + 3.2667x^3 + 4.2x}{x^4 + 4.6667x^2 + 4.2}$$
$$= \frac{0.2844x(x^4 + 11.4846x^2 + 14.7659)}{x^4 + 4.6667x^2 + 4.2}$$
$$= \frac{0.2844x}{(x^4 + 4.6667x^2 + 4.2)/(x^4 + 11.4846x^2 + 14.7659)}$$
$$= \frac{0.2844x}{1 - (6.8179x^2 + 10.5659)/(x^4 + 11.4846x^2 + 14.7659)}$$
$$= \frac{0.2844x}{1 - 6.8179(x^2 + 1.5497)/(x^4 + 11.4846x^2 + 14.7659)}$$
$$= \frac{0.2844x}{1 - 6.8179/[(x^4 + 11.4846x^2 + 14.7659)/(x^2 + 1.5497)]}$$
$$= \frac{0.2844x}{1 - 6.8179/[x^2 + 9.9348 - 0.6304/(x^2 + 1.5497)]}.$$

To evaluate this last formulation, we need one multiply, three divides, and four add/subtracts for a total of nine operations, one less than for the Maclaurin series and two less than if we did not put Eq. (4.12) into continued fraction form. In most cases, there is a greater advantage to continued fractions; the missing powers of x in this example favored evaluation as polynomials.

Only Maple can get a continued fraction from a ratio of polynomials. If we have obtained the Padé approximation to arctan(x) as above, with: `numapprox[padé](arctan(x),x = 0,[5,5]);` and then do

```
numapprox [confracform] (",x);
```

we see

$$\frac{64}{225}x + \frac{1309}{675}\cfrac{1}{x + \cfrac{8743}{2805}\cfrac{1}{x + \cfrac{236196}{1167815}\cfrac{1}{x + \cfrac{1683}{1249}1/x}}}$$

which is not in the same form but is equivalent.

* Acton (1970) is an excellent reference.

A Better Rational Function for e^x

If we start with the third-degree Chebyshev series of Eq. (4.7) and use it to get a Padé-like rational function, we can get a better approximation to the function. We have

$$e^x \approx 1.2661T_0 + 1.1302T_1 + 0.2715T_2 + 0.0443T_3$$

and form

$$1.2661T_0 + 1.1302T_1 + 0.2715T_2 + 0.0443T_3 - P_n(x)/Q_m(x)$$
$$= \frac{(1.2661 + 1.302T_1 + 0.2715T_2 + 0.0443T_3)(1 + b_1T_1) - (a_0 + a_1T_1 + a_2T_2)}{1 + b_1T_1}.$$

where we have taken $n = 2$, $m = 1$.

We expand the numerator; we will set the coefficients of each degree of the T's to zero.

Before we can equate coefficients to zero, we need to resolve the products of Chebyshev polynomials that occur. Recalling that $T_n(x) = \cos n\theta$, we can use the trigonometric identity

$$\cos n\theta \cos m\theta = \frac{1}{2}[\cos(n + m)\theta + \cos(n - m)\theta],$$

$$T_n(x)T_m(x) = \frac{1}{2}[T_{n+m}(x) + T_{|n-m|}(x)].$$

The absolute value of the difference $n - m$ occurs because $\cos(z) = \cos(-z)$. Using this relation we can write the equations

$$a_0 = 1.2661 + \frac{1.1302}{2}b_1,$$

$$a_1 = 1.1302 + \left(\frac{0.2715}{2} + 1.2661\right)b_1,$$

$$a_2 = 0.2715 + \left(\frac{1.1302}{2} + \frac{0.0443}{2}\right)b_1,$$

$$0 = 0.0443 + \frac{0.2715}{2}b_1.$$

Solving, we first get $b_1 = -0.3263$, then we get $a_0 = 1.0817$, $a_1 = 0.6727$, $a_2 = 0.0799$, and

$$e^x \approx \frac{1.0817 + 0.6727T_1 + 0.0799T_2}{1 - 0.3263T_1},$$
$$e^x \approx \frac{1.0018 + 0.6727x + 0.1598x^2}{1 - 0.3263x}. \qquad \textbf{(4.13)}$$

Table 4.5 Comparison of rational approximations [Eq. (4.13)] with Chebyshev series for e^x

x	e^x	Chebyshev	Error	Rational function	Error
−1.0	0.3679	0.3631	0.0048	0.3686	−0.0007
−0.8	0.4493	0.4536	−0.0042	0.4488	0.0006
−0.6	0.5488	0.5534	−0.0046	0.5484	0.0005
−0.4	0.6703	0.6712	−0.0009	0.6707	−0.0004
−0.2	0.8187	0.8154	0.0033	0.8201	−0.0014
0	1.0000	0.9946	0.0054	1.0018	−0.0018
0.2	1.2214	1.2172	0.0042	1.2225	−0.0011
0.4	1.4918	1.4917	0.0001	1.4911	0.0008
0.6	1.8221	1.8267	−0.0046	1.8191	0.0030
0.8	2.2255	2.2307	−0.0051	2.2224	0.0032
1.0	2.7183	2.7121	0.0062	2.7227	−0.0044

The last part of Eq. (4.13) results when we rewrite the T's in powers of x. Table 4.5 shows that we have a better appproximation than the original third-degree Chebyshev series.

Maple can get this same $P_2(x)/Q_1(x)$:

```
with(numapprox):
chebpade(exp(x),x,(2,1]);
```

$$\frac{1.086272879T(0,x) + .6843619105T(1,x) + .08464994161T(2,x)}{T(0,x) - .3181281121T(1,x)}$$

which is more accurate than Eq. (4.13) because Maple reduced it from a Chebyshev series of higher order. We could have had it expressed in powers of x if we inserted the command:

```
with(orthopoly)
```

which would have given

$$\frac{1.001622938 + .6843619105x + .1692998832x^2}{1 - .3181281121x}$$

If we should tabulate the values from this rational function as done in Table 4.5, we would find that the maximum error is −0.0026, about 60% as great.

It was mentioned that the error of a Padé rational function is often less if the numerator is of greater degree than the denominator. Let us test this, using Maple:

```
chebpade(exp(x),x,[1,2]);
```

we get

$$\frac{.9267185059 + .3066786271x}{.9279844930 - .6177257111x + .1440310141x^2}$$

and the maximum error is greater; it is −0.0032.

Minimax Approximations

We have made several improvements over a Maclaurin series approximation to a function: by economizing it, forming a Chebyshev series, developing a Padé rational function, and, last, creating a Chebyshev–Padé rational function.

Still, none of these is "optimal." A theorem, due to Chebyshev, tells us whether a given approximation is optimal. This states that, in order to be optimal, an approximation of degree N is *minimax* if and only if there are at least $N + 2$ maxima and minima and all are of equal magnitude. (For a rational function, N is the sum of the degrees of the numerator and denominator.) Table 4.5 shows five maxima and minima, the correct number, but they are not equal in magnitude. The same is true in Table 4.2 for the Chebyshev series.

A consequence of the minimax theorem is that a bound to the error of the minimax approximation is given by the magnitudes of the smallest and largest errors of an approximation with the correct number of maxima/minima. Thus, we see, from Table 4.5, that the error of a minimax approximation of degree-3 lies between 0.0006 and 0.0044.

Finding a truly minimax approximation is not an easy task;* nonlinear equations are involved. An algorithm due to Remes is the usual way to find it; it begins with an approximation that has the correct number of maxima/minima and, by iterations, converges on the minimax one. The Chebyshev of Eq. (4.7) could be a good starting point, as illustrated by Figure 4.2. The rational function of Eq. (4.13) would work but is far from minimax, as seen in Table 4.5.

Actually, Maple and *Mathematica* will do the work for us. Maple gets a minimax rational approximation to e^x of degrees 2 and 1:

```
numapprox[minimax](exp(x), x = 0..1, [2,1]);
```

$$\frac{1.164275859 + (.8302764291 + .2779023279x)x}{1.164066084 - .3281321682x}$$

We leave it to the student to do this in *Mathematica*.

4.3 Fourier Series

Polynomials are not the only functions that can be used to approximate known functions. Another means for representing known functions are approximations that use sines and cosines, called *Fourier series* after the French mathematician who first proposed, in the early 1800s, that "any function can be represented by an infinite sum of sine and cosine terms."

Fourier used these series in his studies of heat conduction. His belief that any function can be represented in the form of a sum of sine and cosine terms with the proper coefficients, possibly with an infinite number of terms, was disputed by other mathematicians because he did not adequately develop the theory. Actually, the belief is false, for there are

* Ralston is an excellent reference.

functions (mostly esoteric) that do not have a representation as a Fourier series. However, most functions can be so represented.

Representing a function as a trigonometric series is important in solving some partial-differential equations analytically. In this section we will see how to determine the coefficients of a Fourier series.

Because a Fourier series is a sum of sine and/or cosine terms, it will obviously always be a periodic function.

We will only summarize the important theorems concerning Fourier series. Proofs can be found in Conte and de Boor (1980), Fike (1968), Brigham (1974), Ramirez (1985), and Ralston (1965). In the following theorems, $f(x)$ refers to the periodic function being represented or to the periodic extension given by a redefinition. It is essential that $f(x)$ be integrable if we are to compute the coefficients of a Fourier series.

1. $f(x)$ is said to be piecewise continuous on $(0, L)$ if it is continuous on $(0, L)$ except for a finite number of finite discontinuities. If $f(x)$ and/or $f'(x)$ is piecewise continuous on $(0, L)$, $f(x)$ is said to be piecewise smooth. An infinite series is said to converge pointwise to $f(x)$ if the sum of n terms of the series converges to $f(x)$ at the point in $(0, L)$ as $n \to \infty$. An infinite series is said to converge uniformly if it converges pointwise to $f(x)$ at all points in $(0, L)$.
2. If $f(x)$ is continuous and piecewise smooth, its Fourier series converges uniformly to $f(x)$. If $f(x)$ is piecewise smooth, the series converges pointwise to $f(x)$ at all points where $f(x)$ is continuous and converges to the average value where $f(x)$ has a finite discontinuity.
3. If $f(x)$ is piecewise continuous, its Fourier series can be integrated term by term to yield a series that converges pointwise to the integral of $f(x)$.
4. If $f(x)$ is continuous and $f'(x)$ is piecewise smooth, then the Fourier series of $f(x)$ can be differentiated term by term to give a series that converges pointwise to $f'(x)$ wherever $f''(x)$ exists.

The theory of Fourier series is a major topic in mathematics. Most mathematical texts that cover Fourier series at least outline proofs of the preceding theorems.

Any function, $f(x)$, is periodic of period P if it has the same value for any two x-values that differ by P, or

$$f(x) = f(x + P) = f(x + 2P) = \cdots = f(x - P) = f(x - 2P) = \cdots .$$

Figure 4.4 shows such a periodic function. Additional occurrences are shown as dashed on the plot. Observe that the period can be started at any point on the x-axis. Sin(x) and cos(x) are periodic of period 2π; sin($2x$) and cos($2x$) are periodic of period π; sin(nx) and cos(nx) are periodic of period $2\pi/n$.

We now discuss how to find the A's and B's in a Fourier series of the form

$$f(x) \approx \frac{A_0}{2} + \sum_{n=1}^{\infty} [A_n \cos(nx) + B_n \sin(nx)]. \tag{4.14}$$

[We read the symbol "≈" in Eq. (4.14) as "is represented by."] The determination of the coefficients of a Fourier series [when a given function, $f(x)$, can be so represented] is based

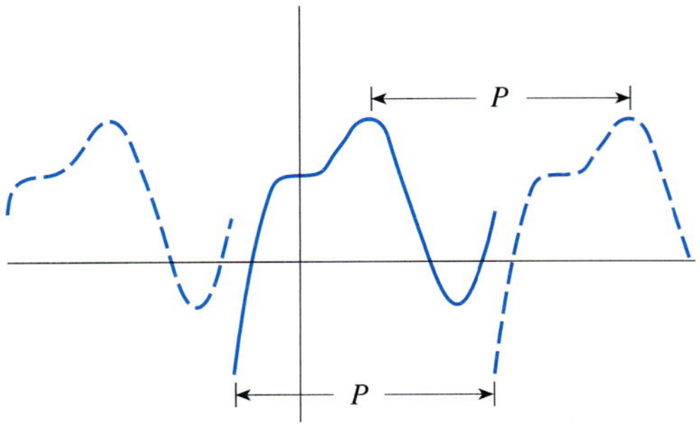

Figure 4.4
Plot of a periodic function of period P

on the *property of orthogonality* for sines and cosines. For integer values of n, m:

$$\int_{-\pi}^{\pi} \sin(nx)\,dx = 0; \tag{4.15}$$

$$\int_{-\pi}^{\pi} \cos(nx)\,dx = \begin{cases} 0, & n \neq 0, \\ 2\pi, & n = 0; \end{cases} \tag{4.16}$$

$$\int_{-\pi}^{\pi} \sin(nx)\cos(mx)\,dx = 0; \tag{4.17}$$

$$\int_{-\pi}^{\pi} \sin(nx)\sin(mx)\,dx = \begin{cases} 0, & n \neq m, \\ \pi, & n = m; \end{cases} \tag{4.18}$$

$$\int_{-\pi}^{\pi} \cos(nx)\cos(mx)\,dx = \begin{cases} 0, & n \neq m, \\ \pi, & n = m. \end{cases} \tag{4.19}$$

Although the term *orthogonal* should not be interpreted geometrically, it is related to the same term used for orthogonal (perpendicular) vectors whose dot product is zero. Many functions, besides sines and cosines, are orthogonal, such as the Chebyshev polynomials that were discussed previously.

To begin, we assume that $f(x)$ is periodic of period 2π and can be represented as in Eq. (4.14). We find the values of A_n and B_n in Eq. (4.14) in the following way.

1. Multiply both sides of Eq. (4.14) by $\cos(0x) = 1$, and integrate term by term between the limits of $-\pi$ and π. (We assume that this is a proper operation; you will find that it works.)

$$\int_{-\pi}^{\pi} f(x)\,dx = \int_{-\pi}^{\pi} \frac{A_0}{2}\,dx + \sum_{n=1}^{\infty} \int_{-\pi}^{\pi} A_n \cos(nx)\,dx + \sum_{n=1}^{\infty} \int_{-\pi}^{\pi} B_n \sin(nx)\,dx \tag{4.20}$$

Because of Eqs. (4.15) and (4.16), every term on the right vanishes except the first, giving

$$\int_{-\pi}^{\pi} f(x)\,dx = \frac{A_0}{2}(2\pi), \qquad \text{or} \qquad A_0 = \frac{1}{\pi}\int_{-\pi}^{\pi} f(x)\,dx. \tag{4.21}$$

Hence, A_0 is found and it is equal to twice the average value of $f(x)$ over one period.

2. Multiply both sides of Eq. (4.14) by $\cos(mx)$, where m is any positive integer, and integrate:

$$\int_{-\pi}^{\pi} \cos(mx)f(x)\,dx = \int_{-\pi}^{\pi} \frac{A_0}{2}\cos(mx)\,dx + \sum_{n=1}^{\infty}\int_{-\pi}^{\pi} A_n\cos(mx)\cos(nx)\,dx + \sum_{n=1}^{\infty}\int_{-\pi}^{\pi} B_n\cos(mx)\sin(nx)\,dx. \tag{4.22}$$

Because of Eqs. (4.16), (4.17), and (4.19), the only nonzero term on the right is when $m = n$ in the first summation, so we get a formula for the A's:

$$A_n = \frac{1}{\pi}\int_{-\pi}^{\pi} f(x)\cos(nx)\,dx, \qquad n = 1, 2, 3, \ldots. \tag{4.23}$$

3. Multiply both sides of Eq. (4.14) by $\sin(mx)$, where m is any positive integer, and integrate:

$$\int_{-\pi}^{\pi} \sin(mx)f(x)\,dx = \int_{-\pi}^{\pi} \frac{A_0}{2}\sin(mx)\,dx + \sum_{n=1}^{\infty}\int_{-\pi}^{\pi} A_n\sin(mx)\cos(nx)\,dx + \sum_{n=1}^{\infty}\int_{-\pi}^{\pi} B_n\sin(mx)\sin(nx)\,dx. \tag{4.24}$$

Because of Eqs. (4.15), (4.17), and (4.18), the only nonzero term on the right is when $m = n$ in the second summation, so we get a formula for the B's:

$$B_n = \frac{1}{\pi}\int_{-\pi}^{\pi} f(x)\sin(nx)\,dx, \qquad n = 1, 2, 3, \ldots. \tag{4.25}$$

By comparing Eqs. (4.21) and (4.23), you now see why Eq. (4.14) had $A_0/2$ as its first term. That makes the formula for all of the A's the same:

$$A_n = \frac{1}{\pi}\int_{-\pi}^{\pi} f(x)\cos(nx)\,dx, \qquad n = 0, 1, 2, \ldots. \tag{4.26}$$

It is obvious that getting the coefficients of Fourier series involves many integrations. We observe that this can be facilitated by a computer algebra system.

The integrations to find the coefficients of a Fourier series can be done numerically, as we explain in Chapter 5. This allows one to get a series that approximates to experimental data, a specially important application. The fast Fourier transform (FFT) is the efficient way to do this.

Fourier Series for Periods Other Than 2π

What if the period of $f(x)$ is not 2π? No problem—we just make a change of variable. If $f(x)$ is periodic of period P, the function can be considered to have one period between

$-P/2$ and $P/2$. The functions $\sin(2\pi x/P)$ and $\cos(2\pi x/P)$ are periodic between $-P/2$ and $P/2$. (When $x = -P/2$, the angle becomes $-\pi$; when $x = P/2$, it is π.) We can repeat the preceding developments for sums of $\cos(2n\pi x/P)$ and $\sin(2n\pi x/P)$, or rescale the preceding results. In any event, the formulas become, for $f(x)$ periodic of period P:

$$A_n = \frac{2}{P}\int_{-P/2}^{P/2} f(x)\cos\left(\frac{n\pi x}{P/2}\right)dx, \qquad n = 0, 1, 2, \ldots, \tag{4.27}$$

$$B_n = \frac{2}{P}\int_{-P/2}^{P/2} f(x)\sin\left(\frac{n\pi x}{P/2}\right)dx, \qquad n = 1, 2, 3, \ldots. \tag{4.28}$$

Because a function that is periodic with period P between $-P/2$ and $P/2$ is also periodic with period P between A and $A + P$, the limits of integration in Eqs. (4.27) and (4.28) can be from 0 to P.

EXAMPLE 4.3 Let $f(x) = x$ be periodic between $-\pi$ and π. (See Figure 4.5.) Find the A's and B's of its Fourier expansion.

For A_0:

$$A_0 = \frac{1}{\pi}\int_{-\pi}^{\pi} f(x)\,dx = \frac{1}{\pi}\int_{-\pi}^{\pi} x\,dx = \left.\frac{x^2}{2\pi}\right]_{-\pi}^{\pi} = 0. \tag{4.29}$$

For the other A's:

$$A_n = \frac{1}{\pi}\int_{-\pi}^{\pi} x\cos(nx)\,dx = \frac{1}{\pi}\left.\left(\frac{\cos(nx)}{n^2} + \frac{x\sin(nx)}{n}\right)\right]_{-\pi}^{\pi} = 0. \tag{4.30}$$

For the B's:

$$\begin{aligned} B_n &= \frac{1}{\pi}\int_{-\pi}^{\pi} x\sin(nx)\,dx = \frac{1}{\pi}\left.\left(\frac{\sin(nx)}{n^2} - \frac{x\cos(nx)}{n}\right)\right]_{-\pi}^{\pi} \\ &= \frac{2(-1)^{n+1}}{n}, \qquad n = 1, 2, 3, \ldots. \end{aligned} \tag{4.31}$$

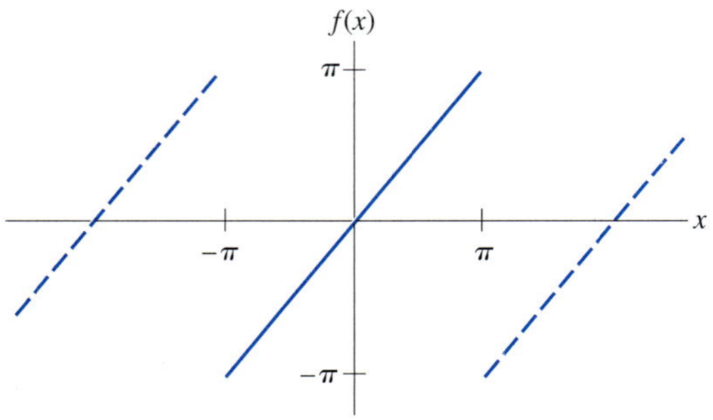

Figure 4.5
Plot of $f(x) = x$, periodic of period 2π

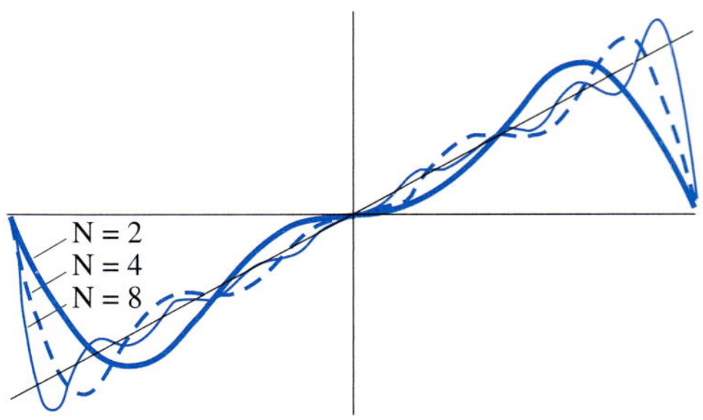

Figure 4.6
Plot of Eq. (4.32) for N = 2, 4, 8

We then have

$$x \approx 2 \sum_{n=1}^{\infty} \frac{(-1)^{n+1}}{n} \sin(nx), \qquad -\pi < x < \pi. \tag{4.32}$$

Figure 4.6 shows how the series approximates to the function when only two, four, or eight terms are used.

EXAMPLE 4.4 Find the Fourier coefficients for $f(x) = |x|$ on $-\pi$ to π:

$$A_0 = \frac{1}{\pi}\int_{-\pi}^{0} (-x)\,dx + \frac{1}{\pi}\int_{0}^{\pi} x\,dx = \frac{2}{\pi}\int_{0}^{\pi} x\,dx = \pi; \tag{4.33}$$

$$\begin{aligned} A_n &= \frac{1}{\pi}\int_{-\pi}^{0} (-x)\cos(nx)\,dx + \frac{1}{\pi}\int_{0}^{\pi} x\cos(nx)\,dx \\ &= \frac{2}{\pi}\left(\frac{\cos(nx)}{n^2} + \frac{x\sin(nx)}{n}\right)\Bigg]_0^{\pi} \\ &= \begin{cases} 0, & n = 2, 4, 6, \ldots, \\ \dfrac{-4}{n^2\pi}, & n = 1, 3, 5, \ldots, \end{cases} \end{aligned} \tag{4.34}$$

$$B_n = \frac{1}{\pi}\int_{-\pi}^{0} (-x)\sin(nx)\,dx + \frac{1}{\pi}\int_{0}^{\pi} x\sin(nx)\,dx = 0. \tag{4.35}$$

Because the definite integrals in Eq. (4.34) are nonzero only for odd values of n, it simplifies to change the index of the summation. The Fourier series is then

$$|x| \approx \frac{\pi}{2} - \frac{4}{\pi}\sum_{n=1}^{\infty} \frac{\cos(2n-1)x)}{(2n-1)^2}. \tag{4.36}$$

Figure 4.7 shows how the series approximates the function when two, four, or eight terms are used.

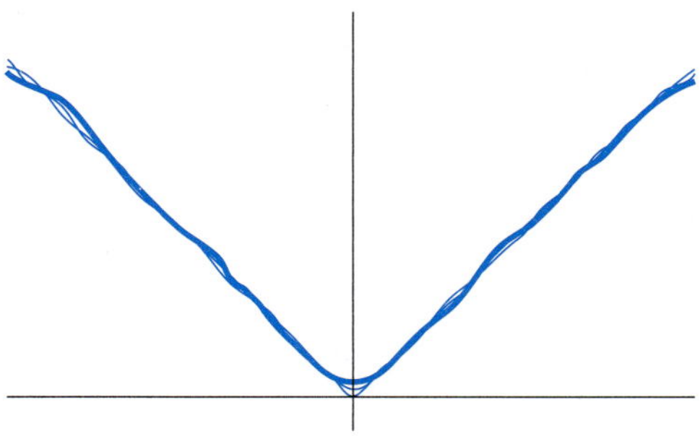

Figure 4.7
Plot of Eq. (4.36) for N = 2, 4, 8

When you compare Eqs. (4.32) and (4.36) and their plots in Figures 4.6 and 4.7, you will notice several differences:

1. The first contains only sine terms, the second only cosines.
2. Equation (4.32) gives a value at both endpoints that is the average of the end values for $f(x)$, where there is a discontinuity.
3. Equation (4.36) gives a closer approximation when only a few terms are used.

Example 4.5 will further examine these points.

EXAMPLE 4.5

Find the Fourier coefficients for $f(x) = x(2 - x) = 2x - x^2$ over the interval $[-2, 2]$ if it is periodic of period 4. Equations (4.27) and (4.28) apply.

$$A_0 = \frac{2}{4}\int_{-2}^{2}(2x - x^2)\,dx = \frac{-8}{3} \tag{4.37}$$

$$A_n = \frac{2}{4}\int_{-2}^{2}(2x - x^2)\cos\left(\frac{n\pi x}{2}\right)dx = \frac{16(-1)^{n+1}}{n^2\pi^2}, \qquad n = 1, 2, 3, \ldots \tag{4.38}$$

$$B_n = \frac{2}{4}\int_{-2}^{2}(2x - x^2)\sin\left(\frac{n\pi x}{2}\right)dx = \frac{8(-1)^{n+1}}{n\pi}, \qquad n = 1, 2, 3, \ldots \tag{4.39}$$

$$x(2 - x) \approx \frac{-4}{3} + \frac{16}{\pi^2}\sum_{n=1}^{\infty}\frac{(-1)^{n+1}}{n^2}\cos\left(\frac{n\pi x}{2}\right) + \frac{8}{\pi}\sum_{n=1}^{\infty}\frac{(-1)^{n+1}}{n}\sin\left(\frac{n\pi x}{2}\right) \tag{4.40}$$

You will notice that both sine and cosine terms occur in the Fourier series and that the discontinuity at the endpoints shows itself in forcing the Fourier series to reach the average value. It should also be clear that the series is the sum of separate series for $2x$ and $-x^2$. Figure 4.8 shows how the series of Eq. (4.40) approximates to the function when 40 terms are used. It is obvious that many more terms are needed to reduce the error to negligible proportions because of the extreme oscillation near the discontinuities, called the *Gibbs phenomenon*. The conclusion is that a Fourier series often involves a lot of computation as well as awkward integrations to give the formula for the coefficients.

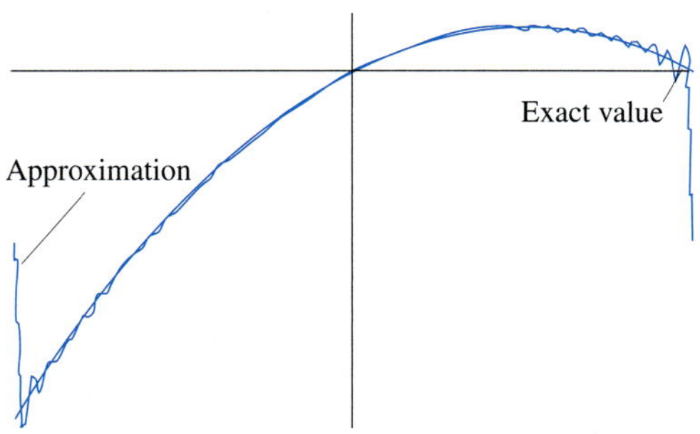

Figure 4.8
Plot of Eq. (4.40) for N = 40

Mathematica has a built-in command to get the Fourier series for a function; the others can get the coefficients by integration, of course. (Maple's `fourier` command gets the Fourier transforms, not the series.)

With *Mathematica*, we must first load a package:

```
<<Calculus'FourierTransform'
```

and, because there is a warning message, do

```
Clear[FourierTrigSeries]; Remove[FourierTrigSeries]
```

and then

```
FourierTrigSeries[x*(2 − x),{x,−2,2},4]
```

which gives the series. However, the sine terms are hard to interpret until we do

```
Collect[%,Pi]
```

which gives a correct result [see Eq. (4.40)], but all the cosine terms are grouped together with a denominator of `Pi`^{2} and the sine terms are similarly grouped with a denominator of `Pi`.

All of the computer algebra systems can do the required integrations to get the coefficients of a Fourier series. If $f(x) = x(2 - x)$ over $[-2, 2]$ and we want A_3, here are the commands that are used:

In Maple, we do

```
2/4*int(x*(2 − x)*cos(3*Pi*x/2), x = −2..2);
```

Mathematica's command is

```
2/4*Integrate[x*(2 − x)*Cos[3*Pi*x/2],{x,−2,2}]
```

With MATLAB, two commands are needed because the first result is symbolic and the integration operation does not permit a multiplier (although the 2/4 could be included in the integrand):

```
a3 = int('x*(2 − x)*cos(3*pi*x/2)',−2, 2)
symmul (a3, '2/4')
```

In all cases, the correct result is obtained.

Fourier Series for Nonperiodic Functions and Half-Range Expansions

The development until now has been for a periodic function. What if $f(x)$ is not periodic? Can we approximate it by a trigonometric series? We assume that we are interested in approximating the function only over a limited interval and we do not care whether the approximation holds outside of that interval. This situation frequently occurs when we want to solve partial-differential equations analytically.

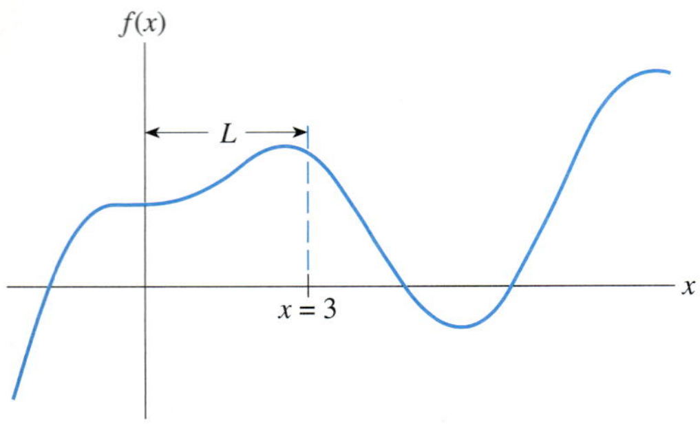

Figure 4.9
A function, $f(x)$, of interest on [0, 3]

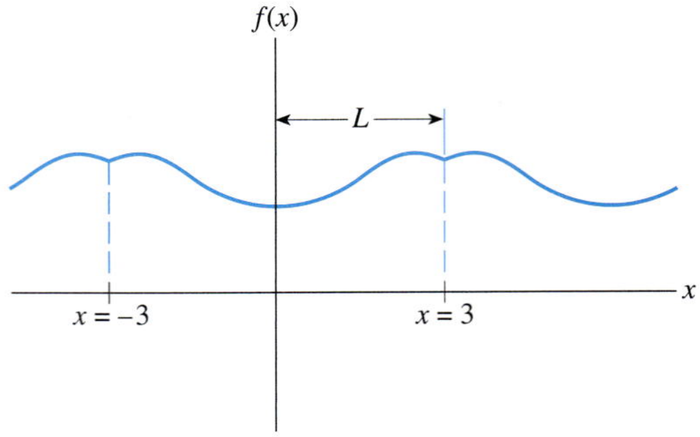

Figure 4.10
Plot of a function reflected about the y-axis, an even function

Suppose we have a function defined for all x-values, but we are only interested in representing it over $(0, L)$.* Figure 4.9 is typical. Because we will ignore the behavior of the function outside of $(0, L)$, we can redefine the behavior outside that interval as we wish. Figures 4.10 and 4.11 show two possible redefinitions.

In the first redefinition, we have reflected the portion of $f(x)$ about the y-axis and have extended it as a periodic function of period $2L$. This creates an *even* periodic function. If we reflect it about the origin and extend it periodically, we create an *odd* periodic function of period $2L$. More formally, we define even and odd functions through these relations:

$$f(x) \text{ is even if } f(-x) = f(x), \tag{4.41}$$

$$f(x) \text{ is odd if } f(-x) = -f(x). \tag{4.42}$$

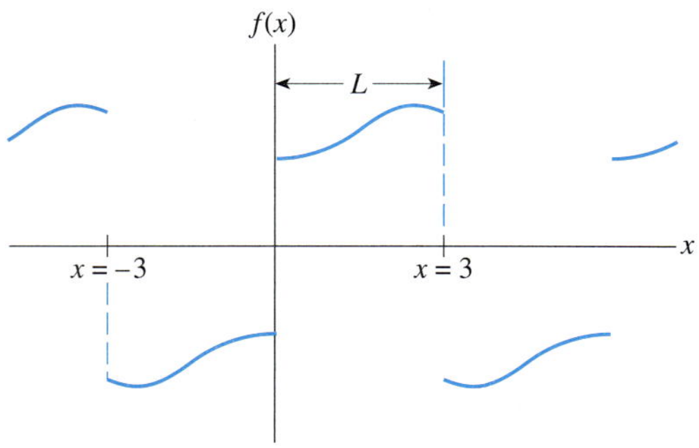

Figure 4.11
Plot of a function reflected about the origin, an odd function

* If the range of interest is $[a, b]$, a simple change of variable can make this $[0, L]$.

It is easy to see that $\cos(Cx)$ is an even function and that $\sin(Cx)$ is an odd function for any real value of C.

There are two important relationships for integrals of even and odd functions. (If you think of the integrals in a geometric interpretation, these relationships are obvious.)

$$\text{If } f(x) \text{ is even, } \int_{-L}^{L} f(x)\,dx = 2\int_{0}^{L} f(x)\,dx. \tag{4.43}$$

$$\text{If } f(x) \text{ is odd, } \int_{-L}^{L} f(x)\,dx = 0. \tag{4.44}$$

It is also easy to show that the product of two even functions is even, that the product of two odd functions is even, and that the product of an even and an odd function is odd. This means that, if $f(x)$ is even, $f(x)\cos(nx)$ is even and $f(x)\sin(nx)$ is odd. Further, if $f(x)$ is odd, $f(x)\cos(nx)$ is odd and $f(x)\sin(nx)$ is even. Because of Eq. (4.41), the Fourier series expansion of an even function will contain only cosine terms (all the B-coefficients are zero). Also, if $f(x)$ is odd, its Fourier expansion will contain only sine terms (all the A-coefficients are zero). These facts are important when we develop the "half-range" expansion of a function.

Therefore, if we want to represent $f(x)$ between 0 and L as a Fourier series and are interested only in approximating it on the interval $(0, L)$, we can redefine f within the interval $(-L, L)$ in two importantly different ways: (1) We can redefine the portion from $-L$ to 0 by reflecting about the y-axis. We then generate an even function. (2) We can reflect the portion between 0 and L about the origin to generate an odd function. Figures 4.10 and 4.11 showed these two possibilities.

Thus two different Fourier series expansions of $f(x)$ on $(0, L)$ are possible, one that has only cosine terms or one that has only sine terms. We get the A's for the even extension of $f(x)$ on $(0, L)$ from

$$A_n = \frac{2}{L}\int_0^L f(x)\cos\left(\frac{n\pi x}{L}\right)dx, \qquad n = 0, 1, 2, \ldots. \tag{4.45}$$

We get the B's for the odd extension of $f(x)$ on $(0, L)$ from

$$B_n = \frac{2}{L}\int_0^L f(x)\sin\left(\frac{n\pi x}{L}\right)dx, \qquad n = 1, 2, 3, \ldots. \tag{4.46}$$

EXAMPLE 4.6 Find the Fourier cosine series expansion of $f(x)$, given that

$$f(x) = \begin{cases} 0, & 0 < x < 1, \\ 1, & 1 < x < 2. \end{cases} \tag{4.47}$$

Figure 4.12 shows the even extension of the function.

Because we are dealing with an even function on $(-2, 2)$, we know that the Fourier series will have only cosine terms. We get the A's with

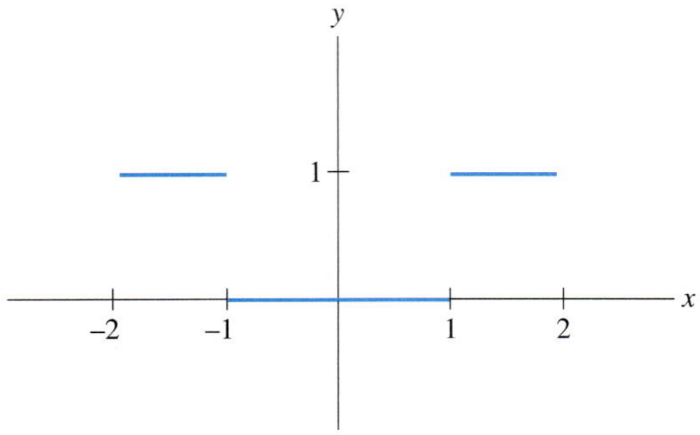

Figure 4.12
Plot of Eq. (4.47) reflected about the y-axis

$$A_0 = \frac{2}{2}\int_1^2 (1)\,dx = 1;$$
$$A_n = \frac{2}{2}\int_1^2 (1)\cos\left(\frac{n\pi x}{2}\right)dx = \begin{cases} 0, & n \text{ even}, \\ \dfrac{2(-1)^{(n+1)/2}}{n\pi}, & n \text{ odd}. \end{cases} \tag{4.48}$$

Then the Fourier cosine series is

$$f(x) \approx \frac{1}{2} + \frac{2}{\pi}\sum_{n=1}^{\infty}\frac{(-1)^n\cos((2n-1)(\pi x/2))}{(2n-1)} \tag{4.49}$$

EXAMPLE 4.7 Find the Fourier sine series expansion for the same function as in Example 4.6. Figure 4.13 shows the odd extension of the function.

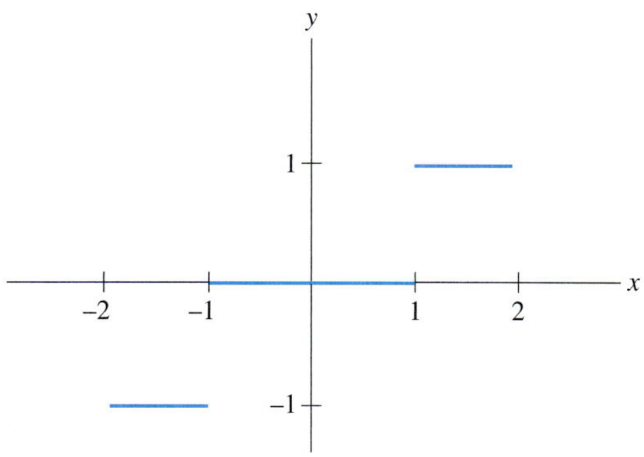

Figure 4.13
Plot of Eq. (4.47) reflected about the origin

We know that all of the A-coefficients will be zero, so we need to compute only the B's:

$$B_n = \frac{2}{2}\int_1^2 (1)\sin\left(\frac{n\pi x}{2}\right) dx$$
$$= \frac{2}{n\pi}\left[-\cos(n\pi) + \cos\left(\frac{n\pi}{2}\right)\right], \qquad n = 1, 2, 3, \ldots. \tag{4.50}$$

The term in brackets gives the sequence 1, −2, 1, 0, 1, −2, 1, 0, ... Because this sequence is awkward to reduce (except by use of the mod function), we simply write

$$f(x) = \frac{2}{\pi}\sum_{n=1}^{\infty}\frac{[\cos(n\pi/2) - \cos(n\pi)]}{n}\sin\left(\frac{n\pi x}{2}\right). \tag{4.51}$$

Summary of Formulas for Computation of Fourier Coefficients

A function that is periodic of period P and meets certain criteria (see below) can be represented by Eq. (4.52):

$$f(x) = \frac{A_0}{2} + \sum_{n=1}^{\infty} A_n \cos\left(\frac{n\pi x}{P/2}\right) + \sum_{n=1}^{\infty} B_n \sin\left(\frac{n\pi x}{P/2}\right). \tag{4.52}$$

The coefficients can be computed with

$$A_n = \frac{2}{P}\int_{-P/2}^{P/2} f(x)\cos\left(\frac{n\pi x}{P/2}\right) dx, \qquad n = 0, 1, 2, \ldots, \tag{4.53}$$

$$B_n = \frac{2}{P}\int_{-P/2}^{P/2} f(x)\sin\left(\frac{n\pi x}{P/2}\right) dx, \qquad n = 1, 2, 3, \ldots. \tag{4.54}$$

(The limits of the integrals can be from a to $a + P$.)

If $f(x)$ is an even function, only the A's will be nonzero. Similarly, if $f(x)$ is odd, only the B's will be nonzero. If $f(x)$ is neither even nor odd, its Fourier series will contain both cosine and sine terms.

Even if $f(x)$ is not periodic, it can be represented on just the interval $(0, L)$ by redefining the function over $(-L, 0)$ by reflecting $f(x)$ about the y-axis or, alternatively, about the origin. The first creates an even function, the second an odd function. The Fourier series of the redefined function will actually represent a periodic function of period $2L$ that is defined for $(-L, L)$.

When L is the half-period, the Fourier series of an even function contains only cosine terms and is called a Fourier cosine series. The A's can be computed by

$$A_n = \frac{2}{L}\int_0^L f(x)\cos\left(\frac{n\pi x}{L}\right) dx, \qquad n = 0, 1, 2, \ldots. \tag{4.55}$$

The Fourier series of an odd function contains only sine terms and is called a Fourier sine series. The B's can be computed by

$$B_n = \frac{2}{L}\int_0^L f(x)\sin\left(\frac{n\pi x}{L}\right) dx, \qquad n = 1, 2, 3, \ldots. \tag{4.56}$$

If $f(x)$ (or its redefined extension) has a finite discontinuity, the Fourier series will converge to the average of the two limiting values at the discontinuity. The Fourier series converges slowly at a point of discontinuity and exhibits more pronounced oscillations (the Gibbs phenomenon) near that point. If $f(x)$ (or the redefined function) has a discontinuity in its first derivative, convergence will be slower at that point.

Exercises

Section 4.1

1. Write a computer program that generates $T_n(x)$ when the value of n is an input quantity.

2. Extend the graphs of several of the Chebyshev polynomials to $[-3, 3]$ and observe that the maximum magnitudes are larger than one outside of $[-1, 1]$.

3. Show that Eq. (4.3) is true for several combinations of n and m. How can you handle the discontinuity at $x = 1$?

4. Graph $T_5(x)$ for x between -1 and 1. Read approximate values for the zeros from the graph. Then use Newton's method to find the values to a precision of ± 0.0001.

▶ **5.** Expand $\cos(6x)$ and compare this to $T_6(x)$. The formula for the cosine of the sum of two angles will help you in this.

▶ **6.** $T_4(x)/8$ has four zeros in $[-1, 1]$. What are their values? Create some other fourth-degree polynomial whose coefficient of x^4 is unity that has different zeros in $[-1, 1]$. Compare the graphs of this with that of $T_4(x)/8$ to verify that it has a larger upper bound to its magnitude within the interval.

7. Verify that the values in Table 4.1 are accurate.

8. For the interval $[0, 1]$, superimpose the graph for the economized polynomial of degree-4 on the graph of the Maclaurin series of the same degree for $f(x) = e^x$. Do it again for those of degree-5.

9. Repeat Exercise 8, but for the interval $[-1, 0]$.

10. Make a table equivalent to Table 4.1 but for the function $f(x) = e^x \cos(x)$.

▶**11.** Extend Eq. (4.6) to include equations for x^{10} and x^{11}.

12. The function $\arctan(x)$ can be represented by this power series:

$$\arctan(x) = x - \frac{x^3}{3} + \frac{x^5}{5} - \frac{x^7}{7} + \frac{x^9}{9} - \cdots.$$

Economize this three times to give a third-degree polynomial. Graph the errors, and compare this graph to the errors of the ninth-degree expansion.

13. Find the first few terms of the Chebyshev series for $\cos(x)$ by rewriting the Maclaurin series in terms of the $T(x)$'s and collecting terms. Convert this to a power series in x. Compare the error of both the Chebyshev series and the power series after truncating each to the fourth degree.

▶**14.** A series expansion for $(1 + x/5)^{1/2}$ is

$$1 + \frac{x}{10} - \frac{x^2}{200} + \frac{x^3}{2000} - \frac{x^4}{16{,}000} + \frac{7x^5}{800{,}000} - \cdots.$$

Convert this to a Chebyshev series, including terms to T_2. What is the maximum error of the truncated Chebyshev series? Compare this to the error of the power series when it is truncated to second degree.

15. Make a table similar to Table 4.2 but for $f(x) = \sin(x)/\exp(x)$. Also do a graph similar to Figure 4.2.

16. Compute the coefficients of Eq. (4.7) from the integration formula for a_i given in Section 4.1.

17. To get the smaller error of a Chebyshev series or an economized power series requires that the approximation be for the interval $[-1, 1]$. Show what change of variable will change $f(x)$ on $[a, b]$ to $f(y)$ on $[-1, 1]$.

▶**18.** The Legendre polynomials (which we will discuss in the next chapter) resemble the Chebyshev polynomials in that they have the same number of zeros within $[-1, 1]$ and the same number of maxima and minima. These Legendre polynomials can be obtained through this recursion formula:

$$L_0(x) = 1, \quad L_1(x) = x,$$
$$(n+1)L_{n+1}(x) = (2n+1)xL_n(x) - nL_{n-1}(x).$$

Compare the graphs of several of these polynomials with Chebyshev polynomials of the same degree. Why are the Legendre polynomials less suited to economizing a power series?

19. Verify Eq. (4.3) after making the substitution from Eq. (4.4). Do this analytically.

20. $\sin(nx)$ is orthogonal over $[-\pi, \pi]$ [see Eq. (4.18)]. Make a change of variable so that it is orthogonal over $[-1, 1]$. What value of n causes this new function to have exactly five minima/maxima on $[-1, 1]$. Compare its graph to that of $T_4(x)$. Then do the same for $\cos(x)$.

Section 4.2

21. Find Padé approximations for these functions, with numerators and denominators each of the third degree:

$$\cos^2(x),\ \sin(x^4 - x),\ xe^x.$$

▶**22.** Compare the errors in $[-1, 1]$ for each of the approximations of Exercise 2 with the errors of the corresponding sixth-degree Maclaurin series.

23. Express the following rational fractions in continued-fraction form. In each part, compare the number of multiplication and division operations with that resulting from evaluating the polynomials by Horner's method (in nested form).

a. $\dfrac{x^2 - 2x + 2}{x^2 + 2x - 2}$.

b. $\dfrac{2x^3 + x^2 + x + 3}{x^2 - x - 4}$.

c. $\dfrac{2x^4 + 45x^3 + 381x^2 + 1353x + 1511}{x^3 + 21x^2 + 157x + 409}$.

24. Convert each of the Padé approximations of Exercise 21 to continued fractions.

25. Estimate the errors of each of the Padé approximations of Exercise 21 by computing the coefficient of the next nonzero term in the numerator. Compare to the actual errors at $x = -1$ and $x = 1$.

26. A Chebyshev series for $\cos(\pi x/4)$ is

$$0.851632 - 0.146437T_2 + 0.00192145T_4 - 9.965 * 10^{-6}T_6.$$

Use this series to develop a Padé-like rational function by the method of Section 4.2 where the function is $R_{3,3}$.

▶**27.** Fike (1968) gives this example of a rational fraction approximation to $\Gamma(1 + x)$ on $[0, 1]$:

$$R_{3,4}(x) = \frac{0.999999 + 0.601781x + 0.186145x^2 + 0.0687440x^3}{1 + 1.17899x - 0.122321x^2 - 0.260996x^3 + 0.060992x^4}.$$

Is this a minimax approximation? If not, what are the bounds of the errors of the $R_{3,4}$ minimax approximation?

28. The rational function of Exercise 27 is $R_{3,4}$. Getting $R_{4,3}$ should have a smaller error. Is this true?

29. The approximations obtained in Exercise 21 are not minimax. However, you can set bounds to the errors of the corresponding minimax approximations from them. What are these bounds?

Section 4.3

▶**30.** Which of these functions is periodic? What is the period if it is periodic?

a. $\sin(2x) + 2\cos(x)$ c. $\sin^3(x)$

b. $e^{-10x}\cos(x)$ d. e^{2ix}, where $i = \sqrt{-1}$

31. Duplicate Example 4.3 and Figures 4.5 and 4.6, but for $f(x) = (x + 1)^2$.

32. Example 4.4 gets the Fourier series for $f(x) = |x|$ between $x = -\pi$ and $x = \pi$. $f(x)$ is also periodic with a period of 2π. Extend the function to the range $[-8, 8]$ and duplicate Figure 4.7 for this larger range.

33. Duplicate the plot of Figure 4.6, but for the extended range of $[-10, 10]$.

▶**34.** Find the Fourier coefficients for $f(x) = x^3$ if it is periodic and one period extends from $x = -1$ to $x = 2$.

35. Find the Fourier coefficients for $f(x) = x^2 - 1$ if it is periodic and one period extends from $x = -1$ to $x = 2$.

36. Show that the Fourier series for

$$f(x) = x^3 + x^2 - 1,$$

between $x = -1$ and $x = 2$, is just the sum of the series in Exercises 34 and 35.

▶**37.** Is the Fourier series for $f(x) * g(x)$ equal to the product of the series for $f(x)$ and $g(x)$?

38. What are the plots of the functions in Exercises 34 and 35 when reflected about the x-axis? When reflected about $x = -1$? When reflected about $x = 2$? Are any of these odd functions?

39. Repeat Exercise 38 but reflect about the y-axis. Are the results even or odd?

40. Suppose we are interested in $f(x) = e^{-x} \sin(2x - 1)$ only in the interval [0, 2]. Sketch the half-range extensions that give

a. An even function.

b. An odd function.

41. Repeat Exercise 40, but for the range $[-1, 3]$.

42. Find the Fourier coefficients for the periodic functions of Exercise 40.

43. Repeat Exercise 42 for the functions of Exercise 41. Is there any relation between these and the series of Exercise 42?

▶44. As Figure 4.7 shows, a finite Fourier series does not match to $f(0) = 0$ for $f(x) = |x|$.

a. How many terms are needed for it to match to within 0.00001?

b. Looking at Figure 4.7 again, a finite Fourier series does not match $f(\pi) = \pi$ at $x = \pi$. How many terms are needed to match within 0.00001?

c. Is the error of a given finite series the same at these two points?

▶45. Figure 4.6 shows how a Fourier series behaves at a discontinuity—it equals the average value of the function. The figure also shows, for $f(x) = x$, that the series matches exactly to $f(x)$ $2N + 1$ times when there are N terms in the series. For this example, what are the x-values where there is match for $N = 4$?

46. Reproduce Figure 4.6, but for $N = 3$ and 5. Based on the figure, it would appear that these two series should match exactly to $f(x)$ at 7 and 11 x-values.

a. Is this true?

b. At what x-values do the series match to $f(x)$?

c. Using this together with the result of Exercise 45, what can you say about the location of points where the series and the function agree?

d. Is the conclusion of part (c) true for other definitions of $f(x)$?

Applied Problems and Projects

APP1. In Section 4.2, the Padé rational functions were developed to approximate $f(x)$ on the interval $[-1, 1]$. If we want to approximate $f(x)$ on a different interval, say, $[a, b]$, a simple linear transformation can change the interval to $[-1, 1]$. What if we want to approximate a function on an interval with one or both endpoints infinite? Devise a transformation for such cases.

APP2. Investigate, for some computer system available to you, how some or all of the following transcendental functions are approximated in Fortran 90. Classify these into Taylor series formulas, Chebyshev polynomials, rational functions, or other types. Which of these are minimax?

a. $\sin(x)$
b. $\cos(x)$
c. $\tan(x)$
d. $\text{atan}(x)$
e. $\exp(x)$
f. $\ln(x)$

APP3. Repeat APP2 for other computer languages: BASIC, Pascal, and C.

APP4. As illustrated by Figure 4.8, the sum of n terms of the Fourier series for a function that has a jump discontinuity has larger oscillations near the discontinuity—the Gibbs phenomenon. For $f(x)$ equal to a square wave, investigate whether the departure of the sum of the series from $f(x)$ for the last "hump" in the curve (the "ear"), decreases when n is increased. What can you conclude about the size of the ear?

APP5. One way to eliminate the Gibbs phenomenon (see APP4) is to abolish the jump discontinuity by subtracting a linear function from $f(x)$. Suppose the linear function is $L(x)$. For $f(x)$ equal to square wave, find the $L(x)$ for which $f(x) - L(x)$ has no jump discontinuities. Compare the accuracy of the sum of 10 terms of the Fourier series for $f(x)$ with that for the sum of 10 terms of the Fourier series for $g(x) = f(x) - L(x)$.

APP6. Another way to ameliorate the problem of the Gibbs phenomenon is to use the so-called Lanczos's factors. Search the literature to find out more about this method. Apply it to obtain an improved approximation to a square wave.

APP7. Chapter 3 described the fitting of functions with polynomials and this chapter describes fitting them with sinusoids (Fourier series). Another possibility is to fit with exponentials, $y(x) = \Sigma c_i \exp(a_i x)$. Is it possible to do this? Under what conditions is it possible? How can the values of c_i and a_i be determined? Specifically, fit a four-term sum to these data and compare to the exact solution, $y = \sin(\pi x/6)$:

x:	1	2	3	4,
y:	$\frac{1}{2}$	$\sqrt{\frac{3}{2}}$	1	$\sqrt{\frac{3}{2}}$.

5

Numerical Differentiation and Integration

The heart of calculus is to find derivatives and integrals of functions that are exploited in many applications. We show in this chapter how derivatives and definite integrals can be computed with a computer program. Of course, computer algebra systems such as MATLAB, Maple, or *Mathematica* can obtain analytic results through their symbolic capabilities.

In this chapter, as in the previous two, we continue to exploit the useful properties of polynomials to develop methods for a computer to do integrations and to find derivatives. Because we can use an interpolating polynomial to approximate a function even if it is known only through a table of values, these methods find application when working with experimental data.

When the function is explicitly known, we can emulate the methods of calculus. But doing so in getting derivatives requires the subtraction of quantities that are nearly equal and that runs into round-off error. However, integration involves only addition, so round off is no problem; of course, we cannot often find the true answer numerically because the analytical value is the limit of the sum of an infinite number of terms. We must be satisfied with approximations for both derivatives and integrals but, for most applications, the numerical answer is adequate.

If we are working with experimental data that are displayed in a table of $[x, f(x)]$ pairs, emulation of calculus is impossible; we must approximate the function behind the data in some way. The polynomial approximations of Chapter 3 are an obvious approach. Still, even if the experimental data are exact, approximating the function with a polynomial is itself inexact. If there is experimental error in the data, there is additional error due to this.

The topics of this chapter are important enough that there are many techniques and these have been implemented in libraries of computer programs in various languages. The history of the methods is rich and goes back more than 300 years; names of famous mathematicians like Newton, Gauss, Lagrange, and Legendre are associated with them.

Contents of This Chapter

5.1 Differentiation with a Computer
Employs the interpolating polynomials of Chapter 3 to derive formulas for getting derivatives. These can be applied to functions known explicitly as well as those whose values are found in a table. Based on a consideration of the error term, a method for getting improved estimates can be found, a procedure called Richardson extrapolation.

5.2 Numerical Integration—The Trapezoidal Rule
Approximates the integrand function with a linear interpolating polynomial to derive a very simple but important formula for numerically integrating functions between given limits. The method can be applied to tabular data. Romberg integration, an extrapolation technique, can improve the accuracy.

5.3 Simpson's Rules
Develops more accurate integration formulas based on approximating the integrand with quadratic or cubic polynomials.

5.4 An Application of Numerical Integration—Fourier Series and Fourier Transforms
Shows how the methods for numerical integration can be used to compute the terms of a Fourier series. When a Fourier series is developed from experimental measurements of periodic phenomena, a *discrete Fourier series* can be obtained. This is a transformation of the data to reveal the fundamental vibrational frequencies of the system.

5.5 Adaptive Integration
Describes a way to reduce the number of function evaluations when Simpson's rule is used. A kind of binary search is used to locate subregions where the size of intervals can be larger. An interesting bookkeeping problem is involved.

5.6 Gaussian Quadrature
Gives the development of an integration method that uses fewer function evaluations by properly selecting the points where the value of the function is computed. The section introduces a representative of *orthogonal polynomials,* the Legendre polynomials.

5.7 Multiple Integrals
Explains how numerical methods can evaluate a multiple integral, with either fixed or variable limits.

5.8 Application of Cubic Splines
Gives the details for using a spline approximation to compute derivatives and integrals.

5.1 Differentiation with a Computer

When you studied calculus, you learned that the derivative of a function, $f(x)$ at $x = a$, is defined as

$$(df/dx)_{x=a} = \lim \frac{f(a + \Delta x) - f(a)}{\Delta x}, \text{ as } \Delta x \rightarrow 0.$$

(This is called a forward-difference approximation. The limit could be approached from the opposite direction, giving a backward-difference approximation.)

It should be clear that a computer can calculate an approximation to the derivative from

$$(df/dx)_{x=a} = \frac{f(a + \Delta x) - f(a)}{\Delta x}$$

if a very small value is used for Δx. What if we do this, recalculating with smaller and smaller values of Δx starting from an initial value that is not small? We should expect to find an optimal value for Δx because round-off errors in the numerator will become great as Δx approaches zero, and these are magnified by the small value in the denominator.

When we try this for $f(x) = e^x \sin(x)$ at $x = 1.9$, starting with $\Delta x = 0.05$ and halving Δx each time, we find that the errors of the approximation decrease as Δx is reduced until about $\Delta x = 0.05/128$. The analytical answer is 4.1653826. Table 5.1 gives the results from a computer program. Notice that each successive error is about 1/2 of the previous error as Δx is halved until Δx gets quite small, at which time round off affects the ratio. At values for Δx smaller than 0.05/128, the error of the approximation increases due to round off. If double precision is used, a more accurate estimate is achieved. In effect, the best value for Δx is when the effects of round-off and truncation errors are balanced.

Table 5.1 Forward-difference approximations for $f(x) = e^x \sin(x)$

Δx	Approximation	Error	Ratio of errors
0.05	4.05010	−0.11528	
0.05/2	4.10955	−0.05583	2.06
0.05/4	4.13795	−0.02743	2.04
0.05/8	4.15176	−0.01362	2.01
0.05/16	4.15863	−0.00675	2.02
0.05/32	4.16199	−0.00389	1.99
0.05/64	4.16382	−0.00156	2.18
0.05/128	**4.16504**	**−0.00034**	**4.67***
0.05/256	4.16504	−0.00034	
0.05/512	4.16504	−0.00034	
0.05/1024	4.16992	0.00454	
0.05/2048	4.17969	0.01430	

* At this point, round-off and truncation errors are in balance, but we still do not achieve six-digit accuracy.

If a backward-difference approximation is used:

$$(df/dx)_{x=a} = \frac{f(a) - f(a - \Delta x)}{\Delta x},$$

similar results are obtained.

MATLAB knows a lot about derivatives. First, it can get the analytical answer to the function of Table 5.1:

```
EDU>> f = 'exp(x)*sin(x)'
f =
exp(x)*sin(x)
EDU>> df = diff(f,'x')
df =
exp(x)*sin(x) + exp(x)*cos(x)
EDU>> numeric(subs(df,1.9,'x'))
ans =
   4.1654
```

Of course it can compute numerically:

```
EDU>> x = [1.9 1.9 1.9 1.9 1.9 1.9 1.9 1.9 1.9];
EDU>> del = [.05 .05/2 .05/4 .05/8 .05/16 .05/32 . . .
        .05/64 .05/128 .05/256];
EDU>> xplus = x + del;
EDU>> f = exp(x).*sin(x);
EDU>> fplus = exp(xplus).*sin(xplus);
EDU>> num = fplus - f;
EDU>> deriv = num./del
      deriv =
         4.0501    4.1096    4.1379    4.1518    4.1586
           4.1620    4.1637    4.1645    4.1650
```

In this, we first created several vectors: the x-values, and values for Δx, $x + \Delta x$, $f(x)$, $f(x + \Delta x)$, and the numerator values. This last was divided by the Δx's to give essentially the same results as in Table 5.1.

You may want use Maple to see how round off causes the results to be less accurate when the precision of the computations is poorer. We found that with a precision of only five digits, the best estimate was 4.1600 at $\Delta x = 0.05/8$. The ratio of errors was again about 2 to 1.

It is not by chance that the errors are about halved each time. Look at this Taylor series where we have used h for Δx:

$$f(x + h) = f(x) + f'(x) * h + f''(\xi) * h^2/2,$$

where the last term is the error. The value of ξ is at some point between x and $x + h$. If we solve this equation for $f'(x)$, we get

$$f'(x) = (f(x + h) - f(x))/h - f''(\xi) * h/2, \tag{5.1}$$

which shows that the errors should be about proportional to h, precisely what Table 5.1 shows. In terms of the order relation, we see that error is $O(h)$. If we repeat this but begin with the Taylor series for $f(x - h)$, it turns out that

$$f'(x) = (f(x) - f(x - h))/h + f''(\zeta) * h/2, \tag{5.2}$$

where ζ is between x and $x - h$, so the two error terms are not identical though both are $O(h)$.

Now, if we add Eqs. (5.2) and (5.1), then divide by 2, we get the *central-difference approximation* to the derivative:

$$f'(x) = (f(x + h) - f(x - h))/(2h) - f'''(\xi)h^2/6. \tag{5.3}$$
$$\text{Error is } O(h^2).$$

We had to extend the two Taylor series by an additional term to get the error because the $f''(x)$ terms cancel.

This shows that using a central-difference approximation is a much preferred way to estimate the derivative; even though we use the same number of computations of the function at each step, we approach the answer much more rapidly. Table 5.2 illustrates this, showing that errors decrease about four fold when Δx is halved [as Eq. (5.3) predicts] and that a more accurate value is obtained.

All of this reminds us that it is best to center the x-value within the points used in the estimate, as we found for interpolation.

What we have found is also in accord with the *mean-value theorem for derivatives*:

$$\frac{f(b) - f(a)}{b - a} = f'(\xi), \quad \text{where } a < \xi < b.$$

The forward-difference approximation will give a value for $f'(x)$ at a point between x and $x + h$; the backward approximation gives a value at a point between $x - h$ and x; the central approximation at a point between $x - h$ and $x + h$. Unless the function behaves wildly near the point x, these three points are close to $x + h/2$, $x - h/2$, and x.

Table 5.2 Central-difference approximations for $f(x) = e^x \sin(x)$

Δx	Approximation	Error	Ratio of errors
0.05	4.15831	−0.00708	
0.05/2	4.16361	−0.00177	4.00
0.05/4	4.16496	−0.00042	4.21
0.05/8	4.16527	−0.00011	3.80
0.05/16	**4.16534**	**−0.00004**	**2.75**
0.05/32	4.16534	−0.00004	
0.05/64	4.16565	−0.00027	

Derivatives from Divided-Difference Tables

There is another way to get derivatives numerically. We can build a table of values for the function, get an interpolating polynomial from appropriate entries, and then differentiate this polynomial. If the x-values are evenly spaced, we could employ a polynomial derived from ordinary differences. If the entries are unevenly spaced, we use divided differences. Because divided differences apply in either case, we do this first. Recall that

$$\begin{aligned} f(x) &= P_n(x) + \text{error} \\ &= f[x_0] + f[x_0, x_1](x - x_0) \\ &\quad + f[x_0, x_1, x_2](x - x_0)(x - x_1) \\ &\quad + \cdots + f[x_0, x_1, \ldots, x_n]\prod_{i=0}^{n-1}(x - x_i) \\ &\quad + \text{error}. \end{aligned} \tag{5.4}$$

If the polynomial is a good match to the function near the x-value where we want the derivative, we should get a good match to the derivative by differentiating Eq. (5.4). Doing this* we get:

$$\begin{aligned} P_n'(x) &= f[x_0, x_1] + f[x_0, x_1, x_2][(x - x_1) + (x - x_0)] + \cdots \\ &\quad + f[x_0, x_1, \ldots, x_n]\sum_{i=0}^{n-1} \frac{(x - x_0)(x - x_1)\cdots(x - x_{n-1})}{(x - x_i)}. \end{aligned} \tag{5.5}$$

To get the error term for Eq. (5.5) we have to differentiate the error term for $P_n(x)$:

$$\text{Error} = (x - x_0)(x - x_1)\cdots(x - x_n)\frac{f^{(n+1)}(\xi)}{(n+1)!}. \tag{5.6}$$

When this error term is differentiated, we will find a sum that has in one of its terms

$$\frac{d}{dx}[f^{(n+1)}(\xi)],$$

which is impossible to evaluate because ξ depends on x in an unknown way. However, if we take $x = x_i$ (where x_i is one of the tabulated points), the difficult term drops out and we get this expression for the error:

Error of the approximation to $f'(x)$, when $x = x_i$, is

$$\text{Error} = \left[\prod_{\substack{j=0 \\ j\neq 1}}^{n}(x_i - x_j)\right]\frac{f^{(n+1)}(\xi)}{(n+1)!}, \qquad \xi \text{ in } [x, x_0, x_n]. \tag{5.7}$$

---

* Recall that the derivative of a product of n factors is a sum of n terms, where each term in this sum is the same n factors but one of the factors is replaced by its derivative in succession. For example,

$$(u * v * w)' = u' * v * w + u * v' * w + u * v * w'.$$

Observe that the error is not zero even when x is a tabulated value, although the interpolating polynomial agrees with $f(x)$ at this point. In fact, the error of the derivative is less at some x-values between the points.

It is not surprising that the next-term rule applies here as it did for interpolating polynomials.

Suppose we have the table given in Table 5.3. The table is for $f(x) = e^x \sin(x)$, so we can compare with the previous computations.

Remember that the first divided differences are computed as

$$f[x_i, x_{i+1}] = (f_{i+1} - f_i)/(x_{i+1} - x_i),$$

which are precisely the forward differences of f_i. Hence, between each successive pairs of the entries in the table, the estimates of $f'(x)$ are just these first differences that are constants. Figure 5.1a shows these estimates superimposed on the curve of analytical values for $f'(x)$. The value of x where $f'(x) = 4.6311$ (the first entry in the table for the first-order difference) is 1.7527, almost exactly halfway between the x-values used to compute it. That is true as well for the other estimates in that column.

If we use two terms of Eq. (5.5) we can compute the estimated derivatives from successive triples of the data. These give linear relations in x as shown in Figure 5.1b. Three terms of Eq. (5.5), which involve groups of four entries from the table, produce the quadratic relations shown in Figure 5.1c. Notice that the estimated derivatives are very close to the analytical when three terms are used.

As we saw in Chapter 3 for interpretation, if we want to estimate the derivative for an x-value near the end of the table, we appear to be severely limited in the degree of interpolating polynomial. We can overcome this limitation by reordering i-values, putting them in reverse order. Our formulas still work correctly, but we must remember to go diagonally upward to get the values for a given value of i.

Because an interpolating polynomial fits better to the function if the x-values used in its construction are such that the x-value for the derivative is centered within them, we should choose the starting point (the i-value) to make this true. (If the x-values are in order, our task is easier.)

Table 5.3 Divided-difference table for $e^x \sin(x)$

i	x_i	f_i	$f[x_i, x_{i+1}]$	$f[x_i \ldots x_{i+2}]$	$f[x_i \ldots x_{i+3}]$	$f[x_i \ldots x_{i+4}]$
0	1.70	5.4283	4.6311	−1.6469	−3.1137	−1.1493
1	1.80	5.8914	4.1371	−3.6708	−4.0331	
2	2.00	6.7188	2.1182	−6.4939		
3	2.35	7.4602	−1.1288			
4	2.50	7.2909				

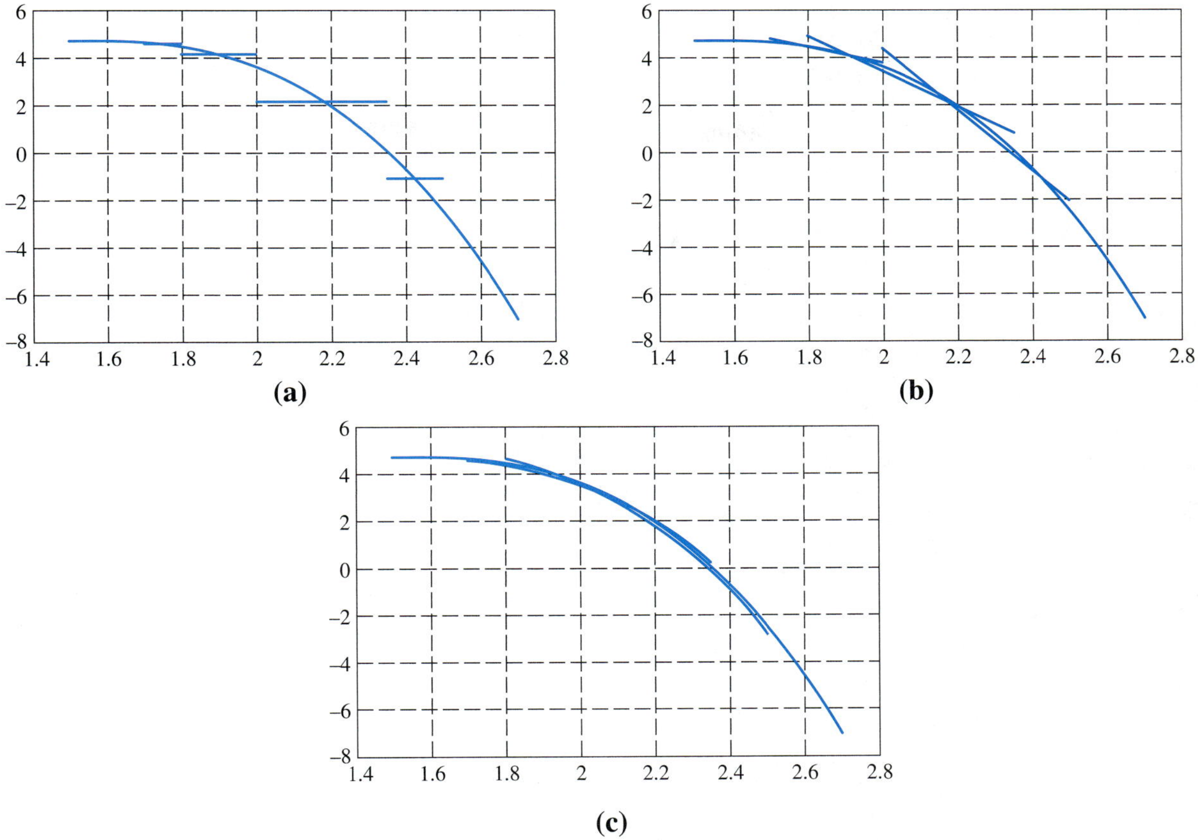

Figure 5.1

An Algorithm to Obtain an Estimate of the Derivative, f′(u), from a Divided-Difference Table given n + 1 data pairs, $(x_i, f_i), i = 0, \ldots, n$:

(Create Table of Divided Differences)
For $i = 0$ To n Step 1 Do
    Set $f(i, 0) = f(i)$ End Do (For i)
For $j = 1$ To n Step 1 Do
    For $i = 0$ To $n - j$ Step 1 Do
        $f(\mathrm{i}, \mathrm{j}) = [f(i + 1, j - 1) - f(i, j - 1)]/[x(i + j) - x(i)]$
    EndDo (For i)
EndDo (For j)

(Now get user inputs)
    u = value to be chosen for $f'(u)$
    Deg = chosen degree of the polynomial

(Compute the derivative)

```
Set PolySum = 0
For j = Deg To 2 Step − 1 Do
    Set Sum = 0
    For k = 0 To j − 1 Step 1 Do
        (p computes the value: [(u − x_0) . . . (u − x_k)])
        Set p = 1
        For ℓ = 0 to j − 1 Step 1 Do
            (evaluates the product: [(u − x_0) . . . (u − x_{j−1})]/(u − x_ℓ)
        If ℓ ≠ k Then
            p = p * [u − x_ℓ] End If
        End Do (For ℓ)
        Set Sum = Sum + f(0, j) * p
    End Do (For k)
    Set PolySum = Polysum + Sum
End Do (For j).
PolySum = PolySum + f(0, 1)

Display PolySum as derivative value at u.
```

Evenly Spaced Data

Even though divided differences can handle any table, it is instructive to see how ordinary differences can estimate the derivative when a table is evenly spaced. Recall from Chapter 3 that we can write an interpolating polynomial in terms of the differences [in this, $s = (x - x_i)/h$, which means that $x = x_i + s * h$]:

$$P_n(x_s) = P_n(s) = f_i + s\Delta f_i + \frac{s(s-1)}{2!}\Delta^2 f_i + \frac{s(s-1)(s-2)}{3!}\Delta^3 f_i$$

$$+ \cdots + \prod_{j=0}^{n-1}(s-j)\,\frac{\Delta^n f_i}{n!} + \text{error};$$

$$\text{Error} = \left[\prod_{j=0}^{n}(s-j)\right]\frac{f^{(n+1)}(\xi)}{(n+1)!}, \qquad \xi \text{ in } [x, x_1, \ldots, x_n].$$

(In this formula, i is the index value where we enter the difference table.)

The derivative of $P_n(s)$ should approximate $f'(x)$. We do exactly the same as we did for the polynomial constructed from a divided-difference table, getting

$$\frac{d}{dx}P_n(s) = \frac{d}{ds}P_n(s)\,\frac{ds}{dx}$$

$$= \frac{1}{h}\left[\Delta f_i + \sum_{j=2}^{n}\left\{\sum_{k=0}^{j-1}\prod_{\substack{\ell=0\\ \ell\neq k}}^{j-1}(s-\ell)\right\}\frac{\Delta^j f_i}{j!}\right]. \qquad (5.8)$$

(The $1/h$ factor comes from $ds/dx = d/dx(x - x_i)/h = 1/h$.)

Again, the error term involves an unknown quantity unless x is one of the tabulated values. When $x = x_i$, $s = 0$. In this case, we get this analog of Eq. (5.7) when an interpolating polynomial of degree n is used:

$$\begin{matrix}\text{Error} \\ (\text{when } x = x_i)\end{matrix} = \frac{(-1)^n h^n}{n+1} f^{(n+1)}(\xi), \quad \xi \text{ in } [x_1, \ldots, x_n]. \tag{5.9}$$

Equation (5.8) is a formula for estimating derivatives from a table of differences that we enter at index value i.

We will illustrate with the data from Table 5.4, which again are for $f(x) = e^x \sin(x)$.

The values of the first order differences, Δf_i, give the forward-difference approximations to f_i' when divided by $\Delta x = 0.2$ (we do not have to divide with divided differences). If we use two terms of Eq. (5.8), we get linear approximations; three terms give quadratics. Plots similar to those in Figure 5.1 result if these are graphed together with $f'(x)$.

Simpler Formulas

Equation (5.8) is awkward to use, but if we stipulate that the x-value where we want the value of the derivative is one of the tabulated values, there is great simplification. To get $f'(x_i)$ we just use

$$f'(x_i) = (1/h)[\Delta f_i - (1/2)\Delta^2 f_i + (1/3)\Delta^3 f_i - \cdots + (1/n)(-1)^{n-1}\Delta^n f_i] \tag{5.10}$$

because at $x = x_i$, $s = 0$. Equation (5.10) is easy to use because the multipliers of the differences are so simple. In addition, the errors are very conviently expressed as "order of." For example, with just one term of Eq. (5.6) we have

$$f'(x_i) = (1/h)[\Delta f_i] - (1/2)hf''(\xi), \quad \text{error is } O(h).$$

With two terms,

$$f'(x_i) = (1/h)[\Delta f_i - (1/2)\Delta^2 f_i] + (1/3)h^2 f^{(3)}(\xi), \quad \text{error is } O(h^2).$$

So, we see that with n terms, the error is $O(h^n)$. We also see that we can estimate the error with the next-term rule. It is easy to show that the use of two terms of Eq. (5.10) gives the central-difference formula for $f'(x_{i+1})$ and that this has an error of $O(h^2)$.

Table 5.4 Differences of $f(x) = e^x \sin(x)$

i	x_i	f_i	Δf_i	$\Delta^2 f_i$	$\Delta^3 f_i$	$\Delta^4 f_i$
1	1.7	5.4283	0.8985	−0.1763	−0.1573	−0.0448
2	1.9	6.3269	0.7223	−0.3336	−0.2020	
3	2.1	7.0491	0.3887	−0.5356		
4	2.3	7.4378	−0.1469			
5	2.5	7.2909				

Higher-Order Derivatives

If we want formulas to estimate the second and higher-order derivatives of tabulated values, we could differentiate Eq. (5.5) or Eq. (5.8). However, we prefer to show you another way to do the job, a technique that has wide applicability. This is the *method of undetermined coefficients.* It is easiest to demonstrate this for a table of evenly spaced x-values.

We begin by getting a formula for the first derivative, one that we have seen already.

Suppose we have three values of $f(x)$ with the x-values differing by a uniform amount, h. We can tabulate these:

x_-	f_-
x_0	f_0
x_+	f_+

where $x_- = x_0 - h$ and $x_+ = x_0 + h$. We want a formula for $f'(x_0)$ in terms of the three function values. The arithmetic is simplified if we translate axes to make $x_0 = 0$, so the values of x becomes $-h$, 0, $+h$.

A second-degree polynomial, $P(x)$, can be fitted to the three points and $P(x)$ is then an approximation to $f(x)$. The derivative of $P(x)$, $P'(x)$, is then an approximation to $f'(x)$. We will want it for $f'(0)$, the derivative at $x = 0$, the center of the x-values:

$$f'(0) \approx P'(0) = A * f_- + B * f_0 + C * f_+, \tag{5.11}$$

where A, B, and C are the unknown coefficients. We write $P(x)$ as the general quadratic

$$P(x) = ax^2 + bx + c.$$

We look at three instances of Eq. (5.11):
Case 1, $P(x) = 1$ implies:

$$c = 1, a = b = 0, f_- = 1, f_0 = 1, f_+ = 1, P'(0) = 0:$$
$$A * (1) + B * (1) + C * (1) = 0.$$

Case 2, $P(x) = x$ implies:

$$b = 1, a = c = 0, f_- = -h, f_0 = 0, f_+ = h, P'(0) = 1:$$
$$A * (-h) + B * (0) + C * (h) = 1.$$

Case 3, $P(x) = x^2$ implies:

$$a = 1, b = c = 0, f_- = h^2, f_0 = 0, f_+ = h^2, P'(0) = 0:$$
$$A * (h^2) + B * (0) + C * (h^2) = 0.$$

In matrix form, these three equations are:

$$\begin{bmatrix} 1 & 1 & 1 \\ -h & 0 & h \\ h^2 & 0 & h^2 \end{bmatrix} \begin{bmatrix} A \\ B \\ C \end{bmatrix} = \begin{bmatrix} 0 \\ 1 \\ 0 \end{bmatrix}. \tag{5.12}$$

We can write the solution to Eq. (5.12) by inspection: From the third equation, $A = -C$; then, from the second, $A = -1/(2h)$ and $C = +1/(2h)$; substituting into the first, $B = 0$. We then have this equation that approximates the derivative:

$$f'(0) = -1/(2h) * f_- + (0) * f_0 + 1/(2h) * f_+ = \frac{f_+ - f_-}{2h},$$

which is the central-difference formula, as we would expect.

We now do the same for the second derivative. The cases are the same, but now, when $P(x) = 1$, $P''(0) = 0$; when $P(x) = x$, $P''(0) = 0$; when $P(x) = x^2$, $P''(0) = 2$. The coefficient matrix is identical to that in Eq. (5.12), so we get

$$\begin{bmatrix} 1 & 1 & 1 \\ -h & 0 & h \\ h^2 & 0 & h^2 \end{bmatrix} \begin{bmatrix} A \\ B \\ C \end{bmatrix} = \begin{bmatrix} 0 \\ 1 \\ 2 \end{bmatrix}. \tag{5.13}$$

Again, the solution is easy: from the second equation, $A = C$; from the third, $A = C = 1/h^2$; from the first, $B = -2/h^2$. The formula for the second derivative is

$$f''(0) = \frac{f_- - 2f_0 + f_+}{h^2}. \tag{5.14}$$

If we want the error term for Eq. (5.14), we work with the Taylor series for $f(x + h)$ and for $f(x - h)$. The result is that the error is $O(h^2)$, which we leave as an exercise.

We also leave as exercises to show that

$$f''(x_0) = \frac{-f_2 + 16f_1 - 30f_0 + 16f_{-1} - f_{-2}}{12h^2} + O(h^4),$$

$$f'''(x_0) = \frac{f_2 - 2f_1 + 2f_{-1} - f_{-2}}{2h^3} + O(h^2),$$

$$f^{(4)}(x_0) = \frac{f_2 - 4f_1 + 6f_0 - 4f_{-1} + f_{-2}}{h^4} + O(h^2).$$

Higher Derivatives with MATLAB

We saw earlier that MATLAB can get the analytical derivatives of a function. It can do the same for higher derivatives. Here we find the second and eighth derivatives of $f(x) = e^x/x$ and evaluate these at $x = 3$.

```
EDU>> f = 'exp(x)/x';
EDU>> df2 = diff(f, 'x', 2)
  df2 =
    exp(x)/x - 2*exp(x)/x^2 + 2*exp(x)/x^3
EDU>> df8 = diff (f, 8)
  df8 =
    exp(x)/x-8*exp(x)/x^2 + 56*exp(x)/x^3 -
```

```
336*exp(x)/x^4 + 1680*exp(x)/x^5 -
  6720*exp(x)/x^6 + 20160*exp(x)/x^7 -
    40320*exp(x)/x^8 + 40320*exp(x)/x^9
```

The expression for the eighth derivative is pretty complicated. We can get the numerical values of these at $x = 3$:

```
EDU>> numeric(subs (df2, 3))
  ans =
    3.7195
EDU>> numeric(subs (df8, 3))
  ans =
    3.7563
```

Extrapolation Techniques

We found earlier that the errors of a central-difference approximation to $f'(x)$ were of $O(h^2)$. In effect, that suggests that the errors are proportional to h^2 although that is true only in the limit as $h \to 0$. Unless h is quite large, we can assume the proportionality. So, from two computations with h being half as large in the second, we can estimate the proportionality factor which we call C. For example, in Table 5.2 we had:

h	**Approximation**
0.05	4.15831
0.025	4.16361

If errors were truly $C(h^2)$, we can write two equations:

$$\text{True value} = 4.15831 + C(0.05^2)$$

$$\text{True value} = 4.16361 + C(0.025^2)$$

from which we can solve for the true value, eliminating the unknown "constant" C, getting:

$$\begin{aligned}\text{True value} &= 4.16361 + (1/3) * (4.16361 - 4.15831)\\ &= 4.16538,\end{aligned}$$

which is very close to the exact value for $f'(1.9)$, 4.165382.

You can easily derive the general formula for improving the estimate, when errors decrease by $O(h^n)$:

$$\underset{\text{estimate}}{\text{Better}} = \underset{\text{accurate}}{\text{more}} + (1/(2^n - 1))(\text{more} - \text{less}), \tag{5.15}$$

where more and less in the last term are the two estimates at h_1 and $h_2 = h_1/2$. "More accurate" is the estimate at the smaller value of h and n is the power of h in the order of the errors.

As a second example, let us apply this to values from Table 5.1 which were from forward-difference approximations. Here the errors are $O(h)$.

h	Approximation
0.05	4.05010
0.025	4.10955

Using Eq. (5.15), we have

$$\text{Better estimate} = 4.10955 + (4.10955 - 4.05010)\,(1/(2^1 - 1))$$
$$= 4.16900,$$

which shows considerable improvement but not as good as from the central differences.

This extrapolation technique applies to any set of computations where the order of the error is known, and we will see later in this chapter that we can apply it to integration methods. Of course it also applies to the computation of higher derivatives, such as from Eq. (5.14).

Richardson Extrapolation

When we compute an extrapolation from two estimates of the derivative using, say, $h = 0.1$ and $h = 0.05$, both of which are of $O(h^2)$, the improved estimate has an error $O(h^4)$ as we show below. If we do another computation of $f'(x)$ at $h = 0.025$ to get a third estimate of $f'(x)$ and use this with the estimate at $h = 0.05$ to extrapolate, we get a second further improved estimate also of error $O(h^4)$. What is the error if we use these two improved estimates to extrapolate again?

Consider the difference between the pair of Taylor series that gave rise to Eq. (5.3) but with more terms:

$$f_i' = (f(x + h) - f(x - h))/(2h) + a_1h^2 + a_2h^4 + a_3h^6 + \cdots . \quad \textbf{(5.16)}$$

(The terms on the right after the first represent the error of the central difference approximation; the odd powers of h drop out through cancellations.)

If we compute a second approximation for f_i' but with h cut in half, we get a better approximation:

$$f_i' = (f(x + h/2) - f(x - h/2))/(h) + a_1h^2/4 + a_2h^4/16 + a_3h^6/64 + \cdots . \quad \textbf{(5.17)}$$

Adding 1/3 of the difference between Eqs. (5.17) and (5.16) to Eq. (5.17) gives Eq. (5.15), but now we see that n will be 4 because the first of the errors terms cancel.

Using the two improved estimates for the derivative, but now adding 1/15 of the difference to the better estimate, results in canceling the next error term; it will be of $O(h^6)$. Continuing in the same fashion gives estimates of $O(h^8)$, $O(h^{10})$, . . . , until there is no change in the improvements. Doing these successive extrapolations is called *Richardson extrapolation.*

Here is an example with $f(x) = x^2 * \cos(x)$ for which $f(1.0) = 0.23913363$. The original values of $f'(1.0)$ are from central differences so they are of $O(h^2)$.

Value of h	$f'(1.0)$	First extrapolations	Second extrapolations	Third extrapolations
0.1	0.226736			
0.05	0.236031	0.239129		
0.025	0.238258	0.239133	0.239134	
0.0125	0.238940	0.239132	0.239132	0.239132

There really was no point in doing the third extrapolation because the second one did not change the value.

The merit of Richardson extrapolation is that we get greatly improved estimates without having to evaluate the function additional times. We can use this technique to extrapolate higher derivatives as well.

An Algorithm to Compute a Richardson Table That Computes the Derivative, f′(x)

Given a function $f(x)$:

Input
    x = value for x
    h = starting value for stepsize h
    MaxStage = maximum number of stages (lines of table)
    Tol = tolerance value for termination
    $d(0, 1) = 0$: the initial value of the table

(Compute lines, (stages) of the table)
    For stage = 1 To MaxStage Step 1 Do
        Set $d(\text{stage}, 1) = [f(x + h) - f(x - h)]/(2h)$.
        For j = 2 to stage Step 1 Do
            Set $d(\text{stage}, j) = d(\text{stage}, j - 1)$
                $+ [d(\text{stage}, j - 1) - d(\text{stage} - 1, j - 1)]/(2^{2j} - 1)$
        EndDo (For j).
        If $|d(\text{stage}, \text{stage}) - d(\text{stage}, \text{stage} - 1)| < \text{Tol}$
        Then Exit EndIf.
        Set $h = h/2$
EndDo (For stage).

On termination, the last computed value is the extrapolated estimate of the derivative.

Extrapolation with Tabulated Values

If we only have an evenly spaced table of $(x, f(x))$ values, as we might have from a set of experiments, we have no way to get new function values where the differences in x are halved. However, if there are enough entries in the table, we may be able to double the Δx's. Table 5.5 is an example.

Table 5.5

i	x_i	f_i
0	2.0	0.123060
1	2.1	0.105706
2	2.2	0.089584
3	2.3	0.074764
4	2.4	0.061277
5	2.5	0.049126
6	2.6	0.038288
7	2.7	0.028722
8	2.8	0.020371
9	2.9	0.013164
10	3.0	0.007026

Suppose we want the derivative at $x = 2.4$. The central difference approximation is -0.12819 from $f(2.3)$ and $f(2.5)$, $h = 0.1$. Now, if we compute the value again, but use the values at $x = 2.2$ and 2.6, where $h = 0.2$, we get -0.12824, a poorer estimate because h is twice as large. However, since we known that both are of $O(h^2)$, we can employ Eq. (5.15) to get an improvement:

$$\begin{aligned} f'(2.4)\ [\text{improved}] &= -0.12819 + (-0.12819 + 0.12824)/3 \\ &= -0.12817. \end{aligned}$$

[The function in Table 5.5 is for $f(x) = e^{-x}\sin(x)$ for, which $f'(2.4) = -0.128171$.]

For convenience, here we collect formulas for computing derivatives.

Formulas for Computing Derivatives

Formulas for the first derivative:

$$f'(x_0) = \frac{f_1 - f_0}{h} + O(h)$$

$$f'(x_0) = \frac{f_1 - f_{-1}}{2h} + O(h^2) \qquad \text{Central difference}$$

$$f'(x_0) = \frac{-f_2 - 4f_1 - 3f_0}{2h} + O(h^2)$$

$$f'(x_0) = \frac{-f_2 + 8f_1 - 8f_{-1} + f_{-2}}{12h} + O(h^4) \qquad \text{Central difference}$$

Formulas for the second derivative:

$$f''(x_0) = \frac{f_2 - 2f_1 + f_0}{h^2} + O(h)$$

$$f''(x_0) = \frac{f_1 - 2f_0 + f_{-1}}{h^2} + O(h^2) \qquad \text{Central difference}$$

$$f''(x_0) = \frac{-f_3 + 4f_2 - 5f_1 + 2f_0}{h^2} + O(h^2)$$

$$f''(x_0) = \frac{-f_2 + 16f_1 - 30f_0 + 16f_{-1} - f_{-2}}{12h^2} + O(h^4) \qquad \text{Central difference}$$

Formulas for the third derivative:

$$f'''(x_0) = \frac{f_3 - 3f_2 + 3f_1 - f_0}{h^3} + O(h)$$

$$f'''(x_0) = \frac{f_2 - 2f_1 + 2f_{-1} - f_{-2}}{2h^3} + O(h^2) \qquad \text{Averaged difference}$$

Formulas for the fourth derivative:

$$f^{iv}(x_0) = \frac{f_4 - 4f_3 + 6f_2 - 4f_1 + f_0}{h^4} + O(h)$$

$$f^{iv}(x_0) = \frac{f_2 - 4f_1 + 6f_0 - 4f_{-1} + f_{-2}}{h^4} + O(h^2) \qquad \text{Central difference}$$

5.2 Numerical Integration—The Trapezoidal Rule

Integral calculus is a most important branch of calculus. It is used to find the velocity of a body when its acceleration is known, to find the distance traveled using the velocity, to compute areas, to predict population growth, and in many other important applications.

In your calculus course, you learned many formulas to get the *indefinite integral* of function $f(x)$, the *antiderivative.* [Given the function, $f(x)$, the antiderivative is a function $F(x)$ such that $F'(x) = f(x)$.] You learned that the *definite integral,*

$$\int_a^b f(x) = F(b) - F(a),$$

can be evaluated from the antiderivative. Still, there are functions that do not have an antiderivative expressible in terms of ordinary functions.

All of our computer algebra systems can find the antiderivative if its table of integrals has it. For example, in Maple,

```
>int(x*sin(x), x);
  sin(x) − x cos(x)
```

but, if the antiderivative in unknown, it just returns $\int f(x)\,dx$:

```
>int(exp(x)/ln(x), x);
```

$$\int \frac{e^x}{\ln(x)}\,dx$$

If we give limits for the integral,

```
>int(x*sin(x), x = 1..2);
  sin(2) - 2 cos(2) - sin(1) + cos(1)
```

Maple gives us $F(b) - F(a)$, which we can evaluate with

```
>evalf(%)
  1.440422421
```

Now we ask, "Is there any way that the definite integral can be found when the antiderivative is unknown?" The answer is "Yes, we can do it numerically."

You learned that the definite integral is the area between the curve of $f(x)$ and the x-axis. That is the principle behind all numerical integration—we divide the distance from $x = a$ to $x = b$ into vertical strips and add the areas of these strips (the strips are often made equal in widths but that is not always required).

The Trapezoidal Rule

When the area between the curve of $f(x)$ and the x-axis is subdivided into strips, one way to draw the strips is to make the top of the strips touch the curve, either at the left corner or the right corner, but that is less accurate than making the top of the strip even with the curve at its midpoint. In effect, these schemes replace the curve for $f(x)$ with a sequence of horizontal lines. We can think of these lines as interpolating polynomials of degree zero.

A much better way is to approximate the curve with a sequence of straight lines; in effect, we slant the top of the strips to match with the curve as best we can. We are approximating the curve with interpolating polynomials of degree-1. The gives us the *trapezoidal rule.* Figure 5.2 illustrates this.

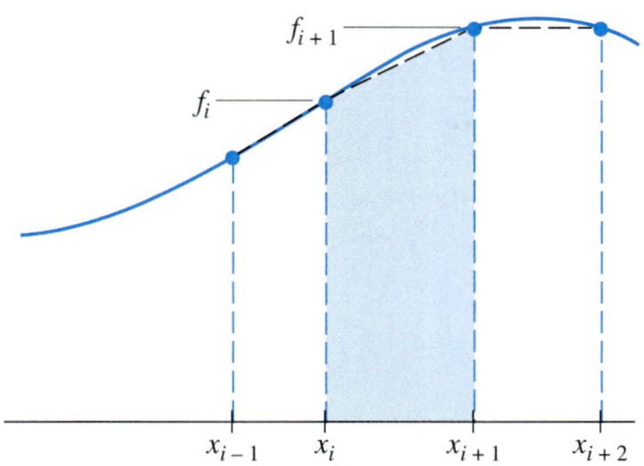

Figure 5.2

From Figure 5.2, it is intuitively clear that the area of the strip from x_i to x_{i+1} gives an approximation to the area under the curve:

$$\int_{x_i}^{x_{i+1}} f(x)\,dx \approx \frac{f_i + f_{i+1}}{2}(x_{i+1} - x_i).$$

We will usually write $h = (x_{i+1} - x_i)$ for the width of the interval.

Another Derivation of the Trapezoidal Rule

An alternative way to obtain the trapezoidal rule is to fit $f(x)$ between pairs of x-values with polynomials of degree-1 and integrate those polynomials. We learned in Chapter 3 that a first-degree Newton–Gregory interpolating polynomial between points x_i and x_{i+1} was

$$f(x) \approx P_1(x) = f_i + s\Delta f_i + \text{error},$$

where $s = (x - x_i)/h$ and the error is given by

$$(h^2/2)(s)(s-1)f''(\xi).$$

We can estimate $\int f(x)$ between the two points by integrating $P_1(x)$:

$$\int_{x_i}^{x_{i+1}} f(x)\,dx \approx \int_{x_i}^{x_{i+1}} P_1(x) = \int_{x_i}^{x_{i+1}} (f_i + s\Delta f_i)\,dx = h\int_0^1 (f_i + s\Delta f_i)\,ds,$$

where we have replaced dx with $h * ds$, and noted that $s = 0$ at x_i and $s = 1$ at x_{i+1}.

Carrying out the integration, we find that

$$\int_{x_i}^{x_{i+1}} f(x)\,dx \approx \int_{x_i}^{x_{i+1}} P_1(x) = h(f_i + (1/2)\Delta f_i) = (h/2)(f_i + f_{i+1}),$$

exactly as we found intuitively. The real reason for this development is to find the error term for one application of trapezoidal integration. We get this by integrating the error term. Doing so, we find

$$\text{Error} = -(1/12)h^3 f'''(\xi) = O(h^3).$$

The Composite Trapezoidal Rule

If we are getting the integral of a known function over a larger span of x-values, say, from $x = a$ to $x = b$, we subdivide $[a, b]$ into n smaller intervals with $\Delta x = h$, apply the rule to each subinterval, and add. This gives the composite trapezoidal rule:

$$\int_a^b f(x)\,dx \approx \sum_{i=0}^{n-1} (h/2)(f_i + f_{i+1}) = (h/2)(f_0 + 2f_1 + 2f_2 + \cdots + 2f_{n-1} + f_n). \qquad \textbf{(5.18)*}$$

---

* In a computer program, you should do $h(f_0/2 + f_1 + f_2 + \cdots + f_{n-1} + f_n/2)$ in order to reduce the number of operations.

The error now is not the *local error* $O(h^3)$ but the *global error,* the sum of n local errors:

$$\text{Global error} = (-1/12)h^3[f''(\xi_1) + f''(\xi_2) + \cdots + f''(\xi_n)].$$

In this equation, each of the ξ_i is somewhere within each subinterval. If $f''(x)$ is continuous in $[a, b]$, there is some point within $[a, b]$ at which the sum of the $f''(\xi_i)$ is equal to $f''(\xi)$, where ξ is in $[a, b]$. We then see that, because $nh = (b - a)$,

$$\text{Global error} = (-1/12)h^3 nf''(\xi) = \frac{-(b-a)}{12} h^2 f''(\xi) = O(h^2).$$

The fact that the global error is $O(h^2)$ while the local error is $O(h^3)$ seems reasonable because, for example, if we double the number of subintervals, we add together twice as many local errors.

EXAMPLE 5.1 Given the values for x and $f(x)$ in Table 5.6, use the trapezoidal rule to estimate the integral from $x = 1.8$ to $x = 3.4$.

Applying the trapezoidal rule:

$$\int_{1.8}^{3.4} f(x)\,dx \approx \frac{0.2}{2}[6.050 + 2(7.389) + 2(9.025) + 2(11.023) + 2(13.464) + 2(16.445) + 2(20.086) + 2(24.533) + 29.964] = 23.9944.$$

The data in Table 5.6 are for $f(x) = e^x$ and the true value is $e^{3.4} - e^{1.8} = 23.9144$. The trapezoidal rule value is off by 0.08; there are three digits of accuracy. How does this compare to the estimated error?

$$\text{Error} = -\frac{1}{12}h^3 nf''(\xi), \qquad 1.8 \le \xi \le 3.4,$$

$$= -\frac{1}{12}(0.2)^3(8)*\begin{Bmatrix} e^{1.8} & (\text{max}) \\ e^{3.4} & (\text{min}) \end{Bmatrix} = \begin{Bmatrix} -0.0323 & (\text{max}) \\ -0.1598 & (\text{min}) \end{Bmatrix}.$$

Table 5.6

x	$f(x)$	x	$f(x)$
1.6	4.953	2.8	16.445
1.8	6.050	3.0	20.086
2.0	7.389	3.2	24.533
2.2	9.025	3.4	29.964
2.4	11.023	3.6	36.598
2.6	13.464	3.8	44.701

Alternatively,

$$\text{Error} = -\frac{1}{12}(0.2)^2(3.4 - 1.8)*\begin{Bmatrix} e^{1.8} & (\max) \\ e^{3.4} & (\min) \end{Bmatrix} = \begin{Bmatrix} -0.0323 \\ -0.1598 \end{Bmatrix}.$$

The actual error was −0.080.

If we had not known the function for which we have tabulated values, we would have estimated $h^2 f''(\xi)$ from the second differences.

An Algorithm for Integration by the Composite Trapezoidal Rule

Given a function $f(x)$:

(Get user inputs).
Input

  a, b = endpoints of interval
  n = number of intervals

(Do the integration)
Set $h = (b - a)/n$.
Set sum = 0
For $i = 1$ to $n - 1$ Step 1 Do
  Set $x = a + h * i$,
  Set sum = sum + 2 * $f(x)$
End Do (For i).
Set sum = sum + $f(a) + f(b)$.
Set ans = sum * $h/2$.

The value of the integral is given by ans.

Unevenly Spaced Data

Data from experimental observations may not be evenly spaced. The trapezoidal rule still applies. Suppose there are five points:

$$\int f(x) \approx \frac{f_0 + f_1}{2}(x_1 - x_0) + \frac{f_1 + f_2}{2}(x_2 - x_1) + \frac{f_2 + f_3}{2}(x_3 - x_2) + \frac{f_3 + f_4}{2}(x_4 - x_3)$$

$$= \sum_{i=0}^{3} \frac{f_i + f_{i+1}}{2}(x_{i+1} - x_i).$$

There is no simple way to express this.

Romberg Integration

We can improve the accuracy of the trapezoidal rule integral by a technique that is similar to Richardson extrapolation. This technique is known as *Romberg integration.*

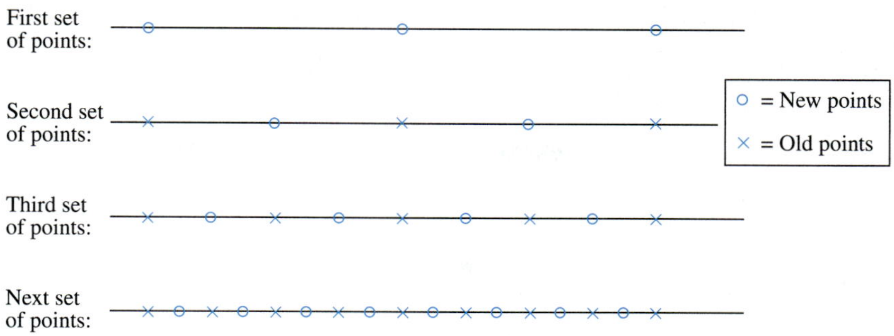

Figure 5.3

Because the integral determined with the trapezoidal method has an error of $O(h^2)$, we can combine two estimates of the integral that have h-values in a 2:1 ratio by Eq. (5.15), which we repeat here:

$$\text{Better estimate} = \text{more accurate} + \frac{1}{2^n - 1}(\text{more accurate} - \text{less accurate}).^* \tag{5.19}$$

When we apply this equation to get the integral of a known function, we begin with an arbitrary value for h in Eq. (5.18). A second estimate is then made with the value of h halved. From these two estimates we extrapolate to get an improved estimate using Eq. (5.19). This has an error of $O(h^4)$.

Obviously, this can be extended to produce a table of successively better estimates. When we find that the values converge, we have the best estimate that we can make in the light of round-off error. As shown before, each new extrapolation has error orders that increase: $O(h^4)$, $O(h^6)$, $O(h^8)$, ...

We can reduce the number of computations because, when h is halved, all of the old points at which the function was evaluated to get Eq. (5.18) appear in the new computation and we thus can avoid repeating the evaluations. Figure 5.3 illustrates this point.

This next example shows how the *Romberg table* appears for the function $f(x) = e^{-x^2}$ integrated between the limits of 0.2 and 1.5. This integral has no closed form solution. It is closely related to the *error function,* a quantity that is so important in statistics and other branches of applied mathematics that values have been tabulated.

EXAMPLE 5.2 Use Romberg integration to find the integral of e^{-x^2} between the limits of $a = 0.2$ and $b = 1.5$. Take the initial subinterval size as $h = (b - a)/2 = 0.65$.

Our first estimate is

$$\text{Integral} = \frac{h}{2}[f(a) + 2f(a + h) + f(b)]$$

* In Eq. (5.19), n is the order of the error. In the first extrapolation, $n = 2$. In successive extrapolations, it is 4, 6, 8, ...

Table 5.7 Romberg table of integrals over interval from 0.2 to 1.5 with an initial h of 0.65

0.66211			
0.65947	0.65859		
0.65898	0.65881	0.65882	
0.65886	0.65882	0.65882	0.65882

$$= \frac{0.65}{2}\,[e^{-0.2^2} + 2e^{-0.85^2} + e^{-1.5^2}]$$

$$= 0.66211.$$

The next estimate uses $h = 0.65/2 = 0.325$:

$$\text{Integral} = \frac{h}{2}\,[f(a) + 2f(a+h) + 2f(a+2h) + 2f(a+3h) + f(b)]$$

$$= \frac{0.325}{2}\,[e^{-0.2^2} + 2e^{-0.525^2} + 2e^{-0.85^2} + 2e^{-1.175^2} + e^{-1.5^2}]$$

$$= 0.65947.$$

Observe that only two new function evaluations appear in the second estimate.

We now extrapolate:

$$\text{Improved} = 0.65947 + \frac{1}{3}\,[0.65947 - 0.66211]$$

$$= 0.65859.$$

Table 5.7 exhibits the calculations when we repeat the estimations, halving the h-value each time.

The Romberg Method for a Tabulated Function

We can apply the Romberg method to integrate a function known only as a table of evenly spaced function values, but now we cannot make h smaller. Instead, we use estimates of the integral with h doubled each time, just as we did to improve the estimates of derivatives in Section 5.1. Here is an example.

EXAMPLE 5.3 Use the Romberg method to get an improved estimate of the integral from $x = 1.8$ to $x = 3.4$ from the data in Table 5.6. In Example 5.1, we found an estimate of 23.9944 when $h = 0.2$. If we now use $h = 0.4$, we compute

$$\text{Integral} = (0.4/2)\,[6.050 + 2(9.025) + 2(13.464) + 2(20.086) + 29.964]$$

$$= 24.2328.$$

Table 5.8 Romberg table for Example 5.3

$h = 0.2$	23.9944	23.9149	23.9147
$h = 0.4$	24.2328	23.9181	
$h = 0.8$	25.1768		

We can extrapolate from these two estimates:

$$\text{Improved} = 23.9944 + (23.9944 - 24.2328)/3 = 23.9149.$$

Now, if we use $h = 0.8$, we get

$$\begin{aligned}\text{Integral} &= (0.8/2)\,[6.050 + 2(13.464) + 29.964]\\ &= 25.1768.\end{aligned}$$

Using this with the estimate when $h = 0.4$:

$$\begin{aligned}\text{Improved} &= 24.2328 + (24.2328 - 25.1768)/3\\ &= 23.9181.\end{aligned}$$

We can use these two improved estimates to extrapolate a second time:

$$\begin{aligned}\text{Further improved} &= 23.9149 + (23.9149 - 23.9181)/15\\ &= 23.9147.\end{aligned}$$

Table 5.8 shows the results. Considering that the function values in Table 5.6 are given only to three decimals, this compares well to the analytical answer of 23.9144 and this is much better than the result from the single estimate with $h = 0.2$, which was 23.9944.

The Romberg method is applicable to a wide class of functions. Smoothness and continuity are not required. However, when $f(x)$ is discontinuous, we should make the evenly spaced points fall on the discontinuities. This can be done if we break the interval into subintervals that are bounded by the discontinuities.

An Algorithm for Romberg Integration

Given a function $f(x)$:

(Get user inputs).
Input
    a, b = endpoints of interval
    Maxstages = number of refinements

(Do the integration)

Set $h = (b - a)/2$.

```
(Do the lines of the table)
Set sum = f(a) + 2 * f(a + h) + f(b).
Set integral(0, 0) = sum * h/2 (first value)
Set distance = 2 * h (distance between added points)
For stage = 1 To Maxstages Step 1 Do
    Set h = h/2.
    Set distance = distance/2.
    For i = 1 To 2^stage Step 1 Do
        Set x = a − h + i * distance.
        Set sum = sum + 2 * f(x).
    End Do (For i).
    Set integral(stage,0) = sum * h/2.
(Now extrapolate)
    For j = 1 To stage Step 1 Do
        Set integral(stage, j) = integral(stage, j − 1) +
            [integral(stage, j − 1) − integral(stage − 1, j − 1)]/(2^j − 1)
    End Do (For j).
End Do (For stage)
```

The last computed value is the estimate of the integral.

An alternative stopping criterion is when two successive computations in a line differ by less than some tolerance value.

5.3 Simpson's Rules

The trapezoidal rule is based on approximating the function with a linear polynomial. We can fit the function better if we approximate it with a quadratic or a cubic interpolating polynomial. Simpson's rules are based on these approximations. There are two of these rules: *Simpson's 1/3 rule* and *Simpson's 3/8 rule*, so-named because the values 1/3 and 3/8 appear in their formulas.

We get the 1/3 rule by integrating the second-degree Newton–Gregory forward polynomial, which fits $f(x)$ at x-values of x_0, x_1, x_2, which are evenly spaced a distance h apart:

$$\int_{x_0}^{x_2} f(x)\,dx \approx \int_{x_0}^{x_2} \left(f_0 + s\Delta f_0 + \frac{s(s-1)}{2}\Delta^2 f_0\right) dx$$

$$= h\int_0^2 \left(f_0 + s\Delta f_0 + \frac{s(s-1)}{2}\Delta^2 f_0\right) ds$$

$$= hf_0 s\Big]_0^2 + h\Delta f_0 \frac{s^2}{2}\Big]_0^2 + h\Delta^2 f_0\left(\frac{s^3}{6} - \frac{s^2}{4}\right)\Big]_0^2$$

$$= h\left(2f_0 + 2\Delta f_0 + \frac{1}{3}\Delta^2 f_0\right) = \frac{h}{3}(f_0 + 4f_1 + f_2).*$$

We get the error by integrating the error of the polynomial:

$$\text{Error} = -\frac{1}{90}h^5 f^{(4)}(\xi), \qquad x_0 < \xi < x_2.$$

It is convenient to think of the strips defined by successive x-values as *panels*. For Simpson's 1/3 rule, there must be an even number of panels.

We get the 3/8 rule similarly, by integrating the third-degree Newton–Gregory interpolating polynomial that fits to four evenly spaced points, and its error term:

$$\int_{x_0}^{x_3} f(x)\,dx \approx \int_{x_0}^{x_3} P_3(x_5)\,dx = \frac{3h}{8}(f_0 + 3f_1 + 3f_2 + f_3).$$

$$\text{Error} = -\frac{3}{80}h^5 f^{(4)}(\xi_1), \qquad x_0 < \xi_1 < x_3.$$

If the number of panels is divisible by 3, the 3/8 rule applies. Observe that the error of the 3/8 rule is actually larger than for the 1/3 rule and both have a local error of $O(h^5)$. The global errors will be $O(h^4)$ for the same reason as with the trapezoidal rule.

You may wonder why we use the 3/8 rule when it has a larger error. One useful application of it is to find the integral from a table of values that has an odd number of panels. Still, the error should be less for a table with an odd number of points by applying the 3/8 rule for the first or last set of three panels and then using the 1/3 rule for the rest. Where the 3/8 rule is used, it is best to choose the panels at one end or the other, or at intermediate points where the function is most nearly straight.

In the example below, we compare the three rules. We will obtain the integral of $\exp(-x^2)$ between $x = 0.2$ and $x = 2.6$ with different values for h, the even spacing between points. This integral has no closed form; it is required to get values for the *error function,* a special function that is important in certain statistical applications and is related to another special function, the *gamma function.*

First, we will use MATLAB to get the true value of the integral:

```
EDU>> f = sym('exp(-x^2)')
  f =
     exp(-x^2)
EDU>> fint = int(f)
  fint =
     1/2*pi^(1/2)*erf(x)
EDU>> fintdef = int(f,.2,2.6)
  fintdef =
```

* This is the way that the rule is usually written and is responsible for its being called the 1/3 Rule. If the coefficient is written $2h/6$, it more closely parallels the trapezoidal rule and the 3/8 rule.

Table 5.9 Comparison of integration methods for the integral of $\exp(-x^2)$ between $x =$ 0.2 and 2.6

Number of panels	Trapezoidal rule		Simpson's 1/3 rule		Simpson's 3/8 rule	
	Value	Error	Value	Error	Value	Erorr
6	0.69378	−0.00513	0.68824	−0.00041	0.68723	−0.00142
12	0.68992	−0.00127	0.68863	−0.00002	0.68860	−0.00005
18	0.68921	−0.00056	0.68865	0.00000	0.68864	−0.00001
24	0.68897	−0.00031	0.68865	0.00000	0.68865	0.00000

```
      1/2*erf(13/5)*pi^(1/2) - 1/2*erf(1/5)*pi^(1/2)
  EDU>> digits(10)
  EDU>> vpa(fintdef)
    ans =
      .6886527145
```

In this, we define the function symbolically, ask for the indefinite integral (which does involve the error function), get the definite integral (but this is not numeric), and finally get the numeric answer with the vpa command.

EXAMPLE 5.4 Find the integral of $\exp(-x^2)$ between $x = 0.2$ and $x = 2.6$. Compare the results at varying values for h with the trapezoidal rule, Simpson's 1/3 rule, and Simpson's 3/8 rule.

Table 5.9 gives the results. With the trapezoidal method, five significant digits of accuracy are not obtained until almost 300 panels have been used. The 1/3 method is better than the 3/8, as we would expect. The ratio of errors when the h-value is halved is close to 2^4 for the 1/3 rule, not quite that for the 3/8 rule (we do not have enough data for a good value), and almost exactly 2^2 for the trapezoidal rule.

Formulas for Integration (Uniform spacing, $\Delta x = h$)

Trapezoidal rule:

$$\int_a^b f(x)\,dx = \frac{h}{2}(f_1 + 2f_2 + 2f_3 + \cdots + 2f_n + f_{n+1}) - \frac{(b-a)}{12}h^2 f''(\xi), \qquad a \le \xi \le b \tag{5.20}$$

Simpson's $\frac{1}{3}$ rule:

$$\int_a^b f(x)\,dx = \frac{h}{3}(f_1 + 4f_2 + 2f_3 + 4f_4 + 2f_5 + \cdots + 4f_n + f_{n+1})$$

$$- \frac{(b-a)}{180} h^4 f^{(4)}(\xi), \qquad a \le \xi \le b \qquad (5.21)$$

(requires an even number of panels)

Simpson's $\frac{3}{8}$ rule:

$$\int_a^b f(x)\,dx = \frac{3h}{8}(f_1 + 3f_2 + 3f_3 + 2f_4 + 3f_5 + 3f_6 + \cdots + 3f_n + f_{n+1})$$

$$- \frac{(b-a)}{80} h^4 f^{(4)}(\xi), \qquad a \le \xi \le b \qquad (5.22)$$

(requires a number of panels divisible by 3)

These formulas, based on approximating the integrand with a polynomial of different degree, are known as *Newton–Cotes formulas*.

It is of interest to see that each of these integration formulas is just the width of the interval, $(b - a)$, times an average value for the function within that interval. That average value is a sum of the weighted values divided by the sum of the weights. For example, if there are six panels (seven points),

Trapezoidal rule: Weights are [1 2 2 2 2 2 1], whose sum is 12 and $(b - a)/12 = h * (1/2)$.

Simpson's 1/3 rule: Weights are [1 4 2 4 2 4 1], whose sum is 18 and $(b - a)/18$ is $h * (1/3)$.

Simpson's 3/8 rule: Weights are [1 3 3 2 3 3 1], whose sum is 16 and $(b - a)/16$ is $h * (3/8)$.

Discontinuous Functions and Improper Integrals

If the function being integrated is discontinuous or whose slope is discontinuous, it is essential that the region be broken up into subintervals bounded by the discontinuities. (It could be that the chosen points within the interval fall at the points of discontinuity and that takes care of this.)

An improper integral is (a) one whose integrand becomes infinite at one or more points on the region of interest, or (b) one with infinity at one or both of the endpoints of the integration. Some improper integrals have a finite value; the integral is said to *converge*. If the limiting value of the integration as we approach the point of singularity is infinite, it is said to *diverge*. It is obvious that none of the integration rules that we have described will work for improper integrals, although we can approximate the answer by gradually closing in on the point of singularity. This is not an easy way to get a good value; there are other integration techniques (called *open formulas*) that we do not discuss here that are better adapted. [*Numerical Recipes* (W.H. Press et al., 1992) is a good reference.] When an improper integral is integrable, often a change of variable will make it proper.

Another problem that is somewhat related is finding the value of the integral for a function that increases exponentially. Formulas that use evenly spaced points will not be adequate. We should use points that are much closer together in the subregion(s) where the slope is great. A plot of the function will reveal this.

Getting Integration Formulas in a Different Way

In Section 5.1, we used the method of undetermined coefficients to get formulas for differentiation. We can use this technique to get formulas for integration. We will illustrate it by starting with the simplest formula.

Suppose we want a formula to estimate the integral of $f(x)$ between $x = x_1$ and $x = x_2$, where $x_2 - x_1 = h$, using only the function values $f(x_1)$ and $f(x_2)$, and is of the form

$$\int_a^b f(x)\,dx = A * f(x_1) + B * f(x_2),$$

where A and B are coefficients to be determined. The two pairs of points, $(x_1, f(x_1))$ and $(x_2, f(x_2))$, permit us to write an interpolating po1ynomial, $P(x)$, of degree-1:

$$f(x) \approx P(x) = ax + b.$$

It simplifies the arithmetic if we translate axes to make $x_1 = 0$ so that $x_2 = h$. There are two cases to consider:

Case 1: $P(x) = 1$ requires $b = 1$, $a = 0$, so

$$\int_0^h P(x)\,dx = \int_0^h (1)\,dx = h = A * P(0) + B * P(h) = A * (1) + B * (1).$$

Case 2: $P(x) = x$ requires $a = 1$, $b = 0$, so

$$\int_0^h P(x)\,dx = \int_0^h (x)\,dx = (h^2)/2 = A * P(0) + B * P(h) = A * (0) + B * (h).$$

We can set up these two equations in matrix form:

$$\begin{bmatrix} 1 & 1 \\ 0 & h \end{bmatrix} \begin{bmatrix} A \\ B \end{bmatrix} = \begin{bmatrix} h \\ h^2/2 \end{bmatrix};$$

whose solution is easy: From the second equation, $B = h/2$; from the first $A + B = h$, so $A = h - B = h - h/2 = h/2$. Our formula is the familiar trapezoidal rule:

$$(h/2) * (f(x_1) + f(x_2)).$$

Now for another formula. If we use three evenly spaced values of $f(x)$, at $x_{-1} = -h$, $x_1 = h$, and the midpoint, $x_0 = 0$ (which we get after translating the axes), the interpolating polynomial, $P(x)$ [which is an approximation to $f(x)$], is now a quadratic,

$$P(x) = ax^2 + bx + c.$$

The formula we desire is

$$\int_{x_{-1}}^{x_1} f(x)\,dx = A * f(x_{-1}) + B * f(x_0) + C * f(x_1).$$

We have three cases for $P(x)$:

Case 1: $P(x) = 1$ requires $c = 1$, $a = b = 0$, so

$$\int_{-h}^{h} P(x)\,dx = \int_{-h}^{h} (1)\,dx = 2h = A * P(-h) + B * P(0) + C * P(h)$$
$$= A * (1) + B * (1) + C * (1).$$

Case 2: $P(x) = x$ requires $b = 1$, $a = c = 0$, so

$$\int_{-h}^{h} P(x)\,dx = \int_{-h}^{h} (x)\,dx = 0 = A * P(-h) + B * P(0) + C * P(h)$$
$$= A * (-h) + B * (0) + C * (h).$$

Case 3: $P(x) = x^2$ requires $a = 1$, $b = c = 0$, so

$$\int_{-h}^{h} P(x)\,dx = \int_{-h}^{h} (x^2)\,dx = 2h^3/3 = A * P(-h) + B * P(0) + C * P(h)$$
$$= A * (h^2) + B * (0) + C * (h^2).$$

The matrix is

$$\begin{bmatrix} 1 & 1 & 1 \\ -h & 0 & h \\ h^2 & 0 & h^2 \end{bmatrix} \begin{bmatrix} A \\ B \\ C \end{bmatrix} = \begin{bmatrix} 2h \\ 0 \\ 2h^3/3 \end{bmatrix},$$

whose solution is easy: From the second equation, $A = C$; from the third, $A = C = h/3$; from the first, $B = 4h/3$, so we get Simpson's 1/3 rule:

$$h/3 * [f(x_{-1}) + 4f(x_0) + f(x_1)].$$

Simpson's 3/8 rule can be derived if one uses four evenly spaced points of $(x, f(x))$. We leave this as an exercise.

5.4 An Application of Numerical Integration—Fourier Series and Fourier Transforms

In Chapter 4, we saw that a Fourier series can approximate functions, even those with discontinuities. The coefficients of the terms of the series are determined by definite integrals. There are functions for which the necessary integrals cannot be found analytically; for these, numerical procedures can be employed.

In this next example, we compare the accuracy of computing Fourier coefficients by the trapezoidal rule and by Simpson's 1/3 rule in a case where the analytical values are possible.

EXAMPLE 5.5 Evaluate the coefficients for the half-range expansions for $f(x) = x$ on [0, 2] numerically and compare to the analytic values. Do this with both 20 intervals and 200 intervals.

For the even extension (the Fourier cosine series), we use Eq. (4.55) to get the A's (all B's are zero):

$$A_n = \left(\frac{2}{2}\right)\int_0^2 x\cos\left(\frac{n\pi x}{2}\right)dx, \qquad n = 0, 1, 2, \ldots.$$

For the even extension (the Fourier sine series), we use Eq. (4.56) to get the B's (all A's are zero):

$$B_n = \left(\frac{2}{2}\right)\int_0^2 x\sin\left(\frac{n\pi x}{2}\right)dx, \qquad n = 1, 2, 3, \ldots.$$

Tables 5.10 and 5.11 show the results. Observe that the accuracy is poorer as the value of n increases.

Discrete Fourier Series

There are a number of applications when measurements of a periodic phenomenon are studied: musical chords, vibrations of structures, shock in automobiles, outputs in electrical and electronic circuits, for example. In analyzing such phenomena, we want to know the frequency spectrum.

When the data are from measurements of the system, we do not know the "function" that generates the information; we only have samples. Most often, this sampling is at successive intervals of time, with Δt being constant. When we fit such data with sine/cosine terms, it is called *Fourier analysis.* Other names are *harmonic analysis* and the

Table 5.10 Comparison of numerical integration with analytica1 results: 20 subdivisions of [0, 2]

	Trapezoidal rule		Simpson's rule		Analytical integration	
n	A_n	B_n	A_n	B_n	A_n	B_n
0	2		2		2	
1	−0.81224	1.27062	−0.81056	1.27324	−0.81057	1.27323
2	0	−0.63138	0	−0.63665	0	−0.63662
3	−0.09175	0.41653	−0.08999	0.42453	−0.09006	0.42441
4	0	−0.30777	0	−0.31860	0	−0.31831
5	−0.03414	0.24142	−0.03219	0.25523	−0.03242	0.25465

Table 5.11 Comparison of numerical integration with analytical results: 200 subdivisions of [0, 2]

	Trapezoidal rule		Simpson's rule		Analytical integration	
n	A_n	B_n	A_n	B_n	A_n	B_n
0	2		2		2	
1	−0.81059	1.27321	−0.81057	1.27324	−0.81057	1.27323
2	0	−0.63657	0	−0.63662	0	−0.63662
3	−0.09008	0.42433	−0.09006	0.42441	−0.09006	0.42441
4	0	−0.31821	0	−0.31831	0	−0.31831
5	−0.03244	0.25452	−0.03242	0.25465	−0.03242	0.25465

finite Fourier transform. This is a "transform" because we change data that are a function of time to a function of frequencies. We form what is called a *discrete Fourier series.*

Why should we want to so transform a set of experimental data? Because knowing which frequencies of a Fourier series are most significant (have the largest coefficients) gives information on the fundamental frequencies of the system. This knowledge is important because an applied periodic external force that includes components of the same frequency as one of these fundamental frequencies causes extremely large disturbances. (Such a periodic force may come from vibrations from rotating machinery, from wind, or from earthquakes.) We normally want to avoid such extreme responses for fear that the system will be damaged.

It is clear from Example 5.5 that the coefficients of a Fourier series can be computed numerically. Example 5.6 demonstrates getting the coefficients from measurements:

EXAMPLE 5.6 An experiment (actually, these are contrived data) showed the displacements given in Table 5.12 when the system was caused to vibrate in its natural modes. The values represent a periodic function on the interval for t of [2, 10] because they repeat themselves after $t = 10$.

We will use trapezoidal integration to find the Fourier series coefficients for the data. Doing so gives these values for the A's and B's:

n	A	B
0	**4.6015**	
1	**1.5004**	**−0.5006**
2	−0.0009	0.0016
3	−0.0017	0.0016
4	0.0008	**4.0011**
5	−0.0017	0.0000
6	−0.0009	0.0022
7	−0.0005	−0.0023
8	−0.0008	0.0009

Table 5.12 Measurements of displacements versus time

t	Displacement	t	Displacement
2.000	3.804	6.250	3.746
2.250	6.503	6.500	5.115
2.500	7.496	6.750	4.156
2.750	6.094	7.000	1.593
3.000	3.003	7.250	−0.941
3.250	−0.105	7.500	−1.821
3.500	−1.598	7.750	−0.329
3.750	−0.721	8.000	2.799
4.000	1.806	8.250	5.907
4.250	4.350	8.500	7.338
4.500	5.255	8.750	6.380
4.750	3.878	9.000	3.709
5.000	0.893	9.250	0.992
5.250	−2.048	9.500	−0.116
5.500	−3.280	9.750	1.047
5.750	−2.088	10.000	3.802
6.000	0.807		

This shows that only A_0, A_1, B_1, and B_4 are important. There would be no amplification of motion from forces that do not include the frequencies corresponding to these.
(Table 5.12 was constructed from

$$f(t) = 2.3 + 1.5\cos(t) - 0.5\sin(t) + 4\sin(4t),$$

plus a small random variation whose values ranged from −0.01 to +0.01. It is the random variations that cause nonzero values for the insignificant A's and B's.)

The Fast Fourier Transform

If we need to do a finite Fourier transform on lots of data, the amount of effort used in carrying out the computations is exorbitant. In the preceding examples, where we reevaluated cosines and sines numerous times, we should have recognized that many of these values are the same. When we evaluate the integrals for a finite Fourier transform, we compute sines and cosines for angles around the origin, as indicated in the figure on the following page.

When we need to find $\cos(nx)$ and $\sin(nx)$, we move around the circle; when $n = 1$, we use each value in turn. For other values of n, we use every nth value, but it is easy to see that these repeat previous values. The *fast Fourier transform* (often written as *FFT*) takes advantage of this fact to avoid the recomputations.

In developing the FFT algorithm, the preferred method is to use an alternative form of the Fourier series. Instead of

$$f(x) \approx \frac{A_0}{2} + \sum_{n=1}^{\infty} [A_n \cos(nx) + B_n \sin(nx)], \qquad (\text{period} = 2\pi), \tag{5.23}$$

we will use an equivalent form in terms of complex exponentials. Utilizing Euler's identity (using i as $\sqrt{-1}$),

$$e^{ijx} = \cos(jx) + i \sin(jx),$$

we can write Eq. (5.52) as

$$\begin{aligned} f(x) &= \sum_{j=0}^{\infty} (c_j e^{ijx} + c_{-j} e^{-ijx}) \\ &= 2c_0 + \sum_{j=1}^{\infty} [(c_j + c_{-j}) \cos(jx) + i(c_j - c_{-j}) \sin(jx)] \\ &= \sum_{j=-\infty}^{\infty} c_j e^{ijx}. \end{aligned} \tag{5.24}$$

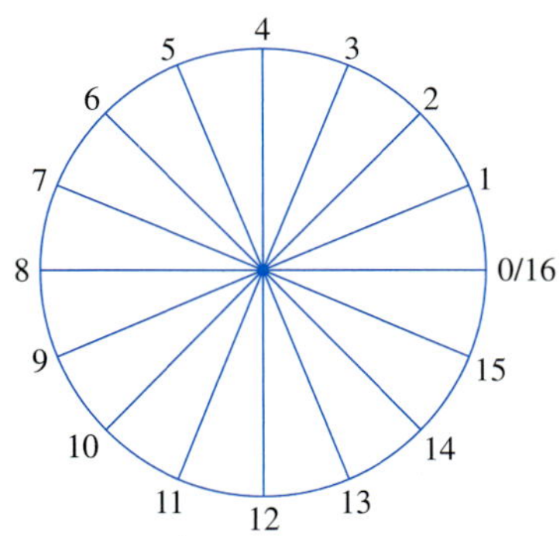

Angles used in computing for 16 points

We can match up the A's and B's of Eq. (5.23) to the c's of (5.24):

$$A_j = c_j + c_{-j}, \qquad B_j = i(c_j - c_{-j}),$$

$$c_j = \frac{A_j - iB_j}{2}, \qquad c_{-j} = \frac{A_j + iB_j}{2}.$$

When $f(x)$ is real valued it is easy to show that $c_0 = \overline{c}_0$ and $c_j = \overline{c}_{-j}$, where the bars represent complex conjugates.

For integers j and k, it is true that

$$\int_0^{2\pi} (e^{ikx})(e^{ijx})\,dx = \int_0^{2\pi} e^{i(k+j)x}\,dx = \begin{Bmatrix} 0 & \text{for } k \neq -j, \\ 2\pi & \text{for } k = -j. \end{Bmatrix}$$

(You can verify the first of these through Euler's identity.) This allows us to evaluate the c's of Eq. (5.24) by the following.

For each fixed k, we get

$$f(x)e^{-ikx} = \sum_{j=-\infty}^{\infty} c_k e^{i(j-k)x},$$

$$\int_0^{2\pi} f(x)e^{-ikx}\,dx = 2\pi c_k, \qquad \text{or}$$

$$c_k = \frac{1}{2\pi}\int_0^{2\pi} f(x)e^{-ikx}\,dx, \qquad k = 0, \pm 1, \pm 2, \ldots.$$

EXAMPLE 5.7 (You should verify each of these.)

1. Let $f(x) = x$; then

$$c_k = \frac{1}{2\pi}\int_0^{2\pi} xe^{ikx}\,dx = -\frac{i}{k}, \qquad k \neq 0.$$

2. Let $f(x) = x(2\pi - x)$; then

$$c_k = \frac{1}{2\pi}\int_0^{2\pi} x(2\pi - x)e^{-ikx}\,dx = -\frac{2}{k^2}, \qquad k \neq 0.$$

3. Let $f(x) = \cos(x)$; then

$$c_k = \frac{1}{2\pi}\int_0^{2\pi} \cos(x)e^{-ikx}\,dx = \begin{cases} \dfrac{1}{2} & \text{for } k = 1 \text{ or } -1, \\ 0 & \text{for all other } k. \end{cases}$$

Note that for Eq. (5.23) this makes $A_1 = 1$ and all the other A_j's $= 0$.

Thus, for a given $f(x)$ that satisfies continuity conditions, we have

$$c_j = \frac{1}{2\pi}\int_0^{2\pi} f(x)e^{-ijx}\,dx, \qquad j = 0, \pm 1, \pm 2, \ldots.$$

The magnitudes of the Fourier series coefficients $|c_j|$ are the *power spectrum of f;* these show the frequencies that are represented in $f(x)$. If we know $f(x)$ in the time domain, we can identify f by computing the c_j's. In getting the Fourier series, we have transformed from the time domain to the frequency domain, an important aspect of wave analysis.

Suppose we have N values for $f(x)$ on the interval $[0, 2\pi]$ at equispaced points, $x_k = 2\pi k/N$, $k = 0, 1, \ldots, N-1$. Because $f(x)$ is periodic, $f_N = f_0, f_{N+1} = f_1$, and so on. Instead of formal analytical integration, we would use a numerical integration method to get the coefficients. Even if $f(x)$ is known at all points in $[0, 2\pi]$, we might prefer to use numerical integration. This would use only certain values of $f(x)$, often those evaluated at uniform intervals. It is also often true that we do not know $f(x)$ everywhere, because we have sampled a continuous signal. In that case, however, it is better to use the discrete Fourier transform, which can be defined as

$$X(n) = \sum_{k=0}^{N-1} x_0(k)e^{-i2\pi nk/N}, \qquad n = 0, 1, 2, \ldots, N-1. \tag{5.25}$$

In Eq. (5.25), we have changed notation to conform more closely to the literature on FFT. $X(n)$ corresponds to the coefficients of N frequency terms, and the $x_0(k)$ are the N values of the signal samples in the time domain. You can think of n as indexing the X-terms and k as indexing the x_0-terms. Equation (5.25) corresponds to a set of N linear equations that we can solve for the unknown $X(n)$. Because the unknowns appear on the left-hand side of Eq. (5.25), this requires only the multiplication of an N-component vector by an $N \times N$ matrix.

It will simplify the notation if we let $W = e^{-i2\pi/N}$, making the right-hand-side terms of Eq. (5.25) become $x_0(k)W^{nk}$. To develop the FFT algorithm, suppose that $N = 4$. We write the four equations for this case:

$$\begin{aligned}
X(0) &= W^0x_0(0) + W^0x_0(1) + W^0x_0(2) + W^0x_0(3),\\
X(1) &= W^0x_0(0) + W^1x_0(1) + W^2x_0(2) + W^3x_0(3),\\
X(2) &= W^0x_0(0) + W^2x_0(1) + W^4x_0(2) + W^6x_0(3),\\
X(3) &= W^0x_0(0) + W^3x_0(1) + W^6x_0(2) + W^9x_0(3).
\end{aligned}$$

In matrix form:

$$\begin{bmatrix} X(0)\\ X(1)\\ X(2)\\ X(3)\end{bmatrix} = \begin{bmatrix} W^0 & W^0 & W^0 & W^0\\ W^0 & W^1 & W^2 & W^3\\ W^0 & W^2 & W^4 & W^6\\ W^0 & W^3 & W^6 & W^9\end{bmatrix} X_0. \tag{5.26}$$

In solving the set of N equations in the form of Eq. (5.26) we will have to make N^2 complex multiplications plus $N(N-1)$ complex additions. Using the FFT, however, greatly reduces the number of such operations. Although there are several variations on the algorithm, we will concentrate on the Cooley–Tukey formulation.

The matrix of Eq. (5.26) can be factored to give an equivalent form for the set of equations. At the same time we will use the fact that $W^0 = 1$ and $W^k = W^{k \bmod (N)}$:

$$\begin{bmatrix} X(0)\\ X(2)\\ X(1)\\ X(3)\end{bmatrix} = \begin{bmatrix} 1 & W^0 & 0 & 0\\ 1 & W^2 & 0 & 0\\ 0 & 0 & 1 & W^1\\ 0 & 0 & 1 & W^3\end{bmatrix}\begin{bmatrix} 1 & 0 & W^0 & 0\\ 0 & 1 & 0 & W^0\\ 1 & 0 & W^2 & 0\\ 0 & 1 & 0 & W^2\end{bmatrix} X_0. \tag{5.27}$$

You should verify that the factored form [Eq. (5.27)] is exactly equivalent to Eq. (5.26) by multiplying out. Note carefully that the elements of the X-vector are scrambled. (The development can be done formally and more generally by representing n and k as binary values, but it will suffice to show the basis for the FFT algorithm by expanding on this simple $N = 4$ case.)

By using the factored form, we now get the values of $X(n)$ by two steps (stages), in each of which we multiply a matrix times a vector. In the first stage, we transform x_0 into x_1 by

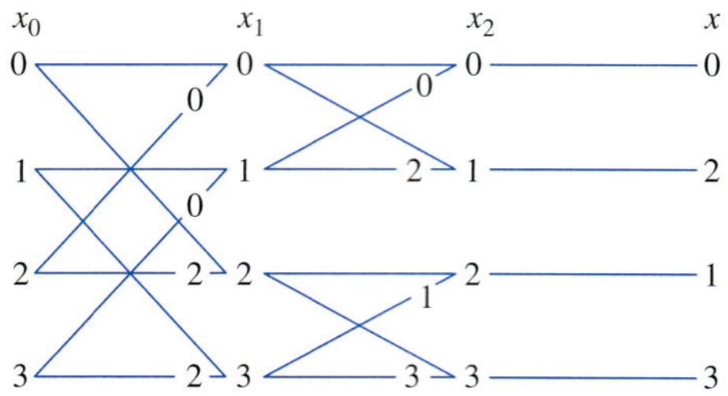

Figure 5.4

multiplying the right matrix of Eq. (5.27) and x_0. In the second stage, we multiply the left matrix and x_1, getting x_2. We get X by unscrambling the components of x_2. By doing the operation in stages, the number of complex multiplications is reduced to $N[\log_2(N)]$. For $N = 4$, this is a reduction by one-half, but for large N it is very significant; if $N = 1024$, there are 10 stages and the reduction in complex multiplies is a hundredfold!

It is convenient to represent the sequence of multiplications of the factored form [Eq. (5.27) or its equivalent for larger N] by flow diagrams. Figure 5.4 is for $N = 4$ and Figure 5.5 is for $N = 16$. Each column holds values of x_{ST}, where the subscript tells which stage is being computed; ST ranges from 1 to 2 for $N = 4$ and from 1 to 4 for $N = 16$. [The number of stages, for N a power of 2, is $\log_2(N)$.] In each stage, we get x-values of the next stage from those of the present stage. Every new x-value is the sum of the two x-values from the previous stage that

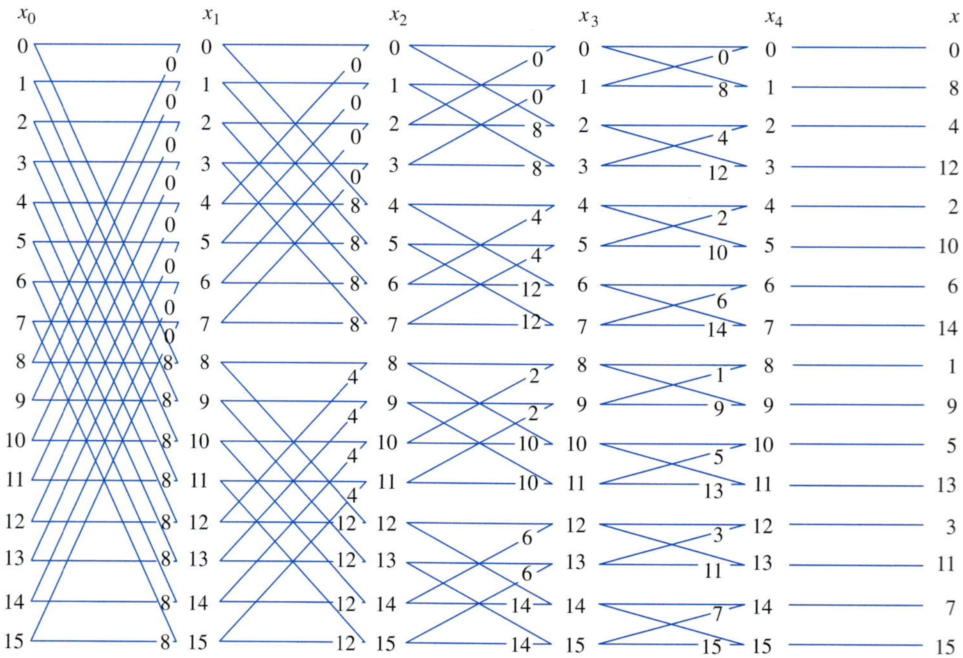

Figure 5.5

connect to it, with one of these multiplied by a power of W. The diagram tells which x_{ST} terms are combined to give an x_{ST+1} term, and the numbers shown within the lines are the powers of W that are used. For example, looking at Figure 5.5 we see that

$$x_2(6) = x_1(2) + W^8 x_1(6),$$
$$x_3(13) = x_2(13) + W^6 x_2(15),$$
$$x_4(9) = x_3(8) + W^9 x_3(9), \qquad \text{and so on.}$$

The last columns in Figures 5.4 and 5.5 indicate how the final x-values are unscrambled to give the X-values. This relationship can be found by expressing the index k of x in the last stage as a binary number and reversing the bits; this gives n in $X(n)$. For example, in Figure 5.5 we see that $x_4(3) = X(12)$ and $x_4(11) = X(13)$. From the bit-reversing rule, we get

$$3 = 0011_2 \rightarrow 1100_2 = 12, \qquad 11 = 1011_2 \rightarrow 1101_2 = 13.$$

Observe also that the bit-reversing rule can give the powers of W that are involved in computing the next stage. For the last stage, the powers are identical to the numbers obtained by bit reversal. At each previous stage, however, only the first half of the powers are employed, but each power is used twice as often. It is of interest to see how we can generate these values. Computer languages that facilitate bit manipulations make this an easy job, but there is a good alternative. Observe how the powers in Figure 5.4 differ from those in Figure 5.5 and how they progress from stage to stage. The following table pinpoints this:

Stage	$N = 4$				$N = 16$															
1:	0	0	2	2	0	0	0	0	0	0	0	0	8	8	8	8	8	8	8	8
2:	0	2	1	3	0	0	0	0	8	8	8	8	4	4	4	4	12	12	12	12
3:					0	0	8	8	4	4	12	12	2	2	10	10	6	6	14	14
4:					0	8	4	12	2	10	6	14	1	9	5	13	3	11	7	15

Can you see what a similar table for $N = 2$ would look like? Its single row would be 0 1. Now we see that the row of powers for the last stage can be divided into two halves, with the numbers in the second half always one greater than the corresponding entry in the first half. The row above is the left half of the current row with each value repeated. This observation leads to the following algorithm:

Algorithm to Generate Powers of W in FFT

For N a power of 2, let $Q = \log_2(N)$

Initialize an array P of length N to all zeros.
Set st = 1.
Repeat
  Double the values of $P(k)$ for $k = 1 .. 2^{st-1}$,

Let each $P(k + 2^{st-1}) = P(k) + 1$ for $k = 1 \,..\, 2^{st-1} - 1$.
Increment stage
Until stage $> Q$.

The successive new values for powers of W are now in array P.

EXAMPLE 5.8 Use the algorithm to generate the powers of W for $N = 8$:

$$Q = \log_2(8) = 3.$$

K	0	1	2	3	4	5	6	7
Initial P array:	0	0	0	0	0	0	0	0
ST = 1, doubled:	0	0	0	0	0	0	0	0
add 1:	0	1	0	0	0	0	0	0
ST = 2, doubled:	0	2	0	0	0	0	0	0
add 1:	0	2	1	3	0	0	0	0
ST = 3, doubled:	0	4	2	6	0	0	0	0
add 1:	0	4	2	6	1	5	3	7

The last row of values corresponds to the bits of the binary numbers 000 to 111, after reversal.

Our discussion has assumed that N is a power of 2; for this case, the economy of the FFT is a maximum. When N is not a power of 2 but can be factored, there are adaptations of the general idea that reduce the number of operations, but they are more than $N\log_2(N)$. See Brigham (1974) for a discussion of this as well as a fuller treatment of the theory behind FFT.

More recently, there has been interest in another transform, called the discrete Hartley transform. A discussion of this transform would parallel our discussion of the Fourier transform. Moreover, it has been shown that this transform can be converted into a fast Hartley transform (FHT) that reduces to $N \log_2(N)$ computations. For a full coverage of the FHT, one should consult Bracewell (1986). The advantages of the FHT over the Fourier transform are its faster and easier computation. Moreover, it is easy to compute the FFT from the Hartley transform. However, the main power of the FHT is that all the computations are done in real arithmetic, so that we can use a language like Pascal that does not have a complex data type. An interesting and easy introduction into the FHT is found in O'Neill (1988).

An Algorithm to Perform a Fast Fourier Transform

Given n data points, (x_i, f_i), $i = 0, \ldots, n - 1$ (with n a power of 2) and x on $[0 \,..\, 2\pi]$:

Set $yr_i = f_i$, $i = 0, \ldots, n - 1$.

Set $yi_i = 0, i = 0, \ldots, n-1$.
Set $c_i = \cos(2i\pi/n), i = 0, \ldots, n-1$. (These are the trigonometric
Set $s_i = \sin(2i\pi/n), i = 0, \ldots, n-1$. values that are used.)
Set numstages $= \log_2(n)$ (The number of stages)
Set $p_i = 0, i = 0, \ldots, n-1$ (Use the previous alogrithm
For stage = 1 To numstages Do to get "bit reversal" values)
    Set $p_i = 2p_i, i = 1, \ldots, 2^{\text{stage-1}}$
    Set $p_{i+2} = p_i + 1, i = 0, \ldots, 2^{\text{stage-1}}$
End Do (For stage)

Set stage = 1 (These values
Set nsets = 1 are for the
Set del = $n/2$ first stage—
Set $k = 0$ k indexes the y-values to be computed.)
Repeat
    For set = 1 To nsets Do
        For $i = 0$ To n/nsets − 1 Do
            Set $j = i$ Mod del + (set − 1) * del * 2 (Indexes old y-values)
            Set $\ell = p_{\text{Int}(k/\text{del})}$ (Indexes c_i, s_i values)
            Set $yyr_k = yr_j + c_\ell * yr_{j+\text{del}} - s_\ell * yi_{j+\text{del}}$.
            Set $yyi_k = yi_j + c_\ell * y_{j+\text{del}} - s_\ell * yr_{j+\text{del}}$.
            Set $k = k + 1$
        End Do (For i).
    End Do (For set).
Set $yr_i = yyr_i, i = 0, \ldots, n-1$. (Reset
Set $yi_i = yyi_i, i = 0, \ldots, n-1$. values
        Set stage = stage + 1. for
        Set nsets = nsets * 2. next
        Set del = del/2. stage.)
        Set $k = 0$.
Until stage > numstages.

When terminated, the A's and B's of the Fourier series are contained in the yr and yi arrays. These must be divided by $n/2$ and should be unscrambled using the p-array values as indices.

Note: If the f_i are complex numbers, set the imaginary parts into array yi.

EXAMPLE 5.9 Use the FFT algorithm to obtain the finite Fourier series coefficients for the same data as in Table 5.12 [These are perturbed values from

$$f(t) = 2.3 + 1.5\cos(t) - 0.5\sin(t) + 4\sin(4t).]$$

A computer program that implements the algorithm gave these results:

n	A_n	B_n
0	**4.6017**	
1	**1.4993**	**−0.4994**
2	0.0017	−0.0010
3	0.0003	−0.0005
4	0.0015	**3.9990**
5	0.0019	0.0009
6	−0.0004	−0.0009
7	−0.0003	−0.0019
8	0.0017	−0.0008
9	−0.0023	0.0019
10	−0.0024	−0.0011
11	0.0003	0.0020
12	0.0008	−0.0033
13	−0.0004	0.0011
14	0.0025	0.0003
15	−0.0005	0.0013
16	−0.0010	

The results are essentially the same as those of Example 5.6, which were computed by the trapezoidal rule.

Observe that we compute exactly as many A's and B's as there are data points. This is not only reasonable (we cannot "manufacture" information) but is in accord with *information theory*.

Information Theory—The Sampling Theorem

In performing a discrete Fourier transform, we work with samples of some function of t, $f(t)$. We normally have data taken at evenly spaced intervals of time. If the interval between samples is D sec, its reciprocal, $1/D$, is called the *sampling rate* (the number of samples per second).

Corresponding to the sampling interval, D, is a critical frequency, called the *Nyquist critical frequency*, f_c, where

$$f_c = \frac{1}{2}D.$$

The reason this is a critical frequency is seen from the following argument. Suppose we sample a sine wave whose frequency is f_c and get a value corresponding to its positive peak amplitude. The next sample will be at the negative peak, the next beyond that at the positive peak, and so on—that is, critical sampling is at a rate of two samples per cycle. We can construct the magnitude of the sine wave from these two samples. If the frequency is less than f_c, we will have more than two samples per cycle and again we can construct the wave correctly. On the other hand, if the frequency is greater than f_c, we have fewer than two samples per cycle and we have inadequate information to determine $f(t)$.

The significance of this theorem is that if the phenomenon described by $f(t)$ has no frequencies greater than f_c, then $f(t)$ is completely determined from samples at the rate $1/D$. Unfortunately, this also means that if there are frequencies in $f(t)$ greater than f_c, all these frequencies are spuriously crowded into the range $[0, f_c]$, causing a distortion of the power spectrum. This distortion is called *aliasing*.

All of this is very clear if we think of the results of an FFT on the samples. If we have N samples of the phenomenon, we certainly cannot determine more than a total of exactly N of the Fourier coefficients, the A's and B's. The last of these will be $A_{n/2}$ (assuming an even number of samples). We see that this corresponds to the Nyquist frequency.

5.5 Adaptive Integration

The trapezoidal rule and Simpson's $\frac{1}{3}$ rule are often used to find the integral of $f(x)$ over a fixed interval $[a, b]$ using a uniform value for Δx. When $f(x)$ is a known function, we can choose the value for $\Delta x = h$ arbitrarily. The problem is that we do not know a priori what value to choose for h to attain a desired accuracy. Romberg-type integration is a way to find the necessary h. We start with two panels, $h = h_1 = (b - a)/2$, and apply one of the formulas. Then we let $h_2 = h_1/2$ and apply the formula again, now with four panels, and compare the results. If the new value is sufficiently close, we terminate and use a Richardson extrapolation to further reduce the error. If the second result is not close enough to the first, we again halve h and repeat the procedure. We continue in this way until the last result is close enough to its predecessor.

We illustrate this obvious procedure with an example.

EXAMPLE 5.10 Integrate $f(x) = 1/x^2$ over the interval $[0.2, 1]$ using Simpson's $\frac{1}{3}$ rule. Use a tolerance value of 0.02 to terminate the halving of $h = \Delta x$. From calculus, we know that the exact answer is 4.0.

We introduce a special notation that will be used throughout this section:

$$S_n[a, b] = \text{the computed value using Simpson's } \tfrac{1}{3} \text{ rule with } \Delta x = h_n \text{ over } [a, b].$$

If we use this notation, the composite Simpson rule becomes

$$I(f) = S_n[a, b] - \frac{(b - a)}{180} h_n^4 f^{(4)}(\xi), \quad a < \xi < b.$$

Using this with $h_1 = (1.0 - 0.2)/2 = 0.4$, we compute $S_1\,[0.2, 1.0]$. We continue halving h, $h_{n+1} = h_n/2$, computing its corresponding $S_{n+1}[a, b]$ until $|S_{n+1} - S_n| < 0.02$, the tolerance value. The following table shows the results:

n	h_n	S_n	$\lvert S_{n+1} - S_n \rvert$
1	0.4	4.948148	
			0.761111
2	0.2	4.187037	
			0.162819
3	0.1	4.024218	
			0.022054
4	0.05	4.002164	
			0.002010
5	0.025	4.000154	

From the table we see that, at $n = 5$, we have met the tolerance criterion, because $|S_5 - S_4| < 0.02$. A Romberg extrapolation gives

$$RS[a, b] = S_5 + \frac{S_5 - S_4}{15} = 4.00002.$$

(We use $RS[a, b]$ to represent the Romberg extrapolation from Simpson's rule.)

The Adaptive Scheme

Using the same value for h throughout the interval may be disadvantageous because the behavior of $f(x)$ may not require such uniformity. Consider Figure 5.6. It is obvious that, in the subinterval $[c, b]$, h can be much larger than in subinterval $[a, c]$, where the curve is much less smooth. We could subdivide the entire interval $[a, b]$ nonuniformly by personal intervention after examining the graph of $f(x)$. We prefer to avoid such intervention.

Adaptive integration automatically allows for different h's on different subintervals of $[a, b]$, choosing values adequate for a specified accuracy. We do not specify where the size

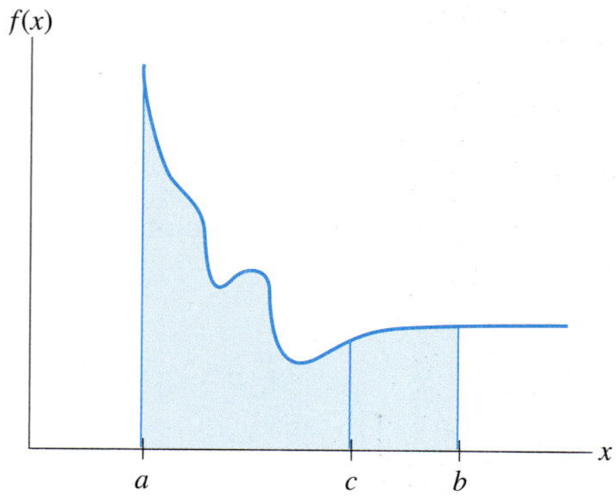

Figure 5.6

change for h occurs; this can occur anywhere within it. We use something like a binary search to locate the point where we should change the size of h. Actually, the total interval $[a, b]$ may be broken into several subintervals, with different values for h within each of them. This depends on the tolerance value, TOL, and the nature of $f(x)$.

To describe this strategy, we repeat the preceding example to find the integral of $f(x) = 1/x^2$ between $x = 0.2$ and $x = 1$. We choose a value for TOL of 0.02, and do the computations in double precision to minimize the effects of round off.

We begin as before by specifying just two subintervals in $[a, b]$. The first computation is a Simpson integration over [0.2, 1] with $h_1 = 0.4$. The result, which we call $S_1[0.2, 1]$, is 4.94814815. The next step is to integrate over each half of [0.2, 1] but with h half as large, $h_2 = 0.2$. We get

$$S_2[0.2, 0.6] = 3.51851852 \quad \text{and} \quad S_2[0.6, 1] = 0.66851852.$$

We now test the accuracy of our initial computations by seeing whether the difference between $S_1[0.2, 1]$ and the sum of $S_2[0.2, 0.6]$ and $S_2[0.6, 1]$ is greater than TOL. (Actually, we compare the magnitude of this difference.)

$$S_1[0.2, 1] - (S_2[0.2, 0.6] + S_2[0.6, 1]) = 0.7611111.$$

Because this result is greater than TOL = 0.02, we must use a smaller value for h.

We continue by applying the strategy to one-half of the original interval. We arbitrarily choose the right half and compute $S_2[0.6, 1]$ with $h = h_2 = (1 - 0.6)/2 = 0.2$, comparing it to $S_3[0.6, 0.8] + S_3[0.8, 1]$ (both of these use $h_3 = h_2/2 = 0.1$). We also halve the value for TOL, getting

$$\begin{aligned} S_2[0.6, 1] - (S_3[0.6, 0.8] + S_3[0.8, 1]) &= 0.66851852 - (0.41678477 + 0.25002572) \\ &= 0.66851852 - 0.66681049 \\ &= 0.001708 \qquad \text{versus TOL} = 0.01. \end{aligned}$$

This passes the test, so we take advantage of the results that we have available and do a Richardson extrapolation to get

$$\begin{aligned} RS[0.6, 1] &= 0.66681049 + \frac{1}{15}(0.66681049 - 0.66851852) \\ &= 0.66669662. \end{aligned}$$

We now move to the next adjacent subinterval, [0.2, 0.6], and repeat the procedure. We compute

$$\begin{aligned} S_2[0.2, 0.6] &= 3.51851852, \qquad \text{with } h_2 = 0.2; \\ S_3[0.2, 0.4] &= 2.52314815; \qquad S_3[0.4, 0.6] = 0.83425926; \\ S_2[0.2, 0.6] - (S_3[0.2, 0.4] + S_3[0.4, 0.6]) &= 0.161111 \qquad \text{versus TOL} = 0.01, \end{aligned}$$

which fails, so we proceed to another level with the right half:

$$\begin{aligned} S_3[0.4, 0.6] &= 0.83425926, \qquad \text{with } h_3 = 0.1; \\ S_4[0.4, 0.5] &= 0.50005144; \qquad S_4[0.5, 0.6] = 0.33334864; \\ S_3[0.4, 0.6] - (S_4[0.4, 0.5] + S_4[0.5, 0.6]) &= 0.000859 \qquad \text{versus TOL} = 0.005, \end{aligned}$$

which passes. We extrapolate:

$$RS[0.4, 0.6] = 0.8333428.$$

The next adjacent interval is [0.2, 0.4]. For this we use TOL = 0.005. We find that this does not meet the criterion, so we next do [0.3, 0.4]. We do meet the TOL level of 0.0025:

$$S_4[0.3, 0.4]\ 0.83356954, \qquad \text{with } h_4 = 0.05;$$

$$S_5[0.3, 0.35] = 0.47620166; \qquad S_5[0.35, 0.4] = 0.35714758;$$

$$S_4[0.3, 0.4] - (S_5[0.3, 0.35] + S_5[0.35, 0.4]) = 0.000220 \qquad \text{versus TOL} = 0.0025,$$

which passes, so

$$RS[0.3, 0.4] = 0.83333492.$$

Our last subinterval is [0.2, 0.3]. We find that we again meet the test. We give only the extrapolated result

$$RS[0.2, 0.3] = 1.666686.$$

Adding all of the *RS*-values gives the final answer:

$$\text{Integral over } [0.2, 1] = 4.00005957.$$

By employing adaptive integration, we reduced the number of function evaluations from 33 to 17.

Bookkeeping and Avoiding Repeating Function Evaluations

It should be obvious that we recomputed many of the values of $f(x)$ in the previous integration. We can avoid these recalculations if we store these computations in such a way as to retrieve them appropriately. We also need to keep track of the current subinterval, the previous subintervals that we return to, and the appropriate value for h and TOL for each subinterval. The mechanism for storing these quantities is a *stack,* a data structure that is a last-in, first-out device that resembles a stack of dishes in a restaurant. Actually, we use just a two-dimensional array of seven columns and as many rows as levels that we wish to accommodate. (Often a large number of levels is provided—say, 200—even though we hardly ever need so many.)

After an initial calculation to get $h_1 = (b - a)/2$, $c = a + h_1$, $f(a)$, $f(c)$, $f(b)$, and $S_1[a, b]$, we store a set of seven values: $a, f(a), f(c), f(b), h$, TOL, $S[a, b]$. We retrieve these values into variables that represent these quantities and continue with the first stage of the computations.

Whenever the test fails after computing for the current subinterval, we store two sets of values in two rows of the seven columns:

First row: $\quad a, f(a), f(d), f(c), h_n$, TOL, $S[a, c]$,

Next row: $\quad c, f(c), f(e), f(b), h_n$, TOL, $S[c, b] \leftarrow$ TOP, where the letters a, d, c, e, b refer to points in the last subinterval that are evenly spaced from left to right in that order. We also use a pointer to the last row stored. It is named TOP to indicate it is the "top" of the stack (even though it points to the last row stored as we normally view an array). Whenever we store a set of values, we add one to TOP; whenever we retrieve a set of values, we subtract one so that TOP always points to the row that is next available for retrieval.

We begin each iteration by retrieving the row of quantities pointed to by TOP (the one labeled "Next row" above). In this way, we can reuse the previously computed function values to get values for computing the rightmost remaining subinterval. (Observe that the next subinterval begins at the c-value for the last subinterval.)

The following algorithm implements the adaptive integration scheme that we have described.

An Algorithm for Computing $I(f) = \int_a^b f(x)\,dx$ with an Adaptive Procedure

Set Value = 0.0.
  Evaluate: $h_1 = (b - a)/2$, $c = a + h_1$, $Fa = f(a)$, $Fc = f(c)$, $F(b) = f(b)$, $Sab = S_1(a, b)$
  Store (a, Fa, Fc, Fb, h_1, Tol, Sab).
  Set top = 1.
Repeat
  Retrieve (a, Fa, Fc, Fb, h_1, Tol, Sab).
  Set top = top − 1.
  Evaluate: $h_2 = h_1/2$, $d = a + h_2$, $e = a + 3h_2$, $Fd = f(d)$,
    $Fe = f(e)$,
    $Sac = S_2(a, c)$, $Scb = S_2(c, b)$, $S_2(a, b) = Sac + Scb$.
    If $|S_2(a, b) - S_1(a, b)| < Tol$ Then
      Compute $RS(a, b)$,
      Value = Value + $RS(a, b)$,
    Else
      $h_1 = h_2$, $Tol = Tol/2$,
      Set top = top + 1,
      Store(a, Fa, Fd, Fc, h_1, Tol, Sac),
      Set top = top + 1,
      Store(c, Fc, Fe, Fb, h_1, Tol, Scb),
Until top = 0.

$I(f)$, the value of the integral, is in variable Value.

5.6 Gaussian Quadrature

Our previous formulas for numerical integration were all predicated on evenly spaced x-values; this means the x-values were predetermined. With a formula of three terms, then, there were three parameters, the coefficients (weighting factors) applied to each of the functional values. A formula with three parameters corresponds to a polynomial of the second degree, one less than the number of parameters. Gauss observed that if we remove the requirement that the function be evaluated at predetermined x-values, a three-term formula will contain six parameters (the three x-values are now unknowns, plus the three weights) and should correspond to an interpolating polynomial of degree-5. Formulas based on this principle are called *Gaussian quadrature formulas*. They can be applied only when $f(x)$ is known explicitly, so that it can be evaluated at any desired value of x.

We will determine the parameters in the simple case of a two-term formula containing four unknown parameters:

$$\int_{-1}^{1} f(t) \approx af(t_1) + bf(t_2).$$

The method is the same as that illustrated in the previous section, by determining unknown parameters. We use an integration interval that is symmetrical about the origin, from -1 to 1 to simplify the arithmetic, and call our variable t. (This notation agrees with that of most authors. As the variable of integration is only a dummy variable, its name is unimportant.) Our formula is to be valid for any polynomial of degree-3; hence it will hold if $f(t) = t^3$, $f(t) = t^2$, $f(t) = t$, and $f(t) = 1$:

$$\begin{aligned}
f(t) &= t^3: & \int_{-1}^{1} t^3\,dt &= 0 = at_1^3 + bt_2^3;\\
f(t) &= t^2: & \int_{-1}^{1} t^2\,dt &= \frac{2}{3} = at_1^2 + bt_2^2;\\
f(t) &= t: & \int_{-1}^{1} t\,dt &= 0 = at_1 + bt_2;\\
f(t) &= 1: & \int_{-1}^{1} dt &= 2 = a + b.
\end{aligned} \tag{5.28}$$

Multiplying the third equation by t_1^2, and subtracting from the first, we have

$$0 = 0 + b[t_2^3 - t_2 t_1^2] = b(t_2)(t_2 - t_1)(t_2 + t_1). \tag{5.29}$$

We can satisfy Eq. (5.29) by either $b = 0$, $t_2 = 0$, $t_1 = t_2$, or $t_1 = -t_2$. Only the last of these possibilities is satisfactory, the others being invalid, or else reduces our formula to only a single term, so we choose $t_1 = -t_2$. We then find that

$$a = b = 1,$$

$$t_2 = -t_1 = \sqrt{\frac{1}{3}} = 0.5773,$$

$$\int_{-1}^{1} f(t)\,dt \approx f(-0.5773) + f(0.5773).$$

It is remarkable that adding these two values of the function gives the exact value for the integral of any cubic polynomial over the interval from -1 to 1.

Suppose our limits of integration are from a to b, and not -1 to 1 for which we derived this formula. To use the tabulated Gaussian quadrature parameters, we must change the interval of integration to $(-1, 1)$ by a change of variable. We replace the given variable by another to which it is linearly related according to the following scheme:

If we let

$$x = \frac{(b-a)t + b + a}{2} \qquad \text{so that } dx = \left(\frac{b-a}{2}\right) dt,$$

then

$$\int_a^b f(x)\,dx = \frac{b-a}{2}\int_{-1}^{1} f\left(\frac{(b-a)t+b+a}{2}\right) dt.$$

EXAMPLE 5.11 Evaluate $I = \int_0^{\pi/2} \sin x\,dx$. (It is not hard to show that $I = 1.0$, so we can readily see the error of our estimate.)

To use the two-term Gaussian formula, we must change the variable of integration to make the limits of integration from -1 to 1.

Let

$$x = \frac{(\pi/2)t + \pi/2}{2}, \qquad \text{so } dx = \frac{\pi}{4}\,dt.$$

Observe that when $t = -1$, $x = 0$; when $t = 1$, $x = \pi/2$. Then

$$I = \frac{\pi}{4}\int_{-1}^{1} \sin\left(\frac{\pi t + \pi}{4}\right) dt.$$

The Gaussian formula calculates the value of the new integral as a weighted sum of two values of the integrand, at $t = -0.5773$ and at $t = 0.5773$. Hence,

$$\begin{aligned} I &= \frac{\pi}{4}[(1.0)(\sin(0.10566\pi)) + (1.0)(\sin(0.39434\pi))] \\ &= 0.99847. \end{aligned}$$

The error is 1.53×10^{-3}.

The power of the Gaussian method derives from the fact that we need only two functional evaluations. If we had used the trapezoidal rule, which also requires only two evaluations, our estimate would have been $(\pi/4)(0.0 + 1.0) = 0.7854$, an answer quite far from the mark. Simpson's $\frac{1}{3}$ rule requires three functional evaluations and gives $I = 1.0023$, with an error of -2.3×10^{-3}, somewhat greater than for Gaussian quadrature.

Gaussian quadrature can be extended beyond two terms. The formula is then given by

$$\int_{-1}^{1} f(t)\,dt = \sum_{i=1}^{n} w_i f(t_i), \qquad \text{for } n \text{ points.} \tag{5.30}$$

This formula is *exact* for functions $f(t)$ that are polynomials of degree $2n - 1$ or less! Moreover, by extending the method we used previously for the 2-point formula, for each n we obtain a system of $2n$ equations:

$$w_1 t_1^k + \cdots + w_n t_n^k = \begin{cases} 0, & \text{for } k = 1, 3, 5, \ldots, 2n - 1; \\ \dfrac{2}{k+1}, & \text{for } k = 0, 2, 4, \ldots, 2n - 2. \end{cases}$$

This approach is obvious. However, this set of equations, obtained by writing $f(t)$ as a succession of polynomials, is not easily solved. We will use an approach that is easier than the methods for a nonlinear system that we used in Chapter 1.

It turns out that the t_i's for a given n are the roots of the nth-degree Legendre polynomial. The Legendre polynomials are defined by recursion:

$$(n + 1)L_{n+1}(x) - (2n + 1)xL_n(x) + nL_{n-1}(x) = 0,$$

$$\text{with } L_0(x) = 1, \qquad L_1(x) = x.$$

Then $L_2(x)$ is

$$L_2(x) = \frac{3xL_1(x) - (1)L_0(x)}{2} = \frac{3}{2}x^2 - \frac{1}{2},$$

whose zeros are $\pm\sqrt{\frac{1}{3}} = \pm 0.5773$, precisely the t-values for the two-term formula.

By using the recursion relation, we find

$$L_3(x) = \frac{5x^3 - 3x}{2},$$

$$L_4(x) = \frac{35x^4 - 30x^2 + 3}{8}, \qquad \text{and so on.}$$

The methods of Chapter 1 allow us to find the roots of these polynomials. After they have been determined, the set of equations analogous to Eq. (5.28) can easily be solved for the weighting factors because the equations are linear with respect to these unknowns.

Table 5.13 lists the zeros of Legendre polynomials up to degree-5, giving values that we need for Gaussian quadrature where the equivalent polynomial is up to degree-9. For example, $L_3(x)$ has zeros at $x = 0$, $+0.77459667$, and -0.77459667.

Before continuing with another example of the use of Gaussian quadrature, it is of interest to summarize the properties ot Legendre polynomials.

1. The Legendre polynomials are *orthogonal* over the interval $[-1, 1]$. That is,

$$\int_{-1}^{1} L_n(x)L_m(x)\,dx \begin{cases} = 0 \text{ if } n \neq m; \\ > 0 \text{ if } n = m. \end{cases}$$

Table 5.13 Values for Gaussian quadrature

Number of terms	Values of t	Weighting factor	Valid up to degree
2	−0.57735027	1.0	3
	0.57735027	1.0	
3	−0.77459667	0.55555555	5
	0.0	0.88888889	
	0.77459667	0.55555555	
4	−0.86113631	0.34785485	7
	−0.33998104	0.65214515	
	0.33998104	0.65214515	
	0.86113631	0.34785485	
5	−0.90617975	0.23692689	9
	−0.53846931	0.47862867	
	0.0	0.56888889	
	0.53846931	0.47862867	
	0.90617975	0.23692689	

This is a property of several other important functions, such as $\{\cos(nx), n = 0, 1, \ldots\}$. Here we have

$$\int_0^{2\pi} \cos(mx)\cos(nx)\,dx \begin{cases} = 0 \text{ if } n \neq m; \\ > 0 \text{ if } n = m. \end{cases}$$

In this case, we say that this function is orthogonal over the interval $[0, 2\pi]$.

2. Any polynomial of degree n can be written as a sum of the Legendre polynomials:

$$P_n(x) = \sum_{i=0}^{n} c_i L_i(x).$$

3. The n roots of $L_n(x) = 0$ lie in the interval $[-1, 1]$.

Using these properties, we are able to show that Eq. (5.30) is exact for polynomials of degree $2n - 1$ or less.

The weighting factors and t-values for Gaussian quadrature have been tabulated. [Love, (1966) gives values for up to 200-term formulas.] We are content to give a few of the values in Table 5.13.

Maple can produce the Legendre polynomials:

```
>with(orthopoly);
>f(x):= P(4,x);
```

$$f(x) = \frac{35}{8}x^4 - \frac{15}{4}x^2 + \frac{3}{8}$$

```
>plot(f(x),x=-1..1);
```

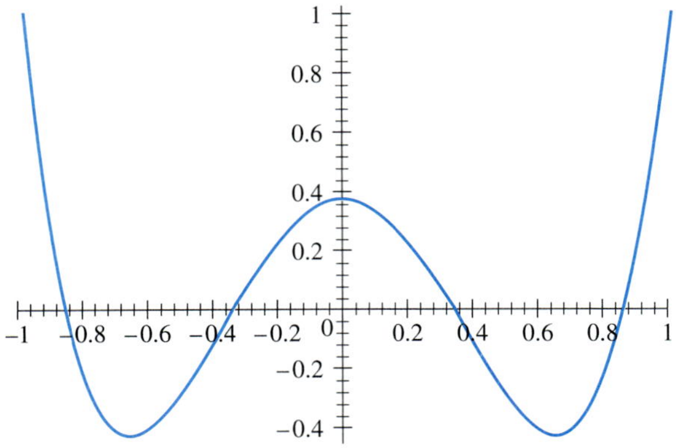

and we see from the plot that the graph crosses the x-axis at the values of the t-values of Table 5.13.

Example 5.12 illustrates the use of the four-term formula.

EXAMPLE 5.12 Repeat Example 5.4, but use the four-term Gaussian formula. Compare to the result of Example 5.4. We are to evaluate

$$I = \int_{0.2}^{2.6} e^{-x}\, dx.$$

We change to variable t for limits $[-1, 1]$:

$$x = \frac{(2.6 - 0.2)t + 2.6 + 0.2}{2} = 1.2t + 1.4.$$

So that

$$\begin{aligned} I &= \frac{2.6 - 0.2}{2} \int_{-1}^{1} e^{-(1.2t+1.4)}\, dt \\ &= 1.2[0.3478\ldots e^{-[1.2(-0.8611\ldots)+1.4]} \\ &\qquad + 0.6521\ldots\ e^{-[1.2(-0.3398\ldots)+1.4]} \\ &\qquad + 0.6521\ldots e^{-[1.2(0.3398\ldots)+1.4]} \\ &\qquad + 0.3478\ldots e^{-[1.2(0.8611\ldots)+1.4]}] \\ &= 0.68833, \text{ whose error is } -0.00032. \end{aligned}$$

This error is less than the error from Simpson's 1/3 Rule with six intervals (its error is -0.00041) and less than the error with the trapezoidal rule with 18 intervals (its error is -0.00056).

Improper Integrals

Because Gaussian quadrature does not use the value of the integrand at the endpoints, it would seem that it could evaluate some improper integrals, those with a singularity at an end of the interval of integration. Analytically, a convergent improper integral is handled by substitutions and taking limits. How does the Gaussian technique work on

$$\int_0^4 \frac{dx}{4 - x} = 4?$$

Using the fourth-order formula with endpoints of [0, 4] gives 3.6127 as a result—not very good. If we add the results for two intervals, [0, 3.9] and [3.9, 4], we get 3.8883. This is better, but still not close. This could be extended. As with the other kinds of numerical integration, when the integrand increases extremely rapidly, we have trouble.

We might hope to evaluate

$$\int_0^\infty \frac{dx}{x^2 + 1} = \pi/2 = 1.5708\ldots$$

if we use a very large number for the upper limit. The four-term formula gets 0.03992 when the interval is [0, 1000]. Adding the results for [1000, 10000], which is only 0.00856, and that for [10000, 100000] which is 0.000085 still does not help. Even though the integrand is very small at large values of x, there is still considerable area under the curve.

5.7 Multiple Integrals

When we need the definite integral of $z = f(x, y)$ over a region defined by limit values for x and y, we do *multiple integration.* In calculus, you learned that a double integral can be evaluated as an iterated integral. So we write

$$\iint_A f(x, y)\, dA = \int_a^b \left(\int_c^d f(x, y)\, dy \right) dx = \int_c^d \left(\int_a^b f(x, y)\, dx \right) dy. \tag{5.31}$$

In Eq. (5.31), the region is the rectangle bounded by the lines

$$x = a, \qquad x = b, \qquad y = c, \qquad y = d.$$

The region does not have to be a rectangle; the limits may not be constants, but we postpone that situation. In computing the iterated integrals, we hold x constant while integrating with respect to y (vice versa in the second case).

We can easily adapt the previous integration formulas to get a multiple integral. Recall that any one of the integration formulas is just a linear combination of values of the function, evaluated at varying values of the independent variable. In other words, a quadrature formula is just a weighted sum of certain functional values. The inner integral is written then as a weighted sum of function values with one variable held constant. We then add together a weighted sum of these sums. If the function is known only at the nodes of a rectangular grid through the region, we are constrained to use these values. The

Newton–Cotes formulas are a convenient set to employ. There is no reason why the same formula must be used in each direction, although it is often particularly convenient to do so.

EXAMPLE 5.13 We illustrate this technique by evaluating the integral of the function of Table 5.14 over the rectangular region bounded by

$$x = 1.5, \qquad x = 3.0, \qquad y = 0.2, \qquad y = 0.6.$$

Let us use the trapezoidal rule in the x-direction and Simpson's $\frac{1}{3}$ rule in the y-direction. (Because the number of panels in the x-direction is not even, Simpson's $\frac{1}{3}$ rule does not apply readily.) It is immaterial which integral we evaluate first. Suppose we start with y being constant:

$$y = 0.2: \qquad \int_{1.5}^{3.0} f(x, y)\,dx = \int_{1.5}^{3.0} f(x, 0.2)\,dx = \frac{h}{2}(f_1 + 2f_2 + 2f_3 + f_4)$$

$$= \frac{0.5}{2}[0.990 + 2(1.568) + 2(2.520) + 4.090]$$

$$= 3.3140;$$

$$y = 0.3: \qquad \int_{1.5}^{3.0} f(x, 0.3)\,dx = \frac{0.5}{2}[1.524 + 2(2.384) + 2(3.800) + 6.136]$$

$$= 5.0070.$$

Similarly, at

$$y = 0.4, \qquad I = 6.6522;$$

$$y = 0.5, \qquad I = 8.2368;$$

$$y = 0.6, \qquad I = 9.7435.$$

We now sum these in the y-direction according to Simpson's rule:

$$f(x, y)\,dx = \frac{0.1}{3}[3.3140 + 4(5.0070) + 2(6.6522) + 4(8.2368) + 9.7435]$$

$$= 2.6446$$

Table 5.14 Tabulation of a function of two variables, $u = f(x, y)$

x \ y	**0.1**	**0.2**	**0.3**	**0.4**	**0.5**	**0.6**
0.5	0.165	0.428	0.687	0.942	1.190	1.431
1.0	0.271	0.640	1.003	1.359	1.703	2.035
1.5	0.447	0.990	1.524	2.045	2.549	3.031
2.0	0.738	1.568	2.384	3.177	3.943	4.672
2.5	1.216	2.520	3.800	5.044	6.241	7.379
3.0	2.005	4.090	6.136	8.122	10.030	11.841
3.5	3.306	6.679	9.986	13.196	16.277	19.198

(In this example, our answer does not check well with the analytical value of 2.5944 because the x-intervals are large. We could improve our estimate somewhat by fitting a higher-degree polynomial than the first to provide the integration formula. We can even use values outside the range of integration for this, using undetermined coefficients to get the formulas.)

The previous example shows that double integration by numerical means reduces to a double summation of weighted function values. The calculations we have just made could be written in the form

$$\begin{aligned}\int f(x, y)\,dx\,dy &= \sum_{j=1}^{m} v_j \sum_{i=1}^{n} w_i f_{ij} \\ &= \frac{\Delta y}{3}\frac{\Delta x}{2}[(f_{1,1} + 2f_{2,1} + 2f_{3,1} + f_{4,1}) \\ &\quad + 4(f_{1,2} + 2f_{2,2} + 2f_{3,2} + f_{4,2}) + \cdots \\ &\quad + (f_{1,5} + 2f_{2,5} + 2f_{3,5} + f_{4,5})].\end{aligned}$$

It is convenient to write this in pictorial operator form, in which the weighting factors are displayed in an array that is a map to the location of the functional values to which they are applied.

$$\int f(x, y)\,dx\,dy = \frac{\Delta y}{3}\frac{\Delta x}{2}\left\{\begin{matrix} 1 & 4 & 2 & 4 & 1 \\ 2 & 8 & 4 & 8 & 2 \\ 2 & 8 & 4 & 8 & 2 \\ 1 & 4 & 2 & 4 & 1 \end{matrix}\right\} f_{i,j}. \tag{5.32}$$

We interpret the numbers in the array of Eq. (5.32) in this manner: We use the values 1, 4, 2, 4, and 1 as weighting factors for functional values in the top row of the portion of Table 5.14 that we integrate over (values were $x = 1.5$ and y varies from 0.2 to 0.6). Similarly, the second column of the array in Eq. (5.32) represents weighting factors that are applied to a column of function values where $y = 0.4$ and x varies from 1.5 to 3.0. Observe that the values in the pictorial operator of Eq. (5.32) follow immediately from the Newton–Cotes coefficients for single-variable integration.

Other combinations of Newton–Cotes formulas give similar results. It is probably easiest for hand calculation to use these pictorial integration operators. Pictorial integration is readily adapted to any desired combination of integration formulas. Except for the difficulty of representation beyond two dimensions, this operator technique also applies to triple and quadruple integrals.

There is an alternative representation to such pictorial operators that is easier to translate into a computer program. We also derive it somewhat differently. Consider the numerical integration formula for one variable

$$\int_{-1}^{1} f(x)\,dx \approx \sum_{i=1}^{n} a_i f(x_i). \tag{5.33}$$

We have seen in Section 5.3 that such formulas can be made exact if $f(x)$ is any polynomial of a certain degree. Assume that Eq. (5.33) holds for polynomials up to degree s.

We now consider the multiple integral formula

$$\int_{-1}^{1}\int_{-1}^{1}\int_{-1}^{1} f(x, y, z)\, dx\, dy\, dz \overset{?}{=} \sum_{i=1}^{n}\sum_{j=1}^{n}\sum_{k=1}^{n} a_i a_j a_k f(x_i, y_j, z_k). \tag{5.34}$$

We wish to show that Eq. (5.34) is exact for all polynomials in x, y, and z up to degree s. Such a polynomial is a linear combination of terms of the form $x^\alpha y^\beta z^\gamma$, where α, β, and γ are nonnegative integers whose sum is equal to s or less. If we can prove that Eq. (5.34) holds for the general term of this form, it will then hold for the polynomial.

To do this we assume that

$$f(x, y, z) = x^\alpha y^\beta z^\gamma.$$

Then, because the limits are constants and the integrand is factorable,

$$\begin{aligned} I &= \int_{-1}^{1}\int_{-1}^{1}\int_{-1}^{1} x^\alpha y^\beta z^\gamma\, dx\, dy\, dz \\ &= \left(\int_{-1}^{1} x^\alpha\, dx\right)\left(\int_{-1}^{1} y^\beta\, dy\right)\left(\int_{-1}^{1} z^\gamma\, dz\right). \end{aligned}$$

Replacing each term according to Eq. (5.34), we get,

$$I = \left(\sum_{i=1}^{n} a_i x_i^\alpha\right)\left(\sum_{j=1}^{n} a_j y_j^\beta\right)\left(\sum_{k=1}^{n} a_k z_k^\gamma\right) = \sum_{i=1}^{n} a_i x_i^\alpha \sum_{j=1}^{n} a_j y_j^\beta \sum_{k=1}^{n} a_k z_k^\gamma. \tag{5.35}$$

We need now an elementary rule about the product of summations. We illustrate it for a simple case. We assert that

$$\begin{aligned} \left(\sum_{i=1}^{3} u_i\right)\left(\sum_{j=1}^{2} v_j\right) &= \sum_{i=1}^{3}\left(\sum_{j=1}^{2} u_i v_j\right) \\ &= \sum_{i=1}^{3}\sum_{j=1}^{2} u_i v_j. \end{aligned}$$

The last equality is purely notational. We prove the first by expanding both sides:

$$\begin{aligned} \left(\sum_{i=1}^{3} u_i\right)\left(\sum_{j=1}^{2} v_j\right) &= \sum_{i=1}^{3} u_i \sum_{j=1}^{2} v_j \\ &= (u_1 + u_2 + u_3)(v_1 + v_2) \\ &= u_1v_1 + u_1v_2 + u_2v_1 + u_2v_2 + u_3v_1 + u_3v_2; \end{aligned}$$

$$\sum_{i=1}^{3}\sum_{j=1}^{2} u_i v_j = (u_1v_1 + u_1v_2) + (u_2v_1 + u_2v_2) + (u_3v_1 + u_3v_2).$$

On removing parentheses, we see the two sides are the same. Using this principle, we can write Eq. (5.35) in the form

$$I = \sum_{i=1}^{n} \sum_{j=1}^{n} \sum_{k=1}^{n} a_i a_j a_k x_i^{\alpha} y_j^{\beta} z_k^{\gamma}, \tag{5.36}$$

which shows that the questioned equality of Eq. (5.34) is valid, and we can write a program for a triple integral by three nested DO loops. The coefficients a_i are chosen from any numerical integration formula. If the three one-variable formulas corresponding to Eq. (5.34) are not identical, an obvious modification of Eq. (5.36) applies. In some cases a change of variable is needed to correspond to Eq. (5.33).

If we are evaluating a multiple integral numerically where the integrand is a known function, our choice of the form of Eq. (5.33) is wider. Of higher efficiency than the Newton–Cotes formulas is Gaussian quadrature. Because it also fits the pattern of Eq. (5.33), the formula of Eq. (5.36) applies. We illustrate this with a simple example.

EXAMPLE 5.14 Evaluate

$$I = \int_0^1 \int_{-1}^0 \int_{-1}^1 yze^x \, dx \, dy \, dz$$

by Gaussian quadrature using a three-term formula for x and two-term formulas for y and z. We first make the changes of variables to adjust the limits for y and z to $(-1, 1)$:

$$y = \frac{1}{2}(u - 1), \qquad dy = \frac{1}{2}\,du;$$

$$z = \frac{1}{2}(v + 1), \qquad dz = \frac{1}{2}\,dv.$$

Our integral becomes

$$I = \frac{1}{16}\int_{-1}^1 \int_{-1}^1 \int_{-1}^1 (u - 1)(v + 1)e^x \, dx \, du \, dv.$$

The two- and three-point Gaussian formulas are, from Section 5.6,

$$\int_{-1}^1 f(x)\,dx = (1)f(-0.5774) + (1)f(0.5774),$$

$$\int_{-1}^1 f(x)\,dx = \left(\frac{5}{9}\right)f(-0.7746) + \left(\frac{8}{9}\right)f(0) + \left(\frac{5}{9}\right)f(0.7746).$$

The integral is then

$$I = \frac{1}{16}\sum_{i=1}^{2} \sum_{j=1}^{2} \sum_{k=1}^{3} a_i a_j b_k (u_i + 1)(v_j - 1)e^{x_k},$$

$$a_1 = 1, \qquad a_2 = 1,$$

$$b_1 = \frac{5}{9}, \qquad b_2 = \frac{8}{9}, \qquad b_3 = \frac{5}{9},$$

and values of u, v, and x as given.

A few representative terms of the sum are

$$I = \frac{1}{16}\left[(1)(1)\left(\frac{5}{9}\right)(-0.5774 + 1)(-0.5774 - 1)e^{-0.7446}\right.$$
$$+ (1)(1)\left(\frac{8}{9}\right)(-0.5774 + 1)(-0.5774 - 1)e^{0}$$
$$+ (1)(1)\left(\frac{5}{9}\right)(-0.5774 + 1)(-0.5774 - 1)e^{0.7746}$$
$$+ (1)(1)\left(\frac{5}{9}\right)(0.5774 + 1)(-0.5774 - 1)e^{-0.7746}$$
$$\left. + \cdots\right].$$

On evaluating, we get $I = -0.58758$. The analytical value is

$$-\frac{1}{4}(e - e^{-1}) = -0.58760.$$

MATLAB can solve Example 5.14:

```
EDU>> int(int(int('y*z*exp(x)','x',-1,1),'y',-1,0),'z',0,1)
ans =
      -1/4*exp(1)+1/4*exp(-1)
EDU>> numeric(ans)
ans =
      -0.5876
```

and both the analytical and numeric results are obtained.

Integrating with Variable Limits

As we said, the region for which we want the integral does not have to be a rectangle. Suppose we want to integrate

$$\iint f(x, y)\, dy\, dx$$

over the region bounded by the lines $x = 0$, $x = 1$, $y = 0$, and the curve $y = x^2 + 1$. The region is sketched in Figure 5.7. If we draw vertical lines spaced at $\Delta x = 0.2$ apart, shown as dashed lines in Figure 5.7, it is obvious that we can approximate the inner integral at constant x-values along any one of the vertical lines (including $x = 0$ and $x = 1$). If we use the trapezoidal rule with five panels for each of these, we get the series of sums

$$S_1 = \frac{h_1}{2}(f_a + 2f_b + 2f_c + 2f_d + 2f_e + f_f),$$

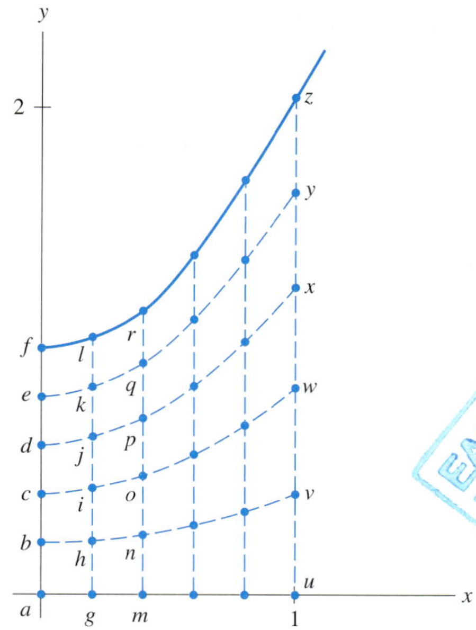

EAGLE LIBRARY
Bolton Institute of Higher Education - EAGLE CAMPUS

Figure 5.7

$$S_2 = \frac{h_2}{2}(f_g + 2f_h + 2f_i + 2f_j + 2f_k + f_l),$$

$$S_3 = \frac{h_3}{2}(f_m + 2f_n + \cdots),$$

$$\vdots$$

$$S_6 = \frac{h_6}{2}(f_u + 2f_v + 2f_w + 2f_x + 2f_y + f_z).$$

The subscripts here indicate the values of the function at the points so labeled in Figure 5.7. The values of the h_i are not equal in each of the equations, but in each they are the vertical distances divided by five. The combination of these sums to give an estimate of the double integral will then be

$$\text{Integral} = \frac{0.2}{2}(S_1 + 2S_2 + 2S_3 + 2S_4 + 2S_5 + S_6).$$

To be even more specific, suppose that $f(x, y) = xy$. Then,

$$S_1 = \frac{1.0/5}{2}(0 + 0 + 0 + 0 + 0 + 0) = 0,$$

$$S_2 = \frac{1.04/5}{2}(0 + 0.0832 + 0.1664 + 0.2496 + 0.3328 + 0.208) = 0.1082,$$

$$S_3 = \frac{1.16/5}{2}(0 + 0.1856 + 0.3712 + 0.5568 + 0.7424 + 0.464) = 0.2691,$$

$$S_4 = \frac{1.36/5}{2}(0 + 0.3264 + 0.6528 + 0.9792 + 1.3056 + 0.816) = 0.5549,$$

$$S_5 = \frac{1.64/5}{2}(0 + 0.5248 + 1.0496 + 1.5744 + 2.0992 + 1.312) = 1.0758,$$

$$S_6 = \frac{2.0/5}{2}(0 + 0.8 + 1.6 + 2.4 + 3.2 + 2.0) = 2.0;$$

$$\text{Integral} = \frac{0.2}{2}(0 + 0.2164 + 0.5382 + 1.1098 + 2.1516 + 2.0)$$

$$= 0.6016 \qquad \text{versus analytical value of } 0.583333.$$

The extension of this to more complicated regions and the adaptation to the use of Simpson's rule should be obvious. If the functions that define the region are not single-valued, we must divide the region into subregions to avoid the problem, but we must also do this when we integrate analytically.

The previous calculations were not very accurate because the trapezoidal rule has relatively large errors. Gaussian quadrature should be an improvement, even using fewer points within the region. Let us use three-point quadrature in the x-direction and four-point quadrature in the y-direction. As in Section 5.6, we must change the limits of integration:

$$\int_0^1 \int_0^{x^2+1} xy\, dy\, dx$$

to

$$\frac{1}{4}\int_{-1}^{1}\int_{-1}^{1} \frac{s+1}{2}\left[\frac{(x^2(s)+1)t + (x^2(s)+1)}{2}\right] dt\, ds$$

in which we make the following substitutions:

$$x = \frac{s+1}{2} \qquad y = \frac{(x^2(s)+1)t + (x^2(s)+1)}{2}.$$

The integral is approximated by the sum

$$\sum_{i=1}^{3}\sum_{j=1}^{4} w_i W_j f(s_i, t_j),$$

where the w_i's, W_j's, s_i's, and t_j's are the values taken from Table 5.13. Using that table, we set $w_1 = 0.55555555$, $w_3 = w_1$, and $w_2 = 0.88888889$; we set $s_1 = -0.77459667$, $s_3 = -s_1$, and $s_2 = 0.0$. The values for the W_j's and t_j's are obtained in the same way. For each fixed i, $i = 1, 2, 3$, let S_i be the corresponding value obtained using Gaussian quadrature for a fixed s_i, where $S_i = \Sigma_{j=1}^{4} (W_j f(s_i, t_j))$.

The following intermediate values are easily verified:

$$S_1 = (0.00279158 + 0.02487506 + 0.05050174 + 0.03741447) = 0.11558285,$$

$$S_2 = (0.01886891 + 0.16813600 + 0.34135240 + 0.25289269) = 0.78125000,$$

$$S_3 = (0.06845742 + 0.61000649 + 1.23844492 + 0.91750833) = 2.83441716.$$

We sum these values as follows:

$$\frac{w_1 S_1 + w_2 S_2 + w_3 S_3}{4} = 0.58333334,$$

which agrees with the exact answer to seven places. In this case, we used only 12 evaluations of the function (exceptionally simple to do here, but usually more costly), compared to the 36 used with the trapezoidal rule.

To keep track of the intermediate computations, it is convenient to use a template such as

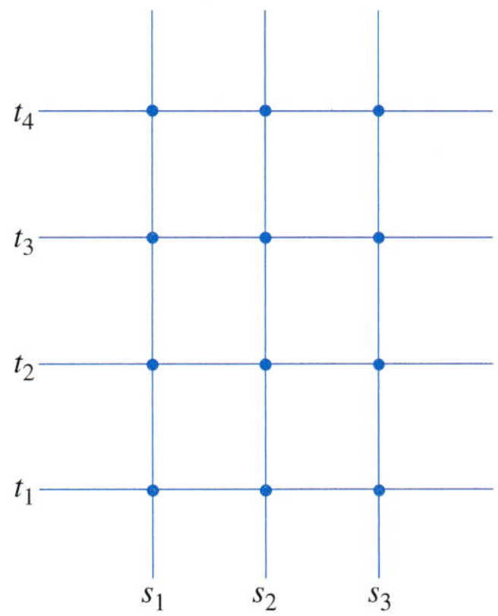

and to compute the S_i's along the verticals. The points (s_i, t_i) within the region are often called *Gauss points.*

MATLAB has no trouble in solving this problem:

```
EDU>> int(int('x*y', 'y', 0, 'x^2+1'), 'x', 0, 1)
   ans =
      7/12
EDU>> numeric (ans)
   ans =
      0.5833
```

Errors in Multiple Integration and Extrapolations

The error term of a one-variable quadrature formula is an additive one just like the other terms in the linear combination (although of special form). It would seem reasonable that it would go through the multiple summations in a similar fashion, so we should expect error terms for multiple integration that are analogous to the one-dimensional case. We illustrate that this is true for double integration using the trapezoidal rule in both directions, with uniform spacings, choosing n intervals in the x-direction and m in the y-direction.

From Section 5.2 we have

$$\text{Error of } \int_a^b f(x)\,dx = -\frac{b-a}{12}h^2 f''(\xi) = O(h^2),$$

$$h = \Delta x = \frac{b-a}{n}.$$

In developing Romberg integration, we observed that the error term could be written as

$$\text{Error} = O(h^2) = Ah^2 + O(h^4) \approx Ah^2 + Bh^4,$$

where A is a constant and the value of B depends on a fourth derivative of the function. Appending this error term to the trapezoidal rule, we get

$$\int_a^b f(x, y)\,dx \bigg|_{y=y_j} = \frac{h}{2}(f_{0,j} + 2f_{1,j} + 2f_{2,j} + \cdots + f_{n,j}) + A_j h^2 + B_j h^4.$$

Summing these in the y-direction and retaining only the error terms, we have

$$\int_c^d \int_a^b f(x, y)\,dx\,dy = \frac{k}{2}\frac{h}{2}\sum_{i=0}^{n}\sum_{j=0}^{m} a_i a_j f_{i,j} + \frac{k}{2}(A_0 + 2A_1 + 2A_2 + \cdots + A_m)h^2$$
$$+ \frac{k}{2}(B_0 + 2B_1 + 2B_2 + \cdots + B_m)h^4 + \overline{A}k^2 + \overline{B}k^4,$$

$$k = \Delta y = \frac{d-c}{m}.$$

In this, $\overline{A}$ and $\overline{B}$ are the coefficients of the error term for y. The coefficients A and B for the error terms in the x-direction may be different for each of the $(m + 1)$ y-values, but each of the sums in parentheses is $2n$ times some average value of A or B, so the error terms become

$$\text{Error} = \frac{k}{2}(nA_{av})h^2 + \frac{k}{2}(nB_{av})h^4 + \overline{A}k^2 + \overline{B}k^4.$$

Because both Δx and Δy are constant, we may take $\Delta y = k = \alpha\Delta x = \alpha h$, where $\alpha = \Delta y/\Delta x$, and the equation can be written, with $nh = (b - a)$,

$$\text{Error} = \left(\frac{b-a}{2}A_{av}\alpha\right)h^2 + \left(\frac{b-a}{2}B_{av}\alpha\right)h^4 + A\alpha^2h^2 + B\alpha^4h^4$$
$$= K_1h^2 + K_2h^4.$$

Here, K_2 will depend on fourth-order partial derivatives. This confirms our expectation that the error term of double integration by numerical means is of the same form as for single integration.

Because this is true, a Romberg integration may be applied to multiple integration, whereby we extrapolate to an $O(h^4)$ estimate from two trapezoidal computations at a 2:1 interval ratio. From two such $O(h^4)$ computations we may extrapolate to one of $O(h^6)$ error.

5.8 Applications of Cubic Splines

In addition to their obvious use for interpolation, splines (Chapter 3) can be used for finding derivatives and integrals of functions, even when the function is known only as a table of values. The smoothness of splines can give improved accuracy in some cases, because of the requirement that each portion have the same first and second derivatives as its neighbor where they join.

For the cubic spline that approximates $f(x)$, we can write, for the interval $x_i \le x \le x_{i+1}$,

$$g(x) = a_i(x - x_i)^3 + b_i(x - x_i)^2 + c_i(x - x_i) + d_i,$$

where the coefficients are determined as in Section 3.3. The method outlined in that section computes S_i and S_{i+1}, the values of the second derivative at each end of the subinterval. From these S-values and the values of $f(x)$, we compute the coefficients of the cubic:

$$a_i = \frac{S_{i+1} - S_i}{6(x_{i+1} - x_i)},$$

$$b_i = \frac{S_i}{2},$$

$$c_i = \frac{f(x_{i+1}) - f(x_i)}{x_{i+1} - x_i} - \frac{2(x_{i+1} - x_i)S_i + (x_{i+1} - x_i)S_{i+1}}{6},$$

$$d_i = f(x_i).$$

Approximating the first and second derivatives is straightforward; we estimate these as the values of the derivatives of the cubic:

$$f'(x) \approx 3a_i(x - x_i)^2 + 2b_i(x - x_i) + c_i, \quad (5.37)$$

$$f''(x) \approx 6a_i(x - x_i) + 2b_i. \quad (5.38)$$

At the $n + 1$ points x_i where the function is known and the spline matches $f(x)$, these formulas are particularly simple:

$$f'(x_i) \approx c_i,$$

$$f''(x_i) \approx 2b_i.$$

(We note that a cubic spline is not useful for approximating derivatives of order higher than second. A higher degree of spline function would be required for these values.)

Approximating the integral of $f(x)$ over the n intervals where $f(x)$ is approximated by the spline is similarly straightforward:

$$\int_{x_1}^{x_{n+1}} f(x)\,dx \approx \int_{x_1}^{x_{n+1}} g(x)\,dx$$

$$= \sum_{i=1}^{n} \left[\frac{a_i}{4}(x - x_i)^4 + \frac{b_i}{3}(x - x_i)^3 + \frac{c_i}{2}(x - x_i)^2 + d_i(x - x_i) \right]_{x_i}^{x_{i+1}}$$

$$= \sum_{i=1}^{n} \left[\frac{a_i}{4}(x_{i+1} - x_i)^4 + \frac{b_i}{3}(x_{i+1} - x_i)^3 + \frac{c_i}{2}(x_{i+1} - x_i)^2 + d_i(x_{i+1} - x_i) \right].$$

If the intervals are all the same size, ($h = x_{i+1} - x_i$), this equation becomes

$$\int_{x_1}^{x_{n+1}} f(x)\,dx \approx \frac{h^4}{4}\sum_{i=1}^{n} a_i + \frac{h^3}{3}\sum_{i=1}^{n} b_i + \frac{h^2}{2}\sum_{i=1}^{n} c_i + h\sum_{i=1}^{n} d_i.$$

We illustrate the use of splines to compute derivatives and integrals by a simple example.

EXAMPLE 5.15 Compute the integral and derivatives of $f(x) = \sin \pi x$ over the interval $0 \le x \le 1$ from the spline that fits at $x = 0, 0.25, 0.5, 0.75$, and 1.0. (See Table 5.15.) We use end condition 1: $S_1 = 0$, $S_5 = 0$. Solving for the coefficients of the cubic spline, we get the results shown in Table 5.16.

The estimated values for $f'(x)$ and $f''(x)$ computed with Eqs. (5.37) and (5.38) are shown in Table 5.17. The errors of these estimates from the exact values ($f'(x) = \pi\cos(\pi x)$ and $f''(x) = -\pi^2 \sin(\pi x)$) are shown in the last two columns.

In general, the cubic spline gives good estimates of the derivatives, the maximum error being 2.5% for the first derivative and 5.0% for the second.

It is of interest to compare these values with estimates of the derivatives from a fourth-degree interpolating polynomial that fits $f(x)$ at the same five points. Table 5.18 exhibits these estimates. For the first derivative, the spline curve gives better results near the ends of the range for $f(x)$; the polynomial gives better results near the midpoint. Both are very good in this example.

Comparison of estimates for the second derivative shows a similar relationship, except for the fourth-degree polynomial, which is very bad at the endpoints.

We readily compute the integral from the cubic spline:

$$\int_0^1 f(x)\,dx \approx \frac{(0.25)^4}{4}(0) + \frac{(0.25)^3}{3}(-12.5376) + \frac{(0.25)^2}{2}(3.1340) + 0.25(2.4142)$$
$$= 0.6362 \qquad (\text{exact} = 0.6366;\ \text{error} = +0.0004).$$

Table 5.15

***i*, point number**	***x***	***f*(*x*)**
1	0	0
2	0.25	0.7071
3	0.50	1.0000
4	0.75	0.7071
5	1.0	0

Table 5.16

i	x	S_i	a_i	b_i	c_i	d_i
1	0	0	−4.8960	0	3.1340	0
2	0.25	−7.3440	−2.0288	−3.6720	2.2164	0.7071
3	0.50	−10.3872	2.0288	−5.1936	0	1.000
4	0.75	−7.3440	4.8960	−3.6720	−2.2164	0.7071

The value for the integral using splines is better than getting it with Simpson's $\frac{1}{3}$ rule using the same panels ($\Delta x = 0.25$), which gives a value of 0.6381. The error there is −0.0015, almost four times greater than from the spline fit.

Observe that the error in the integral is only 0.24%, while the maximum errors in the derivatives are about 2.5% and 5.0%. This is generally true—numerical differentiation, in the words of many authorities, is *basically an unstable process*. We have seen how round-off error is terribly important when a numerical value for the derivative is computed.

Differentiation of "noisy" data encounters a similar problem. If the data being differentiated are from experimental tests, or are observations subject to errors of measurement,

Table 5.17 Estimates of $f'(x)$ and $f''(x)$ from a cubic spline

x	$f'(x)$	$f''(x)$	**Error in** $f'(x)$	**Error in** $f''(x)$
0.00	3.1344	0.0000	0.007146	0.000000
0.05	3.0977	−1.4689	0.005191	−0.075053
0.10	2.9876	−2.9378	0.000275	−0.112090
0.15	2.8039	−4.4067	−0.004766	−0.074028
0.20	2.5469	−5.8756	−0.005287	0.074363
0.25	2.2164	−7.3445	0.005053	0.365600
0.30	1.8340	−7.9529	0.012627	−0.031778
0.35	1.4211	−8.5613	0.005155	−0.232546
0.40	0.9778	−9.1698	−0.007015	−0.216781
0.45	0.5041	−9.7782	−0.012668	0.030114
0.50	−0.0000	−10.3866	0.000000	0.517038
0.55	−0.5041	−9.7782	0.012668	0.030113
0.60	−0.9778	−9.1698	0.007015	−0.216781
0.65	−1.4211	−8.5613	−0.005155	−0.232547
0.70	−1.8340	−7.9529	−0.012628	−0.031778
0.75	−2.2164	−7.3445	−0.005053	0.365598
0.80	−2.5469	−5.8756	0.005287	0.074362
0.85	−2.8039	−4.4067	0.004766	−0.074028
0.90	−2.9876	−2.9378	−0.000275	−0.112088
0.95	−3.0977	−1.4689	−0.005190	−0.075051
1.00	−3.1344	0.0000	−0.007146	0.000003

Table 5.18 Estimates of $f'(x)$ and $f''(x)$ from a polynomial, $P_4(x)$

x	$f'(x)$	$f''(x)$	Error in $f'(x)$	Error in $f''(x)$
0.00	3.0849	1.1505	0.056643	−1.150496
0.05	3.0894	−0.9358	0.013513	−0.608130
0.10	2.9950	−2.8025	−0.007196	−0.247359
0.15	2.8128	−4.4496	−0.013630	−0.031101
0.20	2.5537	−5.8771	−0.012126	0.075874
0.25	2.2288	−7.0849	−0.007321	0.106083
0.30	1.8489	−8.0732	−0.002311	0.088523
0.35	1.4251	−8.8418	0.001151	0.047960
0.40	0.9684	−9.3909	0.002436	0.004319
0.45	0.4897	−9.7203	0.001778	−0.027803
0.50	−0.0000	−9.8301	−0.000000	−0.039509
0.55	−0.4897	−9.7203	−0.001779	−0.027803
0.60	−0.9684	−9.3909	−0.002437	0.004321
0.65	−1.4251	−8.8418	−0.001152	0.047961
0.70	−1.8489	−8.0732	0.002311	0.088524
0.75	−2.2288	−7.0849	0.007320	0.106084
0.80	−2.5537	−5.8771	0.012125	0.075876
0.85	−2.8128	−4.4496	0.013630	−0.031100
0.90	−2.9950	−2.8025	0.007196	−0.247358
0.95	−3.0894	−0.9358	−0.013513	−0.608130
1.00	−3.0849	1.1505	−0.056643	−1.150496

the errors so influence the derivative values calculated by numerical procedures that they may be meaningless. The usual recommendation is to smooth the data first, using methods that are discussed in Chapter 3. Passing a cubic spline through the points and then getting the derivative of this approximation to the data has become quite popular. A least-squares curve may also be used. The strategy involved is straightforward—we don't try to represent the function by one that fits exactly to the data points, because this fits to the errors as well as to the trend of the information. Rather, we approximate with a smoother curve that we hope is closer to the truth than the data themselves. The problem, of course, is how much smoothing should be done. One can go too far and "smooth" beyond the point where only errors are eliminated.

A final situation should be mentioned. Some functions, or data from a series of tests, are inherently "rough." By this we mean that the function values change rapidly; a graph would show sharp local variations. When the derivative values of the function incur rapid changes, a sampling of the information may not reflect them. In this instance, the data indicate a smoother function than actually exists. Unless enough data are at hand to show the local variations, valid values of the derivatives just cannot be obtained. The only solution is more data, especially near the "rough" spots. And then we are beset by problems of accuracy of the data!

Fortunately, this problem does not occur with numerical integration. As you have seen, all the integration formulas add function values together. Because the errors can be positive or negative and the probability for each is the same, errors tend to cancel out. That means that integration is a smoothing process. We assess integration as *inherently stable*. This is generally true of computations that are *global*, in contrast to those that are *local* in nature, such as differentiation.

Exercises

Section 5.1

▶ **1.** Duplicate Table 5.1, but with double precision arithmetic. At what value for Δx is round-off error apparent?

2. Computer algebra systems permit you to use a specified number of digits in the computations. Repeat Exercise 1, but with only three digits of precision.

3. What is the effect of the precision of arithmetic on Table 5.2 where central differences are used?

4. Make a graph for $f(x) = e^{-x/3} * \cos(x)$ from $x = -1$ to $x = 3$.

a. From the graph, predict for what x-value(s) the accuracy of a forward-difference approximation to the derivative with $h = 0.05$ will be most accurate.

b. Confirm your prediction by doing computations.

5. Repeat Exercise 4 but for backward differences.

6. Repeat Exercise 4 but for central differences.

▶ **7.** Make a divided-difference table similar to Table 5.3, but for the function $f(x) = 2x * \cos(2x)$. Use the data in the table to compute $f'(2.0)$

a. Using a forward-difference approximation.

b. Using a backward-difference approximation.

c. Using a central-difference approximation.

8. Find bounds to the errors of each of the computations of Exercise 7 from Eq. (5.7). What are the actual errors?

9. Duplicate Figure 5.1a, b, and c with the function of Exercise 7.

10. Compute a difference table like Table 5.4 but for the same function as in Exercise 7, $f(x) = 2x * \cos(2x)$. Use one, two and three terms of Eq. (5.10) to construct graphs similar to Figure 5.1a, b, and c.

▶ **11.** Compute a value for $f'(0.268)$ from a quadratic interpolating polynomial that fits the table at the three points that should give the most accurate answer. Which points are these?

I	x_i	f_i	$f[x_i, x_{i+1}]$	$f[x_i \cdots x_{i+2}]$	$f[x_i \cdots x_{i+3}]$
0	0.15	0.1761	2.4355	−5.7505	15.3476
1	0.21	0.3222	1.9754	−3.9088	8.7492
2	0.23	0.3617	1.7409	−2.9464	5.9642
3	0.27	0.4314	1.4757	−2.2307	
4	0.32	0.5051	1.2973		
5	0.35	0.5441			

12. The function in Exercise 11 is for $f(x) = 1 + \log_{10}(x)$.

a. What is the error of your answer in Exercise 11?

b. How does this compare with that estimated from the next-term rule?

c. Compute $f'(0.268)$ from other sets of three points and repeat parts (a) and (b) for each of these.

▶ **13.** The differences in the table of Exercise 11 are actually the divided differences of $f(x)$ accurate to six decimal places, even though the function values are shown to only four decimals. Recompute the differences using the tabulated function values and repeat Exercise 12. How much does the rounding affect the errors? Is rounding more important than truncation?

14. Repeat Exercise 11, but this time for $f'(x)$ at $x = 0.21$, 0.22, 0.23, 0.24, 0.25, 0.26, and 0.27. Plot the estimates and compare to a graph of the true values. Make another plot of the errors versus x. At what point is the error smallest?

15. As described in Exercise 13, the differences tabulated in Exercise 11 are based on more accurate function values. Recompute the divided-difference table using the tabulated function values, then repeat Exercise 14. How does rounding change the errors you found in Exercise 14?

16. Use Eq. (5.7) to find bounds for the errors at $x = 0.21$, 0.23, and 0.27 in Exercise 14. Do these bounds bracket the errors found in Exercise 14?

17. Use the next-term rule to estimate the error in Exercise 14. Compare these errors with the actual errors. Are the estimates always larger?

18. Repeat Exercise 17, but with the recomputed table done in Exercises 13 and 15.

▶**19.** The following ordinary difference table is for $f(x) = x + \sin(x)/3$. Use it to find

a. $f'(0.72)$ from a cubic polynomial.
b. $f'(1.33)$ from a quadratic.
c. $f'(0.50)$ from a fourth-degree polynomial.

In each part, choose the best starting i-value.

i	x_i	f_i	Δf_i	$\Delta^2 f_i$	$\Delta^3 f_i$	$\Delta^4 f_i$
0	0.30	0.3985	0.2613	−0.0064	−0.0022	0.0003
1	0.50	0.6598	0.2549	−0.0086	−0.0018	0.0004
2	0.70	0.9147	0.2464	−0.0104	−0.0014	0.0005
3	0.90	1.1611	0.2360	−0.0118	−0.0010	
4	1.10	1.3971	0.2241	−0.0128		
5	1.30	1.6212	0.2113			
6	1.50	1.8325				

20. Use the next-term rule to estimate the errors in Exercise 19. Compare these to the actual errors. Are the estimates always larger?

21. Show that the error of Eq. (5.14) is $O(h^2)$.

▶**22.** Use the method of undetermined coefficients to obtain the formulas for $f''(x)$, $f'''(x)$ and $f^{(4)}(x)$ at x_0 using five evenly spaced points from x_2 to x_{-2}, together with their error terms.

23. Get estimates for the second third and fourth derivatives of $f(x)$ at $x = 0.90$ from the data of the table of Exercise 19. What are the errors?

24. Extrapolate to get $f''(0.90)$ from the table of Exercise 19 as many times as you can. What is the error? How much of this is due to the precision of the data?

25. Show that the first extrapolation for $f'(x_0)$ with h-values differing by 2 to 1 is the same as the formula

$$f_0' = \frac{1}{H}\left(\frac{\Delta f_{-1} + \Delta f_0}{2} - \frac{1}{6}\,\frac{\Delta^3 f_{-2} + \Delta^3 f_{-1}}{2}\right),$$

where H is the smaller of the h's.

26. Can extrapolations similar that of Eq. (5.15) be used for unevenly spaced data? (A Taylor series expansion may be helpful.) If you succeed in getting a formula, use it to estimate a better value for $f'(0.27)$ from the table of Exercise 11. What order of error results?

▶**27.** Apply Richardson extrapolation to get $f'(0.32)$ accurate to five significant figures for $f(x) = \sin^2(x/2)$, starting with $h = 0.1$ and using central differences. When the extrapolations agree to five significant figures, are they that accurate?

28. Repeat Exercise 27, but now for $f''(0.32)$.

29. Can Richardson extrapolation be used with forward differences? If you can do this, repeat Exercise 27 employing forward differences.

30. Create a Richardson table with a computer algebra system. The trick is how to get a display similar to that in Section 5.1.

Section 5.2

▶**31.** The global error of the integral, $\int f(x)\,dx$, between two x-values by the trapezoidal rule is

$$-(1/12)h^3 f''(\xi),$$

where ξ lies inside the two x-values. For these functions and x-values, find the value for ξ:

a. $f(x) = x^3$, $x = [0.2, 0.5]$.
b. $f(x) = e^x$, $x = [-.1, 0.2]$.
c. $f(x) = \sin(x)$, $x = [0, 0.4]$.

32. The global error of the trapezoidal rule is

$$(-(b - a)/12)h^2 f''(\xi),$$

where ξ lies within the range for the integral. Repeat Exercise 31 when the step size, h, is

a. 0.1.
b. 0.01.
c. What are the limiting values as $h \to 0$?

33. Repeat Example 5.1, but now use only four values, for $x = 1.6, 2.2, 2.8$, and 3.4.

34. How small must h be for the trapezoidal rule to attain an error less than 0.001 for

$$\int x^2 \sin(x)\,dx, \text{ between } x = 0.2 \text{ and } 2.8?$$

▶**35.** Use the data in the table to find the integral between $x = 1.0$ and 1.8, using the trapezoidal rule:

a. With $h = 0.1$.
b. With $h = 0.2$.
c. With $h = 0.4$.

x	$f(x)$
1.0	1.543
1.1	1.669
1.2	1.811
1.3	1.971
1.4	2.151
1.5	2.352
1.6	2.577
1.7	2.828
1.8	3.107

36. The function tabulated in Exercise 35 is $\cosh(x)$. What are the errors in parts (a), (b), and (c)? How closely are these proportional to h^2? What errors are present besides the truncation error?

37. Extrapolate from the results of Exercise 35 to get an improved value for the integral (Romberg integration). What is the order of error for this extrapolated answer? How accurate is it?

▶**38.** If the integral of Example 5.1 is wanted correct to five decimal places (error < 0.000005), how small should h be? Recompute the table with this value for h and verify that this gives the desired accuracy.

39. Repeat Exercise 38, but now use Romberg integration. What is the degree of improvement over Exercise 38?

40. Use Romberg integration to evaluate the integral of $f(x) = 1/x$ between $x = 1$ and $x = 3$. Using six significant digits in your computations, continue until there is no change in the fourth decimal place. Is this answer that correct?

Section 5.3

▶**41.** Repeat Exercise 35, but now use Simpson's 1/3 rule.

42. Use the error term for Simpson's 1/3 rule to bound the errors in Exercise 41 for each application of the rule. What are the values for ξ for each value of h?

43. Simpson's 3/8 rule cannot be applied directly to Exercise 41 because the number of panels is not divisible by three. Still, you can use it in combination with the 1/3 rule over two panels. There are several choices of where to use the 1/3 rule. Which of these choices gives the most accurate answer?

44. The function $f(x) = x^2 * \sin(2x)$ is zero at the origin and is zero again at multiples of $\pi/2$.

a. Use Simpson's 1/3 rule to approximate the integral under the first "hump." How large can h be and still attain a value with an error less than 0.001?

b. Repeat part (a) but now get the integral from $x = 0$ to $x = \pi$.

45. Repeat Exercise 44, but now use Simpson's 3/8 rule.

▶**46.** Show that extrapolating once with the trapezoidal rule is equivalent to using Simpson's 1/3 rule with a comparable value for h.

47. Is there an equivalent relation, between extrapolations of the trapezoidal rule and Simpson's 3/8 rule as found in Exercise 46? Find such a relationship if it exists or prove that there is none.

48. Simpson's $\frac{1}{3}$ rule, although based on passing a quadratic through three evenly spaced points, actually gives the exact answer if $f(x)$ is a cubic. The implication is that the area under any cubic between $x = a$ and $x = b$ is identical to the area of a parabola that matches the cubic at $x = a$, $x = b$, and $x = (a + b)/2$. Prove this.

49. Simpson's rules are derived by fitting polynomials of degrees 2 and 3 to the integrand. Obtain a formula that results from fitting a fourth-degree polynomial and its error term. Would this have any advantage over the Simpson's rules?

▶**50.** In solving differential equations, one method finds the integral of the derivative function from a linear sum of past values for the derivative. One example is

$$\int_{x_n}^{x_{n+1}} f(x)\,dx = c_0 f(x_{n-3}) + c_1 f(x_{n-2}) + c_2 f(x_{n-1}) + c_3 f(x_n).$$

What values should be used for the c's?

51. Compute the integral of $f(x) = \sin(x)/x$ between $x = 0$ and $x = 1$ using Simpson's $\frac{1}{3}$ rule with $h = 0.5$ and then with $h = 0.25$. (Remember that the limit of $\sin(x)/x$ at $x = 0$ is 1.) From these two results, extrapolate to get a better result. What is the order of the error after the extrapolation? Compare your answer with the true answer.

▶**52.** Repeat Exercise 51, but use Simpson's 3/8 rule.

53. Prove that *all* integration methods that are based on even-order interpolation formulas (quadratic, quartic, etc.) have a global error order equal to two more than the order of the polynomial, while those based on a polynomial of odd order have a global error just one more than the order of the polynomial.

54. A way to derive integration formulas (as well as formulas for differentiation) is the *symbolic method.* Do research to find out about this method and use it to derive several of the formulas of this chapter.

Section 5.4

▶**55.** Use trapezoidal integration with 24 panels to get the first nine Fourier coefficients for these functions and compare to those from analytical integration:

a. $f(x) = x^3 - 1$ on [0, 3].
b. $f(x) = 2x^2 + 1$ on [−2, 1].
c. $f(x) = e^x \cos(3x)$ on [0, 5].

56. Repeat Exercise 55, but with Simpson's 1/3 rule. How much more accurate are these than the results of Exercise 55?

57. Repeat Exercise 55, but with Simpson's 3/8 rule. Are these less accurate than those from Exercise 57?

58. How many panels would be needed to match to the analytical coefficients to within 0.00001

a. in Exercise 55?
b. in Exercise 56?
c. in Exercise 57?

▶**59.** Verify that Eqs. (5.26) and (5.27) are truly identical.

60. Make a diagram similar to Figure 5.5 for $n = 8$.

61. Use the algorithm given in Section 5.4 that generates the powers of W to be used in an FFT to obtain the values for $n = 16$. These should agree with those in Figure 5.5; do they?

62. Repeat Exercise 61 but now with the bit-reversing rule.

63. Write a procedure in a computer algebra system that does an FFT, with up to 33 pairs of $t, f(t)$ values as an input. Test it by duplicating Example 5.9.

Section 5.5

64. Repeat Example 5.10, but use the trapezoidal rule. At what value for h do the computations terminate? How many function evaluations are required compared to Simpson's 1/3 rule?

65. Repeat Exercise 64, but now use Simpson's 3/8 rule.

66. Solve the problem of Example 5.10 with an adaptive trapezoidal rule. Compare the number of function evaluations with that for Simpson's 1/3 rule.

67. Repeat Exercise 66, but now with Simpson's 3/8 rule.

▶**68.** Use adaptive Simpson's 1/3 rule to obtain the integral of $e^x \cos(2x)$ over the interval [0, $\pi/4$]. Use a value for TOL, the tolerance value, sufficiently small to attain an answer within 0.001 of the exact answer, 0.677312.

69. Repeat Exercise 68, but now use adaptive trapezoidal rule. Compare the number of function evaluations with that used in Exercise 68.

▶**70.** Most programs for adaptive integration will compute the appropriate step size if they use the procedure of Section 5.5. However, in some cases this leads to significant errors. For instance, the integral of $\sin^2(16x)$ between $x = 0$ and $x = \pi/2$ is $\pi/4$, but it is easy to see that the values of $S_1[0, \pi/2]$ and $S_2[0, \pi/2]$ both equal zero, where $h_1 = \pi/4$ and $h_2 = \pi/8$.

How can we solve this problem correctly with the adaptive method of Section 5.5? (It is interesting to know that the HP-15C calculator avoids this error.)

Section 5.6

71. The integral of e^x between 0 and 3 is $(e^3 - e^0) = e^3 - 1 = 19.085537$. How many terms of Gaussian quadrature must be used to obtain the result correct to within 0.001?

72. If Simpson's 1/3 rule were used to get the integral of Exercise 71, how many more function evaluations would be needed?

73. What is the error if the integral of $\sin(x)/x$ over x = [0, 2] is evaluated with a four-term Gaussian formula? How many intervals would be needed with Simpson's 1/3 rule to get the value with the same accuracy?

▶**74.** By using Gaussian formulas of increasing complexity, determine how many terms are needed to evaluate the integral of $x^3 * \sin(x^2)e^{x-3}$ over the interval [−1.5, 2.7] to get accuracy to six significant figures.

75. An n-term Gaussian formula assumes that a polynomial of degree $2n - 1$ is used to fit the function between $x = a$ and $x = b$. Does this mean that the error is the same as for a Newton–Cotes integration formula based on a polynomial of degree $2n - 1$?

▶**76.** Confirm that the values for t in Table 5.13 are correct by getting the zeros of the appropriate Legendre polynomials. Use any method from Chapter 1.

77. Repeat Exercise 76, but get the zeros with a computer algebra system.

78. Two improper integrals are given in Section 5.6 as examples where Gaussian quadrature can be applied. How many terms are needed to get the integrals correct to within 0.0001?

79. Instead of using a Gaussian quadrature formula of higher degree to evaluate an integral, one could break up the interval of integration into subintervals and combine the results from a formula of lower degree. Is there merit to this idea? Find a function where this is of advantage and find another where it is not.

Section 5.7

80. The statement is made in Example 5.13 that "it is immaterial which integral we evaluate first." Confirm that this is true by repeating Example 5.13, but integrate first with respect to y.

▶**81.** Write pictorial operators similar to Eq. (5.32) for

a. Simpson's $\frac{1}{3}$ rule in the x-direction and the trapezoidal rule in the y-direction.

b. Simpson's $\frac{1}{3}$ rule in both directions.

c. Simpson's $\frac{3}{8}$ rule in both directions.

d. What conditions are placed on the number of panels in both directions by parts (a), (b), and (c)?

82. Because Simpson's $\frac{1}{3}$ rule is exact when $f(x)$ is a cubic, evaluation of the following triple integral should be exact. Confirm by evaluating both numerically and analytically. Use Eq. (5.36) adapted for this integral.

$$\int_0^1 \int_0^2 \int_{-1}^0 x^3yz^2 \, dx \, dy \, dz$$

83. Draw a pictorial operator that represents the formula used in Exercise 82. You may want to do this on three widely separated planes, such as

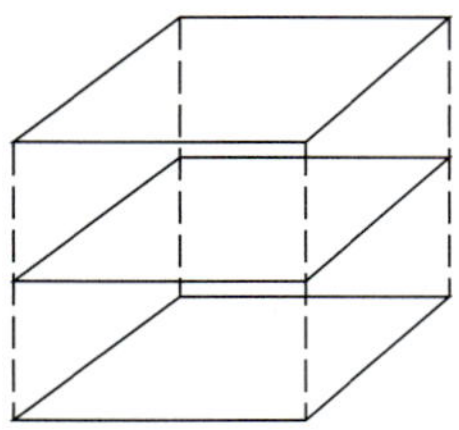

84. Evaluate the following integral, and compare your answers to the analytical solution. Use $h = 0.1$ in both directions in parts (a) and (b),

a. using the trapezoidal rule in both directions.

b. using Simpson's $\frac{1}{3}$ rule in both directions.

c. using Gaussian quadrature, three-term formulas in both directions.

$$\int_{-0.2}^{1.4} \int_{0.4}^{2.6} e^x \sin(2y) \, dy \, dx$$

85. Solve Exercise 84 by performing the trapezoidal rule integrations first with $h = 0.2$ (in both directions), then with $h = 0.1$. and extrapolate. The answer should match part (b) of the exercise. Does it?

▶**86.** Integrate with varying values of Δx and Δy using the trapezoidal rule in both directions, and show that the error decreases about in proportion to h^2:

$$\int_0^1 \int_0^1 (x^2 + y^2) \, dx \, dy.$$

87. Apply Romberg integration to Exercise 86 to get a value of $O(h^6)$.

88. Repeat Exercise 86, but now use Simpson's 1/3 rule. How do errors decrease with h?

89. Extrapolate from two results of Exercise 88. What is the order of the error of the extrapolation?

Section 5.8

▶**90.** The following table is for $f(x) = 1/(x + 2)$. Find values for $f'(x)$ and $f''(x)$ at $x = 1.5$, 2.0, and 2.5 from cubic spline functions that approximate $f(x)$. Compare to the true values to determine the errors. Also compare to derivative values computed from central-difference formulas.

a. Use end condition 1.

b. Use end condition 3.

c. Use end condition 4.

x	**1.0**	**1.5**	**2.0**	**2.5**	**3.0**
$f(x)$	0.333	0.286	0.250	0.222	0.200

91. Plot the values of $f'(x)$ and $f''(x)$ from the cubic splines of Exercise 90 on [1.0, 3.0], and compare to plots of the true values.

92. The comparisons in Exercise 90 may favor the cubic spline because they are based on cubic polynomials, whereas the central-difference formulas are based on quadratics. Repeat Exercise 90, but now use interpolating polynomials of degrees 3 and 4.

93. Repeat Exercise 90, but this time use cubic splines that have the correct slopes at the ends, condition 2.

▶**94.** Integrate sech(x) over [0, 2] by integrating the natural cubic spline curve (end condition 1) that fits at five evenly spaced points on [0, 2]. Compare the result to the analytical value. Also compare to the integral from Simpson's $\frac{1}{3}$ rule.

95. Repeat Exercise 94 using end conditions 2, 3, and 4. For condition 2, use the analytical values for $f'(x)$.

96. Repeat Exercise 94 but now force the values of $f''(x)$ at the ends to the analytical values of the second derivative of sech(x).

Applied Problems and Projects

APP1. When one first hears of Gaussian quadrature, it seems remarkable that just adding the value of the integrand at two points is equivalent to integrating from an interpolating polynomial of degree-3, and that adding a weighted sum of three points is equivalent to using a polynomial of degree-5.

Table 5.13 gives values that determine where to select the points. What if we use values that are slightly incorrect? How much is the approximation of the integral affected if the selected points are off by 1%? By 5%?

APP2. Differential thermal analysis is a specialized technique that can be used to determine transition temperatures and the thermodynamics of chemical reactions. It has special application in the study of minerals and clays. Vold [*Anal. Chem.* 21, 683 (1949)] describes the technique. In this method, the temperature of a sample of the material being studied is compared to the temperature of an inert reference material when both are heated simultaneously under identical conditions. The furnace housing the two materials is normally heated so that its temperature (T_f) increases (approximately) linearly with time (t), and the difference in temperatures (ΔT) between the sample and the reference is recorded. Some typical data are

***t*, min**	**0**	**1**	**2**	**3**	**4**	**5**	**6**	**7**	
ΔT, °F	0.00	0.34	1.86	4.32	8.07	13.12	16.80	18.95	
T_f, °F	86.2	87.8	89.4	91.0	92.7	94.3	95.9	97.5	
t	**8**	**9**	**10**	**11**	**12**	**13**	**14**	**15**	**16**
ΔT	18.07	16.69	15.26	13.86	12.58	11.40	10.33	8.95	6.46
T_f	99.2	100.8	102.3	103.9	105.5	107.1	108.6	110.2	111.8
t	**17**	**18**	**19**	**20**	**21**	**22**	**23**	**24**	**25**
ΔT	4.65	3.37	2.40	1.76	1.26	0.88	0.63	0.42	0.30
T_f	113.5	115.1	116.8	118.4	120.0	121.6	123.2	124.9	126.5

The ΔT values increase to a maximum, then decrease, due to the heat evolved in an exothermic reaction. One item of interest is the time (and furnace temperature) when the reaction is complete. Vold shows that the logarithm of ΔT should decrease linearly after the reaction is over; while the chemical reaction is occurring, the data depart from this linear relation. Vold used a graphical method to find this point. Perform numerical computations to find, from the preceding data, the time and the furnace temperature when the reaction terminates. Compare the merits of doing it graphically or numerically.

APP3. The temperature difference data in APP2 can be used to compute the heat of reaction. To do this, the integral of the values of ΔT is required, from the point where the reaction begins (which is at the point where ΔT becomes nonzero) to the time when the reaction ceases, as found in APP2. Determine the value of the required integral. Which of the methods of this chapter should give the best value for the integral?

APP4. There is a way to integrate numerically called the *midpoint rule*. The estimates the integral of $f(x)$ on the interval $[a, b]$ by this equation:

$$\int_a^b f(x)\,dx \approx (b-a)f\left(\frac{a+b}{2}\right).$$

a. Derive this formula in three different ways.
b. Find its error term.
c. Find at least three functions for which this gives the exact answer. State the condition for this to be true.
d. What is the composite rule for the midpoint rule? What is the error term for it?
e. Outline how adaptive integration would be used for this method.

APP5. The stress developed in a rectangular bar when it is twisted can be computed if one knows the values of a torsion function U that satisfies a certain partial-differential equation. Chapter 8 describes a numerical method that can determine values of U. To compute the stress, it is necessary to integrate $\int\int U\,dx\,dy$ over the rectangular region for which the data given here apply. Determine the stress. (You may be able to simplify the integration because of the symmetry in the data.)

x \ y	**0.0**	**0.2**	**0.4**	**0.6**	**0.8**	**1.0**	**1.2**
0.0	0	0	0	0	0	0	0
0.2	0	2.043	3.048	3.354	3.048	2.043	0
0.4	0	3.123	4.794	5.319	4.794	3.123	0
0.6	0	3.657	5.686	6.335	5.686	3.657	0
0.8	0	3.818	5.960	6.647	5.960	3.818	0
1.0	0	3.657	5.686	6.336	5.686	3.657	0
1.2	0	3.123	4.794	5.319	4.794	3.123	0
1.4	0	2.043	3.048	3.354	3.048	2.043	0
1.6	0	0	0	0	0	0	0

APP6. *Fugacity* is a term used by engineers to describe the available work from an isothermal process. For an ideal gas, the fugacity f is equal to its pressure P, but for real gases,

$$\ln\frac{f}{P} = \int_0^P \frac{C-1}{P}\,dp,$$

where C is the experimentally determined *compressibility factor.* For methane, values of C are

***P*(atm)**	***C***	***P*(atm)**	***C***
1	0.9940	80	0.3429
10	0.9370	120	0.4259
20	0.8683	160	0.5252
40	0.7043	250	0.7468
60	0.4515	400	1.0980

Write a program that reads in the P and C values and uses them to compute and print f corresponding to each pressure given in the table. Assume that the value of C varies linearly between the tabulated values (a more precise assumption would fit a polynomial to the tabulated C values). The value of C approaches 1.0 as P approaches 0.

APP7. The highway patrol uses a radar gun to clock the speed of a motorist. The gun is equipped with a device that records the speed at 4-second intervals as given in the table below.

a. What is the total distance traveled by the car?
b The speed limit is 65 mph. What fraction of the time is he speeding?
c. When do you think the motorist noticed the officer?

Time	0	4	8	12	16	20	24	28	32	36	40
Speed (mph)	64	68	71	74	76	72	64	63	68	73	72

APP8. A cardiod curve is heart-shaped. It can be drawn from the equation

$$r = a(1 - \cos(\theta)).$$

Use a numerical method to compute the length of the curve if $a = 3$ and compare to the analytical answer.

APP9. A variation on APP8 is a lemniscate; the equation is

$$r = a * \sin(2\theta).$$

Draw the curve for $a = 3$. Then repeat APP8 for this curve using Gaussian quadrature.

APP10. Outline a procedure for an adaptive Gaussian quadrature that uses the three-term formula.

6

Numerical Solution of Ordinary Differential Equations

Most problems in the real world are modeled with differential equations because it is easier to see the relationship in terms of a derivative. An obvious example is Newton's Law—$f = M * a$—where the acceleration a is the rate of change of the velocity. Velocity is also a derivative, the rate of change the position, s, of an object of mass, M, when it is acted on by force, f. So we should think of Newton's Law as

$$d^2s/dt^2 = a = f/M,$$

a *second-order ordinary differential equation.* It is *ordinary* because it does not involve partial differentials and *second order* because the order of the derivative is two. The solution to this equation is a function, $s(t)$. This is a particularly easy problem to solve analytically when the acceleration is constant:

$$s(t) = (1/2)\, at^2 + v_0 t + s_0.$$

The solution contains two arbitrary constants, v_0 and s_0, the initial values for the velocity and position. The equation for $s(t)$ allows the computation of a numerical value for s, the position of the object, at any value for time, the independent variable, t.

Many differential equations can be solved analytically and you probably learned how to do this in a previous course. The general analytical solution will include arbitrary constants in a number equal to the order of the equation. If the same number of conditions on the solution are given, these constants can be evaluated.

When all of the conditions on the problem are specified at the same value for the independent variable, the problem is termed an *initial-value problem.* If these are at two different values for the independent variable, usually at the boundaries of some region of interest, it is called a *boundary-value problem.*

This chapter describes techniques for solving ordinary differential equations by numerical methods. To solve the problem numerically, the required number of conditions must be known and these values are used in the numerical solution. We will begin the chapter with a Taylor series method that is not only a good method in itself but serves as the basis for several other methods. We start with first-order initial-value problems and later cover higher-order problems and boundary-value problems.

With an initial-value problem, the numerical solution begins at the initial point and marches from there to increasing values for the independent variable. With a boundary problem, one must march toward the other boundary and match with the condition(s) there. This is not as easy to accomplish. Certain boundary-value problems have a solution only for *characteristic values* for a parameter; these are known as *characteristic-value problems.*

When we attempt to solve a differential equation, we must be sure that there really is a solution and that the solution we get is unique. This requires that $f(x, y)$ in $dy/dx = f(x, y)$ meet the *Lipschitz condition:*

Let $f(x, y)$ be defined and continuous on a region R that contains the point (x_0, y_0). We assume that the region is a closed and bounded rectangle. Then $f(x, y)$ is said to satisfy the Lipschitz condition if:

There is an $L > 0$ so that for all x, y_1, y_2 in R, we have

$$|f(x, y_1) - f(x, y_2)| < L|y_1 - y_2|.$$

For most problems and all examples of this chapter, the condition is met.

There is a similar set of conditions for the solution to a boundary-value problem to exist and be unique. A linear problem of the form

$$\frac{d^2u}{dx^2} = pu' + qu + r, \qquad \text{for } x \text{ on } [a, b],$$

with

$$u(a) = uL, \qquad u(b) = uR,$$

where p, q, and r are functions of x only, has a unique solution if two conditions are met:

$$p, q, \text{ and } r \text{ must be continuous on } [a, b],$$

and

$$q > 0 \text{ on } [a, b].$$

If the problem is nonlinear, more severe conditions apply that involve the partial derivatives of the right-hand side with respect to u and u'.

Contents of This Chapter

6.1 The Taylor-Series Method

Adapts this method from calculus to develop a power series that, if truncated, approximates the solution to a first-order initial-value problem. Unless many terms are used, the solution cannot be carried far beyond the initial point.

6.2 The Euler Method and Its Modifications

Describes a method that is easy to use but is not very precise unless the step size, the intervals for the projection of the solution, is very small. Modifications permit the use of a larger step size or give greater accuracy at the same size of steps. These methods are based on low-order Taylor series.

6.3 Runge–Kutta Methods

Presents methods that are based on more terms of a Taylor series than the Euler methods and are thereby much more accurate. A very widely used method, the Runge–Kutta–Fehlberg method (RKF) allows an estimation of the error as computations are made so the step size can be changed appropriately.

6.4 Multistep Methods

Covers methods that are more efficient than the previous methods, which are called *single-step methods.* They require a number of starting values in addition to the initial value. A Runge–Kutta method is frequently used to get these starting values. A valuable adjunct to a multistep method is to first compute a *predicted* value and then do a second computation to get a *corrected value.* Doing this monitors the accuracy of the computations.

6.5 Higher-Order Equations and Systems

Describes how the methods previously covered can solve an equation of order higher than the first. This is done by converting the equation to a system of first-order problems. Hence, even a system of higher-order problems can be handled.

6.6 Stiff Equations

Discusses a type of problem that poses difficulties in avoiding *instability,* the growth of initial error as the solution proceeds.

6.7 Boundary-Value Problems

Extends the methods previously described to solve a differential equation whose conditions are specified at not just the initial point. This section also describes how the solution can be approximated if the derivatives are replaced by difference quotients, as explained in Chapter 5.

6.8 Characteristic-Value Problems

Shows how that class of boundary-value problems that have a solution only for certain values of a parameter can be solved. These certain values are the *eigenvalues* of the system; eigenvalues and their associated *eigenvectors* are essential matrix-related quantities that have applications in many fields. Two different ways to obtain eigenvalues are described.

6.1 The Taylor-Series Method

As you have seen before, a Taylor series is a way to express (most) functions as a power series. When expanded about the point $x = a$, the coefficients of the powers of $(x - a)$ include the values of the successive derivatives of the function at $x = a$. This means that if we know enough about a function at some point $x = a$, that is, its value and the value of all of its derivatives, we can (usually) write a series that has the same value as the function at all values of x. We will use x_0 to represent $x = a$.

In the present application, we are given the function that is the first derivative of $y(x)$: $y' = f(x, y)$, and an initial value, $y(x_0)$. With this information we can write the Taylor series for $y(x)$ about $x = x_0$. We just differentiate $y'(x) = f(x, y)$ as many times as we desire and evaluate these derivatives at $x = x_0$. The problem is that, when $y'(x)$ involves not just x but the unknown y as well, the higher derivatives may not be easy to come by.

Even so, these higher derivatives can be written in terms of x and the lower derivatives of y. We only want their values at $x = x_0$. Here is an example:

$$\frac{dy}{dx} = -2x - y, \qquad y(0) = -1. \tag{6.1}$$

(This particularly simple example is chosen to illustrate the method so that you can readily check the computational work. The analytical solution,

$$y(x) = -3e^{-x} - 2x + 2$$

is obtained immediately by application of standard methods and will be compared with our numerical results to show the error at any step.)

We develop the relation between y and x by finding the coefficients of the Taylor series in which we expand y about the point $x = x_0$:

$$y(x) = y(x_0) + y'(x_0)(x - x_0) + \frac{y''(x_0)}{2!}(x - x_0)^2 + \frac{y'''(x_0)}{3!}(x - x_0)^3 + \cdots.$$

If we let $x - x_0 = h$, we can write the series as

$$y(x) = y(x_0) + y'(x_0)h + \frac{y''(x_0)}{2}h^2 + \frac{y'''(x_0)}{6}h^3 + \cdots. \tag{6.2}$$

Because $y(x_0)$ is our *initial condition,* the first term is known from the initial condition $y(0) = -1$. (Because the expansion is about the point $x = 0$, our Taylor series is actually the Maclaurin series in this example.)

We get the coefficient of the second term by substituting $x = 0$, $y = -1$ in the equation for the first derivative, Eq. (6.1):

$$y'(x_0) = y'(0) = -2(0) - (-1) = 1.$$

We get the second- and higher-order derivatives by successively differentiating the equation for the first derivative. Each of these derivatives is evaluated corresponding to $x = 0$ to get the various coefficients:

Table 6.1

x	$y(x)$	Anal	Error
0.00000	−1.00000	−1.00000	0.00000
0.10000	−0.91451	−0.91451	0.00000
0.20000	−0.85620	−0.85619	0.00001
0.30000	−0.82251	−0.82245	0.00006
0.40000	−0.81120	−0.81096	0.00024
0.50000	−0.82031	−0.81959	0.00072
0.60000	−0.84820	−0.84643	0.00177

$$y''(x) = -2 - y', \qquad y''(0) = -2 - 1 = -3,$$
$$y'''(x) = -y'', \qquad y'''(0) = 3,$$
$$y^{(4)}(x) = -y''', \qquad y^{(4)}(0) = -3.$$

We then write our series solution for y, letting $x = h$ be the value at which we wish to determine y:

$$y(h) = -1 + 1.0h - 1.5h^2 + 0.5h^3 - 0.125h^4 + \text{error}.$$

Table 6.1 shows how the computed solutions compare to the analytical between $x = 0$ and $x = 0.6$. At the start, the Taylor-series solution agrees well, but beyond $x = 0.3$ they differ increasingly. More terms in the series would extend the range of good agreement.

The error of this computation is given by the next term in the series, evaluated at a point between 0 and x:

$$\text{Error} = (x^5/120)y^{(5)}(\xi), \qquad 0 < \xi < x.$$

We have used the so-called next-term rule before. How good is this estimate of the error at $x = 0.6$? The next term is $(3/120) * (0.6)^5 = 0.00194$, comparing well to the actual error of 0.00177.

We stated earlier that the analytical solution of the example differential equation can be obtained by "the application of standard methods." MATLAB can do this:

```
EDU>> dsolve('Dy = -2*x - y', 'y(0) = -1', 'x')
     ans =
        -2*x + 2 - 3*exp(-x)
```

which is the same as the above with terms in a different order.

Maple can get the Taylor-series solution:

```
> deq := diff(y(x),x) = -2*x - y(x):
> dsolve ({deq, y(0) = -1), y(x), series);
```

$$y(x) = -1 + x - \frac{3}{2}x^2 + \frac{1}{2}x^3 - \frac{1}{8}x^4 + \frac{1}{40}x^5 + O(x^6)$$

which is the series of order 6 and the error order.

When the function that defines $y'(x)$ is not as simple as this, getting the successive derivatives is not as easy. Consider

$$y'(x) = \frac{x}{(y - x^2)}.$$

You will find that the successive derivatives get very messy.

Even though computers are not readily programmed to produce these higher derivatives, computer algebra systems like Maple and *Mathematica* do have the capabilities that we need.

There is another approach—*automatic differentiation.* This is different from the symbolic differentiation that computer algebra systems use. It produces machine code that finds values of the derivatives when dy/dx is defined through a *code list.*

We will not give a thorough explanation, only an example, but L. R. Rall (1981) and Corliss and Chang (1982) are good sources for more information. Here is our example:

Solve $y' = f(x, y) = \dfrac{x}{(y - x^2)}$ using automatic differentiation with $y(0) = 1$.

To do this, we first create a code list, which is just a name for a sequence of statements that define dy/dx, with only a single operation on each line:

```
T1 = x*x
T2 = y - T1
dy/dx = x/T2    [which is f(x, y)].
```

We will use a simplified notation for the terms of the Taylor series:

$$(y)_k = \left(\frac{1}{k!}\right)\left[\frac{d^k y}{dx^k}\right], \qquad k = 0, \ldots, n.$$

And we will use $(x)_0 = x_0$. We then have $(y)_0 = y(x_0)$.

The software for automatic differentiation includes the standard rules for differentiation in recursive form, such as the derivatives of $(u + v)_k$, $(u - v)_k$, $(u * v)_k$, and $(u/v)_k$, plus the elementary functions, including sin, cos, ln, exp, and so on.

In our example, we have $(x)_0 = 0$, $(x)_1 = 1$ (because $dx/dx = 1$), so that $(x)_k = 0$ for all higher derivatives of x. From the initial condition, $(y)_0 = 1$ and from the expression for $y'(x)$, $(y)_1 = 0$. It is not hard to determine that $(y)_2 = 0.5$. The automatic differentiation software develops a recursion formula for the additional coefficients of the Taylor series. This formula is something like this:

$$(y)_k = \alpha_k \sum_{i=1}^{k-1} i(y)_i(y)_{k-1},$$

where the multiplier, α_k, is a complicated function of k.

Similar recursion formulas will be derived by the software for any differential equation that can be compiled into a code list, and these can have any initial condition.

For our example, all the odd-order terms are zero; the even-order terms are:

Order	0	2	4	6	8
Coefficient	1	$\frac{1}{2}$	$\frac{1}{8}$	$\frac{1}{48}$	$\frac{-1}{384}$

Using this in the Taylor series produces $y(0.1) = 1.0050125$, $y(0.2) = 1.0202013$.

The authors are especially grateful to Professor Ramon E. Moore of Ohio State University for calling our attention to this method for solving ordinary differential equations.

While getting the higher derivatives of $y' = x/(y - x^2)$ is awkward by hand, Maple has no trouble. If we want these up to the 22nd power of x, we must first reset the `Order` from its default value, then use the series option of `dsolve`.

```
> Order: = 22:
> deq: = diff(y(x), x) = x/(y(x) - x^2):
> dsolve({deq, y(0) = 1}, y(x), series);
```

$$y(x) = 1 + \frac{1}{2}x^2 + \frac{1}{8}x^4 + \frac{1}{48}x^6 - \frac{1}{384}x^8 - \frac{13}{3840}x^{10} - \frac{47}{46080}x^{12} + \frac{73}{645120}x^{14} + \frac{2447}{10321920}x^{16} + \frac{16811}{185794560}x^{18} - \frac{15551}{3715891200}x^{20} + O(x^{22})$$

The Taylor series is easily applied to a higher-order equation. For example, if we are given

$$y'' = 3 + x - y^2, \qquad y(0) = 1, \qquad y'(0) = -2,$$

we can find the derivative terms in the Taylor series as follows:

$y(0)$, and $y'(0)$ are given by the initial conditions.

$y''(0)$ comes from substitution into the differential equation from $y(0)$ and $y'(0)$.

$y'''(0)$ and higher derivatives are found by differentiating the equation for the previous order of derivative and substituting previously computed values.

6.2 The Euler Method and Its Modifications

The first truly numerical method that we discuss is the Euler method. We can solve the differential equation

$$dy/dx = f(x, y), \qquad y(x_0) = y_0,$$

by using just one term of the Taylor-series method:

$$y(x) = y(x_0) + y'(x_0)(x - x_0) + \text{error},$$
$$\text{error} = (h^2/2)y''(\xi) = O(h^2).$$

This is known as the Euler method. In effect, we project along the tangent line from the starting point, $y(x_0)$. If the increment to x, $(x - x_0) = h$, is small enough, the error will be small. Once we have y at $x_0 + h$, we can repeat to get more y-values:

$$y_{n+1} = y_n + hy'_n + O(h^2).* \tag{6.3}$$

The method is easy to program for we know the formula for $y'(x)$ and a starting value, $y_0 = y(x_0)$.

* This error is just the local error. Over many steps, the global error becomes $O(h)$.

Table 6.2

x_n	y_n	y'_n	hy'_n
0.0	−1.00000	1.00000	0.10000
0.1	−0.90000	0.70000	0.07000
0.2	−0.83000	0.43000	0.04300
0.3	−0.78700	0.18700	0.01870
0.4	−0.76830	−0.03170	

(Analytical answer is −0.81096, error is −0.04266.)

To see this in action, we apply it to the sample equation:

$$\frac{dy}{dx} = -2x - y, \qquad y(0) = -1,$$

where the computation can be done rather simply. It is convenient to arrange the work as in Table 6.2. Here we take $h = 0.1$.

Each of the y_n values is computed using Eq. (6.3), adding hy'_n and y_n of the previous line. Comparing the last result to the analytical answer $y(0.40) = -0.81096$, we see that there is only one-decimal-place accuracy, even though we have advanced the solution only four steps! To gain four-decimal-place accuracy, we must reduce the error by more than 400-fold. Because the global error is about proportional to h, we will need to reduce the step size about 426-fold, to <0.00024.

Improving the Simple Euler Method

The trouble with this most simple method is its lack of accuracy, requiring an extremely small step size. Figure 6.1 suggests how we might improve this method with just a little additional effort.

In the simple Euler method, we use the slope at the beginning of the interval, y'_n, to determine the increment to the function. This technique would be correct only if the function were linear. What we need instead is the correct average slope within the interval. This can be approximated by the mean of the slopes at both ends of the interval.

Suppose we use the arithmetic average of the slopes at the beginning and end of the interval to compute y_{n+1}:

$$y_{n+1} = y_n + h\,\frac{y'_n + y'_{n+1}}{2}. \tag{6.4}$$

This should give us an improved estimate for y at x_{n+1}. However, we are unable to employ Eq. (6.4) directly, because the derivative is a function of both x and y and we cannot evaluate y'_{n+1} with the true value of y_{n+1} unknown. The modified Euler method works around this problem by estimating or "predicting" a value of y_{n+1} by the simple Euler relation, Eq. (6.3). It then uses this value to compute y'_{n+1}, giving an improved estimate

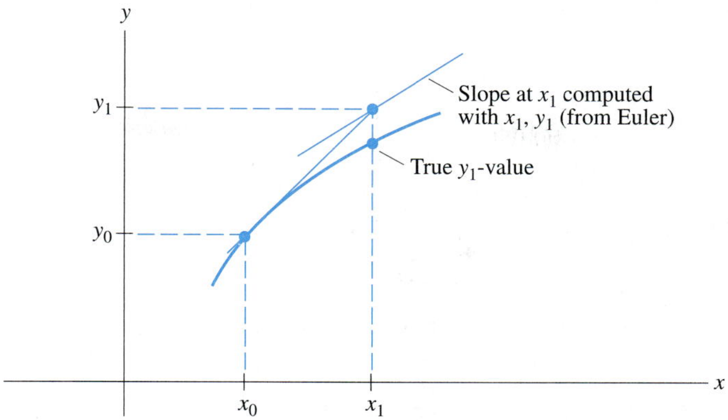

Figure 6.1

(a "corrected" value) for y_{n+1}. Because the "predicted" value for y_{n+1} is not usually very accurate, the value for y'_{n+1} that we compute from it is also inaccurate. One might be tempted to recorrect, using the first "corrected" value to recompute y'_{n+1} to get a better value for y'_{n+1} and repeat this until there is no significant change. However, this is less efficient than using a more powerful method, as we describe in the next section.

Table 6.3 shows the results of this modified Euler method on this same problem, $dy/dx = -2x - y$, $y(0) = -1$.

We can find the error of the modified Euler method by comparing it with the Taylor series:

$$y_{n+1} = y_n + y'_n h + \frac{1}{2} y''_n h^2 + \frac{y'''(\xi)}{6} h^3, \qquad x_n < \xi < x_n + h.$$

Replace the second derivative by the forward-difference approximation for y'', $(y'_{n+1} - y'_n)/h$, which has error of $O(h)$, and write the error term as $O(h^3)$:

$$y_{n+1} = y_n + h\, y'_n + \frac{1}{2}\left[\frac{y'_{n+1} - y'_n}{h} + O(h)\right] h^2 + O(h^3),$$

Table 6.3

x_n	y_n	hy'_n	$y_{n+1,p}$	$hy'_{n+1,p}$	hy'_{av}	$y_{n+1,c}$
0.0	−1.0000	0.1000	−0.9000	0.0700	0.0850	−0.9150
0.1	−0.9150	0.0715	−0.8435	0.0444	0.0579	−0.8571
0.2	−0.8571	0.0457	−0.8114	0.0211	0.0334	−0.8237
0.3	−0.8237	0.0224	−0.8013	0.0001	0.0112	−0.8124
0.4	−0.8124	0.0012	−0.8112	−0.0189	−0.0088	−0.8212
0.5	−0.8212					

[y (0.5) = −0.81959, the analytical value]

$$y_{n+1} = y_n + h\left(y'_n + \frac{1}{2}y'_{n+1} - \frac{1}{2}y'_n\right) + O(h^3),$$

$$y_{n+1} = y_n + h\left(\frac{y'_n + y'_{n+1}}{2}\right) + O(h^3).$$

This shows that the error of one step of the modified Euler method is $O(h^3)$. This is the local error. There is an accumulation of errors from step to step, so that the error over the whole range of application, the so-called global error, is $O(h^2)$. This seems intuitively reasonable, because the number of steps into which the interval is subdivided is proportional to $1/h$; hence the order of error is reduced to $O(h^2)$ on continuing the technique.

Another Way to Improve the Euler Method

The technique that we have called the modified Euler method tries to find a value for the average slope of y between x_n and $x_n + h$ by averaging the slopes at x_n and at x_{n+1}. There are other ways to do this. The *midpoint method* uses the slope at the midpoint of the interval as the average slope. It uses the simple Euler method to estimate y at $x + h/2$ and evaluates y' at the midpoint with this. For some derivative functions this is better than modified Euler and for others it is less accurate; for the example used to construct Tables 6.2 and 6.3, this midpoint method gives precisely the same results.

Propagation of Errors

The errors that we have mentioned for these Euler methods are the truncation errors, those due to truncating the Taylor series on which they are based. There are other errors; round off in particular will enter. It is important to understand that errors made early in the process will also affect the later computations—the early error will be propagated. The analysis of propagated error is not easy. We do it here only for the simple Euler method—this will indicate how such analysis can be accomplished.

We consider the first-order equation $dy/dx = f(x, y)$, $y(x_0) = y_0$. Let

$$Y_n = \text{calculated value at } x_n,$$
$$y_n = \text{true value at } x_n,$$
$$e_n = y_n - Y_n = \text{error in } Y_n;\ y_n = Y_n + e_n.$$

By the Euler algorithm,

$$Y_{n+1} = Y_n + hf(x_n, Y_n).$$

By Taylor series,

$$y_{n+1} = y_n + hf(x_n, y_n) + \frac{h^2}{2}y''(\xi_n), \qquad x_n < \xi_n < x_n + h,$$

$$e_{n+1} = y_{n+1} - Y_{n+1} = y_n - Y_n + h[f(x_n, y_n) - f(x_n, Y_n)] + \frac{h^2}{2}y''(\xi_n) \tag{6.5}$$

$$= e_n + h \frac{f(x_n, y_n) - f(x_n, Y_n)}{y_n - Y_n}(y_n - Y_n) + \frac{h^2}{2} y''(\xi_n)$$

$$= e_n + hf_y(x_n, \eta_n)e_n + \frac{h^2}{2}y''(\xi_n), \qquad \eta_n \text{ between } y_n, Y_n.$$

In Eq. (6.5), we have used the mean-value theorem, imposing continuity and existence conditions on $f(x, y)$ and f_y. We suppose, in addition, that the magnitude of f_y is bounded by the positive constant K in the region of x, y-space in which we are interested.* Hence,

$$e_{n+1} \leq (1 + hK)e_n + \frac{1}{2}h^2y''(\xi_n). \tag{6.6}$$

Here, $y(x_0) = y_0$ is our initial condition, which we assume free of error. Because $Y_0 = y_0$, $e_0 = 0$:

$$e_1 \leq (1 + hK)e_0 + \frac{1}{2}h^2y''(\xi_0) = \frac{1}{2}h^2y''(\xi_0),$$

$$e_2 \leq (1 + hK)\left[\frac{1}{2}h^2y''(\xi_0)\right] + \frac{1}{2}h^2y''(\xi_1) = \frac{1}{2}h^2[(1 + hK)y''(\xi_0) + y''(\xi_1)].$$

Similarly,

$$e_3 \leq \frac{1}{2}h^2[(1 + hK)^2y''(\xi_0) + (1 + hK)y''(\xi_1) + y''(\xi_2)],$$

$$e_n \leq \frac{1}{2}h^2[(1 + hK)^{n-1}y''(\xi_0) + (1 + hK)^{n-2}y''(\xi_1) + \cdots + y''(\xi_{n-1})].$$

If $f_y \leq K$ is positive, the truncation error at every step is propagated to every later step after being amplified by the factor $(1 + hf_y)$ each time. Note that as $h \to 0$, the error at any point is just the sum of all the previous errors. If the f_y are negative and of magnitude such that $|hf_y| < 2$, the errors are propagated with diminishing effect.

We now show that the accumulated error after n steps is $O(h)$; that is, the global error of the simple Euler method is $O(h)$. We assume, in addition, that y'' is bounded, $|y''(x)| < M$, $M > 0$. After taking absolute values, Eq. (6.6) becomes

$$|e_{n+1}| \leq (1 + hK)|e_n| + \frac{1}{2}h^2M.$$

Now we compare to the first-order difference equation:

$$Z_{n+1} = (1 + hK)Z_n + \frac{1}{2}h^2M, \qquad Z_0 = 0. \tag{6.7}$$

* This is essentially the same as the Lipschitz condition, which will guarantee existence and uniqueness of a solution.

Obviously the values of Z_n are at least equal to the magnitudes of $|e_n|$. The solution to Eq. (6.7) is (check by direct substitution)

$$Z_n = \frac{hM}{2K}(1 + hK)^n - \frac{hM}{2K}.$$

The Maclaurin expansion of e^{hk} is

$$e^{hK} = 1 + hK + \frac{(hK)^2}{2} + \frac{(hK)^3}{6} + \cdots,$$

so that

$$1 + hK < e^{hK} \quad (K > 0),$$

$$\begin{aligned} Z_n < \frac{hM}{2k}(e^{hK})^n - \frac{hM}{2K} &= \frac{hM}{2K}(e^{nhK} - 1) \\ &= \frac{hM}{2K}(e^{(x_n - x_0)K} - 1) = O(h). \end{aligned}$$

It follows that the global error e_n is $O(h)$. (This result can be derived without difference equations.)

6.3 Runge–Kutta Methods

The simple Euler method comes from using just one term from the Taylor series for $y(x)$ expanded about $x = x_0$. The modified Euler method can be derived from using two terms:

$$y(x_0 + h) = y(x_0) + y'(x_0) * h + y''(x_0) * h^2/2.$$

If we replace the second derivative with a backward-difference approximation,

$$\begin{aligned} y(x_0 + h) &= y(x_0) + y'(x_0) * h + [(y'(x_0 + h) - y'(x_0))/h] * h^2/2 \\ &= y(x_0) + \frac{y'(x_0) + y'(x_0 + h)}{2} h, \end{aligned}$$

we get the formula for the modified method. What if we use more terms of the Taylor series? Two German mathematicians, Runge and Kutta, developed algorithms from using more than two terms of the series. We will consider only fourth- and fifth-order formulas. The modified Euler method is a second-order Runge–Kutta method.

Second-order Runge–Kutta methods are obtained by using a weighted average of two increments to $y(x_0)$, k_1 and k_2. For the equation $dy/dx = f(x, y)$:

$$\begin{aligned} y_{n+1} &= y_n + ak_1 + bk_2, \\ k_1 &= hf(x_n, y_n), \\ k_2 &= hf(x_n + \alpha h, y_n + \beta k_1). \end{aligned} \tag{6.8}$$

We can think of the values k_1 and k_2 as estimates of the change in y when x advances by h, because they are the product of the change in x and a value for the slope of the curve, dy/dx. The Runge–Kutta methods always use the simple Euler estimate as the first estimate of Δy; the other estimate is taken with x and y stepped up by the fractions α and β of h and of the earlier estimate of Δy, k_1. Our problem is to devise a scheme of choosing the four parameters, a, b, α, β. We do so by making Eq. (6.8) agree as well as possible with the Taylor-series expansion, in which the y-derivatives are written in terms of f, from $dy/dx = f(x, y)$,

$$y_{n+1} = y_n + hf(x_n, y_n) + \frac{h^2}{2} f'(x_n, y_n) + \cdots.$$

An equivalent form, because $df/dx = f_x + f_y\, dy/dx = f_x + f_y f$, is

$$y_{n+1} = y_n + hf_n + h^2\left(\frac{1}{2}f_x + \frac{1}{2}f_y f\right)_n. \tag{6.9}$$

[All the derivatives in Eq. (6.9) are calculated at the point (x_n, y_n).] We now rewrite Eq. (6.9) by substituting the definitions of k_1 and k_2:

$$y_{n+1} = y_n + ahf(x_n, y_n) + bhf[x_n + \alpha h, y_n + \beta hf(x_n, y_n)]. \tag{6.10}$$

To make the last term of Eq. (6.10) comparable to Eq. (6.9), we expand $f(x, y)$ in a Taylor series in terms of x_n, y_n, remembering that f is a function of two variables,* retaining only first derivative terms:

$$f[x_n + \alpha h, y_n + \beta hf(x_n, y_n)] \approx (f + f_x \alpha h + f_y \beta hf)_n. \tag{6.11}$$

On the right side of both Eqs. (6.9) and (6.11) f and its partial derivatives are all to be evaluated at (x_n, y_n).

Substituting from Eq. (6.11) into Eq. (610), we have

$$y_{n+1} = y_n + ahf_n + bh(f + f_x \alpha h + f_y \beta hf)_n,$$

or, rearranging,

$$y_{n+1} = y_n + (a + b)hf_n + h^2(\alpha b f_x + \beta b f_y f)_n. \tag{6.12}$$

Equation (6.12) will be identical to Eq. (6.9) if

$$a + b = 1, \qquad \alpha b = \frac{1}{2}, \qquad \beta b = \frac{1}{2}.$$

---

* Appendix A will remind readers of this expansion.

Note that only three equations need to be satisfied by the four unknowns. We can choose one value arbitrarily (with minor restrictions); hence, we have a set of second-order methods.

One choice can be $a = 0$, $b = 1$; $\alpha = 1/2$, $\beta = 1/2$. This gives the midpoint method. Another choice can be $a = 1/2$, $b = 1/2$; $\alpha = 1$, $\beta = 1$, which give the modified Euler. Still another possibility is $a = 1/3$, $b = 2/3$, $\alpha = 3/4$, $\beta = 3/4$; this is said to give a minimum bound to the error. All of these are special cases of second-order Runge–Kutta methods.

Fourth-order Runge–Kutta methods are most widely used and are derived in similar fashion. Greater complexity results from having to compare terms through h^4, and this gives a set of 11 equations in 13 unknowns. The set of 11 equations can be solved with 2 unknowns being chosen arbitrarily. The most commonly used set of values leads to the procedure:

$$
\begin{aligned}
y_{n+1} &= y_n + \frac{1}{6}(k_1 + 2k_2 + 2k_3 + k_4), \\
k_1 &= hf(x_n, y_n), \\
k_2 &= hf\left(x_n + \frac{1}{2}h, y_n + \frac{1}{2}k_1\right), \\
k_3 &= hf\left(x_n + \frac{1}{2}h, y_n + \frac{1}{2}k_2\right), \\
k_4 &= hf(x_n + h, y_n + k_3).
\end{aligned}
\tag{6.13}
$$

Using Eqs. (6.13) to apply the Runge–Kutta fourth order to the problem, $dy/dx = -2x - y$, $y(0) = -1$ with $h = 0.1$, we obtain the results shown in Table 6.4. The results here are very impressive compared to those given in Table 6.1, where we computed the values using the terms of the Taylor series up to the h^4 term. Table 6.4 agrees to five decimals with the analytical result—illustrating a further gain in accuracy with less effort than with the Taylor-series method of Section 6.1—and it certainly is better than the Euler or modified Euler methods.

Table 6.4

x	y	k_1	k_2	k_3	k_4	k_{avg}
0.0	−1.00000	0.1000	0.0850	0.0858	0.0714	0.0855
0.1	−0.91451	0.0715	0.0579	0.0586	0.0456	0.0584
0.2	−0.85619	0.0456	0.0333	0.0340	0.0222	0.0338
0.3	−0.82246	0.0222	0.0111	0.0117	0.0011	0.0115
0.4	−0.81096	0.0011	−0.0090	−0.0085	−0.0181	−0.0086
0.5	−0.81959	−0.0180	−0.0271	−0.0267	−0.0354	−0.0268
0.6	−0.84644					

(The analytical value of $y(0.6)$ is -0.846434.)

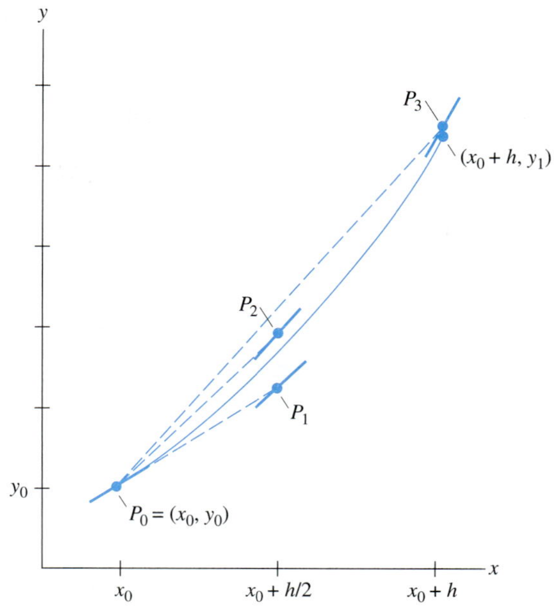

Figure 6.2

Figure 6.2 illustrates the four slope values that are combined in the four *k's* of the Runge–Kutta method.

The local error term for the fourth-order Runge–Kutta method is $O(h^5)$; the global error would be $O(h^4)$. It is computationally more efficient than the modified Euler method because, although four evaluations of the function are required per step rather than two, the steps can be manyfold larger for the same accuracy. The Runge–Kutta techniques have been very popular, especially the fourth-order method just presented. Because going from second to fourth order was so beneficial, we may wonder whether we should use a still higher order of formula. Higher-order (fifth, sixth, and so on) Runge–Kutta formulas have been developed and can be used to advantage in determining a suitable size for h, as we will see. Still, Runge–Kutta methods of order greater than 4 have the disadvantage that the number of function evaluations that are required is greater than the order of the method, while Runge–Kutta methods of order-4 or less require the same number of evaluations as the order.

How Do We Know If the Step-Size Is Right?

One way to determine whether the Runge–Kutta values are sufficiently accurate is to recompute the value at the end of each interval with the step size halved. If only a slight change in the value of y_{n+1} occurs, the results are accepted; if not, the step must be halved again until the results are satisfactory. This procedure is very expensive, however. For instance, to implement Eq. (6.13) this way, we would need an additional seven function evaluations to determine the accuracy of our y_{n+1}. The best case then would require 4 + 6 = 10 function evaluations to go from (x_n, y_n) to (x_{n+1}, y_{n+1}).

A different approach uses two Runge–Kutta methods of different orders. For instance, we could use one fourth-order and one fifth-order method to move from (x_n, y_n) to (x_{n+1}, y_{n+1}). We would then compare our results at y_{n+1}. The Runge–Kutta–Fehlberg method, now one of the most popular of these methods, does just this. Only six functional evaluations (versus ten) are required, and we also have an estimate of the error (the difference of the two y's at $x = x_{n+1}$):

An Algorithm for the Runge–Kutta–Fehlberg Method

Given $y' = f(x, y)$ and $y(x_n) = y_n$, to compute $y(x_{n+1}) = y_{n+1}$ where $x_{n+1} = x_x + h$, evaluate:

$$k_1 = h \cdot f(x_n, y_n),$$

$$k_2 = h \cdot f\left(x_n + \frac{h}{4}, y_n + \frac{k_1}{4}\right),$$

$$k_3 = h \cdot f\left(x_n + \frac{3h}{8}, y_n + \frac{3k_1}{32} + \frac{9k_2}{32}\right),$$

$$k_4 = h \cdot f\left(x_n + \frac{12h}{13}, y_n + \frac{1932k_1}{2197} - \frac{7200k_2}{2197} + \frac{7296k_3}{2197}\right),$$

$$k_5 = h \cdot f\left(x_n + h, y_n + \frac{439k_1}{216} - 8k_2 + \frac{3680k_3}{513} - \frac{845k_4}{4104}\right),$$

$$k_6 = h \cdot f\left(x_n + \frac{h}{2}, y_n - \frac{8k_1}{27} + 2k_2 - \frac{3544k_3}{2565} + \frac{1859k_4}{4104} - \frac{11k_5}{40}\right);$$

$$\hat{y}_{n+1} = y_n + \left(\frac{25k_1}{216} + \frac{1408k_3}{2565} + \frac{2197k_4}{4104} - \frac{k_5}{5}\right), \text{ with global error } O(h^4),$$

$$y_{n+1} = y_n + \left(\frac{16k_1}{135} + \frac{6656k_3}{12825} + \frac{28561k_4}{56430} - \frac{9k_5}{50} + \frac{2k_6}{55}\right),$$

with global error $O(h^5)$;

$$\text{Error, } E = \frac{k_1}{360} - \frac{128k_3}{4275} - \frac{2197k_4}{75240} + \frac{k_5}{50} + \frac{2k_6}{55}.$$

The basis for the Runge–Kutta–Fehlberg scheme is to compute two Runge–Kutta estimates for the new value of y_{n+1} but of different orders of errors. Thus, instead of comparing estimates of y_{n+1} for h and $h/2$, we compare the estimates $\hat{y}_{n+1}$ and y_{n+1} using fourth- and fifth-order (global) Runge–Kutta formulas. Moreover, both equations make use of the same k's, so only six function evaluations are needed versus the previous 11. In addition, one can increase or decrease h depending on the value of the estimated error. As our estimate for the new y_{n+1}, we use the fifth-order (global) estimate.

As an example, we once more solve $dy/dx = -2x - y$, $y(0) = -1$ with $h = 0.1$, using the Runge–Kutta–Fehlberg method:

$$k_1 = 0.1,$$
$$k_2 = 0.0925000,$$
$$k_3 = 0.0889609,$$
$$k_4 = 0.0735157,$$
$$k_5 = 0.0713736,$$
$$k_6 = 0.0853872,$$
$$\hat{y}_1 = -0.914512212, \qquad y_1 = -0.914512251, \qquad \text{Error, } E = -0.000000040.$$

The exact value is $y(0.1) = -0.914512254$. Thus, on the first step, y_1 agrees with the exact answer to eight decimal places with only two additional function evaluations. Moreover, we have the value E to adjust our step size for the next iteration. Of course, we would use the more accurate y_{n+1} for the next step. This algorithm is well documented and implemented in the FORTRAN program, RKF45, of Forsythe, Malcolm, and Moler (1977). MATLAB has two numerical procedures `ode45` and `ode23`. Maple has `rkf45` in its arsenal to get the numerical solution to differential equations.

A summary and comparison of the numerical methods we have studied for solving $y' = f(x, y)$ is presented in Table 6.5.

To see empirically that the global errors of Table 6.5 hold, again consider the example $dy/dx = -2x - y$, $y(0) = -1$. Table 6.6 shows how the errors of $y(0.4)$ decrease as h is halved. The table shows the ratios of errors of successive calculations.

In Table 6.6, we obtain the second row in this way: For a step size of $h = 0.2$, we compute the errors in the values for y at $x = 0.4$ using the three methods indicated at the top of columns two through four. We write down the values of the differences between the computed value and the analytical value. The last three columns represent the ratio between the previous error (larger step size h) and the current. For instance, the 3.3 in the second row is the ratio of 2.11E-01/9.10E-2 for the errors from Euler's method for $h = 0.4$ and $h = 0.2$. We do the same for the modified Euler method and the Runge–Kutta fourth-order method in columns six and seven. We see that as h gets smaller, the last three columns approach the

Table 6.5

Method	Estimate of slope	Global error	Local error	Evaluations per step
Euler	Initial value	$O(h)$	$O(h^2)$	1
Modified Euler	Average, initial and final	$O(h^2)$	$O(h^3)$	2
Midpoint	Midpoint of interval	$O(h^2)$	$O(h^3)$	2
Runge–Kutta (fourth-order)	Weighted average, four values	$O(h^4)$	$O(h^5)$	4
Runge–Kutta–Fehlberg	Weighted average, six values	$O(h^5)$	$O(h^6)$	6

Table 6.6

	Error in value computed at $x = 0.4$			Ratios of successive errors		
h	Euler	Modified Euler	Runge–Kutta 4th	Euler	Modified Euler	Runge–Kutta 4th
0.4000	2.11E-01	2.90E-02	2.40E-04			
0.2000	9.10E-02	6.42E-03	1.27E-05	2.3	4.5	18.9
0.1000	4.27E-02	1.44E-03	7.29E-07	2.1	4.5	17.4
0.0500	2.07E-02	3.48E-04	4.37E-08	2.1	4.1	16.7
0.0250	1.02E-02	8.54E-05	2.76E-09	2.0	4.1	15.8
0.0125	5.06E-03	2.11E-05	1.65E-10	2.0	4.0	16.7

ratios of 2.0, 4.0, and 16.0. This is what we expect, because these three methods are, respectively, $O(h)$, $O(h^2)$, and $O(h^4)$ and because at each stage the step size is halved.

We end this section by showing the Runge–Kutta–Merson method, another fourth-order method even though five different k's must be computed. It can be seen from the formula that the order is given, not by the number of k's, but by the global error.

$$
\begin{aligned}
k_1 &= h \cdot f(x_n, y_n),\\
k_2 &= h \cdot f\left(x_n + \frac{h}{3}, y_n + \frac{k_1}{3}\right),\\
k_3 &= h \cdot f\left(x_n + \frac{h}{3}, y_n + \frac{k_1}{6} + \frac{k_2}{6}\right),\\
k_4 &= h \cdot f\left(x_n + \frac{h}{2}, y_n + \frac{k_1}{8} + \frac{3k_3}{8}\right),\\
k_5 &= h \cdot f\left(x_n + h, y_n + \frac{k_1}{2} - \frac{3k_3}{2} + 2k_4\right);\\
y_{n+1} &= y_n + \frac{(k_1 + 4k_4 + k_5)}{6} + O(h^5);\\
\text{Error, } E &= \frac{1}{30}(2k_1 - 9k_3 + 8k_4 - k_5).
\end{aligned}
$$

As we have already indicated, there are methods that use Runge–Kutta formulas of orders 5, 6, and higher. In fact, the IMSL routine DVERK uses formulas or orders 5 and 6 that were developed by J. H. Verner. In this case, the method uses eight function evaluations. Maple has an option in its procedure for solving differential equations that is called `dverk78`.

Although the Runge–Kutta method has been very popular in the past, it has its limitations in solving certain types of differential equations. However, for a large class of problems the methods presented in this section produce some very stunning results. Also the technique introduced by Fehlberg in comparing two different orders rather than halving step sizes increases the efficiency of the Runge–Kutta methods.

The methods so far discussed are called single-step methods. They use only the information at (x_n, y_n) to get to (x_{n+1}, y_{n+1}). In the next sections, we examine methods that utilize past information from previous points to get (x_{n+1}, y_{n+1}).

Here is the MATLAB solution to our sample problem through its ode45 command, which uses the RKF method with the step size automatically adjusted. We first create an M-file that defines the derivative function:

```
function dydx = deq1(x,y)
dydx = -2*x - y;
```

Now we use the 'ode45' command to get the solution between $x = 0$ and $x = 0.6$ using the RKF method:

```
EDU>> [x,y] = ode45(@deq1, [0,.6], -1)
```

and MATLAB displays a list of the x-values used in the computations followed by the corresponding y-values. Though not apparent here, the procedure uses automatic step-size adjustment. We show only a portion of the whole output; the default of 40 intervals is used. We show the y-values side by side with the x-values. (The solution is much more accurate than four digits.)

```
x =              y =
      0             -1.0000
   0.0150           -0.9853
   0.0300           -0.9713
   0.0450           -0.9580
   0.0600           -0.9453
      .                .
      .                .
      .                .
      .                .
   0.5100           -0.8215
   0.5250           -0.8247
   0.5400           -0.8282
   0.5550           -0.8322
   0.5700           -0.8366
   0.5850           -0.8413
   0.6000           -0.8464
```

6.4 Multistep Methods

Runge–Kutta-type methods (which include Euler and modified Euler as special cases) are called single-step methods because they use only the information from the last step computed. In this, they have the ability to perform the next step with a different step size and are ideal for beginning the solution where only the initial conditions are available. After the solution has begun, however, there is additional information available about the function (and its derivative) if we are wise enough to retain it in the memory of the computer. A multistep method is one that takes advantage of this fact.

The principle behind a multistep method is to utilize the past values of y and/or y' to construct a polynomial that approximates the derivative function, and extrapolate this into the next interval. Most methods use equispaced past values to make the construction of the polynomial easy. The Adams method is typical.* The number of past points that are used sets the degree of the polynomial and is therefore responsible for the truncation error. The order of the method is equal to the power of h in the global error term of the formula, which is also equal to one more than the degree of the polynomial.

To derive the relations for the Adams method, we write the differential equation $dy/dx = f(x, y)$ in the form

$$dy = f(x, y)\, dx,$$

and we integrate between x_n and x_{n+1}:

$$\int_{x_n}^{x_{n+1}} dy = y_{n+1} - y_n = \int_{x_n}^{x_{n+1}} f(x, y)\, dx.$$

To integrate the term on the right, we approximate $f(x, y)$ as a polynomial in x, deriving this by making it fit at several past points. If we use three past points, the approximating polynomial will be a quadratic. If we use four points, it will be a cubic. The more points we use, the better the accuracy (until round off interferes, of course).

You saw in Chapter 3 how interpolating polynomials can be developed. *Mathematica* can do this for us with its `Interpolating Polynomial` function. With this, we can get a quadratic approximation:

$$f(x, y) = \frac{1}{2} h^2(f_n - 2f_{n-1} + f_{n-2})x^2 + \frac{1}{2} h(3f_n - 4f_{n-1} + f_{n-2})x + f_n.$$

Now we again use *Mathematica* to integrate between the limits of $x = x_n$ and $x = x_{n+1}$. The result is a formula for the increment in y:

$$y_{n+1} - y_n = \frac{h}{12}(23f_n - 16f_{n-1} + 5f_{n-2}),$$

and we have the formula to advance y:

$$y_{n+1} - y_n = \frac{h}{12}[23f_n - 16f_{n-1} + 5f_{n-2}] + O(h^4). \qquad \textbf{(6.14)}$$

* This is often called the Adams–Bashford method.

Observe that Eq. (6.14) resembles the single-step formulas of the previous sections in that the increment to y is a weighted sum of the derivatives times the step size, but differs in that past values are used rather than estimates in the forward direction.

EXAMPLE 6.1 We illustrate the use of Eq. (6.14) to calculate $y(0.6)$ for $dy/dx = -2x - y$, $y(0) = -1$. We compute good values for $y(0.2)$ and $y(0.4)$ using a single-step method. In this case we obtain these values using the Runge–Kutta–Fehlberg method with $h = 0.2$. These values are given in Table 6.7.

Then, from Eq. (6.14), we have

$$y(0.6) = -0.81096 + \frac{0.2}{12}[23(0.01096) - 16(0.45619) + 5(1.0)]$$
$$= -0.84508.$$

Comparing our result with the exact solution (−0.84643), we find that the computed value has an error of 0.00135. We can reduce the size of the error by doing the calculations with a smaller step size of 0.1. We use the fifth-order values of the Runge–Kutta–Fehlberg method once again to obtain the values in Table 6.8.

Using Eq. (6.14) again with the values for $f(x, y)$ at $x = 0.3$, $x = 0.4$, $x = 0.5$ from Table 6.8, we recompute $y(0.6)$:

$$y(0.6) = -0.81959 + \frac{0.1}{12}[23(-0.18041) - 16(0.01096) + 5(0.22245)]$$
$$= -0.84636,$$

which has an error of 0.00007.

Adams Fourth-Order Formula

Equation (6.14) is a third-order formula that uses y-values at three past points, x_n, x_{n-1}, and x_{n-2}, to estimate y_{n+1}. Using four past points is equivalent to integrating a cubic interpolating polynomial through four past points. We can use the method of undetermined coefficients to obtain this.

Table 6.7

x	y	y, analytical	$f(x, y)$
0.0	−1.0000000	−1.0000000	1.0000000
0.2	−0.8561921	−0.8561923	0.4561921
0.4	−0.8109599	−0.8109601	0.0109599

Table 6.8

x	y	y, analytical	$f(x, y)$
0.0	−1.00000	−1.00000	1.00000
0.1	−0.91451	−0.91451	0.71451
0.2	−0.85619	−0.85619	0.45619
0.3	−0.82245	−0.82245	0.22245
0.4	−0.81096	−0.81096	0.01096
0.5	−0.81959	−0.81959	−0.18041

We desire a formula of the form

$$\int_{x_n}^{x_{n+1}} f(x)\,dx = c_0 f_{n-3} + c_1 f_{n-2} + c_2 f_{n-1} + c_3 f_n.$$

With four constants, we can make the formula exact when $f(x)$ is any polynomial of degree-3 or less. Accordingly, we replace $f(x)$ successively by x^3, x^2, x, and 1 to evaluate the coefficients.

It is apparent that the formula must be independent of the actual x-values. To simplify the equations, let us shift the origin to the point $x = x_n$; our integral is then taken over the interval from 0 to h, where $h = x_{n+1} - x_n$:

$$\int_0^h f(x)\,dx = c_0 f(-3h) + c_1 f(-2h) + c_2 f(-h) + c_3 f(0).$$

Carrying out the computations by replacing $f(x)$ with the particular polynomials, we have

$$\frac{h^4}{4} = c_0(-3h)^3 + c_1(-2h)^3 + c_2(-h)^3 + c_3(0),$$

$$\frac{h^3}{3} = c_0(-3h)^2 + c_1(-2h)^2 + c_2(-h)^2 + c_3(0),$$

$$\frac{h^2}{2} = c_0(-3h) + c_1(-2h) + c_2(-h) + c_3(0),$$

$$h = c_0(1) + c_1(1) + c_2(1) + c_3(1).$$

The linear system we are to solve is

$$\begin{bmatrix} -27 & -8 & -1 & 0 \\ 9 & 4 & 1 & 0 \\ -3 & -2 & -1 & 0 \\ 1 & 1 & 1 & 1 \end{bmatrix} \begin{bmatrix} c_0 \\ c_1 \\ c_2 \\ c_4 \end{bmatrix} = \begin{bmatrix} 1/4 \\ 1/3 \\ 1/2 \\ 1 \end{bmatrix}$$

whose solution is

$$c_0 = -9/24,\ c_1 = 37/24,\ c_2 = -59/24,\ c_3 = 55/24.$$

The fourth-order Adams formula is then

Table 6.9

Number of points used	Estimate of $y(0.6)$	Error ($h = 0.1$)
3	−0.8463626	0.000072
4	−0.8464420	0.000007

$$y_{n+1} = y_n + \frac{h}{24}[55f_n - 59f_{n-1} + 37f_{n-2} - 9f_{n-3}] + O(h^5). \qquad (6.15)$$

If we repeat Example 6.1 with this fourth-order formula, taking values at $x = 0.2, 0.3, 0.4$, and 0.5, we compute:

$$\begin{aligned} y(0.6) &= -0.81959 + \frac{0.1}{24}[55(-0.18041) - 59(0.01096) \\ &\quad + 37(0.22245) - 9(0.45619)] \\ &= -0.84644. \end{aligned}$$

The error of this computation has been reduced to 0.00001. We summarize the results of these two formulas in Table 6.9.

The Error Term We get the error term for the fourth-order Adams formula by integrating the error of the cubic interpolating polynomial. This turns out to be

$$\text{Error} = \frac{251}{720} h^5 y^{(5)}(\xi),$$

which is $O(h^5)$ as we have used before.

The Adams–Moulton Method

An improvement over the Adams method is the Adams–Moulton method. It uses the Adams method as a *predictor formula,* then applies a *corrector formula,* based on constructing another cubic interpolating formula through four points—the one obtained with the predictor formula and three previously computed points. (You may want to use undetermined coefficients to confirm this.)

Predictor:

$$y_{n+1} = y_n + \frac{h}{24}(55f_n - 59f_{n-1} + 37f_{n-2} - 9f_{n-3}) + \frac{251}{720} h^5 y^{(5)}(\xi_1). \qquad (6.16)$$

Corrector:

$$y_{n+1} = y_n + \frac{h}{24}(9f_{n+1} + 19f_n - 5f_{n-1} + f_{n-2}) - \frac{19}{720}h^5 y^{(5)}(\xi_2). \quad \textbf{(6.17)}$$

We illustrate the Adams–Moulton method using our earlier example, $dy/dx = -2x - y$, $y(0) = -1$. Using Eqs. (6.16) and (6.17) we construct Table 6.10. Here is how the entries in the table were obtained. By the predictor formula of (6.16), we get

$$\begin{aligned} y(0.4) &= -0.8224547 + \frac{0.1}{24}[55(0.2224547) - 59(0.4561923) \\ &\quad + 37(0.7145123) - 9(1.0)] \\ &= -0.8109687. \end{aligned}$$

Then $f(0.4, -0.8109687)$ is computed, to get 0.0109688, and we use the corrector formula of Eq. (6.17) to get

$$\begin{aligned} y(0.4) &= -0.8224547 + \frac{0.1}{24}[9(0.0109688) + 19(0.2224547) \\ &\quad -5(0.4561923) + 0.7145123] \\ &= -0.8109652. \end{aligned}$$

The computations are continued in the same manner to get $y(0.5)$. The corrected value almost agrees to five decimals with the predicted value. Comparing error terms of Eqs. (6.16) and (6.17) and assuming that the two fifth-derivative values are equal, we see that the true value should lie between the predicted and corrected values, with the error in the corrected value being about

$$\frac{19}{251 + 19} \quad \text{or} \quad \frac{1}{14.2}$$

times the difference between the predicted and corrected values. A frequently used criterion for accuracy of the Adams–Moulton method with four starting values is that the corrected value is not in error by more than 1 in the last place if the difference between

Table 6.10

x	y	$f(x, y)$	
0.0	−1.0000000	1.0000000	
0.1	−0.9145122	0.7145123	
0.2	−0.8561923	0.4561923	
0.3	−0.8224547	0.2224547	
0.4	(−0.8109687) predicted		
	(−0.8109652) corrected		(−0.8109601 analytical)
0.5	(−0.8195978) predicted		
	(−0.8195905) corrected		(−0.8195920 analytical)

predicted and corrected values is less than 14 in the last decimal place. If this degree of accuracy is not met, we know that h is too large.

Changing the Step Size

When the predicted and corrected values agree to as many decimals as the desired accuracy, we can save computational effort by increasing the step size. We can conveniently double the step size, after we have seven equispaced values, by omitting every second one. When the difference between predicted and corrected values reaches or exceeds the accuracy criterion, we should decrease step size. If we interpolate two additional y-values with a fourth-degree polynomial, where the error will be $O(h^5)$, consistent with the rest of our work, we can readily halve the step size. Convenient formulas for this are

$$y_{n-1/2} = \frac{1}{128}\,[35y_n + 140y_{n-1} - 70y_{n-2} + 28y_{n-3} - 5y_{n-4}],$$

$$y_{n-3/2} = \frac{1}{128}\,[-5y_n + 60y_{n-1} + 90y_{n-2} - 20y_{n-3} + 3y_{n-4}].$$

Use of these values with y_n, y_{n-1} gives four values of the function at intervals of $\Delta x = h/2$.

The efficiency of Adams–Moulton is about twice that of the Runge–Kutta–Fehlberg and Runge–Kutta methods. Only two function evaluations are needed per step for the former method, whereas six or four are required with the single-step alternatives. All have similar error terms. Change of step size with the multistep methods is considerably more awkward, however.

Stability Considerations

In getting the solution to a differential equation, one must always worry whether the method is *stable*. In a stable method, early errors (due to the imprecision of the method or to an initial value that is slightly incorrect) are damped out as the computations proceed; they do not grow without bound. The opposite is true for an *unstable* method.

In the discussion of the Euler method in Section 6.2, we showed the conditions for stability. This was not a simple task. It is easier to see if a method is stable or unstable by testing it with certain kinds of derivative functions, $y'(x) = f(x, y)$.

Consider this equation:

$$dy/dx = f(x, y) = -2y + 2, \qquad y(0) = -1,$$

whose analytical solution is $y(x) = 1 - 2e^{-2x}$. The curve for $y(x)$ is smooth, starting at $y = -1$, proceeding rapidly upward with a slope of 4, crossing the x-axis at about $x = 0.35$, and approaching the asymptote of $y = 1$ as x increases. By $x = 3$, the y-value is within 0.5% of its limiting values.

Suppose that we use a very simple multistep formula:

$$y_{n+1} = y_{n-1} + 2hf(x_n, y_n), \tag{6.18}$$

which has a truncation error of $(1/6)h^3y'''(\xi)$, smaller than for the simple Euler method, which is $(1/2)h^2y''(\xi)$, particularly with small values for h.

If we apply Eq. (6.18) to $y' = -2y + 2$, $y(0) = -1$, with an h-value of 0.1 we get the results in Table 6.11. (We need starting values at $x = 0$ and $x = 0.1$; these were from the given $y(0) = -1$ and the analytical value at $x = 0.1$.)

Table 6.11 Results from Eq. (6.18)

x	y	Analytical	Error	Rel error
0.20	−0.34502	−0.34064	0.00438	−0.01284
0.30	−0.09946	−0.09762	0.00183	−0.01877
0.40	0.09477	0.10134	0.00658	0.06488
0.50	0.26264	0.26424	0.00160	0.00607
0.60	0.38971	0.39761	0.00790	0.01987
0.70	0.50675	0.50681	0.00005	0.00010
0.80	0.58701	0.59621	0.00920	0.01543
0.90	0.67195	0.66940	−0.00255	−0.00380
1.00	0.71823	0.72933	0.01110	0.01522
1.10	0.78466	0.77839	−0.00626	−0.00805
1.20	0.80437	0.81856	0.01420	0.01734
1.30	0.86291	0.85145	−0.01146	−0.01346
1.40	0.85920	0.87838	0.01918	0.02183
1.50	0.91923	0.90043	−0.01880	−0.02088
1.60	0.89151	0.91848	0.02696	0.02936
1.70	0.96262	0.93325	−0.02937	−0.03147
1.80	0.90646	0.94535	0.03889	0.04114
1.90	1.00004	0.95526	−0.04478	−0.04688
2.00	0.90645	0.96337	0.05692	0.05909
2.10	1.03746	0.97001	−0.06745	−0.06954
2.20	0.89146	0.97545	0.08398	0.08610
2.30	1.08087	0.97990	−0.10098	−0.10305
2.40	0.85911	0.98354	0.12443	0.12651
2.50	1.13723	0.98652	−0.15070	−0.15276
2.60	0.80422	0.98897	0.18474	0.18681
2.70	1.21554	0.99097	−0.22457	−0.22662
2.80	0.71801	0.99260	0.27460	0.27664
2.90	1.32834	0.99394	−0.33439	−0.33643
3.00	0.58667	0.99504	0.40837	0.41041
3.10	1.49367	0.99594	−0.49773	−0.49976
3.20	0.38920	0.99668	0.60747	0.60950
3.30	1.73799	0.99728	−0.74071	−0.74273
3.40	0.09401	0.99777	0.90376	0.90578
3.50	2.10038	0.99818	−1.10221	−1.10422
3.60	−0.34614	0.99851	1.34465	1.34666
3.70	2.63884	0.99878	−1.64006	−1.64207
3.80	−1.00168	0.99900	2.00068	2.00269
3.90	3.43951	0.99918	−2.44033	−2.44234
4.00	−1.97749	0.99933	2.97682	2.97882

Table 6.12 Results from Simple Euler Method

x	y	Analytical	Error	Rel error
0.00	−1.00000	−1.00000	0.00000	0.00000
0.10	−0.60000	−0.63746	−0.03746	0.05877
0.20	−0.28000	−0.34064	−0.06064	0.17802
0.30	−0.02400	−0.09762	−0.07362	0.75416
0.40	0.18080	0.10134	−0.07946	−0.78406
0.50	0.34464	0.26424	−0.08040	−0.30426
0.60	0.47571	0.39761	−0.07810	−0.19642
0.70	0.58057	0.50681	−0.07376	−0.14555
0.80	0.66446	0.59621	−0.06825	−0.11447
0.90	0.73156	0.66940	−0.06216	−0.09286
1.00	0.78525	0.72933	−0.05592	−0.07668
1.10	0.82820	0.77839	−0.04981	−0.06399
1.20	0.86256	0.81856	−0.04400	−0.05375
1.30	0.89005	0.85145	−0.03860	−0.04533
1.40	0.91204	0.87838	−0.03366	−0.03832
1.50	0.92963	0.90043	−0.02921	−0.03244
1.60	0.94371	0.91848	−0.02523	−0.02747
1.70	0.95496	0.93325	−0.02171	−0.02326
1.80	0.96397	0.94535	−0.01862	−0.01969
1.90	0.97118	0.95526	−0.01592	−0.01666
2.00	0.97694	0.96337	−0.01357	−0.01409
2.10	0.98155	0.97001	−0.01154	−0.01190
2.20	0.98524	0.97545	−0.00980	−0.01004
2.30	0.98819	0.97990	−0.00830	−0.00847
2.40	0.99056	0.98354	−0.00701	−0.00713
2.50	0.99244	0.98652	−0.00592	−0.00600
2.60	0.99396	0.98897	−0.00499	−0.00504
2.70	0.99516	0.99097	−0.00420	−0.00424
2.80	0.99613	0.99260	−0.00353	−0.00355
2.90	0.99691	0.99394	−0.00296	−0.00298
3.00	0.99752	0.99504	−0.00248	−0.00249
3.10	0.99802	0.99594	−0.00208	−0.00209
3.20	0.99842	0.99668	−0.00174	−0.00174
3.30	0.99873	0.99728	−0.00145	−0.00146
3.40	0.99899	0.99777	−0.00121	−0.00122
3.50	0.99919	0.99818	−0.00101	−0.00101
3.60	0.99935	0.99851	−0.00084	−0.00085
3.70	0.99948	0.99878	−0.00070	−0.00070
3.80	0.99958	0.99900	−0.00059	−0.00059
3.90	0.99967	0.99918	−0.00049	−0.00049
4.00	0.99973	0.99933	−0.00041	−0.00041

Observe in Table 6.11 that we get good results up to about $x = 0.8$, but from $x = 2$ the computed values are increasingly poor, and as x approaches 4 they are completely useless; they oscillate widely about the asymptotic value for y.

Compare these with the results from a simple Euler computation, also with $h = 0.1$, that are given in Table 6.12. These are much less accurate at small values of x (the magnitudes of the errors from the simple Euler computation between $x = 0.2$ and $x = 0.5$ are on the average nearly 20 times as large).

On the other hand, the Euler results closely resemble the analytical values at larger values for x and do not show the same oscillations.

The method of Eq. (6.18) is unstable while the Euler method is stable.

There is another unstable method but its instability is less apparent. *Milne's method* is a multistep predictor–corrector that uses these equations:

Predictor:

$$y_{n+1} - y_{n-3} = \frac{4h}{3}(2f_n - f_{n-1} + 2f_{n-2}) + \frac{28}{90}h^5 y^{(5)}(\xi_1), \qquad x_{n-3} < \xi_1 < x_{n+1}.$$

Corrector:

$$y_{n+1,c} - y_{n-1} = \frac{h}{3}(f_{n+1} + 4f_n + f_{n-1}) - \frac{h^5}{90}y^{(5)}(\xi_2), \qquad x_{n-1} < \xi_2 < x_{n+1}. \quad \textbf{(6.19)}$$

Observe that the error term after correcting has a multiplier that is less than half that of Adams–Moulton so we should expect very accurate results. However, if we solve the same equation,

$$dy/dx = f(x, y) = -2y + 2, \qquad y(0) = -1,$$

with the formulas of Eq. (6.19), we again observe oscillatory behavior as exhibited in Table 6.13, but the oscillations are slight and do not appear until about $x = 2$ and even at $x = 8$ they are not large but they are increasing in magnitude.

Of course, this demonstration of instability for Milne's method is not entirely satisfactory. We can do this more theoretically. Consider the differential equation

$$dy/dx = Ay,$$

where A is a constant. The general solution is $y = ce^{Ax}$. Suppose now that $y(x_0) = y_0$ is the initial condition; it then follows that the value of c must be $c = y_0 e^{-Ax_0}$. Hence, letting y_n be the value of the function when $x = x_n$, the analytical solution is

$$y_n = y_0 e^{A(x_n - x_0)}.$$

If we solve the differential equation by the method of Milne, we have, from the corrector formula,

$$y_{n+1} = y_{n-1} + \frac{h}{3}(y'_{n+1} + 4y'_n + y'_{n-1}).$$

Letting $y'_n = Ay_n$, from the original differential equation, and rearranging, we get

$$y_{n+1} = y_{n-1} + \frac{h}{3}(Ay_{n+1} + 4Ay_n + Ay_{n-1}),$$

Table 6.13 Results with Milne's method

x	y	Analytical	Error	Rel error
0.40	0.101355	0.101342	−0.000013	−0.000127
0.50	0.264249	0.264241	−0.000008	−0.000029
0.60	0.397630	0.397612	−0.000019	−0.000047
0.70	0.506816	0.506806	−0.000010	−0.000020
0.80	0.596227	0.596207	−0.000020	−0.000033
$\ddots$	$\ddots$	$\ddots$	$\ddots$	$\ddots$
1.80	0.945365	0.945353	−0.000012	−0.000013
1.90	0.955257	0.955258	0.000002	0.000002
2.00	0.963380	0.963369	−0.000011	−0.000012
2.10	0.970006	0.970009	0.000003	0.000003
2.20	0.975456	0.975445	−0.000010	−0.000011
$\ddots$	$\ddots$	$\ddots$	$\ddots$	$\ddots$
3.50	0.998167	0.998176	0.000010	0.000010
3.60	0.998518	0.998507	−0.000011	−0.000011
3.70	0.998767	0.998778	0.000010	0.000010
3.80	0.999010	0.998999	−0.000011	−0.000011
3.90	0.999169	0.999181	0.000011	0.000011
4.00	0.999341	0.999329	−0.000012	−0.000012
$\ddots$	$\ddots$	$\ddots$	$\ddots$	$\ddots$
7.70	0.999968	1.000000	0.000031	0.000031
7.80	1.000032	1.000000	−0.000032	−0.000032
7.90	0.999967	1.000000	0.000033	0.000033
8.00	1.000033	1.000000	−0.000034	−0.000034
8.10	0.999965	1.000000	0.000035	0.000035

$$\left(1 - \frac{hA}{3}\right) y_{n+1} - \frac{4hA}{3} y_n - \left(1 + \frac{hA}{3}\right) y_{n-1} = 0.$$

This a second-order difference equation that has the solution:

$$y_n = C_1 Z_1^n + C_2 Z_2^n,$$

where Z_1, Z_2 are the roots of the quadratic

$$\left(1 - \frac{hA}{3}\right) Z^2 - \frac{4hA}{3} Z - \left(1 + \frac{hA}{3}\right) = 0,$$

which you may check by direct substitution. We can simplify this by letting $hA/3 = r$; the roots of the quadratic are then

$$Z_1 = \frac{2r + \sqrt{3r^2 + 1}}{1 - r},$$

$$Z_2 = \frac{2r - \sqrt{3r^2 + 1}}{1 - r}.$$

What happens if the step size h becomes small? As $h \to 0$, $r \to 0$, and $r^2 \to 0$ even faster. We then can neglect the $3r^2$ terms in comparison to 1 under the radical and get, after dividing the fractions,

$$Z_1 \approx \frac{2r+1}{1-r} = 1 + 3r + O(r^2) = 1 + Ah + O(h^2),$$

$$Z_2 \approx \frac{2r-1}{1-r} = -1 + r + O(r^2) = -\left(1 - \frac{Ah}{3}\right) + O(h^2).$$

We now compare this to the Maclaurin series for the exponential function,

$$e^{hA} = 1 + hA + O(h^2),$$

$$e^{-hA/3} = 1 - \frac{hA}{3} + O(h^2).$$

We see that, for $h \to 0$,

$$Z_1 = e^{hA}, \qquad Z_2 = -e^{-hA/3}.$$

Hence, the Milne solution is represented by

$$y_n = C_1(e^{hA})^n + C_2(e^{-hA/3})^n = C_1 e^{A(x_n - x_0)} + C_2 e^{-A(x_n - x_0)/3}.$$

In this, we have used $x_n - x_0 = nh$. The solution consists of two parts. The first term obviously agrees with the analytical solution. The second term, called a *parasitic term*, will die out as x_n increases if A is a positive constant, but if A is negative, it will grow exponentially with x_n. Note that we get this peculiar behavior independent of h; smaller step size is of no benefit in eliminating the error.

Hamming's Method

The analysis of Milne's method shows that the instability comes from the corrector equation. Hamming describes a way to avoid this instability while still using the Milne predictor with its simplicity. Hammings equations are

Predictor:

$$y_{i+1,p} = y_{i-3} + \frac{4h}{3}(2f_i - f_{i-1} + 2f_{i-2}),$$

which is first modified as

$$y_{i+1,m} = y_{i+1,p} - \frac{112}{121}(y_{i,p} - y_{i,c}),$$

and the modified value is used in the corrector:

$$y_{i+1,c} = \frac{1}{8}[9y_i - y_{i-2} + 3h(f_{i+1,m} + 2f_i - f_{i-1})],$$

The error of this method is not as small as with Milne, but it is a little better than Adams–Moulton.

6.5 Higher-Order Equations and Systems

In the opening portion of this chapter, we pointed out that Newton's law of motion, $f = m * a$, is a differential equation with a being the acceleration, the rate of change of velocity with time. Velocity is itself the derivative of distance with time, dx/dt. So, $f = ma$ is really

$$f = m * d^2x/dt^2,$$

a second-order differential equation.

We can solve this equation numerically by changing it into a pair of first-order equations. We rearrange the equation to put the derivative on the left

$$d^2x/dt^2 = f/m,$$

and then, by letting $dx/dt = y$, a new variable, we have

$$dx/dt = y,$$
$$dy/dt = d^2x/dt^2 = f/m.$$

To solve the original second-order equation for x as a function of time, we need two initial conditions, the starting position, x_0, and the starting velocity, x'_0. So, the equation for dx/dt begins with $x = x_0$, and that for dy/dt begins with $y = y_0 = x'_0$.

Here is another example, a variation on the familiar spring-mass problem. Figure 6.3 shows our system. Mass 1 is a block that rolls along a horizontal surface and whose motion is controlled by the linear spring whose spring constant is k_1. The second mass, m_2, is a wheel of radius r_2 that rolls on the top of mass 1 and is attached to another spring whose spring constant is k_2. The equations of motion for this system are:

$$(m_1 + 0.5m_2)\frac{d^2x_1}{dt^2} - 0.5m_2\frac{d^2x_2}{dt^2} + k_1x_1 = 0,$$

$$-0.5m_2\frac{d^2x_2}{dt^2} + 1.5m_2\frac{d^2x_1}{dt^2} + k_2x_2 = 0.$$

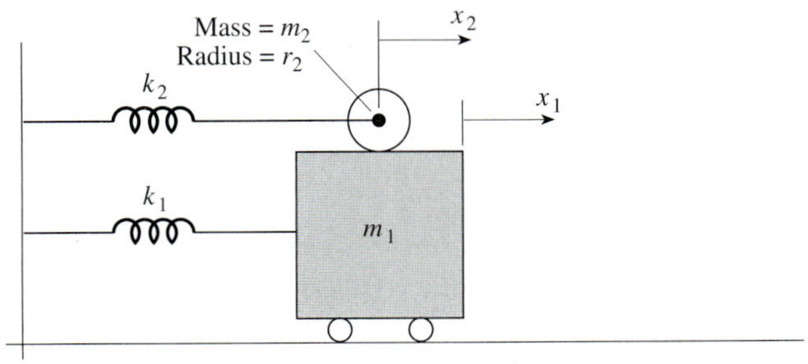

Figure 6.3

These equations make up a system of two second-order equations. To solve this problem numerically, we reduce to a system of four first-order equations by substituting dy/dt for d^2x_1/dt^2 and dz/dt for d^2x_2/dt^2. You should write out these for equations for practice. What are the four initial conditions?

Systems of First-Order Equations

It is clear that all we need to do to solve higher-order equations, even a system of higher-order initial-value problems, is to reduce them to a system of first-order equations. We illustrate how a system of first-order problems can be solved with a pair of equations whose solution at $t = 0.1$ is $x = 0.913936$, $y = -0.909217$.

$$\begin{aligned} \frac{dx}{dt} &= xy + t, \qquad x(0) = 1, \\ \frac{dy}{dt} &= ty + x, \qquad y(0) = -1. \end{aligned} \tag{6.20}$$

Taylor-Series Method

We need the various derivatives $x', x'', x''', \ldots, y', y'', y''', \ldots$, all evaluated at $t = 0$:

$$\begin{aligned} x' &= xy + t, & x'(0) &= (1)(-1) + 0 = -1 \\ y' &= ty + x, & y'(0) &= (0)(-1) + 1 = 1, \\ x'' &= xy' + x'y + 1, & x''(0) &= (1)(1) + (-1)(-1) + 1 = 3, \\ y'' &= y + ty' + x', & y''(0) &= -1 + (0)(1) - 1 = -2, \\ x''' &= x'y' + xy'' + x''y + x'y', & x'''(0) &= -7, \\ y''' &= y' + y' + ty'' + x'', & y'''(0) &= 5, \end{aligned}$$

and so on; and so on;

$$\begin{aligned} x(t) &= 1 - t + \frac{3}{2}t^2 - \frac{7}{6}t^3 + \frac{27}{24}t^4 - \frac{124}{120}t^5 + \cdots, \\ y(t) &= -1 + t - t^2 + \frac{5}{6}t^3 - \frac{13}{24}t^4 + \frac{47}{120}t^5 + \cdots. \end{aligned} \tag{6.21}$$

At $t = 0.1$, $x = 0.9139$ and $y = -0.9092$.

Equations (6.21) are the solution to the set (6.20). Note that we need to alternate between the functions in getting the derivatives; for example, we cannot get $x''(0)$ until $y'(0)$ is known; we cannot get $y'''(0)$ until $x''(0)$ is known. After we have obtained the coefficients of the Taylor-series expansions in Eq. (6.21), we can evaluate x and y at any value of t, but the error will depend on how many terms we employ.

Euler Predictor–Corrector Method (Modified Euler)

We apply the predictor to each equation; then the corrector can be used. Again, note that we work alternately with the two functions.

Take $h = 0.1$. Let p and c subscripts indicate predicted and corrected values, respectively:

$$x_p(0.1) = 1 + 0.1[(1)(-1) + 0] = 0.9,$$
$$y_p(0.1) = -1 + 0.1[(0)(-1) + 1] = -0.9,$$
$$x_c(0.1) = 1 + 0.1\left(\frac{-1 + [(0.9)(-0.9) + 0.1]}{2}\right) = 0.9145,$$
$$y_c(0.1) = -1 + 0.1\left(\frac{1 + [(0.1)(-0.9) + 0.9145]}{2}\right) = -0.9088.$$

In computing $x_c(0.1)$, we used the x_p and y_p. In computing y_c (0.1) after $x_c(0.1)$ is known, we have a choice between x_p and x_c. There is an intuitive feel that one should use x_c, with the idea that one should always use the best available values. This does not always expedite convergence, probably due to compensating errors. Here we have used the best values to date. If we use the corrected values to recompute the value of the derivatives at $h = 0.1$, we can obtain better values. Doing so gives

$$x(0.1) = 0.9135,$$
$$y(0.1) = -0.9089,$$

but this is not as efficient as using a more powerful method. We can now advance the solution another step if desired, by using the computed values at $t = 0.1$ as the starting values. From this point, we can advance one more step, and so on for any value of t. The errors will be the combination of local truncation error at each step plus the propagated error resulting from the use of inexact starting values.

Runge–Kutta–Fehlberg Method

Again there is an alternation between the x and y calculations. In applying this method, one always uses the previous k-value in incrementing the function values and the value of h to increment the independent variable. As in the previous calculations, we alternate between computations for x and for y; for example, we do $k_{1,x}$, then $k_{1,y}$, before doing $k_{2,x}$, and so on.

Keeping in mind that the equations are

$$\frac{dx}{dt} = f(t, x, y) = xy + t, \quad x(0) = 1,$$

$$\frac{dy}{dt} = g(t, x, y) = ty + x, \quad y(0) = -1,$$

the k-values for x and y are

for x:

$$\begin{aligned}
k_{1,x} &= hf(0, 1, -1) \\
&= 0.1[(1)(-1) + 0] \\
&= -0.1; \\
k_{2,x} &= hf(0.025, 0.975, -0.975) \\
&= 0.1[(0.975)(-0.975) + 0.025] \\
&= -0.092562; \\
k_{3,x} &= hf(0.038, 0.965, -0.964) \\
&= 0.1[(0.965)(-0.964) + 0.038] \\
&= -0.089226; \\
k_{4,x} &= hf(0.092, 0.919, -0.915) \\
&= 0.1[(0.919)(-0.915) + 0.092] \\
&= -0.074892; \\
k_{5,x} &= hf(0.1, 0.913, -0.908) \\
&= 0.1[(0.913)(-0.908) + 0.1] \\
&= -0.072904; \\
k_{6,x} &= hf(0.05, 0.954, -0.953) \\
&= 0.1[(0.954)(-0.953) + 0.05] \\
&= -0.085868.
\end{aligned}$$

for y:

$$\begin{aligned}
k_{1,y} &= hg(0, 1, -1) \\
&= 0.1[(0)(-1) + 1] \\
&= 0.1; \\
k_{2,y} &= hg(0.025, 0.975, -0.975) \\
&= 0.1[(0.025)(-0.975) + 0.975] \\
&= 0.095062; \\
k_{3,y} &= hg(0.038, 0.965, -0.964) \\
&= 0.1[(0.038)(-0.964) + 0.965] \\
&= 0.092845; \\
k_{4,y} &= hg(0.092, 0.919, -0.915) \\
&= 0.1[(0.092)(-0.915) + 0.919] \\
&= 0.083461; \\
k_{5,y} &= hg(0.1, 0.913, -0.908) \\
&= 0.1[(0.1)(-0.908) + 0.913] \\
&= 0.082178; \\
k_{6,y} &= hg(0.05, 0.954, -0.953) \\
&= 0.1[(0.05)(-0.953) + 0.954] \\
&= 0.090628.
\end{aligned}$$

Then, using the fifth-order formula, we get

$$\begin{aligned}
x(0.1) &= 1 + (-0.01185 - 0.046307 - 0.037905 + 0.013123 - 0.003122) \\
&= 0.913936; \\
y(0.1) &= -1 + (0.01185 + 0.048185 + 0.042242 - 0.014792 + 0.003296) \\
&= -0.909217.
\end{aligned}$$

Extending the Taylor-series solution even further shows that the Runge–Kutta–Fehlberg values are correct to more than five decimals, whereas the modified Euler values are correct to only three, so $h = 0.1$ may be too large for that method.

Advancing the solution by the Runge–Kutta–Fehlberg method will again involve using the computed values of x and y as the initial values for another step. The errors here will be much less than those for the Euler predictor–corrector method.

Table 6.14

	t	x	x'	t	y	y'
Starting values	0.000	1.0	−1.0	0.00	−1.0	1.0
	0.025	0.9759	−0.9271	0.025	−0.9756	0.9515
	0.050	0.9536	−0.8582	0.050	−0.9524	0.9060
	0.075	0.9330	−0.7929	0.075	−0.9303	0.8632
Predicted	0.10	(0.9139)	(−0.7310)	0.10	(−0.9092)	(0.8230)
Corrected		0.9139			−0.9092	

Adams–Moulton Method

After getting four starting values, we proceed with the algorithm of Eqs. (6.16) and (6.17), again alternately computing x and then y (see Table 6.14.)

In the computations we first get predicted values of x and y:

$$x(0.1) = 0.9330 + \frac{0.025}{24}[55(-0.7929) - 59(-0.8582) + 37(-0.9271) - 9(-1.0)]$$
$$= 0.913937;$$

$$y(0.1) = -0.9303 + \frac{0.025}{24}[55(0.8632) - 59(0.9060) + 37(0.9515) - 9(1.0)]$$
$$= -0.909217.$$

After getting x' and y' at $t = 0.1$, using $x(0.1)$ and $y(0.1)$, we then correct:

$$x(0.1) = 0.9330 + \frac{0.025}{24}[9(-0.7310) + 19(-0.7929) - 5(-0.8582) + (-0.9271)]$$
$$= 0.913936;$$

$$y(0.1) = -0.9303 + \frac{0.025}{24}[9(0.8230) + 19(0.8632) - 5(0.9060) + (0.9515)]$$
$$= -0.909217.$$

The close agreement of predicted and corrected values indicates six-decimal-place accuracy.

In this method, as we advance the solution to larger values of t, the comparison between predictor and corrector values tells us whether the step size needs to be changed.

Our computer algebra systems have no trouble in solving a system of first-order equations. Here is how Maple can solve the same problem that we have used to illustrate the methods:

```
> deqs := {D(x)(t) = x(t)*y(t) + t, D(y)(t) = t*y(t) + x(t)}:
> inits := {x(0) = 1, y(0) = -1}:
> soln := dsolve(deqs union inits, {x(t),y(t)}, numeric,
```

```
        output = array([0, 0.1, 0.2, 0.3, 0.4]));
                          [t, x(t) y(t)]
        0          1.                  -1.
        .1    .91393569117289    -.90921691879919
soln: = .2    .85218609746503    -.83408937511807
        .3    .81063353106742    -.77109331990007
        .4    .78634968913429    -.71735810231063
```

Here, we asked for the solution at x-values between 0 and 0.4 in steps of 0.1 and the results are given in tabular form. MATLAB and *Mathematica* can do so similarly.

6.6 Stiff Equations

Some initial value problems pose significant difficulties for their numerical solution. Acton points out several kinds of such difficulties—one of his examples is Bessel's equation:

$$y'' + y'/x + y = 0, \qquad y(0) = 1, \qquad y'(0) = 0.$$

There is a singularity at the origin, but this is surmounted by the initial value for y ($y = 0$), so that one can replace the equation at $x = 0$ and get a starting value with

$$2y'' + y = 0.$$

There are other difficult situations: The equation may change its form at certain critical points, or it may have a sharp narrow peak that will be missed if too large an interval is used.

One particular difficult case is one that we now discuss—*stiff differential equations*. The word *stiff* comes from an analogy to a spring system where the natural frequency of vibration is very great if the spring constant is large.

When the solution to a differential equation (say, of second order) has a general solution that involves the sum or difference of terms of the form ae^{ct} and be^{dt} where both c and d are negative but c is much smaller than d, the numerical solution can be very unstable even with a very small step size.

An example is the following:

$$\begin{aligned} x' &= 1195x - 1995y, \qquad x(0) = 2, \\ y' &= 1197x - 1997y, \qquad y(0) = -2. \end{aligned} \tag{6.22}$$

The analytical solution of Eq. (6.22) is

$$x(t) = 10e^{-2t} - 8e^{-800t}, \qquad -y(t) = 6e^{-2t} - 8e^{-800t}.$$

Observe that the exponents are all negative and of very different magnitude, qualifying this as a stiff equation. Suppose we solve Eq. (6.22) by the simple Euler method with $h = 0.1$, applying just one step. The iterations are

$$\begin{aligned} x_{i+1} &= x_i + hf(x_i, y_i) = x_i + 0.1(1195x_i - 1995y_i), \\ y_{i+1} &= y_i + hg(x_i, y_i) = y_i + 0.1(1197x_i - 1997y_i). \end{aligned}$$

This gives $x(0.1) = 640$, $y(0.1) = 636$, while the analytical values are $x(0.1) = 8.187$ and $y(0.1) = 4.912$. Such a result is typical (although here exaggerated) for stiff equations.

One solution to this problem is to use an implicit method rather than an explicit one. All the methods so far discussed have been explicit, meaning that new values, x_{i+1} and y_{i+1}, are computed in terms of previous values, x_i and y_i. An implicit method computes the increment only with the new (unknown) values. Suppose that

$$x' = f(x, y) \qquad \text{and} \qquad y' = g(x, y).$$

The implicit form of the Euler method is

$$\begin{aligned} x_{i+1} &= x_i + hf(x_{i+1}, y_{i+1}), \\ y_{i+1} &= y_i + hg(x_{i+1}, y_{i+1}). \end{aligned} \tag{6.23}$$

If the derivative functions $f(x, y)$ and $g(x, y)$ are nonlinear, this is difficult to solve. However, in Eq. (6.22) they are linear. Solving Eq. (6.22) by use of Eq. (6.23) we have

$$\begin{aligned} x_{i+1} &= x_i + 0.1(1195x_{i+1} - 1995y_{i+1}), \\ y_{i+1} &= y_i + 0.1(1197x_{i+1} - 1997y_{i+1}). \end{aligned}$$

The system is linear, so we can write

$$\begin{bmatrix} x_{i+1} \\ y_{i+1} \end{bmatrix} = \begin{bmatrix} (1 - 1195(0.1)) & 1995(0.1) \\ -1197(0.1) & (1 + 1997(0.1)) \end{bmatrix}^{-1} \begin{bmatrix} x_i \\ y_i \end{bmatrix}$$

which has the solution $x(0.1) = 8.23$, $y(0.1) = 4.90$, reasonably close to the analytical values.

In summary, our results for the solution of Eq. (6.22) are

	x(0.1)	**y(0.1)**
Analytical	8.19	4.91
Euler		
Explicit	640	636
Implicit	8.23	4.90

If the step size is very small, we can get good results from the simpler Euler after the first step. With $h = 0.0001$, the table of results becomes

	x(0.0001)	**y(0.0001)**
Analytical	2.61	−1.39
Euler		
Explicit	2.64	−1.36
Implicit	2.60	−1.41

but this would require 1000 steps to reach $t = 0.1$, and round-off errors would be large.

If we anticipate some material from Section 6.8, we can give a better description of stiffness as well as indicate the derivation of the general solution to Eq. (6.22). We rewrite Eq. (6.22) in matrix form:

$$\begin{bmatrix} x \\ y \end{bmatrix}' = A\begin{bmatrix} x \\ y \end{bmatrix}, \qquad \text{where } A = \begin{bmatrix} 1195 & -1995 \\ 1197 & -1997 \end{bmatrix}.$$

The general solution, in matrix form, is

$$\begin{bmatrix} x \\ y \end{bmatrix} = ae^{-2t}v_1 + ce^{-800t}v_2,$$

where

$$v_1 = \begin{bmatrix} 5 \\ 3 \end{bmatrix} \quad \text{and} \quad v_2 = \begin{bmatrix} 1 \\ 1 \end{bmatrix}.$$

You can easily verify that $Av_1 = -2v_1$ and $Av_2 = -800v_2$. This means that v_1 is an eigenvector of A and that -2 is the corresponding eigenvalue. Similarly, v_2 is an eigenvector of A with the corresponding eigenvalue of -800. (In Section 6.8, you will learn additional methods to find the eigenvectors and eigenvalues of a matrix.)

A stiff equation can be defined in terms of the eigenvalues of the matrix A that represents the right-hand sides of the system of differential equations. When the eigenvalues of A have real parts that are negative and differ widely in magnitude as in this example, the system is stiff. In the case of a nonlinear system

$$\begin{bmatrix} x_1 \\ x_2 \\ \vdots \\ x_n \end{bmatrix}' = \begin{bmatrix} f_1(x_1, x_2, \ldots, x_n) \\ f_2(x_1, x_2, \ldots, x_n) \\ \vdots \\ f_n(x_1, x_2, \ldots, x_n) \end{bmatrix},$$

one must consider the Jacobian matrix whose terms are $\partial f_i/\partial x_j$. See Gear (1971) for more information.

6.7 Boundary-Value Problems

As we have seen, a second-order differential equation (or a pair of first-order problems) must have two conditions for its numerical solution. Up until now, we have considered that both of these conditions are given at the start—these are initial-value problems. That is not always the case; the given conditions may be at different points, usually at the endpoints of the region of interest. For equations of order higher than two, more than two conditions are required and these also may be at different x-values. We consider now how such problems can be solved.

Here is an example that describes the temperature distribution within a rod of uniform cross section that conducts heat from one end to the other. Look at Figure 6.4. By concentrating our attention on an element of the rod of length dx located at a distance x from the left end, we can derive the equation that determines the temperature, u, at any point along

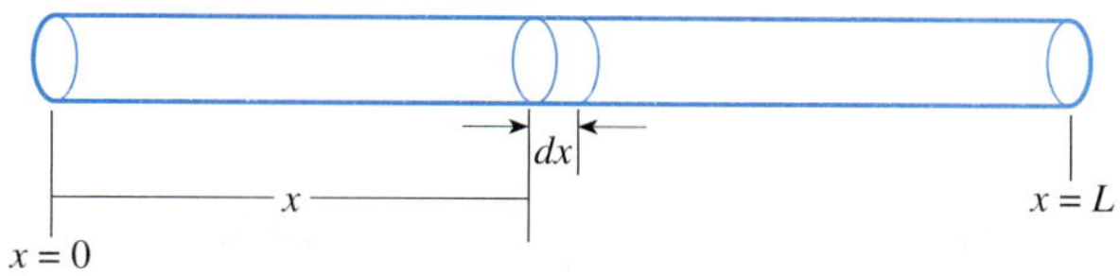

Figure 6.4

the rod. The rod is perfectly insulated around its outer circumference so that heat flows only laterally along the rod. It is well known that heat flows at a rate (measured in calories per second) proportional to the cross-sectional area (A), to a property of the material [k, its thermal conductivity, measured in cal/(sec * cm^2 * (°C/cm))], and to the temperature gradient, du/dx (measured in °C/cm), at point x. We use $u(x)$ for the temperature at point x, with x measured from the left end of the rod. Thus, the rate of flow of heat into the element (at $x = x$) is

$$-kA\left(\frac{du}{dx}\right).$$

The minus sign is required because du/dx expresses how rapidly temperatures increase with x, while the heat always flows from high temperature to low.

The rate at which heat leaves the element is given by a similar equation, but now the temperature gradient must be at the point $x + dx$:

$$-kA\left[\frac{du}{dx} + \frac{d}{dx}\left(\frac{du}{dx}\right)dx\right],$$

in which the gradient term is the gradient at x plus the change in the gradient between x and $x + dx$.

Unless heat is being added to the element (or withdrawn by some means), the rate that heat flows from the element must equal the rate that heat enters, or else the temperature of the element will vary with time. In this chapter, we consider only the case of *steady-state* or *equilibrium* temperatures, so we can equate the rates of heat entering and leaving the element:

$$-kA\left(\frac{du}{dx}\right) = -kA\left[\frac{du}{dx} + \frac{d}{dx}\left(\frac{du}{dx}\right)dx\right].$$

When some common terms on each side of the equation are canceled, we get the very simple relation

$$kA\frac{d}{dx}\left(\frac{du}{dx}\right)dx = kA\frac{d^2u}{dx^2} = 0,$$

where we have written the second derivative in its usual form. For this particularly simple example, the equation for u as a function of x is the solution to

$$\frac{d^2u}{dx^2} = 0,$$

and this is obviously just

$$u = ax + b,$$

a linear relation. This means that the temperatures vary linearly from TL to TR as x goes from 0 to L.

The rod could also lose heat from the outer surface of the element. If this is Q (cal/(sec * cm^2)), the rate of heat flow in must equal the rate leaving the element by conduction along the rod plus the rate at which heat is lost from the surface. This means that:

$$-kA\left(\frac{du}{dx}\right) = -kA\left[\frac{du}{dx} + \frac{d}{dx}\left(\frac{du}{dx}\right)dx\right] + Qp\,dx,$$

where p is the perimeter at point x. (Q might also depend on the difference in temperature within the element and the temperature of the surroundings, but we will ignore that for now.)

If this equation is expanded and common terms are canceled, we get a somewhat more complicated equation whose solution is not obvious:

$$\frac{d^2u}{dx^2} = \frac{Qp}{(kA)}. \tag{6.24}$$

In Eq. (6.24), Q can be a function of x.

The situation may not be quite as simple as this. The cross section could vary along the rod, or k could be a function of x (some kind of composite of materials, possibly). Suppose first that only the cross section varies with x. We will have, then, for the rate of heat leaving the element

$$-k[A + A'\,dx]\left[\frac{du}{dx} + u''\,dx\right],$$

where we have used a prime notation for derivatives with respect to x. Equating the rates in and out as before and canceling common terms results in

$$kAu''\,dx + kA'u'\,dx + kA'u''\,dx^2 = Qp\,dx.$$

We can simplify this further by dropping the term with dx^2 because it goes to zero faster than the terms in dx. After also dividing out dx, this results in a second-order differential equation similar in form to some we have discussed in Section 6.5:

$$kAu'' + kA'u' = Qp. \tag{6.25}$$

The equation can be generalized even more if k also varies along the rod. We leave to the reader as an exercise to show that this results in

$$kAu'' + (kA' + k'A)u' = Qp. \tag{6.26}$$

If the rate of heat loss from the outer surface is proportional to the difference in temperatures between that within the element and the surroundings (u_s), (and this is a common situation), we must substitute for Q:

$$Q = q(u - u_s),$$

giving

$$kAu'' + (kA' + k'A)u' - q * pu = -q * pu_s. \qquad (6.27)$$

This chapter will discuss two ways to solve equations like Eqs. (6.24) to (6.27).

Heat flow has been used in this section as the physical situation that is modeled, but equations of the same form apply to diffusion, certain types of fluid flow, torsion in objects subject to twisting, distribution of voltage, in fact, to any problem where the potential is proportional to the gradient.

The Shooting Method

We can rewrite Eq. (6.27) as

$$A\frac{d^2u}{dx^2} + B\frac{du}{dx} + Cu = D, \qquad (6.28)$$

where the coefficients, A, B, C, and D are functions of x. (Actually, they could also be functions of both x and u, but that makes the problem more difficult to solve. In a temperature-distribution problem, such nonlinearity can be caused if the thermal conductivity, k, is considered to vary with the temperature, u. That is actually true for almost all materials but, as the variation is usually small, it is often neglected and an average value is used.)

To solve Eq. (6.28), we must know two conditions on u or its derivative. If both u and u' are specified at some starting value for x, the problem is an *initial-value problem.* In this section, we consider Eq. (6.28) to have two values of u to be given but these are at two different values for x—this makes it a *boundary-value problem.* In this section, we discuss how the same procedures that apply to an initial-value problem can be adapted.

The strategy is simple: Suppose we know u at $x = a$ (the beginning of a region of interest) and u at $x = b$ (the end of the region). We wish we knew u' at $x = a$; that would make it an initial-value problem. So, why not assume a value for this? Some general knowledge of the situation may indicate a reasonable guess. Or we could blindly select some value. The test of the accuracy of the guess is to see if we get the specified $u(b)$ by solving the problem over the interval $x = a$ to $x = b$. If the initial slope that we assumed is too large, we will often find that the computed value for $u(b)$ is too large. So, we try again with a smaller initial slope. If the new value for $u(b)$ is too small, we have bracketed the correct initial slope. This method is called the *shooting method* because of its resemblance to the problem faced by an artillery officer who is trying to hit a distant target. The right elevation of the gun can be found if two shots are made of which one is short of the target and the other is beyond. That means that an intermediate elevation will come closer.

EXAMPLE 6.2 Solve

$$u'' - \left(1 - \frac{x}{5}\right)u = x, \qquad u(1) = 2, \qquad u(3) = -1.$$

(This is an instance of Eq. (6.28) with $A = 1$, $B = 0$, $C = -(1 - x/5)$, and $D = x$.) Assume that $u'(1) = -1.5$ (which might be a reasonable guess, because u declines

Table 6.15

	Assume $u'(1) = -1.5$		Assume $u'(1) = -3.0$		Assume $u'(1) = -3.4950$	
x	u	u'	u	u'	u	u'
1.00	2.0000	−1.5000	2.0000	−3.0000	2.0000	−3.4950
1.20	1.7614	−0.9886	1.4598	−2.5118	1.3503	−3.0145
1.40	1.6043	−0.4814	0.9921	−2.0719	0.7900	−2.5967
1.60	1.5597	0.0389	0.6192	−1.6598	0.3099	−2.2204
1.80	1.6218	0.5876	0.3275	−1.2580	−0.0997	−1.8671
2.00	1.7976	1.1783	0.1163	−0.8512	−0.4385	−1.5209
2.20	2.0967	1.8227	−0.0118	−0.4259	−0.7076	−1.1679
2.40	2.5309	2.5310	−0.0520	0.0299	−0.9043	−0.7955
2.60	3.1139	3.3116	0.0029	0.5266	−1.0237	−0.3925
2.80	3.8608	4.1706	0.1620	1.0732	−1.0586	0.0511
3.00	4.7876	5.1119	0.4360	1.6773	−1.0000	0.5439

between $x = 1$ and $x = 3$; this number is the average slope over the interval). If we use a program that implements the Runge–Kutta–Fehlberg method, we get the values shown in the first part of Table 6.15.

Because the value for $u(3)$ is 4.7876 rather than the desired -1, we try again with a different initial slope, say $u'(1) = -3.0$, and get the middle part of Table 6.15. The resulting value for $u(3)$ is still too high: 0.4360 rather than -1. We could guess at a third trial for $u'(1)$, but let us interpolate linearly between the first two trials.* Doing so suggests a value for $u'(1)$ of -3.4950. Lo and behold, we get the correct answer for $u(3)$! These results are shown in the third part of Table 6.15.

It was not just by chance that we got the correct solution by interpolating from the first two trials. The problem is *linear* and for linear equations this will always be true. Except for truncation and round-off errors, the exact solution to a linear boundary-value problem by the shooting method is a linear combination of two trial solutions:

Suppose that $x_1(t)$ and $x_2(t)$ are two trial solutions of a boundary-value problem

$$x'' + Fx' + Gx = H, \qquad x(t_0) = A, \qquad x(t_f) = B$$

(where F, G, and H are functions of t only) and both trial solutions begin at the correct value of $x(t_0)$.

We then state that

$$y(t) = \frac{c_1x_1 + c_2x_2}{c_1 + c_2}$$

* If G = guess, and R = result: DR = desired result: G3 = G2 + (DR − R2)(G1 − G2)/(R1 − R2)

is also a solution. We show that this is true, because, since x_1 and x_2 are solutions, it follows that

$$x_1'' + Fx_1' + Gx_1 = H, \quad \text{and} \quad x_2'' + Fx_2' + Gx_2 = H.$$

If we substitute y into the original equations, with

$$y' = \frac{c_1x_1' + c_2x_2'}{c_1 + c_2}, \quad \text{and} \quad y'' = \frac{c_1x_1'' + c_2x_2''}{c_1 + c_2},$$

we get

$$\frac{c_1x_1'' + c_2x_2''}{c_1 + c_2} + \frac{c_1x_1' + c_2x_2'}{c_1 + c_2}F + \frac{c_1x_1 + c_2x_2}{c_1 + c_2}G = \frac{c_1x_1'' + c_1Fx_1' + c_1Gx_1 + c_2x_2'' + c_2Fx_2' + c_2Gx_2}{c_1 + c_2}$$

$$= \frac{c_1H}{c_1 + c_2} + \frac{c_2H}{c_1 + c_2} = H,$$

which shows that y is also a solution that begins at the correct value for $x(t_0)$. The implication of this is that, if c_1 and c_2 are chosen so that $y(t_f) = x(t_f) = B$, $y(t)$ is the correct solution to the boundary-value problem.

It must also be true that $y'(t_0)$ is the correct initial slope and that one can interpolate between every pair of computed values to get correct values for $y(x)$ at intermediate points.

This next example shows that we cannot get the correct solution so readily when the problem is *nonlinear.*

EXAMPLE 6.3 Solve

$$u'' - \left(1 - \frac{x}{5}\right)uu' = x, \qquad u(1) = 2, \qquad u(3) = -1.$$

This resembles Example 6.2 but observe that the coefficient of u' involves u, the dependent variable. This problem is nonlinear and we shall see that it is not as easy to solve. If we again use the Runge–Kutta–Fehlberg method, we get the results summarized in Table 6.16. Here the third trial, which used the interpolated value from the first two trials,

Table 6.16

Assumed value for $u'(1)$	Calculated value for $u(3)$
−1.5	−0.0282
−3.0	−2.0705
−2.2137*	−1.2719
−1.9460*	−0.8932
−2.0215*	−1.0080
−2.0162*	−1.0002
−2.0161*	−1.0000

* Interpolated from two previous values

Table 6.17

x	u	u'
1.0000	2.0000	−2.0161
1.2000	1.5552	−2.4130
1.4000	1.0459	−2.6438
1.6000	0.5318	−2.6352
1.8000	0.0082	−2.3832
2.0000	−0.4272	−1.9472
2.2000	−0.7640	−1.4110
2.4000	−0.9896	−0.8441
2.6000	−1.1022	−0.2848
2.8000	−1.1047	0.2569
3.0000	−1.0000	0.7909

does not give the correct solution. A nonlinear problem requires a kind of search operation. We could interpolate with a quadratic from the results of three trials, an adaptation of Muller's method. Table 6.17 gives the computed values for $u(x)$ between $x = 1$ and $x = 3$ with the final (good) estimate of the initial slope.

The shooting method is often quite laborious, especially with problems of fourth or higher order. With these, the necessity of assuming two or more conditions at the starting point (and matching with the same number of conditions at the end) is slow and tedious.

There are times when it is better to compute "backwards" from $x = b$ to $x = a$. For example, if $u(b)$ and $u'(a)$ are the known boundary values, the technique just described works best if we compute from $x = b$ to $x = a$. Another time that computing backwards would be preferred is in a fourth-order problem where three conditions are given at $x = b$ and only one at $x = a$.

Maple's `dsolve` command works with boundary-value problems. Here is how it can solve Example 6.3.

```
>de2 := diff(u(x),x$2) - (1 - x/5)*u(x)*diff(u(x)x) = x:
>F := dsolve({de2, u(1) = 2, u(3) = -1}, u(x), numeric);
     F := proc(bvp_x . . . end proc
>F(1); F(2); F(3);
        x = 1., u(x) = 2.,∂/∂x u(x) = -2.01607429521390014
   x = 2., u(x) = -.427176163177449108, ∂/∂x u(x) =
     -1.94723020165843686
   x = 3., u(x) = -1.00000000000000022, ∂/∂x u(x) =
     .790910254537530277
>F(1.4); F(2.6);
   x = 1.4, u(x) = 1.04594603838311962, ∂/∂x u(x) =
     -2.64376847138324100
   x = 2.6, u(x) = -1.10221333664797760, ∂/∂x u(x) =
     -.284818239545453100
```

In this, we first defined the second-order equation, then used the dsolve command to get the solution, F, (a "procedure" that is not spelled out). When we asked for values of the solution at $x = 1, 2, 3, 1.4$, and 2.6, Maple displayed results that match to Table 6.17 but with many more digits of precision.

Solution Through a Set of Equations

There is another way to solve boundary-value problems like Example 6.2. We have seen in Chapter 5 that derivatives can be approximated by finite-difference quotients. If we replace the derivatives in a differential equation by such expressions, we convert it into a difference equation whose solution is an approximation to the solution of the differential equation. This method is sometimes preferred over the shooting method, but it really can be used only with linear equations. (If the differential equation is nonlinear, this technique leads to a set of nonlinear equations that are more difficult to solve. Solving such a set of nonlinear equations is best done by iteration, starting with some initial approximation to the solution vector.)

EXAMPLE 6.4 Solve the boundary-value problem of Example 6.2 but use a set of equations obtained by replacing the derivative with a central difference approximation. Divide the region into four equal subintervals and solve the equations, then divide into ten subintervals. Compare both of these solutions to the results of Example 6.2.

When the interval from $x = 1$ to $x = 3$ is subdivided into four subintervals, there are interior points (these are usually called *nodes*) at $x = 1.5$, 2.0, and 2.5. Label the nodes as x_1, x_2, and x_3. The endpoints are x_0 and x_4. We write the difference equation at the three interior nodes. The equation, $u'' - (1 - x/5)u = x$, $u(1) = 2$, $u(3) = -1$, becomes

$$\text{At } x_1: \qquad \frac{(u_0 - 2u_1 + u_2)}{h^2} - \left(1 - \frac{x_1}{5}\right) u_1 = x_1,$$

$$\text{At } x_2: \qquad \frac{(u_1 - 2u_2 + u_3)}{h^2} - \left(1 - \frac{x_2}{5}\right) u_2 = x_2,$$

$$\text{At } x_3: \qquad \frac{(u_2 - 2u_3 + u_4)}{h^2} - \left(1 - \frac{x_3}{5}\right) u_3 = x_3.$$

These equations are all of the form:

$$\text{At } x_i: \qquad \frac{(u_{i-1} - 2u_i + u_{i+1})}{h^2} - \left(1 - \frac{x_i}{5}\right) u_i = x_i,$$

which can be rearranged into:

$$\text{At } x_i: \qquad u_{i-1} - \left[2 + h^2\left(1 - \frac{x_i}{5}\right)\right] u_i + u_{i+1} = h^2 x_i.$$

Substitute $h = 0.5$, substitute the x-values at the nodes, and substitute the u-values at the endpoints and arrange in matrix form, which gives

$$\begin{bmatrix} -2.175 & 1 & 0 \\ 1 & -2.150 & 1 \\ 0 & 1 & -2.125 \end{bmatrix} \begin{bmatrix} u_1 \\ u_2 \\ u_3 \end{bmatrix} = \begin{bmatrix} -1.625 \\ 0.5 \\ 1.625 \end{bmatrix}.$$

Observe that the system is tridiagonal and that this will always be true even when there are many more nodes, because any derivative of u involves only points to the left, to the right, and the central point.

When this system is solved, we get

$$x_1 = 0.552, \qquad x_2 = -0.424 \qquad \text{and} \qquad x_3 = -0.964.$$

If we solve the problem again but with ten subintervals ($h = 0.2$), we must solve a system of nine equations, because there are nine interior nodes where the value of u is unknown. The answers, together with the results from the shooting method for comparison, are

x	**Values from the finite-difference method**	**Values from the shooting method**
1.2	1.351	1.350
1.4	0.792	0.790
1.6	0.311	0.309
1.8	−0.097	−0.100
2.0	−0.436	−0.438
2.2	−0.705	−0.708
2.4	−0.903	−0.904
2.6	−1.022	−1.024
2.8	−1.058	−1.059

There is quite close agreement. It is difficult to say from this which method is more accurate because both are subject to error. We can compare the methods and determine how making the number of subintervals greater increases the accuracy by examining the results for a problem with a known analytical answer.

EXAMPLE 6.5 Compare the accuracy of the finite-difference method with the shooting method on this second-order boundary-value problem:

$$u'' = u, \qquad u(1) = 1.17520, \qquad u(3) = 10.01787,$$

whose analytical solution is $u = \sinh(x)$.

When the problem is solved by finite-difference approximations to the derivatives, the typical equation is

$$u_{i-1} - (2 + h^2)u_i + u_{i+1} = 0.$$

Solving with $h = 1$, $h = 0.5$, and $h = 0.25$, we get the values in Table 6.18. If we solve this with the shooting method (employing Runge–Kutta–Fehlberg), we get Table 6.19.

Table 6.18 Solutions with the finite-difference method

	u-values with		
x	**2 subintervals**	**4 subintervals**	**8 subintervals**
1.25			1.60432
1.50		2.14670	2.13372
1.75			2.79647
2.00	3.73102	3.65488	3.63400
2.25			4.69866
2.50		7.07678	7.05698
2.75			7.79387
error at $x = 2.00$	0.10416	0.02802	0.00714

In both tables, the errors at $x = 2.0$ are shown. This is nearly the maximum error of any of the results.

When the results from the two methods are compared, it is clear that (1) the shooting method is much more accurate at the same number of subintervals, its errors being from 80 to over 500 times smaller; and (2) the errors for the finite-difference method decrease about four times when the number of subintervals is doubled, which is as expected.

The reader should make a similar comparison for other equations.

Derivative Boundary Conditions

The conditions at the boundary often involve the derivative of the dependent variable in addition to its value. A hot object loses heat to its surroundings proportional to the

Table 6.19 Solutions with the shooting method

	u-values with		
x	**2 subintervals**	**4 subintervals**	**8 subintervals**
1.25			1.60192
1.50		2.12931	2.12928
1.75			2.79042
2.00	3.62814	3.62692	3.62686
2.25			4.69117
2.50		7.05025	7.05020
2.75			7.78935
error at $x = 2.00$	0.00128	0.00006	0.00000

Figure 6.5

difference between the temperature at the surface of the object and the temperature of the surroundings. The proportionality constant is called the *heat-transfer coefficient* and is frequently represented by the symbol h. (This can cause confusion because we use h for the size of a subinterval. To avoid this confusion, we shall use a capital letter, H, for the heat-transfer coefficient.) The units of H are cal/sec/cm^2/°C (of temperature difference). In this section we consider a rod that loses heat to the surroundings from one or both ends. Of course, heat could be gained from the surroundings if the surroundings are hotter than the rod.

Names have been given to the various types of boundary conditions. If the value for u is specified at a boundary, it is called a *Dirichlet condition.* This is the type of problem that we have solved before. If the condition is the value of the derivative of u, it is a *Neumann condition.* When a boundary condition involves both u and its derivative, it is called a *mixed condition.*

We now develop the relations when heat is lost from the ends of a rod that conducts heat along the rod but is insulated around its perimeter so that no heat is lost from its lateral surface. First consider the right end of the rod and assume that heat is being lost to the surroundings (implying that the surface is hotter than the surroundings). Figure 6.5 will help to visualize this. At the right end of the rod ($x = x_R$), the temperature is u_R; the temperature of the surroundings is u_{SR}. Heat then is being lost from the rod to the surroundings at a rate [measured in (cal/sec)] of

$$HA(u_R - u_{SR}),$$

where A is the area of the end of the rod. This heat must be supplied by heat flowing from inside the rod to the surface, which is at the rate of

$$-kA\frac{du}{dx},$$

where the minus sign is required because heat flows from high to low temperature. Equating these two rates and solving for du/dx (the gradient) gives (the A's cancel):

$$\frac{du}{dx} = -\left(\frac{H}{k}\right)(u_R - u_{SR}), \qquad \text{at the right end.}$$

Now consider the left end of the rod, at $x = 0$, where $u = u_L$. Assume that the temperature of the surroundings here are at some other temperature, u_{SL}. Here, heat is flowing from right to left, so we have

$$\text{Heat leaving the rod: } -HA(u_L - u_{SL}).$$

For the rate at which heat flows from inside the rod we still have

$$-kA\,\frac{du}{dx},$$

and, after equating and solving for the gradient:

$$\frac{du}{dx} = \left(\frac{H}{k}\right)(u_{\text{L}} - u_{\text{SL}}), \qquad \text{at the left end.}$$

The fact that the signs in the equations for the gradients are not the same can be a source of confusion. Of course, if both ends lose heat to the surroundings, the equilibrium or steady-state temperatures of the rod will just be a linear relation between the two (possibly different) surrounding temperatures. In practical situations of heat distribution in a rod, only one end of the rod loses (or gains) heat to (from) the surroundings, the other end being held at some constant temperature.

A minor problem is presented in the cases under consideration. We need to give consideration to how to approximate the gradient at the end of the rod. One could use a forward difference approximation (at the right end, a backward difference at the left), but that seems inappropriate when central differences are used to approximate the derivatives within the rod. This conflict can be resolved if we imagine that the rod is fictitiously extended by one subinterval at the end of the rod that is losing heat. Doing so permits us to approximate the derivative with a central difference. The "temperature" at this fictitious point is eliminated by using the equation for the gradient. The next example will clarify this.

EXAMPLE 6.6 An insulated rod is 20 cm long and is of uniform cross section. It has its right end held at 100° while its left end loses heat to the surroundings, which are at 20°. The rod has a thermal conductivity, k, of 0.52 cal/(sec * cm * °C), and the heat-transfer coefficient, H, is 0.073 cal/(sec/cm^2/°C). Solve for the steady-state temperatures using the finite-difference method with eight subintervals.

For this example, because the boundary condition at the left end involves both the u-value at the left end and the derivative there, this example has a mixed condition at the left end, whereas it has a Dirichlet condition at the right end.

The equation that applies is Eq. (6.24) with $Q = 0$, because no heat is added at points along the rod:

$$\frac{d^2u}{dx^2} = 0.$$

The typical equation is

$$u_{i-1} - 2u_i + u_{i+1} = 0,$$

and this applies at each node. At the left end we imagine a fictitious point at x_{-1}, and this allows us to write the equation for that node. At the left endpoint, at $x = x_0$, we write an equation for the gradient:

$$\frac{du}{dx} = \left(\frac{H}{k}\right)(u_{\text{L}} - u_{\text{SL}}),$$

or,

$$\frac{(u_1 - u_{-1})}{2h} = \frac{(u_1 - u_{-1})}{(2 * 2.5)}$$
$$= \left(\frac{0.073}{0.52}\right) * (u_0 - 20),$$

which we use to eliminate u_{-1}:

$$u_{-1} = u_1 - (2 * 2.5) * \left[\left(\frac{0.073}{0.52}\right)(u_0 - 20)\right]$$
$$= u_1 - 0.70192u_0 + 14.0385.$$

We will use this last for the equation written at x_0, to give, at that point:

$$u_{-1} - 2u_0 + u_1 = (u_1 - 0.70192u_0 + 14.0385) - 2u_0 + u_1 = 0,$$

or,

$$-2.70192u_0 + 2u_1 = -14.0385,$$

which is the first equation of the set. Here is the augmented matrix for the problem:

$$\begin{bmatrix} -2.70192 & 2 & 0 & 0 & 0 & 0 & 0 & 0 & -14.0385 \\ 1 & -2 & 1 & 0 & 0 & 0 & 0 & 0 & 0 \\ 0 & 1 & -2 & 1 & 0 & 0 & 0 & 0 & 0 \\ 0 & 0 & 1 & -2 & 1 & 0 & 0 & 0 & 0 \\ 0 & 0 & 0 & 1 & -2 & 1 & 0 & 0 & 0 \\ 0 & 0 & 0 & 0 & 1 & -2 & 1 & 0 & 0 \\ 0 & 0 & 0 & 0 & 0 & 1 & -2 & 1 & 0 \\ 0 & 0 & 0 & 0 & 0 & 0 & 1 & -2 & -100 \end{bmatrix}$$

for which the solution is

i:	0	1	2	3	4	5	6	7	(8)
u_i:	41.0103	48.3840	55.7577	63.1314	70.5051	77.8789	85.2526	92.6263	(100)

Observe that the gradient all along the rod is a constant (2.94948°C/cm).

Here is another example that illustrates an important point about derivative boundary conditions.

EXAMPLE 6.7 Solve $u'' = u$, $u'(1) = 1.17520$, $u'(3) = 10.01787$, with the finite-difference method.

This example is identical to that of Example 6.5, except that the boundary conditions are the derivatives of u rather than the values of u. (It has Neumann conditions at both ends.) For this problem, the known solution is $u = \cosh(x) + C$, and the boundary values are values of sinh(1) and sinh(3).

Because the values of u are not given at either end of the interval, we must add fictitious points at both ends; call these u_{LF} and u_{RF}. With four subintervals, ($h = 2/4 = 0.5$), we can write five equations (at each of the three interior nodes plus the two endpoints where u is unknown). We label the nodes from x_0 (at the left end) to x_4 (at the right end). Each equation is of the form:

$$u_{i-1} - 2u_i + u_{i+1} = h^2 u_i, \quad i = 0, 1, 2, 3, 4, \quad h^2 = 0.25,$$

where u_{-1} and u_5 are the fictitious points u_{LF} and u_{RF}.

Doing so gives this augmented matrix:

$$\begin{bmatrix} -2.25 & 1 & 0 & 0 & 0 & -u_{LF} \\ 1 & -2.25 & 1 & 0 & 0 & 0 \\ 0 & 1 & -2.25 & 1 & 0 & 0 \\ 0 & 0 & 1 & -2.25 & 1 & 0 \\ 0 & 0 & 0 & 1 & -2.25 & -u_{RF} \end{bmatrix}$$

There are two more unknowns in this than equations: the unknown fictitious points. However, these can be eliminated by using the derivative conditions at the ends. As before, we use central difference approximation to the derivative:

$$u'(1) = 1.17520 = \frac{(u_1 - u_{LF})}{2h},$$

$$u'(3) = 10.01787 = \frac{(u_{RF} - u_3)}{2h}, \qquad (h = 0.5),$$

which we solve for the fictitious points in terms of nodal points:

$$u_{LF} = u_1 - 1.17520, \qquad u_{RF} = 10.01787 + u_3.$$

Substituting these relations for the fictitious points changes the first and last equations to

$$-2.25u_0 + 2u_1 = 1.17520,$$
$$2u_3 - 2.25u_4 = -10.01787.$$

When the five equations are solved, we get these answers:

x	**Answers**	**cosh(x)**	**Error**
1.0	1.55219	1.54308	−0.00911
1.5	2.33382	2.35241	0.01859
2.0	3.69870	3.76220	0.06350
2.5	5.98870	6.13229	0.14359
3.0	9.77568	10.06770	0.29202

We observe that the accuracy is much poorer than it was in Example 6.5. Take note of the fact that the numerical solution is not identical to the analytical solution; the arbitrary constant is missing (or, we may say, is equal to zero).

Using the Shooting Method

We can solve boundary-value problems where the derivative is involved at one or both end conditions by "shooting." In fact, as this method computes both the dependent variable and its derivative, this is quite natural. Here is how Example 6.7 can be solved by the shooting method.

EXAMPLE 6.8 Solve $u'' = u$, $u'(1) = 1.17520$, $u'(3) = 10.01787$ by the shooting method.

We can begin at either end, but it seems more natural to begin from $x = 1$. To begin the solution, we must guess at a value for $u(1)$—not for the derivative as we have been doing. From this point, we compute values for u and u' by, say, RKF. If the value of $u'(3)$ is not 10.01787, we try again with a guess for $u(1)$. This will probably not give the correct value for $u'(3)$, but, because the problem is linear, we can interpolate to find the proper value to use for $u(1)$. Here are the answers when four subintervals are used:

x	$u(x)$	$u'(x)$	$\cosh(x)$
1.0	1.54319	1.17520	1.54308
1.5	2.35250	2.12932	2.35241
2.0	3.76228	3.62692	3.76220
2.5	7.13236	7.05027	6.13229
3.0	10.06767	10.01790	10.06770

The results are surprisingly accurate even though the subdivision was coarse; the largest error in the $u(x)$ values is 0.00011 at $x = 1$ and the errors are less as x increases. For this example, the shooting method is much more accurate than using finite-difference approximations to the derivative.

Here is an example that has a mixed end condition.

EXAMPLE 6.9 Solve Example 6.6 by the shooting method. We restate the problem:

An insulated rod is 20 cm long and is of uniform cross section. It has its right end held at 100° while its left end loses heat to the surroundings, which are at 20°. The rod has a thermal conductivity, k, of 0.52 cal/(sec * cm * °C), and the heat-transfer coefficient, H, is 0.073 cal/(sec * cm^2 * °C). Use the shooting method with eight subintervals.

The procedure here is similar to that used in Example 6.8 but it is necessary to begin at the right end and solve "backwards." (That is no problem; we just use a negative value for Δx.) Beginning at $x = 0$ would be very difficult because we would have to guess at both $u(0)$ and $u'(0)$.

Finding the correct value for u' at $x = 20$ is not as easy as in the previous example because we must fit to a combination of $u(0)$ and $u'(0)$. Here are the results after finding the correct value for $u'(20)$ by a trial and error technique.

i:	0	1	2	3	4	5	6	7	(8)
u_i:	41.005	48.379	55.754	63.128	70.502	77.877	85.251	92.626	(100)

(The gradient here is 2.94975 throughout.) These values match those of Example 6.6 very closely.

We note that Maple can solve a boundary-value problem with an end condition that involves the derivative.

6.8 Characteristic-Value Problems

Problems in the fields of elasticity and vibration (including applications of the wave equation of modern physics) fall into a special class of boundary-value problems known as *characteristic-value problems*. Some problems of statistics also fall into this class. We discuss only the most elementary forms of characteristic-value problems.

Consider the homogeneous* second-order equation with homogeneous boundary conditions:

$$\frac{d^2u}{dx^2} + k^2u = 0, \qquad u(0) = 0, \qquad u(1) = 0, \tag{6.29}$$

where k^2 is a parameter. (Using k^2 guarantees that the parameter is a positive number.) We first solve this equation nonnumerically to show that there is a solution only for certain particular or "characteristic" values of the parameter. These characteristic values are more often called the *eigenvalues* from the German word. The general solution is

$$u = a\sin(kx) + b\cos(kx),$$

which can easily be verified by substituting into the differential equation. The solution contains the two arbitrary constants a and b because the equation is of second order. The constants a and b are to be determined to make the general solution agree with the boundary conditions.

At $x = 0$, $u = 0 = a\sin(0) + b\cos(0) = b$. Then b must be zero. At $x = 1$, $u = 0 = a\sin(k)$; we may have either $a = 0$ or $\sin(k) = 0$ to satisfy the end condition. However, if $a = 0$, y is everywhere zero—this is called the *trivial solution,* and is usually of no interest. To get a useful solution, we must choose $\sin(k) = 0$, which is true only for certain "characteristic" values:

$$k = \pm n\pi, \qquad n = 1, 2, 3, \ldots .$$

* Homogeneous here means that all terms in the equation are functions of u or its derivatives.

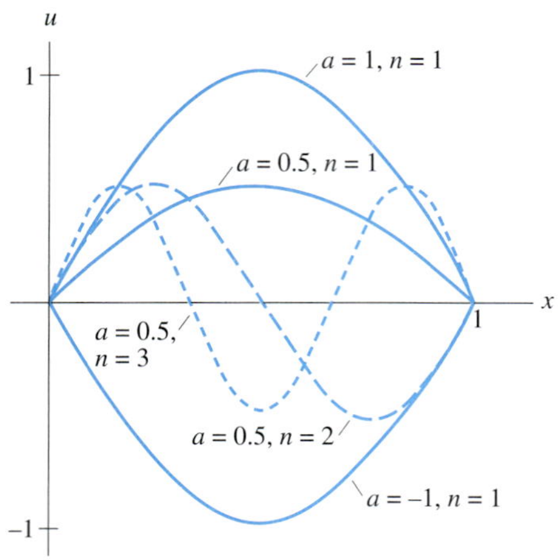

Figure 6.6

These are the eigenvalues for the equation, and the solution to the problem is

$$u = a \sin(n\pi x), \qquad n = 1, 2, 3, \ldots . \tag{6.30}$$

The constant a can have any value, so these solutions are determined only to within a multiplicative constant. Figure 6.6 sketches several of the solutions to Eq. (6.30).

These eigenvalues are the most important information for a characteristic-value problem. In a vibration problem, these give the natural frequencies of the system, which are important because, if the system is subjected to external loads applied at or very near to these frequencies, resonance causes an amplification of the motion and failure is likely.

Corresponding to each eigenvalue is an eigenfunction, $u(x)$, which determines the possible shapes of the elastic curve when the system is at equilibrium. Figure 6.6 shows such eigenfunctions. Often the smallest eigenvalue is of particular interest; at other times, it is the one of largest magnitude.

We can solve Eq. (6.29) numerically, and that is what we concentrate on in this section. We will replace the derivatives in the differential equation with finite-difference approximations, so that we replace the differential equation with difference equations written at all nodes where the value of u is unknown (which are all the nodes of a one-dimensional system except for the endpoints).

EXAMPLE 6.10 Solve Eq. (6.29) with five subintervals. We restate the problem:

$$\frac{d^2u}{dx^2} + k^2 u = 0, \qquad u(0) = 0, \qquad u(1) = 0.$$

The typical equation is

$$\frac{(u_{i-1} - 2u_i + u_{i+1})}{h^2} + k^2 u_i = 0.$$

With five subintervals, $h = 0.2$, and there are four equations because there are four interior nodes. In matrix form these are

$$\begin{bmatrix} 2-0.04k^2 & -1 & 0 & 0 \\ -1 & 2-0.04k^2 & -1 & 0 \\ 0 & -1 & 2-0.04k^2 & -1 \\ 0 & 0 & -1 & 2-0.04k^2 \end{bmatrix} \begin{bmatrix} u_1 \\ u_2 \\ u_3 \\ u_4 \end{bmatrix} = \begin{bmatrix} 0 \\ 0 \\ 0 \\ 0 \end{bmatrix} \tag{6.31}$$

where we have multiplied by -1 for convenience. Observe that this can be written as the matrix equation $(A - \lambda I)u = 0$, where I is the identity matrix and the A matrix is

$$\begin{bmatrix} 2 & -1 & 0 & 0 \\ -1 & 2 & -1 & 0 \\ 0 & -1 & 2 & -1 \\ 0 & 0 & -1 & 2 \end{bmatrix},$$

and $\lambda = 0.04k^2$.

The approximate solution to the characteristic-value problem, Eq. (6.29) is found by solving the system of Eq. (6.31). However, this system is an example of a homogeneous system (the right-hand sides are all equal to zero), and it has a nontrivial solution only if the determinant of the coefficient matrix is zero. Hence, we set

$$\det(A - \lambda I) = 0.$$

Expanding the determinant will give an eighth-degree polynomial in k. (This is *not* the preferred way!) Doing so and getting the zeros of that polynomial gives these values for k:

$$k = \pm 3.09, \qquad k = \pm 5.88, \qquad k = \pm 8.09, \qquad k = \pm 9.51.$$

The analytical values for k are

$$k = \pm 3.14\ (\pm\pi), \qquad k = \pm 7.28\ (\pm 2\pi),$$
$$k = \pm 9.42\ (\pm 3\pi), \qquad k = \pm 12.57\ (\pm 4\pi),$$

and we see that the estimates for k are not very good and get progressively worse. We would need a much smaller subdivision of the interval to get good values. There are other problems with this technique: Expanding the determinant of a matrix of large size is computationally expensive, and solving for the roots of a polynomial of high degree is subject to large round-off errors. The system is very ill-conditioned.*

We normally find the eigenvalues for a characteristic-value problem from $(A - \lambda I)u = 0$ in other ways that are not subject to the same difficulties. We describe these now. For clarity we use small matrices.

The Power Method

The *power method* is an iterative technique. The basis for this is presented below. We illustrate the method through an example.

* One authority says never to use the characteristic polynomial for a matrix larger than 5×5.

EXAMPLE 6.11 Find the eigenvalues (and the eigenvectors) of matrix A:

$$A = \begin{bmatrix} 3 & -1 & 0 \\ -2 & 4 & -3 \\ 0 & -1 & 1 \end{bmatrix}$$

(The eigenvalues of A are 5.47735, 2.44807, and 0.074577, which are found, perhaps, by expanding the determinant of $A - \lambda I$. The eigenvectors are found by solving the equations $Au = \lambda u$ for each value of λ. After normalizing, these vectors are

$$u_1 = [-0.40365, 1, -0.22335],$$
$$u_2 = [1, 0.55193, -0.38115],$$
$$u_3 = [0.31633, 0.92542, 1],$$

where the normalization has been to set the largest component equal to unity.)*

We will find that both the eigenvalues and the eigenvectors are produced by the power method. We begin this by choosing a three-component vector more or less arbitrarily. (There are some choices that don't work but usually the column vector $u = [1, 1, 1]$ is a good starting vector.) We always use a vector with as many components as rows or columns of A.

We repeat these steps:

1. Multiply $A * u$.
2. Normalize the resulting vector by dividing each component by the largest in magnitude.
3. Repeat steps 1 and 2 until the change in the normalizing factor is negligible. At that time, the normalization factor is an eigenvalue and the final vector is an eigenvector.

Step 1, with $u = [1, 1, 1]$:

$$A * u \qquad \text{gives} \qquad [2, -1, 0].$$

Step 2:

Normalizing gives $2 * [1, -.5, 0]$, and u now is $[1, -.5, 0]$.

Repeating, we get

$$A * u = [3.5, -4, .5],$$
$$\text{normalized: } -4 * [-.875, 1, -.125];$$
$$A * u = [-3.625, 6.125, -1.125],$$
$$\text{normalized: } 6.125 * [-.5918, 1, -.1837];$$
$$A * u = [-2.7755, 5.7347, -1.1837],$$
$$\text{normalized: } 5.7347 * [-.4840, 1, -.2064];$$

After 14 iterations, we get

---

* It is more common to set some norm equal to 1.

$$A * u = [-2.21113, 5.47743, -1.22333],$$
$$\text{normalized: } 5.47743 * [-.40368, 1, -.22334].$$

The fourteenth iteration shows a negligible change in the normalizing factor: We have approximated the largest eigenvalue and the corresponding eigenvector. (Twenty iterations will give even better values.) Although not very rapid, the method is extremely simple and easy to program. Any of the computer algebra systems can do this for us.

The Inverse Power Method

The previous example showed how the power method gets the eigenvalue of largest magnitude. What if we want the one of smallest magnitude? All we need to do to get this is to work with the inverse of A. For the matrix A of Example 6.11, its inverse is

$$\begin{bmatrix} 1 & 1 & 3 \\ 2 & 3 & 9 \\ 2 & 3 & 10 \end{bmatrix}.$$

Applying the power method to this matrix gives a value for the normalizing factor of 13.4090 and a vector of [.3163, .9254, 1]. For the original matrix A, the eigenvalue is the reciprocal, 0.07457. The eigenvector that corresponds is the same; no change is needed.

Shifting with the Power Method

As we have seen, the power method may not converge very fast. We can accelerate the convergence as well as get eigenvalues of magnitude intermediate between the largest and smallest by *shifting*. Suppose we wish to determine the eigenvalue that is nearly equal to some number s. If s is subtracted from each of the diagonal elements of A, the resulting matrix has eigenvalues the same as for A but with s subtracted from them. This means that there is an eigenvalue for the shifted matrix that is nearly zero. We now use the inverse power method on this shifted matrix, and the reciprocal of this very small eigenvalue is usually very much larger in magnitude than any other. As shown below, this causes the convergence to be rapid. Observe that if we have some knowledge of what the eigenvalues of A are, we can use this shifted power method to get the value of any of them.

How can we estimate the eigenvalues of a matrix? *Gerschgorin's theorem* can help here. This theorem is especially useful if the matrix has strong diagonal dominance. The first of Gerschgorin's theorems says that the eigenvalues lie in circles whose centers are at a_{ii} with a radius equal to the sum of the magnitudes of the other elements in row i. (Eigenvalues can have complex values, so the circles are in the complex plane.)

Gerschgorin's Theorem We will not give a proof of this theorem,* but only show that it applies in several examples.

---

* Proofs can be found in Ralston (1965) and in Burdern and Faires (2001).

If matrix A is diagonal, its eigenvalues are the diagonal elements:

$$\begin{matrix} 10 & 0 & 0 & & \\ 0 & 7 & 0 & \rightarrow & 4,\ 7,\ 10, \quad \text{which are in} \\ 0 & 0 & 4 & & 4 \pm 0,\ 7 \pm 0,\ \ 10 \pm 0. \end{matrix}$$

If matrix A has small off-diagonal elements:

$$\begin{matrix} 10 & 0.1 & 0.1 & & \\ 0.1 & 7 & 0.1 & \rightarrow & 3.9951, \quad 6.9998, \quad 10.0051, \text{ in} \\ 0.1 & 0.1 & 4 & & 4 \pm 0.2, \quad 7 \pm 0.2, \quad 10 \pm 0.2, \end{matrix}$$

and there is a small change.

When the off-diagonals are larger:

$$\begin{matrix} 10 & 1 & 1 & & \\ 1 & 7 & 1 & \rightarrow & 3.6224, \ 6.8329, \quad 10.5446, \text{ in} \\ 1 & 1 & 4 & & 4 \pm 2, \quad 7 \pm 2, \quad 10 \pm 2, \end{matrix}$$

there is a greater change.

If they are still larger:

$$\begin{matrix} 10 & 2 & 2 & & \\ 2 & 7 & 2 & \rightarrow & 2.8606, \quad 6.2151, \quad 11.9243, \text{ in} \\ 2 & 2 & 4 & & 4 \pm 4, \quad 7 \pm 4, \quad 10 \pm 4, \end{matrix}$$

there is a still greater change, but the theorem holds.

Even in this case, the theorem holds:

$$\begin{matrix} 10 & 4 & 4 & & \\ 4 & 7 & 4 & \rightarrow & 1.0398, \quad 4.4704, \quad 15.4898, \text{ in} \\ 4 & 4 & 4 & & 4 \pm 8, \quad 7 \pm 8, \quad 10 \pm 8. \end{matrix}$$

Whenever the matrix is diagonally dominant or nearly so, shifting by the value of a diagonal element will speed up convergence in the power method.

EXAMPLE 6.12 Given matrix A:

$$\begin{bmatrix} 4 & -1 & 1 \\ 1 & 1 & 1 \\ -2 & 0 & -6 \end{bmatrix}$$

find all of its eigenvalues using the shifted power method.

Gerschgorin's theorem says that there are eigenvalues within -6 ± 2, 1 ± 2, and 4 ± 2. We shift first by -6 and get an eigenvalue equal to -5.76851 (vector = $[-.11574, -.13065, 1]$) using the inverse power method in four iterations; the tolerance on change in the normalization factor was 0.0001. (Getting this largest-magnitude eigenvalue through the regular power method required 23 iterations.) If we repeat but shift by one, the inverse power method gives 1.29915 as an eigenvalue (vector = [.41207, 1, −.11291]) in six iterations. (Using just the inverse power method to get this smallest of the eigenvalues required eight iterations.)

For this 3×3 matrix, we do not have to get the other eigenvalue; the sum of the eigenvalues equals the trace of the matrix. So, if we subtract (−5.76851 + 1.29915) from −1 (the trace) we get the third eigenvalue, 3.46936. (It is always true that the sum of the eigenvalues equals the trace.) The eigenvalues satisfy Gerschgorin's theorem: −5.76851 is in -6 ± 2, 1.29915 is in 1 ± 2, 3.46936 is in 4 ± 2.

Getting the third eigenvalue from the trace does not give us its eigenvector; we can use the shifted inverse power method on the original matrix to find it.

Shifting by 4 in this example runs into a problem; a division by zero is attempted. We overcome this problem by distorting the shift amount slightly. Shifting by 3.9 and employing the inverse power method gives the eigenvalue: 3.46936, and the vector [1, .31936, −.21121] in six iterations. (If a division by zero occurs, it is advisable to distort the shift amount slightly.)

The Basis for the Power Method

The utility of the power method is that it finds the eigenvalue of largest magnitude and its corresponding eigenvector in a simple and straightforward manner. It has the disadvantage that convergence is slow if there is a second eigenvalue of nearly the same magnitude. The following discussion proves this and also shows why some starting vectors are unsuitable.

The method works because the eigenvectors are a set of *basis vectors*. A set of basis vectors is said to *span the space,* meaning that any n-component vector can be written as a unique linear combination of them. Let $v^{(0)}$ be any vector and $x_1, x_2, \ldots, x_n$, be eigenvectors. Then, for a starting vector, $v^{(0)}$,

$$v^{(0)} = c_1x_1 + c_2x_2 + \cdots + c_nx_n.$$

If we multiply $v^{(0)}$ by matrix A, because the x_i are eigenvectors with corresponding eigenvalues λ_i and remembering that $Ax_i = \lambda_i x$, we have,

$$\begin{aligned} v^{(1)} = Av^{(0)} &= c_1Ax_1 + c_2Ax_2 + \cdots + c_nA_nx_n \\ &= c_1\lambda_1x_1 + c_2\lambda_2x_2 + \cdots + c_n\lambda_nx_n, \end{aligned} \tag{6.32}$$

Upon repeated multiplication by A, after m such multiplies, we get,

$$v^{(m)} = A^mv^{(0)} = c_1\lambda_1^mx_1 + c_2\lambda_2^mx_2 + \cdots + c_n\lambda_n^mx_n.$$

Now, if one of the eigenvalues, call it λ_1, is larger than all the rest, it follows that all the coefficients in the last equation become negligibly small in comparison to λ_1^m as m gets large, so

$$A^mv^{(0)} \longrightarrow c_1\lambda_1^mx_1,$$

which is some multiple of eigenvector x_1 with the normalization factor λ_1, provided only that $c_1 \neq 0$. This is the principle behind the power method. Observe that if another of the eigenvalues is exactly of the same magnitude as λ_1, there never will be convergence to a single value. Actually, in this case, the normalization values alternate between two numbers and the eigenvalues are the square root of the product of these values. If another eigenvalue is not equal to λ_1, but is near to it, convergence will be slow. Also, if the starting vector, $v^{(0)}$, is such that the coefficient c_1 in Eq. (6.32) equals zero, the method will not work. (This last will be true if the starting vector is "perpendicular" to the eigenvector that corresponds to λ_1—that is, the dot-product equals zero.) On the other hand, if the starting vector is almost "parallel" to the eigenvector of λ_1, all the other coefficients in Eq. (6.32) will be very small in comparison to c_1 and convergence will be very rapid.

The preceding discussion also shows why shifting and then using the inverse power method can often speed up convergence to the eigenvalue that is near the shift quantity. Here we create, in the shifted matrix, an eigenvalue that is nearly zero, so that using the inverse method makes the reciprocal of this small number much larger than any other eigenvalue.

The power method with its variations is fine for small matrices. However, if a matrix has two eigenvalues of equal magnitude, the method fails in that the successive normalization factors alternate between two numbers. The duplicated eigenvalue in this case is the square root of the product of the alternating normalization factors. If we want all the eigenvalues for a larger matrix, there is a better way.

The QR Method, Part 1—Similarity Transformations

If matrix A is diagonal or upper- or lower-triangular, its eigenvalues are just the elements on the diagonal. This can be proved by expanding the determinant of $(A - \lambda I)$. This suggests that, if we can transform A to upper-triangular, we have its eigenvalues! We have done such a transformation before: The Gaussian elimination method does it. Unfortunately, this transformation changes the eigenvalues!!

There are other transformations that do not change the eigenvalues. These are called *similarity transformations*. For any nonsingular matrix, M, the product $M * A * M^{-1} = B$, transforms A into B, and B has the same eigenvalues as A. The trick is to find matrix M such that A is transformed into a similar upper-triangular matrix from which we can read off the eigenvalues of A from the diagonal of B. The QR technique does this. We first change one of the subdiagonal elements of A to zero; we then continue to do this for all the elements below the diagonal until A has become upper-triangular. The process is slow; many iterations are required, but the procedure does work.

Suppose that A is 4×4. Here is a matrix, Q, also 4×4, that will create a zero in position a_{42}:

$$Q = \begin{bmatrix} 1 & 0 & 0 & 0 \\ 0 & c & 0 & s \\ 0 & 0 & 1 & 0 \\ 0 & -s & 0 & c \end{bmatrix},$$

where

$$d = \sqrt{(a_{22}^2 + a_{42}^2)},$$

$$c = \frac{a_{22}}{d},$$

$$s = \frac{a_{42}}{d}.$$

EXAMPLE 6.13 Given this matrix A, create a zero in position (4, 2) by multiplying by the proper Q matrix.

$$A = \begin{bmatrix} 7 & 8 & 6 & 6 \\ 1 & 6 & -1 & -2 \\ 1 & -2 & 5 & -2 \\ 3 & 4 & 3 & 4 \end{bmatrix}$$

We compute:

$$d = \sqrt{(6^2 + 4^2)} = 7.21110,$$

$$c = \frac{6}{d} = 0.83205,$$

$$s = \frac{4}{d} = 0.55470.$$

The Q matrix is

$$\begin{bmatrix} 1 & 0 & 0 & 0 \\ 0 & .83205 & 0 & .55470 \\ 0 & 0 & 1 & 0 \\ 0 & -.55470 & 0 & .83205 \end{bmatrix}.$$

When we multiply Q by A, we get for $Q * A$:

$$\begin{bmatrix} 7 & 8 & 6 & 6 \\ 2.49615 & 7.21110 & .83205 & .55470 \\ 1 & -2 & 5 & -2 \\ 1.94145 & 0 & 3.05085 & 4.43760 \end{bmatrix}$$

where the element in position (4, 2) is zero, as we wanted. However, we do not yet have a similarity transformation. (The trace has been changed, meaning that the eigenvalues are not the same as those of A.) To get the similarity transformation that is needed, we must now postmultiply by the inverse of Q. Getting the inverse (which is Q^{-1}) is easy in this case because for any Q as defined here, its inverse is just its transpose! (When this is true for a matrix, it is called a *rotation matrix*.) If we now multiply $Q * A * Q^{-1}$, we get

$$\begin{bmatrix} 7 & 9.98460 & 6 & 0.55470 \\ 2.49615 & 6.30769 & 0.83205 & -3.53846 \\ 1 & -2.77350 & 5 & -0.55470 \\ 1.94145 & 2.46154 & 3.05085 & 3.69231 \end{bmatrix},$$

for which the trace is the same as that of the original A and whose eigenvalues are the same. However, it seems that we have not really done what we desired; the element in position (4, 2) is zero no longer! There has been some improvement, though. Observe that the sum of the magnitudes of the off-diagonal elements in row 4 is smaller than in matrix A. This means that 3.69231 is closer to one of the eigenvalues (which will turn out to be 1) than the original value, 4. Also, the element in position (2, 2) (6.30769) is closer to another eigenvalue (which is equal to 7) than the original number, 6.

This suggests that we should continue doing such similarity transformations to reduce all below-diagonal elements to zero. It takes many iterations, but, after doing 111 of these, we get

$$\begin{bmatrix} 10 & 1.5811 & -11.0680 & -3.0000 \\ 0 & 7 & -1.0000 & 0.0000 \\ 0 & 0 & 4 & -3.1623 \\ 0 & 0 & 0 & 1 \end{bmatrix},$$

where the numbers have been rounded to four decimals. (All the below-diagonal elements have a value of 0.00001 or less.) We have found the eigenvalues of A; these are 10, 7, 4, and 1.

The QR Method, Part 2 – Making the Matrix Upper Hessenberg

The trouble with doing such similarity transformations repeatedly is poor efficiency. We can improve the method by first doing a *Householder transformation,* which is a similarity transformation that creates zeros in matrix A for all elements below the "subdiagonal." (This means all elements below the diagonal except for those immediately below the diagonal. We might call such a matrix "almost triangular.") The name for such a matrix is *upper Hessenberg.* The Householder transformation changes matrix A into upper Hessenberg. Once an $n \times n$ matrix has been converted to upper Hessenberg, there are only $n - 1$ elements to reduce, compared to $(n)(n - 1)/2$.

There is another technique that further speeds up the reduction of matrix A to upper-triangular. We can employ shifting (similar to that done in the power method). The easiest way to shift is to do it with the element in the last row and last column.

Here are the steps that we will use:

1. Convert to upper Hessenberg.
2. Shift by a_{nn}, then do similarity transformations for all columns from 1 to $n - 1$.
3. Repeat step 2 until all elements to the left of a_{nn} are essentially zero. An eigenvalue then appears in position a_{nn}.

4. Ignore the last row and column, and repeat steps 2 and 3 until all elements below the diagonal of the original matrix are essentially zero. The eigenvalues then appear on the diagonal.

How do we convert matrix A to upper Hessenberg without changing the eigenvalues? This is best explained through an example.

EXAMPLE 6.14 Convert the same matrix A (as in Example 6.13) to upper Hessenberg.

We recall that A is

$$\begin{bmatrix} 7 & 8 & 6 & 6 \\ 1 & 6 & -1 & -2 \\ 1 & -2 & 5 & -2 \\ 3 & 4 & 3 & 4 \end{bmatrix}.$$

We can create zeros in the first column and rows 3 and 4 by $B * A * B^{-1}$, where

$$B = \begin{bmatrix} 1 & 0 & 0 & 0 \\ 0 & 1 & 0 & 0 \\ 0 & -b_3 & 1 & 0 \\ 0 & -b_4 & 0 & 1 \end{bmatrix}, \qquad B^{-1} = \begin{bmatrix} 1 & 0 & 0 & 0 \\ 0 & 1 & 0 & 0 \\ 0 & b_3 & 1 & 0 \\ 0 & b_4 & 0 & 1 \end{bmatrix},$$

$$b_3 = a_{31}/a_{21} = 1/1 = 1,$$
$$b_4 = a_{41}/a_{21} = 3/1 = 3.$$

Observe that the B matrix is the identity matrix with the two zeros below the diagonal in column 2 replaced with $-b_3$ and $-b_4$, where these values are the elements of column 1 of matrix A that are to be made zero divided by the subdiagonal element in column 1. The inverse of this B matrix is B with the signs changed for the new elements in its column 2.

If we now perform the multiplications $B_1 * A * B_1^{-1}$, we get

$$\begin{bmatrix} 7 & 32 & 6 & 8 \\ 1 & -1 & -1 & -2 \\ 0 & -2 & 6 & 0 \\ 0 & 22 & 6 & 10 \end{bmatrix},$$

which has zeros below the subdiagonal of column 1 and the same eigenvalues as the original matrix A.

We continue this in column 2, where now

$$B_2 = \begin{bmatrix} 1 & 0 & 0 & 0 \\ 0 & 1 & 0 & 0 \\ 0 & 0 & 1 & 0 \\ 0 & 0 & -b_4 & 1 \end{bmatrix}, \qquad \text{with} \qquad b_4 = a_{42}/a_{32} = 22/-2 = -11.$$

Here, B_2^{-1} is the same as B_2 except that the sign of b_4 is changed. Now premultiplying the last matrix by B_2 and postmultiplying by B_2^{-1} gives the lower Hessenberg matrix:

$$B_2 B_1 A B_1^{-1} B_2^{-1} = \begin{bmatrix} 7 & 32 & -60 & 6 \\ 1 & -1 & 21 & -2 \\ 0 & -2 & 6 & 0 \\ 0 & 0 & -38 & 10 \end{bmatrix},$$

which is what was desired.

There is a potential problem with this reduction to the Hessenberg matrix. If the divisor used to create the B matrices is zero or very small, either a division by zero occurs or the round-off error is great. We can avoid these problems by interchanging both rows and columns to put the element of largest magnitude in the subdiagonal position. It is essential to do the interchanges for both rows and columns so that the diagonal elements remain the same.

The QR Method, Part 3—The Steps Combined

If we (1) convert matrix A to upper Hessenberg, and, (2) perform QR operations on this, the final matrix that results is

$$\begin{bmatrix} 10 & 9.8315 & 4.9054 & -3.2668 \\ 0 & 1 & 1.8256 & 2.7199 \\ 0 & 0 & 4 & -1.6958 \\ 0 & 0 & 0 & 7 \end{bmatrix},$$

in which the same eigenvalues appear on the diagonal as when QR operations were done on the original A matrix. However, only seven QR iterations were required after reduction to Hessenberg, compared to 111 if that step is omitted. The other elements are different because row and column interchanges were done in creating the last result.

MATLAB can find the eigenvalues and eigenvectors of a square matrix. Here is an example:

Find the eigenvalues of

$$\begin{bmatrix} 10 & 0 & 0 \\ 1 & -3 & -7 \\ 0 & 2 & 6 \end{bmatrix}.$$

Solution:

We define A in MATLAB:

```
A = [10  0  0; 1  -3  -7; 0  2  6]
A =
```

```
10   0   0
 1  -3  -7
 0   2   6
```

and then do

```
e = eig(A)
e =
    4
   -1
   10
```

If we want both the eigenvalues and eigenvectors:

```
[V, D] = eig(A)
V =
          0         0    0.9977
    -0.7071    0.9615    0.0605
     0.7071   -0.2747    0.0302
D =
     4   0   0
     0  -1   0
     0   0  10
```

where the eigenvectors appear as the columns of V (they are scaled so each has a norm of one) and the eigenvalues are on the diagonal of matrix D. Observe that MATLAB gets all the eigenvectors at once.

Suppose we want to get the eigenvalues of A after its element in row 1, column 2 is changed to one. If that is what we want, we just enter:

```
A(1, 2) = 1;
eig(A)
ans =
   10.0606
   -1.1250
    4.0644
```

MATLAB uses a QR algorithm to get the eigenvalues after converting to Hessenberg form as described. We can also use the characteristic polynomial:

After defining the original matrix (A) in MATLAB, we do

```
EDU>> pp = poly(A)
pp =
   1.0000  -13.0000  25.0000  46.0000
```

which are the coefficients of the cubic

$x^3 - 13x^2 + 25x + 46.$

We get the roots by

```
EDU>> roots(pp)
ans =
    10.0606
     4.0644
    -1.1250
```

which is the same as before, as expected.

Exercises

Section 6.1

1. Use the Taylor series method to get solutions to

$$dy/dx = x + y - xy, \qquad y(0) = 1$$

at $x = 0.1$ and $x = 0.5$. Use terms through x^5.

▶ **2.** The solution to Exercise 1 at $x = 0.5$ is 1.59420. How many terms of a Taylor series must be used to match this?

3. Repeat Exercises 1 and 2 but for

$$y''(x) = x/y, \qquad y(0) = 1, \qquad y'(0) = 1.$$

The correct value for $y(0.5)$ is 1.51676.

4. A spring system has resistance to motion proportional to the square of the velocity, and its motion is described by

$$\frac{d^2x}{dt^2} + 0.1\left(\frac{dx}{dt}\right)^2 + 0.6x = 0.$$

If the spring is released from a point that is a unit distance above its equilibrium point, $x(0) = 1$, $x'(0) = 0$, use the Taylor-series method to write a series expression for the displacement as a function of time, including terms up to t^6.

Section 6.2

5. Repeat Exercise 1, but use the simple Euler method. How small must h be to match to the values of Exercise 1?

▶ **6.** Repeat Exercise 2, but use the simple Euler method. How small must h be?

7. Repeat Exercise 5, but now with the modified Euler method. Comparing to Exercise 5, how much less effort is required?

8. Find the solution to

$$\frac{dy}{dt} = y^2 + t^2, \quad y(1) = 0, \quad \text{at } t = 2,$$

by the modified Euler method, using $h = 0.1$. Repeat with $h = 0.05$. From the two results, estimate the accuracy of the second computation.

9. Solve $y' = \sin(x) + y$, $y(0) = 2$ by the modified Euler method to get $y(0.1)$ and $y(0.5)$. Use a value of h small enough to be sure that you have five digits correct.

▶ **10.** A sky diver jumps from a plane, and during the time before the parachute opens, the air resistance is proportional to the $\frac{3}{2}$ power of the diver's velocity. If it is known that the maximum rate of fall under these conditions is 80 mph, determine the diver's velocity during the first 2 sec of fall using the modified Euler method with $\Delta t = 0.2$. Neglect horizontal drift and assume an initial velocity of zero.

11. Repeat Exercise 8 but use the midpoint method. Are the results the same? If not, which is more accurate?

12. The midpoint method gives results identical to modified Euler for $dy/dx = -2x - xy$, $y(0) = -1$. But for some definitions of dy/dx, it is better; for other definitions, it is worse. What are the conditions on the derivative function that cause

a. The midpoint method to be better?
b. The midpoint method to be poorer?
c. The two methods to give identical results?
d. Give specific examples for parts (a) and (b).

▶ **13.** For some derivative functions, the simple Euler method will have errors that are always positive but for others, the errors will always be negative.

a. What property of the function will determine which kind of error will be experienced?
b. Provide examples for both types of derivative function.

c. When will the errors be positive at first, but then become negative? Give an example where the errors oscillate between positive and negative as the x-values increase.

14. Is the phenomenon of Exercise 13 true for the modified Euler method? If it is, repeat Exercise 13 for this method.

Section 6.3

15. What are the equations that will be used for a second-order Runge–Kutta method if $a = 1/3$, $b = 2/3$, $\alpha = 3/4$ and $\beta = 3/4$. The statement is made that "this is said to give a minimum bound to the error." Test the truth of this statement by comparing this method with modified Euler on the equations of Exercises 1 and 8. Also compare to the midpoint method.

16. What is the equivalent of Eq. (6.10) for a third-order RK method? What then is the equivalent of Eq. (6.12)? Give three different combinations of parameter values that can be employed.

17. Use one set of the parameter values you found in Exercise 16 to solve Exercise 9.

a. How much larger can h be than the value found in Exercise 9?

b. Repeat with the other sets of parameters. Which set is preferred?

18. Solve Exercise 1 with fourth-order Runge–Kutta method. How large can h be to get the correct value at $x = 1.0$, which is 2.19496?

19. Determine y at $x = 1$ for the following equation, using fourth-order Runge–Kutta method with $h = 0.2$. How accurate are the results?

$$dy/dx = 1/(x + y), \qquad y(0) = 2.$$

▶**20.** Using the conditions of Exercise 10, determine how long it takes for the jumper to reach 90% of his or her maximum velocity, by integrating the equation using the Runge–Kutta technique with $\Delta t = 0.5$ until the velocity exceeds this value, and then interpolating. Then use numerical integration on the velocity values to determine the distance the diver falls in attaining $0.9v_{max}$.

21. It is not easy to know the accuracy with which the function has been determined by either the Euler methods or the Runge–Kutta method. A possible way to measure accuracy is to repeat the problem with a smaller step size, and compare results. If the two computations agree to n decimal places, one then assumes the values are correct to that many places. Repeat Exercise 20 with $\Delta t = 0.3$, which should give a global error about one-eighth as large, and by comparing results, determine the accuracy in Exercise 20. (Why do we expect to reduce the error eightfold by this change in Δt?)

22. Solve Exercises 1, 9, and 10 by the Runge–Kutta–Fehlberg method.

23. Using Runge–Kutta–Fehlberg, compare your results to that from fourth-order Runge–Kutta method in Exercise 18.

▶**24.** Solve $y' = 2x^2 - y$, $y(0) = -1$ by the Runge–Kutta–Fehlberg method to $x = 2.0$. How large can h be and still get the solution accurate to 6 significant digits?

25. Add the results from the Runge–Kutta–Fehlberg method to Table 6.6.

26. In the algorithm for the Runge–Kutta–Fehlberg method, an expression for the error is given. Repeat Exercise 19 with the Runge–Kutta–Fehlberg method and compare the actual error to the value from the expression.

Section 6.4

▶**27.** Derive the formula for the second-order Adams method. Use the method of undetermined coefficients.

28. Use the formula of Exercise 27 to get values as in Example 6.1.

29. For the differential equation

$$\frac{dy}{dt} = y - t^2, \quad y(0) = 1,$$

starting values are known:

$$y(0.2) = 1.2186, \qquad y(0.4) = 1{,}4682,$$
$$y(0.6) = 1.7379.$$

Use the Adams method, fitting cubics with the last four (y, t) values and advance the solution to $t = 1.2$. Compare to the analytical solution.

▶**30.** For the equation

$$\frac{dy}{dt} = t^2 - t, \quad y(1) = 0,$$

the analytical solution is easy to find:

$$y = \frac{t^3}{3} - \frac{t^2}{2} + \frac{1}{6}.$$

If we use three points in the Adams method, what error would we expect in the numerical solution? Confirm your expectation by performing the computations.

31. Repeat Exercise 30, but use four points.

32. Solve Exercise 29 with Adams–Moulton fourth order method.

33. For the equation $y' = y * \sin(\pi x)$, $y(0) = 1$, get starting values by RKF for $x = 0.2$, 0.4, and 0.6 and then advance the solution to $x = 1.4$ by Adams–Moulton fourth order method.

34. Get the equivalent of Eqs. (6.16) and (6.17) for a third-order Adams–Moulton method.

35. Derive the interpolation formulas given in Section 6.4 that permit getting additional values to reduce the step size.

▶**36.** Use Eq. (6.18) on this problem

$$dy/dx = 2x + 2, \quad y(1) = 3.$$

a. Is instability indicated?
b. Compare the results with this method to those from the simple Euler method as in Tables 6.11 and 6.12.

37. Use Milne's method on the equation in Exercise in 36. Is there any indication of instability?

38. Parallel the theoretical demonstration of instability with Milne's method with the equation $dy/dx = Ax^n$, where A and n are constants. What do you conclude?

39. What is the error term for Hamming's method? Show that it is a stable method.

Section 6.5

40. The mathematical model of an electrical circuit is given by the equation

$$0.5\frac{d^2Q}{dt^2} + 6\frac{dQ}{dt} + 50Q = 24\sin 10t,$$

with $Q = 0$ and $i = dQ/dt = 0$ at $t = 0$. Express as a pair of first-order equations.

▶**41.** In the theory of beams, it is shown that the radius of curvature at any point is proportional to the bending moment:

$$EI\frac{y''}{\{1 + (y')^2]^{3/2}} = M(x),$$

where y is the deflection of the neutral axis. In the usual approach, $(y')^2$ is neglected in comparison to unity, but if the beam has appreciable curvature, this is invalid. For the cantilever beam for which $y(0) = y'(0) = 0$, express the equation as a pair of simultaneous first-order equations.

42. A cantilever beam is 12 ft long and bears a uniform load of W lb/in. so that $M(x) = W * x^2/2$. Exercise 41 suggests that a simplified version of the differential equation can be used if the curvature of the beam is small. For what value of W, the value of the uniform load, does the simplified equation give a value for the deflection at the end of the beam that is in error by 5%?

▶**43.** Solve the pair of simultaneous equations

$$dx/dt = xy - t, \qquad x(0) = 1,$$
$$dy/dt = x + t, \qquad y(0) = 0,$$

by the modified Euler method from $t = 0$ to $t = 1.0$ in steps of 0.2.

44. Repeat Exercise 43, but with the Runge–Kutta–Fehlberg method. How accurate are these results? How much are the errors less than those of Exercise 43?

45. Use the first results of Exercise 44 to begin the Adams–Moulton method and then advance the solution to $x = 1.0$. Are the results as accurate as with the Runge–Kutta–Fehlberg method?

▶**46.** The motion of the compound spring system as sketched in Figure 6.7 is given by the solution of the pair of simultaneous equations

$$m_1\frac{d^2y_1}{dt^2} = -k_1y_1 - k_2(y_1 - y_2),$$
$$m_2\frac{d^2y_2}{dt^2} = k_2(y_1 - y_2),$$

where y_1 and y_2 are the displacements of the two masses from their equilibrium positions. The initial conditions are

$$y_1(0) = A, \quad y_1'(0) = B, \quad y_2(0) = C, \quad y_2'(0) = D.$$

Express as a set of first-order equations.

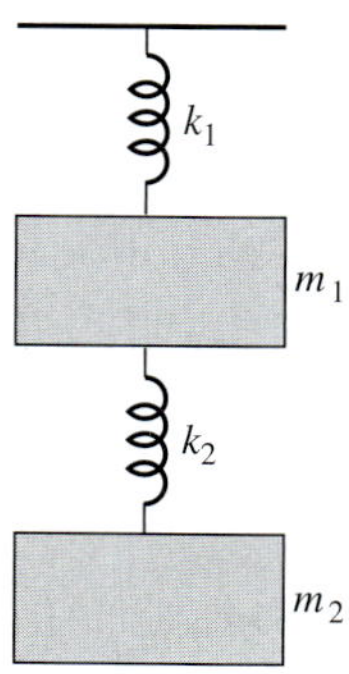

Figure 6.7

47. For the third-order equation

$$y''' + ty' - 2y = t, \quad y(0) = y''(0) = 0, \quad y'(0) = 1,$$

a. Solve for $y(0.2)$, $y(0.4)$, $y(0.6)$ by RKF.

b. Advance the solution to $t = 1.0$ with the Adams–Moulton method.

c. Estimate the accuracy of $y(1.0)$ in part (b).

48. Solve the equation in Exercise 47 by the Taylor-series method. How many terms are needed to be sure that $y(1.0)$ is correct to four significant digits?

49. If some simplifying assumptions are made, the equations of motion of a satellite around a central body are

$$\frac{d^2x}{dt^2} = \frac{-x}{r^3}, \quad \frac{d^2y}{dt^2} = \frac{-y}{r^3},$$

where

$$r = \sqrt{(x^2 + y^2)}, \qquad x(0) = 0.4,$$
$$y(0) = x'(0) = 0, \qquad y'(0) = 2.$$

a. Evaluate $x(t)$ and $y(t)$ from $t = 0$ to $t = 10$ in steps of 0.2. Use any of the single-step methods to do this.

b. Plot the curve for this range of t-values.

c. Estimate the period of the orbit.

Section 6.6

50. Equation 6.22 is for a *stiff* equation. If the coefficients of the equation for x' are changed, for what values is the system no longer stiff?

51. A pair of differential equations has the solution

$$x(t) = e^{-22t} - e^{-t},$$
$$y(t) = e^{-22t} + e^{-t},$$

with initial conditions of $x(0) = 0$, $y(0) = 2$.

a. What are the differential equations?

b. Is that system "stiff"?

c. What are the computed values for $x(0.2)$ and $y(0.2)$ if the equations of part (a) are solved with the simple Euler method, with $h = 0.1$?

d. Repeat part (c), but employing the method of Eq. (6.23). Is this answer closer to the correct value?

e. How small must h be to get the solutions at $t = 0.2$ accurate to four significant digits when using the simple Euler method?

f. Repeat part (e), but now for the method of Eq. (6.23).

▶**52.** When testing a linear system to see if it is "stiff" it is convenient to write it as

$$\begin{bmatrix} x \\ y \end{bmatrix}' = A\begin{bmatrix} x \\ y \end{bmatrix},$$

where the elements of matrix A are the multipliers of x and y in the equations. If the eigenvalues of A are all real and negative and differ widely in magnitude, the system is stiff. (One can get the eigenvalues from the characteristic polynomial as explained in Chapter 2 or with a computer algebra system.)

Suppose that A has these elements:

$$A = \begin{bmatrix} 19 & -20 \\ -20 & 19 \end{bmatrix}.$$

a. What are the eigenvalues of A? Would you call the system stiff?

b. Change the elements of A so that all are positive. What are the eigenvalues of A after this change? Does this make the system "nonstiff"?

53. The definition of a stiff equation as one whose coefficient matrix has negative eigenvalues that "differ widely in magnitude" is rather subjective. Propose an alternate definition of stiffness that is more specific.

Section 6.7

54. Suppose that a rod of length L is made from two dissimilar materials welded together end-to-end. From $x = 0$ to $x = X$, the thermal conductivity is k_1; from $x = X$ to $x = L$, it is k_2. How will the temperatures vary along the rod if $u = 0°$ at $x = 0$ and $u = 100°$ at $x = L$? Assume that Eq. (6.24) applies with $Q = 0$ and that the cross-section is constant.

55. What if k varies with temperature: $k = a + bu + cu^2$? What is the equation that must be solved to determine the temperature distribution along a rod of constant cross section?

56. Solve the boundary value problem

$$d^2x/dt^2 + t\,(dx/dt) - 3x = 3t, \quad x(0) = 1, \quad x(2) = 5$$

by "shooting." (The initial slope is near -1.5.) Use $h = 0.25$ and compare the results from the Runge–Kutta–Fehlberg method and modified Euler methods. Why are the results different? Is it possible to match the Runge–Kutta–Fehlberg method results when the modified Euler method is used? If so, show how this can be accomplished.

57. Repeat Exercise 56, but with smaller values for h. At what h-values with the Runge–Kutta–Fehlberg method are successive computations the same?

58. The boundary-value problem of Exercise 56 is linear. That means that the correct initial slope can be found by interpolating from two trial values. Show that intermediate values from the computations obtained with these two trial values can themselves be interpolated to get correct intermediate values for $x(t)$.

59. If the equation of Exercise 56 is changed only slightly to

$$d^2x/dt^2 + x\,(dx/dt) - 3x = 3t, \quad x(0) = 1, \quad x(2) = 5,$$

it is no longer linear. Solve it by the shooting method using RKF. Do you find that more than two trials are needed to get the solution? What is the correct value for the initial slope? Use a value of h small enough to be sure that the results are correct to five significant digits.

60. Given this boundary-value problem:

$$\frac{d^2y}{d\theta^2} + \frac{y}{4} = 0, \quad y(0) = 0, \quad y(\pi) = 2,$$

which has the solution $y = 2\sin(\theta/2)$,

a. Solve, using finite difference approximations to the derivative with $h = \pi/4$ and tabulate the errors.
b. Solve again by finite differences but with a value of h small enough to reduce the maximum error to 0.5%. Can you predict from part (a) how small h should be?
c. Solve again by the shooting method. Find how large h can be to have maximum error of 0.5%.

61. Solve Exercise 56 though a set of equations where the derivatives are replaced by difference quotients. How small must h be to essentially match to the results of Exercise 56 when RKF was used?

62. Use finite difference approximations to the derivatives to solve Exercise 59. The equations will be nonlinear so they are not as easily solved. One way to approach the solution is to linearize the equations by replacing x in the second term with an approximate value, then using the results to refine this approximation successively. Solve it this way.

63. Solve this boundary-value problem by finite differences, first using $h = 0.2$, then with $h = 0.1$:

$$y'' + xy' - x^2y = 2x^3, \quad y(0) = 1, \quad y(1) = -1.$$

Assuming that errors are proportional to h^2, extrapolate to get an improved answer. Then, using a very small h-value in the shooting method, see if this agrees with your improved answer.

64. Repeat Exercise 60, except with these derivative boundary conditions:

$$y'(0) = 0, \qquad y'(\pi) = 1.$$

In part (a), compare to $y = -2\cos(\theta/2)$.

65. Solve through finite differences with four subintervals:

$$\frac{d^2y}{dx^2} + y = 0, \quad y'(0) + y(0) = 2,$$

$$y'\left(\frac{\pi}{2}\right) + y\left(\frac{\pi}{2}\right) = -1.$$

66. The most general form of boundary condition normally encountered in second-order boundary-value problems is a linear combination of the function and its derivatives at both ends of the region. Solve through finite difference approximations with four subintervals:

$$x'' - tx' + t^2x = t^3,$$
$$x(0) + x'(0) - x(1) + x'(1) = 3,$$
$$x(0) - x'(0) + x(1) - x'(1) = 2.$$

67. Repeat Exercise 63, but use the Runge–Kutta–Fehlberg method. The errors will not be proportional to h^2

68. Repeat Exercise 66, but use the modified Euler method.

69. Can a boundary-value problem be solved with a Taylor-series expansion of the function? If it can, use the Taylor-series technique for several of the above problems. If it cannot be used, provide an argument in support of this.

▶70. In solving a boundary-value problem with finite difference quotients, using smaller values for h improves the accuracy. Can one make h too small?

71. Compare the number of numerical operations used in Example 6.5 to get Tables 6.18 and 6.19.

Section 6.8

72. Consider the characteristic-value problem with k restricted to real values:

$$y'' - k^2y = 0, \quad y(0) = 0, \quad y(1) = 0.$$

a. Show analytically that there is no solution except the trivial solution $y = 0$.
b. Show, by setting up a set of difference equations corresponding to the differential equation with $h = 0.2$, that there are no real values for k for which a solution to the set exists.
c. Show, using the shooting method, that it is impossible to match $y(1) = 0$ for any real value of k [except if $y'(0) = 0$, which gives the trivial solution].

▶73. For the equation

$$y'' - 3y' + 2k^2y = 0, \quad y(0) = 0, \quad y(1) = 0,$$

find the principal eigenvalue and compare to $|k| = 2.46166$,

a. using $h = \frac{1}{2}$.
b. using $h = \frac{1}{3}$.
c. using $h = \frac{1}{4}$.
d. Assuming errors are proportional to h^2, extrapolate from parts (a) and (c) to get an improved estimate.

74. Using the principal eigenvalue, $k = 2.46166$, in Exercise 73, find y as a function of x over [0, 1]. This is the corresponding eigenfunction.

75. Parallel the computations of Exercise 73 to estimate the second eigenvalue. Compare to the analytical value of 4.56773.

76. Find the dominant eigenvalue and the corresponding eigenvector by the power method:

a. $\begin{bmatrix} 3 & 1 \\ 2 & 9 \end{bmatrix}$ b. $\begin{bmatrix} 2 & 3 \\ 6 & 5 \end{bmatrix}$ c. $\begin{bmatrix} 2 & 3 \\ 3 & -2 \end{bmatrix}$

d. $\begin{bmatrix} 6 & 2 & 0 \\ 2 & 4 & 1 \\ 0 & 1 & -1 \end{bmatrix}$ e. $\begin{bmatrix} 1 & 2 & 3 \\ 0 & 1 & 3 \\ 2 & 2 & 1 \end{bmatrix}$

[In part (c), the two eigenvalues are equal but of opposite sign.]

77. For the two matrices

$$A = \begin{bmatrix} -5 & 2 & 1 \\ 1 & -9 & -1 \\ 2 & -1 & 7 \end{bmatrix},$$

$$B = \begin{bmatrix} -4+2i & -1 & -5i \\ -3 & 7+i & -i \\ 2 & -1 & 4-i \end{bmatrix},$$

a. Put bounds on the eigenvalues using Gerschgorin's theorem.
b. Can you tell from part (a) whether either of the matrices is singular?

78. Use the power method or its variations to find all of the eigenvalues and eigenvectors for the matrices of Exercise 77. For matrix B, do you need to use complex arithmetic?

▶79. Get the eigenvalues for matrix A in Exercise 77 from its characteristic polynomial. Then invert the matrix and show that the eigenvalues are reciprocals but the eigenvectors are the same. How do the two characteristic polynomials differ? Can you get the second polynomial directly from the first? Can you do all of this for matrix B?

80. Repeat Exercise 79, but use the power method to get the dominant eigenvalue. Then shift by that amount and get the next one. Finally, get the third from the trace of A.

81. Find three matrices that convert one of the below diagonal elements to zero for matrix A of Exercise 77.

82. Use the matrices of Exercise 81 successively to make one element below the diagonal of A equal to zero, then multiply that product and the inverse of the rotation matrix (which is easy to find because it is just its transpose). We keep the eigenvalues the same because the two multiplications are a similarity transformation.

Repeat this process until all elements below the diagonal are less than 1.0E-4. When this is done, compare the elements now on the diagonal to the eigenvalues of A obtained by iteration. (This will take many steps. You will want to write a short computer program to carry it out.)

▶83. Use similarity transformations to reduce the matrix to upper Hessenberg. (Do no column or row interchanges.)

$$C = \begin{bmatrix} 3 & -1 & 2 & 7 \\ 1 & 2 & 0 & -1 \\ 4 & 2 & 1 & 1 \\ 2 & -1 & -2 & 2 \end{bmatrix}$$

84. Repeat Exercise 83 but with row/column interchanges that maximize the magnitude of the divisors.

85. Repeat Exercise 82 after first converting to upper Hessenberg. How many fewer iterations are needed?

Applied Problems and Projects

APP1. The mass in Figure 6.8 moves horizontally on the frictionless bar. It is connected by a spring to a support located centrally below the bar. The unstretched length of the spring is $L = \sqrt{(10)} = 3.1623$ m (meters); the spring constant is $k = 100$ N/m (newtons per meter); the mass of the block is 3 kg. Let $x(t)$ be the distance from the center of the bar to the location of the block at time t. Clearly the equilibrium position of the block is at $x = 1.0$ m (or $x = -1.0$ m). Let $y_0 = \sqrt{10}$ m (the unstretched length of the spring). This second-order differential equation describes the motion:

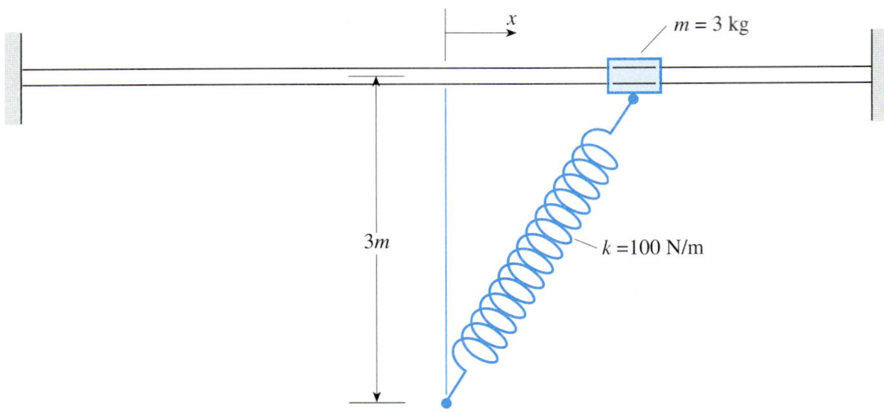

Figure 6.8

$$\frac{d^2x}{dt^2} = -\left(\frac{k}{m}\right)x\left(1 - \frac{y_0}{\sqrt{(x^2 + 9)}}\right).$$

a. Using both single-step and multistep methods, find the position of the block between $t = 0$ and $t = 10$ sec if $x_0 = 1.4$ and the initial velocity is zero.
b. Repeat part (a), but now with the spring stretched more at the start, $x_0 = 2.5$.
c. Use Maple and/or MATLAB to graph the motion for both parts (a) and (b). Compare your graphs to Figure 6.9.

APP2. The equation $y' = 1 + y^2$, $y(0) = 0$ has the solution $y = \tan(x)$. Use modified Euler method to compute values for $x = 0$ to $x = 1.6$ with a value for h small enough to obtain values that differ from the analytical by no more than ± 0.0005. What is the largest h-value to do this? $y(x)$ becomes infinite at $x = \pi/2$. What happens if you try to integrate y' beyond this point? Is there some way you can solve the equation numerically from $x = 0$ to $x = 2$?

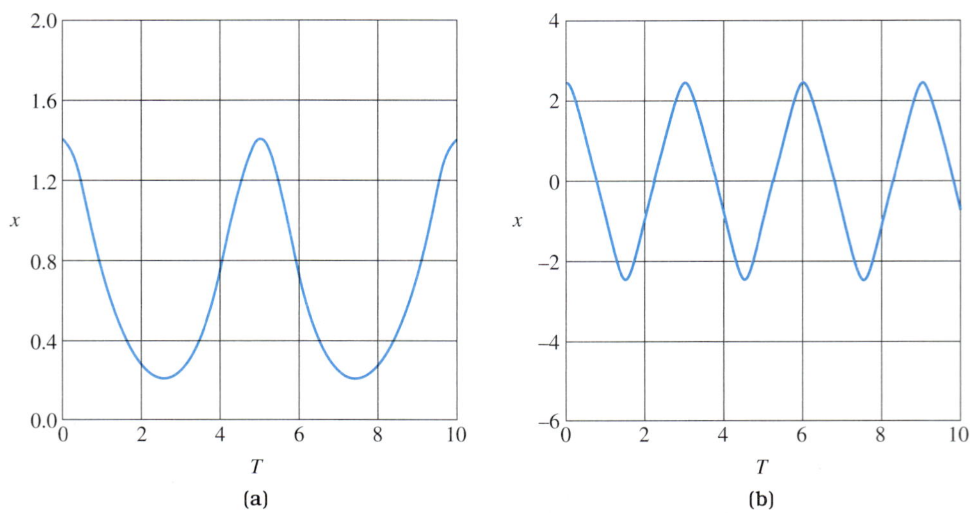

Figure 6.9

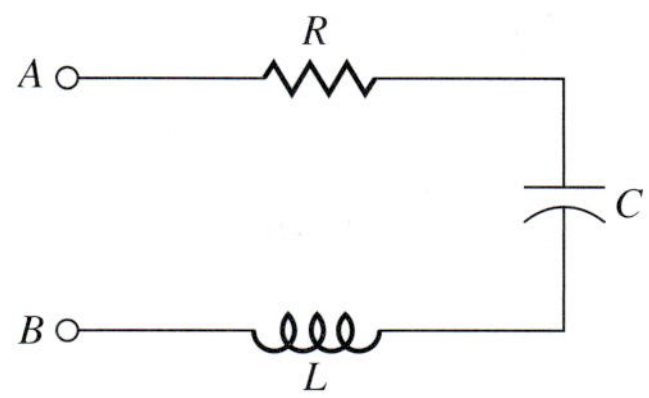

Figure 6.10

APP3. A nonlinear boundary-value problem is more difficult than a linear problem because many trials may be needed to get a good value for the initial slope. From three initial trials it should be possible to use a Muller's-type interpolation. Outline the steps of a program that will do this.

APP4. In an electrical circuit (Figure 6.10) that contains resistance, inductance, and capacitance (and every circuit does), the voltage drop across the resistance is iR (i is current in amperes, R is resistance in ohms), across the inductance it is L, (di/dt) (L is inductance in henries), and across the capacitance it is q/C (q is charge in the capacitor in coulombs, C is capacitance in farads). We then can write, for the voltage, difference between points A and B,

$$V_{AB} = L\frac{di}{dt} + Ri + \frac{q}{C}.$$

Differentiating with respect to t and remembering that $dq/dt = i$, we have a second-order differential equation;

$$L\frac{d^2i}{dt^2} + R\frac{di}{dt} + \frac{1}{C}i = \frac{dV}{dt}.$$

If the voltage V_{AB} (which has previously been 0 V) is suddenly brought to 15 V (let us say, by connecting a battery across the terminals) and maintained steadily at 15 V (so $dV/dt = 0$), current will flow through the circuit. Use an appropriate numerical method to determine how the current varies with time between 0 and 0.1 sec if $C = 1000\ \mu$f, $L = 50$ mH, and $R = 4.7$ ohms; use Δt of 0.002 sec. Also determine how the voltage builds up across the capacitor during this time. You may want to compare the computations with the analytical solution.

APP5. Repeat App 4, but let the voltage source be a 60-Hz sinusoidal input:

$$V_{AB} = 15 \sin(120\pi t).$$

How closely does the voltage across the capacitor resemble a sine wave during the last full cycle of voltage variation?

APP6. After the voltages have stabilized in APP4 (15 V across the capacitor), the battery is shorted so that the capacitor discharges through the resistance and inductor. Follow the current and the capacitor voltages for 0.1 sec, again with $\Delta t = 0.002$ sec. The oscillations of decreasing amplitude are called *damped oscillations*. If the calculations are repeated but with the resistance value increased, the oscillations will be damped out more quickly; at $R = 14.14$ ohms the oscillations should disappear; this is called *critical damping*. Perform numerical computations with values of R increasing from 4.7 to 22 ohms to confirm that critical damping occurs at 14.14 ohms.

APP7. Cooling fins are often welded to objects in which heat is generated to conduct the heat away, thus controlling the temperature. If the fin loses heat by radiation to the surroundings the rate of heat loss from the fin is proportional to the difference in fourth powers of the fin temperature and the surroundings, both measured in absolute degrees. The equation reduces to

$$d^2u/dx^2 = k(u^4 - T^4)$$

where u is the fin temperature, T is the surroundings temperature, and x is the distance along the fin. k is a constant. For a fin of given length L, this is not difficult to solve numerically if $u(0)$ and $u(L)$ are known. Solve for $u(x)$, the distribution of temperature along the fin, if $T = 300$, $u(0) = 450$, $u(20) = 350$, $k = 0.23$, utilizing any of the methods for a boundary-value problem. Use a value for h small enough to get temperatures accurate to 0.1 degree.

APP8. In APP7, suppose the fin is of infinite length and we can assume that $\lim (u(x)) = 0$ as $x \to \infty$. Can this problem be solved numerically? If so, get the solution for $u(x)$ between $x = 0$ and $x = 20$.

APP9. A Foucault pendulum is one free to swing in both the x- and y-directions. It is frequently displayed in science museums to exhibit the rotation of the earth, which causes the pendulum to swing in directions that continuously vary. The equations of motion are

$$\ddot{x} - 2\omega \sin \psi \dot{y} + k^2 x = 0,$$
$$\ddot{y} + 2\omega \sin \psi \dot{x} + k^2 y = 0,$$

when damping is absent (or compensated for). In these equations, the dots over the variable represent differentiation with respect to time. Here ω is the angular velocity of the earth's rotation ($7.29 \times 10^{-5}\ \text{sec}^{-1}$), ψ is the latitude, $k^2 = g/\ell$ where ℓ is the length of the pendulum. How long will it take a 10-m-long pendulum to rotate its plane of swing by 45° at the latitude where you live? How long if located in Quebec, Canada?

APP10. Condon and Odishaw (1967) discuss Duffing's equation for the flux ϕ in a transformer. This nonlinear differential equation is

$$\ddot{\phi} + \omega_0^2 \phi + b\phi^3 = \frac{\omega}{N} E \cos \omega t.$$

In this equation, $E \sin \omega t$ is the sinusoidal source voltage and N is the number of turns in the primary winding, while ω_0 and b are parameters of the transformer design. Make a plot of ϕ versus t (and compare to the source voltage) if $E = 165$, $\omega = 120\pi$, $N = 600$, $\omega_0^2 = 83$, and $b = 0.14$. For approximate calculations, the nonlinear term $b\phi^3$ is sometimes neglected. Evaluate your results to determine whether this makes a significant error in the results.

APP11. Ethylene oxide is an important raw material for the manufacture of organic chemicals. It is produced by reacting ethylene and oxygen together over a silver catalyst. Laboratory studies gave the equation shown.

It is planned to use this process commercially by passing the gaseous mixture through tubes filled with catalyst. The reaction rate varies with pressure, temperature, and concentrations of ethylene and oxygen, according to this equation:

$$r = 1.7 \times 10^6 e^{-9716/T} \left(\frac{P}{14.7} \right) C_E^{0.328} C_O^{0.672},$$

where

r = reaction rate (units of ethylene oxide formed per lb of catalyst per hr),
T = temperature, °K (°C + 273),
P = absolute pressure (lb/in.^{2}),
C_E = concentration of ethylene,
C_O = concentration of oxygen.

Under the planned conditions, the reaction will occur, as the gas flows through the tube, according to the equation

$$\frac{dx}{dL} = 6.42r,$$

where

x = fraction of ethylene converted to ethylene oxide,
L = length of reactor tube (ft).

The reaction is strongly exothermic, so that it is necessary to cool the tubular reactor to prevent overheating. (Excessively high temperatures produce undesirable side reactions.) The reactor will be cooled by surrounding the catalyst tubes with boiling coolant under pressure so that the tube walls are kept at 225°C. This will remove heat proportional to the temperature difference between the gas and the boiling water. Of course, heat is generated by the reaction. The net effect can be expressed by this equation for the temperature change per foot of tube, where B is a design parameter:

$$\frac{dT}{dL} = 24{,}302r - B(T - 225).$$

For preliminary computations, it has been agreed that we can neglect the change in pressure as the gases flow through the tubes; we will use the average pressure of $P = 22$ lb/in.^{2} absolute. We will also neglect the difference between the catalyst temperature (which should be used to find the reaction rate) and the gas temperature. You are to compute the length of tubes required for 65% conversion of ethylene if the inlet temperature is 250°C. Oxygen is consumed in proportion to the ethylene converted; material balances show that the concentrations of ethylene and oxygen vary with x, the fraction of ethylene converted, as follows:

$$C_E = \frac{1 - x}{4 - 0.375x},$$

$$C_O = \frac{1 - 1.125x}{4 - 0.375x}.$$

The design parameter B will be determined by the diameter of tubes that contain the catalyst. (The number of tubes in parallel will be chosen to accommodate the quantities of materials flowing through the reactor.) The tube size will be chosen to control the maximum temperature of the reaction, as set by the minimum allowable value of B. If the tubes are too large in diameter (for which the value of B is small), the temperatures will run wild. If the tubes are *too small* (giving a large value to B), so much heat is lost that the reaction tends to be quenched. In your studies, vary B to find the least value that will keep the maximum temperature below 300°C. Permissible values for the parameter B are from 1.0 to 10.0.

In addition to finding how long the tubes must be, we need to know how the temperature varies with x and with the distance along the tubes. To have some indication of the controllability of the process, you are also asked to determine how much the outlet temperature will change for a 1°C change in the inlet temperature, using the value of B already determined.

APP12. An ecologist has been studying the effects of the environment on the population of field mice. Her research shows that the number of mice born each month is proportional to the number of females in the group and that the fraction of females is normally constant in any group. This implies that the number of births per month is proportional to the total population.

She has located a test plot for further research, which is a restricted area of semiarid land. She has constructed barriers around the plot so mice cannot enter or leave. Under the conditions of the experiment, the food supply is limited, and it is found that the death rate is affected as a result, with mice dying of starvation at a rate proportional to some power of the population. (She also hypothesizes that when the mother is undernourished, the babies have less chance for survival and that starving males tend to attack one another, but these factors are only speculation.)

The net result of this scientific analysis is the following equation, with N being the number of mice at time t (with t expressed in months). The ecologist has come to you for help in solving the equation; her calculus doesn't seem to apply.

$$\frac{dN}{dt} = aN - BN^{1.7}, \qquad \text{with } B \text{ given by Table 6.20.}$$

Table 6.20

t	B	t	B
0	0.0070	5	0.0013
1	0.0036	6	0.0028
2	0.0011	7	0.0043
3	0.0001	8	0.0056
4	0.0004		

As the season progresses, the amount of vegetation varies. The ecologist accounts for this change in the food supply by using a "constant" B that varies with the season.

If 100 mice were initially released into the test plot and if $a = 0.9$, estimate the number of mice as a function of t, for $t = 0$ to $t = 8$.

APP13. A certain chemical company produces a product that is a mixture of two ingredients, A and B. In order to ensure that the product is homogeneous, A and B are fed into a well-mixed tank that holds 100 gal. The desired product must contain two parts of A to one part of B within certain specifications. The normal flows of A and B into the tank are 4 and 2 gal/min. There is no volume change when these are mixed, so the outflow is 6 gal/min and the holding time in the tank is $100/6 = 16.66$ min. Due to an unfortunate accident, the flow of ingredient B is cut off and before this is noticed and corrected, the ratio of A to B in the tank has increased to 10 parts of A to 1 part of B. (There are still 100 gal in the tank.) Set up equations that give the ratio of A to B in the tank as a function of time after the flow of B has been restored to its normal value of 2 gal/min. How long will it take until the output from the tank reaches 2 parts A to 0.99 parts B? How much product is produced (and discarded because it is not up to specification) during this time? How would you suggest that this time to reach specification be reduced?

7

Optimization

The dictionary defines *optimum* as "the best or most favorable degree, quantity, number, etc." In mathematics, we optimize by finding the maximum or minimum of a function. Applications in business are to minimize costs or to maximize profits. In this chapter, we describe methods that usually find the point(s) where a function, $f(x, y, z, \ldots)$, has a minimum value. We find maxima by locating the points where the negative of the function is a minimum.

A function can have several minima and maxima when the range is unrestricted. The smallest of the minima is the global minimum; others are called local minima. The global maximum is the largest of the maxima; others are local maxima. We will often restrict the range and then the maxima/minima can occur at an endpoint of the range or within the range. A function is called *unimodal* when there is exactly one minimum (or maximum) within the range or at an endpoint. Our examples are unimodal.

The chapter begins with a problem that is familiar to students—find the x-value that makes $y = f(x)$ a minimum, the one-dimensional case. We will compare classical analytical methods with purely numerical ones. We then proceed to functions of more than one variable.

Contents of This Chapter

7.1 Finding the Minimum of $y = f(x)$

Begins by pointing out when getting the minimum from $f'(x) = 0$ has problems. A simple search method can be used, but this is less efficient than methods that narrow the interval that encloses the minimum. Once several values for y at some x-values have been computed, interpolation can locate the minimum with less computational cost. Computer algebra systems and spreadsheets can automate the solution.

7.2 Minimizing a Function of Several Variables

Compares the analytical method of setting partial derivatives to zero and solving the resulting system with numerical procedures. These include graphical techniques and searching procedures. A method called *steepest descent* does the searching along lines on which the function decreases most rapidly, but, for some problems, this is less efficient than another searching procedure called *conjugate gradient*. Newton's method can be adapted to locating a minimum.

7.3 Linear Programming

Describes a widely used technique in business applications. This applies when the minimum of a linear function is constrained to lie on the boundaries of a region defined by linear relations. The *simplex method* is most often used to solve these problems, and this can determine the effects of changes in the parameters. Again, computer algebra systems and spreadsheets have facilities for doing this.

7.4 Nonlinear Programming

Is a more difficult problem than one with a linear function subject to linear constraints. A number of ways to solve such problems are discussed.

7.5 Other Optimizations

Briefly describes another problem of importance to the managers of a business who desire to minimize the costs of transporting goods, as well as problems where the values of the independent variables are restricted to integer values or where the values are not known with certainty but only within a range.

7.1 Finding the Minimum of $y = f(x)$

We begin our treatment of optimization by examining ways that we can find a minimum point on the curve of $y = f(x)$, the one-dimensional case. As always in applied mathematics, we wish to solve a problem with the least effort. (So this is itself a minimization problem!)

The problem is not as simple as it might at first seem. A function may not have a minimum point at all, at least not in the normal sense; the function $y = x$ can hardly be said to have a minimum point unless we want to think of $y \to -\infty$ as $x \to -\infty$ to be a minimum. Another example of a function without a minimum is $y = 2/(x^3 - 1)$ (look at its graph to see this).

The function may have several minimum points; we usually want to find the global minimum, the least of all the minima, and that task is often not easy. We might have to locate every one of the many minima and then select the proper one. Consider the graph of $y = 2x - \cos(2x)$ as seen in Figure 7.1. However, this task is simpler if we only desire the global minimum within a restricted range of x-values; the problem is *constrained* in that x must lie in interval $[a, b]$.

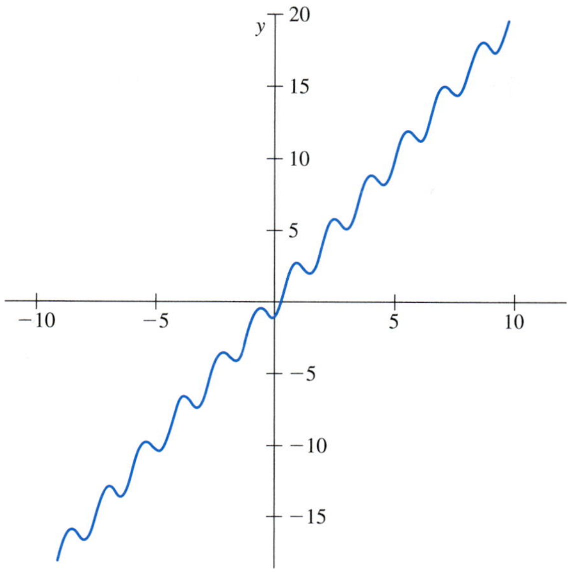

Figure 7.1

The Classical Method—$f'(x) = 0$

It is likely that you first think of locating the minimum point on $y = f(x)$ as a root-finding problem. You say, "Just differentiate to get $f'(x)$ and then locate its zeros." All of us have done this in our calculus course many times. We do have to differentiate between maxima and minima but, of course, examining the value of $f''(x)$ will distinguish between them. This will even tell us if the point is a horizontal inflection. After eliminating all the maxima and horizontal inflections, we arrive at the candidates for the global minimum and we select the right one from the $f(x)$ values at these points.

Actually, we are going to simplify the problem of minimization in 1-D by working with functions that are *unimodal*. This term means that there is exactly one minimum point on $[a, b]$. We will further assume that the function decreases as we move from a toward b and also decreases as we move from b toward a, which eliminates functions whose minimum is at an end point.

Even with these restrictions, there are cases where the analytical method won't work. Figure 7.2 shows two of these.

In Figure 7.2a, the derivative is discontinuous at the minimum point. In Figure 7.2b, there is a discontinuity in $f(x)$ at the minimum point. (Interestingly, if the lines in Figure 7.2a have slopes of -1 and 1, the numerical estimate of the slope at the minimum point from a central difference is zero.)

It is often the case in real-world applications that the equation for $f(x)$ is not known — we can only find a value for the function from an experiment. While we can approximate the function by fitting an equation (probably a polynomial) to data from several experiments, using the

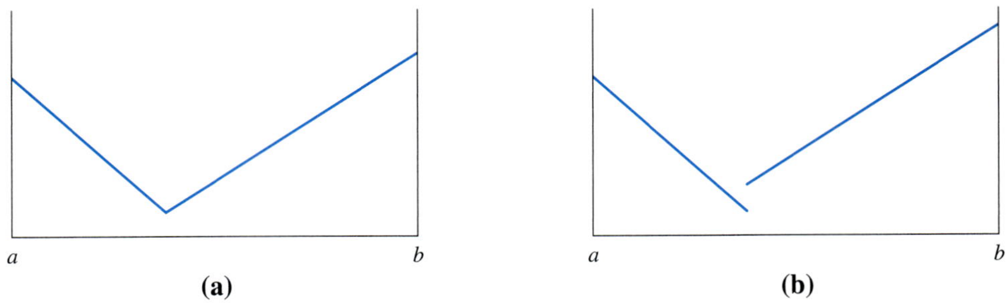

Figure 7.2

classical analytical method to find the minimum point would be terribly expensive. Further, as we saw in an earlier chapter, values for the derivative from such data are apt to be inaccurate.

We argue from this that there is real merit to discovering numerical methods.

Searching for the Minimum

There are several ways that we can use searching methods. If the function is known, we can use a spreadsheet program to list function values at a sequence of x-values. Most spreadsheet programs have a built-in function that will pinpoint the minimum of the list of values. This technique is handy but not very efficient.

A somewhat more efficient search method is what we call the *back-and-forth* technique. In this, one begins at one end of the interval $[a, b]$ and moves toward the other end. Example 7.1 is an illustration.

EXAMPLE 7.1 Find the minimum on $[-3, 1]$ of $f(x) = e^x + 2 - \cos(x)$. Use the back-and-forth method. Begin from $x = -3$ and move toward b ($b = 1$) with $\Delta x = (b - a)/4 = 1$. When the next function value increases, reverse the direction with Δx equal to 1/4 of the previous. Repeat this until $\Delta x < 0.001$.

The successive values are:

At $x = a = -3, f(x)$ is 3.039780, we now begin the search.
With $h = 1$,

$x = -2,\qquad f(x) = 2.551482$

$x = -1,\qquad f(x) = 1.827577$

$x = 0,\qquad f(x) = 2$

We reverse now, $h = -0.25$,

$x = -0.25,\qquad f(x) = 1.809888$

$x = -0.5,\qquad f(x) = 1.728948$

$x = -0.75,\qquad f(x) = 1.740678$

We reverse now, $h = 0.0625$,

$x = -0.6875, \quad f(x) = 1.729997$
$x = -0.625, \quad f(x) = 1.724298$
$x = -0.5625, \quad f(x) = 1.723858$
$x = -0.5, \quad f(x) = 1.728948$

We reverse now, $h = -0.015625$,

$x = -0.515625, \quad f(x) = 1.727143$
$x = -0.53125, \quad f(x) = 1.725695$
$x = -0.546875, \quad f(x) = 1.724602$
$x = -0.5625, \quad f(x) = 1.723858$
$x = -0.578125, \quad f(x) = 1.723460$
$x = -0.59375, \quad f(x) = 1.723404$
$x = -0.609375, \quad f(x) = 1.723685$

We reverse now, $h = 0.00390625$,

$x = -0.6054688, \quad f(x) = 1.723583$
$x = -0.6015625, \quad f(x) = 1.723502$
$x = -0.5976563, \quad f(x) = 1.723443$
$x = -0.59375, \quad f(x) = 1.723404$
$x = -0.5898438, \quad f(x) = 1.723386$
$x = -0.5859375, \quad f(x) = 1.723390$

We reverse now, $h = -0.0009765625$,

$x = -0.5869141, \quad f(x) = 1.723387$
$x = -0.5864258, \quad f(x) = 1.7233882$

Tolerance of 0.001 is met.

We can see several objections to this crude method. We have to compute an extra function value before we know that the direction is to be reversed. Further, some x-values are duplicated after a reversal but we still recompute $f(x)$. (Keeping track of the function value would be very complicated.) We seek an improvement.

One way to improve the efficiency of this crude method is to use three values that bracket the minimum (at $x = -2, -1, 0$, the first three values in Example 7.1) and fit a quadratic polynomial to them, then find the minimum of that. [When $f(x) = ax^2 + bx + c$, $f'(x)$ is $2ax + b$, and this will be zero at $x = -b/2a$.] The easy way to do this is to form the quadratic polynomial from a difference table and find its minimum point. From these three $x, f(x)$ values, we get an estimate of $x_{min} = -0.6923658$ and no additional function evaluations are required.

We can continue from here by successively forming quadratics from three points nearest the minimum. We must compute the function value at the new point. Here is the first set:

$$x = -1, \qquad f(x) = 1.827577;$$
$$x = -0.6923658, \qquad f(x) = 1.730653;$$
$$x = 0, \qquad f(x) = 2.$$

From these points, we find the interpolating quadratic and get its minimum point: $x = -0.6224442$. If we continue, we find the next two estimates of the x-value at the minimum to be

$$-0.5975463 \qquad \text{and} \qquad -0.5878655,$$

which is within 0.0007 of the true $x_{\min}$ of -0.588532744. We have achieved this with only six evaluations of the function rather than the 23 used in the above simple search.

Narrowing the Interval

When we are given a function that has a single minimum point within the interval $[a, b]$, we can say that points a and b bracket or enclose the minimum point. There are ways to narrow that interval and the method known as the *golden section search* is one of the most popular.

The term *golden section* is a number that is said to be the basis for the beautiful architecture of Greek temples. The ratio of the height to the width of the Parthenon is equal to this, a number equal to 0.618034. . . . It is the positive root of the quadratic $r^2 + r - 1 = 0$. Notice that $r^2 = 1 - r = 0.381966$, another number that will be important to us. We will use the symbol s for it.

The bisection method for finding a zero of $f(x)$ can be considered to be a bracketing technique. You recall that we narrow an initial interval that encloses a root [we know that the root is in $[a, b]$ because the sign of $f(a)$ is opposite to that of $f(b)$], by dividing the interval in half successively. Only one intermediate point is enough to narrow the interval.

We now ask, how do we know that a minimum point is somewhere within a given interval? We know at the start that $[a, b]$ is such an interval from our assumptions on $f(x)$. If we know $f(x)$ at one intermediate point, say, x_L, can we say that we know a smaller enclosing interval? No; if $f(x_L)$ is smaller than either $f(a)$ or $f(b)$, it merely confirms our original assumption that $f(x)$ is unimodal. We only can say that the minimum may be between a and x_L, but it could also be between x_L and b.

It takes two intermediate points to narrow the interval that encloses the minimum. Look at Figure 7.3.

In Figure 7.3a, we see that $f(x_L)$ is the least and the minimum is to the left of the two intermediate points. In Figure 7.3b, the two points are the same but the minimum lies between them. With this arrangement of intermediate points, either situation may occur, so we can only conclude that the minimum is bracketed by $[a, x_R]$.

Figure 7.4 shows the opposite—$f(x_R)$ is less than $f(x_L)$. In Figure 7.4a, the minimum is between the points; in Figure 7.4b, it is between x_R and b. Either case is possible, so we can only say that enclosing interval is $[x_L, b]$.

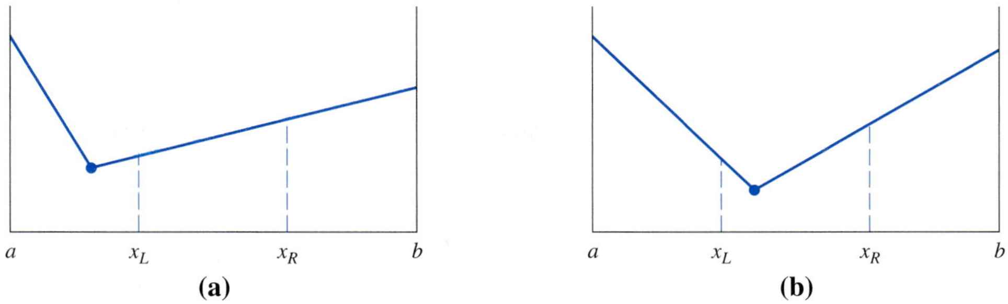

Figure 7.3

What Are the Best Locations for the Intermediate Points?

It is possible to locate the intermediate points anywhere within $[a, b]$, but intuitively one would think they should be placed symmetrically about the midpoint of the interval. Why? The midpoint is the best approximation for the minimum point because the error is not more than $(b - a)/2$, which we know without having to evaluate the function. Putting them at the 1/3 and 2/3 points seems like a good idea. However, this is not such a good choice because it is not clear how one proceeds from them to further narrow the interval.

One often-used choice is based on using the *golden ratio.* As you will see in the following, it provides a clear and excellent way to proceed. Actually, this number is the positive root of the quadratic equation $r^2 + r - 1 = 0$, which is $(\sqrt{5} - 1)/2$. Recall that $r^2 = 1 - r$. It is also related to the numbers in the Fibonacci sequence. This sequence is defined by this recursion formula:

$$F_0 = 1, \qquad F_1 = 1, \qquad F_n = F_{n-2} + F_{n-1},$$

and the first few members are 1, 1, 2, 3, 5, 8, 13, 21, ... As n becomes large, the ratio F_n/F_{n+1} approaches r. (You may want to test this to see how quickly the ratios converge.)

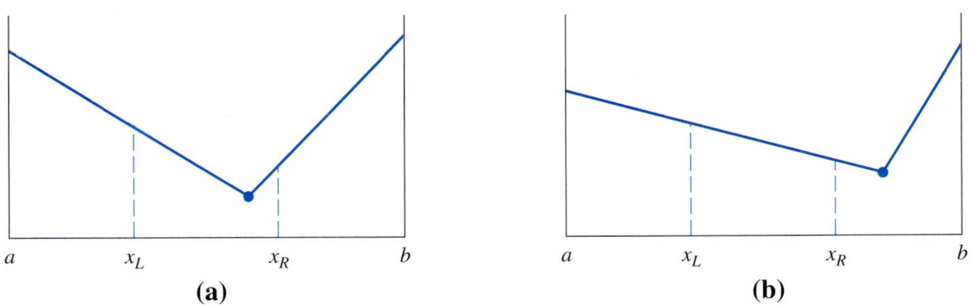

Figure 7.4

(These Fibonacci numbers are also involved in another search method, the *Fibonacci search*. Applied Project 8 asks you to compare that method to the *golden search* that we now describe.)

Using the Golden Mean to Find a Minimum

Another name for the golden section is the *golden mean*. We begin the search by computing the x-values for the two intermediate points:

$$x_L = a + (1 - r) * (b - a), \quad x_R = a + r * (b - a).$$

(We could have written $x_L = a + r^2 * (b - a)$: $r^2 = 1 - r$.) These points are symmetric about the midpoint of $[a, b]$, $(b - a)/2$. One is 0.381966 times $(b - a)$ from a; the other is 0.618034 times $(b - a)$ from a. Next, we compute the function values at these intermediate points, $F_L = f(x_L)$ and $F_R = f(x_R)$.

We compare the two function values and find a new smaller interval in which the minimum lies:

$$\text{If } F_L < F_R, \text{ then the interval is } [a, x_R] \text{ else it is } [x_L, b].$$

We use this to reset the interval. In either case, the new interval is smaller; it is 0.618034 times as large. We redefine point a or b accordingly, redefine either x_L or x_R, and compute a new intermediate point symmetric about the new midpoint. All of this may be clearer from the following box:

An Algorithm for the Golden Section Search

Given $f(x)$ that is unimodal with a minimum in $[a, b]$:

Start:

Compute $x_L = a + (1 - r)(b - a)$, $x_R = a + r(b - a)$,

$(r = 0.618034)$, $F_L = f(x_L)$, $F_R = f(x_R)$.

Continuation:

If $F_R > F_L$, then
$b = x_R$
$x_R = x_L$
$F_R = F_L$
$x_L = a + (1 - r)(b - a)$
$F_L = f(x_L)$
Else
$a = x_L$
$x_L = x_R$
$F_L = F_R$
$x_R = a + r(b - a)$
$F_R = f(x_R)$
and repeat until $x_R - x_L <$ tolerance value.

Notice, because the interval is reduced to 0.618034 times the previous interval, the final interval after n repetitions is

$$\text{Original } (b - a) * 0.618034^n.$$

Here is an example.

EXAMPLE 7.2. Repeat Example 7.1, but now use the golden section search. The function is $f(x) = e^x + 2 - \cos(x)$ and the minimum is within $[-3, 1]$. Continue the search until the intermediate points are within 0.001 of each other. (The correct answer to nine digits is at $x = -0.588532744$.)

The results from a program are

Starting values

X_L	X_R	F_L	F_R	Interval	Width
−1.4721	−0.5279	2.1309	1.7260	[−3.0000, 1.0000]	4.0000
−0.5279	0.0557	1.7260	2.0589	[−1.4721, 1.0000]	2.4721
−0.8885	−0.5279	1.7807	1.7260	[−1.4721, 0.0557]	1.5279
−0.5279	−0.3050	1.7260	1.7833	[−0.8885, 0.0557]	0.9443
−0.6656	−0.5279	1.7274	1.7260	[−0.8885, −0.3050]	0.5836
−0.5279	−0.4427	1.7260	1.7387	[−0.6656, −0.3050]	0.3607
−0.5805	−0.5279	1.7234	1.7260	[−0.6656, −0.4427]	0.2229
−0.6130	−0.5805	1.7238	1.7234	[−0.6656, −0.5279]	0.1378
−0.5805	−0.5604	1.7234	1.7239	[−0.6130, −0.5279]	0.0851
−0.5929	−0.5805	1.7234	1.7234	[−0.6130, −0.5604]	0.0526
−0.6006	−0.5929	1.7235	1.7234	[−0.6130, −0.5805]	0.0325
−0.5929	−0.5882	1.7234	1.7234	[−0.6006, −0.5805]	0.0201
−0.5882	−0.5852	1.7234	1.7234	[−0.5929, −0.5805]	0.0124
−0.5900	−0.5882	1.7234	1.7234	[−0.5929, −0.5852]	0.0077
−0.5882	−0.5870	1.7234	1.7234	[−0.5900, −0.5852]	0.0047
−0.5889	−0.5882	1.7234	1.7234	[−0.5900, −0.5870]	0.0029
−0.5893	−0.5889	1.7234	1.7234	[−0.5900, −0.5882]	0.0018

This tabulation rounds the values to four digits, even though they were computed with about seven digits of precision. The first lines show the start of the computations; two function evaluations were used there. After the blank line, the continuations are shown; only one function evaluation was needed for each step. So, the total number of function evaluations was 18.

The minimum is fairly flat—the computed values for $f(x)$ are the same within seven digits for both x-values in the last line: $x = -0.5888563$ and -0.5881641.

Example 7.2 required 18 evaluations of $f(x)$, while 23 were needed in Example 7.1. This is a significant savings, but we wonder if there can be further improvement.

Parabolic Extrapolations

An improvement will come if we use the first three golden section points to create an interpolating quadratic polynomial and then find where that polynomial has its minimum. Let us do this (as we did with the simple search procedure).

The first three intermediate points computed in Example 7.2 are

$$x = -1.472136, \quad f(x) = 2.130934,$$
$$x = -0.527864, \quad f(x) = 1.725979,$$
$$x = 0.055728, \quad f(x) = 2.058863.$$

from which we can compute the divided-difference table:

−1.472136	2.130934	−0.428854	0.654024.
−0.527864	1.725979	0.570405	
0.055728	2.058863		

Using the procedure from Chapter 3, we find that the quadratic through these points (written in the usual quadratic form) is:

$$P_2(x) = 0.654024x^2 + 0.879194x - 2.00784,$$

whose derivative is zero at

$$x = -0.879194/(2 * 0.654024) = -0.672142.$$

We don't really have to get the quadratic in normal form. Recall that the interpolating polynomial obtained from the divided differences is

$$p_2(x) = a_0 + a_1(x - x_0) + a_2(x - x_0)(x - x_1),$$

from which the derivative is

$$p_2'(x) = a_1 + a_2[(x - x_0) + (x - x_1)].$$

Setting this to zero and solving for x (the minimum point of the parabola) gives

$$x_{\min} = \frac{a_2(x_0 + x_1) - a_1}{2a_2}.$$

Of course, we can get a_1 and a_2 directly from the x and $f(x)$ values without computing the difference table. So, obtaining the estimate of the minimum of our function requires only several arithmetic operations.

This first extrapolation is still quite far from the true x-value at the minimum but it is much closer than the midpoint of the ranges given from the first several steps in the golden section search.

We can continue to construct another quadratic polynomial and repeat this. However, there are now more than three points that may be used to construct a polynomial. Which should we use? We could use four points to construct a cubic interpolating polynomial and find the minimum of the cubic. Our choice is to fit another quadratic to the three points whose function values are least. If this is done successively, we get these x- and f-values for the minimum of the function:

x	$f(x)$
−0.6721	1.72812
−0.5907	1.72339
−0.5892	1.72339
−0.5885	1.72339
−0.5885	1.72339

The repetition of x- and f-values of the last two lines suggests that we have found the minimum point to a precision of four digits. Each of the lines here required exactly one new function evaluation (and some simple arithmetic computations too) but compared to using the golden section, we see a great economy.

There is subtle flaw in the procedure we have described. It can happen that the successive values oscillate. *Brent's method* overcomes this by resorting to a golden mean computation that ends the oscillations. We do not describe this; *Numerical Recipes* [W. H. Press et al., (1992)] is a good reference.

Using MATLAB

MATLAB can readily find the minimum point of $f(x)$ within a given range of x-values. We saw in Chapter 0 how that can be done. Let us repeat Example 7.2. First, though, it is a good idea to plot the function (see Fig. 7.5):

```
EDU>> f = inline('exp(x) + 2 - cos(x)')
EDU>> fplot(f,[-3, 1]); grid on
```

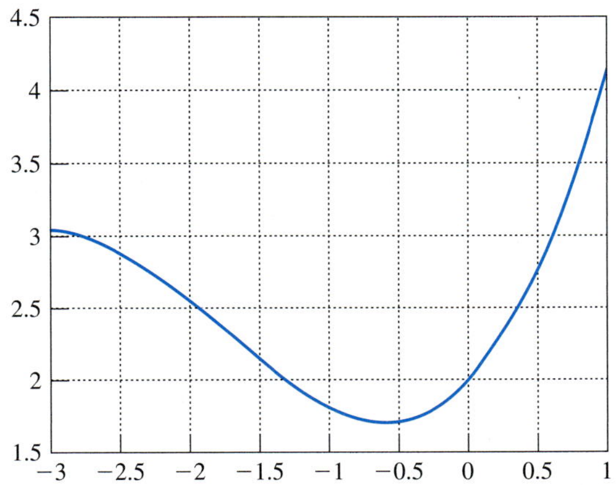

Figure 7.5

Now we ask for the minimum within [−3, 1]:

```
EDU>> fminbnd(f,-3,1,optimset('Display','iter'))

Func-count        x          f(x)      Procedure
     1        -1.47214     2.13093    initial
     2        -0.527864    1.72598    golden
     3         0.0557281   2.05886    golden
     4        -0.672141    1.72812    parabolic
     5        -0.590757    1.72339    parabolic
     6        -0.589208    1.72339    parabolic
     7        -0.588556    1.72339    parabolic
     8        -0.588522    1.72339    parabolic
     9        -0.588489    1.72339    parabolic

Optimization terminated successfully:
the current x satisfies the termination criteria using OPTIONS.TolX
of 1.000000e-004
ans =
  -0.5885
```

We see from this that MATLAB does exactly as we described; three intermediate points are first obtained with the golden mean technique and then it continues with parabolic extrapolations. When the computations stop, the x-value is within 0.00005 of the analytical value.

Using a Spreadsheet Program

We said earlier that one could use a spreadsheet program to set up a sequence of x-values and then use these to get the corresponding $f(x)$-values with the command that locates the minimum of the $f(x)$. That is an inefficient way.

The popular program EXCEL provides a better way. The *Solver* command is an add-in to the standard program. This can be downloaded from the Web site www.solver.com.

To use Excel to find the minimum of the same function as in the examples, we do this:

Choose cell A1 to hold the x-values and enter 0 as a starting value for the minimization.

In cell A3, enter the function: exp(A1) + 2 − cos(A1).

Click on `Tools` and chose `Solver`.

In the dialog box that appears, enter into `Set Target Cell:` the absolute cell reference `$A$3`, into `By changing Cells` the absolute cell reference `$A$1`, into `Subject to the Constraints:` the limits to the range, `$A$1<= 1` and `$A$1>= -3`. Now click on `Solve` then on `OK`.

Doing this produces a value of 0.588491 in cell A1 and 1.72339 in cell A3. These are the x- and $f(x)$ values at the minimum. The values essentially match to those from MATLAB.

Quattro Pro has the same capability. In this the program is called *Optimizer.* These operations will find the minimum for the same example function: $e^x + 2 - \cos(x)$:

Choose cell A1 to hold the x-values and enter 0 as a starting value for the minimization.

Choose cell A3 to hold the function to be minimized, and enter the function

`@exp(A1) + 2 - @cos(A1).`

We see a 2 in cell A3; this is the value of the function at $x = 0$. We are now ready to "optimize."

Click on `Tools/Numeric Tools/Optimizer.`

In the dialog box that appears,

Enter into `Set Solution Cell` the cell reference `$A$3;`
Set to `Min;`
Enter into `Variable Cell(s)` the cell reference `$A$1;`
Click on `Add Constraint;`
Enter into `Cell` the cell reference `$A$1;`
Select >= ;
Enter into `Constant` the value −3;
Then click on `Add Another Constraint;`
Enter into `Cell` the cell reference `$A$1;`
Select <= (the default);
Enter into `Constant` the value 1;
And click `OK`.

This brings us back to the dialog box where we can verify our entries.

We can follow the progress of the minimization by clicking on `Options`. This brings up another dialog box. In this we select `Show Iteration Results`. The defaults in this are what we want: `Target, Forward, Newton.`

Clicking `OK` returns us to the first dialog box. We again click on `OK`.

We are now back to the original spreadsheet. We click again on `Tools/Numeric Tools/Optimizer.`

If we click on `Solve`, we see in cells A1 and A3 the values 1E-06 and 2.000001. Clicking on `Continue Solving` gives us −0.59446 and 1.72349 as the x- and $f(x)$-values. Repeating `Continue Solving`, we see −0.58853 and 1.72349 and these values do not change on further repetition.

The final values match to the analytical. It appears that Quattro Pro is really using the derivative of $f(x)$ for finding the minimum but with the derivative computed numerically. (Can you verify this supposition?)

7.2 Minimizing a Function of Several Variables

When a function depends on the values of more than one independent variable, finding its minimum values, even when constrained to lie within a region, is not easy. While there could be more than three independent variables, many applied problems have at most three, corresponding to our 3-D world. (More than three will be more common in the problems we treat in the next section.) We will restrict our example to functions of the form

$$z = f(x, y)$$

so that the visualization of the function is easier.

The analytical solution to $z = f(x, y)$ is found be solving the pair of equations:

$$\partial f / \partial x = 0 \qquad \text{and} \qquad \partial f / \partial y = 0,$$

and we can tell if the point(s) correspond to a maximum or a minimum or a saddle point by computing the value of this determinant:

$$d(x, y) = \begin{vmatrix} f_{xx} & f_{xy} \\ f_{yx} & f_{yy} \end{vmatrix},$$

where the subscripts indicate the partial derivatives. If $f_{xx} < 0$ and $d(x, y) > 0$ at the point, there is a maximum (but this may be only local, not global). If $f_{xx} > 0$ and $d(x, y) > 0$, the point is a minimum. If $d(x, y) < 0$, it is a saddle point, if $d(x, y) = 0$, the test is inconclusive.

We will illustrate the several ways to minimize $z = f(x, y)$ with this function:

$$z = (x^2 - 2y)^2 + (x - y)^2 + x + 5.$$

We can find where there is a minimum by computing f_x and f_y and setting to zero:

$$f_x = 2(x^2 - 2y)(2x) + 2(x - y)(1) + 1 = 4x^3 - 8xy + 2x - 2y + 1 = 0,$$
$$f_y = 2(x^2 - 2y)(-2) + 2(x - y)(-1) = -4x^2 - 2x + 10y = 0,$$

If we solve $f_y = 0$ for y, we get $y = x(2x + 1)/5$; and substituting this into the equation for $f_x = 0$ and simplifying gives

$$(4x^3 - 12x^2 + 8x + 5)/5 = 0,$$

which has only one real root at $x = -0.380409$. From the equation for y in terms of x, we get $y = -0.0181973$. This is the point of minimum. At this point, $f = 4.78358$, a global minimum.

Finding the Minimum Numerically

If $z = f(x, y)$ has a minimum within a region in the $x - y$ plane, one could locate the minimum by computing $f(x, y)$ at many points within the region and seeing where the value of the function is least. Even when constrained to a small region, this is tedious and not very economical. Still, it may provide a starting point for searching for the minimum and can be used when there are more than two variables.

A somewhat better technique would be to convert this to a sequence of 1-D problems by setting y, say, to a sequence of values and then using the methods of the previous section. This too is not a very good approach and not well adapted to more than two variables.

A variant on this is to solve the equation after setting z equal to a sequence of values. If these new functions of x and y are plotted, we will have a set of contours. Any of the

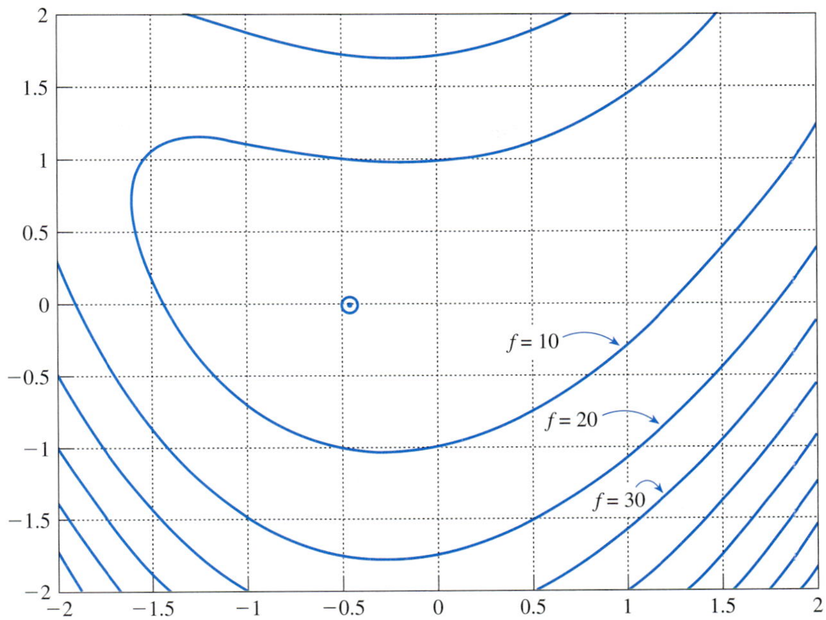

Figure 7.6

computer algebra systems can do this for us. Here are the commands for MATLAB when the function is constrained to lie in the square region whose corners are at (−2, −2) and (2, 2):

```
EDU>> x = - 2:.1:2;
EDU>> y = - 2:.1:2;
EDU>> [X,Y] = meshgrid(x,y);
EDU>> Z = (X.^2 - 2*Y).^2 + (X - Y).^2 + X + 5;
EDU>> contour(x,y,Z); grid on
```

The contour plot looks like Figure 7.6.

We have added the point where $f(x, y)$ is a minimum to the plot. Figure 7.6 is not very helpful because the innermost contour is for $f = 10$, quite far from the minimum of 4.78358. The other contours are at $f = 20, 30, \ldots$.

If we plot contours for values of f near to the minimum, we get Figure 7.7. Observe that the function is quite flat near the minimum point: Even the innermost contour for $f = 5.0$ is not close to the minimum. The other contours are for $f = 5.1$, 5.2, 5.5, and 6.0.

A Simple Search Method

When we have a region in which our function has a minimum point, we can locate it by searching from some starting point within the region. An obvious way to do this is to move from that starting point in the x-direction in small steps until the function stops decreasing.

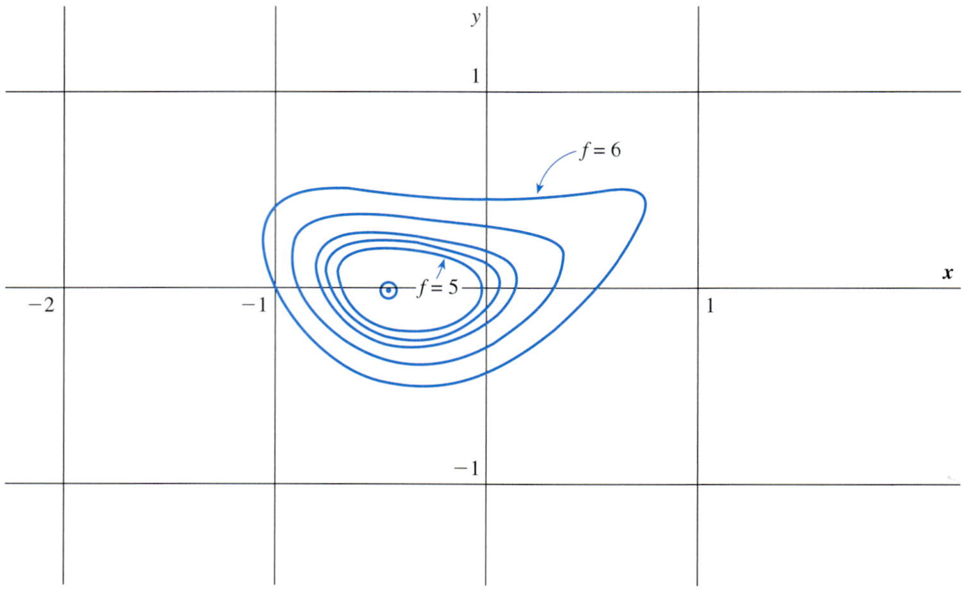

Figure 7.7
Contours for $z = (x^2 - 2y)^2 + (x - y)^2 + x + 5$.
Minimum at $(-0.380\ldots, -0.0182\ldots)$, $f = 4.78358$.

This will happen after we cross a contour and as we approach the other side of it. (It may be that the function does not decrease; if so, we move in the opposite direction. Of course, we could begin in the y-direction.)

This method has been called a *univariant search.* When the function no longer decreases, we begin from the last point and start again, but in the y-direction. After the y-traverse, we do another x-traverse with a smaller step size, going to another y-traverse at the end of this x-traverse. (When the next point at the end of a traverse has the same f-value, it may be that we should use the average of the last two x-values.)

The table shows the results if we do this with the function $f(x, y) = (x^2 - 2y)^2 + (x - y)^2 + x + 5$, starting from $(-1, -1)$ with a step size of $\Delta x = 0.1$. For this problem, we know the answer, $f = 4.78358$ at $x = -0.380409$, $y = -0.0181973$. The table does not complete the tabulation; you may wish to do so. On the second x-traverse, the step size should be reduced, perhaps to 0.05. The process will never be completely finished. When four points are found near the minimum, these can be interpolated.

Observe in the table that the amount of change in f-values at the end of a traverse is only a fraction of that at the start. This itself gives an indication that we are closing in on the minimum point.

There is a problem with such a search method. When the contours are long and narrow and inclined to the axes, it may take many steps to get near the minimum. There are many changes of direction to the search and the approach to the minimum is reached more and more slowly. The difficulty is that we are searching in directions not adapted to the problem. We need a better way.

x	y	$f(x, y)$	
−1.0	−1.0	13.0	
−0.9	−1.0	12.006	
−0.8	−1.0	11.210	
−0.7	−1.0	10.590	
−0.6	−1.0	10.130	
−0.5	−1.0	9.812	
−0.4	−1.0	9.626	
−0.3	−1.0	9.558	
−0.2	−1.0	9.602	(increase, begin y-traverse)
−0.3	−0.9	8.632	
−0.3	−0.8	7.806	
−0.3	−0.7	7.080	
−0.3	−0.6	6.454	
−0.3	−0.5	5.928	
−0.3	−0.4	5.502	
−0.3	−0.3	5.176	
−0.3	−0.2	4.950	
−0.3	−0.1	4.824	
−0.3	0.0	4.798	
−0.3	0.1	4.872	(increase, begin x-traverse)

Finding a Better Search Direction

As we have said, searching in directions parallel to the axes is not usually best. We really want to move in the direction in which the function is decreasing most rapidly. That direction is given by the *gradient,* a vector that points in the direction of most rapid increase in the function values. The gradient of $f(x, y, z)$ is computed by

$$\text{grad}(f) = \nabla f = f_x \boldsymbol{i} + f_y \boldsymbol{j} + f_z \boldsymbol{k}.$$

In this, the subscripts indicate the three partial derivatives at the point (x, y, z) and $\boldsymbol{i}, \boldsymbol{j}$, and $\boldsymbol{k}$ are unit vectors parallel to the axes. Because the gradient vector really points in the direction of most rapid increase in f, we want $-\nabla f$ when we minimize. In the 2-D problem that we are using for examples, we will have

$$-\nabla f = -(f_x \boldsymbol{i} + f_y \boldsymbol{j}).$$

The gradient at any point is perpendicular to the contour curve through that point. (What we call a contour is more commonly called a *level curve* because function has the same value on it.)

What is $-\nabla f$ for the previous example at the point $(-1, -1)$? The function is $z = (x^2 - 2y)^2 + (x - y)^2 + x + 5$. From the previous computations, we know that $f(-1, -1)$ is 13; there is a level curve through this point. We compute the gradient:

$$f_x = [4x^3 - 8xy + 2x - 2y + 1]_{(-1,\,-1)} = (-4 - 8 - 2 + 2 + 1) = -11,$$
$$f_y = [-4x^2 - 2x + 10y]_{(-1,\,-1)} = (-4 + 2 - 10) = -12,$$

so $-\nabla f$ is $11\boldsymbol{i} + 12\boldsymbol{j}$, which points upward from $(-1, -1)$ at an angle of about 47° from the positive x-axis. Let us move along this negative gradient until the function stops decreasing. If we take steps with x-values that differ by 0.2, the y-values will change by $f_y/f_x * \Delta x = 12/11 * 0.2 = 12/55 = 0.21818$. This tabulation shows the results:

x:	-1	-0.8	-0.6	-0.4	-0.2	0.0
y:	-1	$-43/55$	$-31/55$	$-19/55$	$-7/55$	5/55
f:	13	9.0563	6.6133	5.3270	4.8920	5.0413

On the fifth move, we have overshot the minimum along this gradient line. The function value at the fourth step is close to the correct value of $f_{\min} = 4.78353$.

If we search on the negative gradient from this fourth point, we find a minimum at $(-0.16365, -0.08762)$, where f is 4.88294. We then compute the gradient at that point to determine a new search direction. Doing this finds a negative gradient vector that is perpendicular to the former one. This should not be a surprise because we end the gradient search on a contour line and at that point the search vector is tangent to it. The gradient there is perpendicular to the contour and hence to the tangent vector. Eventually we will close in on the true minimum.

A good way to locate the minimum point along a traverse is to determine the linear relation $y = g(x)$ on this vector, substitute this into $f(x, y)$ to reduce it to a function of x only, and then use a method from the previous section. This can be used when there are more than two independent variables.

Figure 7.8 plots the above results superimposed on some of the level curves. Observe that the next gradient vector does not point directly to the minimum point. The major problem with a gradient search is that successive movements are along vectors that

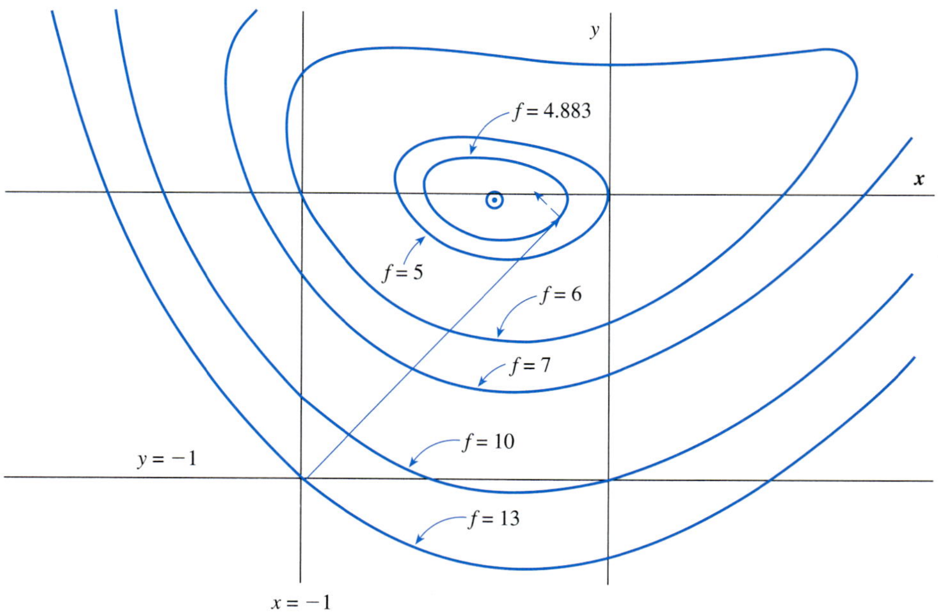

Figure 7.8

are orthogonal, the same as with the univariant search. When the region near the minimum is a long, narrow valley, many right-angled vectors are traversed and these converge on the minimum point only very slowly. The new points that are generated oscillate; they never exactly reach the minimum, but they will come to it within some tolerance value.

Following the negative gradient is called the *method of steepest descent*. The name is appropriate because, at any starting point, we move "downward" in the direction of maximum slope.

Is Steepest Descent the Fastest Way?

The name *steepest descent* might indicate that this is the quickest way to the bottom of a hill. That is not true, as this analogy will show.

Imagine that you are hiking and have followed a path near to the top of a steep hill. As you stop to rest, you notice a small stream nearby that is flowing to the valley below. You realize that the steam is following the negative gradient (the steepest slope) at each point in its journey and that it often winds and curves. The path you traveled does not exactly follow the stream; it takes "short cuts." It is faster to take your path than to meander as the stream does. The reason that the stream has a longer course is because the negative gradient is a local phenomenon.

Sometimes steepest descent is the shortest way. Imagine that you are sitting on the rim of a large circular bowl and look to the bottom. If you slide to the bottom, you will move along the negative gradient and it is the shortest way. Mathematical demonstrations of steepest descent frequently use functions like the bowl; they are spheres, ellipsoids, even paraboloids.

The real world is not so nicely formed. Our hiking analogy resembles it more closely. Still, the method of steepest descent has one advantage: It is sure to find the minimum.

When can we not use this method? If the function is such that the partial derivatives are discontinuous, the gradient will also be discontinuous. If the function is not known as a mathematical relation, such as when function values can only be determined experimentally or through a simulation, we cannot differentiate. Still, by performing more experiments in the neighborhood of the starting point, we can get a numerical estimate. When this is done, the surrounding points are often at the corners of a square region that is centered about the starting point; sometimes points at the midpoints of the edges of the square are included. If these additional function values are expensive to obtain, this may not be a practical way to go.

The Conjugate Gradient Method

For certain objective functions, there is a very rapid way to locate the minimum. It is called the *conjugate gradient* method. When the objective is *quadratic function,* it will find a minimum point in exactly two steps if the function has just two variables, and in exactly three steps if there are three. Each step requires the computation of a number of vectors.

The conjugate gradient method is better than steepest descent in most cases because it takes into account the curvature of the function. This method is important to know: Quadratic functions are not uncommon, and we can approximate other functions as a quadratic function in the neighborhood of a point that we hope is near the minimum.

In this development, we will use many vectors. All vectors will be row vectors and the vector name will be in boldface; if vector $\boldsymbol{w}$ has components u and v, we write $\boldsymbol{w} = [u, v]$. $\boldsymbol{w}^T$ is its transpose, a column vector.

What is a "quadratic function"? If a 2-D function contains only terms in x^2, y^2, $x * y$, x, y, and a constant, it is a quadratic function. A similar definition applies in three dimensions. Any quadratic function can be expressed in a nice way; for $f(x, y)$, this is

$$(1/2)\ [x, y] * H * [x, y]^T + \boldsymbol{b} * [x, y]^T + c,$$

where matrix H is the Hessian matrix* of the function, $[x, y]$ is a row vector, and the components of row vector $\boldsymbol{b}$ are the coefficients of the x- and y-terms; c is the constant term in the equation.

We will use as an example $f(x, y) = x^2 + 2y^2 + xy + 3x$; we can write this as

$$(1/2)\ [x, y] * H * [x, y]^T + \boldsymbol{b} * [x, y]^T + c,$$

where $H = \begin{bmatrix} 2 & 1 \\ 1 & 4 \end{bmatrix}$, $(x, y)^T$ is $\begin{bmatrix} x \\ \hline y \end{bmatrix}$ $\boldsymbol{b} = [3, 0]$, and $c = 0$.

We will illustrate the conjugate gradient method with this function, starting at (0, 0). (It is not difficult to find that $f_{\text{min}} = -18/7$ at $(-12/7, 3/7)$ from $f_x = 2x + y + 3 = 0$, $f_y = 4y + x = 0$.)

We begin the conjugate† gradient method by computing vector $\boldsymbol{x0} = \nabla f(0, 0) = [3, 0]$, the gradient at the starting point. We compute three other vectors from $\boldsymbol{x0}$: $\boldsymbol{q0}, \boldsymbol{v0}, \boldsymbol{x1}$:

Step 1.

a. Compute vector $\boldsymbol{q0}^T = H * \boldsymbol{x0}^T + \boldsymbol{b}^T$:

$$\boldsymbol{q0}^T = \begin{bmatrix} 2 & 1 \\ 1 & 4 \end{bmatrix} * [3, 0]^T + [3, 0]^T = [9, 3]^T, \boldsymbol{q0} = [9, 3].$$

b. Set vector $\boldsymbol{v0} = -\boldsymbol{q0} = [-9, -3]$.

c. Compute multiplier $\alpha 0 = \boldsymbol{v0} * \boldsymbol{v0}^T/(\boldsymbol{v0} * H * \boldsymbol{v0}^T)$:

$$\alpha 0 = [-9, -3] * [-9, -3]^T/([-9, -3] * H * [-9, -3]^T) = 5/14.$$

d. Compute vector $\boldsymbol{x1} = \boldsymbol{x0} + \alpha 0 * \boldsymbol{v0}$:

$$\boldsymbol{x1} = [3, 0] + (5/14) * [-9, -3] = [-3/14, -15/14].$$

---

* The Hessian matrix is formed from the second partial derivatives:

$$H = \begin{bmatrix} f_{xx} & f_{xy} \\ f_{yx} & f_{yy} \end{bmatrix} \text{(but } f_{xy} = f_{yx} \text{ for a quadratic).}$$

† Two vectors, a and b, are conjugate with respect to matrix M if $a * M * b = 0$. (If M is the identity matrix, they are orthogonal.)

Step 2.

We compute two other vectors from $\boldsymbol{x1}$: $\boldsymbol{q1}$, $\boldsymbol{v1}$:

a. Compute vector $\boldsymbol{q1}^T = \mathrm{H} * \boldsymbol{x1}^T + \boldsymbol{b}^T$:

$$= \begin{bmatrix} 2 & 1 \\ 1 & 4 \end{bmatrix} * [-3/14, -15/14]^T + [3, 0]^T = [3/2, -9/2]^T.$$

$\boldsymbol{q1} = [3/2, -9/2]$.

b. Compute multiplier $\beta 0 = \boldsymbol{q1} * H * \boldsymbol{v0}^T/(\boldsymbol{v0} * H * \boldsymbol{v0}^T)$:

$$= \frac{[3/2, -9/2] * \begin{bmatrix} 2 & 1 \\ 1 & 4 \end{bmatrix} * [-9, -3]^T}{[-9, -3] * H * [-9, -3]^T} = 1/4.$$

c. Compute vector $\boldsymbol{v1} = -\boldsymbol{q1} + \beta 0 * \boldsymbol{v0} =$

$$= [-3/2, 9/2] + (1/4) * [-9, -3] = [-15/4, 15/4].$$

d. Compute multiplier $\alpha 1 = -\boldsymbol{q1} * \boldsymbol{v1}^T/(\boldsymbol{v1} * H * \boldsymbol{v1}^T)$:

$$= \frac{[-3/2, 9/2] * [-15/4, 15/4]^T}{[-15/4, 15/4] * H * [-15/4, 15/4]^T} = 2/5.$$

The minimum point is now obtained:

$$\boldsymbol{x}_{\min} = \boldsymbol{x1} + \alpha 1 * \boldsymbol{v1}:$$

$$= [-3/14, -15/14] + (2/5) * [-15/4, 15/4] = [-12/7, 3/7].$$

At this point $f_{\min} = -18/7$, the correct answer.

Vectors $\boldsymbol{v0}$ and $\boldsymbol{v1}$ are conjugates with respect to matrix H:

$$\boldsymbol{v0} * H * \boldsymbol{v1} = [-9, -3] \begin{bmatrix} 2 & 1 \\ 1 & 4 \end{bmatrix} [-15/4, 15/4]^T = 0.$$

This clever way to solve for a minimum applies only to quadratic functions, but we can adapt the conjugate gradient method to functions that are not quadratic by approximating the function with a quadratic. We would fit the quadratic polynomial to the function at the start point and at six adjacent points. This will not exactly get the minimum of the function in two steps so it will require iterations with new quadratic approximations as the minimum is approached.

It is of interest to compare this to the solution by steepest descent. Starting at (0, 0) where the negative gradient is $[-3, 0]$, we move left on the x-axis and find the minimum there at $(-3/2, 0)$, $f = -9/4$. At that point, the negative gradient is [0, 3/2], so we move upward on the line $x = -3/2$ to find the minimum along this vector to be $-81/32$ at $(-3/2, 3/8)$. We again compute the negative gradient and find it to be $[-3/8, 0]$. We move left to the minimum on the vector; it is at $(-27/16, 3/8)$, where $f = -657/256$. This is close to the minimum for the function, but it still differs by 0.2%.

If we continue in this manner, we will find that the moves are along vectors that are parallel to the two axes in turn and we never get exactly to the minimum. Observe that, in this instance, steepest descent is the same as a univariant search.

Newton's Method

We found in Chapter 1 that Newton's method converges quadratically to a zero of a continuous function, $y = f(x)$. This method can be used to find a minimum (or maximum) by finding a zero of the derivative. If the function is quadratic, the solution is found immediately, as this simple example shows:

> Given $y = 2x^2 - x + 4$, what is its minimum value?
> We will do this by finding the zero of dy/dx. The derivative of y is $4x - 1$. (By setting this to zero, we anticipate the answer: $x = 1/4$.) What does Newton's method give, starting from $x_0 = 1$? The iterations are
>
> $$x_{n+1} = x_n - f_0'/f_0''.$$
>
> At $x_0 = 1, f_0' = 4(1) - 1 = 3, f_0'' = 4$, so
>
> $$x_1 = 1 - 3/4 = 1/4, \text{ precisely correct.}$$
>
> It is easy to show that we get the correct answer immediately for any value for x_0. If $y = f(x)$ is not a quadratic, we will have to iterate, but convergence will be quadratic.

How can Newton's method be applied to finding the minimum of a function of more than one variable? We need the equivalents of f' and f''. For f', we use the negative gradient at (x_0, y_0) and for f'', we use the Hessian matrix of partial derivatives computed at the same point. We cannot divide by a matrix, of course, so we will multiply by the inverse of H, H^{-1}:

$$(x, y)_{n+1} = (x, y)_n - H_n^{-1} * (-\nabla f_n).$$

If $z = f(x, y)$ is a quadratic function, we can expect to get to the minimum immediately. Let's see if this is true using the same quadratic function that we used previously:

$$f(x, y) = x^2 + 2y^2 + xy + 3x.$$

(The value we anticipate is $f = -18/7$ at $(-12/7, 3/7)$.

We previously computed the Hessian matrix:

$$H = \begin{bmatrix} 2 & 1 \\ 1 & 4 \end{bmatrix},$$

We need its inverse, which we find to be

$$H^{-1} = \begin{bmatrix} 4/7 & -1/7 \\ -1/7 & 2/7 \end{bmatrix}.$$

If we start from (0, 0), we compute

$$
\begin{aligned}
(x, y)_1 &= (x, y)_0 - H^{-1} * [-\nabla f_0]^T \\
&= [0, 0]^T - \begin{bmatrix} 4/7 & -1/7 \\ -1/7 & 2/7 \end{bmatrix} * [-3, 0]^T \\
&= [-12/7, 3/7]^T, \text{ precisely correct!}
\end{aligned}
$$

When $f(x, y)$ is not a quadratic, we can approximate it with a second-degree polynomial that fits near the starting point and proceed in the same way, except this approximate solution will be inexact. Even so, we will have reached a point nearer to the minimum. We then get another approximation and repeat; this approaches the true minimum as closely as we desire.

Searching Without Using Derivative Values

We really can use only gradient-based searches as described above when the gradient can be obtained as an analytical function. When that is not the case, we can use finite-difference approximations to the derivatives. There are also other approaches.

The univariant search that we have called a "simple search" is a way to minimize without using derivatives, but we really want to move in (almost) the correct direction. There are ways to move more nearly in the right direction, one of which is the *simplex method.* This method begins not from a single point in the region of interest, but from a group of three points. Often, these are chosen to form an isosceles triangle, called a *simplex.* (However, this should not be construed as equivalent to the simplex method for linear programming, discussed in the next section.)

One of the points of the simplex will normally have the largest function value, and obviously we want to move away from that. Call this point p_1. We then locate a new point that is a reflection of point p_1 across the opposite side of the triangle. Dropping p_1 from the set, we have a new triangle that includes the reflected point. We use this triangle to find which point in the current set should be reflected.

Once this process finds that the function value does not decrease at the reflected point, an inward reflection is made, creating the new point within the simplex triangle. The simplex method will eventually close in on the minimum.

Spreadsheets Can Minimize $f(x, y, z, \ldots)$ with Constraints

Both Excel and Quattro Pro can find the minimum or maximum of a function of several variables within a region. The procedure is similar to that described in Section 7.1 for a function of one variable: We choose cells for each of the variables and enter values in them to define a starting point. In another cell, we enter the function to be minimized (or maximized).

The region of interest does not have to be rectangular or polyhedral; we can define its boundaries in terms of the variables by entering the proper relations in other cells. If the region is defined solely by limiting values for the variables themselves, this is not necessary.

In Excel, we invoke `Tools/Solver;` in Quattro Pro, we use `Tools/Numeric Tools/Optimizer.` In dialog boxes, we enter the cell numbers for variables and the function, together with constraints that define the region. Clicking on `Solve` produces the solution and we can get the successive iterations if we want to see them. Options that are available include Gradient, Conjugate, and Newton.

7.3 Linear Programming

A widely used technique for maximizing the profits or minimizing the costs is *linear programming.* It is often used in business to determine those decisions that will increase profitability. It has other business applications, such as finding the optimal schedule for an outside salesman to visit his customers.

The word *programming* here does not mean a computer program in the ordinary sense (although computers are nearly always used to solve the problems). It refers instead to a systematic procedure, one that is based on solving set linear equations. Linear programming is *linear* in that the function whose optimum is sought is a linear combination of two or more (often many) independent variables. The solution is subject to a number of constraints, and these are themselves always a linear combination of the variables. A constraint, for example, might be how a limited resource will be utilized by several competing potential applications.

A Simple Problem

We begin with a simple problem with just two variables, but this will illustrate the method and introduce some of the many special terms of linear programming. The problem is to maximize $f(x_1, x_2) = 5x_1 + 8x_2$,

subject to:

$$x_1 + 3x_2 \le 12,$$
$$3x_1 + 2x_2 \le 15,$$
$$x_1, x_2 \ge 0.$$

Think of a company that is to manufacture two products. The amount of each is measured by x_1 and x_2. $f(x_1, x_2)$ is the *objective function.* This function, $f(x_1, x_2)$, determines the manufacturing profit. The larger the values for x_1 and x_2, the greater the profit. The coefficients are the profit per unit of product.

However, it is not possible to manufacture any desired quantity of these products, for there is a limited amount of two necessary resources. (These might be available employees, critically important parts, machine availability, or the like.) The *constraint relations* show how each of the resources is used up in the manufacturing process. The coefficients

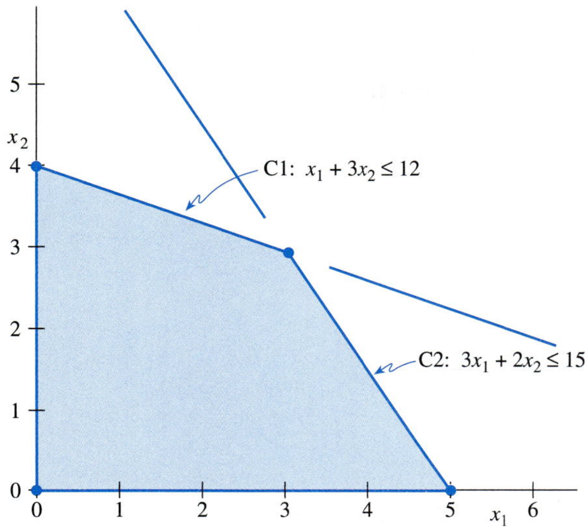

Figure 7.9
The feasible region

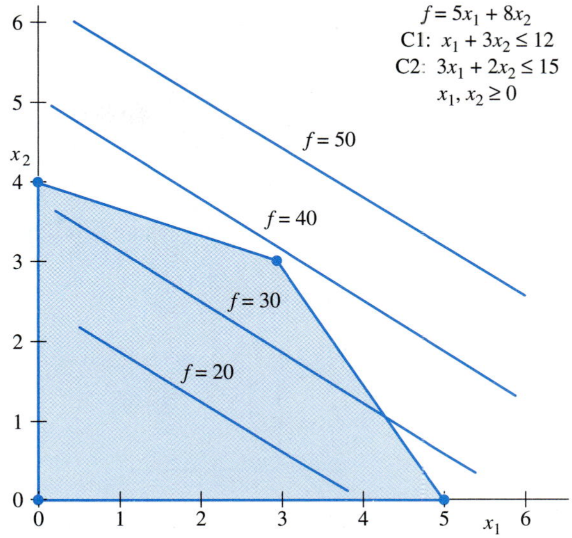

Figure 7.10
Objective function values are superimposed on the feasible region

in the constraint relations represent the required amount of the resource used per unit amount of the product.

Notice that each of the constraints is linear and that the objective function is also a linear combination. The last inequality is a special one; while not a constraint in the same sense as the others, it forces the solution to have only nonnegative values for the variables. This is common because it is impossible to make a neqative quantity of product.

We will first solve this graphically; this will introduce the topic and help to define a number of special terms. A plot of the constraints in Figure 7.9 shows the *feasible region,* the possible production quantities of product 1 and product 2. (We have scaled the numbers to make them small. The actual quantities might be 100 or 1000 times as great.) The region is bordered by the heavy lines.

Observe that the feasible region is bounded by the x_1, x_2 axes (from the nonnegativity condition) and by two intersecting lines. There are four vertices to the polygonal region, including one at the origin and two on the axes.

In Figure 7.10, we redraw the feasible region and superimpose on it a number of lines defined by setting the objection relation equal to several values.

Because the objective function is linear, the lines for $f(x_1, x_2)$ are parallel. The larger the value assigned to the function, the farther from the origin the line lies. Some of the lines do not fall within the feasible region—we cannot achieve that much profit. Some lie within the region but represent choices that give less profit than the maximum. Points on such lines within the region are *feasible solutions.* There is one line (not drawn) that would show the maximum; it would just touch the feasible region. In this example, it will touch at the point (3, 3). A different objective function whose slope is different might touch the region at a different vertex. The important conclusion from this is that the *optimal solution* will always fall at one of the vertices of the region.

The four vertices of the region in Figure 7.10 (we include the origin) are called *basic feasible solutions*. It is then clear that one way to solve this linear programming problem is to find values for x_1 and x_2 at the vertices and from these compute the values for the objective function at each vertex of the region (more commonly called *corner points*), and then select the point where it is a maximum. For our example, these values are

x_1	x_2	$f(x_1, x_2)$
0	0	0
0	4	32
3	3	39
5	0	25

This confirms the fact that the optimal value for the two products is three units and three units, respectively.

Examining Figure 7.10 suggests several other possibilities:

1. If the objective function had different coefficients, the objective function lines might be parallel to one of the constraints and one of these lines will coincide with an edge of the region. In that case, there are multiple optimal solutions. Any combination of choices for x_1 and x_2 that lie on that edge give the same profit.
2. There could be a third constraint and this can have different possible effects:
   a. It could lie totally outside the feasible region and thus not limit the amounts to be produced. We would call this a redundant constraint.
   b. It could coincide with one of the previous constraints. This too is redundant; the region is not affected.
   c. It could lie partially within the region. This would decrease the area of the feasible polygon and might create additional corner points.
3. The graphical method for solving a linear programming problem is fine if there are only two variables. It could be applied (with difficulty) to three variables, but more than three is virtually impossible. We need to find a different way to solve linear programming problems because some applications have hundreds of variables.

The Simplex Method

Even though we have already solved our example, we will use it to introduce the *simplex method,* which is most frequently used for linear programming. We repeat the problem: Maximize

$$f(x_1, x_2) = 5x_1 + 8x_2,$$

subject to

$$x_1 + 3x_2 \leq 12,$$
$$3x_1 + 2x_2 \leq 15,$$
$$x_1, x_2 \geq 0.$$

The simplex method solves the problem through solving a set of equations that represent the constraints. "But our constraints aren't equations, they are inequalities," you say. That is a good observation. We need to change inequalities to equalities. This can be done by a simple device: We add another variable to the constraint, a quantity called a *slack variable.* This measures the amount of the resource not utilized; it takes up the "slack." Call the slack variable for the first constraint x_3, and that for the second, x_4. Our problem then becomes to maximize

$$f(x_1, x_2) = 5x_1 + 8x_2 + 0x_3 + 0x_4$$

subject to

$$x_1 + 3x_2 + x_3 = 12,$$
$$3x_1 + 2x_2 + x_4 = 15,$$
$$x_1, x_2, x_3, x_4 \geq 0.$$

We have expanded the objective function to include the slack variables. They contribute nothing to profits, of course.

In matrix form, the constraint equations are

$$\begin{bmatrix} 1 & 3 & 1 & 0 \\ 3 & 2 & 0 & 1 \end{bmatrix} \begin{bmatrix} x_1 \\ x_2 \\ x_3 \\ x_4 \end{bmatrix} = \begin{bmatrix} 12 \\ 15 \end{bmatrix}.$$

This system is underdetermined; there are only two equations but four variables. Still, we can solve this if we first assign values to two of the variables and move these terms to the right-hand side. Observe that adding the slacks to the system expanded the matrix of constraint coefficients to include an identity matrix.

Let us assign zero to both x_1 and x_2. The system is reduced to

$$\begin{bmatrix} 1 & 0 \\ 0 & 1 \end{bmatrix} \begin{bmatrix} x_3 \\ x_4 \end{bmatrix} = \begin{bmatrix} 12 \\ 15 \end{bmatrix},$$

where the solution is obvious: $x_3 = 12$, $x_4 = 15$. "Of course," you say, "if neither product is made, the entire amount of both of the resources is unused. The slacks measure that." The important result is that we have values for x_1 and x_2 at a corner point, a basic feasible solution to the problem, though surely not the optimum.

In the terminology of linear programming, what we have just done is to cause two variables to leave the system and two to enter. The ones that leave are x_1 and x_2; the ones that enter are x_3 and x_4.

Suppose we allow a new variable to enter the system, replacing one that is already there. So, one of x_3 or x_4 must leave. In effect, we are exchanging a current variable for one not yet in the system.

Which of x_1 or x_2 should we select to enter? Looking at the objective function, we see that the profit from one unit of x_2 is 8, while one unit of x_1 returns only 5; x_2 is the better choice. (You may want to see whether the other choice ends up at the same final answer.) So, x_2 is to enter the system. Now we must decide which of x_3 and x_4 should leave. We answer the question by trying both possibilities:

If x_3 leaves, the variables in the equations are x_2 and x_4, and the system becomes

$$\begin{bmatrix} 3 & 0 \\ 2 & 1 \end{bmatrix} \begin{bmatrix} x_2 \\ x_4 \end{bmatrix} = \begin{bmatrix} 12 \\ 15 \end{bmatrix}, \quad \text{solution: } \begin{aligned} x_2 &= 4, \\ x_4 &= 7. \end{aligned}$$

If x_4 leaves instead, we have

$$\begin{bmatrix} 3 & 0 \\ 2 & 1 \end{bmatrix} \begin{bmatrix} x_2 \\ x_3 \end{bmatrix} = \begin{bmatrix} 12 \\ 15 \end{bmatrix}, \quad \text{solution: } \begin{aligned} x_2 &= 6, \\ x_3 &= -6. \end{aligned}$$

Only the first is acceptable; the second violates the nonnegativity condition. The variables now present are x_2 and x_4. Remembering that x_1 is still zero but now x_2 is 4; we have moved from our initial basic feasible solution, (0, 0) to another basic feasible solution, (0, 4). At this point, the value of the objective function is 32.

We proceed in similar fashion to allow x_1 to enter. x_4 will have to leave. The variables present are the non-slacks, x_1 and x_2. We need to solve:

$$\begin{bmatrix} 1 & 3 \\ 3 & 2 \end{bmatrix} \begin{bmatrix} x_1 \\ x_2 \end{bmatrix} = \begin{bmatrix} 12 \\ 15 \end{bmatrix}, \quad \text{solution: } \begin{aligned} x_1 &= 3, \\ x_2 &= 3. \end{aligned}$$

We have moved to another basic feasible solution (3, 3), where the value of the objective function is 39. In this problem, we know that this must be the optimum point because removing either x_1 or x_2 can only reduce the profit.

(What we have done is to solve for the intersection of the two constraints, a corner point.)

Variations of the Problem

Even this simple example can illustrate how some variants to the problem affect it.

1. What if the lower limit of one of the variables is something other than zero? This would have to be a positive quantity. It also would have to be small enough to lie within the feasible region, or else we would say the problem is *infeasible:* No solution is possible. This is also true if both variables have lower limits other than zero. Having lower limits other than zero will reduce the area of the feasible region. The initial basic feasible solution would still be at one of the lower-limit points. If the nonegativity constraint were replaced by

$$x_1 + x_2 \geq 1,$$

this would chop off a triangle from the lower-left part of the feasible region. We would have to include this inequality in the matrix of constraints. With a greater than or equal relation, the slack variable is subtracted to give the constraint equation.

2. What if additional greater than or equal constraints are included? We just include these with a subtracted slack variable. It is then possible to have a diamond-shaped feasible region.
3. What if the lines for the objective function are parallel to one of the constraints? One of these lines would then coincide with an edge of the region, and any point on this edge is optimal; there is then an infinity of optimal points, all with the same value for the objective function.
4. What if the objective function has a positive slope? (This would mean that one of the products incurs a loss rather than a profit but that, while unlikely, could happen.) The objective function lines would then intersect the constraints. For a region like that of Figure 7.10, the optimum would still occur at a corner point. The simplex method will still find it.
5. What if we want to minimize an objective function? (The coefficients then would represent unit costs rather than unit profits.) The simplex procedure works exactly the same—we just maximize the negative of the objective function.
6. Can we use the simplex method to solve a problem where either the objective function or a constraint is discontinuous? No, the requirement of linearity is absolute.

Another Example

We now present a slightly more complex problem that will show how the simplex method works when there are more than two constraints. It often occurs that there are more constraints than variables. The example still has only two variables, so it could be solved graphically or by computing a list of function values are the corners. Here is our example:

Maximize

$$f(x_1, x_2) = 8x_1 + 9x_2,$$

constraints:

$$\begin{aligned} \text{C1: } & 2x_1 + 4x_2 \le 32, \\ \text{C2: } & 3x_1 + 4x_2 \le 36, \\ \text{C3: } & 6x_1 + 4x_2 \le 60, \\ & x_1, x_2 \ge 0. \end{aligned}$$

We add slacks x_3, x_4, x_5 to the three constraints. In matrix form we have:

$$f = 8x_1 + 9x_2 + 0x_3 + 0x_4 + 0x_5,$$

$$\begin{bmatrix} 2 & 4 & 1 & 0 & 0 \\ 3 & 4 & 0 & 1 & 0 \\ 6 & 4 & 0 & 0 & 1 \end{bmatrix} \begin{bmatrix} x_1 \\ x_2 \\ x_3 \\ x_4 \\ x_5 \end{bmatrix} = \begin{bmatrix} 32 \\ 36 \\ 60 \end{bmatrix},$$

$$x_1, x_2, x_3, x_4, x_5 \geq 0.$$

We begin as is customary with a basic feasible solution at the origin, (0, 0), where $f = 0$. We improve the solution, by bringing in a new variable to replace one of x_3, x_4, or x_5. Our best choice of the variable to bring into the solution is x_2. We need to see which of the current variables is to leave, so we try each in turn.

If x_3 leaves and x_2 enters, the variables in the solution are x_2, x_4, and x_5. We solve:

$$\begin{bmatrix} 4 & 0 & 0 \\ 4 & 1 & 0 \\ 4 & 0 & 1 \end{bmatrix} x = \begin{bmatrix} 32 \\ 36 \\ 60 \end{bmatrix}, \quad \text{whose solution is} \begin{cases} x_2 = 8 \\ x_4 = 4, \\ x_5 = 28 \end{cases}$$

which we can accept; the nonnegativity condition holds.

Let us see if any of the other choices is acceptable. If x_4 leaves instead of x_3, the variables in the solution will be x_2, x_3, and x_5. We solve:

$$\begin{bmatrix} 4 & 1 & 0 \\ 4 & 0 & 0 \\ 4 & 0 & 1 \end{bmatrix} x = \begin{bmatrix} 32 \\ 36 \\ 60 \end{bmatrix}, \quad \text{whose solution is} \begin{cases} x_2 = 9 \\ x_3 = -4. \\ x_5 = 24 \end{cases}$$

This is not acceptable. What if we let x_5 leave instead of x_3? The variables will be x_2, x_3, and x_4. We solve:

$$\begin{bmatrix} 4 & 1 & 0 \\ 4 & 0 & 1 \\ 4 & 0 & 0 \end{bmatrix} x = \begin{bmatrix} 32 \\ 36 \\ 60 \end{bmatrix}, \quad \text{whose solution is} \begin{cases} x_2 = 15 \\ x_3 = -28, \\ x_4 = -24 \end{cases}$$

which is also not acceptable. With variables x_2, x_4, and x_5 in the system, the value for x_2 is 8, x_1 is zero. At (0, 8), f is 72.

We hope to improve the solution by replacing x_3 or x_4. We don't want to put x_5 back in, so we let x_1 replace either x_3 or x_4. If we replace x_3, we will have x_1, x_2, and x_4. We solve:

$$\begin{bmatrix} 2 & 4 & 0 \\ 3 & 4 & 1 \\ 6 & 4 & 0 \end{bmatrix} x = \begin{bmatrix} 32 \\ 36 \\ 60 \end{bmatrix}, \quad \text{whose solution is} \begin{cases} x_1 = 7 \\ x_2 = 4.5, \\ x_4 = -3 \end{cases}$$

which we must reject. We try the other choice, giving the variables as x_1, x_2, and x_3. We solve:

$$\begin{bmatrix} 2 & 4 & 1 \\ 3 & 4 & 0 \\ 6 & 4 & 0 \end{bmatrix} x = \begin{bmatrix} 32 \\ 36 \\ 60 \end{bmatrix}, \quad \text{whose solution is} \begin{cases} x_1 = 8 \\ x_2 = 3. \\ x_3 = 4 \end{cases}$$

We can accept this. So, we have moved from (0, 8) to (8, 3), where the value of f is 91.

Can we improve further? The only possibility is to put x_5 back in, replacing x_3. With variables x_1, x_2, and x_5, we solve:

$$\begin{bmatrix} 2 & 4 & 0 \\ 3 & 4 & 0 \\ 6 & 4 & 1 \end{bmatrix} = \begin{bmatrix} 32 \\ 36 \\ 60 \end{bmatrix}, \quad \text{whose solution is} \begin{cases} x_1 = 4 \\ x_2 = 6 \\ x_5 = 12 \end{cases}.$$

At (4, 6), $f = 86$, and we do not increase the value of f. It seems that the optimum is at (8, 3), where $f = 91$.

There is one more corner point that we could test; it is at (10, 0), where $f = 80$, less than that at other corners.

Are There More Efficient Ways?

We have used a procedure that would most clearly show the basic principle behind the simplex method. This is perhaps not the most efficient. We solved the examples in this way to emphasize that we move from one basic feasible solution to another where the objective function is improved. We did this by replacing one current variable with another. Selecting the variable to enter was easy: We chose the one that would contribute most to the objective function, the one with the larger unit profit. We selected which variable would leave by examining whether the nonnegativity constraints were violated. This examination was done by computing the amounts of the present variables that would remain in the solution when the new variable entered; if any of these were negative, we rejected it.

An alternative procedure sets up a *simplex tableau.* In using this tableau, all of candidates for leaving the basis are tested simultaneously, rather than individually as we have done. The tableau is modified at each iteration by doing the equivalent of a Gauss–Jordan reduction. This may require fewer arithmetic operations but what is happening to the variables is not seen as clearly.

Every linear programming problem has another problem called its *dual* and the solution to the dual problem is the same as for the *primal problem.* The dual may require less effort to solve than the primal, and solving it will be more efficient. We discuss the dual to a primal problem later in this section.

A problem with many variables and many constraints can be solved in the same way as we have described but doing it would be painfully slow. The use of a computer program is essential and there are many available. We can even use the Excel or Quattro Pro spreadsheet programs. Here is how Quattro Pro solves a linear programming problem. We use the last example as an illustration.

Using Quattro Pro

We restate the problem:

$$\text{Maximize } f(x_1, x_2) = 8x_1 + 9x_2,$$

Constraints:

$$\begin{aligned} \text{C1: } 2x_1 + 4x_2 &\le 32, \\ \text{C2: } 3x_1 + 4x_2 &\le 36, \\ \text{C3: } 6x_1 + 4x_2 &\le 60, \\ x_1, x_2 &\ge 0. \end{aligned}$$

After activating Quattro Pro, we decide to use these cells:

Cell A1: holds x_1, set to zero.

Cell A2: holds x_2, set to zero.

Cell A3: holds object function, set to 8*A1 + 9*A2.

Cell A4: holds left-hand side of first constraint, set to 2*A1 + 4*A2.

Cell A5: holds left-hand side of second constraint, set to 3*A1 + 4*A2.

Cell A6: hold left-hand side of third constraint, set to 6*A1 + 4*A2.

The right-hand sides will be entered when constraints are defined. The nonnegativity relations will be set as separate constraints.

We have defined all the necessary parameters. We click on `Tools/Numeric Tools/Optimizer` and see the `Optimizer` input screen. In this we enter:

Solution Cell: A3

Set to Maximize (the default)

Variable Cell(s): A1 . . A2

Click on Add Constraint, in Variable Cell: enter A4, choose ≤ , enter 32 in Constant.

Click on Add Another, in Variable Cell: enter A5, choose ≤ , enter 36 in Constant.

Click on Add Another, in Variable Cell: enter A6, choose ≤ , enter 60 in Constant.

Click on Add Another, in Variable Cell: enter A1, choose ≥ , enter 0 in Constant.

Click on Add Another, in Variable Cell: enter A2, choose ≥ , enter 0 in Constant.

Click OK, which brings us back to the Optimizer input screen. We review our settings and change them if necessary. We click OK and get back to the spreadsheet.

Everything is now in order to get the solution. We again click `Tools/Numeric Tools/Optimizer` and see again the `Optimizer` input screen. We want to see the successive iterations, so we click `Options` and on that screen select `Show Iteration Results`. We click `OK` and are brought back to the `Optimizer` input screen. When we click on `Solve` we see the initial basic feasible solution. By clicking repeatedly on `Continue Solving`, we see this succession of results on the spreadsheet.

The first set of values appears to be the initial basic feasible solution at the origin. (There is some differences from zeros due to the computer's finite precision.) The second set is for a point on one of the edges of the feasible region, but this is not a corner point. The third is near the optimal point and the fourth and subsequent are at (8, 3) where $f = 91$, the optimum.

Cell	Starting Values	Next Values	Third Values	Fourth Values	Fifth Values
A1	0	4.8	8	8	8
A2	1E-06	5.400006	3.000004	3	3
A3	9E-06	87.00006	91.00004	91	91
A4	4E-06	31.20003	28.00002	28	28
A5	4E-06	36.00003	36.00002	36	36
A6	4E-06	50.40003	60.00002	60	60

Using Maple

Maple can solve linear programming problems by the simplex method. Here are the results for the same example:

```
>with(simplex);
  Warning, the protected names maximize and minimize have been
redefined and unprotected
>obj := 8*x + 9*y;
  (the function is echoed)
>constr := {2*x + 4*y <= 32, 3*x* + 4*y <= 36, 6*x + 4*y <= 60};
  (the constraints are echoed)
>maximize(obj, constr, NONNEGATIVE);
  x = 8, y = 3
```

We see the correct value for the coordinates of the optimal point. MATLAB can do similarly but this requires its Optimization Toolbox, which is not a part of the student edition.

The Dual Problem

Every linear programming problem has a counterpart, called its *dual problem.* It is somewhat like a mirror image. One is called the *primal,* the other the *dual*. The optimal values for the objective functions are identical, so solving one is like solving the other.

The coefficients in the objective function of the primal are used as the right-hand sides of the dual constraints; the right-hand sides of the constraints of the primal are used as the coefficients in the objective function of the dual. Rows of the matrix of constraints of the primal become columns in the dual. Maximizing in the primal is equivalent to minimizing in the dual. In the primal, constraints are less than or equal, in the dual they are greater than or equal. This example will make this clearer; it has just two variables and two constraints so that we can graph the solutions:

EXAMPLE 7.3

The primal:	**The dual:**
$\text{Max } f = 5x_1 + 8x_2$	$\text{Min } g = 12y_1 + 10y_2$
$3x_1 + 4x_2 \le 12,$	$3y_1 + 2y_2 \ge 5,$
$2x_1 + 5x_2 \le 10,$	$4y_1 + 5y_2 \ge 8,$
$x_1, x_2 \ge 0.$	$y_1, y_2 \ge 0.$

Figures 7.11 and 7.12 show the graphs. The solutions to each of these were not hard to find: For the primal, $f = 148/7$ at $(x_1, x_2) = (20/7, 6/7)$. For the dual, $g = 148/7$ at $(y_1, y_2) = (9/7, 4/7)$.

It is instructive to put the problems in matrix form:

The primal:

$$\text{Max } [5 \quad 8]\begin{bmatrix} x_1 \\ x_2 \end{bmatrix}$$

$$\text{s.t.} \begin{bmatrix} 3 & 4 \\ 2 & 5 \end{bmatrix} x \le \begin{bmatrix} 12 \\ 10 \end{bmatrix},$$

$$x_1, x_2 \ge 0.$$

The dual:

$$\text{Max } [12 \quad 10]\begin{bmatrix} y_1 \\ y_2 \end{bmatrix}$$

$$\text{s.t.} \begin{bmatrix} 3 & 2 \\ 4 & 5 \end{bmatrix} y \ge \begin{bmatrix} 5 \\ 8 \end{bmatrix},$$

$$y_1, y_2 \ge 0.$$

In more general terms that apply to problems of any size, let row vector $\boldsymbol{c}$ hold the coefficients of the primal objective with column vector $\boldsymbol{x}$ being the variables; let matrix $\mathbf{A}$ hold the coefficients of the constraints with column vector $\boldsymbol{b}$ holding the right-hand sides.

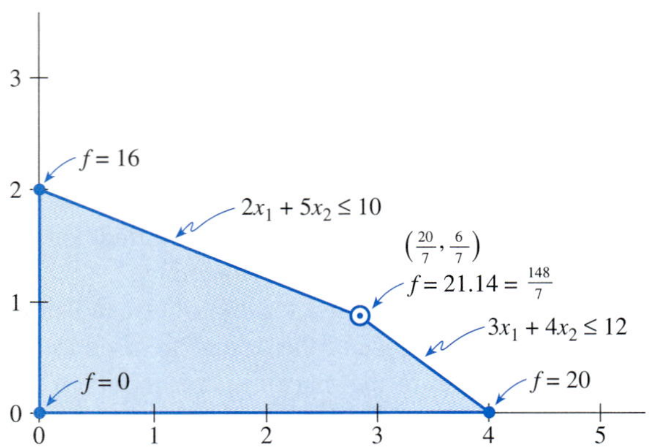

The primal problem

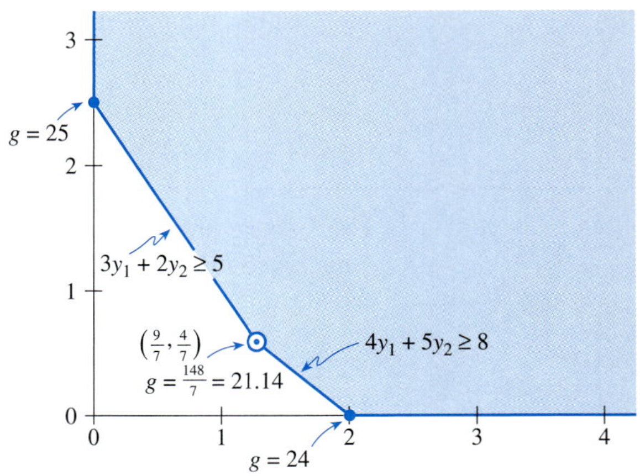

Figure 7.12
The dual problem

For the dual, vector $\boldsymbol{b}^T$ holds the coefficients of the objective that multiplies the variables in column vector $\boldsymbol{y}$; for the constraints of the dual, $\mathbf{A}^T$ is the multiplier of vector $\boldsymbol{y}$ with vector $\boldsymbol{c}^T$ holding the right-hands sides. So we write

The primal:	**The dual:**
Max $\boldsymbol{c}\,\boldsymbol{x}$	Min $\boldsymbol{b}^T \boldsymbol{y}$
s.t. $\mathbf{A}\boldsymbol{x} \le \boldsymbol{b}$	s.t. $\mathbf{A}^T \boldsymbol{y} \ge \boldsymbol{c}^T$
$\boldsymbol{x} \ge 0$	$\boldsymbol{y} \ge 0$

This shows clearly how the primal and dual are related.

If the primal has an equality constraint such as $7x_1 - 2x_2 = 11$, we replace it with two constraints: $7x_1 - 2x_2 \ge 11$ and $7x_1 - 2x_2 \le 11$, but the second is not greater than or equal to, so we change it to $-7x_1 + 2x_2 \ge 11$.

Why do we need to consider the dual to our problem? It is often more efficient to solve the dual, because a primal with three variables and seven constraints becomes a problem with seven variables but only three constraints. If there are seven constraints, we must reduce in seven rows; if there are three, we reduce in only three rows. Fewer rows mean fewer arithmetic operations. Many real-world problems have many more constraints than variables, so working with the dual is greatly preferred.

An example will make this clearer. To make it simple, we will find the dual to the second example of this section. That problem is to maximize

$$f(x_1, x_2) = 8x_1 + 9x_2,$$

constraints:

$$\begin{aligned} \text{C1: } & 2x_1 + 4x_2 \le 32, \\ \text{C2: } & 3x_1 + 4x_2 \le 36, \\ \text{C3: } & 6x_1 + 4x_2 \le 60, \\ & x_1, x_2 \ge 0. \end{aligned}$$

In the dual, the equivalent problem is to minimize

$$g(y_1, y_2, y_3) = 32y_1 + 36y_2 + 60y_3,$$

constraints:

$$\begin{aligned} \text{C1: } & 2y_1 + 3y_2 + 6y_3 \ge 8, \\ \text{C2: } & 4y_1 + 4y_2 + 4y_3 \ge 9, \\ & y_1, y_2, y_3 \ge 0. \end{aligned}$$

To prepare the problem for the simplex method we add slacks; the matrix form of the primal is

$$f = 8x_1 + 9x_2 + 0x_3 + 0x_4 + 0x_5,$$

$$\begin{bmatrix} 2 & 4 & 1 & 0 & 0 \\ 3 & 4 & 0 & 1 & 0 \\ 6 & 4 & 0 & 0 & 1 \end{bmatrix} \begin{bmatrix} x_1 \\ x_2 \\ x_3 \\ x_4 \\ x_5 \end{bmatrix} = \begin{bmatrix} 32 \\ 36 \\ 60 \end{bmatrix},$$

$$x_1, x_2, x_3, x_4, x_5 \geq 0.$$

The matrix form of the dual is

$$g = 32y_1 + 36y_2 + 60y_3 - 0y_4 - 0y_5,$$

$$\begin{bmatrix} 2 & 3 & 6 & -1 & 0 \\ 4 & 4 & 4 & 0 & -1 \end{bmatrix} \begin{bmatrix} y_1 \\ y_2 \\ y_3 \\ y_4 \\ y_5 \end{bmatrix} = \begin{bmatrix} 8 \\ 9 \end{bmatrix},$$

$$y_1, y_2, y_3, y_4, y_5 \geq 0.$$

(Surplus variables are subtracted because the constraints are greater than or equal.)

Sensitivity Analysis

Often, the parameters of a linear programming problem are not known precisely. There could be uncertainty in the coefficients of the objective function, in the right-hand side values for the constraints, or sometimes in the coefficients of the constraints. After all, these numbers are usually obtained from past experience which may now be obsolete.

The selling price of the products may change; this would change the coefficients in the objective function. The amount of available resources may be uncertain (particularly the productivity of workers—for example, a lead employee may become sick); the supplier of a critical component may not be able to complete an order (or possibly he can supply more of a certain product than we at first thought). A machine can break down; other changes are possible.

Changes in selling prices will change the coefficients of the objective function. Changes in the amount of available resources will change the right-hand sides of the constraints. The manager of a business that uses the linear programming (LP) model in making decisions must be concerned about the effect of such changes. (Even the coefficients in one of the constraints can change if a more efficient way to use that resource is found, but that changes the feasible region. This is a less common situation.)

Examining the effect of such variations in the parameters is called *sensitivity analysis* and is an important part of the process. Determining the effect of such uncertainties in the parameters cannot be done in advance; only after the problem has been set up and a solution found can this be done.

One inefficient way to find how a change in a given parameter changes the solution is to solve the problem again with the changed value. That is not practical because a change in some parameters may have little or no effect (that parameter is then an insensitive factor) and we would waste time and effort to find this out. We need a way to know which parameters are most important, and how great a change in their values can be allowed without a significant change in the optimum.

The tableau method for solving an LP problem can be adapted to tell much how much a unit change in any of the factors changes the solution. We have not discussed this method.

The dual problem gives much information about the primal, including how a change in the coefficients of the objective function changes the value at the optimum. We use the problems of Example 7.3 for Example 7.4:

EXAMPLE 7.4

The primal:	**The dual:**
Max $f = 5x_1 + 8x_2$	Min $g = 12y_1 + 10y_2$
$3x_1 + 4x_2 \leq 12,$	$3y_1 + 2y_2 \geq 5,$
$2x_1 + 5x_2 \leq 10,$	$4y_1 + 5y_2 \geq 8,$
$x_1, x_2 \geq 0.$	$y_1, y_2 \geq 0.$

We restate this as matrices and vectors:

The primal:	**The dual:**
Max $[5 \quad 8]\begin{bmatrix} x_1 \\ x_2 \end{bmatrix}$	Min $[12 \quad 10]\begin{bmatrix} y_1 \\ y_2 \end{bmatrix}$
s.t. $\begin{bmatrix} 3 & 4 \\ 2 & 5 \end{bmatrix} x \leq \begin{bmatrix} 12 \\ 10 \end{bmatrix},$	s.t. $\begin{bmatrix} 3 & 2 \\ 4 & 5 \end{bmatrix} y \geq \begin{bmatrix} 5 \\ 8 \end{bmatrix},$
$x_1, x_2 \geq 0.$	$y_1, y_2 \geq 0.$

We already know the solutions: For the primal $f = 148/7$ at $(20/7, 6/7)$; for the dual, $g = 148/7$ at $(9/7, 4/7)$, so we can write

$$\text{Max } f = [5 \quad 8]\begin{bmatrix} 20/7 \\ 6/7 \end{bmatrix} = 148/7, \qquad \text{Min } g = [12 \quad 10]\begin{bmatrix} 9/7 \\ 4/7 \end{bmatrix} = 148/7.$$

What if the objective of the primal were changed from $f = 5x_1 + 8x_2$ to $f = 6x_1 + 8x_2$, a unit change in the first coefficient, c_1? We see from the equation for Max f that the value at the optimal point would increase by 20/7—from 148/7 to 168/7. We can say that $\partial f/\partial c_1$ is equal to the value of x_1 at the optimum. Similarly, $\partial f/\partial c_2$ is equal to the value of x_2 at the optimum.

What if the right-hand side of the first constraint in the primal were changed, from 12 to 13, again a unit change? Look at the equation for Min g in the dual. We see that this will change the solution from 148/7 to 148/7 + 9/7 = 157/7, which is exactly what will be the correspondingly changed solution of the primal! We then see (from the dual) that $\partial f/\partial b_1$ is equal to the value of y_1 at the optimum. The dual shows immediately how the solution to the primal changes if a right-hand side of a constraint is increased.

7.4 Nonlinear Programming

When a problem has nonlinearities, either in the objective function or in the constraints or both, normal linear programming cannot find the answer. In this section, we describe how one can find the optimal values for the variables and the corresponding value of the objective function. We will not give full details of the algorithms but only present examples that illustrate the difficulties and how they are overcome. You will almost always use a prewritten computer program to enter the parameters and find the solution. We will do it with a spreadsheet program.

We describe three situations:

1. The objective function is nonlinear but the constraints are linear.
2. The objective function is linear but some of the constraints are nonlinear.
3. Both the objective function and some of the constraints are nonlinear.

To make the explanation very clear, we will use problems with only two variables. The number of constraints will also be small even though in practice there are often many constraints and many more than two variables. With only two variables, we can show the problem and its solution graphically.

Nonlinear Objective; Linear Constraints

The problem is to maximize

$$f(x_1, x_2) = x_1 * x_2,$$

subject to:

$$2x_1 + x_2 \le 2,$$
$$x_1, x_2 \ge 0.$$

Figure 7.13 shows the feasible region and several curves when the objective function is set equal to certain values.

It is critically important to observe in Figure 7.13 that the optimum is not at a corner point but is on the line representing the single constraint (other than the nonnegativity condition). If the objective function were linear, we would find the optimum at a corner point or everywhere along a constraint line. [If there were a different set of constraints that happened to have a corner point at (1/2, 1), the optimum for $f = xy$ would occur at that corner point. That would be a coincidence; we cannot always assume a corner-point solution.]

A second most important observation is that the curve for the optimal value of the objective function is tangent to the constraint line at the optimal point. This means that they have exactly the same slopes at that point. If there were many variables in the problem, the equivalent observation would be that the gradient of the objective function is a negative multiple of the gradient of the constraint relation that is controlling. (The two gradients will point in opposite directions and may have different magnitudes.)

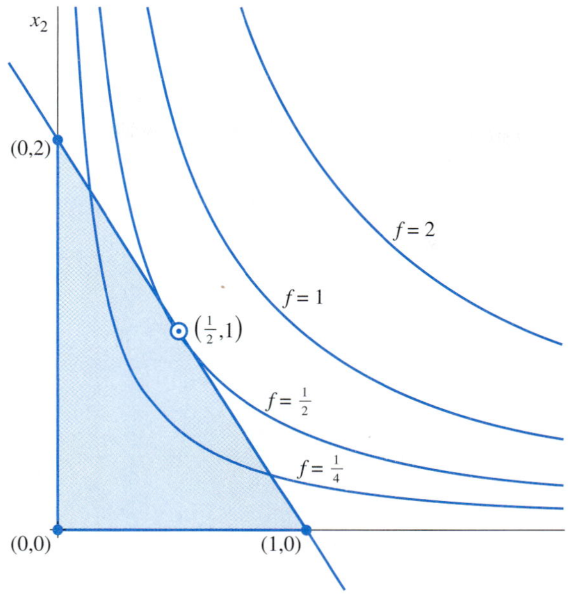

Figure 7.13

If you compute the slope of f at (1/2, 1) you will find that it equals -2 as the graph suggests.

Linear Objective; Nonlinear Constraints

For this example, we will solve graphically: maximize

$$f(x_1, x_2) = x_1 + x_2,$$

subject to:

$$x_1^2 + 4x_2^2 \leq 16,$$
$$x_1 \geq 1,$$
$$x_2 \geq -2.$$

The feasible region is that portion of an ellipse to the right of $x_1 = 1$. Figure 7.14 shows the feasible region and several curves for the objective function.

We see in Figure 7.14 that the optimal value for f is between 4 and 5. We can find the exact value by determining where the ellipse has a slope of -1 in the first quadrant. This turns out to be at $x_1 = \sqrt{12.8} = 3.5777$, $x_2 = (\sqrt{3.2})/2 = 0.8944$. At that point, $f = 4.4721$.

Suppose that we repeat this problem, but minimize instead of maximize. Figure 7.15 shows that this is about -1 at the lower-left corner, $(1, -\sqrt{(15/4)}) = (1, -1.936)$. Substituting the coordinates of the lower-left corner, we compute $f_{\min} = -0.936$.

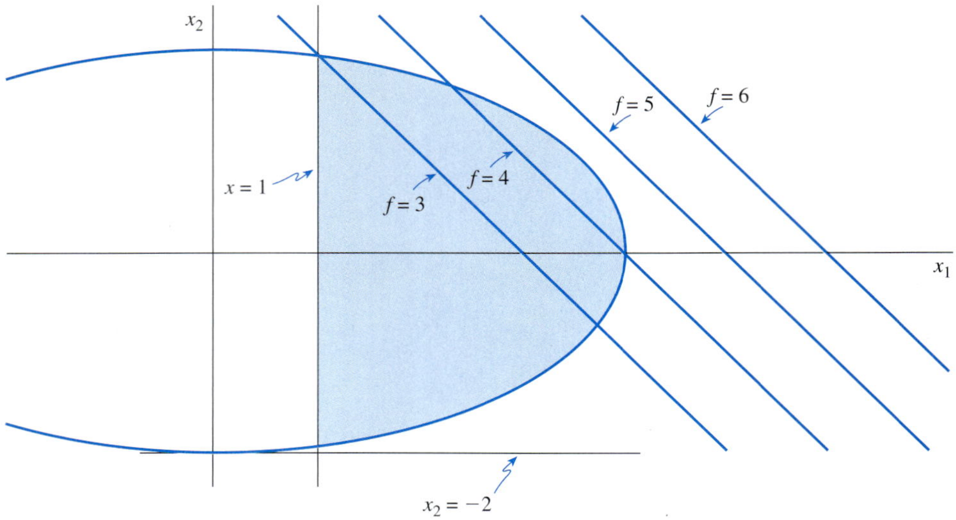

Figure 7.14

Both Objective and Constraints Are Nonlinear

Our next example combines the nonlinear function $f = x_1 * x_2$ with the nonlinear constraints of the last example. Here is our problem: Maximize

$$f(x_1, x_2) = x_1 * x_2,$$

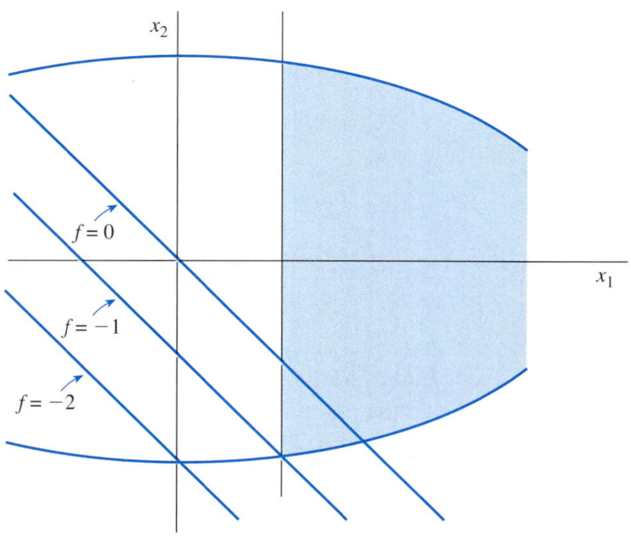

Figure 7.15

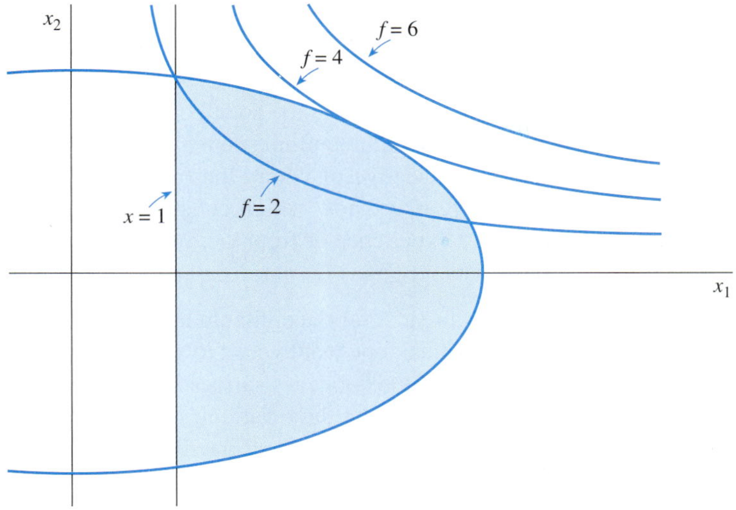

Figure 7.16

subject to:

$$x_1^2 + 4x_2^2 \le 16,$$
$$x_1 \ge 1,$$
$$x_2 \ge -2.$$

We will use the graphical method again. Figure 7.16 shows the familiar feasible region and some of the objective function curves.

The maximum value f occurs at $(2\sqrt{2}, \sqrt{2}) = (2.828, 1.4140)$ where $f = 4$. There is no point in trying to find the minimum. Because the objective is symmetrical about the x_1-axis, we know that it occurs at $(2\sqrt{2}, -\sqrt{2}) = (2.828, -1.4140)$ where $f = -4$.

Conclusions from the Examples

Even though our examples are limited to ones that we can visualize with graphs, there are important conclusions:

1. Finding the optimum for a nonlinear problem is more difficult than when both the objective function and the constraints are linear. Linear programming does not apply.
2. We have confirmed that, if the objective function is nonlinear, it is almost always tangent to one of the constraints at the optimum. This implies that the gradient of the objective function is a negative multiple of the gradient of the constraint(s) that are binding.
3. While this did not occur in our examples, it is possible that the feasible region is concave, not convex as must always be true when all of the constraints are linear (why?). One can imagine situations where the optimum is at a corner point when

both objective and constraints are nonlinear and the gradient condition 2 does not hold; the objective function is not tangent to the region. This would be when the corner point is a cusp.

4. When there are multiple constraints, not all of them may be binding at the optimum.
5. It is not obvious which of the multiple constraints is binding.
6. If there is some knowledge of where the optimum lies, it can speed finding the optimum in that we may know which constraint(s) is binding. (Such knowledge might be from past experience or from the results of the same problem with slightly different constraints.)

In view of conclusion 4, a search for the optimum by searching along the border of the feasible region seems impractical; one would need to search along every one of the edges in a 2-D case, along every one of the boundary surfaces in 3-D, along every one of the "hypersurfaces" when there are more than three variables. Except when there are only two variables, the multivariable methods of Section 7.2 will be involved.

Methods That Underlie Optimization Software

At the outset of the section, we mentioned that, for a real-world problem, one will surely use one of the available optimization programs. There are a number of procedures that this software is based on. We now briefly describe some of them.

1. This is a special case that occurs so infrequently as to be not very useful. If the objective is nonlinear in n variables and we have $(n - 1)$ constraints and all of these are equalities, we can solve the equalities for each variable in terms of just one of them. If these substitutions are used in the objective, we now have a single-variable problem that can be solved analytically or by any numerical method of Section 7.1. If there are more than $(n - 1)$ equality constraints, we solve them in combinations of $(n - 1)$ of them; this requires solving with the substitution obtained from each combination.
2. When the problem has n variables, and $(n - 1)$ inequality constraints, the method of *Lagrange multipliers* can be used. This is covered in most calculus books. The objective function is modified by adding a multiple of the constraint set equal to zero and solving through setting the system of partial derivative equal to zero.
3. *Quadratic programming* is a technique that has a quadratic objective (for two variables, it contains only terms in x_1^2, x_2^2, x_1x_2, x_1, x_2, and a constant term). Constraints can be linear inequalities or equalities. Again, Lagrange multipliers are employed and an equation derived from the gradient is solved.
4. The use of *penalty functions* can convert a constrained problem to an unconstrained function. The objective is expanded to include equality constraints multiplied by a *penalty parameter*. If this is minimized for increasing values of the parameter, the solution converges to the desired result. However, the equations are badly ill-conditioned. A variant of this is the *barrier method,* where the objective is modified to include a multiple of the natural logarithm of the constraint.
5. *Successive linear programming* converts the problem to a linear programming problem by approximating the objective function at a point near to the optimum by a linear combination of the variables and doing the same for the constraint that is

binding. This requires some knowledge of the optimal point. Once that problem has been solved, one iterates from that point to improve the accuracy of the solution. *Successive quadratic programming* is similar, except the objective is approximated by a quadratic function.

6. *Descent methods* move from an initial point toward the optimum by moving in the direction of the gradient at that point (for maximizing) until function values no longer increase, then the gradient is recomputed (perhaps numerically) and movement in a new direction is begun. Eventually, the optimal point is attained. This is somewhat analogous to the univariant search of Section 7.2.

We conclude from this rather superficial review of methods that finding the optimum for a nonlinear objective subject to nonlinear constraints is not easy. It is fortunate that software is available to do the job.

Using a Spreadsheet Program

Both Quattro Pro and Excel can solve for the optimum when either or both of the objective and constraints are nonlinear. Here is how Quattro Pro solves the examples of this section.

Example 1 had a nonliner objective and linear constraints: Maximize

$$f(x_1, x_2) = x_1 * x_2,$$

subject to:

$$2x_1 + x_2 \leq 2,$$
$$x_1, x_2 \geq 0.$$

We choose cell A1 for x_1, cell A2 for x_2, cell A3 for the objective function (A1 * A2), and cell A4 for the left-hand side of the first constraint. We initialize the variables at (1, 1) because, while $f(x, y) = x * y = 0$ at (0, 0), a plot of the function is not a pair of curves but the x and y axes.

We invoke `Tools/Numeric Tools, Optimizer` and enter the parameters: `Solution cell` = A3, select `Max,` set `Variable cell(s)` = A1.. A2, and enter the constraints. To see the successive computations that Quattro Pro uses, we click `Options` and mark `Show Iteration Results.` We return to the optimizer screen and click `OK`; the problem has now been set up.

We again invoke `Tools/Numeric Tools/Optimizer` and click on `Solve,` which begins the optimization process. Clicking on `Continue Solving` repeatedly shows the iterations as new values in the spreadsheet cells. The approach to the optimum is gradual; here are the values that we see

x_1	x_2	f
1	1.000002	1.000002
0.6	0.800002	0.480001
0.5	1.000002	0.500001
0.5	1	0.5

You will find it interesting to plot the successive values along with the constraint line and the objective function for $f = 0.5$. If you do, you will find that the initial movement from the starting point, (1, 1), is perpendicular to the constraint line, then along the constraint line until the optimum is found; it takes only a single step to do this.

A Second Example

The second example that we used has a linear objective but a nonlinear constraint: Maximize

$$f(x_1, x_2) = x_1 + x_2,$$

subject to:

$$x_1^2 + 4x_2^2 \leq 16,$$
$$x_1 \geq 1,$$
$$x_2 \geq -2.$$

We set up the problem on the spreadsheet almost exactly the same as with the previous example using the new parameters and we invoke the solution and display the successive iterates in the same way. Here is what is seen if we begin at (1, 0):

x_1	x_2	f
1	1E−06	1.000001
1.8	0.800002	2.600002
2.2	1.200002	3.400002
2.54356	1.543562	4.087127
2.797915	1.429309	4.227224
3.077707	1.277472	4.355179
3.231592	1.178647	4.410239
3.413433	1.042652	4.456085
3.568482	0.903597	4.472079
3.577687	0.894452	4.472138
3.577979	0.894157	4.472136

and we have reached the correct optimum at point $(\sqrt{12.8}, \sqrt{0.8})$ where f equals the sum of the two coordinate values.

If you plot the successive iterates along with the constraint and the optimum objective curves (here, the objective is just a straight line), you will find that the first significant move is from (1, 0) to (1.8, 0.8), which is within the feasible region. Two additional steps are taken along a line whose slope is 1 (which is perpendicular to the objective) until the constraint curve is reached. From there, additional steps are taken along the border of the feasible region until the optimum is reached. The length of the steps gradually decreases.

Once the constraint curve has been reached, the continuing steps are along the objective curve until the optimum point is found.

If you solve the third example problem with Quattro Pro, starting at (0, 0), you will find similar results, except that after the first step to the constraint $x = 1$, $y = 0$ (a feasible solution), the second step is taken along this edge of the feasible region. The third step is along a line of slope about 0.628 (angle about 32°) to reach the elliptical constraint. Once there, successive steps are taken on the ellipse toward the optimal point. You may want to see how the successive steps are taken from different starting points.

The Excel spreadsheet program can do what we have described for Quattro Pro, except that (0, 0) as a starting point does not work; (1, 1) is satisfactory. Entering constraints is somewhat simpler with this program.

MATLAB has a toolbox, the Optimization Toolbox, that can solve nonlinear programming problems but this is not a part of the student edition. Maple can solve the problem too. Spreadsheet programs are more accessible.

7.5 Other Optimizations

There are many other situations where an optimal result is required. We describe several of these in this section.

The Transportation Problem

The ABC Manufacturing Company has a decision to make. They currently have one manufacturing facility in Mississippi that makes a profitable product. Finished goods are shipped to three distribution warehouses, located in Atlantic City, Chicago, and Los Angeles. Demand has been heavy and is increasing. The managers anticipate that the capacity of the Mississippi plant must be increased if they are to meet the demand over the next four years.

One of their vice presidents, John Adams, points out that the majority of the increase in demand will occur in the western part of the United States, which is supplied by the Los Angeles distribution center. He proposes that they should consider an alternative plan: build a new plant in Mexico to meet the increase in demand, which would save on shipping costs to Los Angeles. Labor costs there will be less, although he admits there will be some concern about product quality.

A major factor as to which alternative should be chosen is the cost of shipping. This is extremely high for the product.

This is a typical *transportation problem:* how the shipping of products from m sources to n distribution points should be allocated to meet the demands at the distribution points with least cost for transporting the goods. One set of constraints is the total amount that is available for shipping from each of the m sources; another set of constraints is the demands at each of the n distribution points. The objective function is to minimize the total of the shipping costs, and this cost depends on which source ships to which location. We must consider each of the possibilities: plants A, B, C, . . . will ship to locations x, y, z, . . . and any plant can ship to any destination.

Adams has determined that the costs of shipping from two manufacturing facilities to the three distribution locations are, in dollars per unit shipped:

	To Atlantic City	**To Chicago**	**To Los Angeles**
From Mississippi	\$ 25	\$ 45	\$110
From Mexico	105	100	33
Units required			
At Atlantic City:	\$ 300		
At Chicago:	400		
At Los Angeles:	600		
Units available			
From Mississippi:	\$ 500		
From Mexico:	1000		

For this simple problem, the solution is quite obvious: Ship 300 units from Mississippi to Atlantic City and the remainder (200) to Chicago. Ship from Mexico enough to fill the rest of the demand at Chicago (200) and the entire demand at Los Angeles (600).

Usually the solution is not so obvious. We can formulate this as a linear programming problem in this way. Think of the table of costs from two sources to three destinations as a 2×3 matrix with the entries as $c_{i,j}$, $i = 1 \ldots 2$, $j = 1 \ldots 3$. Let $x_{i,j}$ be the amounts shipped from source i to destination j, another 2×3 matrix. Let R_j, $j = 1 \ldots 3$ be the requirements at the destinations and A_i, $i = 1 \ldots 2$ be the amounts available at the sources. The problem then can be to minimize

$$\sum_{i=1}^{2} \sum_{j=1}^{3} c_{i,j} x_{i,j},$$

subject to:

$$\sum_{i=1}^{2} x_{i,j} \geq R_j,\ J = 1 \ldots 3,$$

$$\sum_{j=1}^{3} x_{i,j} \leq A_i,\ i = 1 \ldots 2.$$

The objective is just the sum of the dot products of costs times amounts shipped taken by rows and summed by columns. The first constraint is to satisfy the demands, the last is not to ship more than available.

You can easily generalize this to m sources and n destinations. The sums are readily obtained in a spreadsheet.

Such problems are often solved to answer "What If" questions. Mr. Adams might want to consider another possibility: What is the effect if a fourth warehouse were located in Denver?

Integer Programming

Some problems allow only integer values to be valid. This would be the case if we are assigning employees to jobs or selecting from a number of decisions those to implement.

A typical example is the *traveling salesman* problem. He is to start from his home office, visit each of several cities only once, and then return home. He wants to minimize the total distance traveled. One way to find his best route is to visualize two $n \times n$ square arrays, where n is the number of cities to visit (including home). The elements of the first array are the distances between each of the cities taken in pairs. If you construct such an array, you will find it to be symmetric with zeros on the diagonal. The second array will contain the solution. When the optimal route is determined, it will hold ones in locations that represent travel from one city to the next.

One could solve this by finding the total distance for every possible route. If there are many cities, this is expensive; although a program that loops through every possible combination will find the answer. Linear programming can do it, but the real difficulty is defining constraints to assure that only zeros or ones are possible.

The same difficulty appears whenever the solution must have only integer values. One proposed technique is to solve the problem without requiring integers; the integer solution is approximated by rounding the answer. Often this is close enough to the true optimum to be acceptable. This will be the case when the optimum is "flat." If a more accurate answer is needed, the solution when the integer restrictions are ignored can be refined by testing the solutions when each variable in turn is perturbed by one or two.

Stochastic Problems

The word *stochastic* derives from the Greek word meaning "to guess." We guess at a value when the true number is uncertain because it depends on chance: Whether a batter will hit this time up is uncertain, but we can anticipate his success based on past experience, his batting average. It is the same with the parameters in some models of the real world. We may not know the exact value of a parameter, but we do know from past experience a range in which the values will lie and perhaps the frequency of occurrences of values within this range. When this is the case, we are in the field of statistics.

An often mentioned example is the *birthday problem*. In a room of many people, what is the probability that two persons have the same birthday? You may not have heard before that if there are 23 in the room, the chances are that this will occur more than 50% of the time. (We find this by seeing that after one person tells her birthday, the chance that the next will not be born on the same day is 364/365, the chance that the third person will not match the first two is 363/365, and so on. The product of 23 such terms is 0.493 and the probability that at least two were born on the same day is $1 - 0.493 = 0.507$.)

Can we do linear or nonlinear programming when, say, a constraint involves such an uncertain value? One method could be to solve with several values taken within the known range and from these results ascertain the range of results and the frequency of their occurrence within that range. Of course, if the solution is quite insensitive to the value of that uncertain parameter, there will be little effect of the variation and we will probably ignore it.

A technique that is often used in stochastic situations is *simulation,* a topic that we discuss next.

Simulation

The idea of *simulation* for a stochastic problem is to use random numbers to get a value for a parameter whose value is uncertain. There are several ways that a table of random numbers can be constructed; we do not go into this. Most computer languages have a command that produces random numbers, as do spreadsheet programs and some programmable calculators.

We can simulate the birthday problem for a room with n persons by randomly selecting a set of n integers from the set of successive integers from 1 to 365 and seeing if two numbers are the same. By repeating this many times with a different table of random numbers, we estimate the probability that two of the n persons will have the same birthday. Doing this with different values of n will produce a table of values for the probabilities that will compare well to the theoretical.

Obviously, if the stochastic problem can be solved by applying the rules of statistics, we will not do simulation. When the theory does not apply, simulation is an answer.

An Example, Queuing

An example that is often used to illustrate the concept is a *queuing problem,* where customers come at randomly spaced intervals, get served at a single counter for a time that is random in length, and leave. While they are being served, another customer may arrive and have to wait until the first customer leaves, or possibly no new customer arrives. You can imagine that if there is a sudden rush of new customers, a queue of waiting customers will form. How can one find whether a single serving counter is enough to prevent potential customers from not entering the establishment? In other words, if the queue length is often greater than, say, 4, it will be worthwhile to add another serving counter.

If a table is created that indicates at periodic intervals whether a customer enters, how many intervals it takes to serve him, and whether a new customer enters during the next interval, we can find from this the length of the queue of waiting customers. The flip of a coin can decide whether a customer enters during the current period and the length of time for serving a customer can be obtained from a set of random numbers with a distribution based on past experience.

After a sufficient number of periods, one will know the maximum queue length for that simulation operation. Repeating the simulation several times will give a good idea of how often the queue length exceeds four, or some other number.

Simulation Programs

The need to do simulations is important enough that special *simulation languages* have been developed. This avoids having to do it by hand, as suggested in the previous example. There are many such languages, for example, GPSS, SIMSCRIPT, SPICE, QSIM. Some of these have graphical interfaces that facilitate setting up a model of the process being simulated.

MATLAB has *SIMULINK,* an extension for doing modeling and simulation of linear and nonlinear systems. It has a graphical interface to facilitate setting up the simulation model and for observing the results. This is a part of the student edition of MATLAB and we describe a little of it now.

Simulink is activated by typing the word at the command prompt. A graphical representation of various blocks that represent mathematical operations can be connected. Most blocks take an input value and transform it to an output value. A succession of blocks can perform a sequence of mathematical operations, and feedback of values is possible. Many of the operations are numerical integrations of differential equations.

The user manual describes a very simple application: A sine–wave generator sends its output to an integration block as well as to a "Mux" block that combines two inputs. The second input is the output from the integrater. The output of the Mux block goes to a "Scope" block that displays its inputs as a graph. When the model is run, the screen shows the plot of both a sine wave and its integral over a period of time that can be specified by the user.

A more complicated application is also described fully in a demonstration named "thermo"; it is the operation of a thermostat that tries to keep the temperature of a house constant against varying outdoor temperatures. Feedback is involved here. An interesting part is that the cost of heating the house is also computed and displayed.

Exercises

Section 7.1

1. $f(x) = 2x^2 - e^{x/2}$ has a minimum between $x = 0$ and $x = 1$. Use a spreadsheet to create a list of function values between 0 and 1 with x-values spaced by 0.05. Then use the proper spreadsheet function to find the minimum of the values.

2. After you have completed Exercise 1, create a new list of function values for the function between the three x-values nearest the minimum with $\Delta x = 0.005$ and determine a new value for the minimum.

3. Use the three x-values nearest the minimum of the list in Exercise 2 to construct a quadratic interpolating polynomial and find where that polynomial has a minimum. Compare the value with the minimum value of $f(x)$ obtained from $f'(x) = 0$.

▶ **4.** Show that the x-value at the minimum or maximum of any second degree polynomial can be obtained without the need to find a root. If the polynomial is $ax^2 + bx + c$, what relation between the coefficients produces a minimum; what relation produces a maximum?

5. Are there other functions for which a similar result to that of Exercise 4 are obtained? List as many different functions as you can. Is there some general rule?

6. Repeat Exercises 1, 2, and 3, but for $f(x) = x^3 - 4x$. Graph this function and observe that it has a maximum as well as a minimum. How are the x-values for the maximum and the minimum related? Could you tell that this relationship holds without looking at the graph?

7. Propose two numerical methods that can find the minimum for functions with graphs similar to Figure 7.2a and b.

▶ **8.** Use the back- and-forth method to find the minimum of the function of Exercise 1 accurate to ± 0.001.

9. Repeat Exercise 8, but for the function of Exercise 6.

10. Is the back- and-forth method a candidate for Exercise 7? Is it a candidate for both cases of Figure 7.2?

▶**11.** The initial pair of x-value in the golden search procedure narrows the interval that encloses the minimum point to about 61% of the original interval. What pair of x-values will narrow the interval the most? What restrictions should be placed on them? What pair of x-values will narrow the interval the least? What restrictions should be placed on them?

12. Use the golden search procedure to find the minimum of the function in Exercise 1 accurate to ± 0.001. How many iterations are required? Can you know this in advance? How many function evaluations are required?

13. Repeat Exercise 12, but for the function of Exercise 6.

14. Repeat Exercise 12, but for this function:

$$f(x) = (x^2 - x)^2 + x - 5,$$

within the interval $[-2, 2]$.

▶**15.** Find the minimum of Exercise 14 analytically.

16. Repeat Exercises 12, 13, and 14, but get the minimum from a quadratic interpolation polynomial that fits to the first three intermediate points. Refine the answers by employing a new interpolating polynomial that fits to three points closest to the minimum. Repeat with new polynomials until there is no improvement in the answer.

17. Repeat Exercise 16, but with cubics through four points. Compare the efficiency with that of Exercise 16.

18. It is proposed that we extend the idea behind Exercise 16 and 17 as outlined in the steps below. Critique the proposal.

a. Get three intermediate points within the initial interval by Golden Section, fit a quadratic to these and get the x-value at its minimum.

b. Use the four points that are the result of step a to construct an interpolating cubic, and use its minimum x-value to get a fifth point.

c. From these, construct a fourth-degree interpolating polynomial; minimize to get a sixth point.

d. Continue in this fashion until the minimum x-values converge.

19. Use MATLAB (or another computer algebra system) to minimize the functions in Exercises 1, 6, and 14.

20. Repeat Exercise 19, but use Excel or Quattro Pro.

21. The Fibonacci sequence begins with $F_1 = 1$, $F_2 = 1$, and continues with $F_n = F_{n-2} + F_{n-1}$. As n gets large, the ratio of F_n/F_{n+1} approaches the golden mean.

▶ a. How close is the ratio to the golden mean for $n = 20$?

▶ b. How large must n be to approach to within 0.0001?

c. Prove that the limit of F_n/F_{n+1} = golden mean as $n \to \infty$.

Section 7.2

▶**22.** The function $f(x, y) = x^2 + 3y^2 - xy + 3x + 2$ has a minimum within the square whose corners are $(-2, -2)$ and $(0, 0)$. Make a table of f-values evenly spaced within the square spaced apart by 0.2 and find where the smallest value appears. What is the value? How many function evaluations were made?

23. Interpolate from the value found in Exercise 22 together with those at the four neighbors closest to it to get an improved value for the minimum value.

24. Solve for the minimum point in Exercise 22 analytically. What is the f-value? How much do the answers in Exercises 22 and 23 differ from this true minimum?

25. Get contour curves for the function in Exercise 22 for $f = .2, 0, -.2, -.3, -.4$, and $-.45$. Would you say that the minimum is "flat"?

26. Beginning at (0, 0), use a univariant search to find the minimum of Exercise 22. For the initial traverse, use $\Delta x = -0.1$, then use $\Delta y = -0.02$. For each subsequent traverses, decrease the step size five-fold. How many function evaluations are made when the minimum is located to a precision of 0.001? How close is the f-value at that point to the analytical f-value that you obtained in Exercise 24?

▶**27.** Find the minimum point of the function of Exercise 22 by steepest descent starting from (0, 0). Terminate the search when the (x, y) values are within (0.001, 0.001) of the true minimum.

28. Construct a graph similar to Figure 7.8 for the path taken in Exercise 27.

29. We want to solve the problem in Exercise 22 by steepest descent starting from the point on the enclosing square where the gradient has the largest value. Where is that point?

30. The function of Exercise 22 is a quadratic function. That means that the minimum point can be found in two tries with the conjugate gradient method. Confirm that this is true when the starting point is (0, 0). Then repeat this for starting points at the other corners of the square.

▶**31.** The function of Exercise 22 is a quadratic function. That means that the minimum point can be found immediately with Newton's method. Confirm this for starting points from each corner of the enclosing square. What is the Hessian matrix at each of these starting points? What are the inverses of these matrices?

32. If we did not know the function of Exercise 22 explicitly, we could not compute the Hessian matrix equations for the partial derivatives, but we could do so from finite difference approximations. Find approximate values for the derivatives at each corner of the enclosing square from such finite differences, with Δx and $\Delta y = 0.02$. How does the Hessian matrix so determined differ from the true value you found in Exercise 31? Would the approximate values be more accurate if the surrounding points were on lines that are not parallel to the axes?

33. A function that is frequently used to illustrate searching methods for a minimum is Rosenbrock's function:

$$f(x, y) = 100(y - x^2)^2 + (1 - x)^2.$$

Because both terms are never negative, it is clear by inspection that the minimum is zero and occurs at (1, 1). Graph level curves for $f = 1, 2, 5, 15$, and 25 and observe that the "valley" is narrow and curved.

▶**34.** Confirm that the minimum of the function in Exercise 33 is at (1, 1) by constructing a table of function values within the square whose corners are at (0, 0) and (2, 2) with points spaced apart by 0.2. Does round off cause a slight difference in the value of f at (1, 1)? At other points?

35. Some books illustrate searching methods for Rosenbrock's function (see Exercise 33) starting from (−1.2, 1). If steepest descent is used from this starting point, in what direction does one move and to what point before changing direction? What is the direction of the second movement? Superimpose these two movements on the graph of level curves that you made in Exercise 33.

36. Repeat Exercise 35 but use the conjugate gradient method.

▶**37.** Apply Newton's method to find the minimum for the function in Exercise 33 starting from (0, 0).

38. Use a spreadsheet to solve for the minimum of Exercise 22.

39. Use a spreadsheet to solve for the minimum of Exercise 33.

40. A starting simplex for solving for the minimum of the function of Exercise 22 is a triangle with vertices at (0, 0), (0, −.2) and (−.2, 10). What is the f-value at each vertex? If the point where f is greatest is reflected, what new point is chosen for a second simplex? You do not have to exactly reflect, the new point can be anywhere along the line of reflection. How far should the "reflection" go to most greatly reduce the f-value? What is the f-value at that point?

41. Repeat Exercise 40 but with starting simplexes at each corner of the square whose vertices are at (0, 0) and (2, 2). Each of these simplexes has sides parallel to the axes that are 0.2 long. Which starting simplex brings you closest to the optimum point?

Section 7.3

▶**42.** Solve this linear programming problem graphically:

Maximize

$$7x_1 + 3x_2$$

subject to

$$x_1 + x_2 \le 10,$$
$$2x_1 + x_2 \le 12,$$
$$x_1, x_2 \ge 0.$$

43. Repeat Exercise 42 but for these different objective functions. For each, what is the value of the function and at what point does it occur?

a. Max $2x_1 + 3x_2$.
b. Max $11x_1 + 7x_2$.
c. Max $6x_1 + 3x_2$.
d. Min $x_1 + x_2$.
e. Min $x_1 - x_2$.

44. If the constraints of Exercise 42 where changed as follows, how does the solution to Exercise 42 change?

a. First constraint: $x_1 + x_2 \le 15$.
b. Second constraint: $2x_1 + x_2 \le 10$.
c. The nonnegativity constraints:

$$x_1 \ge 0, x_2 \ge 3.$$

▶**45.** If one of these constraints were added to Exercise 42, how would the solution change?

a. $4x_1 - 3x_2 \leq 26$.
b. $4x_1 - 3x_2 \leq 18$.
c. $x_1 + 3x_2 \geq 9$.

46. A region is said to be convex if the line connecting any two points within the region falls entirely within or on the surface of the region. It is said to be concave if some lines cross over a border of the region. Can the feasible region of a linear programming problem ever be concave? Explain why this cannot be or show a case where it is.

47. Write a set of constraints for a linear programming problem with just two variables for which there are no feasible points on either the x- or y-axes.

48. Repeat Exercise 47 for a problem with three variables; there are no feasible points on any of the axes.

49. Can this problem be solved graphically? If you think it can, explain how you would do it. Then get the solution using your plan.

Maximize

$$x_1 + 2x_2 + 3x_3.$$

Subject to:

$$x_1 + 2x_2 \leq 15,$$
$$3x_1 + 4x_3 \leq 21,$$
$$x_1 + x_2 + x_3 \leq 10,$$
$$x_1, x_2, x_3 \geq 0.$$

▶**50.** Write a set of constraints that has no points on either the x-axis or the y-axis; draw the region for the case of two variables. Then write a set of constraints that produces no feasible region and draw these for the case of two variables. Will any combination of both sets of constraints ever result in a feasible region?

51. Under what condition(s) will the optimal point for a problem with just two variables always lie on the x-axis? Under what condition(s) will it always lie on the y-axis? For the case of three variables, what conditions cause the optimum to lie on any one of the three axes?

52. Is it ever possible that the feasible region is composed of two separate regions? If it can, write the constraints for the case of two variables. If it cannot, outline the argument that proves this.

53. Add slack variables to the problem in Exercise 49 to convert the constraints (other than the nonnegativity constraints) to equalities. Then write in matrix form. Finally, use the simplex method to solve.

54. Use either Excel or Quattro Pro to solve the problem in Exercise 49. Do the iterates follow the same path as used in Exercise 53?

55. Use Maple to solve the problem in Exercise 49.

56. Before we can use the simplex method, we must start with a basic feasible solution. In the case of the usual nonnegativity constraints, the origin is such a starting point and it is customary to begin there. If the origin is not a part of the feasible region, how can you find a starting point? Compose a set of constraints that defines such a feasible region and demonstrate that your idea works. Do this for two problems: one with just two variables, and another with four variables.

▶**57.** Formulate this linear programming problem both as the primal and the dual. Solve both problems by the simplex method. (The negative coefficient for x_4 in the objective means that we incur a loss when that is produced, but the second constraint means that we must make 35 units because of certain contractual obligations.

Maximize

$$f = 2x_1 + 3x_2 + x_3 - x_4.$$

Subject to:

$$x_1 + 2x_2 + 3x_3 \leq 120,$$
$$x = 35,$$
$$x_1, x_2, x_3, x_4 \geq 0.$$

58. If the coefficient of x_2 in the objective function in Exercise 57 were changed to 5, how much would the maximum of f change?

59. If the right-hand side of the first constraint in Exercise 57 were changed to 100, how much would the maximum of f change?

▶**60.** Finding the magnitude of changes to the solution of a linear programming problem as done in Exercises 58 and 59 implicitly assumes that the point where the maximum occurs does not change to another feasible solution point. Under what conditions will this assumption not be true? Your explanation should be for both a two-variable problem and for one with more than two variables.

Section 7.4

61. Unlike a linear programming problem, when the program is nonlinear, the feasible region can be concave. Write set of constraints for a two-variable problem that produces a concave feasible region. Plot the region.

62. For the concave region of Exercise 61, find an objective function that is a maximum where it touches the region within the concave portion and not at a corner point. Solve for the optimal point and the maximum function value at that point.

63. Repeat Exercises 61 and 62, but for minimization.

▶**64.** Solve this problem:
Maximize

$$f(x, y) = x^2 + 2y.$$

Subject to:

$$x \geq 1,$$
$$y \geq x - 3,$$
$$x \leq 4,$$
$$y \leq 5 - x.$$
$$x, y \geq 0$$

65. The region of Exercise 64 has four corner points. Find nonlinear objective functions that are a maximum at each of these corner points.

66. Draw the feasible region that is defined by these constraints:

$$y \leq 2x - x^2,$$
$$y \geq x - 4.$$

67. For $f(x, y) = 2x + y$, find

a. Its maximum on the region of Exercise 66.
b. Its minimum on the region of Exercise 66.

68. Repeat Exercise 67, but for $f(x, y) = 2x^2 - y$.

69. Devise a plan whereby a nonlinear problem can be solved graphically when there are three variables. Then pose three problems and solve them by your technique:

a. With a linear objective, nonlinear constraints.
b. With nonlinear objective, linear constraints.
c. With both objective and constraints being nonlinear.

▶**70.** Solve Exercise 64 by approximating the objective with a straight line near the maximum point and use linear programming. Iterate with new approximations to reach the true minimum value of f within 0.002. Do this three times, starting at three different points in the neighborhood of the true minimum.

▶**71.** Solve this problem:
Maximize

$$f(x, y, z) = x^2 + xy + y^2 - 2z.$$

Subject to:

$$x + y + z \leq 10,$$
$$x^2 + y^2 + z^2 \leq 100,$$
$$x, y, z \geq -5.$$

Do this with a spreadsheet. Begin at different starting points. Are there starting points that are invalid?

72. Solve Exercise 64 by drawing contour lines for the objective function, finding one that solves the problem.

73. Use either Quattro Pro or Excel to find the maximum of the third example of Section 7.4:
Maximize

$$f(x_1, x_2) = x_1 * x_2.$$

Subject to:

$$x_1^2 + 4x_2^2 \leq 16,$$
$$x_1 \geq 1,$$
$$x_2 \geq -2.$$

You will find that the maximum value f occurs at $(2\sqrt{2}, \sqrt{2}) = (2.828, 1.4140)$, where $f = 4$. Plot the region and the contour for $f = 4$ and see that it is tangent to the region at the optimal point.

74. If you solve Exercise 73 when asking to see the intermediate steps, you will find that the third step is taken at an angle of about 32°. Why is this so?

Section 7.5

▶**75.** Section 7.5 begins by describing a transportation problem whose solution is obvious. Mr. Adams wants to know if his second alternative, to build a fourth distribution point in Denver, will result in less total shipping cost. If this were done, the shipping costs would be

From Mexico to Denver: \$55
From Mississippi to Denver: \$70
Required amount at Denver: \$200
New required amount at Los Angeles: \$400

What should the decision be? What other factors besides shipping costs should be considered?

76. Formulate the original shipping cost problem of Section 7.5 as a linear programming problem and solve it. Does the solution result in integer quantities? If not, does rounding the results give the same answer as was obtained by inspection?

77. Repeat Exercise 75, but for the alternative of Exercise 76.

▶**78.** A linear programming problem is
Maximize

$$x + 2y,$$

Subject to:

$$5x + 4y \leq 40,$$
$$2x + y \leq 28,$$
$$x, y \geq 0.$$

a. Solve the problem. What are the coordinates of the optimal point and the f-value there? Now, at integer values for x and y nearest to this point, compute f. Does this match to the rounded value of f? If there is discrepancy, is it serious? What if the right-hand sides of the constraints were 100 times as great?

b. Suppose that the x-values are limited to only integer values, but the y-values are not. Can the problem be solved? If it can, how closely does the feasible value in part (a) match to the feasible value in part (b)?

79. Do several trials of simulations of the birthday problem on page 451 by selecting n random integers from the set [1, 2, 3, . . . , 365] for n equal to 23. Be sure that the set of random integers is different for each trial. From the results of the trials, average the number of times that two numbers match. How do the simulation results compare to the theoretical value of 0.507? If the number of trials is increased, is there a better match?

▶80. A barber shop has only one chair. The shop is open from nine in the morning until five in the afternoon, and this period is divided into 15-minute intervals. If the probability that one customer will enter the shop during one time interval is $\frac{1}{3}$ and the probability that two will enter is $\frac{1}{6}$, and it takes 15 minutes to cut a customer's hair, what is the maximum number of customers who must wait? How much time is the barber idle?

Solve this problem by performing trials. You can simulate the arrival of customers by rolling one die; if it comes up a 1 or 2, one customer enters; if it comes up a 3, two customers enter; if it comes up a 4, 5, or 6, no customers enter.

The chore of doing simulations by hand can be made easier if several people work together and their results are pooled.

81. As a variation on Exercise 80, during the noon hour (from 12 to 1), and after 4 P.M. (from 4 to 5), the probability of customers coming into the barber shop is greater: $\frac{1}{2}$ that one will enter, $\frac{1}{3}$ that two will enter. What is now the maximum number who must wait for their haircut? How often is the barber idle during the rush hours?

82. If you have access to one of the specialized simulation languages, use it to answer the questions in Exercise 80 and 81.

83. Use Simulink, a part of the MATLAB student program, to solve several ordinary differential equation problems taken from Chapter 6, including at least one boundary-value problem.

Applied Problems and Projects

APP1. A sales person is headquartered in Kansas City, Kansas, and must visit customers in eight cities: Chicago, Minneapolis, St. Louis, Denver, Omaha, Des Moines, St. Louis, and Oklahoma City.

Look up the distances between all of the cities taken in pairs in a road atlas to construct the distance matrix. Then solve her transportation problem by hand or by writing a computer program that tries each possible way to visit each city exactly once and then return to home with the least distance traveled.

As a variation on this, the salesperson wants to make the trip while traveling only on interstate highways. Does this preclude some potential trips between cities? How can the distance matrix be modified to exclude trips between certain cities? Is the trip longer with this requirement?

As a second variation, some critical requirements require that she visit Chicago before she visits either Omaha or Des Moines, and must visit St. Louis before visiting Oklahoma City. Is there a way to revise the distance matrix to guarantee these exclusions? Is the solution affected by these requirements?

APP2. (*Note:* This is best done as a class project.)
The class divides into five teams. Each team is assigned one of these industries:

a. Petroleum refining
b. Large-scale agriculture

c. Furniture manufacture
d. Freight haulage
e. Confectionary production

Each member of the team is to contact someone in the industry that has been assigned and get answers to these questions:

a. Is linear programming used?
b. If so, how frequently? Are parameters based on experience?
c. If not, how are production quantities decided upon?
d. How are workers assigned jobs?
e. Come up with other questions that the team thinks are pertinent.

APP3. When dice are thrown, it is assumed that they are "fair," meaning that the chance of any of the possible numbers coming up is the same. If this is not true, the dice are said to be "loaded," meaning that some numbers come up more frequently than others. What is the frequency distribution of the total that comes up if two dice are thrown but for one of them, getting a four has a probability of 1/5 rather than 1/6, the other numbers on that die having equal frequencies.

Solve this by simulations, repeating enough times to get a good answer. If you have enough knowledge of statistics, what is the theoretical frequency distribution?

APP4. George Danzig coined the term "linear programming" in 1947. He is given credit for developing the simplex method. Find the answers to the following questions:

a. In what field of science was Danzig an expert?
b. What is the publication where he originally explained the simplex method?
c. Is the explanation he gives in this easy to understand?
d. What references to other related work does he list?

Actually, Danzig is not the earliest to develop the simplex method. Some ten or more years earlier, some Russians introduced the ideas, but this was not well known outside of Russia in 1947. Who are these Russians?

APP5. There are several variations on the simplex method of linear programming. In Section 7.3, a reference is made to the tableau method. Find out what this is and use it to solve the examples of Section 7.3.

APP6. You are to find the minimum of $f(x, y, z)$ within a region that is a cube. If you start at some point on the surface of the cube, you can move closer to the minimum point by going down the gradient until the function value increases.

What point on the surface is the best starting point? Is it where the gradient is greatest? How can you find where this point is located?

Is it preferred to go down the gradient in small steps until the function value increases, or to go down in two larger steps and use three function values (one being at the starting point) to create a quadratic interpolating polynomial to estimate the point where the function is least? (After this, you could use four points to create a cubic polynomial.) Try these schemes on some function whose minimum is within a unit cube centered at the origin.

APP7. If $y = f(x) = ax^2 + bx + c$, a quadratic, and we know that there is a minimum in $[x_1, x_2]$ the minimum can be obtained immediately from a relation involving the coefficients a and b. (What is that relation?) If $y = f(x)$ is not a quadratic but we know that it has a minimum in $[x_1, x_2]$, we have several options:

a. Approximate $f(x)$ with a quadratic from three points in $[x_1, x_2]$ and use the above relation. What are the best choices for these three points?
b. Approximate $f(x)$ with a cubic from four points in $[x_1, x_2]$ and find its minimum. (How would you do this?) Where should the four points be chosen?

c. First approximate $f(x)$ with a quadratic, then use the minimum of that quadratic with the three initial points to construct a cubic and find its minimum.
d. First approximate with a quadratic, get its minimum point, then use this with two of the previous points to construct a second quadratic, and iterate.

Which option is best from the standpoint of least number of arithmetic operations? Test your choice with this function:

$$y = x^4 - 2x^2 + x - 3, [x_1, x_2] = [-3, 0].$$

The function has a second minimum in [0.7, 1.5].

APP8. We have described the golden section search. A Fibonacci search is another way to find the minimum of $y = f(x)$ within $x = [a, b]$. It has the advantage that one knows in advance how many iterations are required to achieve a desired accuracy and hence how many function evaluations are needed. Find information on this method. Is it more economical in using computing power than the golden section search?

APP9. A start-up company, Best Electronics, wants to enter the laptop computer business. They have designs for three models: model A, model B, and model C. To set up the production facility for any model will cost \$20,000. The parts for model A will cost \$126, for model B, \$157, and for model C, \$203.

It is anticipated that sales of model A will be at most 25,000 per year, of model B, 15,000 per year, and of model C, 8,000 per year. The profits from them will be \$65 per unit of model A, \$88 per unit for model B, and \$125 per unit for model C.

Best Electronics will utilize an existing shipping facility that can pack and ship at most 40,000 boxes per year. One box can hold two units of model A but only one unit of model B or C.

Formulate this as a linear programming model and solve for the best production schedule. Observe that the total costs are the sum of the fixed and variable costs.

8

EAGLE LIBRARY
Bolton Institute of Higher
Education - EAGLE CAMPUS

Partial-Differential Equations

The subject of Chapter 6 was ordinary differential equations (ODEs), so called because they involved ordinary derivatives. Some these equations were boundary-value problems where conditions on the problem were specified at the boundaries of some region.

If the region is on a plane or in three-dimensional space, a point in the region has coordinates (x, y) or (x, y, z) and the variation of the dependent function $u = f(x, y, z)$ will be in terms of the space derivatives, $\partial u/dx$, $\partial u/\partial y$, and $\partial u/\partial z$ and/or the corresponding second order derivatives. When a boundary-value problem is defined in terms of these partial derivatives, it is a *partial-differential equation* (PDE). We study PDEs in this chapter.

Types of Partial-Differential Equations

Partial-differential equations (PDEs) are classified as one of three types, with terminology borrowed from the conic sections.

For the second-degree polynomial in x and y,

$$Ax^2 + Bxy + Cy^2 + F = 0,$$

the graph is a quadratic curve, and when

$$B^2 - 4AC < 0, \text{ the curve is an ellipse,}$$
$$B^2 - 4AC = 0, \text{ the curve is a parabola,}$$
$$B^2 - 4AC > 0, \text{ the curve is a hyperbola.}$$

For the general partial-differential equation,

$$A\partial^2 u/\partial x^2 + B\partial^2 u/\partial x\partial y + C\partial^2 u/\partial y^2 + f(x, y, u) = 0,$$

the same terminology is used. If

$$B^2 - 4AC < 0, \quad \text{the equation is elliptic,}$$
$$B^2 - 4AC = 0, \quad \text{the equation is parabolic,}$$
$$B^2 - 4AC > 0, \quad \text{the equation is hyperbolic.}$$

As with the 1-D problems of Chapter 6, the partial-differential equation may have different types of boundary conditions. If the value for u is fixed on some parts of the boundary, it has a *Dirichlet condition* there. If the derivative of u, the *gradient,* is known, it is a *Neumann condition.* (The gradient is always measured along the outward normal.) The condition may be *mixed,* a condition where both the value for u and the gradient is involved. A mixed condition results when heat is lost by conduction or convection to the surroundings.

Elliptic equations describe how a quantity called the *potential* varies within a region. The potential measures the intensity of some quantity (temperature and concentration are "potentials"). The dependent variable, u, that measures the potential at points in the region takes on its equilibrium or *steady-state* value due to values of the potential on the edges or surface of the region. So, elliptic equations are also called potential equations. The general form of an elliptic equation in 2-D is

$$\partial^2 u/\partial x^2 + \partial^2 u/\partial y^2 + f(x, y, u, \partial u/\partial x, \partial u/\partial y) = 0,$$

and we see in comparing with the equations for conic sections that $A = 1$, $B = 0$, and $C = 1$, the values for an ellipse.

How the steady state of the potential is attained from some different starting state is described by a parabolic equation. So, these equations involve time, t, as one of its variables. In effect, we march from the initial state toward the final equilibrium state as time progresses. An important parabolic equation is

$$\partial^2 u/\partial x^2 - (c\rho/k)\, \partial u/\partial t = 0,$$

which tells how temperatures vary with time along a rod subject to certain conditions at its ends. The quantities in $c\rho/k$ are parameters (k = thermal conductivity, ρ = density, c = heat capacity).

Observe that, for this example, $A = 1$, $B = 0$, and $C = 0$, so that $B^2 - 4AC = 0$, the same as for a parabola. This equation and the corresponding ones for 2-D and 3-D regions is then called the *heat equation.* Exactly the same equation but with $c\rho/k$ replaced by $1/D$ describes the molecular diffusion of matter (D is the *diffusion coefficient*), so the equation in this form is called the *diffusion equation.* The ratio $(k/c\rho)$ is sometimes called the *thermal diffusivity.*

The third type of partial-differential equation, hyperbolic equations, is also time-dependent. It tells how waves are propagated; thus it is called the *wave equation.* In 1-D, it shows how a string vibrates. The partial-differential equation for a vibrating string is

$$\partial^2 u/\partial x^2 - (Tg/w)\, \partial^2 u/\partial t^2 = 0,$$

in which T is the tension in the string, g is acceleration of gravity, and w is the weight per unit length. All of these parameters are positive quantities, so we see that, in comparison to the conic-section equation, $A = 1$, $B = 0$, and C is a negative quantity. Therefore, $B^2 - 4AC > 0$, the requirement for a hyperbola. In 2-D, the wave equation describes the propagation of waves.

In this chapter, we discuss the usual techniques for solving partial-differential equations numerically. These methods replace the derivatives with finite-difference quotients. You will see that there are limitations to solving these equations in this way because some regions over which we want to solve the problem do not lend themselves to placing the nodes uniformly. There are ways to overcome this but they are awkward and it is not easy to achieve good accuracy in the solution. To some extent, this chapter is preparation for the next where you will find a more recent way to solve PDEs.

Contents of This Chapter

8.1 Elliptic Equations

In Chapter 6, we described how a boundary problem for an ordinary-differential equation could be solved. We now discuss boundary-value problems where the region of interest is two- or three-dimensional. This makes it a partial-differential equation.

There are two standard forms of elliptic partial-differential equations when the object is two-dimensional:

Laplace's equation: $-\partial/\partial x\,(c_x\partial u/\partial x + c_y\partial u/\partial y) + au = 0.$

Poisson's equation: $-\partial/\partial x\,(c_x\partial u/\partial x + c_y\partial u/\partial y) + au = f(x, y),$

where c_x, c_y, and a are parameters of the system that may depend on u and on the values of x and y. u is the variable whose values within the region we desire, the *potential,* at points (x, y) within the 2-D region. Laplace's equation is often called the *potential equation*.

We will deal with a simplified version where $a = 0$. If $c_x = c_y = c$, a constant, the equations can be rewritten as

$$c(\partial^2 u/\partial x^2 + \partial^2 u/\partial y^2) = 0, \quad \text{or} \quad c(\partial^2 u/\partial x^2 + \partial^2 u/\partial y^2) = f(x, y).$$

There is a special symbol that is often used to represent the sum of the second-order partial derivatives:

$$\nabla^2 u = \partial^2 u/\partial x^2 + \partial^2 y/\partial y^2,$$

and the operator ∇^2 is called the *Laplacian*.

Laplace's equation has many applications besides the steady-state distribution of temperature within an object that we use as our model. We chose this because that situation is easier for most people to visualize.

We derived the equation for temperature distribution within a rod, a one dimensional problem, in Chapter 6. We do this now for a two-dimensional region, a flat plate. Figure 8.1 shows a rectangular slab of uniform thickness τ with an element of size $dx \times dy$. u, the dependent variable, is the temperature within the element. We measure to the location of the element from the lower-left corner of the slab. We consider heat to flow through the element in the direction of positive x and positive y.

The rate at which heat flows into the element in the x-direction is

$$\begin{aligned} -(\text{conductivity})\ (\text{area})\ (\text{temperature gradient}) &= -kA\ \partial u/\partial x, \\ &= -k(\tau dy)\ \partial u/\partial x, \end{aligned}$$

where the derivative is a partial derivative because there are two space dimensions.

Similarly, the rate of heat flow into the element in the y-direction is

$$-k(\tau dx)\ \partial u/\partial y.$$

We equate the rate of heat flow into the element to that leaving plus the rate of flow out of the element from the surface of the slab, Q cal/cm^2 (the system is at steady state). For the rate of heat leaving, we must use the gradients at $x + dx$ and $y + dy$:

$$\text{rate of flow out in } x\text{-direction} = -k(\tau dy)\left[\frac{\partial u}{\partial x} + \frac{\partial^2 u}{\partial x^2}\,dx\right],$$

$$\text{rate of flow out in } y\text{-direction} = -k(\tau dx)\left[\frac{\partial u}{\partial y} + \frac{\partial^2 u}{\partial y^2}\,dy\right],$$

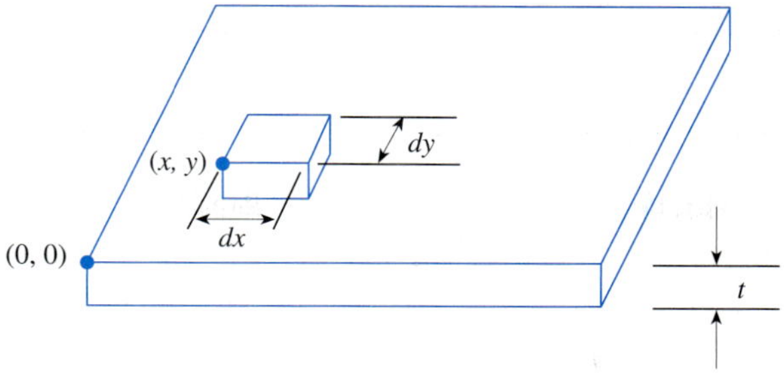

Figure 8.1

so the total flow of heat from the element is

$$-k(\tau\,dy)\left[\frac{\partial u}{\partial x}+\frac{\partial^2 u}{\partial x^2}\,dx\right]-k(\tau\,dx)\left[\frac{\partial u}{\partial y}+\frac{\partial^2 u}{\partial y^2}\,dy\right]+Q(dx\,dy).$$

The sum of the flows into the element must equal the rate at which heat flows from the element plus the heat loss from the surface of the element if the temperature of the element is to remain constant (and we are here considering only the steady-state), so that we have, after some rearrangement:

$$k\tau\left(\frac{\partial^2 u}{\partial x^2}+\frac{\partial^2 u}{\partial y^2}\right)(dx\,dy)=Q(dx\,dy),$$

or

$$\left(\frac{\partial^2 u}{\partial x^2}+\frac{\partial^2 u}{\partial y^2}\right)=\frac{Q}{k\tau} \tag{8.1}$$

If the object under consideration is three-dimensional, a similar development leads to

$$\left(\frac{\partial^2 u}{\partial x^2}+\frac{\partial^2 u}{\partial y^2}+\frac{\partial^2 u}{\partial z^2}\right)=\frac{Q}{k},$$

where now Q is the rate of heat loss per unit volume.

(The loss of heat in the three-dimensional case would have to be through an imbedded "heat-sink," perhaps a cooling coil. It is easier to visualize heat generation within the object, perhaps because there is an electrical current passing through it.)

As we have said, the Laplacian, the sum of the second partial derivatives, is often represented by $\nabla^2 u$, so Eq. (8.1) is frequently seen as

$$\nabla^2 u=\frac{Q}{k\tau}.$$

If the thickness of the plate varies with x and y, a development that parallels that of Section 6.7 gives

$$\tau\,\nabla^2 u+\frac{\partial\tau}{\partial x}\left(\frac{\partial u}{\partial x}\right)+\frac{\partial\tau}{\partial y}\left(\frac{\partial u}{\partial y}\right)=\frac{Q}{k}. \tag{8.2}$$

If both the thickness and the thermal conductivity are variable:

$$k\tau\nabla^2 u + \left(k\frac{\partial\tau}{\partial x} + \tau\frac{\partial k}{\partial x}\right)\left(\frac{\partial u}{\partial x}\right) + \left(k\frac{\partial\tau}{\partial y} + \tau\frac{\partial k}{\partial y}\right)\left(\frac{\partial u}{\partial y}\right) = Q. \qquad \textbf{(8.3)}$$

Solving for the Temperature Within the Slab

The standard way to obtain a solution to Eqs. (8.1), (8.2), and (8.3) is to approximate the derivatives with finite differences. We will use central differences and assume that the elements are all square and of equal size so that nodes are placed uniformly within the slab. This is relatively easy to do if the slab is rectangular and the height and width are in an appropriate ratio. (If this is not true, another technique, the *finite element method,* which we describe in the next chapter, is most often used.) When the nodes are uniformly spaced so that $\Delta x = \Delta y$, we will use the symbol h for that spacing.

A convenient way to write the central difference approximations to the second partial with respect to x is

$$\frac{\partial^2 u}{\partial x^2} = \frac{(uL - 2uO + uR)}{(\Delta x)^2},$$

where uL and uR are temperatures at nodes to the left and to the right, respectively, of a central node whose temperature is uO. The nodes are Δx apart. A similar formula approximates $\partial^2 u/\partial y^2$:

$$\frac{\partial^2 u}{\partial y^2} = \frac{(uA - 2uO + uB)}{(\Delta y)^2},$$

in which uA and uB are at nodes above and below the central node. It is customary to make $\Delta x = \Delta y = h$. So, if we combine these, we get

$$\nabla^2 u = \frac{(uL + uR + uA + uB - 4uO)}{h^2}.$$

Here is an example.

EXAMPLE 8.1 Solve for the steady-state temperatures in a rectangular slab that is 20 cm wide and 10 cm high. All edges are kept at 0° except the right edge, which is at 100°. There is no heat gained or lost from the surface of the slab. Place nodes in the interior spaced 2.5 cm apart (giving an array of nodes in three rows and seven columns) so that there are a total of 21 internal nodes.

Figure 8.2 is a sketch of the slab with the nodes numbered in succession by rows. We could also number them according to their row and column, with node (1, 1) at the upper left and node (3, 7) at the lower right. However, it is better to number them with a single subscript by rows when we are setting up the equations, as we have done in the figure. (In a second example, the alternative numbering system will be preferred.) Let u_i be the temperature at node (i).

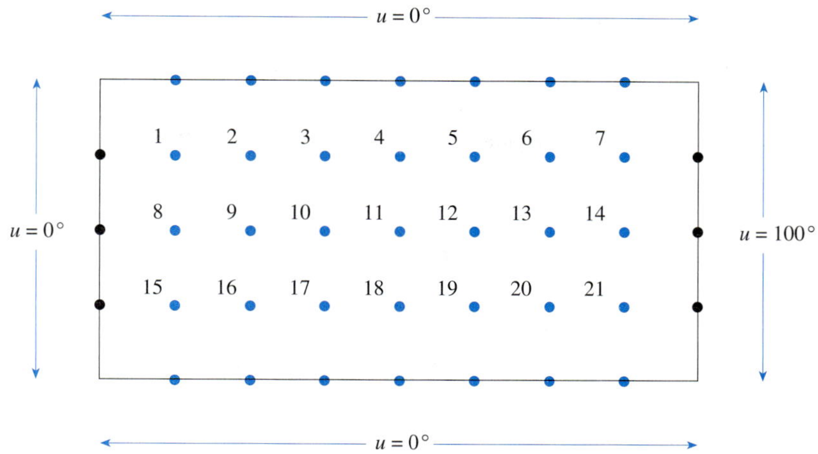

Figure 8.2

The equation that governs this situation is Eq. (8.1) with $Q = 0$:

$$\left(\frac{\partial^2 u}{\partial x^2} + \frac{\partial^2 u}{\partial y^2}\right) = 0. \tag{8.4}$$

We use these approximations for the second-order derivatives at a central node, where the temperature is uO:

$$\frac{\partial^2 u}{\partial x^2} \approx \frac{(uL - 2uO + uR)}{2.5^2};$$

$$\frac{\partial^2 u}{\partial y^2} \approx \frac{(uA - 2uO + uB)}{2.5^2},$$

where uL and uR are nodes to the left and right of the central node. Similarly, nodes uA and uB are nodes above and below the central node. Substituting these into Eq. (8.4) gives

$$\frac{(uL + uR + uA + uB - 4uO)}{6.25} = 0.$$

There is a simple device we can use to remember this approximation to the Laplacian. We call it a "pictorial operator":

$$\nabla^2 u \approx \frac{1}{h^2}\left\{\begin{matrix} & 1 & \\ 1 & -4 & 1 \\ & 1 & \end{matrix}\right\} uO. \tag{8.5}$$

This pictorial operator says: Add the temperatures at the four neighbors to uO, subtract 4 times uO, then divide by h^2, and you have an approximation to the Laplacian.

We can now write the 21 equations for the problem. Because in this example we set the Laplacian for every node equal to zero, we can drop the h^2 term. A node that is adjacent to a boundary will have the boundary value(s) in its equation; this will be subtracted from the right-hand side of that equation before we solve the system. Rather than write out all the equations, we will only show a few of them:

For node 1: $0 + u_2 + 0 + u_8 - 4u_1 = 0$, which, when the nodes are put in order, becomes:

$$-4u_1 + u_2 + u_8 = 0.$$

$$\begin{aligned}
&\text{For node 7:} && u_6 - 4u_7 + u_{14} = -100.\\
&\text{For node 9:} && u_2 + u_8 - 4u_9 + u_{10} + u_{16} = 0.\\
&\text{For node 14:} && u_7 + u_{13} - 4u_{14} + u_{21} = -100.\\
&\text{For node 18:} && u_{11} + u_{17} - 4u_{18} + u_{19} = 0.
\end{aligned}$$

If we write out all 21 equations in matrix form, we get

$$\begin{bmatrix}
-4 & 1 & 0 & 0 & 0 & 0 & 0 & 1 & 0 & 0 & 0 & 0 & 0 & 0 & 0 & 0 & 0 & 0 & 0 & 0 & 0\\
1 & -4 & 1 & 0 & 0 & 0 & 0 & 0 & 1 & 0 & 0 & 0 & 0 & 0 & 0 & 0 & 0 & 0 & 0 & 0 & 0\\
0 & 1 & -4 & 1 & 0 & 0 & 0 & 0 & 0 & 1 & 0 & 0 & 0 & 0 & 0 & 0 & 0 & 0 & 0 & 0 & 0\\
0 & 0 & 1 & -4 & 1 & 0 & 0 & 0 & 0 & 0 & 1 & 0 & 0 & 0 & 0 & 0 & 0 & 0 & 0 & 0 & 0\\
0 & 0 & 0 & 1 & -4 & 1 & 0 & 0 & 0 & 0 & 0 & 1 & 0 & 0 & 0 & 0 & 0 & 0 & 0 & 0 & 0\\
0 & 0 & 0 & 0 & 1 & -4 & 1 & 0 & 0 & 0 & 0 & 0 & 1 & 0 & 0 & 0 & 0 & 0 & 0 & 0 & 0\\
0 & 0 & 0 & 0 & 0 & 1 & -4 & 0 & 0 & 0 & 0 & 0 & 0 & 1 & 0 & 0 & 0 & 0 & 0 & 0 & 0\\
1 & 0 & 0 & 0 & 0 & 0 & 0 & -4 & 1 & 0 & 0 & 0 & 0 & 0 & 1 & 0 & 0 & 0 & 0 & 0 & 0\\
0 & 1 & 0 & 0 & 0 & 0 & 0 & 1 & -4 & 1 & 0 & 0 & 0 & 0 & 0 & 1 & 0 & 0 & 0 & 0 & 0\\
0 & 0 & 1 & 0 & 0 & 0 & 0 & 0 & 1 & -4 & 1 & 0 & 0 & 0 & 0 & 0 & 1 & 0 & 0 & 0 & 0\\
0 & 0 & 0 & 1 & 0 & 0 & 0 & 0 & 0 & 1 & -4 & 1 & 0 & 0 & 0 & 0 & 0 & 1 & 0 & 0 & 0\\
0 & 0 & 0 & 0 & 1 & 0 & 0 & 0 & 0 & 0 & 1 & -4 & 1 & 0 & 0 & 0 & 0 & 0 & 1 & 0 & 0\\
0 & 0 & 0 & 0 & 0 & 1 & 0 & 0 & 0 & 0 & 0 & 1 & -4 & 1 & 0 & 0 & 0 & 0 & 0 & 1 & 0\\
0 & 0 & 0 & 0 & 0 & 0 & 1 & 0 & 0 & 0 & 0 & 0 & 1 & -4 & 0 & 0 & 0 & 0 & 0 & 0 & 1\\
0 & 0 & 0 & 0 & 0 & 0 & 0 & 1 & 0 & 0 & 0 & 0 & 0 & 0 & -4 & 1 & 0 & 0 & 0 & 0 & 0\\
0 & 0 & 0 & 0 & 0 & 0 & 0 & 0 & 1 & 0 & 0 & 0 & 0 & 0 & 1 & -4 & 1 & 0 & 0 & 0 & 0\\
0 & 0 & 0 & 0 & 0 & 0 & 0 & 0 & 0 & 1 & 0 & 0 & 0 & 0 & 0 & 1 & -4 & 1 & 0 & 0 & 0\\
0 & 0 & 0 & 0 & 0 & 0 & 0 & 0 & 0 & 0 & 1 & 0 & 0 & 0 & 0 & 0 & 1 & -4 & 1 & 0 & 0\\
0 & 0 & 0 & 0 & 0 & 0 & 0 & 0 & 0 & 0 & 0 & 1 & 0 & 0 & 0 & 0 & 0 & 1 & -4 & 1 & 0\\
0 & 0 & 0 & 0 & 0 & 0 & 0 & 0 & 0 & 0 & 0 & 0 & 1 & 0 & 0 & 0 & 0 & 0 & 1 & -4 & 1\\
0 & 0 & 0 & 0 & 0 & 0 & 0 & 0 & 0 & 0 & 0 & 0 & 0 & 1 & 0 & 0 & 0 & 0 & 0 & 1 & -4
\end{bmatrix} u = \begin{bmatrix} 0\\ 0\\ 0\\ 0\\ 0\\ 0\\ -100\\ 0\\ 0\\ 0\\ 0\\ 0\\ 0\\ -100\\ 0\\ 0\\ 0\\ 0\\ 0\\ 0\\ -100 \end{bmatrix},$$

and we see that the coefficient matrix is symmetric and banded with a band width of 15. There are modifications of Gaussian elimination that can take advantage of the symmetry and bandedness, and we can use less memory to store the coefficients. You will find that numbering the nodes in a different order can reduce the band width to seven. (An exercise at the end of the chapter asks you to find this preferred ordering.)

When the system of equations is solved by Gaussian elmination, we get these results:

Column	Row 1	Row 2	Row 3
1	0.3530	0.4988	0.3530
2	0.9132	1.2894	0.9132
3	2.0103	2.8323	2.0103
4	4.2957	7.0193	4.2931
5	9.1531	12.6537	9.1531
6	19.6631	27.2893	19.6631
7	43.2101	53.1774	43.2101

Rows 1 and 3 are the same; this is to be expected from the symmetry of boundary conditions at the top and bottom of the region. Nodes near the hot edge are warmer than those farther away.

The accuracy of the solution would be improved if the nodes are closer together; the errors decrease about proportional to h^2, which we anticipate because the central difference approximation to the derivative is of $O(h^2)$. Another way to improve the accuracy is to use a nine-point approximation to the Laplacian. This uses the eight nodes that are adjacent to the central node and has an error of $O(h^6)$. A pictorial operator for this is

$$\nabla^2 u \approx \frac{1}{(6h^2)} \begin{Bmatrix} 1 & 4 & 1 \\ 4 & -20 & 4 \\ 1 & 4 & 1 \end{Bmatrix}. \qquad \textbf{(8.6)}$$

If Example 8.1 is solved using this nine-point formula and with $h = 2.5$ cm, the answers will be within ± 0.0032 of the "analytical" solution (from a series solution given by classical methods for partial differential equations).

Iterative Methods

The difficulty with getting the solution to a problem in the way that was done in the last example is that a very large matrix is needed when the nodal spacing is close. In that example, if $h = 1.25$, the number of equations increases from 21 to 105; if h were 0.625, there would be 465 equations. The coefficient matrix for 465 equations has $465^2 = 216{,}225$ elements! Not only is this an extravagant use of computer memory to store the values but also the solution time may be excessive. However, the matrix is *sparse*, meaning that most of the elements are zero. (Only about 1% of the elements in the last case are nonzero.)

Iterative methods that were discussed in Chapter 2 are an ideal technique for solving a sparse matrix. We do need to arrange the equations so that there is diagonal dominance (and this is readily possible for the problems of this section). We can write the equations in a form useful for iteration from this pictorial operator:

$$uO = \frac{1}{4} \begin{Bmatrix} & 1 & \\ 1 & 0 & 1 \\ & 1 & \end{Bmatrix}, \qquad \textbf{(8.7)}$$

which is, when nodes are specified using row and column subscripts:

$$u_{i,j} = \frac{(u_{i-1,j} + u_{i+1,j} + u_{i,j-1} + u_{i,j+1})}{4}.$$

We can enter the Dirichlet boundary conditions into the equations by substituting these specified values for the boundary nodes that are adjacent to interior nodes.

The name given to this method of solving boundary-value problems is *Liebmann's method*. We illustrate with the same example problem as Example 8.1.

EXAMPLE 8.2 Solve Example 8.1, but now use Liebmann's method. Use $h = 2.5$ cm.

We will designate the temperatures at the nodes by $u_{i,j}$, where i and j are the row and column for the node. Row 1 is at the top; column 1 is at the left and there are three rows and seven columns for interior nodes. The boundary conditions will be stored in row 0 and row 4, and in column 0 and column 8.

Figure 8.3 shows how nodes are numbered for this problem—we use double subscripts to indicate the row and column.

Here is the typical equation for node (i, j):

$$u_{i,j} = \frac{(u_{i,j-1} + u_{i,j+1} + u_{i-1,j} + u_{i+1,j})}{4}, \quad \text{with } i = 1 \ldots 3, \quad j = 1 \ldots 7.$$

It is best to begin the iterations with approximate values for the $u_{i,j}$, but beginning with all values set to zero will also work. Another way to begin the iterations is with all interior node values set to the average of the boundary values. If this is done, 26 iterations give answers that change by less than 0.0001 and that essentially duplicate those of Example 8.1. (If the starting values are all equal to zero, it takes 30 iterations.)

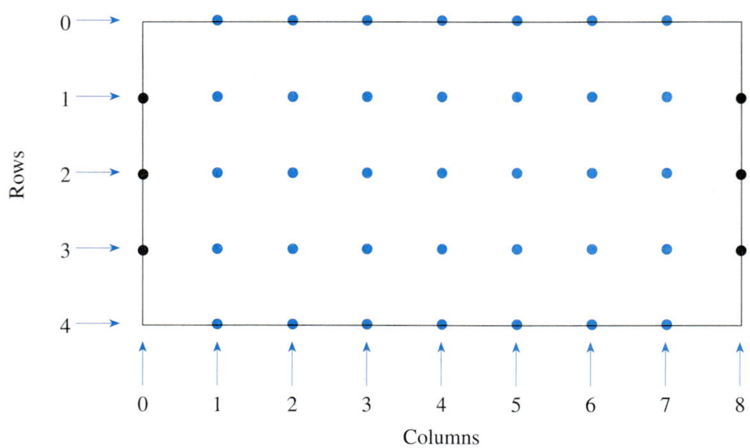

Figure 8.3

Accelerating Convergence in Liebmann's Method

In Chapter 2, it was observed that solving a linear system by iteration can be speeded by applying an *overrelaxation factor* to the process. In the present context, this is called *successive overrelaxation,* abbreviated S.O.R.

To use the S.O.R. techniques, the calculations are made with this formula:

$$u_{i,j} = u_{i,j} + \omega * \frac{(u_{i,j-1} + u_{i,j+1} + u_{i-1,j} + u_{i+1,j} - 4 * u_{i,j})}{4},$$

$$\text{with } i = 1 \ldots 3, \qquad j = 1 \ldots 7,$$

where the $u_{i,j}$ terms on the right are the current values of that variable and the one on the left becomes the new value. The ω-term is called the *overrelaxation factor:*

Solving Example 8.2 with various values for the overrelaxation factor gives these results:

Overrelaxation factor	**Number of iterations**
1.0	26
1.1	22
1.2	18
1.3	15
1.4	18
1.5	21

From this we see that overrelaxation can decrease the number of iterations required by almost one-half.

The optimal value to use for ω, the overrelaxation factor, is not always predictable. There are methods that use the results of the first few iterations to find a good value. For a rectangular region with Dirichlet boundary conditions, there is a formula:

Optimal ω = smaller root of this quadratic equation = 0:

$$\left[\cos\left(\frac{\pi}{p}\right) + \cos\left(\frac{\pi}{q}\right)\right]^2 \omega^2 - 16\omega + 16, \tag{8.8}$$

where p and q are the number of subdivisions of each of the sides. This formula suggests using $\omega = 1.267$ for the previous example. This is about the same as the value $\omega_{\text{opt}} = 1.3$ that was found by trial and error.

Why Does S.O.R. Accelerate Convergence?

We can find the basis for S.O.R. by examining the rate of convergence of iterative methods, both Gauss–Seidel, which we have used on Example 8.2, and the Jacobi method. Both of these techniques can be expressed in the form

$$x^{(n+1)} = Gx^{(n)} = -Bx^{(n)} + b'. \tag{8.9}$$

(Of course, both methods require that matrix A be diagonally dominant, or nearly so.) The two methods differ, and the difference can be expressed through these matrix equations. where A is written as $L + D + U$:

$$\text{Jacobi:} \qquad x^{(n+1)} = -D^{-1}(L + U)x^{(n)} + D^{-1}b, \tag{8.10}$$

$$\text{Gauss–Seidel:} \quad x^{(n+1)} = -(L + D)^{-1}Ux^{(n)} + (L + D)^{-1}b. \tag{8.11}$$

As Eq. (8.9) makes clear, the rate of convergence depends on how matrix B affects the iterations.

We now discuss how matrix B operates in these two methods. If an iterative method converges, $x^{(n+1)}$ will converge to x, where this last is the solution vector. Because it is the solution, it follows that $Ax = b$. Equation (8.9) becomes, for $x^{n+1} = x^n = x$,

$$x = -Bx + b'.$$

Let $e^{(n)}$ be the error in the nth iteration

$$e^{(n)} = x^{(n)} - x.$$

When there is convergence, $e^{(n)} \to 0$, the zero vector, as n gets large. Using Eq. (8.9) it follows that

$$e^{(n+1)} = -Be^{(n)} = B^2e^{(n-1)} = -B^3e^{(n-2)} = \ldots = (-B)^{n+1}e^{(0)}.$$

Now, if $B^n \to 0$, the zero matrix, it is clear that $e^{(n)} \to 0$. To show when this occurs, we need a principle from linear algebra:

> Any square matrix B can be written as UDU^{-1}. If the eigenvalues of B are distinct, then D is a diagonal matrix with the eigenvalues of B on its diagonal. (If some of the eigenvalues of B are repeated, then D may be triangular, but the argument holds in either case.)

From this we write

$$B = UDU^{-1}, \qquad B^2 = UD^2U^{-1}, \qquad B^3 = UD^3U^{-1}, \ldots, \qquad B^n = UD^nU^{-1}.$$

Now, if all the eigenvalues of B (these are on the diagonal of D) have magnitudes less than one, it is clear that D^n will approach the zero matrix and that means that B^n will also. We then see that iterations converge depending on the eigenvalues of matrix B: They must all be less than one in magnitude. Further, the rate of convergence is more rapid if the largest eigenvalue is small. We also see that even if matrix A is not diagonally dominant, there may still be convergence if the eigenvalues of B are less than unity.

This example will clarify the argument.

EXAMPLE 8.3 Compare the rates of convergence for the Jacobi and Gauss–Seidel methods for $Ax = b$, where

$$A = \begin{bmatrix} 6 & -2 & 1 \\ -2 & 7 & 2 \\ 1 & 2 & -5 \end{bmatrix} \qquad b = \begin{bmatrix} 11 \\ 5 \\ -1 \end{bmatrix}.$$

For this example, we have

$$D\begin{bmatrix} 6 & 0 & 0 \\ 0 & 7 & 0 \\ 0 & 0 & -5 \end{bmatrix} \qquad L = \begin{bmatrix} 0 & 0 & 0 \\ -2 & 0 & 0 \\ 1 & 2 & 0 \end{bmatrix} \qquad U = \begin{bmatrix} 0 & -2 & 1 \\ 0 & 0 & 2 \\ 0 & 0 & 0 \end{bmatrix}.$$

and

$$D^{-1} = \begin{bmatrix} 1/6 & 0 & 0 \\ 0 & 1/7 & 0 \\ 0 & 0 & -1/5 \end{bmatrix}.$$

For the Jacobi method, we need to compute the eigenvalues of this B matrix:

$$B = D^{-1}(L + U) = \begin{bmatrix} 1/6 & 0 & 0 \\ 0 & 1/7 & 0 \\ 0 & 0 & 1/5 \end{bmatrix} * \begin{bmatrix} 0 & -2 & 1 \\ -2 & 0 & 2 \\ 1 & 2 & 0 \end{bmatrix} = \begin{bmatrix} 0 & -1/3 & 1/6 \\ -2/7 & 0 & 2/7 \\ -1/5 & -2/5 & 0 \end{bmatrix}.$$

The eigenvalues are $-0.1425 + 0.3366i$, $-0.1425 - 0.3366i$, and 0.2851. The largest in magnitude is 0.3655.

For the Gauss–Seidel method, we need the eigenvalues of this B matrix:

$$B = (L + D)^{-1}U = \begin{bmatrix} 1/6 & 0 & 0 \\ 1/21 & 1/7 & 0 \\ 11/210 & 2/35 & -1/5 \end{bmatrix} * \begin{bmatrix} 0 & -2 & 1 \\ 0 & 0 & 2 \\ 0 & 0 & 0 \end{bmatrix} = \begin{bmatrix} 0 & -1/3 & 1/6 \\ 0 & -2/21 & 1/3 \\ 0 & -11/105 & 1/6 \end{bmatrix},$$

which has these eigenvalues: 0, $0.0357 + 0.1333i$, and $0.0357 - 0.1333i$. The largest in magnitude for the Gauss–Seidel method is 0.1380. We then see that (as expected) the Gauss–Seidel method will converge faster. If we solve this example problem with both methods, starting with [0 0 0] and ending the iterations when the largest change in any element of the solution is less than 0.00001, we find that Gauss–Seidel takes only seven iterations, whereas the Jacobi method takes 12.

We have used overrelaxation (the S.O.R. method) to speed the convergence of the iterations in solving a set of equations by the Gauss–Seidel technique. In view of the last discussion, this must be to reduce the eigenvalue of largest magnitude in the iteration equation. We have used S.O.R. in the following form:

$$x_i^{(n+1)} = x_i^{(n)} + \omega/a_{ii}(b_i - \Sigma a_{ij}x_j^{(n+1)} - \Sigma a_{ij}x_j^{(n)}), \qquad i = 1, 2, \ldots, N,$$

with the first summation from $j = 1$ to $j = i - 1$ and the second from $j = i$ to $j = N$. As shown before, the standard Gauss–Seidel iteration can be expressed in matrix form:

$$x^{(n+1)} = -(L + D)^{-1}Ux^{(n)} + (L + D)^{-1}b, \tag{8.12}$$

which is more convenient for the present purpose. We want the overrelaxation equation to be in a similar form. From $A = L + D + U$, we can write

$$\omega(b - Ax) = \omega(b - (L + D + U)x) = 0.$$

Now, if we add Dx to both sides of this, we get

$$Dx - \omega Lx - \omega Dx - \omega Ux + \omega b = Dx,$$

which can be rearranged into

$$x^{(n+1)} = (D + \omega L)^{-1}[(1 - \omega)D - \omega U]x^{(n)} + \omega(D + \omega L)^{-1}b, \tag{8.13}$$

and this is the S.O.R. form with ω equal to the overrelaxation factor. It is not easy to show in the general case that the eigenvalue of largest magnitude in Eq. (8.13) is smaller than that in Eq. (8.12), but we can do it for a simple example.

EXAMPLE 8.4 Show that overrelaxation will speed the convergence of iterations in solving

$$\begin{bmatrix} 2 & 1 \\ 1 & 3 \end{bmatrix} x = \begin{bmatrix} 6 \\ -2 \end{bmatrix}.$$

For this, the Gauss–Seidel iteration matrix is

$$-(L + D)^{-1}U = \begin{bmatrix} 0 & -1/2 \\ 0 & 1/6 \end{bmatrix},$$

whose eigenvalues are 0 and 1/6.

For the overrelaxation equation, the iteration matrix is

$$(D + \omega L)^{-1}[(1 - \omega)D - \omega U] = \begin{bmatrix} 1 - \omega & -\omega/2 \\ -\omega(1 - \omega)/3 & (\omega^2/6 - \omega + 1) \end{bmatrix}. \tag{8.14}$$

We want the eigenvalues of this, which are, of course, functions of ω. We know that, for any matrix, the product of its eigenvalues equals its determinant (why?), so we set

$$\lambda_1 * \lambda_2 = \text{det(iteration matrix)} = (\omega - 1)^2.$$

To get the smallest possible value for λ_1 and λ_2, we set them equal, so $\lambda_1 = \lambda_2 = (\omega - 1)$. We also know that, for any matrix, the sum of its eigenvalues equals its trace, so

$$\lambda_1 + \lambda_2 = 2(\omega - 1) = \frac{\omega^2}{6} - 2\omega + 2,$$

which has a solution $\omega = 1.045549$. Substituting this value of ω into Eq. (8.14) gives

$$\begin{bmatrix} -0.0455 & -0.5228 \\ 0.0159 & 0.1366 \end{bmatrix},$$

whose eigenvalues are $0.0456 \pm 0.0047i$, whose magnitudes are smaller than the largest for the Gauss–Seidel matrix, which is $1/6 = 0.16667$.

Poisson's Equation

The previous examples were for an equation known as *Laplace's equation:*

$$\nabla^2 u = 0.$$

If the right-hand side is nonzero, we have *Poisson's equation:*

$$\nabla^2 u = R,$$

where R can be a function of position in the region (x, y). To solve a Poisson equation, we need to make only a minor modification to the methods described for Laplace's equation.

EXAMPLE 8.5 Solve for the *torsion function,* ϕ, in a bar of rectangular cross section, whose dimensions are 6 in. × 8 in. (The tangential stresses are proportional to the partial derivatives of the torsion function when the bar is twisted.) The equation for ϕ is

$$\nabla^2\phi = -2, \qquad \text{with } \phi = 0 \text{ on the outer boundary of the bar's cross section.}$$

If we subdivide the cross section of the bar into 1-in. squares, there will be 35 interior nodes at the corners of these squares ($h = 1$). If we use the iterative technique, the equation for ϕ is

$$\phi_{i,j} = \frac{(\phi_{i,j-1} + \phi_{i,j+1} + \phi_{i-1,j} + \phi_{i+1,j} + 2)}{4}, \qquad i = 1 \ldots 7, \qquad j = 1 \ldots 5.$$

Convergence will be hastened if we employ overrelaxation. Equation (8.8) predicts ω_{opt} to be 1.383. Using overrelaxation with this value for ω converges in 13 iterations to the values in Table 8.1.

If overrelaxation is not employed, it takes 25 iterations to get the values of Table 8.1. Again, overrelaxation cuts the number of iterations about in half.

Derivative Boundary Conditions

Just as we saw in Section 6.7 for a one-dimensional problem, two-dimensional problems may have derivative boundary conditions. These may be of either Neumann or mixed type. We can define a more universal type of boundary conditions by the relation:

$$Au + B = Cu', \qquad \text{where } A, B, \text{ and } C \text{ are constants.}$$

If $C = 0$, we have a Dirichlet condition: $u = -B/A$. If $A = 0$, the condition is Neumann: $u' = B/C$. If none of the constants is zero, it is mixed condition. This relation can match a boundary condition for heat loss from the surface:

$$-ku' = H(u - u_s)$$

Table 8.1 Torsion function at interior nodes for Example 8.5

2.042	3.047	3.353	3.047	2.043
3.123	4.794	5.319	4.794	3.123
3.657	5.686	7.335	5.686	3.657
3.818	5.960	7.647	5.960	3.818
3.657	5.686	7.335	5.686	3.657
3.123	4.794	5.319	4.794	3.123
2.042	3.048	3.354	3.048	2.043

by taking $A = H$, $B = -H * u_s$, $C = -k$.

Here is an example that shows how this universal type of boundary conditions can be handled.

EXAMPLE 8.6 Find the steady-state temperatures in a slab that is 5 cm × 9 cm and is 0.5 cm thick. Everywhere within the slab, heat is being generated at the rate of 0.6 cal/sec/cm^3. The two 5-cm edges are held at 20° while heat is lost from the bottom 9-cm edge at a rate such that $\partial u/\partial y = 15$. The top edge exchanges heat with the surroundings according to $-k\,\partial u/\partial y = H * (uO - u_s)$, where k, the thermal conductivity, is 0.16; H, the heat transfer coefficient, is 0.073; and u_s, the temperature of the surroundings, is 25°. (uO in this case is the temperature of a node on the top edge.) No heat is gained or lost from the surfaces of the slab. Place nodes within the slab (and on the edges) at a distance 1 cm apart so that there are a total of 60 nodes.

Figure 8.4 illustrates the problem. In Figure 8.4, rows of fictitious nodes are shown above and below the top and bottom nodes in the slab. These are needed because there are derivative boundary conditions on the top and bottom edges.

The Dirichlet conditions on the left and the right will be handled by initializing the entire array of nodal temperatures to 20°, and omitting these left- and right-edge nodes from the iterations that find new values for the nodal temperatures.

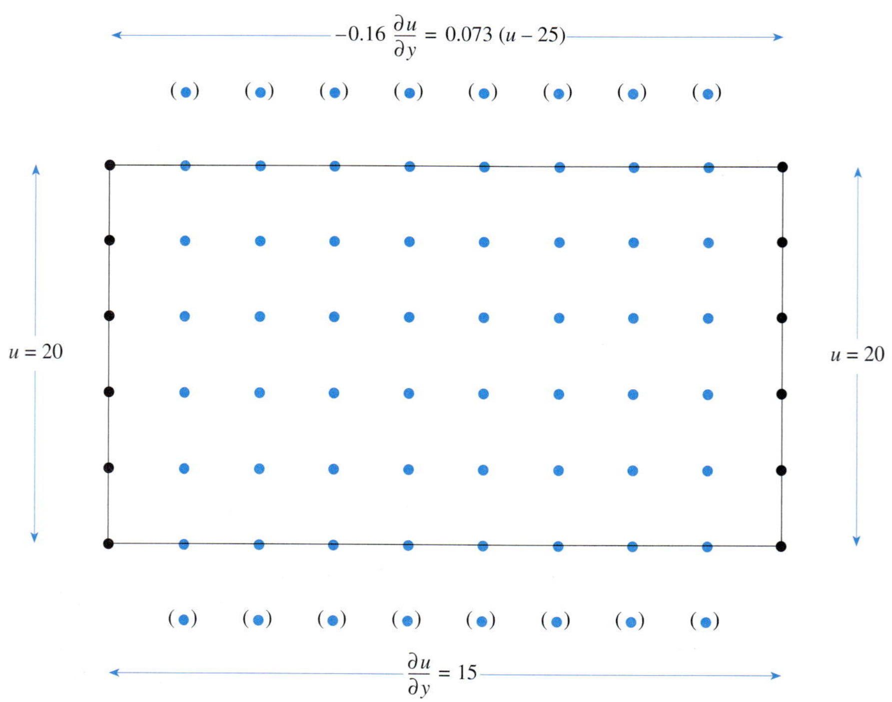

Figure 8.4

(These edge nodes are the uL or uR in the formula:

$$uO = \frac{(uL + uR + uA + uB)}{4} - \frac{Q * h^2}{kt}$$

For computations along the bottom edge (row 6), where $\partial u/\partial y = 15$, the gradient, $\partial u/\partial y$, will be computed by

$$\frac{\partial u}{\partial y} = \frac{(uA - uF)}{2h} = 15, \qquad uF = uA - 2 * h * 15,$$

where uA is at a node in the fifth row and uF is a fictitious node. (Take note of the fact that, if the gradient here is positive, heat flows in the negative y-direction, so heat is being lost as specified.) The equation for computing temperatures along the bottom edge is then

$$uO = \frac{(uL + uR + uA + uB)}{4} - \frac{Q * h^2}{kt}, \qquad \text{with } uB = uF.$$

For computations along the top edge where the relation is

$$-k\frac{\partial u}{\partial y} = H * (uO - uS),$$

temperatures will be computed using a fictitious node above uO, uF, from

$$uO = \frac{(uL + uR + uA + uB)}{4} - \frac{Q * h^2}{kt},$$

where $uA = uF$, and, because

$$\partial u/\partial y = \frac{(uF - uB)}{2h},$$

where uB is a node in the second row, we have

$$-k\frac{\partial u}{\partial y} = \frac{-k * (uF - uB)}{2h} = H * (uO - uS),$$

which gives

$$uA = uF = uB - \left(\frac{2 * H * h}{k}\right) * uO + \left(\frac{2 * H * h}{k}\right) * uS.$$

When these replacements are included in a program and overrelaxation is employed ($\omega = 1.57$), the results after 28 iterations are as shown in Table 8.2.

Table 8.2 Temperatures after 28 iterations for Example 8.6

20.000	73.510	107.915	128.859	138.826	138.826	128.859	107.915	73.510	20.000
20.000	90.195	137.476	167.733	180.743	180.743	167.733	137.476	90.195	20.000
20.000	99.793	155.061	189.855	207.669	207.669	189.855	155.061	99.794	20.000
20.000	103.918	163.119	200.956	219.410	219.410	200.956	163.119	103.919	20.000
20.000	102.762	162.539	201.442	220.604	220.604	201.442	162.539	102.762	20.000
20.000	94.589	152.834	191.669	210.959	210.959	191.669	152.834	94.589	20.000

The Alternating Direction Implicit (A.D.I.) Method

When the partial-differential equations of this chapter are solved (using the finite-difference method), the resulting coefficient matrix is sparse. The sparseness increases as the number of nodes increases: If there are 21 nodes, 81% of the values are zeros; if there are 105 nodes, 96% are zeros; for a $30 \times 30 \times 30$ three-dimensional system, only 0.012% of the $729 * 10^6$ values are nonzero!

The coefficient matrices are not only sparse in two- and three-dimensional problems. They are also *banded,* meaning that the nonzero values fall along diagonal bands within the matrix. There are solution methods that take advantage of this banding, but, because the location of the bands depends strongly on the number of nodes in rows and columns, it is not simple to accomplish. Only for a tridiagonal coefficient matrix is getting the solution straightforward.

One way around the difficulty, as we have shown, is iteration. This is an effective way to decrease the amount of memory needed to store the nonzero coefficients and to (usually) speed up the solution process. However, as we saw in Section 6.7, the system of equations for the one-dimensional case always has a tridiagonal coefficient matrix, and, for this, neither the computational time nor the storage requirements is excessive. We ask "Is there a way to get a tridiagonal coefficient matrix when the region has two or three dimensions?" The answer to this question is yes, and the technique to achieve this is called the *alternating direction implicit method,* usually abbreviated to the *A.D.I. method.*

The trick to get a tridiagonal coefficient matrix for computing the temperatures in a slab is this: First make a traverse of the nodes across the rows and consider the values above and below each node to be constants. These "constants" go on the right-hand sides of the equations, of course. (We know that these "constant" values really do vary, but we will handle that variation in the next step.) After all the nodes have been given new values with the horizontal traverse, we now make a traverse of the nodes by columns, assuming for this step that the values at nodes to the right and left are constants. There is an obvious bias in these computations, but the bias in the horizontal traverse is balanced by the opposing bias of the second step. If the object is three-dimensional, three passes are used: first in the x-direction, then in the y-direction, and finally in the z-direction.

A.D.I. is particularly useful in three-dimensional problems but it is easier to explain with a two-dimensional example. When we attack Laplace's equation in two dimensions, we write the equations as

$$\nabla^2 u = \frac{(uL - 2uO + uR)}{(\Delta x)^2} + \frac{(uA - 2uO + uB)}{(\Delta y)^2} = 0,$$

where, as before, uL, uR, uA, and uB stand for temperatures at the left, right, above, and below the central node, respectively, where it is uO. When, as is customary, $\Delta x = \Delta y$, the denominators can be canceled. The row-wise equations for the $(k + 1)$ iteration are

$$(uL - 2uO + uR)^{(k+1)} = -(uA - 2uO + uB)^{(k)}, \tag{8.15}$$

where the right-hand nodal values are the constants for the equations. When we work column-wise, the equations are for the $(k + 2)$ iteration

$$(uA - 2uO + uB)^{(k+2)} = -(uL - 2uO + uR)^{(k+1)}, \tag{8.16}$$

where, again, the right-hand nodal values are the constants.

We can speed up the convergence of the iterations by introducing an acceleration factor, ρ, to make Eq. (8.15) become

$$uO^{(k+1)} = uO^{(k)} + \rho(uL - 2uO + uR)^{(k+1)} + \rho(uA - 2uO + uB)^{(k)},$$

and Eq. (8.16) becomes

$$uO^{(k+2)} = uO^{(k+1)} + \rho(uA - 2uO + uB)^{(k+2)} + \rho(uL - 2uO + uR)^{(k+1)},$$

where the last terms in both use the values from the previous traverse.

Rearranging further, we get the tridiagonal systems

$$-uL^{(k+1)} + \left(\frac{1}{\rho} + 2\right) uO^{(k+1)} - uR^{(k+1)} = \left[uA + \left(\frac{1}{\rho} - 2\right) uO + uB\right]^{(k)}, \tag{8.17}$$

and

$$-uA^{(k+2)} + \left(\frac{1}{\rho} + 2\right) uO^{(k+2)} - uB^{(k+2)} = \left[uL + \left(\frac{1}{\rho} - 2\right) uO + uR\right]^{(k+1)}, \tag{8.18}$$

for the horizontal and vertical traverses, respectively.

In writing a program for the A.D.I. method, we must take note of the fact that the coefficient matrices for the two traverses are not identical because the boundary values enter differently. Here is a deliberately simple example that illustrates the procedure.

EXAMPLE 8.7 A rectangular plate is 6 in. × 8 in. The top edge (an 8-in. edge) is held at 100°, the right edge at 50°, and the other two edges at 0°. Use the A.D.I. method to find the steady-state temperatures at nodes spaced 1 in. apart within the plate.

There are 5 * 7 = 35 interior nodes, so there are 35 equations in each set (the horizontal and vertical traverses). With $\rho = 0.9$, and starting with all interior values set to 0°, the values of Table 8.3 result after 28 iterations, which is when the maximum change in any of the values is less than 0.001. (If we begin with the interior nodes set to the average of the boundary values, these values are reached in 24 iterations with $\rho = 1.1$.)

For this particular example, the number of nodes is small enough that Liebmann's method with overrelaxation could be used. That method is somewhat more efficient because it requires only 15 iterations to attain the same accuracy.

Table 8.3 Temperatures at interior nodes for Example 8.7

48.523	67.828	74.203	78.669	79.341	77.984	71.464
27.262	44.122	53.644	58.803	61.178	61.132	57.873
17.404	28.754	37.982	42.188	45.434	47.495	48.804
9.599	17.508	23.344	27.534	30.878	34.518	40.209
4.484	8.336	11.352	13.728	17.024	19.492	27.425

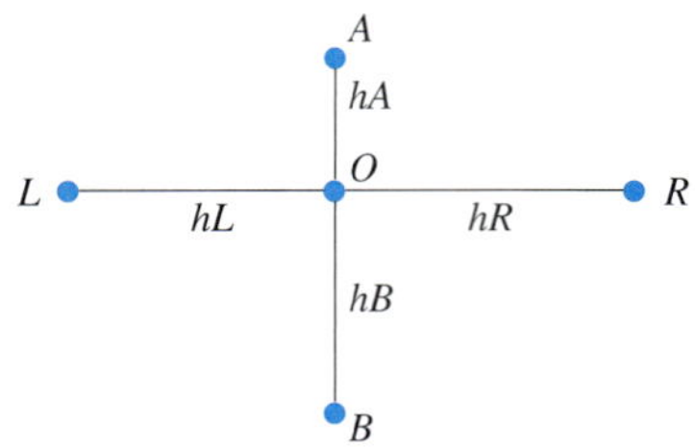

Figure 8.5

Nodes Spaced Nonuniformly

All of the examples that we have used so far have had regions where the nodes can be spaced uniformly. That is not always the case. There are three reasons why we may need a nonuniform spacing:

1. A rectangular region may have width and length incompatible with a uniform spacing.
2. The region may be nonrectangular.
3. We may want nodes closer together in some areas to improve the accuracy where the dependent variable is changing rapidly.

(If the region is three-dimensional, analogous cases apply.)

For case 2, we may be able to change the coordinate system and use an appropriate redefinition of the Laplacian. In any case, we can approximate it for a set of nodes not uniformly spaced. Consider Figure 8.5.

Figure 8.5 illustrates a situation where the four nodes around the central node have different spacing. As shown in the figure, the distances to points L, R, A, and B from point O, the central node, are hL, hR, hA, and hB. These points are nodes to the left, right, above, and below the central node, respectively. The u-values at these points are uL, uR, uA, and uB. Approximate the first derivatives between points L and O and between points O and R with:

$$\left(\frac{\partial u}{\partial x}\right)_{L,\,O} = \frac{(uO - uL)}{hL}, \qquad \left(\frac{\partial u}{\partial x}\right)_{O,\,R} = \frac{(uR - uO)}{hR}.$$

These can be interpreted as central difference approximations at points halfway between points L and O and halfway between O and R. We then approximate the second derivative with:

$$\begin{aligned}\frac{\partial^2 u}{\partial x^2} &= \frac{\left[\left(\frac{\partial u}{\partial x}\right)_{O,\,R} - \left(\frac{\partial u}{\partial x}\right)_{L,\,O}\right]}{\left[\frac{(hL + hR)}{2}\right]} \\ &= \frac{2}{(hL + hR)}\left[\frac{uL}{hL} - \frac{(hL + hR)}{(hL * hR)/uO} + \frac{uR}{hR}\right],\end{aligned} \tag{8.19}$$

but this is not a central difference approximation at exactly point O. We can use it to approximate the second derivative there but doing so incurs an error of $O(h)$. We can do the same to approximate $\partial^2 u/\partial y^2$ by using the points in a vertical line.

Using Eq. (8.19) is not the best way to handle the problem, however. The *finite-element method* (*FEM*)* is much to be preferred and we describe this in the next chapter. In FEM, the region is divided into subregions and these can be other than squares, usually triangles in 2-D. The subregions, which have common vertices, can be of varying sizes. A boundary that is not straight is approximated by a sequence of straight lines that can be very short where the boundary is sharply curved.

8.2 Parabolic Equations

The second class of partial-differential equations is usually called the *diffusion equation* or the *heat equation* because the typical examples are the molecular diffusion of matter and the flow of heat within regions. We will use heat flow as our example, similarly to Section 8.1. In contrast to that for an elliptic PDE, the situation is not the steady state but is time dependent; temperatures vary with time.

We begin with the 1-D case, but we will extend the treatment to 2-D and 3-D. For l-D, we think of heat flowing along a rod. (If the temperatures do reach a steady state, these will be the same as those found by the method of Section 8.1.)

Figure 8.6 shows a rod of length L with an element of length dx in the interior. No heat leaves or enters the rod through its circumference (it may be insulated) but flows only along the rod. As described in Chapter 6, heat flows into the element from the left at a rate, measured in cal/sec, of

$$-kA\left(\frac{du}{dx}\right).$$

The minus sign is required because du/dx expresses how rapidly temperatures increase with x, whereas the heat always flows from high temperature to low.

The rate at which heat leaves the element is given by a similar equation, but now the temperature gradient must be at the point $x + dx$:

$$-kA\left[\frac{du}{dx}+\frac{d}{dx}\left(\frac{du}{dx}\right)dx\right],$$

in which the gradient term is the gradient at x plus the change in the gradient between x and $x + dx$.

These two relations are precisely those of Section 6.7. Now, however, we do not assume that these two rates are equal, but that their difference is the rate at which heat is stored

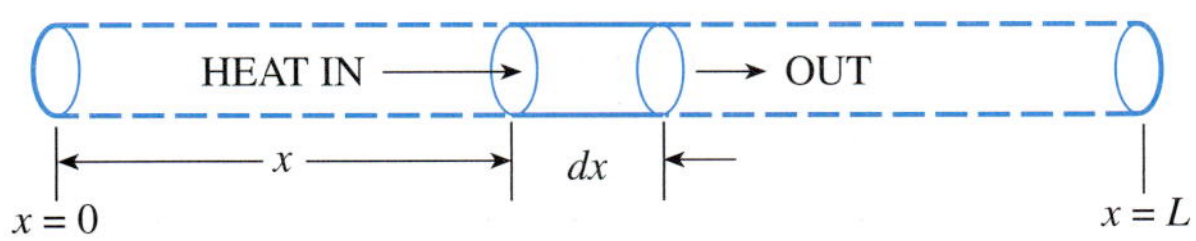

Figure 8.6

* The abbreviation FEA is sometimes used, from *finite-element analysis*.

within the element. This heat that is stored within the element raises its temperature. The rate of increase in the amount of heat that is stored is related to the rate of change in temperature of the element by an equation that involves the volume of the element ($A * dx$, measured in cm^3), the density of the material (ρ, measured in cal/gm), and a property of the material called the heat capacity, [c, measured in cal/(gm * °C)]:

$$\text{rate of increase of heat stored} = c\rho(A\,dx)\frac{du}{dt}.$$

We equate this increase in the rate of heat storage to the difference between the rates at which heat enters and leaves:

$$-kA\left(\frac{\partial u}{\partial x}\right) - \left(-kA\left[\frac{\partial u}{\partial x} + \frac{\partial}{\partial x}\left(\frac{\partial u}{\partial x}\right)dx\right]\right) = c\rho\,(A\,dx)\frac{\partial u}{\partial t}, \tag{8.20}$$

where the derivatives are now partial derivatives because there are two independent variables, x and t. We can simplify Eq. (8.20) to

$$k\left(\frac{\partial^2 u}{\partial x^2}\right) = c\rho\frac{\partial u}{\partial t}. \tag{8.21}$$

If the region is a slab or a three-dimensional object, we have the analogous equation

$$k\,\nabla^2 u = c\rho\frac{\partial u}{\partial t}, \tag{8.22}$$

in which the Laplacian appears.

It may be that the material is not homogeneous and its thermal properties may vary with position. Also, there could be heat generation within the element equal to Q cal/(sec * cm^3). In this more general case we have, in three dimensions,

$$\frac{\partial}{\partial x}\left[k(x, y, z)\frac{\partial u}{\partial x}\right] + \frac{\partial}{\partial y}\left[k(x, y, z)\frac{\partial u}{\partial y}\right] + \frac{\partial}{\partial z}\left[k(x, y, z)\frac{\partial u}{\partial z}\right] + Q(x, y, z)$$
$$= c(x, y, z) * \rho(x, y, z)\frac{\partial u}{\partial t}.$$

Our illustrations will stay with the simpler cases represented by Eqs. (8.21) and (8.22).

In order to solve these equations for unsteady-state heat flow (and they apply as well to diffusion or to any problem where the potential is proportional to the gradient), we need to make the solution agree with specified conditions along the boundary of the region of interest. In addition, because the problems are time dependent, we must begin with specified initial conditions (at $t = 0$) at all points within the region. We might think of these problems as both boundary-value problems with respect to the space variables and as initial-value problems with respect to time.

Solving the Heat Equation

We describe three different ways to solve for temperatures as they vary with time along a rod, the one-dimensional case. All three techniques are similar in that they replace the space derivative with a central difference. They differ in that different finite-difference quotients are used for the time derivative. We begin with what is called the *explicit method.* We use this forward approximation for the time derivative:

$$\frac{\partial u}{\partial t} \approx \frac{u_i^{j+1} - u_i^j}{\Delta t} \qquad \text{(at point } x_i \text{ and time } t_j\text{)}, \tag{8.23}$$

where we use subscripts to indicate the location and superscripts to indicate the time.* For the derivative with respect to x, we use (at point x_i and time t_j):

$$\frac{\partial^2 u}{\partial x^2} \approx \frac{u_{i+1}^j - 2u_i^j + u_{i-1}^j}{(\Delta x)^2}. \tag{8.24}$$

Observe that we are using a forward difference in Eq. (8.23) but a central difference in Eq. (8.24). From the discussion in Chapter 3, we know that the first has an error of order $O(\Delta t)$, whereas the second has an error of order $O(\Delta x)^2$. This difference in orders has an important consequence, as will be seen.

Substituting these approximations into Eq. (8.21) and solving for u_i^{j+1}, we get

$$u_i^{j+1} = r * (u_{i+1}^j + u_{i-1}^j) + (1 - 2r) * u_i^j, \tag{8.25}$$

where

$$r = \frac{k\,\Delta t}{c\rho(\Delta x)^2}.$$

Equation (8.25) is a way that we can march through time one Δt at a time. For $t = t_1$, we have the u's at t_0 from the initial conditions. At each subsequent time interval, we have the values for the previous time from the last computations. We apply the equation at each point along the rod where the temperature is unknown. (If an end condition involves a temperature gradient, that endpoint is included.)

The use of Eq. (8.25) to compute temperatures as a function of position and time is called the *explicit method* because each subsequent computation is explicitly given from the previous u-values.

An example will clarify the procedure.

EXAMPLE 8.8 Solve for the temperatures as a function of time within a large steel plate that is 2 cm thick. For steel, $k = 0.13$ cal/(sec * cm * °C), $c = 0.11$ cal/(g * °C), and $\rho = 7.8$ g/cm^3. Because the plate is large, neglect lateral flow of heat and consider only the flow perpendicular to the faces of the plate.

Initially, the temperatures within the plate, measured from the top face (where $x = 0$) to the bottom (where $x = 2$) are given by this relation:

$$u(x) = 100x, \quad 0 \le x \le 1; \qquad u(x) = 200 - 100x, \quad 1 \le x \le 2.$$

* The x_i are locations of evenly spaced nodes. The t_j are times spaced apart by Δt.

The boundary conditions, both at $x = 0$ and at $x = 2$, are $u = 0°$. Use $\Delta x = 0.25$ so there are eight subdivisions. Number the interior nodes from 1 to 7 so that node 0 is on the top face and node 8 is at the bottom.

The value that we use for Δt depends on the value that we choose for r; the ratio $(k\Delta t)/[c\rho(\Delta x)^2]$. Let us use $r = 0.5$ for a first trial. Doing so greatly simplifies Eq. (8.25). It becomes

$$u_i^{j+1} = 0.5(u_{i+1}^j + u_{i-1}^j). \tag{8.26}$$

(We shall compare the results of this first trial to other trials with different values for r.) With $r = 0.5$, the value of Δt is $rc\rho(\Delta x)^2/k = 0.5(0.11)(7.8)(0.25)^2/0.13 = 0.206$ sec.

We use Eq. (8.26) to compute temperatures at each node for several time steps. When this is done, the results shown in Table 8.4 are obtained. Because the values are symmetrical about the center of the rod, only those for the top half are tabulated, and the values for $x = 0$, which are all $u = 0$, are omitted. Table 8.4 also shows values from the "analytical" solution at $x = 0.5$ and at $x = 1$ from the series solution given by a classical method for solving the problem.

It is apparent from the conditions for this example that the temperatures will eventually reach the steady-state temperatures; at $t = \infty$, u will be 0° everywhere. The values in Table 8.4 are certainly approaching this equilibrium temperature. (All temperatures are within 0.1 of 0.0 after 85 time steps.)

Table 8.4 Computed and analytical temperatures for Example 8.8

		***x* value**					
		0.25	**0.50**		**0.75**	**1.00**	
Time steps	***t***	**(computed)**	**(comp)**	**(anal)**	**(computed)**	**(comp)**	**(anal)**
0	0	25.00	50.00	50.00	75.00	100.00	100.00
1	0.206	25.00	50.00	49.58	75.00	75.00	80.06
2	0.413	25.00	50.00	47.49	62.50	75.00	71.80
3	0.619	25.00	43.75	44.68	62.50	62.50	65.46
4	0.825	21.88	43.75	41.71	53.13	62.50	60.11
5	1.031	21.88	37.50	38.79	53.13	53.13	55.42
6	1.237	18.75	37.50	35.99	45.31	53.13	51.18
7	1.444	18.75	32.03	33.37	45.31	45.31	47.33
8	1.650	16.02	32.03	30.91	38.67	45.31	43.79
9	1.856	16.02	27.34	28.63	38.67	38.67	40.52
10	2.062	13.67	27.34	26.51	33.01	38.67	37.51
11	2.269	13.67	23.34	24.55	33.01	33.01	34.72
12	2.475	11.67	23.34	22.73	28.17	33.01	32.15
13	2.681	11.67	19.92	21.04	28.17	28.17	29.76
14	2.887	9.96	19.92	19.48	24.05	28.17	27.55

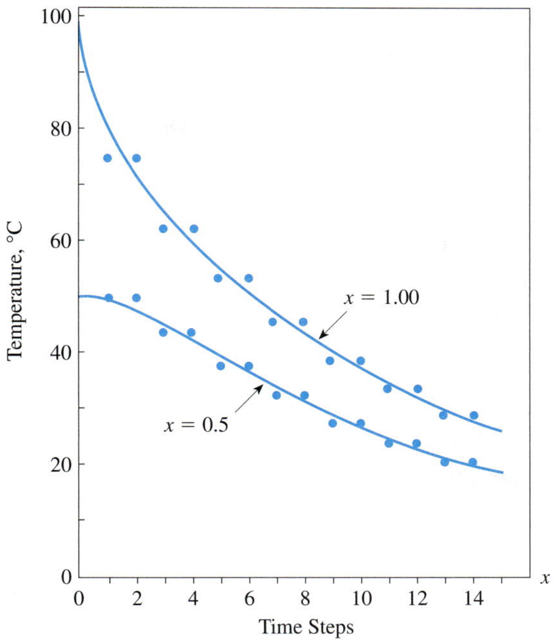

Figure 8.7

The computed values generally follow the analytical but oscillate above and below successive values. This is shown more clearly in Figure 8.7, where the computed temperatures at the center node and at $x = 0.5$ cm are plotted. The curves represent the analytical solution. If the computations are repeated but with two other values of r ($r = 0.4$ and $r = 0.6$), we find an interesting phenomenon. Of course, the values of Δt will change as well. With the smaller value for r, 0.4, the computed results are much more accurate, and the differences from the analytical values are about half as great during the early time steps and become only one-tenth as great after ten time steps. We would expect somewhat better agreement because the time steps are smaller, but the improvement is much greater than this change would cause.

On the other hand, using a value of 0.6 for r results in extremely large errors. In fact, after only eight time steps, some of the calculated values for u are negative, a patently impossible result. Figure 8.8 illustrates this quite vividly. The open circles in the figure are results with $r = 0.6$; the solid points are for $r = 0.4$. The explanation for this behavior is *instability*. The maximum value for r to avoid instability (which is particularly evident for $r = 0.6$) is $r = 0.5$. The oscillation of points about the analytical curve in Figure 8.7 shows incipient instability. Even this value for r is too large if the boundary conditions involve a gradient.

The Crank–Nicolson Method

The reason why there as instability when r is greater than 0.5 in the explicit method is the difference in orders of the finite-difference approximations for the spatial derivative and

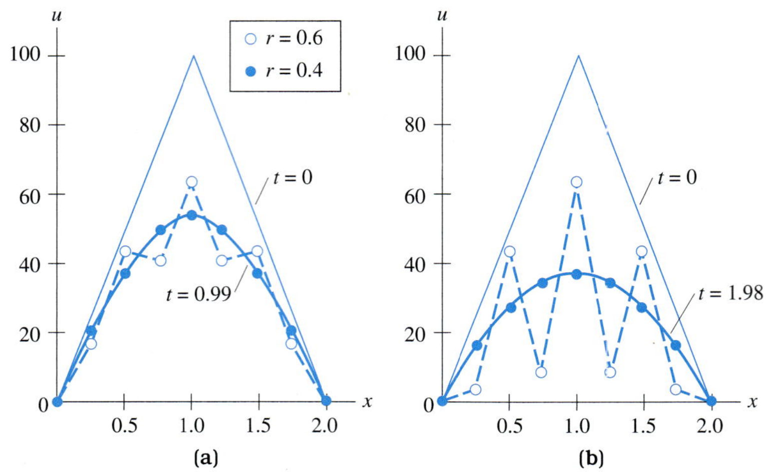

Figure 8.8

the time derivative. The *Crank–Nicolson method* is a technique that makes these finite-difference approximations of the same order.

The difference quotient for the time derivative, $(u_i^{j+1} - u_i^j)/\Delta t$, can be considered a central-difference approximation at the midpoint of the time step. If we do take this as a central-difference approximation, we will need to equate it to a central-difference approximation of the spatial derivative at the same halfway point in the time step, and this we can hope to obtain by averaging two approximations for $\partial^2 u/\partial x^2$, one computed at the start and the other at the end of the time step. So, we write, for

$$\frac{\partial u}{\partial t} = \frac{k}{c\rho}\frac{\partial^2 u}{\partial x^2},$$

this approximation:

$$\frac{u_i^{j+1} - u_i^j}{\Delta t} = \frac{1}{2}\frac{k}{c\rho}\left[\frac{u_{i+1}^j - 2u_i^j + u_{i-1}^j}{(\Delta x)^2} + \frac{u_{i+1}^{j+1} - 2u_i^{j+1} + u_{i-1}^{j+1}}{(\Delta x)^2}\right],$$

which we solve for the u-values at the end of the time step to give

$$-ru_{i+1}^{j+1} + (2 + r)u_i^{j+1} - ru_{i-1}^{j+1} = ru_{i+1}^j + (2 - r)u_i^j + ru_{i-1}^j,$$
$$r = \frac{k\Delta t}{c\rho(\Delta x)^2}. \tag{8.27}$$

Equation (8.27) is the Crank–Nicolson formula, and using it involves solving a set of simultaneous equations, because the equation for u_i^{j+1} includes two adjacent u-values at $t = t^{j+1}$. Hence, this is an *implicit method*. Fortunately, the coefficient matrix is tridiagonal. A most important advantage of the method is that it is stable for any value for r, although smaller values usually give better accuracy. This next example illustrates the method.

EXAMPLE 8.9 Solve Example 8.8, but now use the Crank–Nicolson method. Compare the results with $r = 0.5$ and with $r = 1.0$ to the analytical values.

Employing Eq. (8.27) gives the results shown in Table 8.5 for the centerline temperatures with $r = 0.5$ and in Table 8.6 for the centerline temperatures with $r = 1.0$. The error columns are the differences between the computed temperatures and those from the series solution. In Table 8.5, these range from 2.0% to 2.7% of the analytical values, whereas in Table 8.6, they range from 1.0% to 2.5%. One would expect the errors with $r = 0.5$ to be smaller, but this is not the case. Both sets of computations are more accurate than those in Table 8.4, where the explicit method was used with $r = 0.5$.

The Theta Method – A Generalization

In the Crank–Nicolson method, we interpret the finite-difference approximation to the time derivative as a central difference at the midpoint of the time interval. In the *theta method,* we make a more general statement by interpreting this approximation to apply at some other point within the time interval. If we interpret it to apply at a fraction θ of Δt, we then equate the time-derivative approximation to a weighted average of the spatial derivatives at the beginning and end of the time interval, giving this relation:

$$\frac{u_i^{j+1} - u_i^j}{\Delta t} = \left(\frac{k}{c\rho}\right)\left[\frac{(1-\theta)(u_{i+1}^j - 2u_i^j + u_{i-1}^j)}{(\Delta x)^2} + \frac{\theta\,(u_{i+1}^{j+1} - 2u_i^{j+1} + u_{i-1}^{j+1})}{(\Delta x)^2}\right].$$

Observe that using $\theta = 0.5$ gives the Crank–Nicolson method, whereas using $\theta = 0$ gives the explicit method. If we use $\theta = 1$, the theta method is often called the *implicit method.* For $\theta = 1$, the analog of Eq. (8.27) is

$$-ru_{i+1}^{j+1} + (1 + 2r)u_i^{j+1} - ru_{i-1}^{j+1} = u_i^j, \qquad r = \frac{k\Delta t}{c\rho(\Delta x)^2}.$$

Table 8.5 Centerline temperatures with Crank–Nicolson Method, $r = 0.5$

Time steps	t	u-values	Error
0	0	100.00	—
1	0.206	82.32	2.26
2	0.413	73.48	1.68
3	0.619	66.86	1.40
4	0.825	61.34	1.23
5	1.031	56.52	1.10
6	1.237	52.21	1.03
7	1.444	48.30	0.97
8	1.650	44.71	0.92
9	1.856	41.40	0.88
10	2.062	38.36	0.85

Table 8.6 Centerline temperatures with Crank–Nicolson Method, $r = 1.0$

Time steps	t	u-values	Error
0	0	100.00	—
1	0.413	71.13	0.67
2	0.825	61.53	1.42
3	1.237	51.97	0.79
4	1.650	44.67	0.88
5	2.062	38.29	0.78
6	2.475	32.88	0.73
7	2.887	28.23	0.68
8	3.300	24.23	0.61

For any value of θ, the typical equation is

$$-ru_{i+1}^{j+1} + (1 + 2r\theta)u_i^{j+1} - ru_{i-1}^{j+1}$$
$$= r(1 - \theta)u_{i+1}^j + [1 - 2r(1 - \theta)]u_i^j + r(1 - \theta)u_{i-1}^j. \quad (8.28)$$

What value is best for θ? Burnett (1987) suggests that $\theta = \frac{2}{3}$ is nearly optimal, but he points out that a case can be made for using $\theta = 0.878$. This next example compares the use of these two values.

EXAMPLE 8.10 Solve Example 8.8 by the theta method with $\theta = \frac{2}{3}$, 0.878, and 1.0. Compare these to results from the Crank–Nicolson and explicit methods.

Using Eq. (8.28), computations were made for ten time steps. Table 8.7 shows how the values at the centerline, $x = 1.0$ differ from the analytical values. It is interesting to observe that, for this problem, the Crank–Nicolson results ($\theta = 0.5$) have smaller errors than those with larger values for θ. Even the results from the explicit method ($\theta = 0$) are better than those with $\theta = 1.0$ (although the explicit values oscillate around the analytical). This suggests that there is an optimal value for θ less than $\frac{2}{3}$ and greater than zero. We leave this determination as an exercise, as well as the comparison at other values for x. We also leave as an exercise to find if there is an optimal value in other problems.

Stability Considerations

We have seen in our examples that when the ratio $k\Delta t/c\rho(\Delta x)^2$ is greater than 0.5, the explicit method is unstable. Crank–Nicolson and the implicit methods do not have such a limitation. We now look at this more analytically. We also discuss the *convergence* of the methods.

Table 8.7 Comparisons of results from the theta method, $r = 0.5$

	Errors in computed centerline temperatures θ-value				
Time steps	**2/3**	**0.878**	**1.0**	**0.5**	**0.0**
1	3.57	4.88	5.51	2.26	5.06
2	2.48	3.55	4.15	1.68	3.20
3	1.98	2.79	3.28	1.40	2.96
4	1.71	2.37	2.78	1.23	2.39
5	1.53	2.11	2.46	1.10	2.29
6	1.43	1.97	2.29	1.03	1.95
7	1.35	1.86	2.16	0.97	2.02
8	1.30	1.80	2.09	0.92	1.52
9	1.27	1.76	2.04	0.88	1.85
10	1.23	1.72	2.00	0.85	1.16

By *convergence,* we mean that the results of the method approach the analytical values as Δt and Δx both approach zero. By *stability,* we mean that errors made at one stage of the calculations do not cause increasingly large errors as the computations are continued, but rather will eventually damp out.

We will first discuss convergence, limiting ourselves to the simple case of the unsteady-state heat-flow equation in one dimension:*

$$\frac{\partial U}{\partial t} = \frac{k}{c\rho}\frac{\partial^2 U}{\partial x^2}. \tag{8.29}$$

We will use the symbol U to represent the exact solution to Eq. (8.29), and u to represent the numerical solution. At the moment, we assume that u is free of round-off errors, so the only difference between U and u is the error made by replacing Eq. (8.29) by the difference equation. Let $e_i^j = U_i^j - u_i^j$, at the point $x = x_i$, $t = t_j$. By the explicit method, Eq. (8.29) becomes

$$u_i^{j+1} = r(u_{i+1}^j + u_{i-1}^j) + (1 - 2r)u_i^j, \tag{8.30}$$

where $r = k\,\Delta t/c\rho(\Delta x)^2$. Substituting $u = U - e$ into Eq. (8.30), we get

$$e_i^{j+1} = r(e_{i+1}^j + e_{i-1}^j) + (1 - 2r)e_i^j - r(U_{i+1}^j + U_{i-1}^j) - (1 - 2r)U_i^j + U_i^{j+1}. \tag{8.31}$$

By using Taylor-series expansions, we have

$$\begin{aligned}
U_{i+1}^j &= U_i^j + \left(\frac{\partial U}{\partial x}\right)_{i,j}\Delta x + \frac{(\Delta x)^2}{2}\frac{\partial^2 U(\xi_1, t_j)}{\partial x^2}, \qquad x_i < \xi_1 < x_{i+1},\\
U_{i-1}^j &= U_i^j - \left(\frac{\partial U}{\partial x}\right)_{i,j}\Delta x + \frac{(\Delta x)^2}{2}\frac{\partial^2 U(\xi_2, t_j)}{\partial x^2}, \qquad x_{i-1} < \xi_2 < x_i,\\
U_i^{j+1} &= U_i^j + \Delta t\,\frac{\partial U(x_i, \eta)}{\partial t}, \qquad t_j < \eta < t_{j+1}.
\end{aligned}$$

Substituting these into Eq. (8.31) and simplifying, remembering that $r(\Delta x)^2 = k\,\Delta t/c\rho$, we get

$$e_i^{j+1} = r(e_{i+1}^j + e_{i-1}^j) + (1 - 2r)e_i^j + \Delta t\left[\frac{\partial U(x_i, \eta)}{\partial t} - \frac{k}{c\rho}\frac{\partial^2 U(\xi, t_j)}{\partial x^2}\right],$$
$$t_j \le \eta \le t_{j+1}, \qquad x_{i-1} \le \xi \le x_{i+1}. \tag{8.32}$$

* We could have treated the simpler equation $\partial U/\partial T = \partial^2 U/\partial X^2$ without loss of generality, because with the change of variables—$X = \sqrt{c\rho}\,x$, $T = kt$—the two equations are seen to be identical.

Let E^j be the magnitude of the maximum error in the row of calculations for $t = t_j$, and let $M > 0$ be an upper bound for the magnitude of the expression in brackets in Eq. (8.32). If $r \le \frac{1}{2}$, all the coefficients in Eq. (8.32) are positive (or zero) and we may write the inequality

$$|e_i^{j+1}| \le 2rE^j + (1 - 2r)E^j + M\Delta t = E^j + M\Delta t.$$

This is true for all the e_i^{j+1} at $t = t_{j+1}$, so

$$E^{j+1} \le E^j + M\,\Delta t.$$

This is true at each time step,

$$E^{j+1} \le E^j + M\,\Delta t \le E^{j-1} + 2M\,\Delta t \le \cdots \le E^0 + (j+1)M\,\Delta t = E^0 + Mt_{j+1} = Mt_{j+1},$$

because E^0, the errors at $t = 0$, is zero, as U is given by the initial conditions.

Now, as $\Delta x \to 0$, $\Delta t \to 0$ if $k\,\Delta t/c\rho(\Delta x)^2 \le \frac{1}{2}$, and $M \to 0$, because, as both Δx and Δt get smaller,

$$\left[\frac{\partial U(x_i, \eta)}{\partial t} - \frac{k}{c\rho} * \frac{\partial^2 U(\xi, t_j)}{\partial x^2}\right] \to \left(\frac{\partial U}{\partial t} - \frac{k}{c\rho} * \frac{\partial^2 U}{\partial x^2}\right)_{i,j} = 0.$$

This last is by virtue of Eq. (8.29), of course. Consequently, we have shown that the explicit method is convergent for $r \le \frac{1}{2}$, because the errors approach zero as Δt and Δx are made smaller.

For the solution to the heat-flow equation by the Crank–Nicolson method, the analysis of convergence may be made by similar methods. The treatment is more complicated, but it can be shown that each E^{j+1} is no greater than a finite multiple of E^j plus a term that vanishes as both Δx and Δt become small, and this is independent of r. Hence, because the initial errors are zero, the finite-difference solution approaches the analytical solution as $\Delta t \to 0$ and $\Delta x \to 0$, requiring only that r stay finite. This is also true for the θ method whenever $0.5 \le \theta \le 1$.

We begin our discussion of stability with a numerical example. Because the heat-flow equation is linear, if two solutions are known, their sum is also a solution. We are interested in what happens to errors made in one line of the computations as the calculations are continued, and because of the additivity feature, the effect of a succession of errors is just the sum of the effects of the individual errors. We follow, then, a single error,* which most likely occurred due to round off. If this single error does not grow in magnitude, we will call the method *stable,* because then the cumulative effect of all errors affects the later calculations no more than a linear combination of the previous errors would. (Because round-off errors are both positive and negative, we can expect some cancellation.)

Table 8.8 illustrates the principle. We have calculated for the simple case where the boundary conditions are fixed, so that the errors at the endpoints are zero. We assume that a single

---

* A computation made assuming that each of the interior points has an error equal to e at $t = t_1$ demonstrates the effect more rapidly.

Table 8.8 Propagation of errors—explicit method

t	Endpoint x_1	x_2	x_3	x_4	Endpoint x_5
t_0	0	0	0	0	0
t_1	0	e	0	0	0
t_2	0	0	$0.50e$	0	0
t_3	0	$0.25e$	0	$0.25e$	0
t_4	0	0	$0.25e$	0	0
t_5	0	$0.125e$	0	$0.125e$	0
t_6	0	0	$0.125e$	0	0
t_7	0	$0.062e$	0	$0.062e$	0
t_8	0	0	$0.062e$	0	0

error of size e occurs at $t = t_1$ and $x = x_2$. The explicit method, $k\Delta t/c\rho(\Delta x)^2 = \frac{1}{2}$, was used. The original error quite obviously dies out. As an exercise, it is left to the student to show that with $r > 0.5$, errors have an increasingly large effect on later computations. Table 8.9 shows that errors damp out for the Crank–Nicolson method with $r = 1$ even more rapidly than in the explicit method with $r = 0.5$. The errors with the implicit method also die out with $r = 1$, more rapidly than with the explicit method but less rapidly than with Crank–Nicolson.

A More Analytical Argument

To discuss stability in a more analytical sense, we need some material from linear algebra. In Chapter 6, we discussed eigenvalues and eigenvectors of a matrix. We recall that for the matrix A and vector x, if

$$Ax = \lambda x,$$

then the scalar λ is an eigenvalue of A and x is the corresponding eigenvector. If the N eigenvalues of the $N \times N$ matrix A are all different, then the corresponding N eigenvectors

Table 8.9 Propagation of errors—Crank–Nicolson method

t	x_1	x_2	x_3	x_4	x_5
t_0	0	0	0	0	0
t_1	0	e	0	0	0
t_2	0	$0.071e$	$0.286e$	$0.071e$	0
t_3	0	$0.092e$	$0.082e$	$0.092e$	0
t_4	0	$0.036e$	$0.064e$	$0.036e$	0
t_5	0	$0.024e$	$0.030e$	$0.024e$	0
t_6	0	$0.012e$	$0.018e$	$0.012e$	0
t_7	0	$0.007e$	$0.009e$	$0.007e$	0
t_8	0	$0.004e$	$0.005e$	$0.004e$	0

are linearly independent, and any N-component vector can be written uniquely in terms of them.

Consider the unsteady-state heat-flow problem with fixed boundary conditions. Suppose we subdivide into $N + 1$ subintervals so there are N unknown values of the temperature being calculated at each time step. Think of these N values as the components of a vector. Our algorithm for the explicit method (Eq. 8.25) can be written as the matrix equation*

$$\begin{bmatrix} u_1^{j+1} \\ u_2^{j+1} \\ \vdots \\ u_N^{j+1} \end{bmatrix} = \begin{bmatrix} (1-2r) & r & & \\ r & (1-2r) & r & \\ & & \vdots & \\ & & r & (1-2r) \end{bmatrix} \begin{bmatrix} u_1^{j} \\ u_2^{j} \\ \vdots \\ u_N^{j} \end{bmatrix}, \tag{8.33}$$

or

$$u^{j+1} = Au^j,$$

where A represents the coefficient matrix and u^j and u^{j+1} are the vectors whose N components are the successive calculated values of temperature. The components of u^0 are the initial values from which we begin our solution. The successive rows of our calculations are

$$\begin{aligned} u^1 &= Au^0, \\ u^2 &= Au^1 = A^2u^0, \\ &\vdots \\ u^j &= Au^{j-1} = A^2u^{j-2} = \cdots = A^ju^0. \end{aligned}$$

(Here the superscripts on the A's are exponents; on the vectors they indicate time.)

Suppose that errors are introduced into u^0, so that it becomes $\overline{u}^0$. We will follow the effects of this error through the calculations. The successive lines of calculation are now

$$\overline{u}^j = A\overline{u}^{j-1} = \cdots = A^j\overline{u}^0.$$

Let us define the vector e^j as $u^j - \overline{u}^j$ so that e^j represents the errors in u^j caused by the errors in $\overline{u}^0$. We have

$$e^j = u^j - \overline{u}^j = A^ju^0 - A^j\overline{u}^0 = A^je^0. \tag{8.34}$$

This shows that errors are propagated by using the same algorithm as that by which the temperatures are calculated, as was implicitly assumed earlier in this section.

---

* A change of variable is required to give boundary conditions of $u = 0$ at each end. This can always be done for fixed end conditions.

Now the N eigenvalues of A are distinct (see below) so that its N eigenvectors x_1, $x_2, \ldots, x_N$ are independent, and

$$\begin{aligned} Ax_1 &= \lambda_1 x_1, \\ Ax_2 &= \lambda_2 x_2, \\ &\vdots \\ Ax_N &= \lambda_N x_N. \end{aligned}$$

We now write the error vector e^0 as a linear combination of the x_i:

$$e^0 = c_1 x_1 + c_2 x_2 + \cdots + c_N x_N,$$

where the c's are constants. Then e^1 is, in terms of the x_i,

$$e^1 = Ae^0 = \sum_{i=1}^{N} Ac_i x_i = \sum_{i=1}^{N} c_i A x_i = \sum_{i=1}^{N} c_i \lambda_i x_i,$$

and for e^2,

$$e^2 = Ae^1 = \sum_{i=1}^{N} Ac_i \lambda_i x_i = \sum_{i=1}^{N} c_i \lambda_i^2 x_i.$$

(Again, the superscripts on vectors indicate time; on λ they are exponents.) After j steps, Eq. (8.34) can be written

$$e^j = \sum_{i=1}^{N} c_i \lambda_i^j x_i.$$

If the magnitudes of all of the eigenvalues are less than or equal to unity, errors will not grow as the computations proceed; that is, the computational scheme is stable. This then is the analytical condition for stability: that the largest eigenvalue of the coefficient matrix for the algorithm be one or less in magnitude.

The eigenvalues of matrix A (Eq. 8.33) can be shown to be

$$1 - 4r \sin^2 \frac{n\pi}{2(N+1)}, \qquad n = 1, 2, \ldots, N$$

(note that they are all distinct). We will have stability for the explicit scheme if

$$-1 \le 1 - 4r \sin^2 \frac{n\pi}{2(N+1)} \le 1.$$

The limiting value of r is given by

$$-1 \le 1 - 4r \sin^2 \frac{n\pi}{2(N+1)}$$

$$r \le \frac{\dfrac{1}{2}}{\sin^2 \left(\dfrac{n\pi}{2(N+1)} \right)}.$$

Hence, if $r \le \frac{1}{2}$, the explicit scheme is stable.

The Crank–Nicolson scheme, in matrix form, is

$$\begin{bmatrix} (2+2r) & -r & & \\ -r & (2+2r) & -r & \\ & & \vdots & \\ & & -r & (2+2r) \end{bmatrix} \begin{bmatrix} u_1^{j+1} \\ u_2^{j+1} \\ \vdots \\ u_N^{j+1} \end{bmatrix} = \begin{bmatrix} (2-2r) & r & & \\ r & (2-2r) & r & \\ & & \vdots & \\ & & r & (2-2r) \end{bmatrix} \begin{bmatrix} u_1^{j} \\ u_2^{j} \\ \vdots \\ u_N^{j} \end{bmatrix},$$

or

$$Au^{j+1} = Bu^j.$$

We can write

$$u^{j+1} = (A^{-1}B)u^j,$$

so that stability is given by the magnitudes of the eigenvalues of $A^{-1}B$. These are

$$\frac{2 - 4r\sin^2\left(\dfrac{n\pi}{2(N-1)}\right)}{2 + 4r\sin^2\left(\dfrac{n\pi}{2(N-1)}\right)}, \qquad n = 1, 2, \ldots, N.$$

Clearly, all the eigenvalues are no greater than one in magnitude for any positive value of r. A similar argument shows that the implicit method is also unconditionally stable.

The Heat Equation in Two or Three Dimensions

In dimensions greater than one, the equation that we are to solve is

$$\frac{\partial u}{\partial t} = \frac{k}{c\rho}\nabla^2 u. \tag{8.35}$$

We will apply finite-difference approximations to the derivatives as we did in 1-D. We show how a typical example is solved.

Suppose we have a rectangular region whose edges fit to evenly spaced nodes. If we replace the right-hand side of Eq. (8.35) with central-difference approximations, where $\Delta x = \Delta y = h$, and $r = k\,\Delta t/(c\rho h^2)$, the explicit scheme becomes

$$u_{i,j}^{k+1} - u_{i,j}^{k} = r(u_{i+1,j}^{k} - 2u_{i,j}^{k} + u_{i-1,j}^{k} + u_{i,j+1}^{k} - 2u_{i,j}^{k} + u_{i,j-1}^{k})$$

or

$$u_{i,j}^{k+1} = r(u_{i+1,j}^{k} + u_{i-1,j}^{k} + u_{i,j+1}^{k} + u_{i,j-1}^{k}) + (1-4r)u_{i,j}^{k}.$$

In this scheme, stability requires that the value of r be $\frac{1}{4}$ or less in the simple case of Dirichlet boundary conditions. (Note that this corresponds again to the numerical value that gives a particularly simple formula.) In the more general case with $\Delta x \neq \Delta y$, the criterion is

$$\frac{k\,\Delta t}{c\rho[(\Delta x)^2 + (\Delta y)^2]} \le \frac{1}{8}.$$

The analogous equation in three dimensions, with equal grid spacing each way, has the coefficient $(1 - 6r)$, and $r \le \frac{1}{6}$ is required for convergence and stability.

The difficulty with the use of the explicit scheme is that the restrictions on Δt require inordinately many rows of calculations. One then looks for a method in which Δt can be made larger without loss of stability. In one dimension, the Crank–Nicolson method was such a method. In the 2-D case, using averages of central-difference approximations to give $\partial^2 u/\partial x^2$ and $\partial^2 u/\partial y^2$ at the midvalue of time, we get

$$\begin{aligned} u_{i,j}^{k+1} - u_{i,j}^{k} = \frac{r}{2}\,[&u_{i+1,j}^{k+1} - 2u_{i,j}^{k+1} + u_{i-1,j}^{k+1} + u_{i+1,j}^{k} - 2u_{i,j}^{k} + u_{i-1,j}^{k} \\ &+ u_{i,j+1}^{k+1} - 2u_{i,j}^{k+1} + u_{i,j-1}^{k+1} + u_{i,j+1}^{k} - 2u_{i,j}^{k} + u_{i,j-1}^{k}]. \end{aligned}$$

The problem now is that a set of $(M)(N)$ simultaneous equations must be solved at each time step, where M is the number of unknown values in the x-direction and N in the y-direction. Furthermore, the coefficient matrix is no longer tridiagonal, so the solution to each set of equations is slower and memory space to store the elements of the matrix may be exorbitant.

The advantage of a tridiagonal matrix is retained in the alternating direction implicit scheme (A.D.I.) proposed by Peaceman and Rachford (1955). It is widely used in modern computer programs for the solution of parabolic partial-differential equations. We discussed the A.D.I. method in Section 8.1 applied to elliptic equations. For parabolic equations, we approximate $\nabla^2 u$ by adding a central-difference approximation to $\partial^2 u/\partial x^2$ written at the beginning of the time interval to a similar expression for $\partial^2 u/\partial y^2$ written at the end of the time interval. We will use subscripts L, R, A, and B to indicate nodes to the left, right, above, and below the central node, respectively, where $u = u_0$. We then have

$$u_0^{j+1} - u_0^{j} = r[u_L^{j} - 2u_0^{j} + u_R^{j}] + r[u_A^{j+1} - 2u_0^{j+1} + u_B^{j+1}], \tag{8.36}$$

where $r = k\,\Delta t/c\rho\Delta^2$ and $\Delta = \Delta x = \Delta y$. The obvious bias in this formula is balanced by reversing the order of the second derivative approximations in the next time span:

$$u_0^{j+2} - u_0^{j+1} = r[u_L^{j+2} - 2u_0^{j+2} + u_R^{j+2}] + r[u_A^{j+1} - 2u_0^{j+1} + u_B^{j+1}]. \tag{8.37}$$

Observe that in using Eq. (8.36), we make a vertical traverse through the nodes, computing new values for each column of nodes. Similarly, in using Eq. (8.37) we make a horizontal traverse, computing new values row by row. In effect, we consider u_L and u_R as fixed when we do a vertical traverse; we consider u_A and u_B as fixed for horizontal traverses.

EXAMPLE 8.11 A square plate of steel is 8 in. wide and 6 in. high. Initialty, all points on the plate are at 50°. The edges are suddenly brought to the temperatures shown in Figure 8.9 and held at

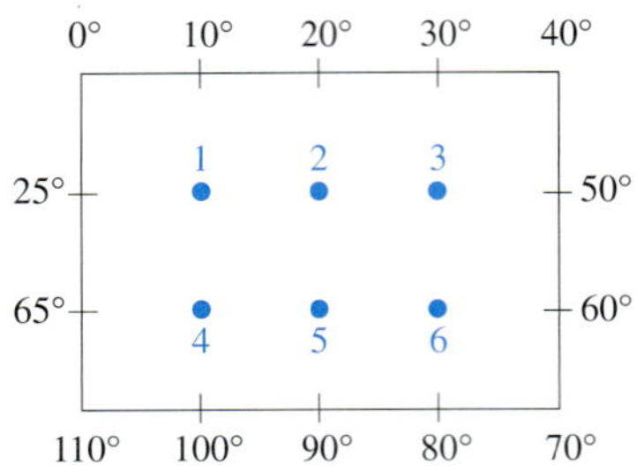

Figure 8.9

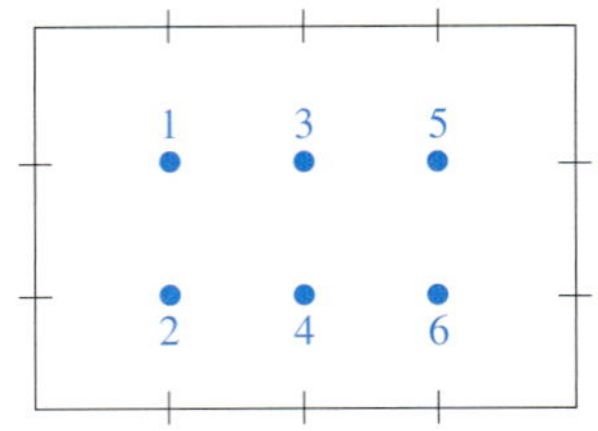

Figure 8.10

these temperatures. Trace the history of temperatures at nodes spaced 2 in. apart using the A.D.I. method, assuming that heat flows only in the x- and y-directions.

Figure 8.9 shows a numbering system for the internal nodes, all of which start at 50°, as well as the temperatures at boundary nodes.

Using Eq. (8.36), the typical equation for a vertical traverse is

$$-ru_A^{j+1} + (1 + 2r)u_0^{j+1} - ru_B^{j+1} = (ru_L + (1 - 2r)u_0 + ru_R)^j.$$

If we use this equation and the numbering system of Figure 8.9 to set up the equations for a vertical traverse, we do not get the tridiagonal system that we desire, but we do if the nodes are renumbered as shown in Figure 8.10. To keep track of the different numbering systems, we will use v for temperatures when a vertical traverse is made (numbered as in Fig. 8.10) and u when a horizontal traverse is made (numbered as in Fig 8.9).

This is the set of equations for a vertical traverse:

$$\begin{bmatrix} (1+2r) & -r & & & & \\ -r & (1+2r) & & & & \\ \hdashline & & (1+2r) & -r & & \\ & & -r & (1+2r) & & \\ \hdashline & & & & (1+2r) & -r \\ & & & & -r & (1+2r) \end{bmatrix} \begin{Bmatrix} v_1 \\ v_2 \\ \hdashline v_3 \\ v_4 \\ \hdashline v_5 \\ v_6 \end{Bmatrix} = \begin{Bmatrix} r(25) + (1-2r)u_1 + ru_2 + r(10) \\ r(65) + (1-2r)u_4 + ru_5 + r(100) \\ \hdashline ru_1 + (1-2r)u_2 + ru_3 + r(20) \\ ru_4 + (1-2r)u_5 + ru_6 + r(90) \\ \hdashline ru_2 + (1-2r)u_3 + r(50) + r(30) \\ ru_5 + (1-2r)u_6 + r(60) + r(80) \end{Bmatrix}.$$

When we apply Eq. (8.36) to get a set of equations for a horizontal traverse, we get (the dashed lines show they break into subsets)

$$\begin{bmatrix} (1+2r) & -r & & & & \\ -r & (1+2r) & -r & & & \\ & -r & (1+2r) & & & \\ \hdashline & & & (1+2r) & -r & \\ & & & -r & (1+2r) & -r \\ & & & & -r & (1+2r) \end{bmatrix} \begin{Bmatrix} u_1 \\ u_2 \\ u_3 \\ \hdashline u_4 \\ u_5 \\ u_6 \end{Bmatrix} = \begin{Bmatrix} r(10) + (1-2r)v_1 + rv_2 + r(25) \\ r(20) + (1-2r)v_3 + rv_4 \\ r(30) + (1-2r)v_5 + rv_6 + r(50) \\ \hdashline rv_1 + (1-2r)v_2 + r(100) + r(65) \\ rv_3 + (1-2r)v_4 + r(90) \\ rv_5 + (1-2r)v_6 + r(80) + r(60) \end{Bmatrix}.$$

A value must be specified for r. Small r's give better accuracy but smaller Δt's, so more time steps are required to compute the history. If we take $r = 1$, Δt is 26.4 sec.

The first vertical traverse gives results for $t = 26.4$ sec. We get the first set of v's from

$$\begin{bmatrix} 3 & -1 & & & & \\ -1 & 3 & & & & \\ & & 3 & -1 & & \\ & & -1 & 3 & & \\ & & & & 3 & -1 \\ & & & & -1 & 3 \end{bmatrix} \begin{Bmatrix} v_1 \\ v_2 \\ v_3 \\ v_4 \\ v_5 \\ v_6 \end{Bmatrix} = \begin{Bmatrix} 25 - (1)50 + 50 + 10 \\ 65 - (1)50 + 50 + 100 \\ 50 - (1)50 + 50 + 20 \\ 50 - (1)50 + 50 + 90 \\ 50 - (1)50 + 50 + 30 \\ 50 - (1)50 + 60 + 80 \end{Bmatrix} = \begin{Bmatrix} 35 \\ 165 \\ 70 \\ 140 \\ 80 \\ 140 \end{Bmatrix}.$$

Solving, we get these values:

$$\{33.75 \quad 66.25 \quad 43.75 \quad 61.25 \quad 47.5 \quad 62.5\}.$$

These values are used to build the right-hand sides for the next computations, a horizontal traverse, getting these equations for $t = 52.8$ sec:

$$\begin{bmatrix} 3 & -1 & & & & \\ -1 & 3 & -1 & & & \\ & -1 & 3 & & & \\ & & & 3 & -1 & \\ & & & -1 & 3 & -1 \\ & & & & -1 & 3 \end{bmatrix} \begin{Bmatrix} u_1 \\ u_2 \\ u_3 \\ u_4 \\ u_5 \\ u_6 \end{Bmatrix} = \begin{Bmatrix} 10 - (1)33.75 + 66.25 + 25 \\ 20 - (1)43.75 + 61.25 \\ 30 - (1)47.5 + 62.5 + 50 \\ 33.75 - (1)66.25 + 100 + 65 \\ 43.75 - (1)61.25 + 90 \\ 47.5 - (1)62.5 + 80 + 60 \end{Bmatrix} = \begin{Bmatrix} 67.5 \\ 37.5 \\ 95 \\ 132.5 \\ 72.5 \\ 125 \end{Bmatrix},$$

which have the solution (a set of u's)

$$\{35.595 \quad 39.286 \quad 44.762 \quad 66.786 \quad 67.857 \quad 64.286\}.$$

We continue by alternating between vertical and horizontal traverses to get the results shown in Table 8.10. This also shows the steady-state temperatures that are reached after a long time. The steady-state temperatures could have been computed by the methods of Section 8.1. We observe that the A.D.I. algorithm for steady-state temperatures is essentially identical to what we have seen here.

The compensation of errors produced by this alternation of direction gives a scheme that is convergent and stable for all values of r, although accuracy requires that r not be too large. The 3-D analog alternates three ways, returning to each of the three formulas after every third step. [Unfortunately, the 3-D case is not stable for all fixed values of $r > 0$. A variant due to Douglas (1962) is unconditionally stable, however.] When the nodes are renumbered, in each case tridiagonal coefficient matrices result.

Note that the equations can be broken up into two independent subsets, corresponding to the nodes in each column or row. (See the first set of equations of Example 8.11.) This is always true in the A.D.I. method; each row gives a set independent of the equations from the other rows. For columns, the same thing occurs. For very large problems, this is important,

Table 8.10 Results for Example 8.11 using the A.D.I. method

AT START, TEMPS ARE				
0.0000	10.0000	20.0000	30.0000	40.0000
25.0000	50.0000	50.0000	50.0000	50.0000
65.0000	50.0000	50.0000	50.0000	60.0000
110.0000	100.0000	90.0000	80.0000	70.0000
AFTER ITERATION 1 TIME = 26.4 – VALUES ARE				
0.0000	10.0000	20.0000	30.0000	40.0000
25.0000	33.7500	43.7500	47.5000	50.0000
65.0000	66.2500	61.2500	62.5000	60.0000
110.0000	100.0000	90.0000	80.0000	70.0000
AFTER ITERATION 2 TIME = 52.8 – VALUES ARE				
0.0000	10.0000	20.0000	30.0000	40.0000
25.0000	35.5952	69.2857	44.7619	50.0000
65.0000	66.7857	67.8571	64.2857	60.0000
110.0000	100.0000	90.0000	80.0000	70.0000
AFTER ITERATION 3 TIME = 79.2 – VALUES ARE				
0.0000	10.0000	20.0000	30.0000	40.0000
25.0000	35.2679	42.0536	45.8929	50.0000
65.0000	67.1131	65.0893	63.1548	60.0000
110.0000	100.0000	90.0000	80.0000	70.0000
AFTER ITERATION 4 TIME = 105.6 – VALUES ARE				
0.0000	10.0000	20.0000	30.0000	40.0000
25.0000	36.2443	41.8878	46.3832	50.0000
65.0000	66.1366	65.2551	62.6644	60.0000
110.0000	100.0000	90.0000	80.0000	70.0000
STEADY-STATE TEMPERATURES:				
0.0000	10.0000	20.0000	30.0000	40.0000
25.0000	35.8427	41.8323	46.1760	50.0000
65.0000	66.5383	65.3106	62.8716	60.0000
110.0000	100.0000	90.0000	80.0000	70.0000

because it permits the ready overlay of main memory in solving the independent sets. Observe also that each subset can be solved at the same time by parallel processors.

Regions Not Fitted with a Uniform Grid

As discussed in Section 8.1, it is possible to place nodes unevenly and approximate the space derivatives differently, as in Eq. (8.19). Or we might use a different coordinate system (polar or spherical coordinates, for example). However, the most frequently used procedure in such a case is the finite-element method of Chapter 9.

8.3 Hyperbolic Equations

The third class of partial-differential equations, the hyperbolic, is time dependent. They describe vibrations within objects and especially how waves are propagated. Because of this, they are called *wave equations.*

The simplest of the wave equations is that for a vibrating string, the 1-D situation. Another example is that of waves traveling along the length of a long, narrow trough. In 2-D, you might imagine a drum head that is set to vibrating by the musician. The 3-D case is harder to visualize; one could think of a cherry suspended within a bowl of transparent gelatin that moves when the container is tapped with a spoon. In all cases, we want to model the motion and, in the real world, that motion decreases with time due to frictional forces that oppose the motion.

The Vibrating String

We can develop the 1-D wave equation, an example of hyperbolic partial-differential equations, by considering the oscillations of a taut string stretched between two fixed endpoints. Figure 8.11 shows the string with displacements from the straight line between the endpoints greatly exaggerated. The figure shows an element of the string of length dx between points A and B. We use u for the displacements, measured perpendicularly from the straight line between the ends of the string. We focus our attention on the element of the string in Figure 8.11. It is shown enlarged in Figure 8.12, which also shows the angles, α_A and α_B, between the ends of element and the horizontal. (The bending of the element between points A and B is exaggerated as are the displacements.) The figure also indicates that the tension in the stretched string is a force, T. Taking the upward direction as positive, we can write, for the upward forces at each end of the element (these are the vertical components of the tensions).

$$\text{Upward force at point } A = -T\sin(\alpha_A),$$
$$\text{Upward force at point } B = T\sin(\alpha_B).$$

Remembering that Figure 8.12 has displacements and angles greatly exaggerated, the tangents of these angles are essentially equal to the sines. We then can write

$$\text{Upward force at point } A = -T\tan(\alpha_A) = -T\left(\frac{\partial u}{\partial x}\right)_A,$$

$$\text{Upward force at point } B = T\tan(\alpha_B) = T\left(\frac{\partial u}{\partial x}\right)_B = T\left[\left(\frac{\partial u}{\partial x}\right)_A + \frac{\partial}{\partial x}\left(\frac{\partial u}{\partial x}\right)dx\right].$$

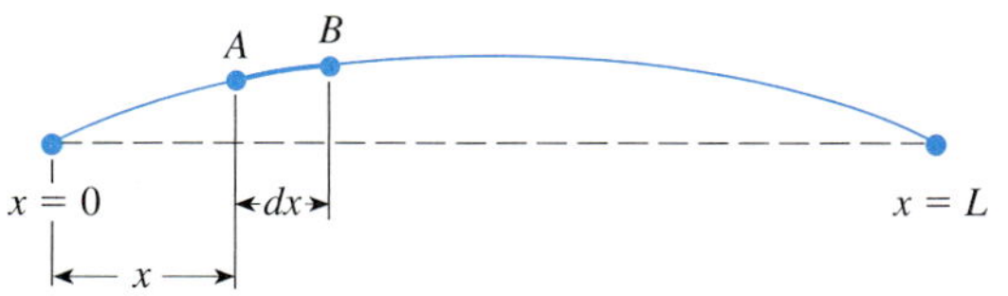

Figure 8.11

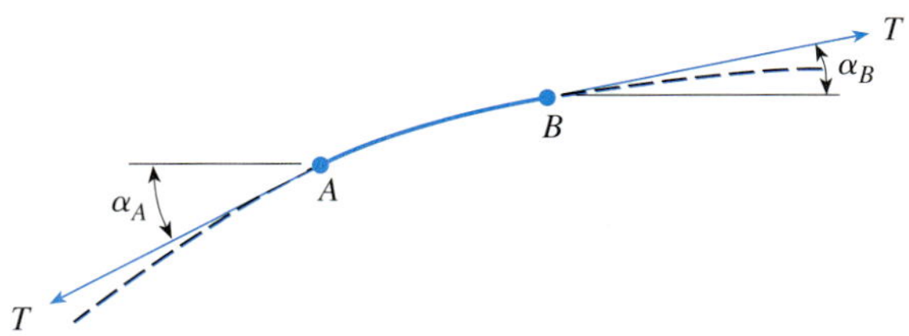

Figure 8.12

The net force acting on the element then is

$$T\left(\frac{\partial^2 u}{\partial x^2}\right) dx.$$

Now, using Newton's law, we equate the force to mass × acceleration (in the vertical direction). Our simplifying assumptions permit us to use $w\ dx$ as the weight (w is the weight per unit length), so

$$\left(T\frac{\partial^2 u}{\partial x^2}\right) dx = \left(w\frac{dx}{g}\right)\left(\frac{\partial^2 u}{\partial t^2}\right) \quad \text{or} \quad \frac{\partial^2 u}{\partial t^2} = \left(\frac{Tg}{w}\right)\frac{\partial^2 u}{\partial x^2}. \tag{8.38}$$

As pointed out in Section 8.1. when Eq. (8.38) is compared to the general form of second-order partial-differential equations, we see that $A = 1$, $B = 0$, and $C = -Tg/w$, and so this falls in the class of hyperbolic equations.

If we have a stretched membrane (like a drum head) instead of a string, the governing equation is

$$\frac{\partial^2 u}{\partial t^2} = \left(\frac{Tg}{w}\right)\nabla^2 u. \tag{8.39}$$

The solution to Eq. (8.38) or Eq. (8.39) must satisfy given boundary conditions along the boundary of the region of interest as well as given initial conditions at $t = 0$. Because the problem is of second order with respect to t, these initial conditions must include both the initial velocity and the initial displacements at all points within the region.

Solving the Vibrating String Problem

We can solve Eq. (8.38) numerically by replacing the derivatives with finite-difference approximations, preferring to use central differences in both cases. If we do this, we get

$$\frac{Tg}{w} * \frac{u_{i+1}^j - 2u_i^j + u_{i-1}^j}{(\Delta x)^2} = \frac{u_i^{j+1} - 2u_i^j + u_i^{j-1}}{(\Delta t)^2}$$

where the subscripts indicate x-values and the superscripts indicate t-values.* (If the boundary conditions involve derivatives, we will approximate them with central differences in the way that we are accustomed.) If we solve for the displacement at the end of the current time step, u_i^{j+1}, we get

$$u_i^{j+1} = \frac{Tg(\Delta t)^2}{w(\Delta x)^2}(u_{i+1}^j + u_{i-1}^j) - u_i^{j-1} + 2\left(1 - \frac{Tg(\Delta t)^2}{w(\Delta x)^2}\right)u_i^j.$$

If we make $Tg(\Delta t)^2/w(\Delta x)^2$ equal to 1, the maximum value that avoids instability, there is considerable simplification:

$$u_i^{j+1} = u_{i+1}^j + u_{i-1}^j - u_i^{j-1}, \qquad \Delta t = \frac{\Delta x}{\sqrt{(Tg/w)}}. \tag{8.40}$$

---

* We again assume evenly spaced nodes and evenly spaced time intervals.

Equation (8.40) shows how one can march through time: To get the new value for u at node i, we add the two u-values last computed at nodes to the right and left and subtract the value at node i at the time step before that. That is fine for the second time step; we have the initial u-values (at $t = 0$) and those for step 1 (at $t = \Delta t$). We also have the necessary information for all subsequent computations. But how do we get the value for the first time step? We seem to need the values of u one time step before the start!

That really is no problem if we recognize that the oscillation of the vibrating string is a periodic function and that the "starting point" is just an arbitrary instant of time at which we happen to know the displacement and the velocity. That suggests that we can get the u-values at $t = -\Delta t$ from the specified initial velocities. If we use a central-difference approximation:

$$\frac{u_i^1 - u_i^{-1}}{2\Delta t} = \frac{\partial u}{\partial t} \qquad \text{at } x_i \text{ and } t = 0,$$

$\partial u/\partial t$ at $t = 0$ is known; it is one of the initial conditions, call it $g(x)$. So we can write

$$u_i^{-1} = u_i^1 - 2g(x)\,\Delta t. \tag{8.41}$$

If we substitute Eq. (8.41) into Eq. (8.40), we have (but for $t = 0$ only),

$$u_i^1 = \frac{1}{2}\,(u_{i+1}^0 + u_{i-1}^0) + g(x)\,\Delta t. \tag{8.42}$$

Our procedure then is to use Eq. (8.42) for the first time step, then use Eq. (8.40) to march on through time after that first step.* As we will see, Eq. (8.40) is not only stable but also can give exact answers. It is interesting that using a value for $Tg(\Delta t)^2/w(\Delta x)^2$ less than 1, while stable, gives results that are less accurate.

An example will illustrate the technique.

EXAMPLE 8.12 A banjo string is 80 cm long and weighs 1.0 gm. It is stretched with a tension of 40,000 g. At a point 20 cm from one end it is pulled 0.6 cm from the equilibrium position and then released. Find the displacements along the string as a function of time. Use $\Delta x = 10$ cm. How long does it take to complete one cycle of motion? From this, compute the frequency of the vibrations.

If Eq. (8.42) is used to begin the calculations and Eq. (8.40) thereafter, the results are as shown in Table 8.11. The initial velocities are zero because the string is just released after being displaced. Observe that the displacements are reproduced every 16 time steps.

Figure 8.13 illustrates how the displacements change with time; it also shows that, after 16 Δt's, the original u-values are reproduced, which will be true for every 16 time steps. Because the original displacements are reproduced every 16 time steps, we can compute the frequency of the vibrations. Each time step is

$$\Delta t = \sqrt{\frac{w}{Tg}} * \Delta x = \sqrt{\frac{1.0/80}{40000 * 980}} * 10 = 0.000179 \text{ sec},$$

* There is a more accurate way to start the computations that we discuss a little later.

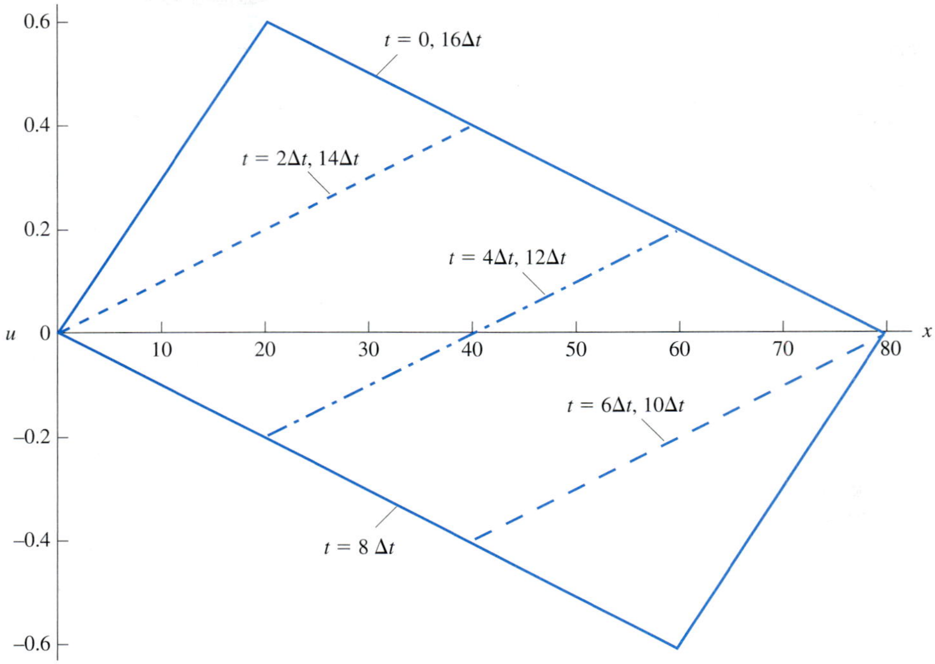

Figure 8.13

and the frequency is

$$f = \frac{1}{16 * 0.000179} = 350 \text{ hertz.}$$

The standard formula from physics is

$$f = \left(\frac{1}{2L}\right)\sqrt{\frac{Tg}{w}} = \left(\frac{1}{160}\right)\sqrt{\frac{40000 * 980}{1.0/80}} = 350 \text{ hertz,}$$

precisely the same!

It seems remarkable that we get exactly the correct frequency, but what about the accuracy of the displacements? We will find that these too are precisely correct, as the next discussion shows. It is also apparent that the computations are stable when $Tg(\Delta t)^2/w(\Delta x)^2$ equals 1.

The D'Alembert Solution

The simple vibrating string problem is one where the analytical solution is readily obtained. This analytical solution is called the *D'Alembert solution*. Consider this expression for $u(x, t)$:

$$u(x, t) = F(x + ct) + G(x - ct), \tag{8.43}$$

where F and G are arbitrary functions.

Table 8.11 Results for vibrating string example

Time steps	*u*-values at *x* = 0	10	20	30	40	50	60	70	80
0	0.00	0.30	0.60	0.50	0.40	0.30	0.20	0.10	0.00
1	0.00	0.30	0.40	0.50	0.40	0.30	0.20	0.10	0.00
2	0.00	0.10	0.20	0.30	0.40	0.30	0.20	0.10	0.00
3	0.00	−0.10	0.00	0.10	0.20	0.30	0.20	0.10	0.00
4	0.00	−0.10	−0.20	−0.10	0.00	0.10	0.20	0.10	0.00
5	0.00	−0.10	−0.20	−0.30	−0.20	−0.10	0.00	0.10	0.00
6	0.00	−0.10	−0.20	−0.30	−0.40	−0.30	−0.20	−0.10	0.00
7	0.00	−0.10	−0.20	−0.30	−0.40	−0.50	−0.40	−0.30	0.00
8	0.00	−0.10	−0.20	−0.30	−0.40	−0.50	−0.60	−0.30	0.00
9	0.00	−0.10	−0.20	−0.30	−0.40	−0.50	−0.40	−0.30	0.00
10	0.00	−0.10	−0.20	−0.30	−0.40	−0.30	−0.20	−0.10	0.00
11	0.00	−0.10	−0.20	−0.30	−0.20	−0.10	0.00	0.10	0.00
12	0.00	−0.10	−0.20	−0.10	0.00	0.10	0.20	0.10	0.00
13	0.00	−0.10	0.00	0.10	0.20	0.30	0.20	0.10	0.00
14	0.00	0.10	0.20	0.30	0.40	0.30	0.20	0.10	0.00
15	0.00	0.30	0.40	0.50	0.40	0.30	0.20	0.10	0.00
16	0.00	0.30	0.60	0.50	0.40	0.30	0.20	0.10	0.00
17	0.00	0.30	0.40	0.50	0.40	0.30	0.20	0.10	0.00
18	0.00	0.10	0.20	0.30	0.40	0.30	0.20	0.10	0.00
19	0.00	−0.10	0.00	0.10	0.20	0.30	0.20	0.10	0.00
20	0.00	−0.10	−0.20	−0.10	0.00	0.10	0.20	0.10	0.00

If we substitute this into the vibrating string equation, which we repeat,

$$\frac{\partial^2 u}{\partial t^2} = \frac{Tg}{w}\frac{\partial^2 u}{\partial x^2}, \tag{8.44}$$

we find that the partial-differential equation is satisfied, because

$$\begin{aligned} \frac{\partial u}{\partial t} &= F'\frac{\partial(x+ct)}{\partial t} + G'\frac{\partial(x-ct)}{\partial t} = cF' - cG', \\ \frac{\partial^2 u}{\partial t^2} &= c^2F'' + c^2G''; \end{aligned} \tag{8.45}$$

$$\begin{aligned} \frac{\partial u}{\partial x} &= F'\frac{\partial(x+ct)}{\partial x} + G'\frac{\partial(x-ct)}{\partial x} = F' + G', \\ \frac{\partial^2 u}{\partial x^2} &= F'' + G''. \end{aligned} \tag{8.46}$$

In Eqs. (8.45) and (8.46), the primes indicate derivatives of the arbitrary functions. Now, substituting these expressions for the second partials into Eq. (8.44), we see that the equation for the vibrating string is satisfied when $c^2 = (Tg/w)$. This means that we can get the solution to Eq. (8.44) if we can find functions F and G that satisfy the initial conditions and the boundary conditions. That too is not difficult. Suppose we are given the initial conditions

$$u(x, 0) = f(x), \qquad \frac{\partial u}{\partial t}(x, 0) = g(x).$$

The combination

$$u(x, t) = \left(\frac{1}{2}\right)[f(x + ct) + f(x - ct)] + \left(\frac{1}{2c}\right)\int_{x-ct}^{x+ct} g(v)\, dv \tag{8.47}$$

is of the same form as Eq. (8.43). It certainly fulfills the boundary conditions, for substituting $t = 0$ in Eq. (8.47) gives $u(x, 0) = f(x)$ and differentiating with respect to t gives

$$\frac{1}{2}[f' * c + f' * (-c)] = 0$$

for the first term of Eq. (8.47), and

$$\left(\frac{1}{2c}\right)\{(c)[g(x + ct)] - (-c)[g(x - ct)]\} = g(x)$$

(when $t = 0$) for the second term.

We have thus shown that the solution to the vibrating string problem is exactly that given by Eq. (8.47). Now we ask "Does the simple algorithm of Eq. (8.40) match Eq. (8.47) for the example problem?" We can show that the answer to the question is yes in the following way.

First, for $Tg(\Delta t)^2/w(\Delta x)^2$ equal to 1, $\Delta x = c\Delta t$. Recalling that u_i^j represents the u-value at $x = x_i = i\Delta x$ and at $t = t_j = j\Delta t$, we see that $ct_j = cj\Delta t = j\Delta x$. If we write $u(x_i, t_j)$ using our subscript/superscript notation, it becomes

$$\begin{aligned} u_i^j &= F(x_i + ct_j) + G(x_i - ct_j) = F(i\Delta x + j\Delta x) + G(i\Delta x - j\Delta x) \\ &= F[(i + j)\, \Delta x] + G[(i - j)\, \Delta x]. \end{aligned} \tag{8.48}$$

Now let us use Eq. (8.48) to write each term on the right-hand side of Eq. (8.40), the algorithm that we used in the example.

$$\begin{aligned} u_{i+1}^j &= F[(i + 1 + j)\, \Delta x] + G[(i + 1 - j)\, \Delta x], \\ u_{i-1}^j &= F[(i - 1 + j)\, \Delta x] + G[(i - 1 - j)\, \Delta x], \\ u_i^{j-1} &= F[(i + j - 1)\, \Delta x] + G[(i - j + 1)\, \Delta x]. \end{aligned}$$

In the example, both F and G are linear functions of x, so that $F(a) + F(b) = F(a + b)$, and the same is true for G. If we combine these terms in Eq. (8.40),

$$\begin{aligned} u_{i+1}^j + u_{i-1}^j - u_i^{j-1} &= F[(i + 1 + j)]\, \Delta x + (i - 1 + j)\, \Delta x - (i + j - 1)\, \Delta x] \\ &\quad + G[(i + 1 - j)\, \Delta x + (i - 1 - j)\, \Delta x - (i - j + 1)\, \Delta x] \\ &= F\{[i + (j + 1)]\, \Delta x\} + G\{[i - (j + 1)]\, \Delta x\} \\ &= u_i^{j+1}, \end{aligned}$$

and the validity of Eq. (8.40) is proved. The important implication from this is that, if we have correct values for the u's at two successive time steps, all subsequent computed values will be correct.

When the Initial Velocity Is Not Zero

Example 8.12 had the string starting with zero velocity. What if the initial velocity is not zero? Equation (8.42) was a very simple way to begin the computations, but it gave correct results only because $g(x)$ was zero in Eq. (8.47). This next example shows that Eq. (8.42) is inadequate when $g(x) \neq 0$ and that there is a better way to begin.

EXAMPLE 8.13 A string is 9 units long. Initially, it is in its equilibrium position (just a straight line between the supports). It is set into motion by striking it so that it has an initial velocity given by $\partial u/\partial t = 3\sin(\pi x/L)$. Take $\Delta x = 1$ unit and let $c^2 = Tg/w = 4$. When the ratio $c^2(\Delta t)^2/(\Delta x)^2 = 1$, the value of Δt is 0.5 time units. Find the displacements at the end of one Δt.

Because $\Delta x = 1$ and the length is 9, the string is divided into nine intervals; there are eight interior nodes. We are to compute the u-values at $t = \Delta t = 0.5$.

As we have seen, Eq. (8.42) is one way to get these starting values. However, looking at Eq. (8.47), we see that there is an alternative technique. If we substitute $t = \Delta t$ in that equation and remember that $c\Delta t = \Delta x$, we get for $u(x_i, \Delta t)$

$$\begin{aligned} u(x_i, \Delta t) &= \frac{1}{2}\,[f(x_i + \Delta x) + f(x_i - \Delta x)] + \left(\frac{1}{2c}\right)\int_{x-\Delta x}^{x+\Delta x} g(v)\,dv \\ &= \frac{1}{2}\,[u_{i+1}^0 + u_{i-1}^0] + \left(\frac{1}{2c}\right)\int_{x-\Delta x}^{x+\Delta x} g(v)\,dv. \end{aligned} \tag{8.49}$$

Equation (8.49) differs from Eq. (8.42) only in the last term. If $g(x)$ = a constant, the last terms are equal, but if $g(x)$ is not constant, we should do the integration in Eq. (8.49). Table 8.12 compares the results of both techniques and also gives the answers from the analytical solution. Only values for x between 1 and 4 are given as the displacements for the right half of the string are the same as for the left half. Simpson's rule was used to do

Table 8.12 Comparison of ways to begin the wave equation at $t = \Delta t$ with $\Delta x = 1$

	u = values from		
x	Eq. (8.42)	Eq. (8.49)	Analytical
1	0.5130	0.5027	0.50267
2	0.9642	0.9448	0.94472
3	1.2990	1.2729	1.27282
4	1.4772	1.4475	1.44740

the integrations. We see from the tabulated results that the values using Eq. (8.49) are almost exactly the same as the analytical values (they are the same within one in the fourth decimal place) but that the results from Eq. (8.42) are less accurate (they each differ by 2.0% from the analytical). We could improve the accuracy with Eq. (8.42) by decreasing the size of Δx (and reducing Δt correspondingly). By making $\Delta x = 0.5$, the errors are reduced fourfold as expected.

Stability of the Solution

We have said that the numerical solution of the vibrating string problem is stable if this ratio is not greater than 1:

$$\frac{Tg\,(\Delta t)^2}{w\,(\Delta x)^2} \leq 1.$$

Because we ordinarily set that ratio equal to 1, it is sufficient to demonstrate stability for that scheme.

For this demonstration, assume that all computations are correct up to a certain point in time, but then an error of size 1 occurs. If the method is stable, that error will not increase. Table 8.13 traces how this single error is propagated. It is allowable to think only of the effect of this single error because for a linear problem that this is, the *principal of superposition* says that we can add together the effects of each of the errors. Equation (8.40) was used and the ends of the string are specified so they are free of error.

Table 8.13 Propagation of single error in numerical solution to wave equation

Initially error-free values	**0.0**	**0.0**	**0.0**	**0.0**	**0.0**	**0.0**	**0.0**
	0.0	0.0	0.0	0.0	0.0	0.0	0.0
Error made here ⟶	0.0	0.0	1.0	0.0	0.0	0.0	0.0
	0.0	1.0	0.0	1.0	0.0	0.0	0.0
	0.0	0.0	1.0	0.0	1.0	0.0	0.0
	0.0	0.0	0.0	1.0	0.0	1.0	0.0
	0.0	0.0	0.0	0.0	1.0	0.0	0.0
	0.0	0.0	0.0	0.0	0.0	0.0	0.0
	0.0	0.0	0.0	0.0	−1.0	0.0	0.0
	0.0	0.0	0.0	−1.0	0.0	−1.0	0.0
	0.0	0.0	−1.0	0.0	−1.0	0.0	0.0
	0.0	−1.0	0.0	−1.0	0.0	0.0	0.0
	0.0	0.0	−1.0	0.0	0.0	0.0	0.0
	0.0	0.0	0.0	0.0	0.0	0.0	0.0
	0.0	0.0	1.0	0.0	0.0	0.0	0.0
	0.0	1.0	0.0	1.0	0.0	0.0	0.0
	0.0	0.0	1.0	0.0	1.0	0.0	0.0

The Wave Equation in Two Dimensions

The finite-difference method can be applied to hyperbolic partial-differential equations in two or more space dimensions. A typical problem is the vibrating membrane. Consider a thin, flexible membrane stretched over a rectangular frame and set to vibrating. As we have seen, the equation is

$$\frac{\partial^2 u}{\partial t^2} = \frac{Tg}{w}\left(\frac{\partial^2 u}{\partial x^2} + \frac{\partial^2 u}{\partial y^2}\right),$$

in which u is the displacement, t is the time, x and y are the space coordinates, T is the uniform tension per unit length, g is the acceleration of gravity, and w is the weight per unit area. For simplification, let $Tg/w = c^2$. Replacing each derivative by its central-difference approximation, and using $h = \Delta x = \Delta y$, gives (we recognize the Laplacian on the right-hand side)

$$\frac{u_{i,j}^{k+1} - 2u_{i,j}^{k} + u_{i,j}^{k-1}}{(\Delta t)^2} = c^2\,\frac{u_{i+1,j}^{k} + u_{i-1,j}^{k} + u_{i,j+1}^{k} + u_{i,j-1}^{k} - 4u_{i,j}^{k}}{h^2}. \tag{8.50}$$

Solving for the displacement at time t_{k+1}, we obtain

$$u_{i,j}^{k+1} = \frac{c^2(\Delta t)^2}{h^2}\left\{\begin{matrix} & 1 & \\ 1 & 0 & 1 \\ & 1 & \end{matrix}\right\} u_{i,j}^{k} - u_{i,j}^{k-1} + \left(2 - 4\frac{c^2(\Delta t)^2}{h^2}\right) u_{i,j}^{k}. \tag{8.51}$$

In Eqs. (8.50) and (8.51), we use superscripts to denote the time. If we let $c^2(\Delta t)^2/h^2 = \frac{1}{2}$, the last term vanishes and we get

$$u_{i,j}^{k+1} = \frac{1}{2}\left\{\begin{matrix} & 1 & \\ 1 & 0 & 1 \\ & 1 & \end{matrix}\right\} u_{i,j}^{k} - u_{i,j}^{k-1}. \tag{8.52}$$

For the first time step, we get displacements from Eq. (8.53), which is obtained by approximating $\partial u/\partial t$ at $t = 0$ by a central-difference approximation involving $u_{i,j}^{1}$ and $u_{i,j}^{-1}$.

$$u_{i,j}^{1} = \frac{1}{4}\left\{\begin{matrix} & 1 & \\ 1 & 0 & 1 \\ & 1 & \end{matrix}\right\} u_{i,j}^{0} + (\Delta t)g(x_i, y_j). \tag{8.53}$$

In Eq. (8.53), $g(x, y)$ is the initial velocity.

It should not surprise us to learn that this ratio $c^2(\Delta t)^2/h^2 = \frac{1}{2}$ is the maximum value for stability, in view of our previous experience with explicit methods. However, in contrast with the wave equation in one space dimension, we do not get exact answers from the numerical procedure of Eq. (8.52), and we further observe that we must use smaller time steps in relation to the size of the space interval. Therefore, we advance in time more slowly. However, the numerical method is straightforward, as the following example will show.

EXAMPLE 8.14 A membrane for which $c^2 = Tg/w = 3$ is streched over a square frame that occupies the region $0 \le x \le 2$, $0 \le y \le 2$, in the xy-plane. It is given an initial displacement described by

$$u = x(2 - x)y(2 - y),$$

and has an initial velocity of zero. Find how the displacement varies with time.

We divide the region with $h = \Delta x = \Delta y = \frac{1}{2}$, obtaining nine interior nodes. Initial displacements are calculated from the initial conditions: $u^0(x, y) = x(2 - x)y(2 - y)$; Δt is taken at its maximum value for stability, $h/(\sqrt{2}\, c) = 0.2041$. The values at the end of one time step are given by

$$u^1_{i,j} = \frac{1}{4}\left\{\begin{matrix} & 1 & \\ 1 & 0 & 1 \\ & 1 & \end{matrix}\right\} u^0_{i,j},$$

Table 8.14 Displacements of a vibrating membrane—finite-difference method: $\Delta t = h/(\sqrt{2}c)$

	Grid location								
t	**(0.5, 0.5)**	**(1.0, 0.5)**	**(1.5, 0.5)**	**(0.5, 1.0)**	**(1.0, 1.0)**	**(1.5, 1.0)**	**(0.5, 1.5)**	**(1.0, 1.5)**	**(1.5, 1.5)**
0	0.5625 (0.5625)	0.750 (0.750)	0.5625	0.750	1.000 (1.000)	0.750	0.5625	0.750	0.5625
0.204	0.375 (0.380)	0.531 (0.536)	0.375	0.531	0.750 (0.755)	0.531	0.375	0.531	0.375
0.408	−0.031 (−0.044)	0.000 (−0.009)	−0.031	0.000	0.062 (0.083)	0.000	−0.031	0.000	−0.031
0.612	−0.375 (−0.352)	−0.531 (−0.539)	−0.375	−0.531	−0.750 (−0.813)	−0.531	−0.375	−0.531	−0.375
0.816	−0.500 (−0.502)	−0.750 (−0.746)	−0.500	−0.750	−1.125 (−1.114)	−0.750	−0.500	−0.750	−0.500
1.021	−0.375 (−0.407)	−0.531 (−0.535)	−0.375	−0.531	−0.750 (−0.691)	−0.531	−0.375	−0.531	−0.375
1.225	−0.031 (−0.015)	0.000 (0.008)	−0.031	0.000	0.062 (0.030)	0.000	−0.031	0.000	−0.031
1.429	0.375 (0.410)	0.531 (0.534)	0.375	0.531	0.750 (0.688)	0.531	0.375	0.531	0.375

Note: Analytical values are in parentheses.

because $g(x, y)$ in Eq. (8.53) is everywhere zero. For succeeding time steps, Eq. (8.52) is used. Table 8.14 gives the results of our calculations. Also shown in Table 8.14 (in parentheses) are analytical values, computed from the double infinite series:

$$u(x, y, t) = \sum_{m=1}^{\infty} \sum_{n=1}^{\infty} B_{mn} \sin\frac{m\pi x}{a} \sin\frac{n\pi y}{b} \cos\left(c\pi t \sqrt{\frac{m^2}{a^2} - \frac{n^2}{b^2}}\right),$$

$$B_{mn} = \frac{16a^2b^2A}{\pi^6 m^3 n^3}(1 - \cos m\pi)(1 - \cos n\pi),$$

which gives the displacement of a membrane fastened to a rectangular framework, $0 \leq x \leq a$, $0 \leq y \leq b$, with initial displacements of $Ax(a - x)y(b - y)$.

We observe that the finite-difference results do not agree exactly with the analytical calculations. The finite-difference values are symmetrical with respect to position and repeat themselves with a regular frequency. The very regularity of the values itself indicates that the finite-difference computations are in error, because they predict that the membrane could emit a musical note. We know from experience that a drum does not give a musical tone when struck; therefore, the vibrations do not have a cyclic pattern of constant frequency, as exhibited by our numerical results.

Decreasing the ratio of $c^2(\Delta t)^2/h^2$ and using Eq. (8.51) gives little or no improvement in the average accuracy; to approach closely to the analytical results, $h = \Delta x = \Delta y$ must be made smaller. When this is done, Δt will need to decrease in proportion, requiring many time steps and leading to many repetitions of the algorithm and extravagant use of computer time. One remedy is the use of implicit methods, which allow the use of larger ratios of $c^2(\Delta t)^2/h^2$. However, with many nodes, this requires large, sparse matrices similar to the Crank–Nicolson method for parabolic equations in two space dimensions. A.D.I. methods have been used for hyperbolic equations—tridiagonal systems result. We do not discuss these methods.

As with other types of partial-differential equations, if the region is not rectangular or if we desire nodes closer together in some parts of the region, it is much preferred to employ the finite-element method, discussed in the next chapter.

Exercises

Section 8.1

1. Show that Eq. (8.2) results if the thickness of the slab varies with position (x, y).

2. Show that Eq. (8.3) applies if both thickness and thermal conductivity vary with position in a slab.

▶ 3. The mixed second derivative $\partial^2 u/(\partial x\, \partial y)$ can be considered as

$$\frac{\partial}{\partial x}\left(\frac{\partial u}{\partial y}\right) = \frac{\partial^2 u}{\partial x\, \partial y} = \frac{\partial}{\partial y}\left(\frac{\partial u}{\partial x}\right).$$

If the nodes are spaced apart a distance h in both the x- and y-directions, show that this derivative can be represented by the pictorial operator

$$\frac{1}{4h^2}\begin{Bmatrix} -1 & & 1 \\ & & \\ 1 & & -1 \end{Bmatrix} + O(h^2).$$

4. What ordering of nodes in Example 8.1 will reduce the band width of the coefficient matrix to seven? Can this be done in more than one way? Can it be reduced to less than seven?

▶ **5.** If d^2u/dx^2 is represented as this fourth-order central-difference formula

$$\frac{d^2u}{dx^2} = \frac{-u_{i+2} + 16u_{i+1} - 30u_i + 16u_{i-1} - u_{i-2}}{12h^2},$$

find the fourth-order operator for the Laplacian. (This requires the function to have a continuous sixth derivative.)

6. Derive the nine-point approximation for the Laplacian of Eq. (8.6).

7. Solve Example 8.1 using the nine-point approximation to the Laplacian. What is the band width of the coefficient matrix if numbered as in Figure 8.2? What ordering of nodes will give the minimum band width? Is this the same as the preferred ordering of Exercise 4?

8. The coefficient matrix of Example 8.1 is bonded and symmetric. If it is solved taking advantage of this structure rather than as it is shown, how many fewer arithmetic operations will be needed to get the solution?

9. A rectangular plate of constant thickness has heat flow only in the x- and y-directions (k is constant). If the top and bottom edges are perfectly insulated and the left edge is at 100° and the right edge at 200°, it is obvious that there is no heat flow except in the x-direction and that temperatures vary linearly with x and are constant along vertical lines.

a. Show that such a temperature distribution satisfies both Eqs. (8.5) and (8.6).

b. Show that the temperatures also satisfy the relation derived in Exercise 5. How should nodes adjacent to the edges be handled?

10. What is the operator equivalent to Eq. (8.7) for the nine-point formula?

▶**11.** Solve for the steady-state temperatures in the plate of the figure when the edge temperatures are as shown. The plate is 10 cm × 8 cm, and the nodal spacing is 2 cm.

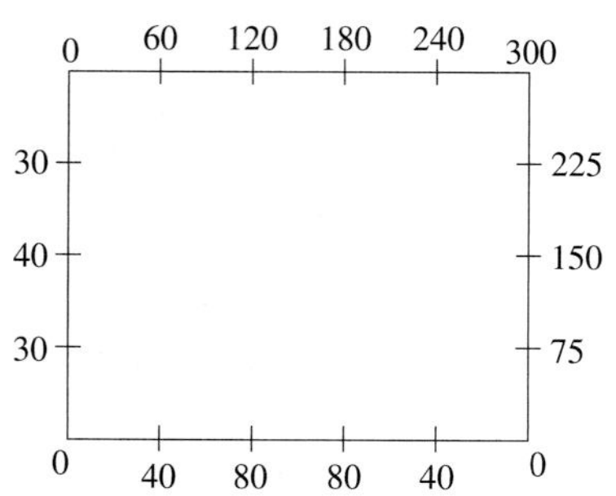

12. Repeat Exercise 11, but with the nine-point formula. Get the solution both by Gaussian elimination and by iteration. How many iterations does it take to reach the solution with a maximum error of 0.001 at any node?

13. The region on which we solve Laplace's equation does not have to be rectangular. We can apply the methods of Section 8.1 to any region where the nodes fall on the boundary. Solve for the steady-state temperatures at the eight interior points of this figure.

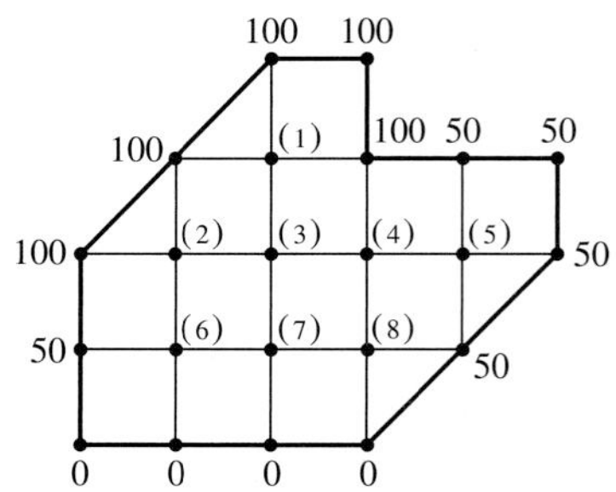

▶**14.** Solve Exercise 11 by Liebmann's method with all elements of the initial u-vector equal to zero. Then repeat with all elements equal to 300°, the upper bound to the steady-state temperatures. Repeat again with the initial values all equal to the arithmetic average of the boundary temperatures. Compare the number of iterations needed to reach a given tolerance for convergence in each case. What is the effect of the tolerance value that is used?

15. Repeat Exercise 14, but now use overrelaxation with the factor given in Eq. (8.8).

16. Find the torsion function ϕ for a 2 in. × 2 in. square bar.

a. Subdivide the region into nine equal squares, so that there are four interior nodes. Because of symmetry, all of the nodes will have equal ϕ-values.

b. Repeat, but subdivide into 36 equal squares with 25 interior nodes. Use the results of part (a) to get starting values for iteration.

17. Solve

$$\nabla^2 u = 2 + x^2 + y^2$$

over a hollow square bar, 5 in. in outside dimension and with walls 2 in. thick (so that the inner square hole is 1 in. on a side). The origin for x and y is the center of the object. On the inner and outer surfaces, $u = 0$.

18. Solve

$$\nabla^2 u = 2 + x^2 + y^2$$

over a hollow square bar whose outside width is 5 in. There is an inner concentric square hole of width 2 in.

(so that the thickness of the wall is 1.5 in.). The origin for x and y is the center of the object. On the outer and inner surfaces, $u = 0$. Space nodes 0.5 in. apart.

19. Can Exercise 18 be solved by iterations as well as by elimination? Repeat it using a method other than the one you used in solving Exercise 18. Which method would be preferred if nodes are spaced very closely together, say, at 0.01 in.?

20. Repeat Exercise 17 but use overrelaxation. Find the optimal overrelaxation factor experimentally. Does this match to that from Eq. (8.8)?

▶21. Solve for the steady-state temperatures in the region of Exercise 13, except now the plate is insulated along the edge where the temperatures were zero. All temperatures on the other edges are as shown in the figure.

22. Solve a modification of Example 8.1, where along every edge there is an outward gradient of -15°C/cm. Is it possible to get a unique solution?

23. Solve Exercise 11 by the A.D.I. method using $\rho = 1.0$. Begin with the initial values equal to the arithmetic average of the boundary temperatures. Compare the number of iterations needed to those required with Liebmann's method (Exercise 14) and with those using S.O.R. with the optimal overrelaxation factor (Exercise 15).

24. Repeat Exercise 16 but now use the A.D.I. method. Vary the value of ρ to find the optimal value experimentally.

▶25. A cube is 7 cm along each edge. Two opposite faces are held at 100°, the other four faces are held at 0°. Find the interior temperatures at the nodes of a 1 cm network. Use the A.D.I. method

26. Repeat Exercise 25, but now the two opposite edges have a mixed condition: The outward normal gradient equals $0.25(u - 18)$, where u is the surface temperature.

Section 8.2

▶27. Suppose that the rod sketched in Figure 8.6 is tapered, with the diameter varying linearly from 2 in. at the left end to 1.25 in. at the right end; the rod is 14 in. long and is made of steel. If 200 BTU/hr of heat flows from left to right (the flow is the same at each x-value along the rod—steady state), what are the values of the gradient at

a. The left end?
b. The right end?
c. $x = 3$ in.?

28. Solve for the temperatures at $t = 2.06$ sec in the 2-cm thick steel slab of Example 8.8 if the initial temperatures are given by

$$u(x, t) = 100 \sin\left(\frac{\pi x}{2}\right).$$

Use the explicit method with $\Delta x = 0.25$ cm. Compare to the analytical solution: $100e^{-0.3738t} \sin(\pi x/2)$.

29. Repeat Exercise 28, but now with Crank–Nicolson.

30. Repeat Exercise 28, but now with the theta method:

a. $\theta = 2/3$.
b. $\theta = 0.878$.
c. $\theta = 1.0$.

▶31. Solve for the temperatures in a cylindrical copper rod that is 8 in. long and whose curved outer surface is insulated so that heat flows only in one direction. The initial temperature is linear from 0°C at one end to 100°C at the other, when suddenly the hot end is brought to 0°C and the cold end is brought to 100°C. Use $\Delta x = 1$ in. and an appropriate value of Δt so that $k\,\Delta t/c\rho(\Delta x)^2 = \frac{1}{2}$. Look up values for k, c, and ρ in a handbook. Carry out the solution for 10 time steps.

32. Repeat Exercise 31, but with $\Delta x = 0.5$ in., and compare the temperature at points 1 in., 3 in., and 6 in. from the cold end with those of the previous exercise. You will need to compute more time steps to match the 10 steps done previously.

You will find it instructive to graph the temperatures for both sets of computations.

33. Repeat Exercise 31 but with $\Delta x = 1.0$ and Δt such that the ratio $k\Delta t/c\rho(\Delta x^2) = 1/4$. Compare the results with both Exercises 31 and 32.

▶34. A rectangular plate 3 in. × 4 in. is initially at 50°. At $t = 0$, one 3-in. edge is suddenly raised to 100°, and one 4-in. edge is suddenly cooled to 0°. The temperature on these two edges is held constant at these temperatures. The other two edges are perfectly insulated. Use a 1 in. grid to subdivide the plate and write the A.D.I. equations for each of the six nodes where unknown temperatures are involved. Use $r = 2$, and solve the equations for four time steps.

35. A cube of aluminum is 4 in. on each side. Heat flows in all three directions. Three adjacent faces lose heat by conduction to a flowing fluid; the other faces are held at a constant temperature different from that of the fluid. Set up the equations that can be solved for the temperature at nodes using the explicit method with a 1-in. spacing between all nodes. How many time steps are needed to reach 15.12 sec using the maximum r-value for stability? (Look up the properties of aluminum in a handbook). How many equations must be solved at each time step?

36. Repeat Exercise 35 for Crank–Nicolson with $r = 1$.

37. Repeat Exercise 35 for the implicit method with $r = 1$.

38. Repeat Exercise 35 for the A.D.I. method with $r = 1$.

▶**39.** Demonstrate that the explicit method is unstable with $r = 0.6$ by performing computation similar to that of Table 8.8

40. Demonstrate that the explicit method is stable if $r = 0.25$ by performing computations similar to that of Table 8.8. Do the errors damp out as rapidly?

▶**41.** Suppose that the end conditions are not $u =$ a constant as in Table 8.8 but rather $u_x = 0$. Demonstrate by performing calculations similar to those in Table 8.8 that the explicit method is still stable for $r = 0.5$ but that the errors damp out much more slowly. Observe that the errors at a later stage become a linear combination of earlier errors.

42. Demonstrate by performing calculations similar to those in Table 8.8 that the Crank–Nicolson method is stable even if $r = 10$. You will need to solve a system of equations in this exercise.

43. Compute the largest eigenvalue of the coefficient matrix in Eq. (8.33) for $r = 0.5$, then for $r = 0.6$. Do you find that the statements in the text relative to eigenvalues are confirmed?

▶**44.** Starting with the matrix form of the implicit method. show that for $A^{-1}B$ none of the eigenvalues exceed 1 in magnitude.

Section 8.3

45. Classify the following as elliptic, parabolic, or hyperbolic.

a. $(Tw_x)_x = p * g$.

b. $(xu_x)_x + u_y = \dfrac{(2 + x + y)}{(1 - x)}$.

c. $kU_{tt} + mU_{xt} - (au_x)_x + bU = f(x, t)$.

d. $(TW_x)_x - k^2W_t = 0,\ W(0) = 0,\ W(L) = 0$.

▶**46.** For what values of x and y is this equation elliptic, parabolic, hyperbolic?

$$(1 + y)u_{xx} + 2(1 - x)u_{xy} - (1 - y)u_{yy} = f(x, y).$$

47. Divide the (x, y)-plane into regions where this equation is elliptic, parabolic, hyperbolic:

$$x^3u_{xx} - 2x^2yu_{xy} + xu_{yy} = x^2 - u_x + u_y.$$

48. What would be the equivalent of Eq. (8.38) if the weight per unit length of the string is not constant but varies, $w = W(x)$?

49. If the banjo string of Example 8.12 is tightened or shortened (as by holding it down on a fret with a finger), the pitch of the sound is higher. What would be the frequency of the sound if the tension is made 42,500 gm and the effective length is 65 cm? Compare your answer to the analytical value that is given by

$$f = (1/2L)\sqrt{(Tg/w)}.$$

50. A vibrating string has $Tg/w = 4\ \text{cm}^2/\text{sec}^2$ and is 48 cm long. Divide the length into subintervals so that $\Delta x = L/8$. Find the displacement for $t = 0$ to $t = L$ if both ends are fixed and the initial conditions are

▶a. $y = x(x - L)/L^2$, $y_t = 0$. (y_t is the velocity.)

b. the string is displaced $+2$ units at $L/4$ and -1 unit at $5L/8$, $y_t = 0$.

▶c. $y = 0$, $y_t = x(L - x)/L^2$. (Use Eq. (8.42).)

▶d. the string is displaced 1 unit at $L/2$, $y_t = -y$.

e. Compare part (a) to the analytical solution,

$$y = \frac{8}{\pi^3}\sum_{n=1}^{\infty}\frac{1}{(2n-1)^3}\sin\left[(2n-1)\frac{\pi x}{L}\right]\cos\left[(4n-2)\frac{\pi t}{L}\right].$$

51. The function u satisfies the equation

$$u_{xx} = u_{tt},$$

with boundary conditions of $u = 0$ at $x = 0$ and $u = 0$ at $x = 1$, and with initial conditions

$$u = \sin(\pi x), \qquad u_t = 0, \qquad \text{for } 0 \le x \le 1.$$

Solve by the finite-difference method and show that the results are the same as the analytical solution,

$$u(x, t) = \sin(\pi x)\cos(\pi t).$$

52. The ends of the vibrating string do not have to be fixed. Solve the equation $u_{xx} = u_{tt}$ with $y(x, 0) = 0$, $y_t(x, 0) = 0$ for $0 \le x \le 1$, and end conditions of

$$y(0, t) = 0, \qquad y(1, t) = \sin\left(\frac{\pi t}{4}\right), \qquad y_x(1, t) = 0.$$

53. If the initial velocity of a vibrating string is not zero, Eq. (8.42) is an inaccurate way to start the solution, so parts (c) and (d) of Exercise 50 are not exact. Repeat these computations, but use Eq. (8.49) employing Simpson's $\frac{1}{3}$ rule. How much difference does this make in the answers?

54. Repeat Exercise 53, but now use more points around x_i. Does this change the answers to Exercise 53?

▶**55.** A string that weighs w lb/ft is tightly stretched between $x = 0$ and $x = L$ and is initially at rest. Each point is given an initial velocity of

$$y_t(x, 0) = v_0 \sin^3\left(\frac{\pi x}{L}\right).$$

The analytical solution is

$$y(x, t) = \frac{v_0 L}{12a\pi}\left(9 \sin\frac{\pi x}{L} \sin\frac{a\pi t}{L} - \sin\frac{3\pi x}{L} \sin\frac{3a\pi t}{L}\right),$$

where $a = \sqrt{Tg/w}$, with T the tension and g the acceleration due to gravity. When $L = 3$ ft, $w = 0.02$ lb/ft, and $T = 5$ lb, with $v_0 = 1$ ft/sec, the analytical formula predicts $y = 0.081$ in. at the midpoint when $t = 0.01$ sec. Solve the problem numerically to confirm this. Does your solution agree with the analytical solution at other values of x and t?

56. Solve the vibrating membrane problem of Example 8.14 with different initial conditions:

$$u(x, y) = 0, \qquad u_t(x, y) = x^2(2 - x)y^2(2 - y).$$

57. Repeat Exercise 56 with the initial conditions reversed:

$$u(x, y) = x^2(2 - x)\, y^2(2 - y), \qquad u_t(x, y) = 0.$$

▶**58.** A membrane is stretched over a frame that occupies the region in the xy-plane bounded by

$$x = 0, \qquad x = 3, \qquad y = 0, \qquad y = 2.$$

At $t = 0$, the point on the membrane at (1, 1) is lifted 1 unit above the xy-plane and then released. If $T = 6$ lb/in. and $w = 0.55$ lb/in.^{2}, find the displacement of the point (2, 1) as a function of time.

59. How do the vibrations of Exercise 58 change if $w = 0.055$ with other parameters remaining the same?

60. The frame holding the membrane of Exercise 58 is distorted by lifting the corner at (3, 2) 1 unit above the xy-plane. (The members of the frame elongate so that the corner moves vertically.) The membrane is set to vibrating in the same way as in Exercise 58. Follow the vibrations through time. [Assume that the rest positions of points on the membrane lie on the two planes defined by the adjacent edges that meet at (0, 0) and at (3, 2).]

Applied Problems and Projects

APP1. A classic problem in elliptic partial-differential equations is to solve $\nabla^2 u = 0$ on a region defined by $0 \le x \le \pi$, $0 \le y \le \infty$, with boundary condition of $u = 0$ at $x = 0$, at $x = \pi$, and at $y = \infty$. The boundary at $y = 0$ is held at $u = F(x)$. This can be quite readily solved by the method of separation of variables, to give the series solution

$$u = \sum_{n=1}^{\infty} B_n e^{-ny} \sin nx,$$

with

$$B_n = 2\int_0^{\pi} F(x) \sin nx\, dx.$$

Solve this equation numerically for various definitions of $F(x)$. (You will need to redefine the region so that $0 \le y \le M$, where M is large enough that changes in u with y at $y = M$ are negligible.) Compare your results to the series solution. You might try

$$F(x) = 100 \sin(x); \qquad F(x) = 4x(\pi - x)/\pi^2; \qquad F(x) = 100(\pi - |2x - \pi|).$$

APP2. The equation

$$2\frac{\partial^2 u}{\partial x^2} + \frac{\partial^2 u}{\partial y^2} - \frac{\partial u}{\partial x} = 2$$

is an elliptic equation. Solve it on the unit square, subject to $u = 0$ on the boundaries. Approximate the first derivative by a central-difference approximation. Investigate the effect of size of Δx on the results, to determine at what size reducing it does not have further effect.

APP3. If you write out the equations for Example 8.1, you will find that the coefficient matrix is symmetric and banded. How can you take advantage of this in solving the equations by Gaussian elimination? Would Gauss–Jordan be preferred? Is the matrix still symmetric and banded if the nodes are numbered by columns?

APP4. A symmetric banded coefficient matrix of width b can be stored in an $n \times (b + 1)/2$ array. Develop an algorithm for reducing the coefficient matrix by Gaussian elimination. Test it with program using a system of width 5. How many fewer operations are needed compared to elimination when the matrix is not compressed ($n \times 5$ versus $n \times 3$)?

APP5. If we want to improve the accuracy of the solution to Example 8.6, there are several alternative strategies, including

a. Recompute with nodes more closely spaced but still in a uniform grid.
b. Use a higher-order approximation, such as Eq. (8.6).
c. Add additional nodes only near the right and left sides because the gradient is large there (see Table 8.2) and errors will be greater.

Discuss the pros and cons of each of these choices. Be sure to consider how boundary conditions will be handled. In part (c), how should equations be written where the nodal spacing changes?

APP6. Solve Example 8.1 by S.O.R. with different values for ω. What value is optimal? How do the starting values that are used affect this?

APP7. A vibrating string, with a damping force-opposing its motion that is proportional to the velocity, follows the equation

$$\frac{\partial^2 y}{\partial t^2} = \frac{Tg}{w} * \frac{\partial^2 y}{\partial x^2} - B\frac{\partial y}{\partial t}.$$

where B is the magnitude of the damping force. Solve the problem if the length of the string is 5 ft with $T = 24$ lb, $w = 0.1$ lb/ft, and $B = 2.0$. Initial conditions are

$$y(x)\big|_{t=0} = \frac{x}{3}, \qquad 0 \le x < 3,$$

$$y(x)\big|_{t=0} = \frac{5}{2} - \frac{x}{2}, \qquad 3 \le x \le 5,$$

$$\left.\frac{\partial y}{\partial t}\right|_{t=0} = x(x - 5).$$

Compute a few points of the solution by difference equations.

APP8. When steel is forged, billets are heated in a furnace until the metal is of the proper temperature, between 2000°F and 2300°F. It can then be formed by the forging press into rough shapes that are later given their final finishing operations. To produce a certain machine part, a billet of size 4 in. × 4 in. × 20 in. is heated in a furnace whose temperature is maintained at 2350°F. You have been requested to estimate how long it will take all parts of the billet to reach a temperature above 2000°F. Heat transfers to the surface of the billet at a very high rate, principally through radiation. It has been suggested that you can solve the problem by assuming that the surface temperature becomes 2250°F instantaneously and remains at that temperature. Using this assumption, find the required heating time.

Because the steel piece is relatively long compared to its width and thickness, it may not introduce significant error to calculate as if it were infinitely long. This will simplify the problem, permitting a two-dimensional treatment rather than a three-dimensional one. Such a calculation should also give a more conservative estimate of heating time. Compare the estimates from two- and three-dimensional approaches.

APP9. After you have calculated the answers to APP8, your results have been challenged on the basis of assuming constant surface temperature of the steel. Radiation of heat flows according to the equation

$$q = E\sigma(u_F^4 - u_S^4) \text{ Btu/(hr * ft}^2)$$

where E = emissivity (use 0.80), σ is the Stefan–Boltzmann constant (0.171×10^{-8} Btu/(hr * ft^2 * °R^4), u_F and u_S are the furnace and surface absolute temperatures, respectively (°F + 460°).

The heat radiating to the surface must also flow into the interior of the billet by conduction, so

$$q = -k\frac{\partial u}{\partial x},$$

where k is the thermal conductivity of steel (use 26.2 Btu/(hr * ft^3 * (°F/ft)) and $(\partial u/\partial x)$ is the temperature gradient at the surface in a direction normal to the surface. Solve the problem with this boundary condition, and compare your solution to that of APP8. (Observe that this is now a nonlinear problem. Think carefully how your solution can cope with it.)

APP10. A horizontal elastic rod is initially undeformed and is at rest. One end, at $x = 0$, is fixed, and the other end, at $x = L$ (when $t = 0$), is pulled with a steady force of F lb/ft^2. It can be shown that the displacements $y(x, t)$ of points originally at the point x are given by

$$\frac{\partial^2 y}{\partial t^2} = a^2\frac{\partial^2 y}{\partial x^2}, \qquad y(0, t) = 0, \qquad \left.\frac{\partial y}{\partial t}\right|_{x=L} = \frac{F}{E}.$$

$$y(x, 0) = 0, \qquad \left.\frac{\partial y}{\partial t}\right|_{t=0} = 0.$$

where $a^2 = Eg/\rho$; E = Young's modulus (lb/ft^2); g = acceleration of gravity; ρ = density (lb/ft^3). Find y versus t for the midpoint of a 2-ft-long piece of rubber for which $E = 1.8 \times 10^6$ and $\rho = 70$ if $F/E = 0.7$.

APP11. A circular membrane, when set to vibrating, obeys the equation (in polar coordinates)

$$\frac{1}{r}\frac{\partial}{\partial r}\left(r\frac{\partial u}{\partial r}\right) + \frac{1}{r^2}\frac{\partial^2 u}{\partial \theta^2} = \frac{w}{Tg}\frac{\partial^2 u}{\partial t^2}.$$

A 3-ft-diameter kettledrum is started to vibrating by depressing the center $\frac{1}{2}$ in. If $w = 0.072$ lb/ ft^2 and $T = 80$ lb/ft, find how the displacements at 6 in. and 12 in. from the center vary with time. The problem can be solved in polar coordinates, or it can be solved in rectangular coordinates using the method of Eq. (8.19) to approximate $\nabla^2 u$ near the boundaries.

APP12. A flexible chain hangs freely, as shown in Figure 8.14. For small disturbances from its equilibrium position (hanging vertically), the equation of motion is

$$x\frac{\partial^2 y}{\partial x^2} + \frac{\partial y}{\partial x} = \frac{1}{g}\frac{\partial^2 y}{\partial t^2}.$$

In this equation, x is the distance from the end of the chain, y is the displacement from the equilibrium position, t is the time, and g is the acceleration of gravity. A 10-ft-long chain is originally hanging freely. It is set into motion by striking it sharply at its midpoint, imparting a velocity there of 1 ft/sec. Find how the chain moves as a result of the blow. If you find you need additional information at $t = 0$, make reasonable assumptions.

APP13. Shipment of liquefied natural gas by refrigerated tankers to industrial nations may become an important means of supplying the world's energy needs. It must be stored at the receiving port, however.

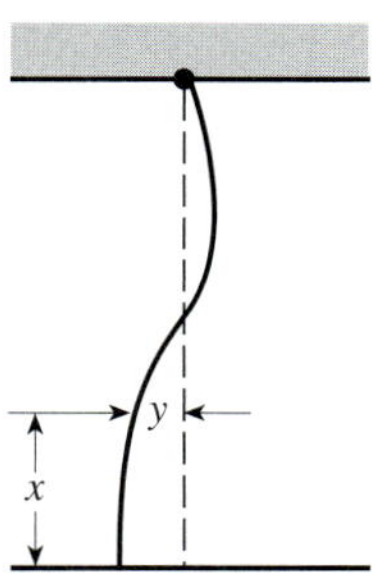

Figure 8.14

[A. R. Duffy and his coworkers (1967) discuss the storage of liquefied natural gas in underground tanks.] A commercial design, based on experimental verification of its feasibility, contemplated a prestressed concrete tank 270 ft in diameter and 61 ft deep, holding some 600,000 bbl of liquefied gas at −258°F. Convection currents in the liquid were shown to keep the temperature uniform at this value, the boiling point of the liquid.

Important considerations of the design are the rate of heat gained from the surroundings (causing evaporation of the liquid gas) and variation of temperatures in the earth below the tank (relating to the safety of the tank, which could be affected by possible settling or frost-heaving.)

The tank itself is to be made of concrete 6 in. thick, covered with 8 in. of insulation (on the liquid side). (A sealing barrier keeps the insulation free of liquid, otherwise, its insulating capacity would be impaired.) The experimental tests showed that there is a very small temperature drop through the concrete: 12°F. This observed 12°F temperature difference seems reasonable in light of the relatively high thermal conductivity of concrete. We expect then that most of the temperature drop occurs in the insulation or in the earth below the tank.

Because the commercial-design tank is very large, if we are interested in ground temperatures near the center of the tank (where penetration of cold will be a maximum), it should be satisfactory to consider heat flowing in only one dimension, in a direction directly downward from the base of the tank. Making this simplifying assumption, compute how long it will take for the temperature to decrease to 32°F (freezing point of water) at a point 8 ft away from the tank wall. The necessary thermal data are

	Insulation	**Concrete**	**Earth**
Thermal conductivity (Btu/(hr * ft * °F))	0.013	0.90	2.6
Density (lb/ft^3)	2.0	150	132
Specific heat (Btu/(lb * °F))	0.195	0.200	0.200

Assume the following initial conditions: temperature of liquid, −258°F; temperature of insulation, −258°F to 72°F (inner surface to outer); temperature of concrete, 72°F to 60°F; temperature of earth, 60°F.

APP14. XYZ Metallurgical has a problem. A slab of steel, 6 ft long, 12 in. wide, and 3 in. thick, must be heat treated and it is a rush job. Unfortunately, their large furnace is down for repairs and the only furnace that can be used will hold just three feet of the slab. It has been proposed that it would be possible to use this furnace if the three feet of the slab that protrude from the furnace are well insulated. (See the figure.) The heat treating requires that all of the slab be held between 950°F and 900°F for at least an hour. The portion that is outside the furnace is covered with a 1 in. thickness of insulation whose thermal conductivity, k, is 0.027 Btu/(hr * ft * °F). Even though you are a new employee, the manager has asked you to determine three things:

(1) Is one inch of this insulation sufficient for all of the slab to reach 900°F with the furnace at 950°F?
(2) If it is, how long will it take for the end of the slab to reach that temperature?
(3) If one inch is insufficient, how much of this same insulation should be used?

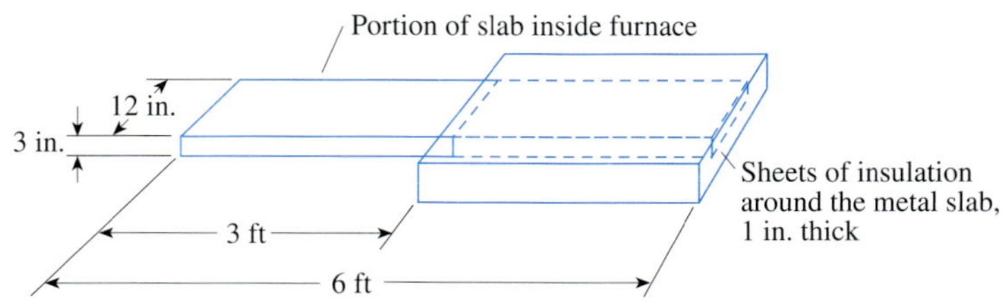

9

Finite-Element Analysis

This chapter remedies the major problem when a partial-differential equation is solved through replacing the derivatives with finite-difference quotients. In that technique, nodes must be in rectangular arrays. In finite-element analysis (often abbreviated FEA), the topic of this chapter, nodes can be spaced in any desired orientation so that a region of any shape can be accommodated. The method is also called the finite-element method (FEM).

In particular, curved boundaries can be approximated by closely spaced nodes. It is not difficult to place modes closer together in subregions where the function is changing rapidly, thus improving the accuracy. A program to carry out FEA is not as simple as for the finite-difference method but software is available to define the region, set up the equations for all types of boundary conditions, and then get the solution. We will describe one of these programs, that from MATLAB in its PDE Toolbox.* This program is most user-friendly—a graphical user interface even lets the user draw a 2-D region on the computer screen.

The basis of FEA is to break up the region of interest into small subregions, the elements. With a 2-D region, elements can be triangles (the most common) or rectangles, even "triangles" or "rectangles" with curved sides. In 3-D, they may be pyramids or bricks. Once the region and its elements are defined, the equations for the system are set up and solved. The equation must, of course, incorporate the boundary conditions, which can be of any type.

The problems that can be solved with FEA include all three types of partial-differential equations, and other problems such as eigenvalue problems, which we do not discuss.

In this chapter, we develop the background for finite elements from a branch of mathematics called the *calculus of variations,* which offers three solution methods that do not use finite elements.

* This toolbox is not a part of the student edition.

Contents of This Chapter

9.1 Mathematical Background

Finite-element analysis is based on some elegant mathematics. We begin the discussion with the *Rayleigh–Ritz method* for solving boundary-value problems. The method comes from that part of mathematics called the *calculus of variations*.

In the Rayleigh–Ritz method, we solve a boundary-value problem by approximating the solution with a finite linear combination of *basis* functions. (We define basis functions and the requirements that are placed on them a little later.) In the calculus of variations, we seek to minimize a special class of functions called *functionals*. The usual form for a functional in problems with one independent variable is

$$I[y] = \int_a^b F\left(x, y, \frac{dy}{dx}\right) dx. \tag{9.1}$$

Observe that $I[y]$ is not a function of x because x disappears when the definite integral is evaluated. The argument y of $I[y]$ is not a simple variable but a function, $y = y(x)$. The square brackets in $I[y]$ emphasize this fact. A functional can be thought of as a "function of functions." The value of the right-hand side of Eq. (9.1) will change as the function $y(x)$ is varied, but when $y(x)$ is fixed, it evaluates to a scalar quantity (a constant). We seek the $y(x)$ that minimizes $I[y]$.

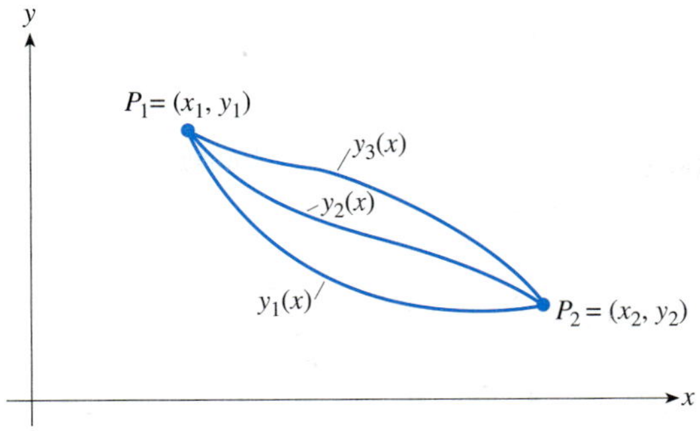

Figure 9.1

Let us illustrate this concept by a very simple example where the solution is obvious in advance—find the function $y(x)$ that minimizes the distance between two points. Although we know what $y(x)$ must be, let's pretend we don't. Figure 9.1 suggests that we are to choose from among the set of curves $y_i(x)$ of which $y_1(x)$, $y_2(x)$, and $y_3(x)$ are representative. In this simple case, the functional is the integral of the distance along any of these curves:

$$I[y] = \int_{x_1}^{x_2} \sqrt{(dx)^2 + (dy)^2} = \int_{x_1}^{x_2} \sqrt{1 + \left(\frac{dy}{dx}\right)^2}\, dx.$$

To minimize $I[y]$, just as in calculus, we set its derivative to zero. There are certain restrictions on all the curves $y_i(x)$. Obviously, each must pass through the points (x_1, y_1) and (x_2, y_2). In addition, for the optimal trajectory, the Euler–Lagrange equation must be satisfied:

$$\frac{d}{dx}\left[\frac{\partial}{\partial y'} F(x, y, y')\right] = \frac{\partial}{\partial y} F(x, y, y'). \qquad \textbf{(9.2)}$$

Applying this to the functional for shortest distance, we have

$$F(x, y, y') = (1 + (y')^2)^{1/2},$$

$$\frac{\partial F}{\partial y} = 0,$$

$$\frac{\partial F}{\partial y'} = \frac{1}{2}(1 + (y')^2)^{-1/2}(2y'),$$

$$\frac{d}{dx}\frac{\partial F}{\partial y'} = \frac{d}{dx}\left(\frac{y'}{\sqrt{1 + (y')^2}}\right) = \frac{\partial F}{\partial y} = 0.$$

[The last comes from Eq. (9.2).]

From this, it follows that

$$\frac{y'}{\sqrt{1 + (y')^2}} = c.$$

Solving for y' gives

$$y' = \sqrt{\frac{c^2}{1 - c^2}} = \text{a constant} = b,$$

and, on integrating,

$$y = bx + a.$$

As stated, $y(x)$ must pass through P_1 and P_2; this condition is used to evaluate the constants a and b.

Let us advance to a less trivial case. Consider this second-order linear boundary-value problem over $[a, b]$:*

$$y'' + Q(x)y = F(x), \qquad y(a) = y_0, \qquad y(b) = y_n. \tag{9.3}$$

(An equation that has y = constant at the endpoints is said to be subject to Dirichlet conditions.) It turns out that the functional that corresponds to Eq. (9.3) is

$$I[u] = \int_a^b \left[\left(\frac{du}{dx} \right)^2 - Qu^2 + 2Fu \right] dx. \tag{9.4}$$

(If the boundary equations involve a derivative of y, the functional must be modified.)

We can transform Eq. (9.4) to Eq. (9.3) through the Euler–Lagrange conditions, so optimizing Eq. (9.4) gives the solution to Eq. (9.3). Observe carefully the benefit of operating with the functional rather than the original equation: We now have only first-order instead of second-order derivatives. This not only simplifies the mathematics but also permits us to find solutions even when there are discontinuities that cause y not to have sufficiently high derivatives.

If we know the solution to our differential equation, substituting it for u in Eq. (9.4) will make $I[u]$ a minimum. If the solution isn't known, perhaps we can approximate it by some (almost) arbitrary function and see whether we can minimize the functional by a suitable choice of the parameters of the approximation. The Rayleigh–Ritz method is based on this idea. We let $u(x)$, which is the approximation to $y(x)$ (the exact solution), be a sum:

$$u(x) = c_0v_0(x) + c_1v_1(x) + \cdots + c_nv_n(x) = \sum_{i=0}^{n} c_iv_i(x). \tag{9.5}$$

There are two conditions on the v's in Eq. (9.5): They must be chosen such that $u(x)$ meets the boundary conditions, and the individual v's must be linearly independent (meaning that no one v can be obtained by a linear combination of the others). We call the v's *trial functions;* the c's and v's are to be chosen to make $u(x)$ a good approximation to the true solution to Eq. (9.3).

If we have some prior knowledge of the true function, $y(x)$, we may be able to choose the v's to closely resemble $y(x)$. Most often we lack such knowledge, and the usual choice then is to use polynomials. We must find a way of getting values for the c's to force $u(x)$ to be close to $y(x)$. We will use the functional of Eq. (9.4) to do this.

---

* This equation is a prototype of many equations in applied mathematics. Equations for heat conduction, elasticity, electrostatics, and so on in a one-dimensional situation are of this form.

If we substitute $u(x)$ as defined by Eq. (9.5) into the functional, Eq. (9.4), we get

$$I(c_0, c_1, \ldots, c_n) = \int_a^b \left[\left(\frac{d}{dx} \Sigma\, c_i v_i \right)^2 - Q(\Sigma\, c_i v_i)^2 + 2F \Sigma\, c_i v_i \right] dx. \tag{9.6}$$

We observe that I is an ordinary function of the unknown c's after this substitution, as reflected in our notation. To minimize I, we take its partial derivatives with respect to each unknown c and set to zero, resulting in a set of equations in the c's that we can solve. This will define $u(x)$ in Eq. (9.5).

We now substitute the $u(x)$ of Eq. (9.5) into the functional. If we partially differentiate with respect to, say, c_i where this is one of the unknown c's, we will get

$$\frac{\partial I}{\partial c_i} = \int_a^b 2 \left(\frac{du}{dx} \right) \frac{\partial}{\partial c_i} \left(\frac{du}{dx} \right) dx - \int_a^b 2Qu \left(\frac{\partial u}{\partial c_i} \right) dx + \int_a^b 2F \left(\frac{\partial u}{\partial c_i} \right) dx, \tag{9.7}$$

where we have broken the integral into three parts.

An example will clarify the procedure.

EXAMPLE 9.1 Solve the equation $y'' + y = 3x^2$, with boundary points (0, 0) and (2, 3.5). (Here $Q = 1$ and $F = 3x^2$.) Use polynomial trial functions up to degree 3. If we define $u(x)$ as

$$u(x) = \frac{7x}{4} + c_2(x)(x - 2) + c_3(x^2)(x - 2), \tag{9.8}$$

we have linearly independent v's. The boundary conditions are met by the first term, and because the other terms are zero at the boundaries, $u(x)$ also meets the boundary conditions. [It is customary to match the boundary conditions with the initial term(s) of $u(x)$ and then make the succeeding terms equal zero at the boundaries, as we have done here.]

Examination of Eq. (9.7) shows that we need these quantities:

$$\frac{du}{dx} = \frac{7}{4} + c_2(2x - 2) + c_3(3x^2 - 4x),$$

$$\frac{\partial}{\partial c_2}\left(\frac{du}{dx}\right) = 2x - 2, \qquad \frac{\partial}{\partial c_3}\left(\frac{du}{dx}\right) = 3x^2 - 4x, \tag{9.9}$$

$$\frac{\partial u}{\partial c_2} = x(x - 2), \qquad \frac{\partial u}{\partial c_3} = x^2(x - 2).$$

We now substitute from Eq. (9.9) into Eq. (9.7). Note that we have two equations, one for the partial with respect to c_2 and the other from the partial with respect to c_3. The results from this step are:

$$\begin{aligned}\frac{\partial I}{\partial c_2}:\quad 0 = &\int_0^2 2\left[\frac{7}{4} + c_2(2x - 2) + c_3(3x^2 - 4x)\right](2x - 2)\, dx \\ &- \int_0^2 2(1)\left[\frac{7x}{4} + c_2(x^2 - 2x) + c_3(x^3 - 2x^2)\right](x^2 - 2x)\, dx \\ &+ 2\int_0^2 (3x^2)(x^2 - 2x)\, dx,\end{aligned} \tag{9.10}$$

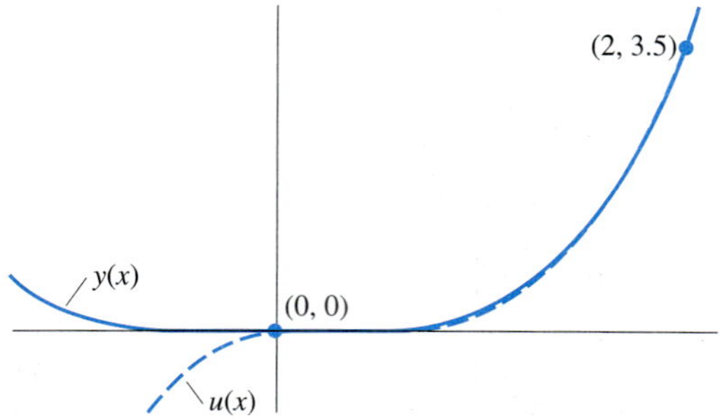

Figure 9.2

$$\frac{\partial I}{\partial c_3}:\quad 0 = \int_0^2 2\left[\frac{7}{4} + c_2(2x - 2) + c_3(3x^2 - 4x)\right](3x^2 - 4x)\,dx$$
$$- \int_0^2 2(1)\left[\frac{7x}{4} + c_2(x^2 - 2x) + c_3(x^3 - 2x^2)\right](x^3 - 2x^2)\,dx \quad \textbf{(9.11)}$$
$$+ 2\int_0^2 (3x^2)(x^3 - 2x^2)\,dx.$$

We now carry out the integrations. Although there are quite a few of them, all are quite simple in our example. With a more complicated $Q(x)$ and $F(x)$, this might require numerical integrations. The result of this step is the pair of equations

$$\frac{16}{5}c_2 + \frac{16}{5}c_3 = \frac{74}{15}, \quad \textbf{(9.12)}$$
$$\frac{16}{5}c_2 + \frac{128}{21}c_3 = \frac{36}{5},$$

which we solve to get the coefficients in our $u(x)$. On expanding, we find that

$$u(x) = \left(\frac{119}{152}\right)x^3 - \left(\frac{46}{57}\right)x^2 + \left(\frac{53}{228}\right)x. \quad \textbf{(9.13)}$$

Figure 9.2 shows that our $u(x)$ agrees well with the exact solution, which is $6\cos(x) + 3(x^2 - 2)$, over the interval $[0, 2]$. Table 9.1 compares computed values and the error of $u(x)$. The largest errors occur near the middle of the interval.

The Collocation Method

There are other ways to approximate $y(x)$ in Example 9.1. The *collocation method* is what is called a "residual method." We begin by defining the residual, $R(x)$, as equal to the left-hand side of Eq. (9.3) minus the right-hand side:

$$R(x) = y'' + Qy - F. \quad \textbf{(9.14)}$$

Table 9.1

x	$y(x)$	$u(x)$	Error	x	$y(x)$	$u(x)$	Error
0.00	0.000	0.000	0.000	1.10	0.352	0.321	0.031
0.10	0.000	0.016	−0.016	1.20	0.494	0.470	0.024
0.20	0.000	0.020	−0.020	1.30	0.675	0.658	0.017
0.30	0.002	0.018	−0.016	1.40	0.900	0.892	0.008
0.40	0.006	0.014	−0.008	1.50	1.174	1.175	−0.001
0.50	0.015	0.012	−0.003	1.60	1.505	1.513	−0.008
0.60	0.032	0.018	0.014	1.70	1.897	1.909	−0.012
0.70	0.059	0.036	0.023	1.80	2.357	2.370	−0.013
0.80	0.100	0.070	0.030	1.90	2.890	2.898	−0.008
0.90	0.160	0.126	0.034	2.00	3.500	3.500	0.000
1.00	0.242	0.208	0.034				

We approximate $y(x)$ again with $u(x)$ equal to a sum of trial functions, usually chosen as linearly independent polynomials, just as for the Rayleigh–Ritz method. We substitute $u(x)$ into $R(x)$ and attempt to make $R(x) = 0$ by a suitable choice of the coefficients in $u(x)$. Of course, normally we cannot do this everywhere in the interval $[a, b]$, so we select several points at which we make $R(x) = 0$. [The number of points where we do this must equal the number of unknown coefficients in $u(x)$.] An example will clarify the procedure.

EXAMPLE 9.2 Solve the same equation as in Example 9.1, but this time use collocation.

The equation we are to solve is

$$y'' + y = 3x^2, \qquad y(0) = 0, \qquad y(2) = 3.5. \tag{9.15}$$

We take $u(x)$ as before to satisfy the boundary conditions:

$$u(x) = \frac{7x}{4} + c_2(x)(x - 2) + c_3(x^2)(x - 2). \tag{9.16}$$

The residual is, after substituting $u(x)$ for $y(x)$.

$$R(x) = u'' + u - 3x^2, \tag{9.17}$$

which becomes, when we differentiate u twice to get u'',

$$R(x) = c_2(2) + c_3(6x - 4) + \frac{7x}{4} + c_2(x^2 - 2x) + c_3(x^3 - 2x^2) - 3x^2. \tag{9.18}$$

Because there are two unknown constants, we can force $R(x)$ to be zero at two points in $[0, 2]$. We do not know which two points will be the best choices, so we arbitrarily take them as $x = 0.7$ and $x = 1.3$. (These points are more or less equally spaced in the interval.)

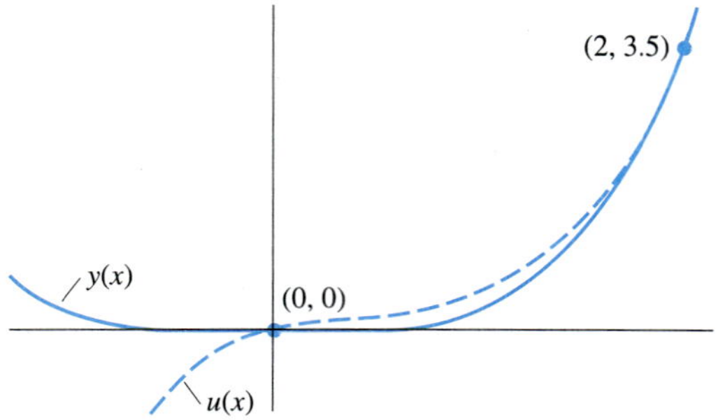

Figure 9.3

Setting $R(x) = 0$ for these choices gives a pair of equations in the c's:

$$\text{From } x = 0.7: \quad \frac{1090c_2 - 437c_3 - 245}{1000} = 0,$$
$$\text{From } x = 1.3: \quad \frac{1090c^2 + 2617c_3 - 2795}{1000} = 0. \tag{9.19}$$

When these are solved for the c's, we get, for $u(x)$,

$$u(x) = \left(\frac{425}{509}\right)x^3 - \left(\frac{61607}{55481}\right)x^2 + \left(\frac{140023}{221924}\right)x, \tag{9.20}$$

in which the coefficients are quite different than in Eq. (9.13). Figure 9.3 shows that this approximation is not as good as that obtained by the Rayleigh–Ritz technique. (But the amount of arithmetic is certainly less! We could improve the approximation by using more terms in $u(x)$.) Table 9.2 compares the approximation with the exact solution.

Table 9.2

x	$y(x)$	$u(x)$	Error	x	$y(x)$	$u(x)$	Error
0.00	0.00	0.00	0.00	1.10	0.35	0.46	−0.11
0.10	0.00	0.05	−0.05	1.20	0.49	0.60	−0.11
0.20	0.00	0.09	−0.09	1.30	0.67	0.78	−0.10
0.30	0.00	0.11	−0.11	1.40	0.90	1.00	−0.10
0.40	0.01	0.13	−0.12	1.50	1.17	1.27	−0.09
0.50	0.02	0.14	−0.13	1.60	1.50	1.59	−0.08
0.60	0.03	0.16	−0.13	1.70	1.90	1.97	−0.07
0.70	0.06	0.18	−0.12	1.80	2.36	2.41	−0.05
0.80	0.10	0.22	−0.12	1.90	2.89	2.92	−0.03
0.90	0.16	0.28	−0.12	2.00	3.50	3.50	0.00
1.00	0.24	0.36	−0.11				

The Galerkin Method

The Galerkin method is widely used, especially in the very popular technique that we will describe in Section 9.2. It is important to know the Galerkin method because of its widespread application.

Like collocation, Galerkin is a "residual method" that uses the $R(x)$ of Eq. (9.14), except that now we multiply $R(x)$ by weighting functions, $W_i(x)$. The $W_i(x)$ can be chosen in many ways, but Galerkin showed that using the individual trial functions, v_i, of Eq. (9.5) is an especially good choice.

Once we have selected the v's for Eq. (9.5), we compute the unknown coefficients by setting the integral over $[a, b]$ of the weighted residual to zero:

$$\int_a^b W_i(x)R(x)\,dx = 0, \qquad i = 0, 1, \ldots, n, \tag{9.21}$$

where $W_i(x) = v_i$. (Observe that using Dirac delta functions for the $W_i(x)$ gives the collocation method.)

Let us use the Galerkin method on the same example as before.

EXAMPLE 9.3 Solve

$$y'' + y = 3x^2, \qquad y(0) = 0, \qquad y(2) = 3.5$$

by the Galerkin method. Use the same $u(x)$ as before:

$$u(x) = \frac{7x}{4} + c_2(x)(x-2) + c_3(x^2)(x-2),$$

so that $v_2 = x(x-2)$ and $v_3 = x^2(x-2)$.

The residual is

$$R(x) = y'' + y - 3x^2,$$

which becomes, after substituting u'' and u for y'' and y, respectively,

$$R(x) = c_2(2) + c_3(6x-4) + \frac{7x}{4} + c_2(x)(x-2) + c_3(x^2)(x-2) - 3x^2. \tag{9.22}$$

We now carry out two integrations (because there are two unknown c's):

$$\text{Using } v_2 \text{ as a } W_i\text{:} \qquad \int_0^2 [x(x-2)] * R(x)\,dx = 0,$$

$$\text{Using } v_3 \text{ as a } W_i\text{:} \qquad \int_0^2 [x^2(x-2)] * R(x)\,dx = 0,$$

which gives two equations in the c's:

$$\begin{aligned} -\frac{24c_2 + 24c_3 - 37}{15} &= 0, \\ -\frac{2(84c_2 + 160c_3 - 180)}{105} &= 0. \end{aligned} \tag{9.23}$$

Solving Eqs. (9.23) for c_2 and c_3 gives

$$u(x) = \left(\frac{101}{152}\right)x^3 - \left(\frac{103}{228}\right)x^2 - \left(\frac{1}{128}\right)x. \tag{9.24}$$

Although Eq. (9.24) looks different from Eq. (9.13), between $x = 0$ and $x = 2$ it gives values for $u(x)$ that have errors about twice those from the Rayleigh–Ritz technique. Equation (9.24) differs from the analytical solution by little more than does the Rayleigh–Ritz equation. [The maximum error of Eq. (9.24) is 0.058; for Eq. (9.13), it is 0.034.]

Although the Rayleigh–Ritz method is slightly more accurate in this example, the Galerkin method is much easier and we never have to find the variational form.

9.2 Finite Elements for Ordinary-Differential Equations

The disadvantages of the methods of the previous section are twofold: Finding good trial functions [the v's in Eq. (9.5)] is not easy, and polynomials [the usual choice when we have no prior knowledge of the behavior of $y(x)$] may interpolate poorly. [We can think of $u(x)$ as an interpolation function between the boundary conditions that also obeys the differential equation.] This is especially true when the interval $[a, b]$ is large.

The remedy to this problem is based on the observation in Chapters 3 and 5 that a function can be approximated by even low-degree polynomials if the polynomial fits the function at values that are closely spaced. We then hope that we can get the solution to a boundary-value problem by applying the Galerkin method to subintervals of $[a, b]$, the boundaries of the equation. It turns out that our hope is fulfilled.

The method that we now describe is called *finite-element analysis (FEA),* also called the *finite-element method (FEM).* The strategy is as follows:

1. Subdivide $[a, b]$ into n subintervals, called *elements,* that join at $x_1, x_2, \ldots, x_{n-1}$. Add to this array $x_0 = a$ and $x_n = b$. We call the x_i the *nodes* of the interval. Number the elements from 1 to n where element (i) runs from x_{i-1} to x_i. The x_i need not be evenly spaced.
2. Apply the Galerkin method to each element separately to interpolate (subject to the differential equation) between the end nodal values, $u(x_{i-1})$ and $u(x_i)$, where these u's are approximations to the $y(x_i)$'s that are the true solution to the differential equation. [These nodal values are actually the c's in our adaptation of Eq. (9.5), the equation for $u(x)$.]
3. Use a low-degree polynomial for $u(x)$. Our development will use a first-degree polynomial, although quadratics or cubics are often used. (The development for these higher-degree polynomials parallels what we will do but is more complicated.)
4. The result of applying Galerkin to element (i) is a pair of equations in which the unknowns are the nodal values at the ends of element (i), the c's. When we have done this for each element, we have equations that involve all the nodal values, which we

combine to give a set of equations that we can solve for the unknown nodal values. (The process of combining the separate *element equations* is called *assembling the system.*)

5. These equations are adjusted for the boundary conditions and solved to get approximations to $y(x)$ at the nodes; we get intermediate values for $y(x)$ by linear interpolation.

We now begin the development. Although it involves several steps, each step is straightforward. The differential equation that we will solve is

$$y'' + Q(x)y = F(x) \qquad \text{subject to boundary conditions at } x = a \text{ and } x = b. \tag{9.25}$$

(We will specify the boundary conditions later.)

Step 1 Subdivide $[a, b]$ into n elements, as discussed. Focus attention on element (i) that runs between x_{i-1} and x_i. To simplify the notation, call the left node L and the right node R.

Step 2 Write $u(x)$ for element (i):

$$\begin{aligned} u(x) &= c_L N_L + c_R N_R = c_L \frac{x-R}{L-R} + c_R \frac{x-L}{R-L} \\ &= c_L \frac{x-R}{-h_i} + c_R \frac{x-L}{h_i}, \end{aligned} \tag{9.26}$$

Recognize that the N's in Eq. (9.26) are really first-degree Lagrangian polynomials. When we use such linear interpolation, the shape functions are often called *hat functions*. (*Chapeau functions,* from the French, is another name.) The reason for this name will become apparent.

Figure 9.4 sketches N_L and N_R within element (i). Because the values of the N's vary (from unity to zero) as x goes from x_L to x_R, they are functions of x. Note also that the c's in Eq. (9.26) are independent of x.

The reason that our N's are called "hat functions" is clear when we look at a sketch of the N's for several adjacent elements in Figure 9.5. Observe that we combine the $N_R(x)$ and $N_L(x)$ of Figure 9.4 that join at x_i into a quantity that we call N_i.

Step 3 Apply the Galerkin method to element (i). The residual is

$$R(x) = y'' + Qy - F = u'' + Qu - F, \tag{9.27}$$

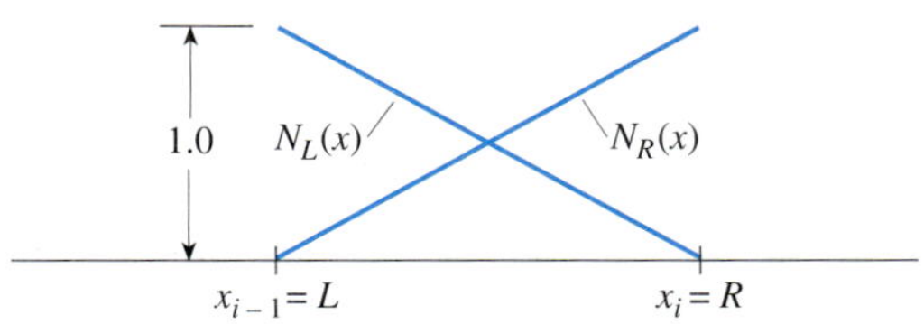

Figure 9.4
N_L and N_R within element (i)

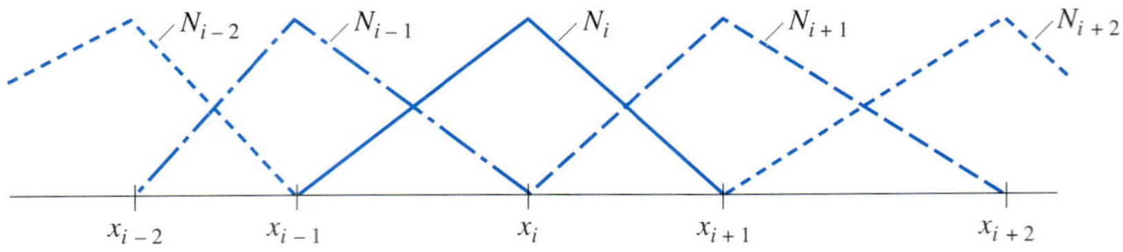

Figure 9.5

where we have substituted $u(x)$ for $y(x)$. The Galerkin method sets the integral of R weighted with each of the N's (over the length of the element) to zero:

$$\int_L^R N_L R(x)\,dx = 0,$$
$$\int_L^R N_R R(x)\,dx = 0. \qquad (9.28)$$

Now expand Eq. (9.28):

$$\int_L^R u'' N_L\,dx + \int_L^R QuN_L\,dx - \int_L^R FN_L\,dx = 0. \qquad (9.29)$$

$$\int_L^R u'' N_R\,dx + \int_L^R QuN_R\,dx - \int_L^R FN_R\,dx = 0. \qquad (9.30)$$

Step 4 Transform Eqs. (9.29) and (9.30) by applying integration by parts* to the first integral. In the second integral, we will take Q out from the integrand as Q_{av}, an average value within the element. We also take F outside the third integral. When this is done, Eq. (9.29) becomes

$$-\int_L^R \left(\frac{du}{dx}\right)\left(\frac{dN_L}{dx}\right) dx + Q_{av}\int_L^R uN_L\,dx - F_{av}\int_L^R N_L\,dx + N_L \left.\frac{du}{dx}\right|_{x=R} - N_L \left.\frac{du}{dx}\right|_{x=L} = 0. \qquad (9.31)$$

In the last two terms of Eq. (9.31), $N_L = 1$ at L and is zero at R, so the equation can be simplified:

$$-\int_L^R \left(\frac{du}{dx}\right)\left(\frac{dN_L}{dx}\right) dx + Q_{av}\int_L^R uN_L\,dx - F_{av}\int_L^R N_L\,dx - \left.\frac{du}{dx}\right|_{x=L} = 0. \qquad (9.32)$$

Doing similarly with Eq. (9.30) gives

$$-\int_L^R \left(\frac{du}{dx}\right)\left(\frac{dN_R}{dx}\right) dx + Q_{av}\int_L^R uN_R\,dx - F_{av}\int_L^R N_R\,dx + \left.\frac{du}{dx}\right|_{x=R} = 0. \qquad (9.33)$$

---

* From $d(UV) = U\,dV + V\,dU$, we have

$$\int_a^b U\,dV = -\int_a^b V\,dU + UV\Big|_a^b, \quad \text{so} \quad \int_L^R u'' N_i\,dx = -\int_L^R u'(dN_i/dx) + N_i u'\Big|_L^R.$$

Step 5 Change signs in Eqs. (9.32) and (9.33); substitute from Eq. (9.26) for u, du/dx, dN_L/dx, and dN_R/dx; and carry out the integrations. We show this separately for each term in Eq. (9.32):

$$\int_L^R \left(\frac{du}{dx}\right)\left(\frac{dN_L}{dx}\right) dx = \int_L^R \left[\frac{-c_L}{h_i} + \frac{c_R}{h_i}\right]\left(\frac{-1}{h_i}\right) dx = \left(\frac{c_L}{h_i^2} - \frac{c_R}{h_i^2}\right)\int_L^R dx \quad (9.34a)$$
$$= \left(\frac{1}{h_i}\right) c_L - \left(\frac{1}{h_i}\right) c_R.$$

$$-Q_{\text{av}}\int_L^R (c_L N_L + c_R N_R) N_L\, dx = -c_L Q_{\text{av}}\int_L^R N_L^2\, dx - c_R Q_{\text{av}}\int_L^R N_R N_L\, dx$$
$$= -c_L Q_{\text{av}}\int_L^R \left(\frac{x-R}{h_i}\right)^2 dx$$
$$-c_R Q_{\text{av}}\int_L^R \left(\frac{x-L}{h_i}\right)\left(\frac{x-R}{-h_i}\right) dx \quad (9.34b)$$
$$= -\left(Q_{\text{av}} * \frac{h_i}{3}\right) c_L - \left(Q_{\text{av}} * \frac{h_i}{6}\right) c_R.$$

$$F_{\text{av}}\int_L^R N_L\, dx = F_{\text{av}}\int_L^R \frac{(x-R)}{-h_i}\, dx = F_{\text{av}} * \frac{h_i}{2}. \quad (9.34c)$$

Doing the same with Eq. (9.33) gives

$$\int_L^R \left(\frac{du}{dx}\right)\left(\frac{dN_R}{dx}\right) dx = -\left(\frac{1}{h_i}\right) c_L + \left(\frac{1}{h_i}\right) c_R. \quad (9.35a)$$

$$-Q_{\text{av}}\int_L^R (c_L N_L + c_R N_R) N_R\, dx = -\left(Q_{\text{av}} * \frac{h_i}{6}\right) c_L - \left(Q_{\text{av}} * \frac{h_i}{3}\right) c_R. \quad (9.35b)$$

$$F_{\text{av}}\int_L^R N_R\, dx = F_{\text{av}} * \frac{h_i}{2}. \quad (9.35c)$$

Step 6 Substitute the result of step 5 [Eqs. (9.34) and (9.35) into Eqs. (9.32) and (9.33), and rearrange to give two linear equations in the unknown c_L and c_R:

$$\left(\frac{1}{h_i} - \frac{Q_{\text{av}}h_i}{3}\right) c_L + \left(\frac{-1}{h_i} - \frac{Q_{\text{av}}h_i}{6}\right) c_R = \frac{-F_{\text{av}}h_i}{2} - \left.\frac{du}{dx}\right|_{x=L},$$
$$\left(\frac{-1}{h_i} - \frac{Q_{\text{av}}h_i}{6}\right) c_L + \left(\frac{1}{h_i} - \frac{Q_{\text{av}}h_i}{3}\right) c_R = \frac{-F_{\text{av}}h_i}{2} + \left.\frac{du}{dx}\right|_{x=R}. \quad (9.36)$$

We call the pair of equations in (9.36) the *element equations*. We can do the same for each element to get n such pairs.

Step 7 Combine (assemble) all the element equations together to form a system of linear equations for the problem. We now recognize that point R in element (i) is precisely the same as point L in element ($i + 1$). Renumber the c's as $c_0, c_1, \ldots, c_n$. Also notice that the gradient (du/dx) must be the same on either side of the join of the elements—that is, $(du/dx)_{x=R}$ in element (i) equals $(du/dx)_{x=L}$ in element ($i + 1$). This means that these terms cancel when we do the assembling except in the first and last equations. (On rare occasions this is not true, but in that case the difference in the two gradients is a known value.)

The result of this step is this set of $n + 1$ equations (numbered from 0 to n),

$$[K]\{c\} = \{b\}, \tag{9.37}$$

where the diagonal elements of $[K]$ are

$$\left(\frac{1}{h_1} - Q_{\text{av},1} * \frac{h_1}{3}\right) \quad \text{in row } 0,$$

$$\left(\frac{1}{h_i} - Q_{\text{av},i} * \frac{h_i}{3}\right) + \left(\frac{1}{h_{i+1}} - Q_{\text{av},i+1} * \frac{h_{i+1}}{3}\right) \quad \text{in rows 1 to } n-1,$$

$$\left(\frac{1}{h_n} - Q_{\text{av},n} * \frac{h_n}{3}\right) \quad \text{in row } n;$$

and elements above and to the left of the diagonal in rows 1 to n are

$$\left(\frac{-1}{h_i} - Q_{\text{av},i} * \frac{h_i}{6}\right).$$

The elements of $\{c\}$ are c_i, $i = 0$ to n.

The elements of $\{b\}$ are

$$-F_{\text{av},1} * \frac{h_1}{2} - \left(\frac{du}{dx}\right)_{x=a} \quad \text{in row } 0,$$

$$-F_{\text{av},i} * \frac{h_i}{2} - F_{\text{av},i+1} * \frac{h_{i+1}}{2} \quad \text{in rows 1 to } n-1,$$

$$-F_{\text{av},n} * \frac{h_n}{2} + \left(\frac{du}{dx}\right)_{x=b} \quad \text{in row } n.$$

In the preceding equations, $Q_{\text{av},i}$ and $F_{\text{av},i}$ are values of Q and F at the midpoints of element (i).

Step 8 Adjust the set of equations from step 6 for the boundary conditions. We will handle two cases: Case (1), a Dirichlet condition is specified—$y(a)$ = constant [and/or $y(b)$ = constant]. Case (2), a Neumann condition is specified—dy/dx = constant at $x = a$ and/or $x = b$. (If $Q = 0$, we cannot have a Neumann condition at both ends, because the solution would be known only to within an additive constant.) [We leave case (3), mixed conditions, as an exercise; it is a modification of case (2).]

Case (1): Dirichlet condition. In this case, c is known at the end node. Suppose this is $y(a) = A$. Then the equation in row 0 is redundant, and so we remove it from the set of equations of step 6. In the next row, we move $k_{10} * A$ to the right-hand side (subtracting this from the element computed in step 6). If the condition is $y(b) = B$, we do the same but with the last and next to last equations.

Case (2): Neumann condition. In this case, c is not known at the end node. Suppose the condition is $dy/dx = A$ at $x = a$. We retain the equation in row 0 and substitute the given value of dy/dx into the right-hand side. If the condition is $dy/dx = B$ at $x = b$, we do the same with the last equation.

Step 9 Solve the set of equations for the unknown c's after adjusting, in step 8, for the boundary conditions. The c's are approximations to $y(x)$ at the nodes. If intermediate values of y are needed between the nodes, we obtain them by linear interpolation.

Examples will clarify the procedure.

EXAMPLE 9.4 Solve $y'' + y = 3x^2$, $y(0) = 0$, $y(2) = 3.5$. (We solved this same equation in Section 9.1.) Subdivide into seven elements that join at $x = 0.4, 0.7, 0.9, 1.1, 1.3$, and 1.6.

Table 9.3 shows the values we need to build the system of equations.

The augmented matrix of the set of equations from step 6 is

$$\left[\begin{array}{cccccccc|c} 2.367 & -2.567 & 0.000 & 0.000 & 0.000 & 0.000 & 0.000 & 0.000 & -0.024 \\ -2.567 & 5.600 & -3.383 & 0.000 & 0.000 & 0.000 & 0.000 & 0.000 & -0.160 \\ 0.000 & -3.383 & 8.167 & -5.033 & 0.000 & 0.000 & 0.000 & 0.000 & -0.328 \\ 0.000 & 0.000 & -5.033 & 9.867 & -5.033 & 0.000 & 0.000 & 0.000 & -0.492 \\ 0.000 & 0.000 & 0.000 & -5.033 & 9.867 & -5.033 & 0.000 & 0.000 & -0.732 \\ 0.000 & 0.000 & 0.000 & 0.000 & -5.033 & 8.167 & -3.383 & 0.000 & -1.378 \\ 0.000 & 0.000 & 0.000 & 0.000 & 0.000 & -3.383 & 5.600 & -2.567 & -2.890 \\ 0.000 & 0.000 & 0.000 & 0.000 & 0.000 & 0.000 & -2.567 & 2.367 & -1.944 \end{array}\right]. \tag{9.38}$$

To adjust for the boundary conditions, we eliminate the first and last equations and subtract $(0)(-2.567) = 0$ from the right-hand side of the top row and subtract $(3.50)(-2.567) = -8.9845$ from the right-hand side of the bottom row to get

$$\left[\begin{array}{ccccccc} 5.600 & -3.383 & 0.000 & 0.000 & 0.000 & 0.000 & -0.160 \\ -3.383 & 8.167 & -5.033 & 0.000 & 0.000 & 0.000 & -0.328 \\ 0.000 & -5.033 & 9.867 & -5.033 & 0.000 & 0.000 & -0.492 \\ 0.000 & 0.000 & -5.033 & 9.867 & -5.033 & 0.000 & -0.732 \\ 0.000 & 0.000 & 0.000 & -5.033 & 8.167 & -3.383 & -1.378 \\ 0.000 & 0.000 & 0.000 & 0.000 & -3.383 & 5.600 & 6.094 \end{array}\right]. \tag{9.39}$$

We have shown Eqs. (9.38) and (9.39) in their full form but observe that the system is tridiagonal (and symmetric, too). It would have been better to store these as 8×4 arrays, so

Table 9.3

Element	L	R	Midpoint	h_i	Q_{av}	F_{av}
1	0	0.4	0.2	0.4	1	0.12
2	0.4	0.7	0.55	0.3	1	0.9075
3	0.7	0.9	0.8	0.2	1	1.92
4	0.9	1.1	1.0	0.2	1	3
5	1.1	1.3	1.2	0.2	1	4.32
6	1.3	1.6	1.45	0.3	1	6.3075
7	1.6	2.0	1.8	0.4	1	9.72

Table 9.4

x	$u(x)$	**Anal.**	**Error**
0.000	0.0000	0.0000	0
0.400	−0.0024	0.0064	0.0088
0.700	0.0433	0.0591	0.0158
0.900	0.1371	0.1597	0.0226
1.100	0.3232	0.3516	0.0284
1.300	0.6419	0.6750	0.0331
1.600	1.4759	1.5048	0.0289
2.000	3.5000	3.5031	0.0031

that Eq. (9.39) would be

$$\left[\begin{array}{cccc:c} 0 & 5.600 & -3.383 & & -0.160 \\ -3.383 & 8.167 & -5.033 & & -0.328 \\ -5.033 & 9.867 & -5.033 & & -0.492 \\ -5.033 & 9.867 & -5.033 & & -0.732 \\ -5.033 & 8.167 & -3.383 & & -1.378 \\ -3.383 & 5.600 & 0 & & 6.094 \end{array}\right].$$

When the system of Eq. (9.39) is solved, we get the solution as shown in Table 9.4. The table also shows the analytical solution and the errors of our computation. The table indicates that closer spacing of nodes near $x = 1$ would give better answers.

This next example solves a boundary-value problem, which has Neumann conditions at the ends of the region.

EXAMPLE 9.5 Solve $y'' - (x + 1)y = e^{-x}(x^2 - x + 2)$ subject to Neumann conditions of

$$y'(2) = 0, \qquad y'(4) = -0.036631.$$

Use four elements of equal lengths. Compare to the analytical solution

$$y(x) = e^{-x}(x - 1).$$

Table 9.5 gives values that we need to set up the equations.

Table 9.5

Element	L	R	**Midpoint**	h_i	Q_{av}	F_{av}
1	2	2.5	2.25	0.5	−3.25	−0.5072
2	2.5	3.0	2.75	0.5	−3.75	−0.4355
3	3.0	3.5	3.25	0.5	−4.25	−0.3611
4	3.5	4.0	3.75	0.5	−4.75	−0.2896

The initial matrix of equations is

$$\left[\begin{array}{ccccc|c} 2.542 & -1.729 & 0.000 & 0.000 & 0.000 & 0.127 \\ -1.729 & 5.167 & -1.688 & 0.000 & 0.000 & 0.236 \\ 0.000 & -1.688 & 5.333 & -1.646 & 0.000 & 0.199 \\ 0.000 & 0.000 & -1.646 & 5.500 & -1.604 & 0.163 \\ 0.000 & 0.000 & 0.000 & -1.604 & 2.792 & 0.072 \end{array}\right].$$

After adjusting for boundary conditions, we get

$$\left[\begin{array}{ccccc|c} 2.542 & -1.729 & 0.000 & 0.000 & 0.000 & 0.127 \\ -1.729 & 5.167 & -1.688 & 0.000 & 0.000 & 0.236 \\ 0.000 & -1.688 & 5.333 & -1.646 & 0.000 & 0.199 \\ 0.000 & 0.000 & -1.646 & 5.500 & -1.604 & 0.163 \\ 0.000 & 0.000 & 0.000 & -1.604 & 2.792 & 0.036 \end{array}\right],$$

and the solution is

x	$u(x)$	**Anal.**	**Error**
2.000	0.1334	0.1353	1.896E-03
2.500	0.1228	0.1231	3.230E-04
3.000	0.0996	0.0996	−2.031E-05
3.500	0.0758	0.0755	−3.280E-04
4.000	0.0564	0.0549	−1.432E-03

We again observe that the system is tridiagonal and symmetric. Will this be true for mixed-boundary conditions?

Other Kinds of Elements

We have used the "hat" functions because they are the simplest kind of element for a 1-D problem. This may not always be adequate. This sketch shows why.

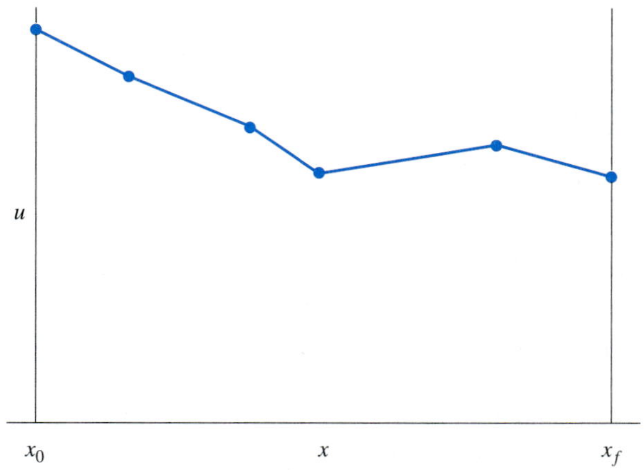

The dots in the figure represent the computed u-values at the nodes and the straight lines that connect the dots are the supposed intermediate values (because the value of u is assumed to vary linearly within the elements). This certainly does not correctly describe the function $u(x)$! (Of course, if the elements are much smaller, the broken line would be a better approximation.) How can we remedy this defect in the procedure?

The obvious way is to use a shape function that better approximates the true function. A quadratic shape function would involve three nodes: two at the ends and an intermediate one. This will force the solution to behave like a parabola that passes through the three nodes. A cubic shape function might be used; it would involve four nodes. Using such higher-order shape functions adds some modest complications to the development of the procedure, but the general approach is identical to what we have shown. When such higher-order shape functions are used, the integration of the analogs of Eqs. (9.31) or (9.5) is usually done by numerical methods. The resulting system of equations is no longer tridiagonal, but the nonzero elements of the coefficient matrix are still clustered about the main diagonal.

There is still a flaw with such higher-order elements in the 1-D problem—the curve for u is not continuous in slope at the juncture of the elements. This flaw could be eliminated if the shape function were splines, but this is not often done because that complicates the procedure significantly.

Sometimes the user of finite-element analysis wants to know the *flux* in addition to the u-values. In one dimension, the flux is $k\partial/\partial x$. (In the sketch, this is proportional to the slopes.) With the linear hat function, the flux values are discontinuous and have larger errors than $u(x)$. Higher-order shape functions help to overcome this.

Convergence Rates

A numerical analyst is always greatly concerned about the accuracy of the numerical solutions. For finite-element-method procedures, the question is "How do the errors decrease when we put nodes closer together?" It can be shown that, with linear elements, errors are of order $O(h^2)$, where h is a measure of the nodal spacing. Quadratic elements give an $O(h^3)$ accuracy; higher orders than two give even better accuracy as the mesh is refined. As we have said, the rate of decrease is a limit value that is achieved only as the h-value gets very small. (The rate of decrease in the errors with quadratic or higher-order shape functions also depends on the integration method used in formulating the system of equations.) Also, a very interesting phenomenon has been observed in studies of the effect of smaller h-values on accuracy—errors may not always decrease uniformly as the spacing is made closer. As a mesh is gradually refined, anomalous behavior can occur.

It is frequently the case that nodes are not uniformly spaced—in fact, this is one of the major advantages of the finite-element method; we can put nodes closer together where the solution $u(x)$ varies most rapidly to get better accuracy in that subregion. This imposes a problem about how best to define "h" in the order of convergence. We shall not pursue this but only remark that if a mesh is refined to improve the accuracy of the numerical solution, we must refine it everywhere, not just in selected parts of the region.

The errors in the flux do not decrease as rapidly with smaller spacing of the nodes. For a linear shape function, errors decrease as $O(h)$.

Burnett (1987) is an excellent reference.

9.3 Finite Elements for Partial-Differential Equations

When an elliptic partial-differential equation is solved by replacing derivatives with finite-difference approximations, there are serious difficulties if the region is irregular. Analytical methods are also very awkward to apply in such cases.

The finite-element method has no such problems. As we saw in Section 9.2 for one-dimensional boundary-value problems, nodes can be placed wherever the problem solver desires with the finite-element method. This is also true for two- and three-dimensional regions. They can be placed along any boundary so as to approximate it closely. It is the method of choice for solving elliptic partial-differential equations on regions of arbitrary shape.

Although setting up the equations that solve partial-differential equations is no easy task, computer programs are available that do so. It is important to understand how this method works, although this text cannot give everything that today's scientists and engineers might want to know. Our treatment will give a basic knowledge.

The introduction to finite elements in Sections 9.1 and 9.2 is important background for what we shall do here. Recall that two ways of applying variational methods to subdivisions of the region of interest were presented: Rayleigh–Ritz and Galerkin. The first of these minimized the functional for the problem by setting partial derivatives to zero; the second by setting integrals of a weighted residual to zero. The two methods are equivalent for most problems, and both can be used for elliptic equations. We choose the former, in part to provide variety from the presentation in Section 9.2.

The elliptic equation that we will solve in this section is

$$u_{xx} + u_{yy} + Q(x, y)u = F(x, y) \tag{9.40}$$

on region R that is bounded by curve L, with boundary conditions

$$u(x, y) = u_0 \quad \text{on } L_1, \qquad \frac{\partial u}{\partial n} = \alpha u + \beta \quad \text{on } L_2,$$

where $\partial u/\partial n$ is the outward normal gradient.

Observe that we have Dirichlet conditions on some parts of the boundary and mixed boundary conditions on other parts. For our notation, we will use $u(x, y)$ as the exact solution to Eq. (9.40) and $v(x, y)$ as our approximation to $u(x, y)$. Although the finite-element method is most often used when the region is three dimensional, we will simplify the development by doing it in only two dimensions.

Here is our plan of attack:

Step 1. Find the functional that corresponds to the partial-differential equation. This is well known for a large class of problems.

Step 2. Subdivide the region into subregions (elements). Although many kinds of elements can be used, our treatment will consider only triangular elements. The elements must span the entire region and approximate the boundary relatively closely. Every node (the vertices of our triangular elements) and every side of the triangles must be common with adjacent elements except for sides on the boundaries.

Step 3. Write an interpolating relation that gives values for the dependent variable within an element based on the values at the nodes (the vertices of the triangles). We will use linear interpolation from the three nodal values for the element. We will write the interpolation function as the sum of three terms; each term involves a quantity c_i, the value of $v(x, y)$ at a node.

Step 4. Substitute the interpolating relation into the functional, and set the partial derivatives of the functional with respect to each c to zero. This gives three equations, with the c's as unknowns for each element.

Step 5. Combine together (assemble) the element equations of step 4 to get a set of system equations. Adjust these for the boundary conditions of the problem, then solve. This will give the values for the unknown nodal values, the c's, that are approximations to $u(x, y)$ at the nodes. We can get approximations to $u(x, y)$ at intermediate points in the region by using the interpolating relations.

We will discuss each of these five steps in turn. We will provide simple examples to illustrate some of them.

Step 1. Find the Functional

For Eq. (9.40) the functional is well known:

$$I[u] = \iint_{\text{Region}} \left[\left(\frac{\partial u}{\partial x} \right)^2 + \left(\frac{\partial u}{\partial y} \right)^2 - Qu^2 + 2Fu \right] dx\, dy - \oint_{L_2} [\alpha u^2 + 2\beta u] dL. \tag{9.41}$$

It is possible to develop Eq. (9.41) using the Galerkin technique. Workers in the field of structural analysis usually derive it from the principle of virtual work. We will take it as a given.

Step 2. Subdivide the Region

As stipulated, we will use triangular elements, which will be defined by our choice of nodes. The placement of nodes is, in part, an art. In general, we place nodes close together in subregions where the solution is expected to vary rapidly. It is advantageous to make the sides run in the direction of the largest gradient. Along the curved parts of the boundary, nodes should be placed so that a side of the triangle closely approximates the boundary.

Some of these recommendations depend on knowing the nature of the solution in advance. Often, however, a better placement for the nodes can be accomplished after some preliminary computations or after preliminary trials using the finite-element method with nodes placed arbitrarily.

The chore of defining the nodes' coordinates is facilitated by computer programs that allow the user to place nodes with a pointing device on a graphical display of the region. These programs even permit rotating 3-D regions or looking at cross sections. Once the nodes have been located, the program connects them to create the elements.

Computer routines are available that can divide any given planar region into triangles automatically, but they usually do not have the expertise of an experienced engineer.

Step 3. Write the Interpolating Relations

This part of the development is longer than the previous one. As stated, we will use a linear relation. Figure 9.6a is a sketch of typical element (i) whose nodes are numbered r, s, and t in counterclockwise direction. The nodal values are c_r, c_s, and c_t as indicated in Figure 9.6b. The shaded triangle shows how $v(x, y)$ varies within the element.

Within typical element (i), we write

$$v(x, y) = N_r c_r + N_s c_s + N_t c_t = \sum_{j=r,s,t} N_j c_j \tag{9.42}$$

$$= (N_r \;\; N_s \;\; N_t) \begin{Bmatrix} c_r \\ c_s \\ c_t \end{Bmatrix} = (N)\{c\},$$

where the N's (called *shape functions*) will be defined so that $v(x, y)$ at an interior point is a linear interpolation from the nodal values, the c's. We have shown in Eq. (9.42) that $v(x, y)$ can be expressed as the product of vectors (N) and $\{c\}$. (We use parentheses to enclose a row vector and curly brackets to enclose a column vector in this section.) Vector and matrix notation will be useful. We will indicate matrix M by $[M]$.

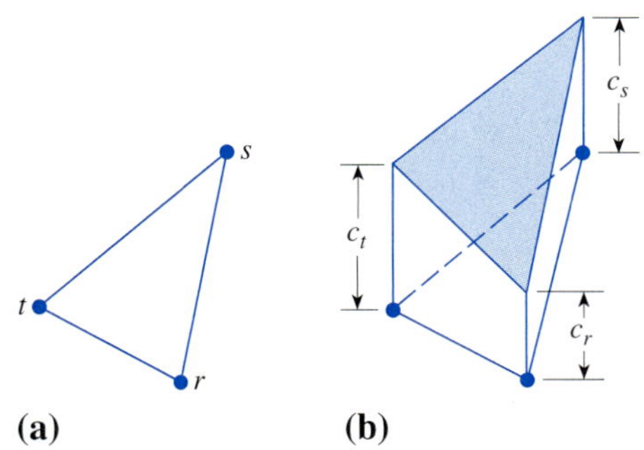

Figure 9.6

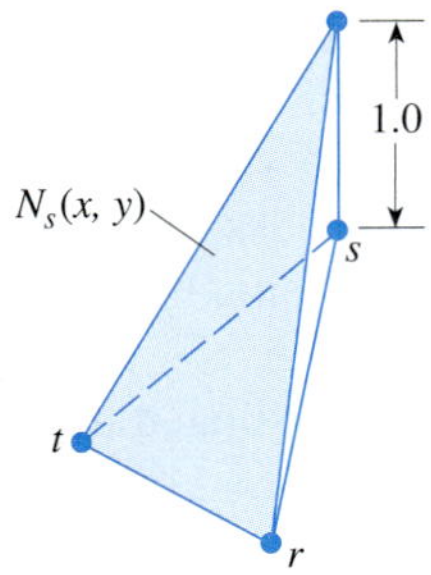

Figure 9.7

Figure 9.6b suggests that $v(x, y)$ lies on the plane above the element that passes through the nodal values. Equation (9.42) does not define $v(x, y)$ outside of element (i); there will be similar expressions for the other elements, but their N's and c's will differ.

A sketch of the entire region would not show $v(x, y)$ as a plane. Instead, it would be a surface composed of planar facets, each in a plane above an element. $v(x, y)$ for the entire region is continuous, but $v'(x, y)$ is not. (This is one of the flaws in our choice of element. Some other element definitions do not have this flaw.)

Another name for the N's of Eq. (9.42) is *pyramid function.* The reason for this name is illustrated in Figure 9.7, where N_s of Figure 9.6 is drawn. Its height at node s is unity and zero at the other nodes. It looks like an unsymmetrical pyramid whose base is the element with its apex directly above node s. The other two N's are similar. It is obvious that the N's are functions of x and y and that the c's are independent of x and y. We now develop expressions for the N's.

Because $v(x, y)$ varies linearly with position within the element, an alternative way to write the linear relation is

$$v(x, y) = a_1 + a_2x + a_3y = (1 \quad x \quad y)\{a\}, \tag{9.43}$$

which must agree with the nodal values when $(x, y) = (x_j, y_j)$, $j = r, s, t$. Hence

$$\begin{aligned} v \text{ at } r: &\quad c_r = a_1 + a_2x_r + a_3y_r, \\ v \text{ at } s: &\quad c_s = a_1 + a_2x_s + a_3y_s, \\ v \text{ at } t: &\quad c_t = a_1 + a_2x_t + a_3y_t. \end{aligned}$$

This is a system of equations

$$[M]\{a\} = \{c\} \qquad \text{(curly brackets show a column vector)}, \tag{9.44}$$

where

$$[M] = \begin{bmatrix} 1 & x_r & y_r \\ 1 & x_s & y_s \\ 1 & x_t & y_t \end{bmatrix}, \qquad \{a\} = \begin{Bmatrix} a_1 \\ a_2 \\ a_3 \end{Bmatrix}, \qquad \{c\} = \begin{Bmatrix} c_r \\ c_s \\ c_t \end{Bmatrix}.$$

Solving for $\{a\}$:

$$\{a\} = [M^{-1}]\{c\}.$$

The inverse of M is not difficult to find

$$[M^{-1}] = \frac{1}{2(\text{Area})}\begin{bmatrix} (x_s y_t - x_t y_s) & (x_t y_r - x_r y_t) & (x_r y_s - x_s y_r) \\ (y_s - y_t) & (y_t - y_r) & (y_r - y_s) \\ (x_t - x_s) & (x_r - x_t) & (x_s - x_r) \end{bmatrix}, \tag{9.45}$$

with 2(Area) = det(M). The value of the determinant is the sum of the elements in row 1 of Eq. (9.45) within the brackets. Area is the area of the triangular element.* You should verify that $[M^{-1}][M] = [I]$ to ensure that Eq. (9.45) truly gives the inverse matrix.

To apply the interpolating function to the minimization of the quadratic functional, Eq. (9.41), we prefer to write $v(x, y)$ in terms of the shape functions of Eq. (9.42). This task is easy. We have, from Eqs. (9.42) and (9.43),

$$v(x, y) = a_1 + a_2 x + a_3 y = (1 \quad x \quad y)\{a\}$$
$$= (1 \quad x \quad y)[M^{-1}]\{c\}.$$

However, in terms of N (from Eq. 9.42),

$$v(x, y) = (N)\{c\}. \tag{9.46}$$

Comparing the two expressions, we have

$$(N) = (1 \quad x \quad y)[M^{-1}], \tag{9.47}$$

where M^{-1} is given by Eq. (9.45). Observe carefully that Eq. (9.47) says that each N is a linear function of x and y of the form

$$N_j = A_j + B_j x + C_j y, \qquad j = r, s, t \tag{9.48}$$

and that the coefficients are in column j of $[M^{-1}]$.

We have found the expressions for the N's. Before we go on, we digress to show an example that will clarify this step.

EXAMPLE 9.6 For the triangular element shown in Figure 9.8 with nodes r, s, and t in counterclockwise order, find $\{a\}$, $\{N\}$, and $v(0.8, 0.4)$.

Node	x	y	c
r	0	0	100
s	2	0	200
t	0	1	300

Before we do any computations, we can find $v(0.8, 0.4)$ by inspection. (See Fig. 9.8.) Point 1 is at (0, 0.4), so v there is 180 by linear interpolation between nodes r and t. Similarly, v at point (2) is 240. The point (0.8, 0.4) is $\frac{2}{3}$ of the distance from points 1 and 2,

* That Area $= \frac{1}{2}$ det (M) is shown in most books on vectors where the cross product is explained.

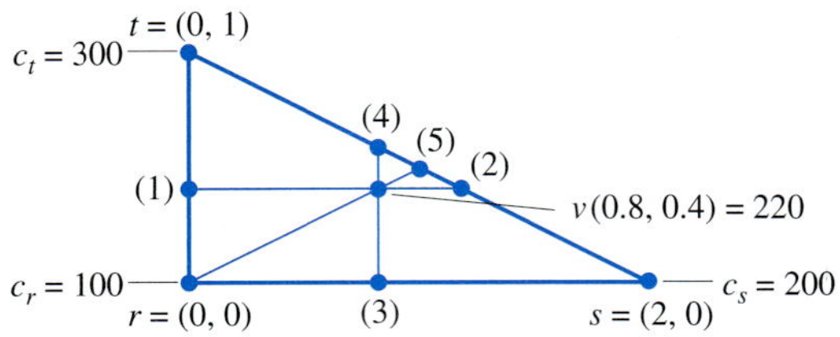

Figure 9.8

so $v(0.8, 0.4) = 180 + \frac{2}{3}(240 - 180) = 220$. We get the same result by interpolating between points 3 and 4, and between node r and (5).

To get $\{a\}$ we first compute $[M^{-1}]$:

$$[M] = \begin{bmatrix} 1 & 0 & 0 \\ 1 & 2 & 0 \\ 1 & 0 & 1 \end{bmatrix}, \qquad [M^{-1}] = \begin{bmatrix} 1 & 0 & 0 \\ -0.5 & 0.5 & 0 \\ -1 & 0 & 1 \end{bmatrix}.$$

Then we compute

$$\{a\} = [M^{-1}]\{c\} = \begin{bmatrix} 1 & 0 & 0 \\ -0.5 & 0.5 & 0 \\ -1 & 0 & 1 \end{bmatrix} \begin{Bmatrix} 100 \\ 200 \\ 300 \end{Bmatrix} = \begin{Bmatrix} 100 \\ 50 \\ 200 \end{Bmatrix},$$

giving $v(x, y) = 100 + 50x + 200y$. (You should confirm that this gives the correct values at each of the nodes.) If we substitute $x = 0.8$, $y = 0.4$, we get $v = 220$, as we should.

From Eq. (9.47),

$$(N) = (1 \quad x \quad y)[M^{-1}] = (1 - 0.5x - y, \quad 0.5x, \quad y).$$

In other words, we have

$$N_r = 1 - 0.5x - y,$$
$$N_s = 0.5x,$$
$$N_t = y.$$

(You should confirm that these also have the proper values at each of the nodes.) It is important to notice that the coefficients of the N's [the A_i, B_i, and C_i of Eq. (9.48)] can be read directly from the columns of $[M^{-1}]$.

In what follows we will need the partial derivatives of the N's with respect to x and to y. From $N_j = A_j + B_jx + C_jy$, we see that these are constants that can be read from rows 2 and 3 of $[M^{-1}]$ in column j.

At this point we know how to write $v(x, y)$ within the single triangular element (i) as $v(x, y) = (N^{(i)})\{c^{(i)}\}$. (The superscripts (i) tell which element is being considered whenever this is necessary.) We now stipulate that $(N^{(i)}) \equiv (0)$ everywhere outside of element (i). Therefore we can write

$$v(x, y) = \sum_{i=\text{all elements}} (N^{(i)})\{c^{(i)}\}.$$

This is a mathematical statement of the previous observation that $v(x, y)$ is a surface composed of joined planar facets.

We are now ready for step 4 of our plan. This too is lengthy but each portion is easy.

Step 4. Substitute $v(x, y)$ into the Functional and Minimize

We continue to work with typical element (i) whose nodes are r, s, and t. Repeating Eq. (9.46), our $v(x, y)$ is

$$v(x, y) = (N)\{c\} = N_r c_r + N_s c_s + N_t c_t,$$

where the N's are given by Eqs. (9.47) and (9.45). Recall that each N is $A_j + B_j x + C_j y$ with the coefficients given by the elements in column j of $[M^{-1}]$.

Our objective is to develop a set of three equations for element (i), which is, in matrix form,

$$[K]\{c\} = \{b\},$$

and which is a prototype of similar equations for all other elements.

When we substitute $v(x, y)$ for element (i) into the functional of Eq. (9.41), we get

$$I(c_r, c_s, c_t) = \iint_{\text{element }(i)} \left[\left(\frac{\partial v}{\partial x}\right)^2 + \left(\frac{\partial v}{\partial y}\right)^2 - Qv^2 + 2Fv\right] dx\,dy - \oint_{L_2} [\alpha v^2 + 2\beta v^2]\, dL. \tag{9.49}$$

[I is now an ordinary function of the c's. The integral is only over the area of element (i) because the N's that define $v(x, y)$ in element (i) are zero outside of (i). The last term appears only if element (i) has a side on the boundary. Actually, we will postpone handling this last term for now and handle it as an adjustment to the equations after they have been developed.]

We minimize I by setting the three partials (with respect to each of the three c's) to zero. We now develop expressions for these partials. First consider $\partial I/\partial c_r$.

For the first term in the integrand:

$$\frac{\partial}{\partial c_r}\left[\left(\frac{\partial v}{\partial x}\right)^2\right] = 2\left(\frac{\partial v}{\partial x}\right)\left(\frac{\partial}{\partial c_r}\left(\frac{\partial v}{\partial x}\right)\right).$$

However,

$$\frac{\partial v}{\partial x} = \left(\frac{\partial N_r}{\partial x}\right) c_r + \left(\frac{\partial N_s}{\partial x}\right) c_s + \left(\frac{\partial N_t}{\partial x}\right) c_t$$
$$= B_r c_r + B_s c_s + B_t c_t,$$

by virtue of Eq. (9.48). (The B's come from row 2 of $[M^{-1}]$.)

Also, $\partial/\partial c_r(\partial v/\partial x) = B_r$ because c_s and c_t are independent of c_r. Hence

$$\frac{\partial}{\partial c_r}\left[\left(\frac{\partial v}{\partial x}\right)^2\right] = 2(B_r^2 + B_rB_s + B_rB_t),$$

The result for the second term is similar:

$$\frac{\partial}{\partial c_r}\left[\left(\frac{\partial v}{\partial y}\right)^2\right] = 2(C_r^2 + C_rC_s + C_rC_t),$$

where the C's come from row 3 of $[M^{-1}]$.

We next consider the Q term. Q is independent of c_r, so

$$\frac{\partial}{\partial c_r}(-Qv^2) = -Q\left[2v\left(\frac{\partial v}{\partial c_r}\right)\right] - 2Q(N_rc_r + N_sc_s + N_tc_t)(N_r).$$

Finally we work with the F term. F is independent of c_r, so

$$\frac{\partial}{\partial c_r}(2Fv) = 2F\left(\frac{\partial v}{\partial c_r}\right) = 2F(N_r).$$

Putting all this together, we have

$$\begin{aligned}\frac{\partial I}{\partial c_r} = 0 = &\iint_{(i)} 2[B_r^2c_r + B_rB_sc_s + B_rB_tc_t]\,dx\,dy \\ &+ \iint_{(i)} 2[C_r^2c_r + C_rC_sc_s + C_rC_tc_t]\,dx\,dy \\ &- \iint_{(i)} 2Q[N_r^2c_r + N_rN_sc_s + N_rN_tc_t]\,dx\,dy \\ &+ \iint_{(i)} 2FN_r\,dx\,dy.\end{aligned} \tag{9.50}$$

Equation (9.50) really is a formulation with the c's unknown:

$$K_{rr}c_r + K_{rs}c_s + K_{rt}c_t = b_r, \tag{9.51}$$

where

$$K_{rr} = \iint_{(i)} 2B_r^2\,dx\,dy + \iint_{(i)} 2C_r^2\,dx\,dy - \iint_{(i)} 2QN_r^2\,dx\,dy,$$

$$K_{rs} = \iint_{(i)} 2B_rB_s\,dx\,dy + \iint_{(i)} 2C_rC_s\,dx\,dy - \iint_{(i)} 2QN_rN_s\,dx\,dy,$$

$$K_{rt} = \iint_{(i)} 2B_rB_t\,dx\,dy + \iint_{(i)} 2C_rC_t\,dx\,dy - \iint_{(i)} 2QN_rN_t\,dx\,dy,$$

$$b_r = -\iint_{(i)} 2FN_r\,dx\,dy.$$

[Remember, we postpone handling the last part of Eq. (9.49).]

Now we recognize that the B's and C's of Eq. (9.51) are constants, so we can bring them out from under the integral sign. If we use average values for Q and F within the element, we can also bring these out as their average values. (The best average value to use is the value of Q and F at the centroid of the triangular element.)

This means that we have to evaluate these five integrals:

$$I_1: \quad \iint_{(i)} dx\,dy$$

$$I_2: \quad \iint_{(i)} N_r^2\,dx\,dy \qquad I_3: \quad \iint_{(i)} N_r N_s\,dx\,dy \qquad I_4: \quad \iint_{(i)} N_r N_t\,dx\,dy$$

$$I_5: \quad \iint_{(i)} N_r\,dx\,dy.$$

The first of these is easy: I_1 = Area of the element, which we already know from having computed $[M^{-1}]$. The other integrals are laborious to compute directly, but there is a useful formula for the integral of the product of powers of linear functions over a triangle:

$$\iint_{\text{(triangle)}} N_r^{\ell} N_s^{m} N_t^{n}\ dx\,dy = \frac{2\ell!m!n!}{(\ell + m + n + 2)!}\text{(Area)}.$$

Using this with the proper values for the exponents, ℓ, m, and n, gives

$$I_2 = \frac{\text{(Area)}}{6},$$

$$I_3 = \frac{\text{(Area)}}{12},$$

$$I_4 = \frac{\text{(Area)}}{12},$$

$$I_5 = \frac{\text{(Area)}}{3}.$$

The terms in Eq. (9.51) are then

$$K_{rr}c_r + K_{rs}c_s + K_{rt}c_t + b_r, \qquad \textbf{(9.52)}$$

where

$$K_{rr} = 2\text{(Area)}\left(B_r^2 + C_r^2 - \frac{Q_{\text{av}}}{6}\right),$$

$$K_{rs} = 2\text{(Area)}\left(B_r B_s + C_r C_s - \frac{Q_{\text{av}}}{12}\right),$$

$$K_{rt} = 2\text{(Area)}\left(B_r B_t + C_r C_t - \frac{Q_{\text{av}}}{12}\right),$$

$$b_r = -2\text{(Area)}\left(\frac{F_{\text{av}}}{3}\right).$$

If we do the same with $\partial I/\partial c_s$ and $\partial I/\partial c_t$, we get two more equations in the c's for element (i). All together we have three equations, which we call the *element equations.* We simplify these equations somewhat by omitting the common factor of 2 for each of them to get

$$[K]\begin{Bmatrix} c_r \\ c_s \\ c_t \end{Bmatrix} = \begin{Bmatrix} b_r \\ b_s \\ b_t \end{Bmatrix},$$

where

$$\text{(diagonals)} \qquad K_{jj} = \text{Area}\left[B_j^2 + C_j^2 - \frac{Q_{\text{av}}}{6}\right], \qquad j = r, s, t;$$

$$\text{(off-diagonals)} \qquad K_{jk} = \text{Area}\left[B_jB_k + C_jC_k - \frac{Q_{\text{av}}}{12}\right] \qquad \begin{cases} j \neq k, \\ j = r, s, t, \\ k = r, s, t; \end{cases}$$

$$\text{(rhs)} \qquad b_j = -\text{Area}\left[\frac{F_{\text{av}}}{3}\right], \qquad j = r, s, t.$$

Observe that $[K]$ is symmetrical: $K_{ij} = K_{ji}$.

Here is an example to clarify the formation of the element equations.

EXAMPLE 9.7 Find the element equations for the element of Example 9.6 if $Q(x, y) = (xy)/2$ and $F(x, y) = x + y$.

The nodes are $(x, y) = (0, 0)$, $(2, 0)$, and $(0, 1)$. We had, for $[M^{-1}]$,

$$[M^{-1}] = \begin{bmatrix} 1 & 0 & 0 \\ -0.5 & 0.5 & 0 \\ -1 & 0 & 1 \end{bmatrix}.$$

Area = 1. Centroid is at $x = (0 + 2 + 0)/3 = \frac{2}{3}$, $y = (0 + 0 + 1)/3 = \frac{1}{3}$. $Q_{\text{av}} = \left(\frac{2}{3}\right)\left(\frac{1}{3}\right)/2 = \frac{1}{9}$. $F_{\text{av}} = \frac{2}{3} + \frac{1}{3} = 1$.

Using Eq. (9.52), we find that the element equations are

$$\begin{bmatrix} 1.2315 & -0.2592 & -1.0092 \\ -0.2592 & 0.2315 & -0.0093 \\ -1.0092 & -0.0093 & 0.9815 \end{bmatrix}\begin{Bmatrix} c_r \\ c_s \\ c_t \end{Bmatrix} = \begin{Bmatrix} -0.3333 \\ -0.3333 \\ -0.3333 \end{Bmatrix}.$$

We are now ready for step 5 of the plan.

Step 5. Assemble the Equations, Adjust for Boundary Conditions, Solve

There are three separate operations in step 5: (i) assemble the equations, (ii) adjust for boundary conditions, and (iii) solve the equations.

(i) Do the Assembly As we have seen, there are three equations for every element. However, some or all nodes of element (i) are shared with other elements; the c-value for a shared node then appears in the equations of all elements that share the node. Combining all of the element equations will create a global system coefficient matrix with as many rows and columns as there are nodes in the system. We combine (assemble) the system matrix in the following way.

Suppose there are n nodes in the system. Number the nodes in order, from 1 to n. Associate the number of each node with the row and column of every element matrix where the c for that node appears on the diagonal. Also associate the node numbers with the rows and columns of the system matrix in the same way.

We get the entry in row (i) and column (j) of the system matrix by adding the values from row (i) of every element matrix that has row (i), then adding these in the columns where the column-node numbers match. We also add the b_i's from these rows to get the b_i of the system matrix. An example will clarify this operation.

EXAMPLE 9.8 Suppose there are five nodes that define three elements, as shown in Figure 9.9 with the element matrices of Eq. (9.53a, b, c) below. Construct the system matrix without adjusting for boundary conditions.

$$\text{Element [1]} \quad \begin{matrix} (1) \rightarrow \\ (2) \rightarrow \\ (4) \rightarrow \end{matrix} \begin{bmatrix} K_{11} & K_{12} & K_{13} \\ K_{21} & K_{22} & K_{23} \\ K_{31} & K_{32} & K_{33} \end{bmatrix} \begin{Bmatrix} c_1 \\ c_2 \\ c_4 \end{Bmatrix} \begin{matrix} = \\ = \\ = \end{matrix} \begin{Bmatrix} b_1 \\ b_2 \\ b_3 \end{Bmatrix} \tag{9.53a}$$
$$\begin{matrix} \uparrow & \uparrow & \uparrow \\ (1) & (2) & (4) \end{matrix}$$

$$\text{Element [2]} \quad \begin{matrix} (2) \rightarrow \\ (3) \rightarrow \\ (4) \rightarrow \end{matrix} \begin{bmatrix} K_{11} & K_{12} & K_{13} \\ K_{21} & K_{22} & K_{23} \\ K_{31} & K_{32} & K_{33} \end{bmatrix} \begin{Bmatrix} c_2 \\ c_3 \\ c_4 \end{Bmatrix} \begin{matrix} = \\ = \\ = \end{matrix} \begin{Bmatrix} b_1 \\ b_2 \\ b_3 \end{Bmatrix} \tag{9.53b}$$
$$\begin{matrix} \uparrow & \uparrow & \uparrow \\ (2) & (3) & (4) \end{matrix}$$

$$\text{Element [3]} \quad \begin{matrix} (4) \rightarrow \\ (5) \rightarrow \\ (1) \rightarrow \end{matrix} \begin{bmatrix} K_{11} & K_{12} & K_{13} \\ K_{21} & K_{22} & K_{23} \\ K_{31} & K_{32} & K_{33} \end{bmatrix} \begin{Bmatrix} c_4 \\ c_5 \\ c_1 \end{Bmatrix} \begin{matrix} = \\ = \\ = \end{matrix} \begin{Bmatrix} b_1 \\ b_2 \\ b_3 \end{Bmatrix} \tag{9.53c}$$
$$\begin{matrix} \uparrow & \uparrow & \uparrow \\ (4) & (5) & (1) \end{matrix}$$

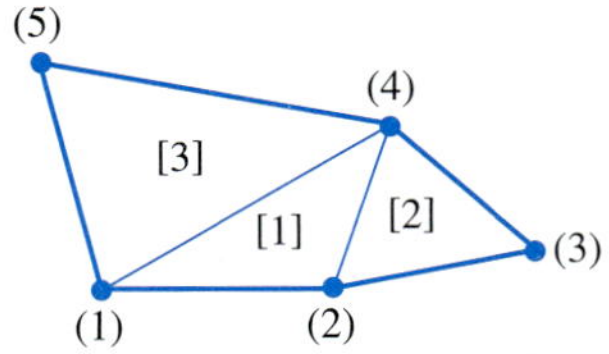

Figure 9.9

[The rows and columns of Eqs. (9.53) could have been in a different order, although we always go counterclockwise around the element in selecting the nodes.]

We construct the system matrix as follows, where the superscripts indicate the element number that provides the value:

For row 1	For row 2
col 1: $K_{11}^{[1]} + K_{33}^{[3]}$	col 1: $K_{21}^{[1]}$
col 2: $K_{12}^{[1]}$	col 2: $K_{22}^{[1]} + K_{11}^{[2]}$
col 3: 0	col 3: $K_{12}^{[2]}$
col 4: $K_{13}^{[1]} + K_{31}^{[3]}$	col 4: $K_{23}^{[1]} + K_{13}^{[2]}$
col 5: $K_{32}^{[3]}$	col 5: 0
b: $b_1^{[1]} + b_3^{[3]}$	b: $b_2^{[1]} + b_1^{[2]}$
and so forth.	

[A zero appears in column 3 of row 1 because node (3) is not in any element that includes node (1). A zero appears in column 5 of row 2 because node (5) is not in any element that includes node (2).]

Once the system matrix has been assembled, we make the adjustments for boundary conditions.

(ii) Adjust for Boundary Conditions There are two types of boundary conditions: non-Dirichlet conditions on some parts of the boundary (L_2) and Dirichlet conditions on other parts (L_1). We will always select nodes such that only one of the two types of conditions pertains to any side of the element. Hence there will always be a node at the point where the two types join. These two types of boundary conditions require two separate adjustments. We prefer to apply the adjustment for a boundary condition that involves the outward normal derivative to the system equations first and then do the adjustment for Dirichlet conditions.

Adjusting for Non-Dirichlet Conditions Non-Dirichlet conditions (those that involve the outward normal derivative) are associated, not with the nodes, but with sides of the triangular elements, sides that correspond to part of L_2 of Eq. (9.41). Consider an element that has a non-Dirichlet condition on one side that lies between nodes r and s. The effect of the boundary condition on the equations comes from differentiating the last term of Eq. (9.49) with respect to the c's. However, if we take α and β out from the integrand as average values, we see from Eq. (9.49) that they are of the same form as the Q and F terms except they are line integrals rather than area integrals. That similarity lets us immediately write the result of the differentiation with respect to c_r as

$$-2\alpha_{\text{av}} \oint (N_r^2 c_r + N_r N_s c_s)\, dL - 2\beta_{\text{av}} \oint N_r\, dL, \qquad (9.54)$$

where the line integrals are along the side between nodes r and s. It is important to note that we have not included c_t because node t is not on the side we are considering. (The average values of α and β should be taken at the midpoint of the side.) When we integrate, we have

$$-2\alpha_{\text{av}} L\left(\frac{c_r}{3} + \frac{c_s}{6}\right) - 2\beta_{\text{av}}\frac{L}{2}. \tag{9.55}$$

(It is easy to evaluate the integrals when we remember that the N's are linear from 1 to 0 between the two nodes.)

Precisely the same relations result when we differentiate with respect to node c_s, except the roles of r and s are interchanged in Eqs. (9.54) and (9.55). The net result would be to add $2\beta L/2$ to the right-hand sides of the rows for c_r and c_s. We also would subtract the multipliers of c_r and c_s in Eq. (9.55) from the coefficients for c_r and c_s in row r. The similar equations from the partials with respect to c_s provide subtractions from the coefficients in row s.

Recall, however, that we canceled a 2 factor when we constructed the element equations and so we must do so here. We make this adjustment to the element equations for every element that has a derivative condition on a side.

Adjusting for Dirichlet Conditions For every node that appears on the boundary where there is a Dirichlet condition, the u-value is specified. We insert this known value in place of the c of that node in every equation where it appears and transpose to the right-hand side. (Actually, if the node number is m, all entries in column m of the matrix are multiplied by the value and subtracted from the right-hand side of the corresponding row.) We also remove the row corresponding to the number of the known node from the set of equations. (The column for this node has already been "removed" by being transferred to the right-hand side.)

Removing the rows for those nodes with a Dirichlet condition is simplified in a computer program if these rows are at the top or the bottom of the matrix. There are other ways to handle Dirichlet conditions that avoid having to remove the rows.

This completes our construction of the system equations.

(iii) Getting the Solution We solve the system in the usual way, perhaps preferring an iterative procedure if the system is large.

An example, intentionally simple, follows.

EXAMPLE 9.9 The region shown in Figure 9.10 has four nodes. It is divided into just two elements. The values for u are specified at nodes (3) and (4), and the outward normal gradient is specified on three sides as indicated. The equation we are to solve is

$$u_{xx} + u_{yy} - \left(\frac{y}{10}\right)u = \frac{x}{4} + y - 12.$$

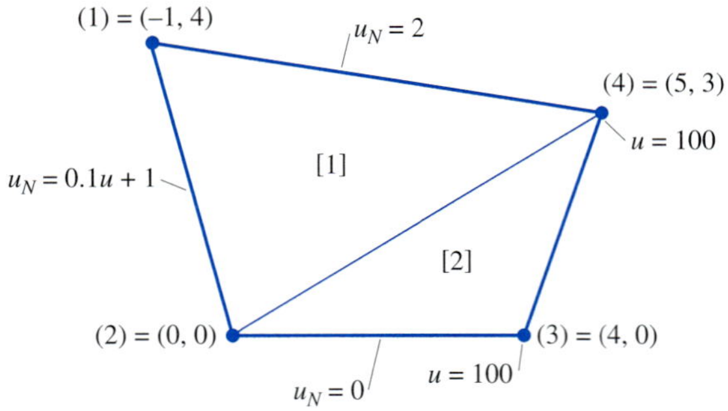

Figure 9.10

Find the solution by the finite-element method. The element-matrix inverses are:

***M* inverse for element 1 area is 11.5**

$$\begin{bmatrix} 1.000 & 0.000 & 0.000 \\ -0.043 & 0.174 & -0.130 \\ -0.261 & 0.043 & 0.217 \end{bmatrix}$$
(2) (4) (1)

$Q_{av} = -0.233 \quad F_{av} = -9.333$

***M* inverse for element 2 area is 6**

$$\begin{bmatrix} 1.000 & 0.000 & 0.000 \\ -0.250 & 0.250 & 0.000 \\ 0.083 & -0.417 & 0.333 \end{bmatrix}$$
(2) (3) (4)

$Q_{av} = -0.1 \quad F_{av} = -10.25$

From these we get these element equations:

Element equations for element 1

$$\begin{matrix} 2 \rightarrow \\ 4 \rightarrow \\ 1 \rightarrow \end{matrix} \begin{bmatrix} 1.2516 & 0.0062 & -0.3633 & 35.7778 \\ 0.0062 & 0.8168 & 0.0714 & 35.7778 \\ -0.3633 & 0.0714 & 1.1864 & 35.7778 \end{bmatrix}$$
(2) (4) (1)

Element equations for element 2

$$\begin{matrix} 2 \rightarrow \\ 3 \rightarrow \\ 4 \rightarrow \end{matrix} \begin{bmatrix} 0.5167 & -0.5333 & 0.2167 & 20.5000 \\ -0.5333 & 1.5167 & -0.7833 & 20.5000 \\ 0.2167 & -0.7833 & 0.7667 & 20.5000 \end{bmatrix}$$
(2) (3) (4)

These equations assemble to give this unadjusted system matrix:

$$\begin{bmatrix} 1.186 & -0.363 & 0.000 & 0.071 & 35.778 \\ -0.363 & 1.768 & -0.533 & 0.223 & 56.278 \\ 0.000 & -0.533 & 1.517 & -0.783 & 20.500 \\ 0.071 & 0.223 & -0.783 & 1.583 & 56.278 \end{bmatrix}$$

We need the lengths of sides 4–1, 1–2, and 2–3. They are

$$\text{Side 4–1:}\quad 6.083, \qquad \text{Side 1–2:}\quad 4.123, \qquad \text{Side 2–3:}\quad 4.$$

We now adjust for the derivative conditions to get the modified system:

$$\begin{bmatrix} 1.049 & -0.432 & 0.000 & 0.003 & 43.922 \\ -0.432 & 1.631 & -0.533 & 0.154 & 64.422 \\ 0.000 & -0.533 & 1.517 & -0.783 & 20.500 \\ 0.003 & 0.154 & -0.783 & 1.446 & 64.422 \end{bmatrix}.$$

The second adjustment is for the known values at nodes (3) and (4), giving

$$\begin{bmatrix} 1.049 & -0.432 & 43.650 \\ -0.432 & 1.631 & 102.339 \end{bmatrix}$$

which we solve to get these estimates of u_1 and u_2:

$$u_1 = 75.71, \qquad u_2 = 82.80.$$

Solving an Elliptic Problem with MATLAB

MATLAB's professional version has a toolbox, the *Partial Differential Equation Toolbox* (not included in the student version that we use) that can solve all three types of partial-differential equations. We describe here how it solves an elliptic problem.

This illustrative example solves Laplace's equation to get the temperature distribution on a region that is a rectangle whose width is twice its height and that has a quarter circle removed from its upper-right corner:

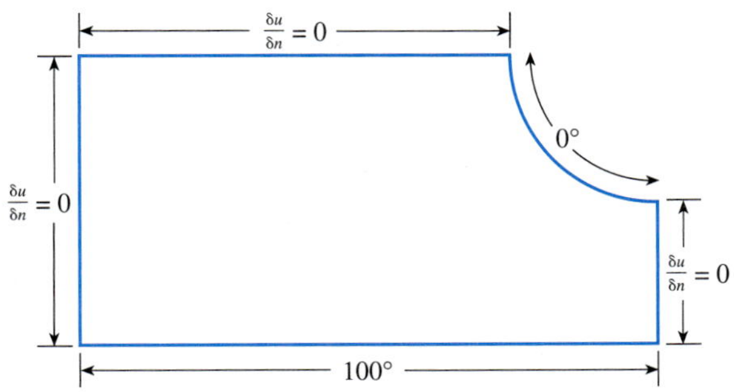

Along the base of the figure, the temperature is 100°; on the arc it is 0° (Dirichlet conditions). All other edges are insulated (Neumann condition, $\partial u/\partial n = 0$). We desire the steady-state temperature distribution.

We will obtain the solution through six steps:

1. Draw the region.
2. Save the region as an M-file.
3. Set boundary conditions.
4. Create a mesh of triangular elements.
5. Define the type of equation to be solved.
6. Solve the problem.

1. Draw the Region The Toolbox provides two different ways to do this: with a Graphical User Interface (GUI), or through commands typed into the command screen.

We will use the second way. Our region will be composed of two parts: the rectangle and a circle that is subtracted from it. We begin with the rectangle. We use the command:

```
>> pderect([-1 1 -.5 .5])
```

where the parameter is a vector of the x-coordinates followed by the y-coordinates of two opposite corners. [The corners are at $(-1, -0.5)$ and $(1, 0.5)$.] After the command is entered, we see the rectangle in a separate window that we will call the "figure window."

This window has a menu bar as well as another bar that has icons; these icons are quick ways to call for many menu commands.

We now create the circle. We go back to the command window and enter

```
>> pdecirc(1, 0.5, 0.5)
```

This superimposes a circle on the rectangle with center at $x = 1$, $y = 0.5$, and radius of 0.5, which we can view in the figure window. The figure window has a box labeled "Set formula" that reads `R1 + C1`, which we change to `R1 − C1` by clicking the box and using the keyboard to make the change. The figure window does not yet reflect the change but it is in effect.

From now on, each step is done in the figure window.

2. Save the Region It is always good to save the description of the region. This permits one to retrieve it at a later time. Saving is done by invoking `File/Save As` in the figure window. We give it a name, say, FIGA, and it is added to the list of M-files.

3. Set Boundary Conditions We invoke `Boundary/Boundary Mode` and see the region displayed (the distorted rectangle is now seen). Its outline is red with arrows indicating a counterclockwise ordering. We establish the boundary conditions by double-clicking on a boundary and then entering parameters into a dialog box.

We begin with the base of the figure. After double-clicking on the base, we see the dialog box. Select `Dirichlet` (actually, this is the default), and make the value of $t = 100$. Click OK and we are returned to the figure window. We double-click on the arc and select `Dirichlet`, and make $t = 0$ (both are default values).

We need to establish a Neumann condition on each of the other sides of the rectangle. This is easy to do: double-click on the side, select `Neumann`, set $g = 0$, $q = 0$ (default values), click `OK`.

The boundary conditions are now established.

4. Create a Mesh of Triangular Elements Clicking on `Mesh/Initialize Mesh` in the menu bar creates a coarse mesh of triangular elements:

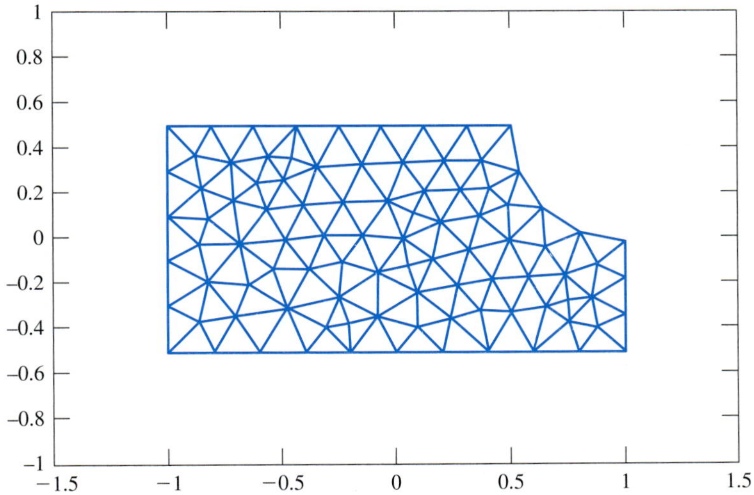

We can refine this mesh with `Mesh/Refine Mesh` but we stay with the current mesh for now.

5. Define the Type of Equation to Be Solved We do this by clicking on `PDE/PDE Specification` in the menu bar. In the dialog box that appears, we select `Elliptic` (the default) and set $c = 1, a = 0, f = 0$. We then click `OK` to finish this step. (These parameters are for our equation, $\nabla^2 u = 0$.)

6. Solve the Problem Clicking `Solve/Solve PDE` in the menu bar gets the solution. The software in the toolbox sets up the equations, assembles these, adjusts for boundary conditions, and solves the system of equations.

We see a display of the region with colors indicating the temperature in each element. On the right of this is a vertical bar that shows how colors and temperatures are related. Our figure here is not in color but the output actually indicates the temperatures within each element by colors that vary from bright red (100°) to bright blue (0°). Because we use a coarse mesh, it is easy to see the temperature of each individual element by its color. This would be difficult with a fine mesh.

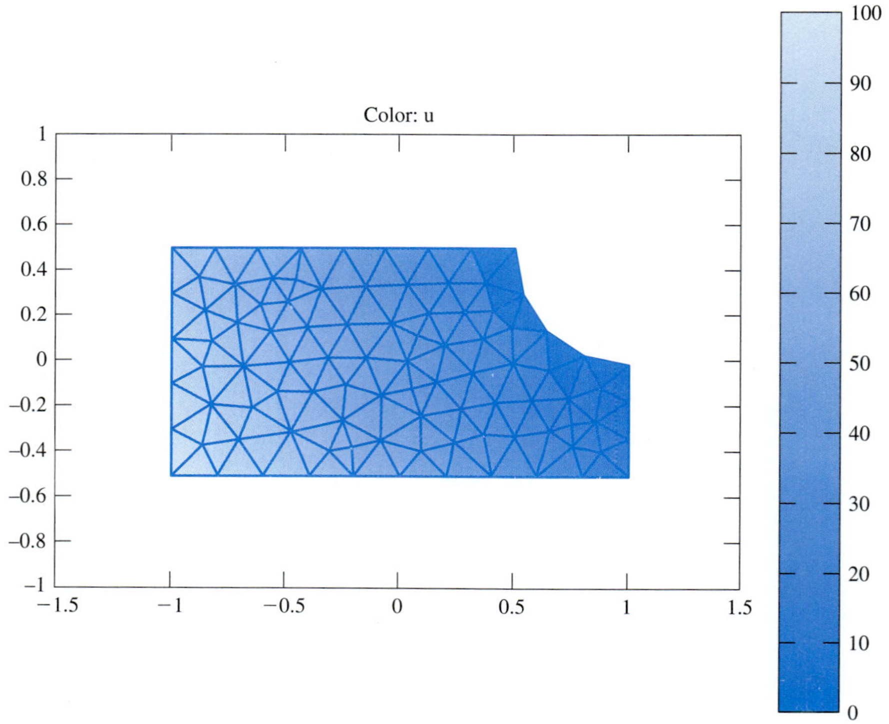

Another way to see the solution is to get the isotherms. Selecting only `Contour` in the `Plot Parameters` dialog box gives a plot of isotherms within the region, with $\Delta u = 5°$. This is shown in the next figure. On the computer screen, these isotherms are colored to indicate the temperatures.

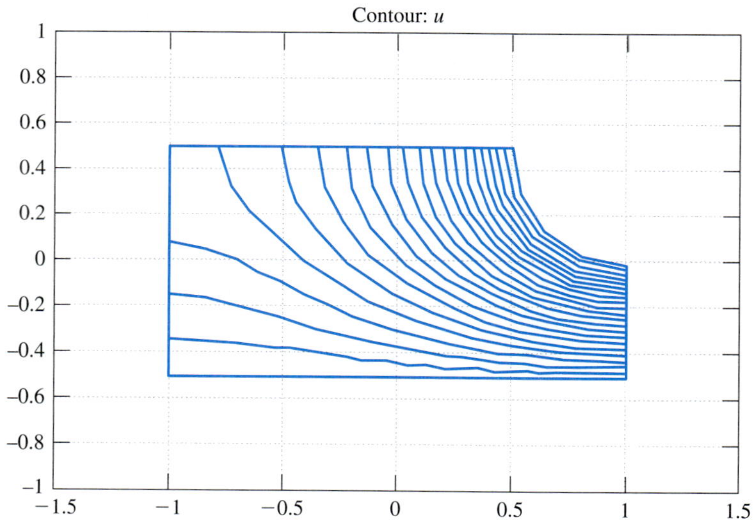

The Heat Equation

As we have just seen, the finite-element method is often preferred for solving boundary-value problems. It is also the preferred method for solving the heat equation when the region of interest is not regular. You should know something about this application of finite elements, but we do not give a full treatment.

Consider the heat-flow equation in two dimensions with heat generation given by $F(x, y)$:

$$\frac{\partial u}{\partial t} = \frac{k}{c\rho}\left(\frac{\partial^2 u}{\partial x^2} + \frac{\partial^2 u}{\partial y^2}\right) + F(x, y), \tag{9.56}$$

which is subject to initial conditions at $t = 0$ and boundary conditions that may be Dirichlet or may involve the outward normal gradient. Although this is really a three-variable problem (in x, y, and t), it is customary to approximate the time derivative with a finite difference and apply finite elements only to the spatial region. Doing so, we can rewrite Eq. (9.56) as

$$\frac{u^{m+1} - u^m}{\Delta t} = \frac{k}{c\rho}\left(\frac{\partial^2 u}{\partial x^2} + \frac{\partial^2 u}{\partial y^2}\right) + F(x, y), \tag{9.57}$$

where we have used a forward difference as in the explicit method. (We might prefer Crank–Nicolson or the implicit method, but we will keep things simple.)

To apply finite elements to the region, we do exactly as described previously—cover the region with joined elements, write element equations for the right-hand side of Eq. (9.56), assemble these, adjust for boundary conditions, and solve. However, we must also consider the time variable. We do so by considering Eq. (9.57) to apply at a fixed point in time, t_m. Because we know the values of u everywhere within the region at $t = t_0$, we surely know the initial nodal values. We then can solve Eq. (9.57) for the u-values at $t = t_0 + \Delta t$, where the size of Δt is chosen small enough to ensure stability.

We will use the Galerkin procedure to derive the element equations to provide some variety from the above. In this procedure you will remember that we integrate the residual weighted with each of the shape functions and set them to zero. (The integrations are done over the element area.) If we stay with linear triangular elements, there are three shape functions, N_r, N_s, and N_t, where the subscripts denote the three vertices (nodes) of the element taken in counterclockwise order.

The residual for Eq. (9.56) is

$$\text{Residual} = u_t - \alpha(u_{xx} + u_{yy}) - F, \tag{9.58}$$

where we have used the subscript notation for derivatives and have abbreviated $k/c\rho$ with α.

As stated, we will use linear triangular elements; within each element we approximate u with

$$u(x, y) \approx v(x, y) = N_r c_r + N_s c_s + N_t c_t. \tag{9.59}$$

This means that Galerkin integrals are

$$\iint N_j[v_t - \alpha(v_{xx} + v_{yy}) + F]\, dx\, dy = 0, \qquad j = r, s, t. \tag{9.60}$$

If we apply integration by parts (as we did in Section 9.2) to the second derivatives of Eq. (9.60), we can reduce the order of these derivatives. Doing so and replacing v from Eq. (9.59) gives a set of three equations for each element, which we write in matrix form:

$$[C]\left\{\frac{\partial c}{\partial t}\right\} + [K]\{c\} = \{b\}. \tag{9.61}$$

The components of $\{c\}$ are the nodal temperatures of the element, of course; those of $\{\partial c/\partial t\}$ are the time derivatives. The components of the matrices of Eq. (9.61) are

$$\begin{aligned} C_{i,j} &- \iint N_i N_j \, dx\, dy, \qquad i,j = r, s, t \\ K_{ij} &= \alpha_{\text{av}} \iint \left[\left(\frac{\partial N_i}{\partial x}\right)\left(\frac{\partial N_j}{\partial x}\right) + \left(\frac{\partial N_i}{\partial y}\right)\left(\frac{N_j}{\partial y}\right)\right] dx\, dy, \qquad i,j = r, s, t. \\ b_i &= \iint F N_i \; dx\, dy + \oint u_N N_i \; dL, \qquad i = r, s, t. \end{aligned} \tag{9.62}$$

In Eq. (9.62), the line integral in the b's is present only along a side of an element on the boundary of the region where the outward normal gradient u_N is specified in a boundary condition.

From the development of Eq. (9.52), we know how to evaluate all of the integrals of Eq. (9.62) when the elements are triangles. [See, for example, Burnett (1987) for the evaluations for other types of elements.]

As stated, we will use a finite-difference approximation for $\partial c/\partial t$. If this is a forward difference as suggested, we get the explicit formula

$$\frac{1}{\Delta t}[C]\{c^{m+1}\} = \frac{1}{\Delta t}[C]\{c^m\} - [K]\{c^m\} + \{b\}, \tag{9.63}$$

where all the c's on the right are nodal temperatures at $t = t_m$ and the nodal temperatures on the left in $\{c^{m+1}\}$ are at $t = t_{m+1}$.

We can put Eq. (9.63) into a more familiar iterating form by multiplying through by $\Delta t[C]^{-1}$:

$$\{c^{m+1}\} = \{c^m\} - \Delta t[C]^{-1}[K]\{c^m\} + \Delta t[C]^{-1}\{b\}. \tag{9.64}$$

[We can make Eq. (9.64) more compact by combining the multipliers of $\{c^m\}$.]

In principle, we have solved the heat-flow problem by finite elements. We construct the equations for every element from Eq. (9.64) and assemble them to get the global matrix, then adjust for boundary conditions just as before. This gives a set of equations in the unknown nodal values that we use to step forward in time from the initial point. With the explicit method illustrated here, each time step is just a matrix multiplication of the current nodal temperatures (and a vector addition) to get the next set of values. If we had used an implicit method such as Crank–Nicolson, we would have had to solve a set of equations at each step, but, unfortunately, they are not tridiagonal. We might hope for some equivalent to the A.D.I. method, but A.D.I. requires that the nodes be uniformly spaced. The conclusion is that the finite-element method in two or three dimensions is a problem that is expensive to solve. In one dimension, however, the system is tridiagonal, so that situation is not bad.

Solving a Parabolic Problem with MATLAB

The Partial Differential Equation Toolbox can solve all types of partial-differential equations. We show here how it can solve the heat equation. In the previous description of solving an elliptic problem with the toolbox, the solution is the steady-state distribution of temperatures. This is not reached instantaneously; the progress of the solution from an initial state to the steady state can be found by solving the heat equation:

$$\partial u/\partial t = k/c\rho\, \nabla^2 u.$$

MATLAB's generic form of a parabolic equation is

$$\mathbf{d}*\partial u/\partial t - \nabla(\mathbf{c}*\nabla u) + \mathbf{a}u = \mathbf{f},$$

where we have used boldface to pinpoint the parameters. For our equation, we want **d** = 1, **c** = k/cρ (the thermal diffusivity), **a** = 0, and **f** = 0.

Let us see how the steady state is approached as time advances for the same region and boundary conditions as before. We will take the initial temperatures within the region as 0°.

The procedure is almost exactly the same as before, only step 5 is different:

1. Define the region.
2. Define the boundary conditions.
3. Enter the values for the parameters of the equation.
4. Establish a mesh of triangular elements.
5. Enter values for the initial values for u and a list of times for which the solution is computed.
6. Solve the problem and display the results

1. Define the Region We saved the region with the file name `FIGA` so all we have to do is enter this file name as a command.

2. Define Boundary Conditions We could have saved the previous set of conditions as an M-file, but we neglected to do that so we do it again. Because several of the boundaries have the same Neumann condition, it is advantageous to do `Edit/Select All`, set the conditions to Neumann with $\partial u/\partial n = 0$, and reset the two with Dirichlet conditions afterward. If we save this with the filename 'FIG_BC,' we can do steps 1 and 2 from that file.

3. Enter Values of Equation Parameters From `PDE/PDE Specifications`, we select `Parabolic`, and make d = 1, c = 1, a = 0, and f = 0 to match our equation.

4. Initialize the Mesh The easiest way to do this is with the triangular-shaped icon in the toolbar. We see the same mesh as before.

5. Enter Initial Temperature and List of Times This is done through the `Solve/Parameters/Solve Parameters` combination. We enter `u0 = 0` (the default), and enter into the time field 0:0.1:0.1 to obtain the solution after one-tenth of a second. (We will revise this after seeing this solution to find the temperatures within the object after 0.2, 0.4, 0.8 and 10.0 seconds.)

6. Solve the Equation We have many options here. Clicking on the = icon gives a color image similar to that from our elliptical example, except the temperatures are lower. Getting the isotherms is a better way to see the temperature distribution. This is accompished by `Plot/Parameters` and then choosing only `Contour`.

We repeated step 6 with different ending times to see how the isotherms change over time. At $t = 10.0$, the temperatures are essentially at steady state. (Smaller values for **c** in the equation delay the time to reach equilibrium.)

The figures show the isotherms for the sequence of ending times. By counting the number of isotherms, we estimate the temperature at the origin (0, 0) to be

t:	0.1	0.2	0.4	0.8	10.0
temp:	28°	43°	57°	66°	68.6°

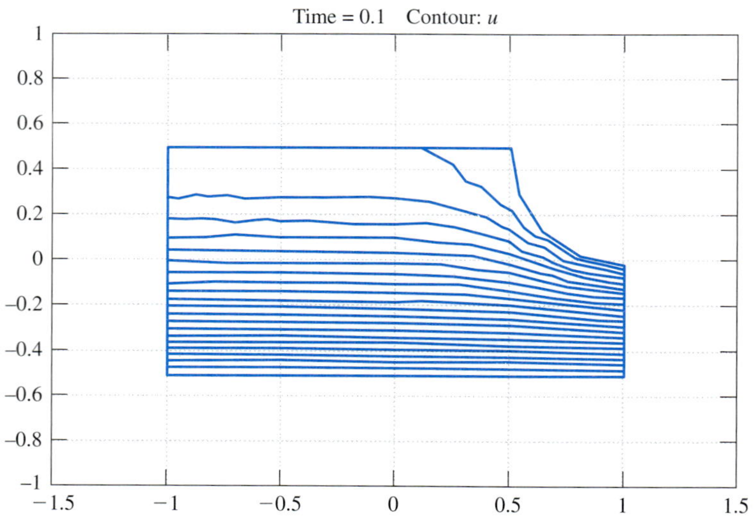

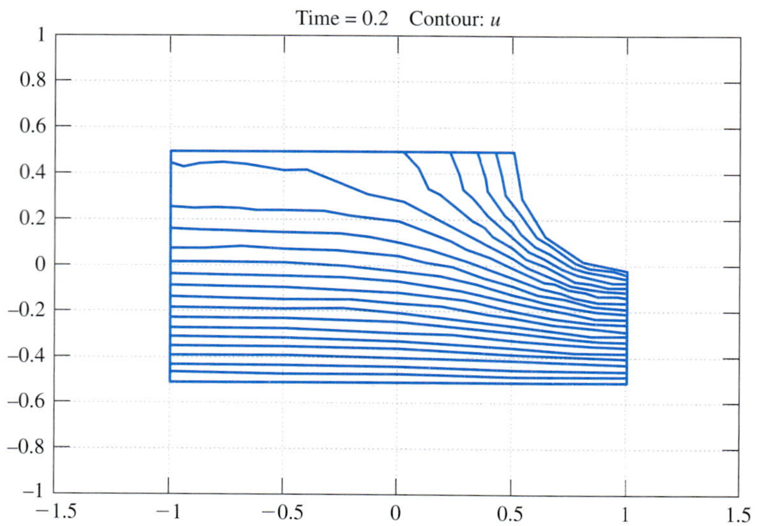

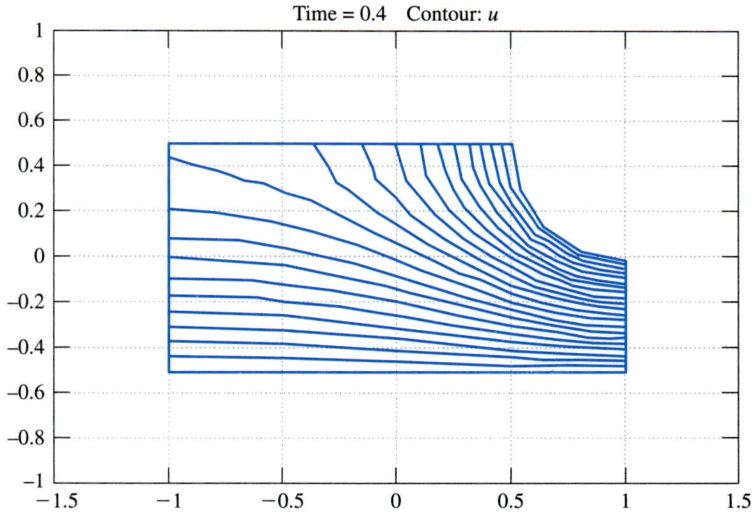
Time = 0.4 Contour: u
1
0.8
0.6
0.4
0.2
0
–0.2
–0.4
–0.6
–0.8
–1
−1.5
−1
−0.5
0
0.5
1
1.5

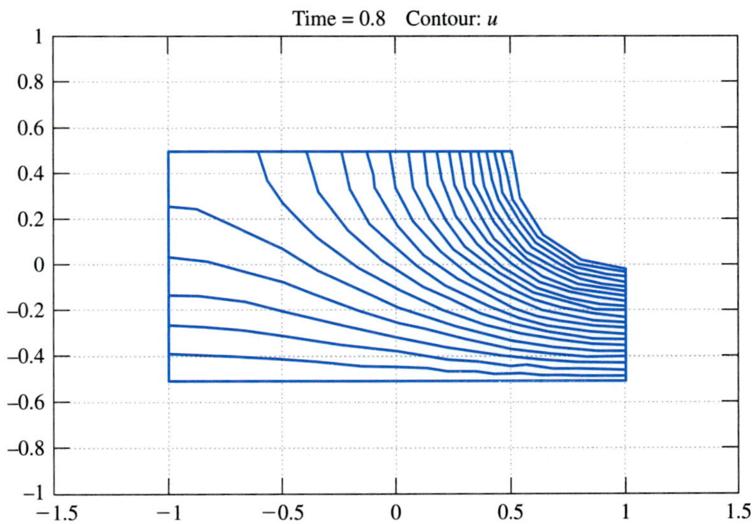
Time = 0.8 Contour: u
1
0.8
0.6
0.4
0.2
0
–0.2
–0.4
–0.6
–0.8
–1
−1.5
−1
−0.5
0
0.5
1
1.5

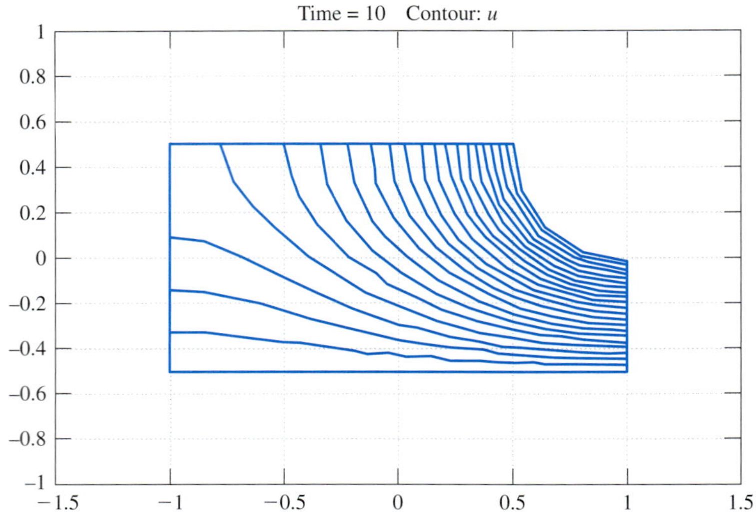

The Wave Equation

We will only outline how finite elements are applied to the wave equation, because this topic is too complex for full coverage here. Just as for the heat equation, finite elements are used for the space region and finite differences for time derivatives. We will develop only the vibrating string case (one dimension); two or three space dimensions are handled analogously but are harder to follow.

The equation that is usually solved is a more general case of the simple wave equation we have been discussing. In engineering applications, damping forces that serve to decrease the amplitude of the vibrations are important, and external forces that excite the system are usually involved. We therefore use, for a 1-D case, this equation for the displacement of points on the vibrating string, $y(x, t)$:

$$\frac{\partial}{\partial x}\left[T(x)\frac{\partial y}{\partial x}\right] - h(x)\frac{\partial y}{\partial t} + F(x, t) = \frac{w(x)}{g}\frac{\partial^2 y}{\partial t^2}. \tag{9.65}$$

Here T represents the tension, which is allowed to vary with x; h represents a damping coefficient that opposes motion in proportion to the velocity; F is the external force; and w/g is the mass density. There are boundary conditions (at $x = a$ and $x = b$) as well as initial conditions that specify initial displacements and velocities.

The approach is essentially identical to that used for unsteady-state heat flow: Apply finite elements to x and finite differences to the time derivatives. We will use linear one-dimensional elements, so we subdivide $[a, b]$ into portions (elements) that join at points that we call nodes. Within each element, we approximate $y(x, t)$ with $v(x, t)$,

$$y(x, t) \approx v(x, t) = N_L c_L + N_R c_R, \tag{9.66}$$

where c_L and c_R are the approximations to the displacements at the nodes at the left and right ends of a typical linear element. The N's are shape functions (in this 1-D case, we have called them "hat functions").

By using the Galerkin procedure, we can get this integral equation, which we will eventually transform into the element equations:

$$\int_L^R N_i\left(\frac{W}{g}\right) y_{tt}\,dx + \int_L^R N_i'(T)y_x\,dx + \int_L^R N_i(h)y_t\,dx$$
$$= \int_L^R N_i(F)\,dx + N_i(T)y_x\Big]_{x=R} - N_i(T)y_x\Big]_{x=L}, \qquad i = \mathrm{L, R}. \tag{9.67}$$

In Eq. (9.67) we have used subscript notation for the partial derivatives of y with respect to t and x and primes to represent the derivatives of the N's with respect to x (because the N's are functions of x only).

We now use Eq. (9.66) to find substitutions for y and its derivatives:

$$\begin{aligned} y(x,t) &\approx N_{\mathrm{L}}c_{\mathrm{L}} + N_{\mathrm{R}}c_{\mathrm{R}},\\ \frac{\partial y}{\partial x} &\approx N_{\mathrm{L}}'c_{\mathrm{L}} + N_{\mathrm{R}}'c_{\mathrm{R}},\\ \frac{\partial y}{\partial t} &\approx N_{\mathrm{L}}\dot{c}_{\mathrm{L}} + N_{\mathrm{R}}\dot{c}_{\mathrm{R}},\\ \frac{\partial^2 y}{\partial t^2} &\approx N_{\mathrm{L}}\ddot{c}_{\mathrm{L}} + N_{\mathrm{R}}\ddot{c}_{\mathrm{R}}. \end{aligned} \tag{9.68}$$

Here we employ the dot notation for time derivatives. (The c's vary with time, of course, but the N's do not.)

We now substitute from Eqs. (9.68) into Eq. (9.67) to get a pair of equations for each element (we write them in matrix form):

$$[M]\{\ddot{c}\} + [C]\{\dot{c}\} + [K]\{c\} = \{b\},$$

$$\left.\begin{aligned} M_{ij} &= \int_L^R N_i\left(\frac{w}{g}\right) N_j\,dx,\\ C_{ij} &= \int_L^R N_i(h)N_j\,dx,\\ K_{ij} &= \int_L^R N_i'(T)N_j'\,dx, \end{aligned}\right\} \quad i, j = \mathrm{L, R}, \tag{9.69}$$

$$b_i = \int_L^R N_i(F)\,dx + N_i(T)\frac{\partial y}{\partial x}\Big]_{x=R} - N_i(T)\frac{\partial y}{\partial x}\Big]_{x=L}, \qquad i = \mathrm{L, R}.$$

We will replace the time derivatives with finite differences, selecting central differences because they worked so well in the finite-difference solution to the simple wave equation. Thus we get

$$[M]\frac{\{c^{m+1} - 2c^m + c^{m-1}\}}{(\Delta t)^2} + [C]\frac{\{c^{m+1} - c^{m-1}\}}{2\Delta t} + [K]\{c^m\} = \{b^m\}. \quad \textbf{(9.70)}$$

Now we solve Eq. (9.70) for $\{c^{m+1}\}$:

$$\left(\frac{1}{(\Delta t)^2}[M] + \frac{1}{2\Delta t}[C]\right)\{c^{m+1}\} = \left(\frac{2}{(\Delta t)^2}[M] - [K]\right)\{c^m\} - \left(\frac{1}{(\Delta t)^2}[M] - \frac{1}{2\Delta t}[C]\right)\{c^{m-1}\} + \{b^m\}.$$

(9.71)

Notice that we need two previous sets of displacements to advance to the new time, t_{m+1}. We faced this identical problem when we solved the simple wave equation with finite differences, and we solve it in the same way. We use the initial velocities (given as one of initial conditions) to get $\{c^{-1}\}$to start the solution:

$$\{c^{-1}\} = \{c^1\} - 2\Delta t\{g(x)\}, \quad \textbf{(9.72)}$$

where $\{g(x)\}$ is the vector of initial velocities. [In view of our earlier work, we expect improved results if we use a weighted average of the g-values if the $g(x)$'s are not constants.]

We have not specifically developed the formulas for the components of the matrices and vector of Eqs. (9.69), but they are identical to those we derived when we applied finite elements to boundary-value problems in Section 9.2 because we will take out w, h, T, and F as average values within the elements. So we just copy from Section 9.2:

$$\begin{aligned}
M_{11} &= M_{22} = \left(\frac{w}{g}\right)\frac{\Delta}{3}, & M_{12} &= M_{21} = \left(\frac{w}{g}\right)\frac{\Delta}{6},\\
C_{11} &= C_{22} = h\frac{\Delta}{3}, & C_{12} &= C_{21} = h\frac{\Delta}{6},\\
K_{11} &= K_{22} = \frac{T}{\Delta}, & K_{12} &= K_{21} = -\frac{T}{\Delta},\\
b_1 &= F\frac{\Delta}{2} - \left[T\frac{\partial y}{\partial x}\right]_{x=L} & b_2 &= F\frac{\Delta}{2} + \left[T\frac{\partial y}{\partial x}\right]_{x=R}.
\end{aligned} \quad \textbf{(9.73)}$$

In this set, Δ represents the length of the element.

We now have everything we need to construct the element equations. Except for the end elements (and then only if the boundary conditions involve the gradient), the gradient terms in Eqs. (9.73) cancel between adjacent elements. Assembly in this case is very simple because there are always two elements that share each node (except at the ends).

What advantage is there to finite elements over finite differences? The major one is that we can use nodes that are unevenly spaced without having to modify the procedure. The advantage becomes really significant in two- and three-dimensional situations, but the other side of the coin is that solving the equations for each time step is not easy.

Solving the Wave Equation with MATLAB

The wave equation is a hyperbolic partial-differential equation. Lets see how MATLAB's PDE Toolbox handles an example. We will solve Example 8.14 by FEM. (The vibrating string problem can be solved with `pdep`, available in the student edition.)

The steps in the procedure are identical to those for a parabolic equation except for step five:

1. Define the region.
2. Define the boundary conditions.
3. Enter the values for the parameters of the equation.
4. Establish a mesh of triangular elements.
5. Enter the initial values for u, $\partial u/\partial t$, and a list of times for which the solution is computed.
6. Solve the problem and display the results.

Example 8.14 finds the displacements of a square flexible membrane that has an initial displacement but zero initial velocity. We will put the center of the square at the origin rather than a corner. This changes the initial displacement function to $(1 - x^2)(1 - y^2)$.

1. We draw the square with `pderect ([−1 1 −1 1])` and we see the square in the figure window. It is labeled `SQ1`.
2. All boundaries are at $u = 0$. Doing `Boundary/Boundary Mode` shows the region in red. This means that the Dirichlet conditions with $u = 0$ are automatically supplied. (We can verify this by double-clicking on a side.)
3. We do `PDE/PDE Specification` and fill in the dialog box to have $c = 1$, $a = 0$, $f = 0$, and $d = 1$.
4. Clicking on the triangular icon creates a coarse mesh of triangles. We will stay with this coarse mesh to make it easier to see how the individual elements change with time. A finer mesh would give a more accurate solution.
5. We do `Solve/Parameters` and fill in the dialog box with `Time = 0:0.2:1` and `u(t0) = (1 − x.^2).*(1 − y.^2).`
6. We are now ready for the solution. For this problem, seeing the results as a "movie" is best. So we do `Plot/Parameters` and select only `Height (3-D Plot)` and `Animation` in the dialog box. When we click on `Plot,` we see the membrane go from its initial bubblelike position to its mirror image on the other side of the (x, y) plane and back again repeatedly. The animation repeats itself several times. This figure shows the final position that is reached after one second.

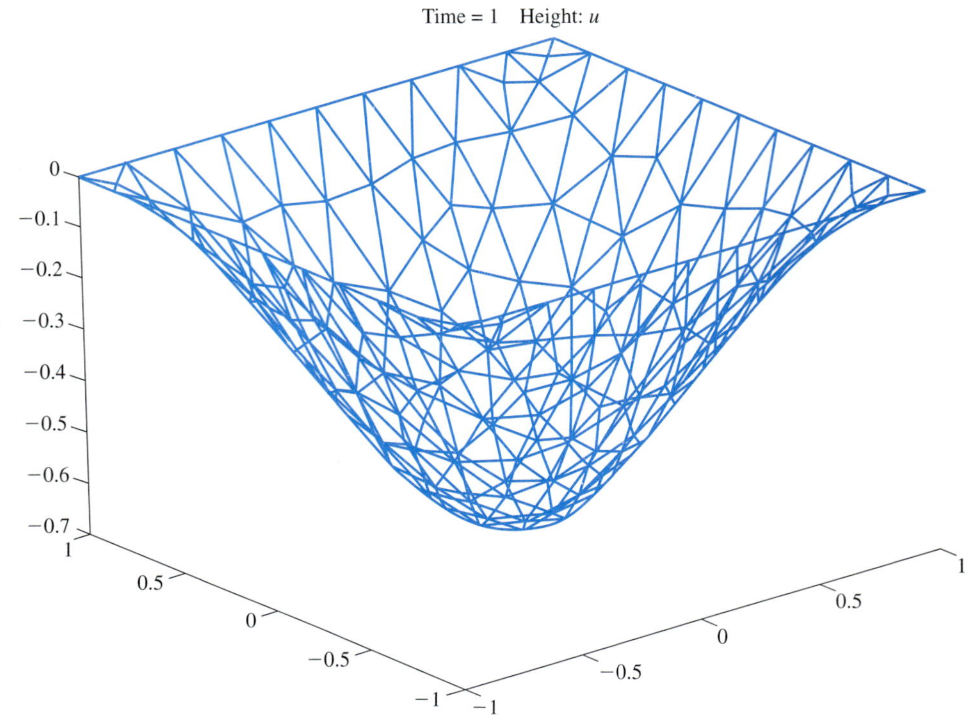

Exercises

Section 9.1

1. Show that the integrand of Eq. (9.4) is equivalent to Eq. (9.3) if the Euler–Lagrange condition is used. This means that Eq. (9.4) is the functional for any second-order boundary-value problem of the form

$$y'' + Q(x)y = F(x),$$

subject to Dirichlet boundary conditions

$$y(a) = A, \qquad y(b) = B,$$

where A and B are constants.

▶ 2. Use the Rayleigh–Ritz method to approximate the solution of

$$y'' = 3x + 1, \qquad y(0) = 0, \qquad y(1) = 0,$$

using a quadratic in x as the approximating function. Compare to the analytical solution by graphing the approximation and the analytical solution.

3. Repeat Exercise 2, but this time, for the approximating function, use

$$ax(x - 1) + bx^2(x - 1).$$

Show that this reproduces the analytical solution.

4. Another approximating function that meets the boundary condition of Exercise 3 is

$$ax(x - 1) + bx(x - 1)^2.$$

Use this to solve by the Rayleigh–Ritz technique.

5. Suppose that the boundary conditions in Exercise 3 are $y(0) = 1$, $y(1) = 3$. Modify the procedure of Exercise 3 to get a solution.

▶ 6. Solve Exercise 2 by collocation, setting the residual to zero at $x = \frac{1}{3}$ and $x = \frac{2}{3}$. Compare this solution to that from Exercise 2.

7. Repeat Exercise 6, except now use different points within [0, 1] for setting the residual to zero. Are some pairs of points better than others?

8. Repeat Exercise 3, but now use collocation. Does it matter where within [0, 1] you set the residual to zero?

9. Use Galerkin's technique to solve Exercise 2. Is the same solution obtained?

10. Repeat Exercise 3, but now use Galerkin.

Section 9.2

11. Suppose that, in Eq. (9.25), $Q(x) = \sin(x)$ and $F(x) = x^2 + 2$. For an element that occupies [0.33, 0.45],
   - ▶a. Find N_L and N_R of Eq. (9.26).
   - b. Write out the integrals of Eq. (9.28).
   - c. Write out the element equations (9.36).
   - ▶d. Compute the correct average values for Q and F.

12. Repeat Exercise 11 for two adjacent elements. These occupy [0.21, 0.33) and [0.45, 0.71].

13. Assemble the three pairs of element equations of Exercises 11 and 12 to form a set of four equations with the nodal values at $x = 0.21$, $x = 0.33$, $x = 0.45$, and $x = 0.71$ as unknowns.

▶14. Solve by the finite-element method:

$$y'' + xy = x^3 - \frac{4}{x^3}, \qquad y(1) = -1, \qquad y(2) = 3.$$

Put nodes at $x = 1.2$, 1.5, and 1.75 well as at the ends of [1, 2]. Compare your solution to the analytical solution, which is $y = x^2 - 2/x$.

15. Repeat Exercise 14, except for the end condition at $x = 1$ of $y'(1) = 4$.

16. Repeat Exercise 14, but with more nodes. Place added nodes at $x = 1.1$, 1.3, 1.4, 1.65, and 1.9. Compare the errors with those of Exercise 14.

Section 9.3

17. Confirm that Eq. (9.45) is in fact the inverse of matrix M in Eq. (9.44).

18. Find M^{-1}, a, N, and $u(x, y)$ for these triangular elements:
   - a. Nodes: (1.2, 3.1), (−0.2, 4), (−2, −3); u-values at these nodes: 5, 20, 7; point where u is to be determined: (−1, 0)
   - b. Nodes: (20, 40), (50, 10), (5, 10); u-values at these nodes: 12.5, 6.2, 10.1; point where u is to be determined: (20, 20)
   - c. Nodes: (12.1, 11.3), (8.6, 9.3), (13.2, 9.3); u-values at these nodes: 121, 215, 67; point where u is to be determined: (10.6, 9.6)

19. Confirm that the sum of the entries in the first row of M^{-1} is equal to twice the area for each of the elements in Exercise 18.

▶20. Find the element equations for the element in part (c) of Exercise 18 if $Q = x^2y$ and $F = -x/y$ (these refer to Eq. 9.40). There are no derivative conditions on any of the element boundaries.

21. Solve Example 8.1 (Chapter 8) by finite elements. Place nodes at each corner and at the midpoints of the top and bottom edges, also at points 9, 12, and 14. Draw triangular elements whose vertices are at these nodes. Compare the answers at each node to those obtained with finite-difference approximations to the derivatives.

22. In Exercise 21, the temperatures in the top half of the slab are the same as those in the bottom half because of symmetry in the boundary conditions. Solve the problem for the top half only of the slab with the same nodes as in Exercise 21. (Along the horizontal midline, the gradient will be zero).

▶23. For a triangular element that has nodes at points (1.2, 3.2), (4.3, 2.7), and (2.4, 4.1), find the components of each matrix in the element equations [Eqs. (9.61) and (9.62)] if the material is aluminum.

24. For heat flow in one dimension, the governing equation is

$$k(\partial^2 u/\partial x^2) = c\rho(\partial u/\partial t).$$

Repeat the development of the analog of Eq. (9.62) for this case.

25. Use the equations that you derived in Exercise 24 to solve Exercise 31 of Chapter 8. Place the nodes exactly as those used in the finite-difference solution. Are the resulting equations the same?

26. Use finite elements to solve Exercise 34 of Chapter 8. Place interior nodes at three arbitrarily selected points (but do not make these symmetrical). Create triangular elements with these nodes and the four corner points. Set up the element equations, assemble, and solve for four time steps. Use the resulting nodal temperatures to estimate the same set of temperatures that were computed by finite differences. Compare the two methods of solving the problem.

27. Solve Example 8.6 (Chapter 8) by finite elements. Place nodes strategically along the edges and within the slab so there are a total of 14 or 15 nodes. Use triangular elements. Compare the solution to that obtained with finite-difference approximations. (You may want to take

advantage of symmetry in the boundary conditions to solve the problem with fewer elements.)

▶28. Rederive Eq. (9.64), but now for the Crank–Nicolson method.

29. Repeat Exercise 28, but now for the theta method.

30. Set up the finite-element equations for advancing the solution to part (a) of Exercise 50 of Chapter 8.

▶31. Set up the finite-element equations for starting the solution to part (a) of Exercise 50 of Chapter 8. Do this first for the analog of Eq. (8.42) and then for the analog of Eq. (8.49).

▶32. If we were to solve part (c) of Exercise 50 of Chapter 8, would there be an advantage to using shorter elements near the middle of the string where the displacements depart more from linearity?

33. Solve, using finite elements, Example 8.14, except with initial conditions of

$$u(x, y) = 0, \qquad u_t(x, y) = x^2(2 - x)y^2(2 - y).$$

34. Repeat Exercise 33, but with these initial conditions:

$$u(x, y) = x^2(2 - x)y^2(2 - y), \qquad u_t(x, y) = 0.$$

35. Solve Exercise 58 of Chapter 8 using finite elements. Where do you think interior nodes should be placed if there are

a. 6 of them?
b. 12 of them?

Compare the solutions from these two cases to that from the finite-difference method.

36. Solve Exercise 60 of Chapter 8 by finite elements, placing five interior nodes at points that you think are best. Justify your choice of nodal positions.

▶37. Using the isotherm plots from the MATLAB solution to a parabolic equation, count the isotherms (there are 20 curves) to see how the temperature at the upper-left corner varies with time. Plot these. Can you find an equation that fits?

Applied Problems and Projects

APP1. Use the Internet to find software that solves both ordinary- and partial-differential equations. Can you find any that use the finite-element method? (*Hint*: Try http://gams.nist.gov/ and search the topic: partial differential equations.)

APP2. Write a computer program that uses finite elements to solve the vibrating string problem. Test it by solving Example 8.13.

APP3. Repeat APP2, but now for the heat equation, Eq. (9.56). Test it by solving Exercises 26 and 27.

APP4. Write a computer program (using your favorite language) to solve a two-dimensional elliptic partial-differential equation. Allow for both Dirichlet and non-Dirichlet boundary conditions. Have the program read in the required data from a file. Provide function procedures to compute the values for $f(x, y)$ and $q(x, y)$. Here is a suggested data structure:

NN = the total number of nodes

NK = the number of boundary nodes with Dirichlet conditions. (NN − NK = number of nodes whose values are not specified, that is, the interior nodes and those boundary nodes whose values are not specified.)

VX (NN) = an array to hold the x-values for all nodes in the order that nodes are numbered. There is an advantage if the nodes whose u-values are specified are numbered so as to follow those nodes where the u-values must be computed.

VY (NN) = an array to hold the corresponding y-values for all nodes

M (NE, 4, 3) = an array to hold the element matrices. The first subscript indicates the element number. The second and third subscripts indicate the row and column of the matrix. The fourth row holds the node numbers for nodes in this element in counterclockwise order. There is an advantage if the unspecified nodes come before the nodes whose u-values are known.

UU (NN) = an array to hold unknown and known u-values at nodes in order of the node number. Zeros may be used as fillers for unknown u-values.

AE (NE) = an array to hold areas of the elements

F(NE) = an array to hold average f-values for each element

Q(NE) = an array to hold average q-values for each element

A(NN, NN + 1) = the system matrix

Here is what your logic might look like:

1. Read in NN, NE, NU.
2. Read in (x, y) values for the nodes, storing in VX and VY.
3. Read in node numbers for each element in turn (nodes should be in counterclockwise order), storing in the fourth row of the element matrices.
4. Read in the unknown and known u-values for each node.
5. Compute average values for f and q in each element. (You may prefer to evaluate these at the centroid of the element.) Store in F and Q.
6. Read in the known u-values, storing in UK.
7. Compute the area for each element and its inverse [Eq. (9.49), the area from the first row elements].
8. Find the element equations and add the appropriate values to the system matrix.
9. Adjust the system matrix for non-Dirichlet boundary conditions. (You may want to have the user input the a and b values for these and the node numbers at the ends of the element boundary where this applies. Alternatively, these could have been read in with the other parts of the data.)
10. Adjust the system matrix for Dirichlet conditions using values from the UU array.
11. Solve the system.
12. Display the u-values for each node.

APP5. Write and test a program that solves the vibrating membrane problem using the finite-element method.

APP6. In developing the element equations, a number of integrals must be evaluated [see Eq. (9.51)]. For triangular elements, these are very easy to get: Each is just the area divided by a number. These simple triangular elements that we have discussed are called *C^0-linear elements.*

Other types of elements besides these simple triangles are sometimes useful. For example, connecting the nodes with lines that form quadrilateral elements can cut the number of elements almost in half. For these, the integrals are not so readily evaluated.

Even if we stay with triangular elements, the accuracy of the solution is improved if we add one node within each of the three sides. Such additional nodes can even permit the "triangle" to have curved sides. Such a more elaborate triangular element is called a *C^0-quadratic element*. This idea can be extended to add more than three nodes to the triangle, and additional nodes are sometimes added to quadrilateral elements.

For all of these more elaborate elements, the shape functions no longer have a "flat top" like that sketched in Figure 9.7. The normal procedure for these is to employ Gaussian quadrature in which a weighted sum of the integrand at certain points, called *Gauss-points*, approximates the integral quite well.

For a square region with opposite corners at $(-1, -1)$ and $(1, 1)$, these Gauss-points are at $x = \pm\sqrt{3}/3$, $y = \pm\sqrt{3}/3$, as given in Table 5.13. For a region that is a triangle with vertices at (0, 0), (1,0), (0, 1), there are three Gauss-points at $\left(\frac{1}{6}, \frac{1}{6}\right)$, $\left(\frac{2}{3}, \frac{1}{6}\right)$, and $\left(\frac{1}{6}, \frac{2}{3}\right)$, each weighted with $\frac{1}{3}$. For elements that do not conform to these basic cases, they must be mapped to coincide with them. Where are the Gauss-points for

a. A triangle whose vertices are $(-1, 3)$, $(7, 1)$, and $(2, 7)$?
b. A quadrilateral whose vertices are $(1, 2)$, $(5, -1)$, $(6, 3)$, $(3, 5)$?

APP7. Use MATLAB's PDE Toolbox to solve several of the examples of Chapters 8 and 9. Define the regions both with the mouse on the graphical user interface and also by using commands.

APP8. There are other software packages that let you solve engineering and scientific problems with FEA. Two of these are ALGOR and MSC/Nastran. Find information on these and compare their capabilities with that of MATLAB's PDE Toolbox. The Internet is a good place to get some information. Your library may have books on them, too.

APP9. Search for information on finite elements with a Web browser. Write a report on what you find.

APPENDIX A

Some Basic Information from Calculus

Because a number of results and theorems from the calculus are frequently used in the text, we collect here a number of these items for ready reference, and to refresh the student's memory.

Open and Closed Intervals

For the open interval $a < x < b$, we use the notation (a, b), and for the closed interval $a \leq x \leq b$, we use the notation $[a, b]$.

Continuous Functions

If a real-valued function is defined on the interval (a, b), it is said to be *continuous* at a point x_0 in that interval if for every $\epsilon > 0$ there exists a positive nonzero number δ such that $|f(x) - f(x_0)| < \epsilon$ whenever $|x - x_0| < \delta$ and $a < x < b$. In simple terms, we can meet any criterion of matching the value of $f(x_0)$ (the criterion is the quantity ϵ) by choosing x near enough to x_0, without having to make x equal to x_0, when the function is continuous.

If a function is continuous for all x-values in an interval, it is said to be continuous on the interval. A function that is continuous on a closed interval $[a, b]$ will assume a maximum value and a minimum value at points in the interval (perhaps the endpoints). It will also assume any value between the maximum and the minimum at some point in the interval.

Similar statements can be made about a function of two or more variables. We then refer to a domain in the space of the several variables instead of to an interval.

Sums of Values of Continuous Functions

When x is in $[a, b]$, the value of a continuous function $f(x)$ must be no greater than the maximum and no less than the minimum value of $f(x)$ on $[a, b]$. The sum of n such values must be bounded by $(n)(m)$ and $(n)(M)$, where m and M are the minimum and maximum values. Consequently, the sum is n times some intermediate value of the function. Hence,

$$\sum_{i=1}^{n} f(\xi_i) = nf(\xi) \quad \text{if } a \le \xi_i \le b, \quad i = 1, 2, \ldots, n, \qquad a \le \xi \le b.$$

Similarly, it is obvious that

$$c_1 f(\xi_1) + c_2 f(\xi_2) = (c_1 + c_2) f(\xi), \qquad \xi_1, \xi_2, \xi \text{ in } [a, b],$$

for the continuous function f when c_1 and c_2 are both equal to or greater than one. If the coefficients are positive fractions, dividing by the smaller gives

$$c_1 f(\xi_i) + c_2 f(\xi_2) = c_1\left[f(\xi_1) + \frac{c_2}{c_1} f(\xi_2)\right] = c_1\left(1 + \frac{c_2}{c_1}\right) f(\xi) = (c_1 + c_2) f(\xi),$$

so the rule holds for fractions as well. If c_1 and c_2 are of unlike sign, this rule does not hold unless the values of $f(\xi_1)$ and $f(\xi_2)$ are narrowly restricted.

Mean-Value Theorem for Derivatives

When $f(x)$ is continuous on the closed interval $[a, b]$, then at some point ξ in the interior of the interval

$$f'(\xi) = \frac{f(b) - f(a)}{b - a}, \qquad a < \xi < b,$$

provided, of course, that $f'(x)$ exists at all interior points. Geometrically, this means that the curve has at one or more interior points a tangent parallel to the secant line connecting the ends of the curve (Fig. A.1).

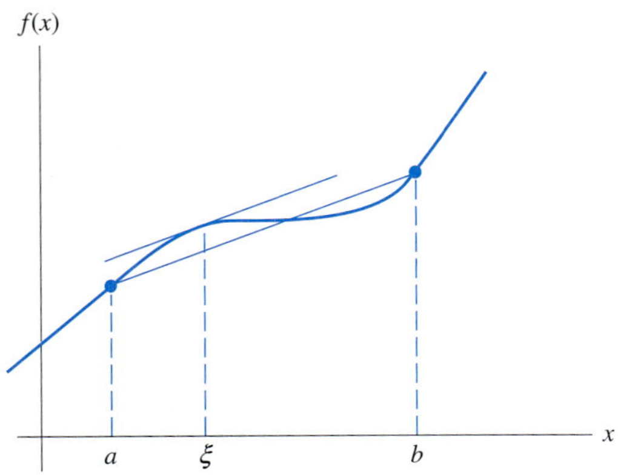

Figure A.1

Mean-Value Theorems for Integrals

If $f(x)$ is continuous and integrable on $[a, b]$, then

$$\int_a^b f(x)\,dx = (b - a)f(\xi), \qquad a < \xi < b.$$

This says, in effect, that the value of the integral is an average value of the function times the length of the interval. Because the average value lies between the maximum and minimum values, there is some point ξ at which $f(x)$ assumes this average value.

If $f(x)$ and $g(x)$ are continuous and integrable on $[a, b]$, and if $g(x)$ does not change sign on $[a, b]$, then

$$\int_a^b f(x)g(x)\,dx = f(\xi)\int_a^b g(x)\,dx, \qquad a < \xi < b.$$

Note that the previous statement is a special case $[g(x) = 1]$ of this last theorem, which is called the *second theorem of the mean for integrals.*

Taylor Series

If a function $f(x)$ can be represented by a power series on the interval $(-a, a)$, then the function has derivatives of all orders on that interval and the power series is

$$f(x) = f(0) + f'(0)x + \frac{f''(0)}{2!}x^2 + \frac{f'''(0)}{3!}x^3 + \cdots.$$

The preceding power-series expansion of $f(x)$ about the origin is called a *Maclaurin series.* Note that if the series exists, it is unique and any method of developing the coefficients gives this same series.

If the expansion is about the point $x = a$, we have the *Taylor series*

$$f(x) = f(a) + f'(a)(x - a) + \frac{f''(a)}{2!}(x - a)^2 + \frac{f'''(a)}{3!}(x - a)^3 + \cdots.$$

We frequently represent a function by a polynomial approximation, which we can regard as a truncated Taylor series. Usually, we cannot represent a function exactly by this means, so we are interested in the error. Taylor's formula with a remainder gives us the error term. The remainder term is usually derived in elementary calculus texts in the form of an integral:

$$f(x) = f(a) + f'(a)(x - a) + \frac{f''(a)}{2!}(x - a)^2 + \cdots$$
$$+ \frac{f^{(n)}(a)}{n!}(x - a)^n + \int_a^x \frac{(x - t)^n}{n!} f^{(n+1)}(t)\,dt.$$

Because $(x - t)$ does not change sign as t varies from a to x, the second theorem of the mean allows us to write the remainder term as

$$\text{Remainder of Taylor series} = \frac{(x-a)^{n+1}}{(n+1)!} f^{(n+1)}(\xi), \qquad \xi \text{ in } [a, x].$$

The derivative form is the more useful for our purposes. It is occasionally useful to express a Taylor series in a notation that shows how the function behaves at a distance h from a fixed point a. If we call $x = a + h$ in the preceding series, so that $x - a = h$, we get

$$f(a+h) = f(a) + f'(a)h + \frac{f''(a)}{2!}h^2 + \cdots \frac{f^{(n)}(a)}{n!}h^n + \frac{f^{(n+1)}(\xi)}{(n+1)!}h^{n+1}.$$

Taylor Series for Functions of Two Variables

For a function of two variables, $f(x, y)$, the rate of change of the function can be due to changes in either x or y. The derivatives of f can be expressed in terms of the partial derivatives. For the expansion in the neighborhood of the point (a, b),

$$\begin{aligned} f(x, y) &= f(a, b) + f_x(a, b)(x-a) + f_y(a, b)(y-b) \\ &\quad + \frac{1}{2!}[f_{xx}(a, b)(x-a)^2 + 2f_{xy}(a, b)(x-a)(y-b) + f_{yy}(a, b)(y-b)^2] \\ &\quad + \cdots. \end{aligned}$$

Descartes' Rule of Signs

Let $p(x)$ be a polynomial with real coefficients and consider the equation $p(x) = 0$. Descartes' rule of signs is a simple method for giving us an estimate of the number of real roots of this equation on both sides of $x = 0$. The rule states that

1. The number of positive real roots is equal to the number of variations in the signs of the coefficients of $p(x)$ or is less than that number by an even integer.
2. The number of negative real roots is determined the same way, but for $p(-x)$. Here also the number of negative roots is equal to the number of variations in the signs of the coefficients of $p(-x)$ or is less than that number by an even integer.

For example, the polynomial equation $p(x) = x^6 - 3x^5 + 2x^4 - 6x^3 - x^2 + 4x - 1 = 0$ will have 5, 3, or 1 positive and 1 negative real root. We can assume then that the number of real roots are at least 2 but can be as many as 6! (There are actually 3 positive, 1 negative, and 2 complex roots.)

APPENDIX B

Software Resources

Many who use this book will want to write programs to carry out the algorithms, but there are many excellent software packages available that professionals prefer to use. The advantage is that the software packages are both reliable and robust. Here is a partial list of software sources and computer algebra systems, organized alphabetically. There is a wealth of information about these and other products on the Internet. Just typing the name of the product or resource into a Web search engine will provide a list of up-to-date sites (plus other sites that use the same words in their name).

DERIVE is a computer algebra system (CAS) that first appeared in 1988, about the same time as Mathematica. DERIVE has the advantage of being menu-driven rather than command-driven. The earlier versions of DERIVE were developed by Soft Warehouse, but in 1999 Texas Instruments took over the product and continues its support. The most recent version is DERIVE 5; it provides both symbolic and numeric operations and can display 2-D graphs and 3-D surfaces. *Source*: www.education.ti.com/derive

GAMS (Guide to Available Mathematical Software) contains over 9000 software modules from over 90 packages, such as IMSL, NAG, BLAS, and EISPACK. Some of these are proprietary. GAMS is a software repository that includes abstracts, documentation, as well source code. GAMS is a project of the National Institute of Standards and Technology (NIST) that "studies techniques to provide scientists and engineers with improved access to reusable software components." *Source:* www.gams.nist.gov/

IBM's ESSL (Engineering and Scientific Subroutine Library) consists of routines that are designed for parallel processors. These are callable from several programming languages. The packages include routines for numerical quadrature, interpolation, random number generation, FFT, linear systems, and eigenvalue problems. The focus of ESSL has been on vector mainframes and RS/6000 processors. *Source:* www.rs6000.ibm.com/software/apps/essl.html

IMSL (International Mathematical and Statistical Library) is a library of hundreds of subroutines available to writers of programs in C, C++, Fortran, or Java, on UNIX, Windows, or Linux. IMSL is owned by Visual Numerics. *Source:* www.vni.com/products/imsl/

LAPACK is a library of Fortran 77 subroutines for solving systems of linear equations, least-squares solutions to linear systems, eigenvalue problems, and singular value decompositions. Its original goal was to make EISPACK and LINPACK run more efficiently on vector and parallel processors. LAPACK makes use of the package Basic Linear Algebra Subprograms (BLAS). *Source:* www.netlib.org/lapack/

Maple is a powerful CAS that performs both symbolic and numerical computations. In addition, it provides excellent and easy-to-use two-dimensional and three-dimensional color graphics. The software runs on PCs, Macs, workstations, and mainframes. It has an impressive collection of tools for solving differential equations, including the traditional Euler and RK4 procedures. There is a student version of the package as well. Its Web site can offer examples of a variety of applications. Its most current version is Version 8. *Source:* www.mapleapps.com/

Mathcad is a CAS that is different from other computer algebra systems in that one can use standard mathematics notation (such as an integral sign) to formulate the problem. It can solve problems both numerically and symbolically. It has graphics capabilities and excellent tutorial support. The current version is Mathcad 2001i. *Source:*www.mathsoft.com/

Mathematica continues to be one of the best-known software packages for doing a wide variety of mathematical problems. It has excellent 2-D and 3-D graphing capabilities; it provides both symbolic and numeric computations. There are excellent *Mathematica* tutorials that can be downloaded from the Web as well as other product supports. (See: http://library.wolfram.com/tutorials/) Stephen Wolfram is associated with *Mathematica* and he has written extensively for it. *Source:* www.wolfram.com/

MATLAB is a very popular and powerful CAS, which has specialized toolboxes for applications such as simulations, optimization, and partial-differential equations. Its newsletter contains articles about new applications of MATLAB. (We have used version 6, release 13 extensively in this book.) Cleve Moler, who has done much in numerical computing, is associated with this product; he is the author of articles in the MATLAB newsletter. *Source:* www.mathworks.com/

NAG (Numerical Algorithms Group) is a not-for-profit company that first started providing mathematical software in the early 1970s. Although it first began with Fortran, it now provides support for users of C, C++, Fortran 90, Java, and other compilers. *Source:* www.nag.com/

Netlib is a collection of mathematical software, papers, and databases. It has been a popular site on the Internet, with 182 million hits by the end of October 2002. Their Web site has a list of topics to choose from. *Source:* www.netlib.org/

Numerical Recipes, a book from Cambridge University Press, is a collection of over 300 numerical routines. There are versions for C, C++, Fortran 77/90, Basic, and Pascal. In addition, the source code is available on tape, diskette, and CD. The book discusses the algorithm as well as gives the code. *Source:* www.cup.org/ (then search on Numerical Recipes).

Solver is a software product from Frontline systems. It is an optimizer for Microsoft Excel using linear, quadratic, and mixed-integer programming, nonlinear optimization, and global optimization. Solver is also incorporated in other spreadsheets such as Lotus and Quattro Pro. The Web site also offers a short but useful tutorial on how to use the product. Many analysts use spreadsheets in solving numerical problems or computer algebra systems instead of other software. Some of the CAS products allow for importing data from Excel. *Source:* www.solver.com/

Answers to Selected Exercises

Chapter 0

6. You could write an expression that gives L as a function of angle c, but there is a better alternative. Think of the projection of the ladder onto ground level. This is identical to Figure 0.1b, so we know that the critical angle is the same, $c = 0.4677$ radians. We compute the length of the tipped ladder as the hypotenuse of a right triangle with sides equal to 33.42 ft and 6 ft 7 in.: 34.06 ft, about 7.7 in. longer.

8. One way would be to do it graphically. The ladder cuts off a circular segment when the bottom is placed against the circumference of the well; it cuts another circular segment that is exactly the same at the top. A rectangular well whose width equals the distance between the bases of the two segments is an equivalent problem. Draw this rectangular well. Cut out a ladder of the correct width and place it on the drawing. Cut off the end so the top of the ladder is exactly even with the ground and measure it.

A more analytical way would be to consider it to be a trigonometry problem. Let H = depth of the well, D = its diameter, W = width of the ladder, V = width of its rails, L = its length, and A = angle of inclination from the vertical. Using these variables, we can write

$$\tan(A) = \frac{2\sqrt{(D/2)^2 - (W/2)^2} - V * \cos(A)}{H - V * \sin(A)}$$

and

$$L = \frac{H - V * \sin(A)}{\cos(A)}$$

Substitute in the given values for H, D, W, and V. Then, MATLAB solves the first equation for $A = 0.34965$ radians. From the second equation, $L = 181.558$ in.

13. There are no values that you can enter from the keyboard that correspond to the inequalities. However, if ε is a value slightly less than eps, and if $X = Y = 1 + \varepsilon$ and Z is exactly 1, all inequalities hold.

15. a. 0.9999907

b. 1.000054

c. 1.00099

18. $x + y = [1.14, 2.65]$. Width is sum of widths.

$x - y + z = [2.22, 7.18]$. Width is sum.

$x * z = [0, 8.763]$. Width not obviously related.

$y/z = [-\infty, \infty]$. Zero is within both y and z.

25. Parallel processing applies when step $n + 1$ does not depend on the completion of step n. The different processing units that work in parallel could be a group of different single-processor computers connected in a distributed network.

Distributed computing applies when the same problem must be solved with different parameters. Of course, the individual steps in the solutions might benefit from parallel processing.

31. When computed term by term, a polynomial of degree n requires $(n^2 + n)/2$ multiplies and n adds. $(1 + 2 + 3 + 4 + \ldots = (n^2 + n)/2)$. The total is $(n^2 + 3n)/2$.

If computed with nested multiplication, the nth-degree polynomial requires n multiplies and n adds, a total of $2n$.

The ratio of numbers of operations is $(n + 3)/4$. As n gets large, this approaches $n/4$.

33. If the numerator is of degree n and the denominator is of degree d, $(n^2 + 3n)/2 + (d^2 + 3d)/2$ multiplies and adds are required if evaluated term by term (see answer to Exercise 31) plus one divide, a total of $n^2/2 + d^2/2 + 3n/2 + 3d/2 + 1$. If $n = d$, this total is $n^2 + 3n + 1$.

For a function of degree n in the numerator and degree d in the denominator, the number of multiplies and adds is $2n + 2d$ plus one more for the division. If both numerator and denominator are of degree n, the total is $4n + 1$. As n gets large, the ratio of operations with term-by-term evaluations to the operations when nested approaches $n/4$.

Chapter 1

3. From the graphs, there is an intersection at about (1.125, 0.425). Using $f(x) = x^3 - 1 - \cos(x)$ and the starting interval [0, 2], bisection finds the solution, $x = 1.12657$, in 17 iterations when tolerance on change in x-value is 1E-5.

7. We solve $(b - a)/2^n = 10^{-4}$ for n. This is

$$n = \frac{\ln(b - a) + 4\ln(5)}{\ln(2)} + 4.$$

9. The two solutions are $x = -5.7591$ and $x = -3.6689$. The tolerance was set at 1E−5:

Regula falsi gets the first root starting from $[-6, -4]$ in 13 iterations; it gets the second from $[-4, -2]$ in 23 iterations.

Bisection gets the first root starting from $[-6, -4]$ in 17 iterations; it gets the second from $[-4, -2]$ in 17 iterations.

The secant method gets the first root starting from $[-6, -4]$ in 4 iterations; it gets the second from $[-4, -2]$ in 3 iterations.

14. Let $f(x) = x^2 - N = 0$, so $f'(x) = 2x$. Then,

$$x_1 = x_0 - \frac{x_0^2 - N}{2x_0} = \frac{x_0^2 + N}{2x_0} = \frac{x_0 + N/x_0}{2}.$$

20. Two equations result from the conditions:

$$x + y = 20; \sqrt{xy} = 9.$$

Solve either for y, substitute in the other, get

$$x^2 - 20x + 81 = 0.$$

Use the quadratic formula to get $x = 14.358899$ and 5.6411011. Corresponding to these, $y = 5.6411011, 14.358899$.

24. Continued synthetic division by $(x - a)$ does not get $P^{(n)}(a)$ but the remainder is $P^{(n)}(a)/n!$, which is true as well for $n = 0$ and 1.

28. The convergence is quadratic. Starting from $x_0 = 5$, we get

x_n	5	4.55	4.25	4.0792	4.0113	4.00028	4.00000
Correct digits	0	1	1	1	2	4	8?
Ratio of errors		0.55	0.454	0.317	0.143	0.025	?

Applying Newton's method to $P'(x)$ to find the triple root results in only linear convergence. Quadratic convergence will be obtained if we apply it to $P''(x)$.

30. a. Starting from $x_0 = 2.1$, convergence is to $x = 2.01$ quadratically.

b. Starting from $x_0 = 1.9$, convergence is to $x = 1.99$ quadratically.

c. Starting from $x_0 = 2.0$ fails, $f'(2.0) =$ zero.

d. Starting from $x_0 = 2.02$ flies of off to large values, $f'(2.02)$ is really zero but round off causes this to be missed.

32. Muller's method in self-starting mode does get the root nearest zero. However, if there are two distinct roots equally distant from zero, it tends to favor the negative one. If these two roots of equal magnitude have a third one that is close to one of these, it favors the root with a neighbor.

37. The relations of (a), (b), and (c) all can be derived from $x^3 = 4$. Only the relation in (b) converges to $x = 1.5874$ starting from $x_0 = 1$; the others diverge.

40. a. $(x - 1)/(x^2 - 2x)$ converges slowly to $x = -0.80193$ from $x_0 = -1$. Does not converge to the other roots.

b. $\sqrt{((2x^2 + x + 1)/x)}$ converges to $x = 2.24697$ from $x_0 = 1$. Does not converge to the other roots.

c. $\sqrt{(x^3 - x + 1)/2)}$ converges to $x = 0.554969$ from $x_0 = 0$. Does not converge to the other roots.

42. a. The division gives a nonzero remainder, -3, so $x^2 + x + 1$ is not a factor.

b. Division by $x^2 + 2x + 3$ gives a zero remainder; it is a factor.

50. Rearranging the second equation to $x = \sqrt{2 - y^2 + x}$ and using the first as it stands does converge from (1, 1) to give the solution, $x = 1.990759$, $y = 0.1662412$, in 12 iterations.

Chapter 2

3. a. $v1 * v2 = \begin{bmatrix} -6 & -8 & 2 \\ 9 & 12 & -3 \\ 12 & 16 & -4 \end{bmatrix}$, $v3 * v2 = \begin{bmatrix} 12 & 16 & -4 \\ 6 & 8 & -2 \\ -9 & -12 & 3 \end{bmatrix}$,

$v2 * v1 = [2] = 2$, $v2 * v3 = [23] = 23$.

b. $A * v1 = \begin{bmatrix} -4 \\ 11 \\ -4 \end{bmatrix}$, $C * v1 = \begin{bmatrix} 3 \\ -7 \\ -14 \end{bmatrix}$, $D * v1 = \begin{bmatrix} 6 \\ 17 \\ 25 \end{bmatrix}$,

$A * v3 = \begin{bmatrix} -4 \\ 11 \\ -4 \end{bmatrix}$, $C * v3 = \begin{bmatrix} 1 \\ 12 \\ 17 \end{bmatrix}$, $D * v3 = \begin{bmatrix} -13 \\ 6 \\ -28 \end{bmatrix}$,

c. $v2 * A = [20, \ 12, \ 8]$, $v2 * B = [-1, \ 4]$,

$v2 * C = [9, \ 3, \ 1]$, $v2 * D = [-2, \ 15, \ 3]$.

d. $v1 * v1^T = \begin{bmatrix} 4 & -6 & -8 \\ -6 & 9 & 12 \\ -8 & 12 & 16 \end{bmatrix}$, $v2 * v2^T = [26]$,

$v3 * v3^T = \begin{bmatrix} 16 & 8 & -12 \\ 8 & 4 & -6 \\ -12 & -6 & 9 \end{bmatrix}$,

$v1^T * v1 = [29]$, $v2^T * v2 = \begin{bmatrix} 9 & 12 & -3 \\ 12 & 16 & -4 \\ -3 & -4 & 1 \end{bmatrix}$, $v3^T * v3 = [29]$.

5. a. For A: $x^2 + 8x - 47$.
For B: $x^3 + x^2 - 18x - 30$.

b. For A: $[-11.9373, \ 3.9373]$.
For B: $[4.4927, \ -3.6765, \ -1.8163]$.

c. $A * v = [7.9817, -7.7698]^T$ is not a multiple of v, so v is not an eigenvector.

9c. $$\begin{bmatrix} 1 & 0 & 0 & 0 & 0 \\ 0 & 1 & 0 & 0 & 0 \\ 0 & 0 & 0 & 0 & 1 \\ 0 & 0 & 0 & 1 & 0 \\ 0 & 0 & 1 & 0 & 0 \end{bmatrix} * A * \begin{bmatrix} 1 & 0 & 0 & 0 & 0 \\ 0 & 0 & 0 & 1 & 0 \\ 0 & 0 & 1 & 0 & 0 \\ 0 & 1 & 0 & 0 & 0 \\ 0 & 0 & 0 & 0 & 1 \end{bmatrix} = \begin{bmatrix} 3 & -1 & -2 & 5 & 0 \\ -2 & -5 & 4 & 3 & 3 \\ -5 & -3 & 3 & 5 & 4 \\ 2 & 2 & 4 & -3 & 0 \\ 5 & 3 & -1 & 2 & -6 \end{bmatrix}.$$

The result is A with the requested interchanges.

15. Solution is $x_1 = 3.2099$, $x_2 = -0.23457$, $x_3 = 0.71605$. No interchanges were required.

21. For a system of n equations, one right-hand side:

In column 1, n divides to put a 1 on the diagonal, n multiplies for each of $(n - 1)$ rows and the same number of subtracts to reduce in that column. (The 1 on the diagonal does not have to be computed nor the zeros below the diagonal.)

In column 2, $(n - 1)$ divides to put a 1 on the diagonal, $(n - 1)$ multiplies for each of $(n - 1)$ rows and the same number of subtracts to reduce in that column.

In column 3, $(n - 2)$ divides to put a 1 on the diagonal, $(n - 2)$ multiplies for each of $(n - 1)$ rows and the same number of subtracts to reduce in that column.

So, in column i, $(n - i)$ divides to put a 1 on the diagonal, $(n - i)$ multiplies for each of $(n - 1)$ rows and the same number of subtracts to reduce in that column.

No operations are needed to do back-substitution.

Total operations:

$$\begin{aligned}&\Sigma i + (n-1)\,\Sigma i + 2(n-1)\,\Sigma i \text{ for } i \text{ from 1 to } n,\\ &= n\,(n+1)/2 + 2(n-1)(\,n)(n+1)/2\\ &= n^2/2 + n/2 + n^3 + n^2 - n^2 - n\\ &= n^3 + n^2/2 - n/2 = O(n^3).\end{aligned}$$

28. The number of comparisons to find the pivot row is the same in both cases. If no interchanges are required, using an order vector is actually slower due to the overhead of setting up the vector. In the worst case, interchanges will occur in $(n - 1)$ columns. For this situation, using the vector requires only $(n - 1)$ numbers to be interchanged; not using it requires $(n + 1)$ in column 1, (n) in columns 2, $(n - 1)$ in column 3, . . . or $(n + 1) + (n) + (n - 1) + \ldots + 3$. This computes to $2(n - 1) + (n^2 - n)/2$. The difference in these totals is $n^2/2 + n/2 + 1$ and twice this is the number of add/subtract times saved by using the order vector.

31. The LU equivalent of the coefficient matrix is

$$\begin{bmatrix} 8 & -3 & 2 \\ -0.125 & 9.625 & 2.250 \\ 0.125 & -0.1688 & 4.1299 \end{bmatrix},$$

where the U matrix has ones on its diagonal. Rows were interchanged. Using this to solve with the given right-hand sides gives

a. $[-0.3711,\quad 0.3585,\quad 0.5220]^T$.

b. $[1.1635,\quad 0.1132,\quad -0.9843]^T$.

34. a. The solution is
[46.3415, 85.3859, 95.1220, 95.1220, 85.3659, 46.3415].

b. Reduction: for each of $(n - 1)$ rows, two multiplies, two subtracts;

Back-substitution: in row n, one divide, in rows $(n - 1)$ to 1, one multiply, one subtract, one divide.

Total: $4\,(n - 1) + 1 + 3(n - 1) = 7n - 6$, much less than Gaussian elimination when not compacted.

36. For column one of L: $l_{i1} = a_{i1}$.

For row one of U: $u_{1j} = a_{1j}/a_{11}$.

Alternate now between columns of L and rows of U:

For column i of L $(2 \le i \le n,\ i \le j \le n)$:
$l_{i,j} = a_{ji} - \Sigma l_{jk} * u_{ki},\ k = 1\ .\ .\ (j - 1).$
For row i of U $(2 \le i \le (n - 1),\ i + 1 \le j \le n)$:
$u_{ij} = (a_{ij} - \Sigma l_{ik} * u_{kj})/l_{ii},\ k = 1\ .\ .\ (i - 1).$

41. a. det $(A) = -142$, not singular.

b. det $(B) = 0$, singular, also lu(B) has zero $B_{4,4}$.

c. det $(C) = -108$, not singular.

45. a. det (H_4) = 1.65E–7. (A zero determinant means singular.)

b. [1.11, 0.228, 1.95, 0.797].

c. [0.988, 1.42, −0.428, 2.10].

Answers are poor because round-off effect is great when the matrix is nearly singular.

50. The determinant is 35. When $A_{3,3}$ is changed, it is −5. The changed matrix is more nearly singular. In fact, if $A_{3,3} = -3.75$, it is singular.

54. Both Gaussian elimination and Gauss–Jordan get the same result:

$$A^{-1} = \begin{bmatrix} 0.1429 & 0.1429 & -0.1429 \\ -0.4000 & 0.6000 & -0.2000 \\ 0.0857 & 0.0857 & 0.1143 \end{bmatrix}.$$

Gaussian: 25 multiplies/divides, 11 adds/subtracts; total = 36.
Gauss-Jordan: 29 multiplies/divides, 15 adds/subtracts; total = 44.

58. a. 1-norm = 17.74; 2-norm = 9.9776, ∞-norm = 8.12.

b. 1-norm = 17; 2-norm = 9.3274, ∞-norm = 7.

61. a. 1-norm = 21, 2-norm = 14.4721, ∞-norm = 20.

b. 1-norm = 18, 2-norm = 14.7774, ∞-norm = 21.1.

Even though the norms are nearly the same, the determinants are very different: −170 versus 515.133.

65. Norms of H_4:

1-norm = 2.0833.

2-norm = 1.5002.

∞-norm = 2.0833.

fro-norm = 1.5097.

71. Condition numbers:

For matrix of Exercise 67: 30,697

For matrix of with $A_{3,2}$ changed: 9.8201

The determinants are very different:

−0.0305 and −90.8807

76. Any multiple of the identity matrix, $a * I$, has a condition number of 1 because its eigenvalues are all equal to a and for its inverse, they are all equal to $1/a$. So, the product of the largest of these is unity.

Changing any element of $a * I$ increases the condition number because at least one of the eigenvalues of the matrix and its inverse are greater than one.

The zero matrix has a condition number of infinity.

79. We switch rows 2 and 3 to make diagonally dominant. Then Jacobi takes 34 iterations to get the solution from [0, 0, 0]:

[−0.14332, −1.37459, 0.71987].

80. The same answer as in Exercise 79 is obtained in 13 iterations.

88. When doing row i, all elements to the left of the diagonal will become zero; we do not have to specifically calculate them. So, we reassign one of the processors from this set, say, PROCESSOR $(i, i-1)$

to replace PROCESSOR $(i, n + 1)$. The n^2 processors are adequate to perform the back-substitution phase.

Chapter 3

3. Equation for Exercise 1: $-1.7833x^2 + 20.9067x - 48.4395$.

Equation for Exercise 2: $-1.7667x^2 + 20.7533x - 48.1910$.

6. Interpolating polynomial is $1.4762x^2 + 0.2429x + 1$. At $x = 1.3$, this gives 3.8095; true value is 3.6693. The error is 0.1402; bounds to error are 0.0595, 0.4396.

10. For n points, there are n terms; each term requires $2n - 2$ subtractions, $2n - 3$ multiplies, and 1 divide. We then use n adds to get the interpolate. The total number of operations is then $n(2n - 2 + 2n - 3 + 1) + n = 4n^2 - 3n$.

If we have n processors working in parallel, each processor can compute each term at the same time; $(2n - 2 + 2n - 3 + 1)$ operations are required. We then add these terms. The $(n - 1)$ adds to do this require $\sqrt{N}$ addition-times where $N = \sqrt{N}$ rounded up.

With $2n$ processors, all the numerators and denominators can be computed in parallel; each term then requires only half as many subtract and multiply times.

15. All the second-order differences are the same—they equal 1. That means that $f(x)$ is a quadratic polynomial. $P_2(x) = x^2 - 4x + 3$.

17. a. The third differences are nearly zero; they are all less than 0.0005, meaning that a third-degree polynomial will fit to the desired precision.

b. A second-degree polynomial will fit quite well; the second differences are all less than 0.0016.

c. Fitting a quadratic to three points near the center of the range estimates $f(1.2)$ as 0.1831 (compare to 0.1823), $f(1.5) = 0.4054$ (compare to 0.4055), and $f(1.25) = 0.2234$ (compare to 0.2231).

d. Divided differences of order n are the ordinary differences of order n divided by $n!h^n$.

21. The best choice of points should be $x = 1.25$, 1.3, and 1.35. A quadratic from these gives estimates for $f(1.2) = 0.1822$ (compare to 0.1823), $f(1.4) = 0.3362$ (compare to 0.3365), $f(1.45) = 0.3707$ (compare to 0.3716), and $f(1.5) = 0.4063$ (compare to 0.4055). However, choosing $x = 1.35$, 1.40, and 1.45 gives estimates that match equally well.

26. a. For divided differences, each entry in a column takes one subtract and one divide. There are six first differences, five second differences, and four third differences: $(2)(6 + 5 + 4) = 30$.

b. For ordinary differences, there are the same number of subtracts but no divides: $(1)(6 + 5 + 4) = 15$.

31. a. $y(0.54) = 0.1664$.

b. This should be a better choice because 0.54 is better centered, but the same value for $y(0.54)$ is obtained.

c. A fourth-degree polynomial through the central five points.

36.

$$\begin{bmatrix} 1.220 & 0.490 & 0.000 & 0.000 & -1.326 \\ 0.490 & 1.240 & 0.130 & 0.000 & -1.715 \\ 0.000 & 0.130 & 0.620 & 0.180 & -0.829 \\ 0.000 & 0.000 & 0.180 & 2.440 & -1.865 \end{bmatrix}$$

40. a. Equation (3.9) together with Eq. (3.10) show this directly; the a_i, b_i, c_i, and d_i have the same values throughout the region of fit.

b. It is obvious that, if the coefficients are not all zero, $S_0 = 0$ is not equal to $p''(x_0)$, and $S_n = 0$ is not equal to $p''(x_n)$.

44. A spline curve using end condition 1 deviates most from the function in the first segment, at $x =$ 0.155; the deviation is 0.858. The equation for x in $[-1, 1]$ is

$$0.278x^3 + 0.836x^2 - 1.279x + 0.164.$$

47.

$$[u^4, u^3, u^2, u, 1] \begin{bmatrix} 1 & -4 & 6 & -4 & 1 \\ -4 & 12 & -12 & 4 & 0 \\ 6 & -12 & 6 & 0 & 0 \\ -4 & 4 & 0 & 0 & 0 \\ 1 & 0 & 0 & 0 & 0 \end{bmatrix} [p_0, p_1, p_2, p_3, p_4]^T$$

52.

$$dx/du = -(3/6)\,(1 - u)^2 x_{i-1} + (1/6)\,(6u^2 - 6u)x_i + (1/6)\,(-9u^2 + 6u + 3)x_{i+1} + (3/6)\,(u^2)x_{i+2};\ \text{at } u = 0,\ dx/du = (3/6)\,(x_{i+1} - x_{i-1}).$$

The expression for dy/du is similar, so

$$dy/dx = (y_{i+1} - y_{i-1})/(x_{i+1} - x_{i-1}),$$ which is the slope between points adjacent to p_i.

55. For both Bezier and B-spline curves, changing a single point changes the curve only within the intervals where that point enters the equations. Its influence is localized, in contrast to a cubic spline, where changing any one point affects the entire curve.

59. The same value is obtained: $f(1.6, 0.33) = 1.841$.

62. Because z is linear in x, it is preferred to fit only to y-values. Choose points where y is in [0.2, 0.7] and for $x = 2.5$ and $x = 3.1$. The interpolate from this is $z = 4.5163$. Adding a ninth point ($y = 0.9$) does not change the result.

67. The second normal equations of Eq. (3.25) is

$$a\Sigma x_i + bN = \Sigma Y_i.$$

If this is divided by N, we get

$$a\Sigma x_i/N + b = \Sigma Y_i/N,$$

which proves the assertion.

70. Making $y(4) = 5$ changes the equation the most [part (a)]. The changes in part (b), $y(4) = 0$, and in part (c); $y(4) = 4$, cause the same lesser changes because these added points are the same distance from the line. The equations are

a. $9 - 1.5x$.

b. $7.333 - 1.5x$.

c. $8.667 - 1.5x$.

The original equation is $8 - 1.5x$.

75. $\ln(F) = 3.4083 + 0.49101 * \ln(P)$, or $F = 30.214\, P^{0.49101}$.

77. Fitting polynomials of degrees 3, 4, 5, 6, and 7 gives

Degree	3	4	5	6	7
σ^2	21.14	25.95	2.080	2.674	1.484

The optimal degree is 5.

Chapter 4

5. Write $\cos(6x)$ as

$$\begin{aligned}\cos(3x + 3x) &= \cos(3x) * \cos(3x) - \sin(3x) * \sin(3x)\\ &= 2\cos^2(3x) - 1\\ &= 2[4\cos^3(x) - 3\cos(x)]^2 - 1\\ &= 32\cos^6(x) - 48\cos^4(x) + 18\cos^2(x) - 1.\end{aligned}$$

6. The zeros of $T_4(x)/8 = x^4 - x^2 + 1/8$ are at ±0.923870, ±0.382683. The maximum magnitude on $[-1, 1]$ is 1/4, reached three times, once within the interval and twice at the endpoints.

Comparing the graph of $T_4(x)/8$ to that for $P_4(x)$, that has zeros at ±0.2 and ±0.6 (equally spaced within $[-1, 1]$), we see that the maximum magnitudes are less within the interval but at the endpoints the magnitude is much greater: 0.6144 compared to 0.125.

11. $x^{10} = (1/512)(126T_0 + 210T_2 + 120T_4 + 45T_6 + 10T_8 + T_{10})$.

$x^{11} = (1/1024)(462T_1 + 330T_3 + 165T_5 + 55T_7 + 11T_9 + T_{11})$.

14. The Chebyshev series of degree-2 is

$$\begin{aligned}&0.99748T_0(x) + 0.10038T_1(x) - 0.002532T_2(x)\\ &= 1.000001 + 0.10038x - 0.005064x^2.\end{aligned}$$

Maximum errors: For the Chebyshev series, -0.000130 at $x = -1$, for the truncated Maclaurin series $= -0.000573$ at $x = -1$. The Chebyshev series has a smaller error by a factor of 4.4.

18. Chebyshev polynomials have all their maxima/minima equal to 1 in magnitude in $[-1, 1]$. All Legendre polynomials have maxima/minima equal to 1 at $x = -1$ or $x = +1$ but, their intermediate maxima/minima are less than 1 in magnitude.

22. For $\cos^2(x)$: Maclaurin is $1 - x^2 + x^4/3 - x^5/120 - 2x^6/45$, Padé is $(1 - 2x^2/3)/(1 + x^2/3)$.

At $[-1, 1]$, Maclaurin errors are [0.003038, 0.003038], Padé errors are [0.04193, 0.04193], much larger. The series fits well throughout $[-1, 1]$; the Padé only through $[-0.5, 0.5]$.

For $\sin(x^4 - x)$: Maclaurin is $-x + x^3/6 + x^4 - x^5/120 - x^6/2$,

Padé is

$$(-x + 0.04738x^2 - 0.1676x^3)/(1 - 0.04738x + 0.3343x^2 + 0.9921x^3).$$

Both are poor approximations; the series fits well only within $[-0.6, 0.6]$; Padé fits well only within $[-0.4, 0.4]$.

At $[-1, 1]$, Maclaurin errors are $[-0.4324, 0.3417]$ Padé errors are $[-2.2098, 0.49154]$, much larger.

In contrast to these, the Padé approximation for xe^x is a better approximation. The series is

$$x + x^2 + x^3/2 + x^4/6 + x^5/24 + x^6/120.$$

Padé is

$$(60x + 24x^2 + 3x^3)/(60 - 36x + 9x^2 - x^3).$$

Both fit well throughout $[-1, 1]$; errors for series are

$$[-0.00121, 0.00162], \text{ for Pade } [0.000451, -0.000468].$$

27. The expression is not minimax. If it were, the error curve would have nine equal maxima/minima on [0, 1].

30. a. Periodic, period = 2π.

b. Not periodic.

c. Periodic, period = 2π.

d. Periodic, period = π.

34. The expressions for the A's and B's are complicated. The first few coefficients are

$A_0 = 5/2$

$A_1 = -2.02270$ $B_1 = -0.40528$

$A_2 = 0.92811$ $B_2 = 0.93071$

$A_3 = 0.15198$ $B_3 = -0.90655$

$A_4 = -0.64531$ $B_4 = 0.27386$

37. No, it is true only for $f(x)$ or $g(x)$ equal to a constant.

44. The match to $f(0) = 0$ within 0.00001 requires 31,347 terms (single precision). Some other results:

Terms	100	1000	10,000	20,000	30,000
Error	3.183E−3	3.184E−4	3.190E−5	1.599E−5	1.068E−5

The match to $f(\pi) = \pi$ gives similar results until the number of terms exceeds about 1600 with single precision, but from then on, the error does not decrease. With double precision, the match to within 0.00001 occurs at 31,831 terms. The conclusion seems to be that the same error is obtained at both $x = 0$ and $x = \pi$.

45. The Fourier matches the function at zero and at ±0.68969, ±1.3773, ±2.0584, ±2.7150.

Chapter 5

1. Round off does not show until $\Delta x = 0.05/2^{24}$.

7. The divided-difference table is

x	$f(x)$	1-diff	2-diff	3-diff	4-diff
1.7	−3.2871	0.5878	8.2698	−0.8764	−3.5842
1.8	−3.2283	3.0688	7.7001	−3.7438	
2.0	−2.6146	7.3039	5.0795		
2.35	−0.0582	9.8436			
2.5	1.4183				

The true value of $f'(2.0)$ is 4.7471.

a. Forward difference gives 7.3039.

b. Backward difference gives 3.0688.

c. The central difference requires evenly spaced points, but doing $(f_+ - f_-)/(x_+ - x_-)$ gives 5.7638; the average of parts (a) and (b) is 5.1864.

11. The best points to use are at $x = 0.23, 0.27$, and 0.32. The quadratic through these points is

$$P_2(x) = -2.9833x^2 + 3.2342x - 0.2243,$$
$$P_2'(x) = -5.9666x + 3.2342,$$
$$P_2'(0.268) = 1.6352.$$

13. The recomputed table is

x	$f(x)$	1-diff	2-diff	3-diff
0.15	0.1761	2.4350	−5.7500	15.6253
0.21	0.3222	1.9750	−3.8750	8.1060
0.23	0.3617	1.7425	−2.9833	6.7359
0.27	0.4314	1.4740	−2.1750	
0.32	0.5051	1.3000		
0.35	0.5441			

$f'(0.242) = 1.9750 - 3.8750\,(0.032 + 0.012) = 1.8045$. The error is -0.0099. Truncation causes a greater error than does rounding.

19. a. 1.2505.

b. 1.0792.

c. 1.2925.

22. For $f'(x)$: Multiplier is $1/h$,
coefficients are $[1/12, -2/3, 0, 2/3, -1/12]$.

For $f''(x)$: Multiplier is $1/h^2$,
coefficients are $[-1/12, 4/3, -5/2, 4/3, -1/12]$.

For $f'''(x)$: Multiplier is $1/h^3$,
coefficients are $[-1/2, 1, 0, -1, 1/2]$.

For $f^{(4)}(x)$: Multiplier is $1/h^4$,
coefficients are $[1, -4, 6, -4, 1]$.

27. Using double precision, the Richardson table is

0.157021273		
0.157217754	0.157283248	
0.157266897	0.157283278	0.157283280

Exact value $= 0.157283$; the estimate agrees to six places.

31. a. Analytical value $= 0.015225$, trapezoidal rule gives 0.01995, error is -0.004725, $\xi = 0.35$.

b. Analytical value $= 0.316565$, trapezoidal rule gives 0.318936, error is -0.002371, $\xi = 0.0524$.

c. Analytical value $= 0.078939$, trapezoidal rule gives 0.077884, error is 0.001055, $\xi = 0.1992$.

For each part, the value of ξ is near the midpoint.

35. a. 1.7684.

b. 1.7728.

c. 1.7904.

38. With 1431 intervals ($h = 0.00112$), value is 23.914454, error $= -2.5$ E–6.

41. $h = 0.1$: 1.76693.
$h = 0.2$: 1.76693.
$h = 0.4$: 1.76720.

46. For n, an even integer, let T_h, T_{2h}, be trapezoidal rule integrals with step sizes h and $2h$. It is easy to show that

$$T_h - T_{2h} = (h/2)\,(-f_0 + 2f_1 - 2f_2 + \ldots - f_n), \text{ from which}$$
$$T_h + (1/3)\,(T_h - T_{2h}) = (h/3)\,(f_0 + 4f_1 + 2f_3 + 4f_3 + \ldots + f_n),$$

which is Simpson's 1/3 rule.

50. $c_0 = -9h/24$, $c_1 = 37h/24$, $c_2 = -59h/24$, $c_3 = 55h/24$.

52. With 12 intervals, integral = 0.946083, error = −1.0E−7.

55. (a)

	Anal.	**Trap.**		**Anal.**	**Trap.**
A_0	11.5000	11.5234			
A_1	4.1035	4.1270	B_1	−7.2882	−7.2391
A_2	1.0259	1.0496	B_2	−4.1339	−4.0353
A_3	0.4560	0.4801	B_3	−2.8164	−2.6677
A_4	0.2565	0.2812	B_4	−2.1282	−1.9283

(b)

	Anal.	**Trap.**		**Anal.**	**Trap.**
A_0	6.0000	6.0104			
A_1	−2.5659	−2.5616	B_1	0.6245	0.6390
A_2	0.5990	0.5748	B_2	−0.8723	−0.8705
A_3	0.2026	0.2134	B_3	0.6366	0.6036
A_4	−0.4705	−0.4375	B_4	−0.1400	−0.1082

(c)

	Anal.	**Trap.**		**Anal.**	**Trap.**
A_0	7.0315	6.4430			
A_1	5.8064	5.2159	B_1	−8.5227	−8.7310
A_2	−8.1319	−8.7287	B_2	−19.9476	−20.3658
A_3	−21.3088	−21.9165	B_3	2.5787	1.9479
A_4	−10.0939	−10.7172	B_4	8.3339	7.4857

59. Multiply the matrices, add exponents of (W^i) (W^j), write $W^0 = 1$, write W^n as $W^{n \bmod 4}$, then unscramble the rows.

68. For any value of TOL ≥ 0.002, the same result is obtained after five iterations, 0.6773188, which has an error of 7E–6. The analytical answer is 0.677312.

70. Break the interval into subintervals: $[0, 1]$, $[1, \pi/2]$.

74. Correct value is -0.700943. Even five terms in the Gaussian formula is not enough. Simpson's 1/3 rule attains five digits of accuracy with 400 intervals. The result from an extrapolated Simpson's rule gets this in seven levels, using 128 intervals.

76. The values are readily confirmed.

81. a.

$$\frac{\Delta x}{3}\frac{\Delta y}{2}\begin{Bmatrix} 1 & 2 & 2 & 1 \\ 4 & 8 & 8 & 4 \\ 2 & 4 & 4 & 2 \\ 4 & 8 & 8 & 4 \\ 1 & 2 & 2 & 1 \end{Bmatrix}.$$

b.

$$\frac{\Delta x}{3}\frac{\Delta y}{3}\begin{Bmatrix} 1 & 4 & 2 & 4 & 1 \\ 4 & 16 & 8 & 16 & 4 \\ 2 & 8 & 4 & 8 & 2 \\ 4 & 16 & 8 & 16 & 4 \\ 1 & 4 & 2 & 4 & 1 \end{Bmatrix}.$$

c.

$$\frac{3\Delta x}{8}\frac{3\Delta y}{8}\begin{Bmatrix} 1 & 3 & 3 & 2 & 3 & 3 & 1 \\ 3 & 9 & 9 & 6 & 9 & 9 & 3 \\ 3 & 9 & 9 & 6 & 9 & 9 & 3 \\ 2 & 6 & 6 & 4 & 6 & 6 & 2 \\ 3 & 9 & 9 & 3 & 9 & 9 & 3 \\ 3 & 9 & 9 & 3 & 9 & 9 & 3 \\ 1 & 3 & 3 & 2 & 3 & 3 & 1 \end{Bmatrix}.$$

d. for *a*: Any number in the *y*-direction, even number in the *x*-direction.

for *b*: Even number in both directions.

for *c*: Divisible by 3 in both directions.

86. Analytical value = 2/3.

Δx	Δy	Integral	Error	Error/h^2
0.5	0.5	0.75	−0.0833	−0.3333
0.25	0.5	0.71875	−0.05208	−0.3333*
0.5	0.25	0.71875	−0.05208	−0.3333*
0.25	0.25	0.6875	−0.0208	−0.3333
0.125	0.125	0.6719	−0.0052	−0.3333

*Using the average of the squares of the *h*-values.

90.

	x	End Condition 1	3	4	Exact value	Central diff. ($h = 0.1$)
$f'(x)$:	1.5	−0.0841	−0.0823	−0.0819	−0.0816	−0.0817
	2.0	−0.0627	−0.0630	−0.0632	−0.0625	−0.0625
	2.5	−0.0489	−0.0497	−0.0494	−0.0494	−0.0494
$f''(x)$:	1.5	0.0596	0.0467	0.0440	0.0466	0.0466
	2.0	0.0257	0.0307	0.0310	0.0313	0.0313
	2.5	0.0296	0.0227	0.0240	0.0219	0.0220

94. Value = 1.29919; Simpson's rule: 1.30160; exact: 1.30176.

Chapter 6

2. The correct answer is 1.59420. Eight terms of the Taylor series gives this result, seven terms gives 1.58421, six terms gives 1.59418.

6. With $h = 1/2^{15}$, single precision gives 1.59419; double precision (rounded) gives 1.59420.

10. Equation is $dv/dt = 32.2 - cv^{3/2}$, $v(0) = 0$. At 80 mi/hr (117.333 ft/sec) $dv/dt = 0$, giving $c =$ 0.025335.

t:	0.2	0.4	0.6	0.8	1.0	1.2
v:	6.3986	12.6822	18.7997	24.7198	30.4209	35.8883
t:	1.4	1.6	1.8	2.0		
v:	41.1127	46.0888	50.8149	55.2919		

13. a. The concavity of $y(x)$; if concave upward, the simple Euler method will have positive errors; the computed values lag behind the true values. If concave downward, the errors will be negative.

b. Example 1: $dy/dx = e^x$ always has positive errors,

Example 2: $dy/dx = -e^x$ always has negative errors.

c. When concavity changes from upward to downward and repeats. Example: $y = \exp(x^4) - \exp(x^2)$.

20. Interpolating linearly between $v(6.0)$ and $v(6.5)$, $v = 105.60$ ft/sec at $t = 6.36$ sec. Distance traveled is about 435 ft.

24. If the answer is rounded, $h = 0.25$ gives 3.32332; all digits are correct.

27. The equations, in matrix form, are

$$\begin{bmatrix} 1 & 1 & 1 \\ 0 & -h & -2h \\ 0 & h^2 & 4h^2 \end{bmatrix} \begin{bmatrix} c_0 \\ c_1 \\ c_2 \end{bmatrix} = \begin{bmatrix} h \\ h^2/2 \\ h^3/3 \end{bmatrix},$$

which has the solution $c_0 = 23h/12$, $c_1 = -16h/12$, $c_2 = 5h/12$.

30.

t:	1.6	1.8	2.0
y:	0.25200	0.49067	0.8333

Computed values are the same as the analytical; y is a cubic polynomial.

36. Eq. (6.18) gets exactly the analytical values for $y(10)$, (which is 120), with $h = 0.2$ or even $h = 1.0$. This is because the derivative function is linear. The modified Euler method also gets the analytical result with $h = 0.2$ and with $h = 1$. The Euler method with $h = 0.2$ gives $y(10) = 118.2$.

41. Let $y' = z$ so that $y'' = z'$. Then we have

$$y' = z, \quad y(0) = 0;$$
$$EIz' = M(1 + z^2)^{3/2}, z(0) = 0.$$

43. At $t = 1.0$, $x = 1.25689$, $y = 1.56012$. If the solution is extended beyond $t = 2$, the x-values increase rapidly and cause overflow near $t = 2.35$.

46. Let $y_1' = u$, $y_2' = v$, then

$$y_1' = u, v_1(0) = \mathrm{A},$$
$$m_1u' = -k_1y_1 - k_2(y_1 - y_2), u(0) = \mathrm{B},$$
$$y_2' = v, y_2(0) = \mathrm{C},$$
$$m_2v' = k_2(y_1 - y_2), v(0) = \mathrm{D}.$$

52. The eigenvalues are -1 and 39; they differ in magnitude but are not both negative. When all the elements of the matrix are positive, the eigenvalues are exactly the same. In contrast, the eigenvalues for the matrix of Eq. (6.22) are -2 and -800, showing that it is very stiff.

60. a.

0	**y**	**% error**
0	0	0
$\pi/4$	0.77015	0.625
$\pi/2$	1.42153	0.518
$3\pi/4$	1.85370	0.321
π	2	0

b. With $h = \pi/5$, largest error is 0.404%.

c. Shooting has a maximum error $< 0.5\%$, with $h = \pi/2$.

70. If h is too small, round-off errors can distort the solution. It can also increase the size of the system of equations beyond the capacity of the computer to solve them.

73. The exact answer is 2.46166.

a. $(h = 1/2)$: $k = 2.0000$,

b. $(h = 1/3)$: $k = 2.25895$,

c. $(h = 1/4)$: $k = 2.34774$,

d. Extrapolated: $k = 2.46366$.

79. Characteristic polynomial is $x^3 + 7x^2 - 58x - 319$; roots are $7.2024, -9.5783, -4.6241$. The eigenvalues of A^{-1} are the reciprocals: $0.1308, -0.1044, -0.2163$. They have the same eigenvectors; the vector corresponding to the first of the eigenvalues is $[-0.0723, 0.0570, -0.9958]$.

For A^{-1}, the polynomial is $1/319(319x^3 + 58x^2 - 7x - 1)$. The coefficients are the negatives of that for A, in reverse order, and scaled by $1/\det(A) = 1/319$.

83. The upper Hessenberg matrix:

$$\begin{bmatrix} 3 & 21 & -2.375 & 7 \\ 1 & 0 & 0.625 & -1 \\ 0 & 8 & -2.125 & 5 \\ 0 & 0 & -5.8281 & 7.125 \end{bmatrix}.$$

Chapter 7

4. If $f(x) = ax^2 + bx + c$, $f_x = 2ax + b = 0$ gives $x_{\min} = -b/2a$ and $f_{\min} = c - b^2/(4a)$. The maximum, if there is one, is at the same x-value.

8. With $f(x) = 2x^2 - e^{x/2}$, and starting from $x = 0$ with $\Delta x = 0.1$, f-values increase at $x = 0.2$. Reversing with $\Delta x = -0.01$, they decrease but increase at $x = 0.12$. Reversing again with $\Delta x = 0.001$, they decrease but stop decreasing at $x = 0.134$. The f-value at $x = 0.133$ is the same, -1.03338. Interpolating, we arrive at $x = 0.1335$, $f = -1.03338$. This compares well to the exact answer, $f = -1.03338$ at $x = 0.133637$.

11. The most narrowing occurs when the two x-values are at the midpoint $\pm\,\varepsilon$, where ε is the smallest value not zero. The least narrowing occurs when each x-value is within ε of the endpoints.

15. For $f(x) = (x^2 - x)^2 + x - 5$, $f_x = 2(x^2 - x)(2x - 1) + 1$. Setting $f_x = 0$ and solving, we get -0.260689 where f is -5.15268.

21. a. With $n = 20$, the ratio is 0.618034, correct to six digits.

b. We get 0.6181818 with $n = 9$, in error by only 0.000148.

22. The least f-value is -0.44 at $(-1.6, -0.2)$. The analytical value is -0.454545 at $(-1.6363, -0.2727)$. The table required 441 computations of the function.

27. Steepest descent from $(0, 0)$ turns out to be a univariant search. We begin along the negative x-axis (the negative gradient) and stop at $(-3/2, 0)$. The negative gradient there points downward; we move to $(-3/2, -1/4)$. The next movement is parallel to the x-axis, to $(-13/8, -1/4)$. Eventually, we will arrive at $(-1.63636, -0.272727)$, where $f = -0.454545$.

31. The analytical answer is $f = -5/11$ at $(-18/11, -3/11)$. Starting from $(0, 0)$:

$$H = \begin{bmatrix} 2 & -1 \\ -1 & 6 \end{bmatrix}, \quad H^{-1} = 1/11 * \begin{bmatrix} 6 & 1 \\ 1 & 2 \end{bmatrix}, \quad \nabla f = [3, \quad 0]^T$$

$$x_{\min} = [0, \quad 0]^T - H^{-1} * \nabla f = 1/11 * [-18, \quad -3]^T.$$

Starting from $(-2, 0)$: H and H^{-1} are the same.

$$\nabla f = [-1, \quad 2]^T, x_{\min} = [-2, \quad 0]^T - H^{-1} * \nabla f = 1/11 * [-18, \quad -3]^T.$$

Starting from $(-2, -2)$: H and H^{-1} are the same.

$$\nabla f = [1, \quad -10]^T, x_{\min} = [-2, \quad -2]^T - H^{-1} * \nabla f = 1/11 * [-18, \quad -3]^T.$$

Starting from $(0, -2)$: H and H^{-1} are the same.

$$\nabla f = [5, \quad -12]^T, x_{\min} = [0, \quad -2]^T - H^{-1} * \nabla f = 1/11 * [-18, \quad -3]^T.$$

We arrive at the exact answer from each corner of the square.

34. At the minimum point, (1, 1), the exact value for the minimum, $f = 0$, is obtained. There is no round-off error here because all quantities in the computation of f are integers. At all other points in the table, single precision gives exact values because no term has more than 5 significant digits.

37. We need these quantities:

$$f_x = 400x^3 + 2x(1 - 200y) - 2,\ f_y = 200y - 200x^2,$$
$$f_{xx} = 1200x^2 - 2(200y - 1),\ f_{yy} = 200,\ f_{xy} = f_{yx} = -400x.$$

From these,

$$H = \begin{bmatrix} 2 & 0 \\ -0 & 200 \end{bmatrix},\quad H^{-1} = \begin{bmatrix} 1/2 & 0 \\ 0 & 1/200 \end{bmatrix}.$$

At (0, 0),

$$\nabla f = [-2,\quad 0]^T,\quad \text{so } H^{-1} * [-2,\quad 0]^T = [-1,\quad 0]^T.$$

This gives

$$x_1 = [0,\quad 0]^T - [-1,\quad 0]^T = [1,\quad 0]^T.$$

At (1, 0),

$$\nabla f = [400,\ -200]^T \quad \text{and} \quad H^{-1} = \begin{bmatrix} 1/402 & 1/201 \\ 1/201 & 601/40200 \end{bmatrix}$$

giving

$$x_2 = [-1,\quad 0]^T - H^{-1} * \nabla f = [1,\quad 0]^T - [0,\quad -1]^T = [1,\quad 1]^T;$$

so we get exactly the correct answer in two steps.

42. There are four corner points at which $f(x, y)$ is

$(x, y) =$	(0, 0)	(0, 10)	(6, 0)	(2, 8)	
$f(x, y) =$	0	30	42	38	max is 42 at (6, 0).

45. a. This constraint does not intersect the feasible region, so it is redundant; no effect.

b. This constraint cuts the feasible region and produces two new corners, at (9/2, 0) and (27/5, 6/5). At these points, the function has values 31.5 and 41.4. The optimum is reduced to 41.4.

c. This constraint also cuts the original feasible region. There are four corner points:

$(x, y) =$	(0, 3)	(0, 10)	(2, 8)	(27/5, 6/5)	
$f(x, y) =$	9	30	38	41.4	max is 41.4 at (27/5, 6/5).

50. There are many possible combinations of constraints.

No corner on axes: $x + y \le 10,\ x, y \ge 2$.

No feasible region: $x + y \le 4,\ x + y \ge 8$.

No combination gives a feasible region.

57. The primal has the solution:

$$x_1 = 120, x_2 = 0, x_3 = 0, x_4 = 35, f = 205.$$

To construct the dual, all constraints must be $\leq$ so we rewrite the equality as two constraints, $x_4 \leq 35, -x_4 \leq -35$.

The dual then is

$$\begin{aligned} &\text{Min } g(y) = 120y_1 + 35y_2 - 35y_3, \text{ subject to} \\ &2y_1 \geq 3, \\ &3y_1 \geq 1, \\ &y_2 - y_3 \geq -1, \\ &y_1 \geq 2, \end{aligned}$$

which has the solution $y_1 = 2, y_2 = 0, y_3 = 1, g = 205$.

60. For two variables: (1) If the magnitude of the slope of the objective function becomes greater than or less than the slopes of the constraints that define the optimal point, (2) if the magnitude of the slopes of the constraints that define the optimal become greater than or less than the slope of the objective function.

For three variables: similar except we are dealing with planes.

64. The objective function defines a parabola; the constraints define a fesible region with four vertices, at (1, 0), (1, 4), (4, 1), and (3, 0). The solution is $f = 18$ at (4, 1), the corner point where the parabola touches the feasible region.

70. The optimum of Exercise 64 is at (4, 1). A straight line that has the same values as the objective function at $x = 3$ and at $x = 4$ is $y = 15 - 7x/2$. This linear objective touches the feasible region at (4, l) where $f = 18$ (the same as for the nonlinear objective). Fitting to other straight lines near $x = 4$ will have the same result.

71. Starting from (0, 0, 0) we find the solution is at $(6\frac{2}{3}, 6\frac{2}{3}, -3\frac{1}{3})$ where $f = 140$. The same result is obtained with other starting values unless these are all negative or are all greater than $16\frac{2}{3}$ where a different solution is found.

75. The solution is again obvious. Ship these amounts:

From/To	Atlantic City	Chicago	Los Angeles	Denver
Mississippi	300	200		
Mexico	0	200	400	200
Cost to ship	$7,500	$29,000	$13,200	$11,000

where the total shipping cost is $60,700, which exceeds the costs of the original configuration by $4,400. This scheme is only better if the costs to supply customers from Denver and to establish that facility are reduced by more than $4,400.

78. a. The solution is $f(x, y) = 96/9 = 10.667$ at $(56/9 = 6.222, 20/9 = 2.222)$. The nearest point with integer coordinates is (6, 2) where f is 10. The rounded value for the solution to the original problem is 11.

b. If x can only have integer va1ues, the feasible region is defined by a sequence of points at these x-values. The objective function with x restricted to integers will match to the feasible region at

(6, 17/7 = 2.29) where the objective has a value of 74/7 = 10.5714, not much different from the value with x unrestricted.

80. The shop is open for 32 15-minute periods. To simplify, assume that customers enter only at the start of a period. The problem can be solved by setting up these variables:

B shows if the barber is busy of not, a Boolean variable.

A, the number of customers who enter together.

Q, the length of the queue, the number who must wait.

P, the period, which varies from 1 to 32.

Use a random number function to generate random integers from 1 to 6, of which two are selected to represent one customer entering, one to represent two entering, and three to represent none. Begin with $B = 0$ (not busy), $Q = 0$ (no customers waiting). Then, for each period in turn, get the value for A:

If $B = 0$, and

$A = 0$ and $Q = 0$, go to next period.

$A = 1$ and $Q = 0$ set $B = 1$, go to next.

$A = 1$ and $Q \neq 0$ set $B = 1$, go to next.

$A = 2$ set $B = 1$, $Q = Q + 1$, go to next.

If $B = 1$, and

$A = 0$ and $Q = 0$ set $B = 0$, go to next.

$A = 1$ and $Q = 0$, go to next.

$A = 1$ and $Q \neq 0$, go to next.

$A = 2$ and $Q \neq 0$, set $Q = Q + 1$, go to next.

The results will ordinarily be different for each trial when the random numbers are different. For one trial, we found that one customer arrived in 13 periods (expected value is $10\frac{2}{3}$, two arrived in 6 periods (expected value is $5\frac{1}{3}$, so this was a good day. The barber was idle for 7 periods and the maximum length of the queue (number waiting) was 2. He served 25 customers. The maximum number he could serve in a day is 32 and he would never be idle.

Chapter 8

3.

$$u_y = (u_{i,\, j+1} - u_{i,\, j-1})/2h.$$

$$u_{xy} = \frac{(u_{i+1,j+1} - u_{i+1,j-1})/2h - (u_{i-1,j+1} - u_{i-1,j-1})/2h}{2h}$$

$$= \frac{u_{i+1,j+1} + u_{i-1,j-1} - u_{i+1,j-1} - u_{i-1,j+1}}{4h^2},$$

which is the same as the given operator.

5.

$$\frac{1}{12h^2}\left\{\begin{matrix} & & -1 & & \\ & & 16 & & \\ -1 & 16 & -60 & 16 & -1 \\ & & 16 & & \\ & & -1 & & \end{matrix}\right\} u_{i,j}$$

11. Interior temperatures:

64.21	105.20	146.65	186.41
61.63	89.94	114.99	134.00
52.39	77.93	89.38	84.59

14. Interior temperatures, with a tolerance of 0.00001:

64.21	105.20	146.65	186.41
61.63	89.94	114.99	134.00
52.39	77.93	89.38	84.59

With initial values all equal to zero, 31 iterations were needed. With initial values all equal to 300, 32 iterations were needed. With initial values all equal to 93.89 (the average of the boundary temperatures), 27 iterations were needed. The final values are not exactly the same for these three cases.

21. Values at interior points, laid out as in the figure:

	93.40		
82.13	73.62	67.56	54.39
54.91	51.37	42.23	
36.13	34.72		

25. There are six "layers" of nodes; each layer has 6 * 6 = 36 nodes; the total number of nodes is 6 * 36 = 216, so there are 216 equations. There are three sets of these, one for each direction (x, y, z). Even though each system is tridiagonal, getting a convergent solution is not done quickly.

27. Using $k = 2.156$ Btu/(hr * in^2 * (°F/in))

a. -29.53 °F/in

b. -75.59 °F/in

c. -34.91 °F/in

31. With units of Btu, lb, in., sec, °F: $k = 0.00517$, $c = 0.0919$, $\rho = 0.322$. With $\Delta x = 1$ in., $\Delta t = 2.862$ sec. Using $r = 0.5$, at $t = 28.62$:

x:	0	1	2	3	4	5	6	7	8
u:	100	85.94	73.44	60.94	50.00	39.06	26.56	14.06	0

34.

34.722	38.589	50.744	70.106	100.00
29.644	33.296	45.066	65.376	100.00

19.495	21.816	30.152	49.058	100.00
0.000	0.000	0.000	0.000	

These values are within 3.5° of the steady-state values.

39. After 22 time steps, a single error grows to become larger than the original error and then continues to grow by a factor of 1.0485 at each succeeding time step.

41. After seven time steps, the maximum error has decreased to 0.219 times the original error. As time increases, the maximum error decreases by a factor of 0.875 for two time steps and this factor gets smaller as time progresses.

44.

N:	3	3	3	3
r:	0.5	1.0	2.0	3.0
Eigenvalue:	0.7735	0.6306	0.4605	0.3627
N:	4	4	4	4
r:	0.5	1.0	2.0	3.0
Eigenvalue:	0.8396	0.7236	0.5669	0.4660

46. The discriminant is $4(1 - x)^2 + 4(1 + y)(1 - y)$. When set to zero, this describes a hyperbola whose center is at (1, 0) and whose vertices are at (1, 1) and (1, −1). The equation is parabolic at points on this curve. Above the upper branch and below the lower branch, it is elliptic. Between the two branches, it is hyperbolic.

50. a. $\Delta t = 3$ sec. Displacements versus time:

t	$x = 0$	6	12	18	24	30	36	42	48
0.00	0.00	−0.11	−0.19	−0.23	−0.25	−0.23	−0.19	−0.11	0.00
3.00	0.00	−0.09	−0.17	−0.22	−0.23	−0.22	−0.17	−0.09	0.00
6.00	0.00	−0.06	−0.13	−0.17	−0.19	−0.17	−0.13	−0.06	0.00
			. . . . .						
45.00	0.00	−0.09	−0.17	−0.22	−0.23	−0.22	−0.17	−0.09	0.00
48.00	0.00	−0.11	−0.19	−0.23	−0.25	−0.23	−0.19	−0.11	0.00

c. $\Delta t = 3$ sec.

t	$x = 0$	6	12	18	24	30	36	42	48
0.00	0.00	0.00	0.00	0.00	0.00	0.00	0.00	0.00	0.00
3.00	0.00	0.33	0.56	0.70	0.75	0.70	0.56	0.33	0.00
6.00	0.00	0.56	1.03	1.31	1.41	1.31	1.03	0.56	0.00
			. . . . .						
48.00	0.00	0.00	0.00	0.00	0.00	0.00	0.00	0.00	0.00
51.00	0.00	0.33	0.56	0.70	0.75	0.70	0.56	0.33	0.00

d. $\Delta t = 3$ sec.

t	$x = 0$	6	12	18	24	30	36	42	48
0.00	0.00	0.25	0.50	0.75	1.00	0.75	0.50	0.25	0.00
3.00	0.00	−0.50	−1.00	−1.50	−2.25	−1.50	−1.00	−0.50	0.00
6.00	0.00	−1.25	−2.50	−4.00	−4.00	−4.00	−2.50	−1.25	0.00
			. . . . .						
45.00	0.00	1.00	2.00	3.00	3.75	3.00	2.00	1.00	0.00
48.00	0.00	0.25	0.50	0.75	1.00	0.75	0.50	0.25	0.00

55. With $\Delta x = 0.3$, $\Delta t = 0.003344$ sec. After three time steps ($t = 0.01003$), $y(1.5) = 0.0067334$ ft $=$ 0.0808 in. (same as analytical). Other values agree with the series solution.

58. Assuming that the initial displacements form a pyramid with flat faces whose peak is at (1, 1). Using $\Delta x = \Delta y = 0.5$, Δt is 0.00544 sec. There appears to be no repetitive pattern. The initial displacements are

0.000	0.000	0.000	0.000	0.000	0.000	0.000
0.000	0.500	0.500	0.500	0.500	0.250	0.000
0.000	0.500	1.000	0.750	0.500	0.250	0.000
0.000	0.500	0.500	0.500	0.500	0.250	0.000
0.000	0.000	0.000	0.000	0.000	0.000	0.000

Some values for the node at (2, 1):

Steps:	0	1	2	4	6	8	10	14
$u(2, 1)$:	0.500	0.500	0.250	−0.234	−0.625	0.313	0.897	−0.932

Chapter 9

2. Let $u(x) = c(x)(x - 1)$. The Rayleigh–Ritz integral gives $2c/3 + 0 = 2(5/12)$, so $c = 5/4$. Some values:

x:	0	0.2	0.4	0.6	0.8	1.0
u:	0	−0.200	−0.300	−0.300	−0.200	0
Analytical:	0	−0.176	−0.288	−0.312	−0.224	0

6. $R(x) = y'' - 3x - 1$. If $u = cx(x - 1)$, $u'' = 2c$. Since there is only one constant, set $R = 0$ at $x =$ 1/2. We then have $2c - 3(1/2) - 1 = 0$, giving $c = 5/4$. This is identical to the answer of Exercise 2.

11. a. $N_L = (x - 0.45)/(-0.12)$, $N_R = (x - 0.33)/0.12$.

d. The best averages when the functions are nonlinear are the integrals over the element boundaries divided by the width of the interval. This gives $F_{av} = 2.153$ and $Q_{av} = 0.3800$. However, these differ little from the values at the midpoint of the interval: 2.1521 and 0.3802.

14.

x:	1.0	1.2	1.5	1.75	2
$u(x)$:	−1	−0.2307	0.9174	1.9197	3
Analytical:	−1	−0.2267	0.9167	1.9196	3

20. The augmented matrix is

$$\begin{bmatrix} -974.54 & -488.12 & -488.72 & 1.738 \\ -488.12 & -975.41 & -487.86 & 1.738 \\ -488.72 & -487.85 & -974.81 & 1.738 \end{bmatrix}.$$

23. The element equations are formed from

$$c_{i,j} = 0.2825 \text{ if } i = j,\ 0.1412 \text{ if } i \neq j,$$

$$[K] = \frac{k * A}{c\rho}\begin{bmatrix} 0.489 & 0.089 & -0.573 \\ 0.089 & 0.196 & -0.285 \\ -0.573 & -0.285 & 0.857 \end{bmatrix}, \text{ (A is area, 1.695);}$$

$$b_i = 0.565\,F_{av}.$$

28. $(2 + r)\{c^{m+1}\} = (2 - r)(c^m) + 2r[K^{-1}]\{b\}.$

31. The element equations for $t = t_1$ are

$$\begin{bmatrix} 2 & 1 \\ 1 & 2 \end{bmatrix}\{c\}^1 = (1/2)\begin{bmatrix} -2 & 8 \\ 8 & -2 \end{bmatrix}\{c\}^0.$$

32. Yes, closer nodes are helpful when the function is nonlinear.

37. The temperature difference between successive contour lines is 100/21 = 4.76°. The initial temperature at the corner is 100°. Interpolating between contours:

Time	0	0.1	0.2	0.4	0.8	10.0
Contour		19.17	16.05	10.10	5.38	3.64
Temperature	100.0	91.3	76.4	48.1	25.6	17.3

The plot shows an S-shaped curve. Fitting to least-squares polynomials of degrees 3 and 4 gets matches to the points within 3.7° and 1.0°. A better fit will result from the use of an equation of the form $1/y = a + be^{-x}$ or the so-called Gompertz relation: $y = a * b^{c^x}$.

References

Acton, F. S. (1970). *Numerical Methods That Work.* New York: Harper and Row.

Aki, S. G. (1989). *The Design and Analysis of Parallel Algorithms.* Englewood Cliffs, NJ: Prentice-Hall.

Allaire, P. W. (1985). *Basics of the Finite Element Method.* Dubuque, IA: Brown.

Allen, D. N. (1954). *Relation Methods.* New York: McGraw-Hill.

Anderson, E., Z. Gai, C. Bishof, J. Demmel, and J. Dongarra et al. (1996). *LAPACK Users' Guide.* 2nd ed. Philadelphia: SIAM.

Andrews, G., and R. Olsson (1993). *The SR Programming Language.* Redwood City, CA: Benjamin Cummings.

Andrews, Larry C. (1985). *Elementary Partial Differential Equations with Boundary Value Problems.* Philadelphia: Saunders College Publishing.

Arney, David C. (1987). *The Student Edition of DERIVE, Manual.* Reading, MA: Addison-Wesley—Benjamin Cummings.

Atkinson, Kendall E. (1989). *An Introduction to Numerical Analysis.* 2nd ed. New York: Wiley.

Bartels, Richard, J. Beatty, and B. Barsky (1987). *An Introduction to Splines for Use in Computer Graphics and Geometric Modeling.* Los Altos, CA: Morgan Kaufmann.

Bertsekas, Dimitri, and John Tsitsiklis (1989): *Parallel and Distributed Computation: Numerical Methods.* Englewood Cliffs, NJ: Prentice-Hall.

Birkhöff, Garrett, Richard Varga, and David Young (1962). Alternating direction implicit methods. *Advances in Computers* 3:187–273.

Boisvert, R. (1994) NIST's GAMS: A "card catalog" for the computer user. *SIAM NEWS* 27.

Borse, G. J. (1997). *Numerical Methods with MATLAB.* Boston: ITP.

Bracewell, Ronald N. (1986). *The Hartley Transform.* New York: Oxford University Press.

Brigham, E. Oron (1974). *The Fast Fourier Transform.* Englewood Cliffs, NJ: Prentice-Hall.

Burden, Richard L., and J. Douglas Faires (2001). *Numerical Analysis.* 7th ed. Pacific Grove, CA: Brooks/Cole.

Burnett, David S. (1987). *Finite Element Analysis: From Concepts to Applications.* Reading, MA: Addison-Wesley.

Campbell, Leon, and Laizi Jacchia (1941). *The Story of Variable Stars.* Philadelphia: Blakiston.

Carnahan, Brice (1964). *Radiation Induced Cracking of Pentanes and Dimethylbutanes.* Ph.D. dissertation, University of Michigan.

Carnahan, Brice, et al. (1969). *Applied Numerical Methods.* New York: Wiley.

Carslaw, H. S., and J. C. Jaeger (1959). *Conduction of Heat in Solids.* 2nd ed. London: Oxford University Press.

Chandy, K. M., and S. Taylor (1992). *An Introduction to Parallel Programming.* Boston: Jones and Bartlett.

Chapman, Stephen J. (2000). *MATLAB Programming for Engineers.* Pacific Grove, CA: Brooks/Cole.

Chapra, Stephen C., and Raymond P. Canale (2002). *Numerical Methods for Engineers.* 4th ed. New York: McGraw-Hill.

Char, B., K. Geddes, G. Gonnet, B. Leong, M. Monagan, and S. Watt (1992). *First Leaves: A Tutorial Introduction to Maple V.* New York: Springer-Verlag.

Cheney, Ward, and David Kincaid (1999). *Numerical Mathematics and Computing.* 4th ed. Pacific Grove, CA: Brooks/Cole.

Condon, Edward, and Hugh Odishaw, eds. (1967). *Handbook of Physics.* New York: McGraw-Hill.

Conte, S. D., and C. de Boor (1980). *Elementary Numerical Analysis.* 3rd ed. New York: McGraw-Hill.

Cooley, J. W., and J. W. Tukey (1965). An algorithm for the machine calculations of complex Fourier series. *Mathematics of Computation* 19:297–301.

Corliss, G., and Y. F. Chang (1982). Solving ordinary differential equations using Taylor series. *ACM Transactions on Mathematical Software* 8:114–144.

Corliss, Robert M. (2002). *Essential Maple 7.* 2nd ed. New York: Springer-Verlag.

Crow, Frank (1987). Origins of a teapot. *IEEE Computer Graphics and Applications* 7(1):8–19.

Datta, B. N. (1995). *Numerical Linear Algebra and Applications.* Pacific Grove, CA: Brooks/Cole.

Davis, Alan J. (1980). *The Finite Element Method.* Oxford: Clarendon Press.

Davis, Phillip J., and Phillip Rabinowitz (1967). *Numerical Integration.* Waltham, MA: Blaisdell.

de Boor, C. (1978). *A Practical Guide to Splines.* New York: Springer-Verlag.

De Santis, R., F. Gironi, and L. Marelli (1976). Vector-liquid equilibrium from a hard-sphere equation of state. *Industrial and Engineering Chemistry Fundamentals* 15(3):183–189.

Dongarra, J., I. Duff, D. Sorensen, and H. van der Vorst (1991). *Solving Linear Systems on Vector and Shared Memory Computers.* Philadelphia: SIAM.

Dongarra, J. J., J. R. Bunch, C. B. Moler, and G. W Stewart. (1979). *LINPACK User's Guide.* Philadelphia: SIAM.

Douglas, J. (1962). Alternating direction methods for three space variables. *Numerical Mathematics* 4:41–63.

Duffy, A. R., J. E. Sorenson, and R. E. Mesloh (1967). Heat transfer characteristics of belowground LNG storage. *Chemical Engineering Progress* 63(6):55–61.

Edgar, Thomas F., David M. Himmelblau, and Leon S. Lasdon (2001). *Optimization of Chemical Processes.* 2nd ed. New York: McGraw-Hill.

Etter, D. M. (1993). *Quattro Pro—A Software Tool for Engineers and Scientists.* Redwood City, CA: Benjamin/Cummings.

Fausett, Laurene (2002). *Numerical Methods Using MathCad.* Englewood Cliffs, NJ: Prentice-Hall.

Fike, C. T. (1968). *Computer Evaluation of Mathematical Functions.* Englewood Cliffs, NJ: Prentice-Hall.

Fletcher, R. (1987). *Practical Methods of Optimization.* 2nd ed. New York: Wiley.

Forsythe, G. E., M. A. Malcolm, and C. B. Moler (1977). *Computer Methods for Mathematical Computation.* Englewood Cliffs, NJ: Prentice-Hall.

Forsythe, G. E., and C. B. Moler (1967). *Computer Solution of Linear Algebraic Systems.* Englewood Cliffs, NJ: Prentice-Hall.

Fox, L. (1965). *An Introduction to Numerical Linear Analysis.* New York: Oxford University Press.

Gear, C. W. (1967). The numerical integration of ordinary differential equations. *Mathematics of Computation* 21:146–156.

Gear, C. W. (1971). *Numerical Initial Value Problems in Ordinary Differential Equations.* Englewood Cliffs, NJ: Prentice-Hall.

Gockenbach, Mark S. (2002). *Partial Differential Equations: Analytical and Numerical Methods.* Philadelphia: SIAM.

Hageman, L. A., and D. M. Young (1981). *Applied Iterative Methods.* New York: Academic Press.

Hamming, R. W. (1971). *Introduction to Applied Numerical Analysis.* New York: McGraw-Hill.

Hamming, R. W. (1973). *Numerical Methods for Scientists and Engineers.* 2nd ed. New York: McGraw-Hill.

Harrington, Steven (1987). *Computer Graphics: A Programming Approach.* New York: McGraw-Hill.

Heath, Michael T. (2002). *Scientific Computing, An Introductory Survey.* 2nd ed. New York: McGraw-Hill.

Henrici, Peter H. (1964). *Elements of Numerical Analysis.* New York: Wiley.

Higham, Desmond J., and Nicholas J. Higham (2000). *MATLAB Guide.* Philadelphia: SIAM.

Hillier, Frederick S., and Gerald J. Liebermann (1974). *Operations Research.* 2nd ed. San Francisco: Holden-Day.

Housholder, Alston S. (1970). *The Numerical Treatment of a Single Nonlinear Equation.* New York: McGraw-Hill.

IEEE Standard for Binary Floating-Point Arithmetic (1985). Institute of Electrical and Electronics Engineers, Inc., New York.

JaJa, J. (1992). *An Introduction to Parallel Algorithms.* Reading, MA: Addison-Wesley.

Jones, B. (1982). A note on the T transformation. *Nonlinear Analysis, Theory, Methods and Applications* 6:303–305.

Kahaner, D., C. Moler, and S. Nash (1989). *Numerical Methods and Software.* Englewood Cliffs, NJ: Prentice-Hall.

Kaplan, Wilfrid (1999). *Maxima and Minima with Applications: Practical Optimization and Duality.* New York: Wiley.

Lee, Peter. and Geoffrey Duffy (1976). Relationships between velocity profiles and drag reduction in turbulent fiber suspension flow. *Journal of the American Institute of Chemical Engineering* 22(4): 750–753.

Love, Carl H. (1966). *Abscissas and Weights for Gaussian Quadrature.* National Bureau of Standards, Monograph 98.

Luenberger, David G. (1973). *Introduction to Linear and Nonlinear Programming*. Reading, MA: Addison-Wesley.

Maron, Melvin J., and Robert J. Lopez (1991). *Numerical Analysis: A Practical Approach.* 3rd ed. Belmont, CA: Wadsworth.

Moskowitz, Herbert, and Gordon P. Wright (1979). *Operations Research Techniques for Management*. Englewood Cliffs. NJ: Prentice-Hall.

Muller, D.E. (1956). A method of solving algebraic equations using an automatic computer. *Math Tables and Other Aids to Computation* 10:208–215.

Nash, Stephen G., and Ariela Sofer (1996). *Linear and Nonlinear Programming.* New York: McGraw-Hill.

O'Neill, Mark A. (1988). Faster than fast Fourier. *BYTE* 13(4):293–300.

Orvis, William J. (1987). *1-2-3 for Scientists and Engineers.* San Francisco: Sybex.

Peaceman, D. W., and H. H. Rachford (1955). The numerical solution of parabolic and elliptic differential equations. *Journal of the Society for Industrial and Applied Mathematics* 3:28–41.

Penrod, E. B., and K. V. Prasanna (1962). Design of a flat-plate collector for a solar earth heat pump. *Solar Energy* 6(1):9–22.

Pinsky, Mark A. (1991). *Partial Differential Equations and Boundary Value Problems with Applications.* 2nd ed. New York: McGraw-Hill.

Pizer, Stephen J. (1975). *Numerical Computing and Mathematical Analysis.* Chicago: Science Research Associates.

Pokorny, C., and C. Gerald (1989). *Computer Graphics: The Principles Behind the Art and Science.* Irvine, CA: Franklin, Beedle, and Associates.

Polak, Elijah (1997). *Opimization, Algorithms and Consistent Approximations.* New York: Springer-Verlag.

Pratap, Rudra (2002). *Getting Started with MATLAB: A Quick Introduction for Scientists and Engineers.* New York: Oxford University Press.

Prenter, P. M. (1975). *Splines and Variational Methods.* New York: Wiley.

Press, W., B. Flannery, S. Teudolsky, and W. Vetterling (1992). *Numerical Recipes in C: The Art of Scientific Computing.* 2nd ed. New York: Cambridge University Press.

Press, W., B. Flannery, S. Teudolsky, and W. Vetterling (1992). *Numerical Recipes in FORTRAN: The Art of Scientific Computing.* 2nd ed. New York: Cambridge University Press.

Press, W., B. Flannery, S. Teudolsky, and W. Vetterling (1996). *Numerical Recipes in FORTRAN 90: The Art of Parallel Scientific Computing.* 2nd ed. New York: Cambridge University Press.

Rall, L. B. (1981). *Automatic Differentiation: Techniques and Applications.* Springer-Verlag.

Ralston, Anthony (1965). *A First Course in Numerical Analysis.* New York: McGraw-Hill.

Ramirez, Robert W. (1985). *The FFT, Fundamentals and Concepts.* Englewood Cliffs, NJ: Prentice-Hall.

Rao, Sigiresu S. (2002). *Applied Numerical Methods for Engineers and Scientists.* Englewood Cliffs, NJ: Prentice-Hall.

Rice, John R. (1983). *Numerical Methods, Software, and Analysis.* New York: McGraw-Hill.

Richtmyer, R. D. (1957). *Difference Methods for Initial Value Problems.* New York: Wiley Interscience.

Sabot, G. W., ed (1995). *High Performance Computing:* Reading, MA: Addison-Wesley.

Sedgwick, R. (1992). *Algorithms in C ++.* Reading, MA: Addison-Wesley. [Other versions available: *in C, in Pascal.*]

Shampine, L., and R. Allen (1973). *Numerical Computing.* Philadelphia: Saunders.

Smith, G. D. (1978). *Numerical Solution of Partial Differential Equations.* 2nd ed. London: Oxford University Press.

Stallings, William (1990). *Computer Organization and Architecture.* New York: Macmillan.

Stewart, G. W. (1973). *Introduction to Matrix Computations.* New York: Academic Press.

Stoer, J., and R. Burlirsch (1993). *Introduction to Numerical Analysis.* 2nd ed. New York: Springer-Verlag.

Traub, J. F. (1964). *Iterative Methods for the Solution of Equations.* Englewood Cliffs, NJ: Prentice-Hall.

Van Loan, C. F. (1997). *Introduction to Scientific Computing: A Matrix-Vector Approach Using MATLAB.* Englewood Cliffs, NJ: Prentice-Hall.

Varga, Richard (1959). *p*-Cyclic matrices: A generalization of the Young–Frankel successive over relaxation scheme. *Pacific Journal of Mathematics* 9:617–628.

Vichnevetsky, R. (1981). *Computer Methods for Partial Differential Equations.* Vol. 1, *Elliptic Equations and the Finite Element Method.* Englewood Cliffs, NJ: Prentice-Hall.

Walker, D. W., and J. J. Dongarra. (1996). MPI: A Standard Message-Passing Interface. *SIAM NEWS* 29(1).

Waser, S., and M. J. Flynn (1982). *Introduction to Arithmetic for Digital Systems Designers.* New York: Holt, Rinehart and Winston.

Wilkinson, J. H. (1963). *Rounding Errors in Algebraic Processes.* Englewood Cliffs, NJ: Prentice-Hall.

Wilkinson, J. H. (1965). *The Algebraic Eigenvalue Problem.* London: Oxford University Press.

Wolfram, Stephen (1999). *The Mathematica Book:* 4th ed. Wolfram Media/Cambridge University Press.

Index